P9-CAA-997

# Advanced Engineering Economics

# ADVANCED ENGINEERING ECONOMICS

**Chan S. Park**
*Auburn University*

**Gunter P. Sharp-Bette**
*Georgia Institute of Technology*

JOHN WILEY & SONS, INC.
New York • Chichester • Brisbane • Toronto • Singapore

***Library of Congress Cataloging in Publication Data:***

Park, Chan S.
Advanced engineering economics/Chan S. Park, Gunter P. Sharp-Bette.
p. cm.
Includes bibliographical references.
ISBN 0-471-79989-0
1. Engineering economy. I. Sharp-Bette, Gunter P. II. Title.
TA177.4.P368 1990
658.15—dc20

89-24851
CIP

Printed in the United States of America

10 9 8 7 6 5 4 3 2 1

*Dedicated* To Our Families
—Kim (Inkyung) and Maria

# ABOUT THE AUTHORS

## CHAN S. PARK, Ph.D., P.E.

Dr. Chan S. Park is Professor of Industrial Engineering at Auburn University. At present, he is on leave as a Visiting Professor of Industrial and Operations Engineering at the University of Michigan (1990). He is a graduate of Hanyang University (B.S.), Purdue University (M.S.I.E.), and the Georgia Institute of Technology (Ph.D.). He is a licensed professional engineer in the state of Florida.

Dr. Park is a senior member of the Institute of Industrial Engineers. In 1984–1985 he served as Director of its Engineering Economy Division. He was also a member of the board of directors of the American Society of Engineering Education from 1982 through 1985. He is currently Manuscript Editor for *The Engineering Economist.*

Dr. Park has written numerous technical papers on economic decision analysis and computer applications. He received the 1987 Alfred V. Bodine/SME Award from the Society of Manufacturing Engineers for directing a best paper on machine tool economics. Dr. Park is the author of *Interactive Microcomputer Graphics* (1985) and *Modern Engineering Economic Analysis* (1991), both from Addison-Wesley.

## GUNTER P. SHARP-BETTE, Ph.D., P.E.

Dr. Gunter P. Sharp-Bette is on the faculty of the School of Industrial and Systems Engineering at the Georgia Institute of Technology. He holds degrees from Georgia Tech (B.I.E., Ph.D.) and Stanford University (M.S.) and is a licensed professional engineer in the state of Georgia.

Dr. Sharp is active in several professional societies, including the Institute of Industrial Engineers, the Operations Research Society of America, and The Institute of Management Sciences. He has written numerous technical papers on economic decision analysis, transport development, and material handling systems analysis. He was Manuscript Editor of *The Engineering Economist* during the years 1979 through 1983.

Dr. Sharp has extensive national and international experience in the fields of feasibility studies, economic development, transport planning, and material handling systems. He was project director of a four-year, nine-university research program that evaluated improvements in freight transportation systems according to their potential for economic development. Since 1982 Dr. Sharp has been Program Director of Warehousing Systems Research in the Material Handling Research Center at Georgia Tech.

# PREFACE

Project evaluation and selection techniques constitute an important body of knowledge for those concerned with making individual, corporate, and public investment decisions. During the past twenty years a number of important and useful concepts and techniques have been developed and are reported primarily in journals and specialized texts. The purpose of this book is to present these concepts and techniques in an integrated framework built on traditional engineering economics principles.

The book is intended for first-year graduate students in engineering, business, operations research, and systems analysis. For a complete understanding of the material, the student should have a background in calculus, introductory linear programming, and introductory probability and statistics and has some familiarity with simulation. For students not familiar with these background topics, the relevant sections can be skipped without loss of continuity, or supplementary material can be provided by the instructor. A first course in engineering economics or business finance is helpful but not necessary, since the treatment of basic topics here is fairly complete.

Although the text is oriented primarily toward students, the style of presentation will also make it useful to practitioners who have some background in engineering economics. Wherever possible, intuitive interpretations are given to complement the analytical explanations.

The material is presented in four parts.

Part One: Basic Concepts and Techniques in Economic Analysis, Chapters 1–5.

Part Two: Deterministic Analysis, Chapters 6–8.

Part Three: Stochastic Analysis, Chapters 9–13.

Part Four: Special Topics in Engineering Economic Analysis, Chapter 14–16.

A brief description of each chapter follows.

Chapter 1, Accounting Income and Cash Flow, introduces the concept of cash flow and distinguishes it from accounting income. There is a brief presentation of the relationships among the balance sheet, the income statement, and the funds flow statement.

Chapter 2, Interest and Equivalence, includes discrete compounding, continuous compounding, continuous cash flows, and equivalence. *Inflation* is included in this early chapter, since it is directly related to the expected earning power and opportunity cost of invested funds. The treatment of *noncomparable payment and compounding periods* is more general than that in most texts. *Sensitivity analysis* is introduced early in the text with an example of home ownership analysis during inflation.

Chapter 3, Transform Techniques in Cash Flow Modeling, covers $z$-transforms for discrete cash flows and Laplace transforms for continuous cash flows. These powerful modeling techniques allow us to deal with more complex cash flow profiles, such as step, ramp, decay, and growth, with relatively little computational effort. Applications to *profit margin analysis* and *declining loan balance* illustrate the techniques. This material is not usually found in engineering economics texts. (This chapter can be skipped without loss of continuity.)

Chapter 4, Depreciation and Corporate Taxation, introduces these important topics early. Besides the traditional depreciation methods (straight-line, declining balance, and sum-of-the-years' digits), the chapter covers the 1986 tax law changes and the *generalized cash flow method.* The last item is a powerful analysis technique that is very useful for sensitivity analysis and for situations in which revenues cannot be associated with an investment. It is demonstrated here by measuring the present value loss of the depreciation tax shield owing to inflation.

Chapter 5, Selecting a Minimum Attractive Rate of Return, begins with sources and costs of funds. The emphasis is on understanding the issues and judging when an approximation provides good results. The use of a weighted-average cost of capital is demonstrated with an example showing the difference in application between *net equity flows* (with an equity interest rate) and *after-tax composite flows* (with a tax-adjusted weighted-average rate. Computing a weighted-average cost of capital under different conditions of capital structure, growth, and risk is discussed to give the reader a view of this complex issue.

Chapter 6, Measures of Investment Worth—Single Project, begins Part Two of the book. The chapter presents a variety of investment criteria: present value, annual equivalent, internal rate of return, modified internal rate of return, benefit–cost ratio, and others. These (more than ten) criteria are analyzed and compared with respect to an individual project. They are shown to be equivalent, except for the payback period. The chapter also covers the distinctions between pure and mixed investments, Descartes' rule and the *Norstrom criterion* for number of roots, and the *project balance* concept.

Chapter 7, Decision Rules for Selecting among Multiple Alternatives, extends the concept of Chapter 6 to project selection in the case of unlimited borrowing and lending at a fixed rate. After a section on formulating mutually exclusive alternatives, the ten equivalent criteria are separated into *four groups,* based on the *ordering to be used with incremental analysis.* A single comprehensive example demonstrates the use of criteria from each of the four groups. Not surprisingly, present value is the easiest criterion to apply in terms of computational effort and ordering. A section on comparing projects with unequal lives emphasizes the need for clear assumptions about period of use, salvage values, and replacement of shorter-lived assets.

Chapter 8, Deterministic Capital Budgeting Models, includes the present value model, pure capital rationing, Weingartner's horizon model, Bernhard's general model, integer programming, multiple objectives, and a case study of a dividend–terminal-wealth model. Some of this material is new. The progression is from the most restrictive to the most general models. When possible, a *single example with variations* is used to explain the relationships among the different models and their assumptions.

Chapter 9, Utility Theory, begins Part Three. The chapter is a brief treatment of the axioms of choice and of risk attitudes. A section covers empirical procedures and results. An important operational method, *mean–variance analysis,* is related to the utility concept.

Chapter 10, Measures of Investment Worth under Risk—Single Project, is an extensive treatment of investment criteria when used in a stochastic framework. Included are statistical moments of net present value derived from *sums, products, quotients,* and *powers* of other random variables. This material is rarely seen outside the statistics literature, despite its obvious applications to profit analysis. Also included are statistical moments based on uncertain timing and statistical distributions of net present value using two moments and using four moments (Pearson's system of curves). A section covers beta-function estimators and Hillier's technique for correlated cash flows.

Chapter 11, Methods for Comparing Risky Projects, begins with mean–variance and mean–semivariance, and then moves on to stochastic dominance. Again, a *single example with variations* is used to demonstrate first-degree, second-degree, and third-degree dominance. Third-degree dominance, though extremely useful, is rarely covered in textbooks. A brief treatment of portfolio theory is followed by a section on discrete capital rationing under risk and a *multiperiod index model* for project portfolio selection. This index model is new, and it is demonstrated by a case study. Uncertainty resolution deals with time-dependent criteria of investment worth.

Chapter 12, Risk Simulation, begins with an overview of simulation and input probability distributions and then covers sampling procedures for independent random variables. A section on *sampling for dependent random variables* brings this important topic within the grasp of the student and the practitioner. Included are sampling based on regression, discriminant sampling in the absence of data, and the normal transformation method. After a section on output analysis, there is a comprehensive example demonstrating use of a computer routine.

Chapter 13, Decision Tree Analysis, contains a brief treatment of decision trees, covering the expected value criterion, the expected utility criterion, the value of perfect information, sampling with imperfect information, and sensitivity analysis.

Chapter 14, Evaluation of Public Investments, is the first of the special topics in Part Four. The chapter begins with a discussion of the nature of different benefits and costs and then covers estimation methods. An example in the context of a mass transit system illustrates some of the concepts. Risk and uncertainty are handled by methods for obtaining the *distribution of B/C ratios* and incremental $B/C$ ratios. In some cases these distributions are exact; this method was previously accessible only in the literature.

Chapter 15, Economic Analysis in Public Utilities, is a brief treatment of capital costs, the revenue requirement method, flow-through accounting, and the normalizing method. The equivalence between the present value method and the revenue requirement method is shown.

Chapter 16, Procedures for Replacement Analysis, covers specialized concepts and techniques for this application. Obsolescence and technology changes are explained with several real examples. The basic principles of sunk costs, outsider point of view, and economic life are explained. Infinite planning period methods include *closed-form expressions* for geometric cost changes. The generalized cash flow method is used to obtain the *after-tax economic life* of an asset and also to provide the year-by-year costs for the finite planning period and dynamic programming methods.

The *interest tables* in Appendix A are accurate to *six* or more *significant digits*. This enables the instructor to demonstrate equivalence and alternative solution approaches with greater precision. They also include values for multiples of 6 (e.g., 36, 42, 48, 72), allowing direct use of the tables for many problems with monthly compounding. Continuous compounding tables are excluded; we are of the opinion that a person using continuous compounding would most likely be modeling continuous cash flows, and then the techniques of Chapter 3 would be used. The statistical tables of Appendix B give the normal distribution, and the chi-square distribution.

Each chapter contains numerous examples demonstrating the concepts and techniques, and there is a problem section at the end of each chapter. (A solution manual is planned.) Case studies are included to demonstrate the application of principles in a more complex or real-life setting.

We have tested most of the material in our graduate classes at

Auburn University and Georgia Tech. The material in the text is sufficient for a two-quarter course. Much of the subject matter can also be covered in a one-quarter or one-semester course by speeding up the review part (for students with an earlier first course) and by selectively deleting some of the material. Alternatively, the book can be used for independent study of selected topics following a formal course.

We would like to thank the many individuals who have encouraged us in writing this book, including Charity Robey, Bill Stenquist, Jerry Thuesen, Ed Unger, and John White. We also wish to thank the reviewers, Ken Case, M. Jeya Chandra, and Jane Fraser, for their many helpful comments. We are grateful to George Prueitt for his assistance in developing the interest tables and Kyung-Il Choe for his assistance in reviewing Chapters 10 and 11. The expert editorial assistance provided by Mrs. Mary Joan Mykytka is gratefully acknowledged. Special thanks for their untiring devotion in typing the manuscript are extended to Carmella Bell, Joene Owen, Pat Hicks, Anita Race, and Barbara Vickery.

CHAN S. PARK
GUNTER P. SHARP-BETTE

*Auburn, Alabama*
*Atlanta, Georgia*

# CONTENTS

## PART ONE
## BASIC CONCEPTS AND TECHNIQUES IN ECONOMIC ANALYSIS

## PART TWO
## DETERMINISTIC ANALYSIS

## PART THREE
## STOCHASTIC ANALYSIS

## PART FOUR
## SPECIAL TOPICS IN ENGINEERING ECONOMIC ANALYSIS

# PART ONE

# BASIC CONCEPTS AND TECHNIQUES IN ECONOMIC ANALYSIS

# 1

# Accounting Income and Cash Flow

## 1.1 WHAT IS INVESTMENT?

International Business Machines (IBM) invests a hundred million dollars in a supercomputer model. The president of the company may invest hundreds of thousands of dollars by purchasing shares of a newly formed company that manufactures microcomputer chips. A plant manager in the company may invest several thousand dollars in a time-sharing resort apartment. The marketing manager considers the investment of some bonus money in dog racing at a local track. The marketing manager's secretary may invest savings in a federally insured savings and loan association. The secretary's 12-year-old son buys savings certificates with the money earned through delivering the local paper; he hopes to buy a car. The secretary's 3-year-old daughter deposits all the pennies she finds in a piggy bank in the hope that enough money will accumulate to buy her favorite candy.

What we just described are investment activities from the perspectives of the investors—an analogy similar to that introduced by Sharpe in his definition of investment [2]. A deposit in a savings account at a bank is an investment in the eyes of the depositor. Even cash stored in a piggy bank can be viewed as an investment—one yielding a penny for every penny invested, assuming no fire or theft. Each example involves the sacrifice of something now for the prospect of something later. And this, in the general sense, is investment. In examining individual examples such as these, Professor Sharpe further notes that "We see two different factors involved, *time* and *risk*. The sacrifice takes place in the present and is certain. The reward comes later, if at all, and the magnitude may be uncertain." In some cases, the element of time predominates (e.g., government bonds). In others, risk (uncertain outcome) is the dominant factor (e.g., betting on a dog race). In yet others, both are important (e.g., supercomputer development).

The investment examples given are often classified in two categories: financial investments and real investments. A *financial investment* is one in which the investor allocates his or her resources to some form of financial instrument, such as stocks or bonds. *Real investments,* on the other hand, are represented by physical assets such as a new plant and equipment. This book is about evaluating such investment options, considering both time and risk. We must emphasize that we are primarily concerned with real (project) investments as contrasted with financial investments.

Economic evaluation of real investments is variously referred to as economic analysis, engineering economy, and economic decision analysis. In particular, the application of economic analysis techniques in the comparison of engineering design alternatives is referred to as *engineering economy.* The widespread use of the engineering economy principle in nonengineering areas, however, has brought about use of the more general term *economic analysis.*

Any contemplated or proposed investment requires evaluation to ensure, with reasonable confidence, that the expected benefits of the project will exceed costs. Whatever methodology is used, certain basic principles remain unchanged. For example, these principles should be applicable whether the investment is for a new firm, the expansion of an existing facility, or the formation of a new corporate entity. Our goal is to identify such principles, formulate a logical decision framework for economic evaluation, and examine a general analytical methodology for assessing the worth of any proposed capital project.

## 1.2 THE CORPORATE INVESTMENT FRAMEWORK

The general principles we present should be equally applicable to personal investments and corporate investments, but we will place our emphasis on real investments by a typical firm. For that reason, we first need to define the basic objective of the firm and review some of its functions.

### 1.2.1 The Objective of the Firm

We will assume that the objective of any firm is to maximize its value to its stockholders. The market price of a firm's stock represents the value of that particular firm. The value is governed by present and prospective earnings per share; the timing, duration, and risk of these earnings; and other factors that affect the market price of the stock. Thus, the market value is an index of a company's progress and reflects the market's assessment of how well the firm is managed for the benefit of its stockholders. Therefore, we can view maximizing the stockholders' wealth as equivalent to maximizing the market value of a share of stock.

It is certainly possible to have objectives other than maximization of the stockholders' wealth. Frequently, maximizing profits is regarded as the proper objective of the firm, but this objective is not as inclusive as that of maximizing stockholders' wealth. For example, a firm could always raise total profits by issuing stock and using the proceeds to invest in U.S. savings bonds. The firm's total profits could also be increased by investing the dividends that would normally be paid to stockholders in some other ventures—or, at the very least, by investing the dividends in a risk-free savings account. Such a policy would surely decrease the market value of the firm's stock.

Earnings-per-share data are the key financial statistics for most investors. These ratios represent, on a per-share basis, the equity of the holder of common stock in a company's current profits. Maximization of earnings per share, however, cannot be a fully appropriate objective of the firm. The shortcoming of this objective is that it does not reflect either the timing and duration of expected returns or the risk and uncertainty of the prospective earning streams. For example, suppose two companies have the same expected future earnings per share. If the earning stream of one company is subject to considerably more uncertainty than the earning stream of the other, the market price per share of its stock may be less.

### 1.2.2 The Functions of the Firm

As mentioned earlier, we are primarily concerned with economic analysis from the perspective of economic decisions at the company level. At this level, a firm must make three major decisions: (1) the investment decision, (2) the financing decision, and (3) the dividend decision. Each must be considered with the objective of the firm in mind. Ideally, an optimal combination of the three decisions will maximize the value of the firm to its stockholders.

***The Investment Decision.*** Capital budgeting, a major aspect of the investment decision, is the allocation of capital to investment projects whose benefits are to be realized in the future. How the capital is to be allocated is one of the most important and most difficult decisions that any firm must make. Because the future benefits, in most cases, are not known with certainty, investment decisions are always based on predictions about the future—often the distant future. Furthermore, investment decisions frequently require judgmental estimates about future events. This lack of certainty about the future makes capital budgeting decisions one of the most challenging tasks.

The capital budgeting process embraces a rather broad and diverse class of activities in allocating capital resources to competing investment projects. Essentially, these activities include the administration and organization of a capital expenditure program, the development of new investment opportunities, the estimation of future cash flows of investment projects, and the review of the investment program. We are concerned primarily, however, with the analytical techniques used during the decision-making phase of the capital budgeting process.

***The Financing Decision.*** The second major decision of the firm is the financing decision. For any firm to make capital investments, it must have a source of funds. In general, these funds are obtained from several sources: the sales of stocks, retained earnings, the sale of bonds, and short-term borrowing from a financial institution. Normally these financing sources are grouped into two classes, equity financing referring to the first two and debt financing referring to the second two.

The firm obtains equity funds by selling to investors shares of common stock, which represent ownership of portions of the firm. (There are many variations in the ways of offering stock; one such is preferred stock.) The firm may also secure capital by retaining a portion of the net profits that it generates for any operating year. This method of obtaining capital is called *retained earnings*. Because the net earnings of the firm are themselves the outcome of a prior equity capital investment by the firm, they belong to the equity investors of the firm. In addition, the firm can obtain capital by borrowing from financial institutions or by selling bonds to investors. Unlike equity financing, this debt financing does not change the ownership attributes of the firm.

***The Dividend Decision.*** The third important decision of the firm is its dividend policy. Investors buy stocks in the expectation of receiving returns on their invested capital. These returns on stockholders' capital are called *dividends*. The dividend decision includes the percentage of earnings paid to stockholders in cash dividends, the stability of absolute dividends over time, and repurchase of stock. The amount of dividends paid to stockholders has an effect on the amount of money available for equity financing. Thus, the value of a dividend to investors must be balanced against the cost of losing these earnings as a means of equity financing. Consequently, a balance must be established between dividend payments and investment needs.

For the purposes of our discussion, the functions of the firm were divided into three decisions. In practice, these functions are rarely separable—they are interrelated. The decision to invest in a new capital project, for example, involves financing of the investment. The financing decision then affects the dividend decision, because any retained earnings used in internal financing represent dividends forgone by the stockholders. With the objective of maximizing stockholders' wealth, the firm should ideally strive for an optimal combination of the three decisions.

The market price rises and falls depending on (1) how the company is doing at a particular time, (2) what is happening to other stock prices, and (3) most important, how investors expect the company to do in the future. Certainly, the company's decisions to invest in various projects and the actual performance of these projects can be major forces driving the market value of the company's stock.

### 1.2.3 The Analysis Framework

We will not attempt in this book to set down an optimum approach to the entire decision area but will focus on problems related to investment decisions. As already mentioned, in practice, the investment decisions are not always independent of the source of financing. For convenience in economic analysis, however, the two activities are usually separated. This approach is followed in this book—first the selection procedures for investment projects are analyzed, and then the choice of financing sources is considered. After this, the appropriate modifications are made to the investment models to allow a formally fixed quantity, the capital budget, to become a decision variable.

### 1.2.4 Accounting Information

During the capital budgeting process, we often need to seek accounting information to determine the effect on the company's financial condition of undertaking an investment project. In practice, we rarely make a capital investment decision on the basis of the profitability of the project alone but rather consider the project in the context of its impact on the financial strength and position of the company.

Accounting is generally the source of much of the past financial data that are needed in making estimates of future financial conditions. Accounting is also a major source of data for postaudit analyses that might be made regarding the actual performance of an investment project compared with the results predicted during the investment decision process. To use accounting information in a proper perspective, we need to understand accounting principles and the significance of accounting data.

The principles of accounting are built on a foundation of basic concepts. These concepts are so basic that most accountants do not consciously refer to them. Nor do accountants fully agree on how many such concepts there are or how they should be defined. There is fairly general support, however, for the following list. (See [1] for a complete discussion.)

1. Monetary measurement: Accounting records only facts that can be expressed in monetary terms.
2. Business entity: Accounts are kept for business entities as distinguished from the person(s) associated with those entities.
3. Going concern: Accounting assumes that a business will continue indefinitely and that it is not about to be sold.
4. Cost: Assets are ordinarily entered on the accounting records at the price paid to acquire them, and this cost, rather than the market value, is the basis for subsequent accounting for the assets.
5. Dual aspect: The amount of assets equals the amount of equities. There are two types of equities: liabilities, which are the claims of creditors, and owners' equity, which is the claim of the owners of the business. Accounting systems are set up in such a way that a record is made of two aspects of each event that changes assets and equities equally.
6. Conservatism: An asset is recorded at the lower of two reasonably possible values, or an event is recorded in such a way that the owners' equity is lower than it would otherwise be.
7. The accrual concept: Income is measured as the difference between revenues and expenses rather than as the difference between cash receipts and disbursements.
8. The realization concept: Revenue is recognized at the time it is realized.
9. The consistency concept: Events of the same character are treated in the same fashion from one period to another.
10. The materiality concept: Events that do not have a significant effect on the accounts are not recorded.

These concepts will be explained further through the financial statements in the subsequent sections. For this purpose, three financial statements will be reviewed: (1) the balance sheet, (2) the income statement, and (3) the funds flow statement. These financial statements report on a company's position at a point in time and on its operations over some past period. Their real value, however, lies in helping us to predict the company's future earnings and dividends, as well as its cash flows.

## 1.3 THE BALANCE SHEET

The balance sheet is a statement of the financial position of a company as of a reporting date. The balance sheet shows the sources from which current operating funds have been obtained (liabilities and owners' equity) and the types of property and property rights in which these funds are currently locked up (assets).

### 1.3.1 Reporting Format

The example shown in Table 1.1 represents the 19x0 and 19x1 comparative balance sheet accounts of Sunset Inc., a manufacturer of power surge suppressors for personal computers. This example will be used in subsequent discussions. (Balance sheets do not all follow precisely the same format or use the same account titles.)

***Accounting Equation.*** The dollar amounts recorded in the various accounts in the balance sheet must conform to the fundamental accounting equation

$$\text{Assets} = \text{Liabilities} + \text{Net worth} \quad \text{or} \quad \text{Assets} - \text{Liabilities} = \text{Net worth}$$

and a typical layout of the balance sheet looks like this:

| | Assets | Liabilities and net worth | |
|---|---|---|---|
| Assets | Current assets | Current liabilities | Liabilities |
| | | Long-term liabilities | |
| | Fixed assets<br>Other assets | Paid-in capital<br>Retained earnings | Net worth |

The asset aside of the balance sheet indicates how the funds provided to the company have been invested as of the date of the balance sheet. The several liability items show the amount of those funds obtained from trade creditors (accounts payable), from lenders (long-term debt), and from other creditors.

**Table 1.1** ***Comparative Balance Sheet***

**SUNSET INC.**
BALANCE SHEET
*December 31*

| | 19x1 | 19x0 | Changes |
|---|---|---|---|
| ***Assets*** | | | |
| Current assets | | | |
| Cash | $ 7,500 | $ 5,500 | +$2,000 |
| Marketable securities | 3,000 | 5,000 | − 2,000 |
| Accounts receivable | 23,700 | 19,500 | + 4,200 |
| Inventories | 37,700 | 39,800 | − 2,100 |
| Prepaid expenses | 2,000 | 1,500 | + 500 |
| Deferred charges | 2,500 | 3,000 | − 500 |
| Total current assets | 76,400 | 74,300 | + 2,100 |
| Fixed Assets | | | |
| Plant and equipment | 154,000 | 145,000 | + 9,000 |
| Less accumulated depreciation | (70,000) | (50,000) | + 2,000 |
| Total fixed assets | 84,000 | 95,000 | −11,000 |
| Total assets | $160,400 | $169,300 | − 8,900 |
| ***Liabilities and Owners' Equity*** | | | |
| Current liabilities | | | |
| Accounts payable | 10,000 | 26,000 | −16,000 |
| Wages payable | 16,000 | 15,000 | + 1,000 |
| Accrued taxes | 2,000 | 3,500 | − 1,500 |
| Total current liabilities | 28,000 | 44,500 | −16,500 |
| Other liabilities | | | |
| First mortgage bonds | 20,000 | 22,000 | − 2,000 |
| Debentures 8% | 10,000 | 10,000 | 0 |
| Total other liabilities | 30,000 | 32,000 | − 2,000 |
| Total liabilities | $58,000 | $76,500 | −18,500 |
| Owners' equity | | | |
| Preferred stock, 6%, $100 par value | 10,000 | 10,000 | 0 |
| Common stock, $4 par value (10,000 shares) | 40,000 | 40,000 | 0 |
| Capital surplus | 10,000 | 10,000 | 0 |
| Retained earnings | 42,400 | 32,800 | + 9,600 |
| Total owners' equity | 102,400 | 92,800 | + 9,600 |
| Total liabilities and equity | $160,400 | $169,300 | − 8,900 |

The owners' equity section shows the funds supplied by the owners. If the business is a corporation, the owners are stockholders and their contribution consists of two principal parts: funds directly obtained by selling stocks (paid-in capital) and funds that owners provided by permitting earnings to remain in the business (retained earnings). The capital obtained from various sources has been invested according to the management's best judgment; a certain fraction is invested in buildings, another fraction is in inventories, another fraction is in the form of cash, and so on. Therefore, the asset side of the balance sheet reflects the result of these management judgments as of the reporting date. We will look at some of these balance sheet items in more detail.

***Assets.*** The *assets* of a company consist of all its property: what it owns and what is owed to it. The assets are generally divided into three principal categories and are listed in decreasing order of liquidity, where liquidity refers to the ease with which an asset can be converted to cash.

*Current assets* are listed first. These assets can be converted to cash or its equivalent in less than one year. For example, cash, the most liquid asset, is generally listed at the top of the current assets column on a balance sheet. Other current assets in order of decreasing liquidity would be marketable securities, accounts receivable and inventories, prepaid expenses, and deferred charges. Prepaid expenses might include insurance premiums and rent, whereas deferred charges represent expenses for materials and services whose benefits will extend into the future.

*Fixed assets* are listed second. These assets are relatively permanent and are not converted to cash or its equivalent in the normal business operating cycle. The major types of fixed assets are land, buildings, factory machinery, equipment, tools, office equipment, furniture, automobiles, trucks, and other vehicles.

*Other assets,* if any, are listed third. Assets in this category include investments made in other companies (to control their operations) and intangible assets such as goodwill, copyrights, and franchises.

***Liabilities and Net Worth (Owners' Equity).*** The *liabilities* of a company disclose where the funds were obtained to acquire the assets and to operate the business. Liabilities are claims of creditors that take precedence over the rights of the owners to the assets. *Net worth* refers to the portion of the assets of a company that is provided by the investors (owners).

Liabilities are separated into current and other liabilities. *Current liabilities* are debts that are payable within one year. They include accounts and notes payable within one year, accrued expenses (wages, salaries, interest, rent, taxes, etc., owed but not yet due for payment), advance payments, and deposits from customers. *Other liabilities* include long-term liabilities such as bonds, mortgages, and long-term notes, which are due and payable more than one year in the future.

Net worth (or owners' equity) is the liability of the company to its owners. This represents the amount that is available to the owners after all other debts have been paid. It generally consists of preferred and common stock, capital surplus, and retained earnings. *Capital surplus* is the amount of money received from the sale of stock in excess of the face value of the stock. *Retained earnings* are the accumulated net operating income after taxes less the dividends paid to stockholders. In the following sections, we will look at each balance sheet item in detail.

### 1.3.2 Cash versus Other Assets

Although the assets are all stated in terms of dollars, only cash amounts represent actual money. In Table 1.1, receivables are bills that others owe to Sunset Inc.; inventories consist of raw materials, work in process, and finished goods available for sale; and fixed assets consist of Sunset's plant and equipment. Sunset can write checks for a total of $7,500 as of December 31, 19x1 (while current liabilities of $44,500 are due within a year). The noncash assets should produce cash flows eventually, but they do not represent cash in hand.

### 1.3.3 Liabilities versus Stockholders' Equity

From the basic accounting equation, we note that the stockholders' equity, or net worth, is a residual amount:

$$\begin{aligned} &\text{Assets} - \text{Liabilities} &&= \text{Stockholders' equity} \\ &\$160{,}400 - \$58{,}000 &&= \$102{,}400 \end{aligned}$$

Suppose assets decline in value—for example, suppose some of the accounts receivable are written off as bed debts. Liabilities remain constant, so the value of the net worth must decline. On the other hand, if asset values rise, the benefits will accrue exclusively to the stockholders.

Recall that the owners' equity (or stockholders' equity) account consists of two parts: (1) the amount contributed by the stockholders and (2) retained earnings. The amount contributed by the stockholders is called *invested capital* and is further divided into three accounts: preferred stock, common stock, and capital surplus (paid-in capital). Preferred stock has preference over common stock in the receipt of dividends and in the right to assets in the event of liquidation. There is no substantial difference in the accounting treatment of these two types of stock. Accountants generally assign a *par value* (book value) to both preferred and common stock; Sunset's preferred stock has a par value of $100 (100 shares), and the common stock has a par value of $4 (10,000 shares) as of December 31, 19x1.

#### *Example 1.1*

Suppose Sunset were to sell 10,000 additional shares (common stock of $4 par) at a market price of $10 per share. The company would raise $100,000, and the cash accounts would go up by this amount. Of the total, $40,000 would be added to common stock and $60,000 to *capital surplus*. Thus, after the sale, common stock accounts would show $40,000 + $40,000 = $80,000, capital surplus would show $10,000 + $60,000 = $70,000, and there would be 20,000 shares (common stock) outstanding. □

### 1.3.4 Inventory Valuation

Sunset uses the first in, first out (FIFO) method to determine the inventory value $37,700 shown on its balance sheet. This inventory valuation method assumes that the first goods acquired are the first goods sold. In other words, it assumes that each sale is made out of the oldest goods in stock; the ending inventory therefore consists of the most recently acquired goods. This FIFO valuation method may be adopted by any business regardless of whether the physical flow of merchandise is actually oldest units first.

Sunset Inc. could use the last in, first out (LIFO) method. This method assumes that the most recently acquired goods are sold first and that the ending inventory consists of "old" merchandise acquired in the earliest purchases. Under the LIFO method, the cost of goods sold tends to reflect the cost of the items most recently purchased. Advocates of the LIFO approach contend that matching of current costs of merchandise against current sales prices, regardless of which physical units of merchandise are being delivered to customers, actually reflects business pricing decisions.

If this contention is correct, the conventional FIFO system can result in the reporting of false "inventory profits" during periods of rising prices. Thus, the inventory valuation method can have a significant effect on financial statements. In selecting a method, the company should consider the probable effect on the balance sheet, on the income statement, on the amount of taxable income, and on such business decisions as pricing goods.

#### *Example 1.2*

Assume that 100 units priced at $10 each are in inventory at the beginning of the year, that during the year 200 units are purchased at $11 each and another 200 units are purchased at $12 each, and that 100 units remain in inventory at the end of the year. What are the amount of the ending inventory and the cost of the 400 units that were sold?

| | Units | Unit Cost | Total Cost |
|---|---|---|---|
| Beginning inventory | 100 | $10 | $1,000 |
| First purchase | 200 | 11 | 2,200 |
| Second purchase | 200 | 12 | 2,400 |
| Goods available for sale | 500 | $11.20 | $5,600 |
| Goods sold during the year | 400 | | |
| Ending inventory | 100 | | |

Under FIFO, the 100 units in inventory at the end of the year are priced at $12 each, or $1,200, and the cost of goods sold is $10(100) + $11(200) + $12(100) = $4,400.

Under LIFO, the 100 units of ending inventory are priced at $10 each, or $1,000, and the cost of goods sold is $12(200) + $11(200) = $4,600.

An *average* cost method can be another acceptable practice. Under this pricing method, the average cost of the 500 units available for sale is [$10(100) + $11(200) + $12(200)]/500 = $11.20. The 100 units in inventory are priced at $11.20 each, or $1,120, and the cost of goods sold is $11.20(400) = $4,480. □

### 1.3.5 Depreciation

With the exception of land, most fixed assets have a limited useful life; that is, they will be of use to the company over a limited number of future accounting periods. A fraction of the cost of an asset is properly chargeable as an expense in each of the accounting periods in which the asset is used by the company. The accounting process for this gradual conversion of fixed assets into expense is called *depreciation*. There are several ways to depreciate an asset, the straight-line method (charging an equal fraction of the cost each year) being the simplest. (The depreciation methods will be introduced in Chapter 4.) Obviously, the depreciation method chosen for financial reporting purposes will affect both the balance sheet and the income statement.

In the case of Sunset, we interpret the data as follows: the company has invested $154,000 in plant and equipment as of the reporting date; the portion of the total investment charged to operations for all periods to date is $70,000, and the portion of the total investment that remains to be charged to operations of future periods is $84,000. Therefore, the amount shown as accumulated depreciation on the balance sheet does *not* represent the accumulation of any cash; it is merely that portion of the assets' original cost that has already been charged against income.

### 1.3.6 Working Capital

In examining the balance sheet, the analyst usually makes a basic distinction between the working capital items and the other items. *Working capital* is defined as current assets minus current liabilities.

| Current assets − Current liabilities = Working capital |
|---|

Therefore, working capital is often considered to be funds available for future operations, because it is the excess left after satisfying all current obligations.

As of December 31, 19x1, Sunset's working capital is

$$\$76,400 - \$28,000 = \$48,400$$

As mentioned, current assets are normally converted into cash within a year or a single operating cycle, and current liabilities are payable within a year. Thus, items in the current section correspond to reasonably recent transactions and are more likely to reflect current market conditions than items in the other sections, which may correspond to transactions that occurred many years previously. In general, the company's financial condition is stronger if working capital increases and weaker if it decreases.

## 1.4 THE INCOME STATEMENT

The balance sheet just discussed shows the financial condition of the company on the *reporting date* only and does not indicate whether the company is making or losing money. The income statement (or profit-or-loss statement) is an important financial statement that summarizes the revenue items, the expense items, and the difference between them (net income) for an accounting period.

| Revenue − Expenses = Net income (or loss) |
|---|

### 1.4.1 Methods of Reporting Income

An accounting method is a set of rules that we use to decide when and how to record revenue and expenses in our book and how to prepare the income statement for our accounting period. The two principal methods of accounting are the cash basis method and the accrual basis method.

***Cash Basis Method.*** If we record on a cash basis, we include as gross revenue all revenue that we received during the year in cash or its equivalent (e.g., savings interest). We deduct all expenses made during the year in cash or its equivalent, if deduction of such expenses is allowed by law.

***Accrual Basis Method.*** If we report revenue when it is earned, whether it is actually received or not, and we report expenses when they are incurred, whether they are paid or not, we are using the accrual method of accounting.

The tax law does not prescribe a uniform method of accounting to be used by all companies. It expects companies to adopt the forms and methods of accounting suitable for them. If it is necessary to include an inventory, however, *the accrual method must be used for purchases and sales* by the tax law. Therefore, most manufacturing and merchandising businesses use the accrual basis in recording revenues and expenses.

### 1.4.2 Reporting Format

As shown in Table 1.2, the income statement presents the details of this procedure according to generally accepted accounting principles. Although many variations exist in preparing an income statement, a common income statement might include any of the following [1].

1. *Revenues,* the gross sales for the period. Gross sales are the total invoice price of the goods shipped (or services rendered) plus the cash sales made during the period. Sales returns and allowances represent the sales value

**Table 1.2** ***The Income Statement***

**SUNSET INC.**
INCOME STATEMENT AND RETAINED EARNINGS
*(For Year Ended December 31, 19x1)*

| | | |
|---|---|---|
| ***Income statement*** | | |
| Net sales | | |
| Sales and other operating revenue | | $303,000 |
| Less sales return and allowances | | (3,000) |
| | | 300,000 |
| Cost of goods sold | | |
| Labor | 120,000 | |
| Materials | 60,000 | |
| Overhead | 8,000 | |
| Depreciation | 20,000 | |
| Total | | (208,000) |
| Gross profit | | 92,000 |
| Operating expenses | | |
| Selling | 15,720 | |
| General administration | 29,000 | |
| Lease payments | 14,000 | |
| Total | 58,720 | (58,720) |
| Net operating profit | | 33,280 |
| Nonoperating revenues | | 0 |
| Nonoperating expenses | | |
| Interest payments | | (5,200) |
| Net income before taxes | | 28,080 |
| Income taxes (30%) | | (8,424) |
| Net income | | $19,656 |
| ***Statement of retained earnings*** | | |
| Cash dividends | | |
| Preferred stock (per share, $6) | | 600 |
| Common stock (per share, $.95) | | 9,456 |
| Total dividends | | $10,056 |
| Retained earnings | | |
| Beginning of year (1/1/19x1) | | 32,800 |
| Current year | | 9,600 |
| End of year | | $42,400 |
| Earnings per share of common stock | | |
| Net applicable income, (19,656 − 600)/10,000 | | $1.91 |

of goods that were returned by the customer or for which the customer was given a credit for any reason. The amount could have been subtracted from the gross sales figures directly.

2. *Cost of sales,* the manufacturing or purchase costs of the goods sold during the period. These include direct labor, material, depreciation, and other manufacturing expenses.
3. *Gross profit* or *margin,* the difference between revenues and cost of sales; item 1 less item 2.
4. *Operating expenses,* the selling, administration, and general expenses associated with operating the company's principal business activity during the period.
5. *Net operating profit,* item 3 less item 4.
6. *Nonoperating revenues,* revenue derived from sources other than operations during the period, such as interest on the temporary investment of excess cash.
7. *Nonoperating expenses,* expenses not directly related to the principal business activity and the financial costs of borrowed money.
8. *Income taxes,* the income tax expense, based on item 5 + item 6 − item 7.
9. *Net income,* item 5 + item 6 − item 7 − item 8.

As a part of the income statement, many companies report changes in the common equity accounts between balance sheet dates under the heading of the *statement of retained earnings.* (This could be a separate statement.) If this optional format is followed, the additional entries to the income statement would be

10. *Retained earnings* at the beginning of the year.
11. *Cash dividends* on common stock.
12. *Retained earnings* at the end of the year, item 10 less item 11.
13. *Earnings per share (EPS),* the net income after tax divided by the number of outstanding shares of common stock.
14. *Dividends per share (DPS),* the amount of dividends paid to holders of common stock divided by the number of outstanding shares of common stock.

We will address many measurement issues related to accounting income by using Sunset's income statement in Table 1.2.

### 1.4.3 Measurement of Revenue

Revenues are measured by using the accrual concept; that is, revenues accrue in the accounting period in which they are earned, which is not necessarily the same as the period in which the cash is received. In accounting, a careful distinction is made between revenues and cash receipts, as illustrated in the following example.

### *Example 1.3*

The following tabulation shows various types of sales transactions and classifies the effect of each on cash receipts and sales revenue for "this year."

| Transaction | Amount | "This Year" Cash Receipts | Sales Revenue |
|---|---|---|---|
| Cash sales made this year | $500 | $500 | $500 |
| Credit sales made last year but cash received this year | 300 | 300 | 0 |
| Credit sales made this year and cash received this year | 100 | 100 | 100 |
| Credit sales made this year but cash received next year | 200 | 0 | 200 |
| Total | | $900 | $800 |

Note that in this illustration the total cash receipts do not equal the total sales revenue for the period. The totals would be equal in a particular accounting period only if (1) the company made all its sales for cash or (2) the amount of cash collected from credit customers in an accounting period happened to equal the amount of credit sales made during that period. □

### 1.4.4 Measurement of Expenses

A similar distinction is made between expenditures and expenses. An *expenditure* takes place when an asset or service is acquired. The expenditure may be by cash, by another asset, or by incurring a liability. Over the entire life of a business, most expenditures made by the business become *expenses,* and there are no expenses that are not represented by an expenditure. As we shall see, however, there is no necessary correspondence between expense and expenditure in any accounting period. Four types of items must be considered in distinguishing between the amounts that are properly considered to be expenses of an accounting period and the expenditures [1].

1. Expenditures this year that are also expenses of this year.
2. Expenditures made prior to this year that become expenses during this year. These appear as assets on the balance sheet at the beginning of this year. First, there are inventories of products; these become expenses when the products are sold. Second, there are prepaid expenses and deferred charges. These represent services or other assets purchased prior to this year but not yet used when the year begins. They become expenses in the year in which the services are used or the assets are consumed (e.g., insurance protection, prepaid rents, and prepaid taxes). Third, there are long-lived fixed assets (except land) that have a limited useful life; that is, they do not last forever. They are purchased with the expectation that they will be used in the operation of the business in future periods, and they will become expenses in these future periods (e.g., depreciation expenses).

3. Expenditures made this year that will become expenses in future years. These appear as prepaid expenses and become assets on the balance sheet at the end of this year.
4. Expenses of this year that will be paid for in a future year. On the balance sheet at the end of this year, these will appear as liabilities (e.g., deferred charges).

### *Example 1.4*

In 19x1, $10,000 of diesel fuel for a lift truck was purchased for cash. This was an expenditure of $10,000, the exchange of one asset for another. If only $8,000 worth of diesel fuel was consumed in 19x1, your book entry for 19x1 should record an expense of $8,000 rather than $10,000. □

### *Example 1.5*

A company purchased three years of insurance protection worth $2,400 on December 31, 19x1. The $2,400 appears as an asset on the balance sheet of December 31, 19x1 as a prepaid expense. In 19x2, $800 becomes an expense and $1,600 remains as an asset on the balance sheet of December 31, 19x2. In 19x3, $800 more becomes an expense, and so forth. □

### 1.4.5 Retained Earnings, Cash Dividends, and Earnings per Share

According to Table 1.2, Sunset Inc. earned $19,656 during 19x1, paid out $10,056 in dividends, and put $9,600 back into the business. Thus, the balance sheet item "retained earnings" increased from $32,800 at the end of 19x0 to $42,400 at the end of 19x1.

Note that the retained earnings account on the balance sheet represents a claim against assets. Further, companies retain earnings primarily to expand the business; this means investing in the plant and equipment, in inventories, and so on, not in a bank account. Thus, retained earnings as reported on the balance sheet do *not* represent the amount of cash that the company has and are not "available" for the payment of dividends or anything else.

In analyzing the income statement, investors pay particular attention to the ratio called *earnings per share* (*EPS*). This is computed by dividing *net income applicable* to the *common stock* by the number of shares of common stock outstanding. If the company has only common stock, the net income used in this ratio is the same as the net income shown on the income statement. If the company has senior securities such as preferred stocks, those that have a claim on net income ahead of the claim of holders of common stock, the income figure used in the calculation of EPS is the amount that remains after the claims of the senior securities have been deducted from net income.

### *Example 1.6*

Sunset Inc. in 19x1 had a net income of $19,656. On January 1, it had 100 shares of $6 preferred stock outstanding (which means that each share of this stock is entitled to an annual dividend of \$6 before any dividends can be paid to holders of common stock) and 10,000 shares of common stock. The *EPS* were (\$19,656 − \$600)/10,000 = \$1.91 per share. □

## 1.5 THE FUNDS FLOW STATEMENT

As a company like Sunset Inc. goes about its business, it makes sales, which lead to a reduction of inventories, to an increase in cash, and, if the sales price exceeds the cost of the item sold, to a profit. These business transactions cause the balance sheet to change, and they are also reflected in the income statement. The *funds flow statement* summarizes these business transactions of the period

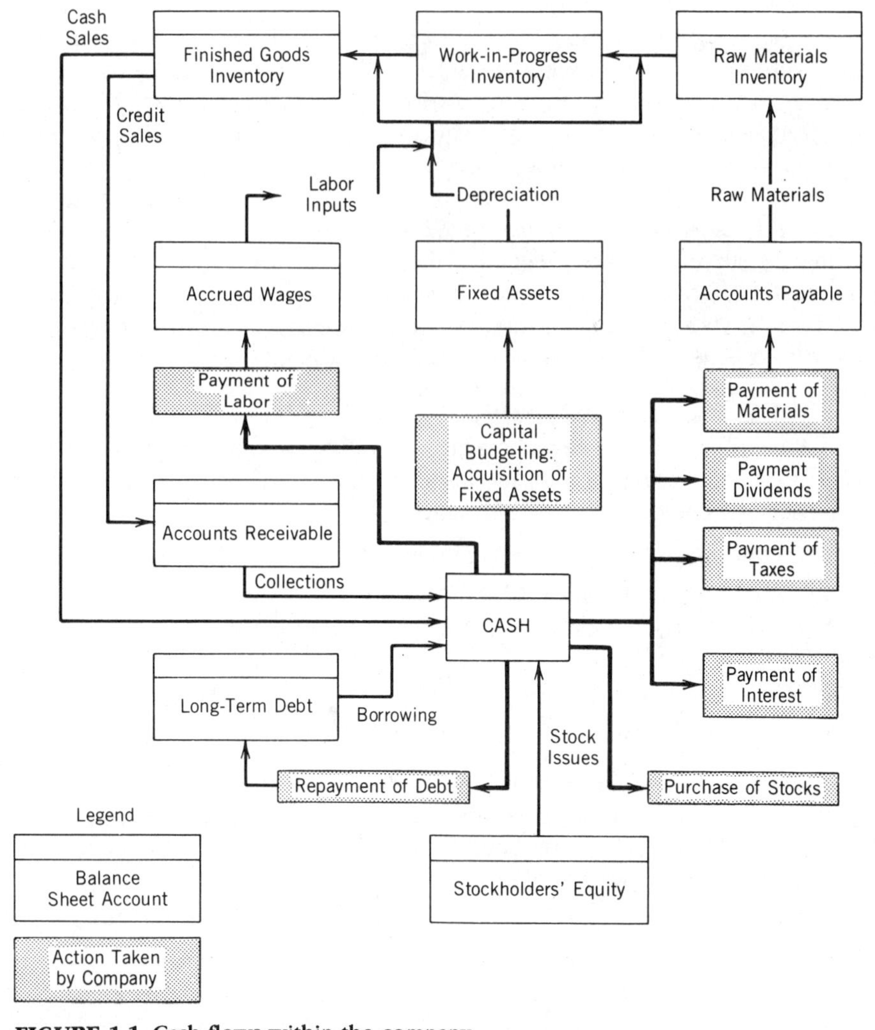


**FIGURE 1.1** Cash flows within the company.

from a different view; it details how the company acquired funds (sources) and how the funds were used (uses).

### 1.5.1 The Cash Flow Cycle

To understand how business transactions influence the financial statement, we need to understand the cash flow cycle within a company. As shown in Figure 1.1, the flow of funds in a typical company may be visualized as a continuous process. For the going concern, there are really no starting and ending points. For every use of funds, however, there must be an offsetting source. In Figure 1.1, rectangles represent balance sheet accounts and shaded boxes represent actions taken by the company.

A finished product is produced with a variety of inputs—raw material, net fixed assets (plant and equipment), and labor. These inputs are ultimately paid for in cash. The product is then sold either for cash or on credit, thus reducing the inventory. A credit sale creates an account receivable, which, when collected, becomes cash. The cash account, the focal point in Figure 1.1, fluctuates over time with the production schedule, sales, collection of receivables, capital expenditures, and financing. Current asset accounts such as the finished goods inventory and receivables fluctuate with sales, production schedules, and management policies for receivables and inventories.

Funds may be defined in several different ways depending on the purpose of the analysis. Although they are sometimes defined as *cash,* funds are treated by many analysts as *working capital*—the most common approach by far. The graphic representation of funds flow in Figure 1.1, when converted into numerical form, is shown in annual reports as the *statement of changes in financial position,* often called *the sources and uses of funds statement.*

### 1.5.2 Basic Relationship

We will first look at how a management decision might affect the balance sheet accounts. As an example, suppose a company wished to buy a new plant. The company could do one of five things.

1. It could borrow, thus increasing long-term debt.
2. It could sell more stock, thus increasing the capital stock.
3. It could forgo dividends, which would show up as an increase in retained earnings.
4. It could use available cash, thus decreasing working capital.
5. It could sell some of its existing fixed assets, thus decreasing fixed assets.

The first three possibilities would be reflected on the balance sheet as an increase in equities; the last two would show up as a decrease in assets.

Looking at the other side of the coin, what uses could the company make of funds that it acquired? It could use the funds in one of five ways.

1. It could add new fixed assets by buying equipment.
2. It could add to working capital.
3. It could pay off existing debt.

4. It could pay additional dividends to the stockholders, which decreases retained earnings.
5. It could repay capital to the stockholders.

The first three possibilities would show up as an increase in assets and the fourth and fifth as a decrease in equities. We can therefore generalize as follows.

| Sources of funds are indicated by increases in equities and decreases in assets. Uses of funds are indicated by increases in assets and decreases in equities. |
|---|

With the dual-aspect accounting concept, total sources of funds must equal total uses of funds. The following relationships therefore exist.

| Sources | | Uses |
|---|---|---|
| Increase in equities + Decreases in assets | = | Increase in assets + Decreases in equities |

### 1.5.3 Funds Statement on a Cash Basis

The starting point in preparing a statement of changes in financial position on a *cash basis* is to determine the change in each balance sheet item and to record it as either a source or a use of funds under the rules previously defined. Table 1.3 shows the changes that occurred in Sunset's balance sheet accounts during the calender year 19x1, with each change designated as a source or a use.

Sources of funds that increase cash are

1. A net decrease in any asset other than cash or fixed assets.
2. A gross decrease in fixed assets.
3. A net increase in any liability.
4. Proceeds from the sale of stock.
5. Funds provided by operations.

In referring to Table 1.3, we will look at how Sunset Inc. obtained and used the funds for the year 19x1. First, from the income statement in Table 1.2, we can determine the amount of funds derived from operation. Because depreciation and deferred income taxes (if any) are not cash outflow from the company, we add these two items to net income to obtain the amount of funds derived from operation.

$$\begin{aligned}\text{Funds provided from operation} &= \text{Net income} + \text{Depreciation} \\ &\quad + \text{Deferred taxes} \\ &= \$19{,}656 + \$20{,}000 + \$0 \\ &= \$39{,}656\end{aligned}$$

**Table 1.3** ***Funds Statement Based on Cash***

| SUNSET INC.<br>CASH FLOW STATEMENT<br>*(December 31, 19x0 to December 31, 19x1)* | | |
|---|---|---|
| Sources of cash | | |
| Cash provided from operations | | |
| Net income after tax | | $19,656 |
| Add noncash expense: Depreciation | | 20,000 |
| Total cash from operation | | 39,656 |
| Adjustments to convert to cash basis | | |
| Decrease in marketable securities | 2,000 | |
| Decrease in inventories | 2,100 | |
| Decrease in deferred charges | 500 | |
| Increase in wages payable | 1,000 | |
| Total adjustments | | 5,600 |
| Total sources of cash | | $45,256 |
| Uses of cash | | |
| Cash dividends | | 10,056 |
| Acquisition of equipment | | 9,000 |
| Reduction of long-term debt | | 2,000 |
| Adjustments to convert to cash basis | | |
| Increase in accounts receivable | 4,200 | |
| Increase in prepaid expenses | 500 | |
| Decrease in accounts payable | 16,000 | |
| Decreases in accrued taxes | 1,500 | |
| Total adjustments for expenses | | 22,200 |
| Total uses of cash | | $43,256 |
| Net increase in cash flow | | $2,000 |

Note that depreciation is not a source of funds. It is a noncash expense that simply reduces the company's tax obligation.

Since we know that this income statement revenue is based on the accrued accounting method, it is further adjusted to the amount of sales-related cash inflows. For example, we examine the $2,100 inventories decrease in Table 1.3, which is treated as a source of cash. By definition, the amount of ending inventories is calculated by

$$\text{Ending Inventories} = \text{Beginning inventories} + \text{New purchases} - \text{Units used}$$

Here, the ending inventories being smaller than the beginning inventories would mean that Sunset used old inventories (in addition to new purchases) to produce the goods during the year. Since these old inventories were purchased in the previous year, the actual inventory cash expenses during the year should be less than the inventory (material) expenses reported in the income statement. Similarly, we may treat the changes in the remaining accounts (such as marketable securities, deferred charges, and wages payable) as sources of cash.

Uses of funds include

1. A net increase in any asset other than cash or fixed assets.
2. A gross increase in fixed assets.
3. A net decrease in any liability.
4. A repurchase of stock.
5. Cash dividends.

To compute gross changes in fixed assets, we add depreciation for the period to net fixed assets at the ending financial statement date, and subtract from this amount the net fixed assets at the beginning financial statement date. For the situation of Sunset Inc., we compute the gross increase in fixed assets as follows.

$$\text{Gross increase in fixed assets} = 84{,}000 + 20{,}000 - 95{,}000$$
$$= \$9{,}000$$

First, the $4,200 accounts receivable increase during the year means that the year's revenues exceeded collections. Therefore, this amount has the same effect as the use of cash, as reflected in Table 1.3. If accounts receivable had decreased during the year, this would mean that the amount of actual cash collection had exceeded the sales revenue for the period, and, thus, the amount of the decrease should therefore be added to sales revenue (or treated as source of cash). The other remaining uses of funds are summarized in Table 1.3. When we subtract the total uses of funds other than cash in Table 1.3 from the total sources, the difference should equal the actual change in cash between the two balance sheet statements.

### 1.5.4 Funds Statement as Working Capital

The *working-capital-based* funds statement in Table 1.4 is essentially the same as the cash-based funds statement. The only difference is that net changes in current assets and liabilities are grouped into one account, working capital. The basic elements listed in the funds statement are

| Sources | | Uses |
|---|---|---|
| Funds provided by operation<br>Proceeds from sale of fixed assets<br>Additional long-term debt | = | New plant and equipment<br>Cash dividends<br>Increase in working capital |

**Table 1.4** ***Cash Flow Statement Showing Working Capital Position***

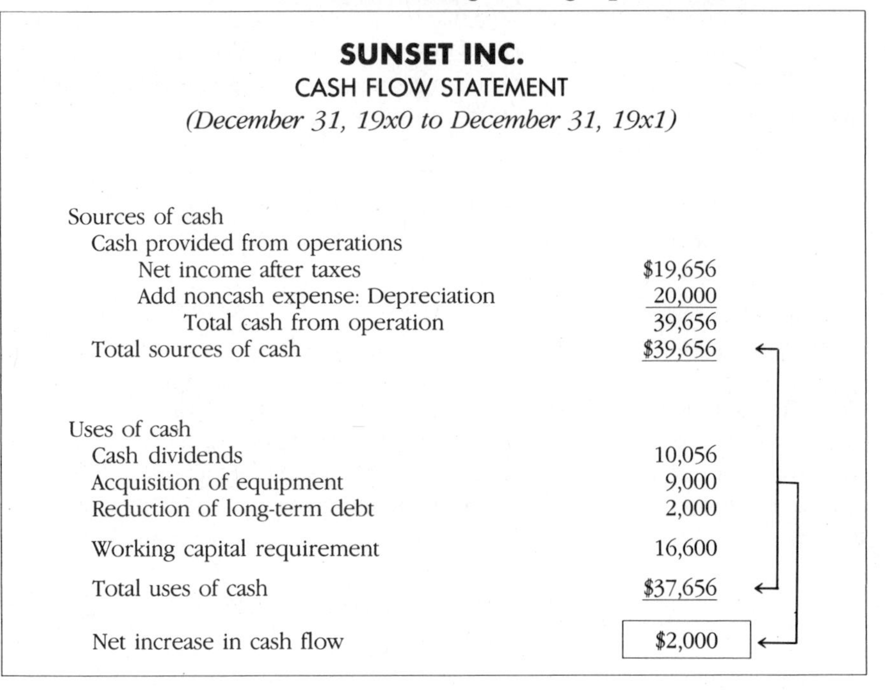

**SUNSET INC.**
CASH FLOW STATEMENT
*(December 31, 19x0 to December 31, 19x1)*

| | |
|---|---|
| Sources of cash | |
| Cash provided from operations | |
| Net income after taxes | $19,656 |
| Add noncash expense: Depreciation | 20,000 |
| Total cash from operation | 39,656 |
| Total sources of cash | $39,656 |
| Uses of cash | |
| Cash dividends | 10,056 |
| Acquisition of equipment | 9,000 |
| Reduction of long-term debt | 2,000 |
| Working capital requirement | 16,600 |
| Total uses of cash | $37,656 |
| Net increase in cash flow | $2,000 |

To illustrate, we consider the adjustment entries to convert to cash basis in Table 1.3. We note that these adjustment entries are all related to either current assets or current liabilities accounts. By definition, working capital is the difference between current assets and current liabilities. This implies that we can determine the net change in working capital requirement by

$$\text{Working capital requirement} = \text{Change in current asset} - \text{Change in current liabilities}$$

Using either Table 1.1 or Table 1.3, we first list all entries related to the current assets (except cash) and the current liabilities by indicating an increase by "+" sign and a decrease by "−" sign, respectively.

| | | | |
|---|---|---|---|
| Marketable securities | −2,000 | Wages payable | + 1,000 |
| Inventories | −2,100 | Accounts payable | −16,000 |
| Deferred charges | − 500 | Accrued taxes | − 1,500 |
| Accounts receivable | +4,200 | | |
| Prepaid expense | + 500 | | |
| Net change in current assets: | + 100 | Net change in current liabilities: | −16,500 |

The net change in working capital is then calculated as

$$\begin{aligned}\text{Net change in working capital} &= 100 - (-16{,}500)\\ &= \$16{,}600\end{aligned}$$

With this net change being positive, Sunset has a net *requirement* of working capital in the amount of $16,600 that has to be expended during the year. In this situation the working capital requirement appears as uses of cash in the cash flow statement. If this amount were negative, there would have been a cash inflow from working capital *release,* which could add to the sources of cash.

## 1.6 TAX ACCOUNTING AND BUSINESS ACCOUNTING

Up to this point, we have discussed how business transactions affect the company's financial statement. Now we will look in detail at how the company's net income (accounting income) can be affected by the choice of depreciation method used in reporting income to stockholders. We will also address the question of whether the use of cash flows or net income is more appropriate in investment analysis.

### 1.6.1 Deferred Income Taxes

Many companies use one depreciation method for tax purposes and another for financial reporting purposes. It is common practice to use an accelerated depreciation method to compute the actual tax liability and a straight-line method for financial reporting. (Different depreciation methods do not change the company's overall tax obligation; they simply cause the annual tax obligation to vary.) The result is that the income tax liability reported on an income statement can be different from the taxes actually paid. The difference between the computed taxes and the actual tax liability is called *deferred taxes.* Thus, the balance sheet item "deferred taxes" represents the cumulative taxes the company has deferred over the years, and it appears in the liability section of the balance sheet. Sunset uses the same depreciation method for calculating its tax liability and for computing net income to report to its stockholders, so it has no such balance sheet entry.

### 1.6.2. Computing Deferred Income Taxes

As we have discussed, a company's cash inflows (sources of funds) are *approximately* equal to its net income plus depreciation plus any taxes it has deferred, and all three of these quantities depend on the depreciation methods the company uses for tax and reporting purposes. Therefore, we will first look at how the depreciation method can affect the reported accounting income.

The current tax law allows companies to use either the prescribed rapid depreciation method, commonly referred to as the modified accelerated cost recovery system (MACRS), or the optional straight-line method when depreciating their assets for tax purposes. Most companies use the rapid method when calculating their taxes because depreciation is a tax-deductible expense and a larger depreciation lowers the current tax liability. However, larger depreciation expenses also reduce a company's net income (accounting income).

A generally accepted accounting principle states that a company should depreciate its assets for reporting purposes by using the method that most accurately reflects the decline in the value of its assets over time. For most companies, this method is straight-line depreciation. A few companies use rapid depreciation for both tax and reporting purposes, but it is much more common to use rapid depreciation for calculating taxes and straight-line depreciation for reporting income to investors.

### *Example 1.7*

To illustrate the concept, consider Table 1.5, which shows before-tax income, taxes, and net income for a company that has sales of $200,000 a year, costs equal to 60% of sales, and a single asset that costs $100,000. This asset has a 5-year useful life, and it will have a salvage value of zero at the end of the 5 years. If the company uses a 3-year property class rate under rapid depreciation for tax purposes and the straight-line method for financial reporting purpose, the annual depreciation allowed each year would be

| | Year End | | | | |
|---|---|---|---|---|---|
| **Item** | 1 | 2 | 3 | 4 | 5 |
| MACRS rate* | 33% | 45% | 15% | 7% | 0 |
| MACRS depreciation | 33,000 | 45,000 | 15,000 | 7,000 | |
| Straight-line (SL) rate | 20% | 20% | 20% | 20% | 20% |
| SL depreciation | 20,000 | 20,000 | 20,000 | 20,000 | 20,000 |

*Rounded to the nearest full percentage.

In Table 1.5 the upper section shows the calculation of actual taxes owed by the MACRS method, and the lower section shows the calculation of taxes and net income for reporting purposes by the straight-line method. Notice that after-tax (A/T) incomes are larger in the early years under the straight-line method. For example, using MACRS, in year 1 the company reports to the Internal Revenue Service (IRS) $33,000 in depreciation, pays $86,800 in income taxes, and has a net income of $130,200. It reports to stockholders only $20,000 depreciation, $92,000 taxes, and $138,000 net income. In general, the company would prefer to report to its stockholders an income of $138,000 rather than $130,200.

Although the company reports taxes of $92,000, its tax bill is only $86,800. The company does not actually pay the difference of $5,200 in reported taxes in year 1. The $5,200 difference is reported on the income statement and the balance sheet as deferred taxes; that is, the company has deferred payment of these taxes until a later date by using an accelerated depreciation method for calculating them.

**Table 1.5** ***Effects of Depreciation on Accounting Income***

| | Operating Year | | | | |
|---|---|---|---|---|---|
| Item | 1 | 2 | 3 | 4 | 5 |
| *Case 1: Tax calculations (thousands of dollars)* | | | | | |
| Revenues | $500 | $500 | $500 | $500 | $500 |
| Less | | | | | |
| Labor and materials | 200 | 200 | 200 | 200 | 200 |
| Depreciation (MACRS)* | 33 | 45 | 15 | 7 | 0 |
| Operating expenses | 50 | 50 | 50 | 50 | 50 |
| Net income B/T | 217 | 205 | 235 | 243 | 250 |
| Less | | | | | |
| Income taxes (40%) | 86.8 | 82 | 94 | 97.2 | 100 |
| Net income A/T | $130.2 | $123.0 | $141 | $145.8 | $150 |
| Earnings per share | 1.30 | 1.23 | 1.41 | 1.46 | 1.50 |
| *Case 2: Calculations for stockholder reporting (thousands of dollars)* | | | | | |
| Revenues | $500 | $500 | $500 | $500 | $500 |
| Less | | | | | |
| Labor and materials | 200 | 200 | 200 | 200 | 200 |
| Depreciation (SL) | 20 | 20 | 20 | 20 | 20 |
| Operating expenses | 50 | 50 | 50 | 50 | 50 |
| Net income B/T | 230 | 230 | 230 | 230 | 230 |
| Less | | | | | |
| Income taxes (40%) | 92 | 92 | 92 | 92 | 92 |
| Net income A/T | $138 | $138 | $138 | $138 | $138 |
| Earning per share | 1.38 | 1.38 | 1.38 | 1.38 | 1.38 |
| Deferred income taxes | 5.2 | 10.0 | −2.0 | −5.2 | −8 |

*Rounded to the nearest full percentage. (For the 3-year MACRS property, the applicable percentages are 33.33%, 44.44%, 14.82%, and 7.41%).

NOTE: B/T = before tax; A/T = after tax.

Notice also that when MACRS depreciation is used, the net income fluctuates for several years and then is stable at a level above the net income reported to stockholders. Income reported to stockholders under straight-line depreciation is stable over the entire period. Companies prefer stable or growing earnings to fluctuating earnings. □

### 1.6.3 Estimating Cash Flows from Income Statement

The cash flows that accrue to the company are calculated in Table 1.6 on the basis of the income statement and balance sheet (with no working capital requirement). Notice that cash flows, which are equal to net income plus depreciation plus deferred taxes, are the same regardless of the depreciation method used. This will always be so because net income increases by (1 − tax

**Table 1.6** *Effects of Depreciation on Cash Flows*

| | Operating Year | | | | |
|---|---|---|---|---|---|
| Item | 1 | 2 | 3 | 4 | 5 |
| *Case 1: Tax calculations (thousands of dollars)* | | | | | |
| Net income | $130.2 | $123 | $141 | $145.8 | $150 |
| Depreciation (MACRS) | 33 | 45 | 15 | 7 | 0 |
| Cash flow | $163.2 | $168 | $156 | $152.8 | $150 |
| *Case 2: Calculations for stockholder reporting (thousands of dollars)* | | | | | |
| Net income | $138 | $138 | $138 | $138 | $138 |
| Depreciation (SL) | 20 | 20 | 20 | 20 | 20 |
| Deferred income taxes | 5.2 | 10 | −2 | −5.2 | −8 |
| Cash flow* | $162.2 | $168 | $156 | $152.8 | $150 |

*Cash flow = Net income + Depreciation + Deferred taxes.

rate) (net change in depreciation amount) and deferred taxes increase by (tax rate) (net change in depreciation amount), which together exactly offset the decrease in depreciation cash flows.

### 1.6.4 Use of Cash Flows in Evaluating Investments

Managers and financial analysts are primarily concerned with the stream of cash flows that accrue to a company from its operations. As we have seen, net income (accounting income) and cash flows are rarely, if ever, the same. Changes in income can occur without any corresponding changes in cash flows. During a period of investment in plant and inventories, a company can even experience a decrease in cash at the same time that income is increasing.

When the company makes sales on credit, the sales appear as accounting income. However, the company has not yet received the cash, it cannot spend the cash, and the ultimate collection of the cash is uncertain. For the purposes of investment analysis, we are more interested in the point when the cash is to be received. At this time the company reaches a new decision point. The cash may be returned to the stockholders by payment of a dividend, or it may be used to retire debt, to increase the working capital, or to acquire new assets.

We use the cash flows of an investment in evaluating its economic value. One advantage of using the cash flow approach is that it avoids difficult problems in the measurement of corporate income that necessarily accompany the accrual method of accounting. These problems include the following.

1. In what time period should revenue be recognized?
2. What expenses should be treated as investments and therefore capitalized and depreciated over several time periods?

3. What method of depreciation should be used in measuring income as reported to management and shareholders?
4. Should LIFO, FIFO, or some other method be used to measure inventory flow?

There are disagreements about the answers to these questions. Different approaches may lead to different measures of income. If income is used to evaluate investment worth, investments may look good or bad, depending on how income is measured. The use of cash flows for investment evaluation minimizes many of these complications.

## 1.7 INVESTMENT PROJECT AND ITS CASH FLOWS

Projects are usually started with an investment, which is a cash flow from the company to the project. Cash outflows that move from the company to the project are arbitrarily given a negative sign. Thus, the net cash flow −$1,000 indicates that $1,000 flows from the company to the project at a particular point in time. A project may require several years of investment before it starts to generate any revenue. When the project is in full operation, periodic cash inflow from the project to the company is assumed to be generated by product sales. Even though the cash flows may actually take place throughout the year, in cash flow analysis each flow is usually assumed to occur at the end of the year. In this section, we will look at how we can estimate a cash flow series for a typical investment project.

### 1.7.1 The Cash Flow Statement

Suppose that the company could introduce a new power regulator, Model X, with the expected cash flow for the year shown in Table 1.7. To do this, the company has to make an investment of $35,000 at the end of 19x0. The statement in the table indicates that the project is expected to generate $54,000 in net funds from its first year of operation. The disposal of an old machine, because new equipment has been purchased, adds $3,000, and proceeds from a new short-term loan add another $10,000. This short-term loan is secured at the beginning of the year at an interest rate of 10%, and both interest and principal are payable at the end of the year. Therefore, the total cash inflow to the company during the year is $77,000.

In the same period, $10,000 of this net cash inflow is a cash outflow to repay the principal of the loan, $30,000 is a cash outflow for other new equipment purchased, $12,000 is retained in the project as an increase in its working capital, and the balance of $25,000 is the net cash return from the project to Sunset. Thus, the net cash inflow to the company from the power regulator model X project would be $25,000 for 19x1. Figure 1.2 illustrates the general nature of composition of cash flow generation and applications.

**Table 1.7** ***Procedure for Computing the Project Cash Flow***

| *Net Operating Income Statement* | | |
|---|---|---|
| Operating revenues (sales) | | $200,000 |
| Less | | |
| Cost of goods sold | | |
| Labor | 40,000 | |
| Materials | 30,000 | |
| Overhead | 20,000 | |
| Depreciation | 10,000 | |
| Total | $100,000 | |
| Operating cost | 10,000 | |
| Interest paid on debt | 1,000 | (111,000) |
| Taxable income | | 89,000 |
| Income tax | | 35,000 |
| Net income from operations | | 54,000 |

| *Project Cash Flow Statement—Power Regulator Model-X for the Year Ending December 31, 19x1* | |
|---|---|
| Sources of cash | |
| Net income from operations | 54,000 |
| Add: Noncash expense depreciation | 10,000 |
| Net cash provided from operations | 64,000 |
| Add | |
| Sale of old equipment replaced by new equipment (net proceeds after tax adjustment) | 3,000 |
| New borrowed money for the project | 10,000 |
| Total cash generated by the project | $77,000 |
| Uses of cash | |
| Repayment of principal | 10,000 |
| Purchase of new equipment | 30,000 |
| Working capital requirement for the project during the period | 12,000 |
| Total cash disbursed for the project | $52,000 |
| Net cash inflow to the company | $25,000 |

## *Example 1.8*

A chemical processing firm is planning to add a duplicate polyethylene plant at another of its locations. The financial information for the first project year is provided as follows.

| | | |
|---|---|---|
| Sales | | $1,500,000 |
| Manufacturing cost | | |
| Direct material | $150,000 | |
| Direct labor | 200,00 | |
| Overhead | 100,000 | |
| Depreciation | 200,000 | |

| | |
|---|---:|
| Interest payment | 20,000 |
| Operating Expenses | 150,000 |
| Taxable income | 680,000 |
| Income taxes | 272,000 |
| Net income after tax | 408,000 |
| Equipment purchase | 400,000 |
| Increase in inventories | 100,000 |
| Decrease in accounts receivable | 20,000 |
| Increase in wages payable | 30,000 |
| Decrease in notes payable | 40,000 |

Compute the net cash flow from this project during the first year.

- The first step is to calculate the net change in working capital, using (+) for increase and (−) for decrease.

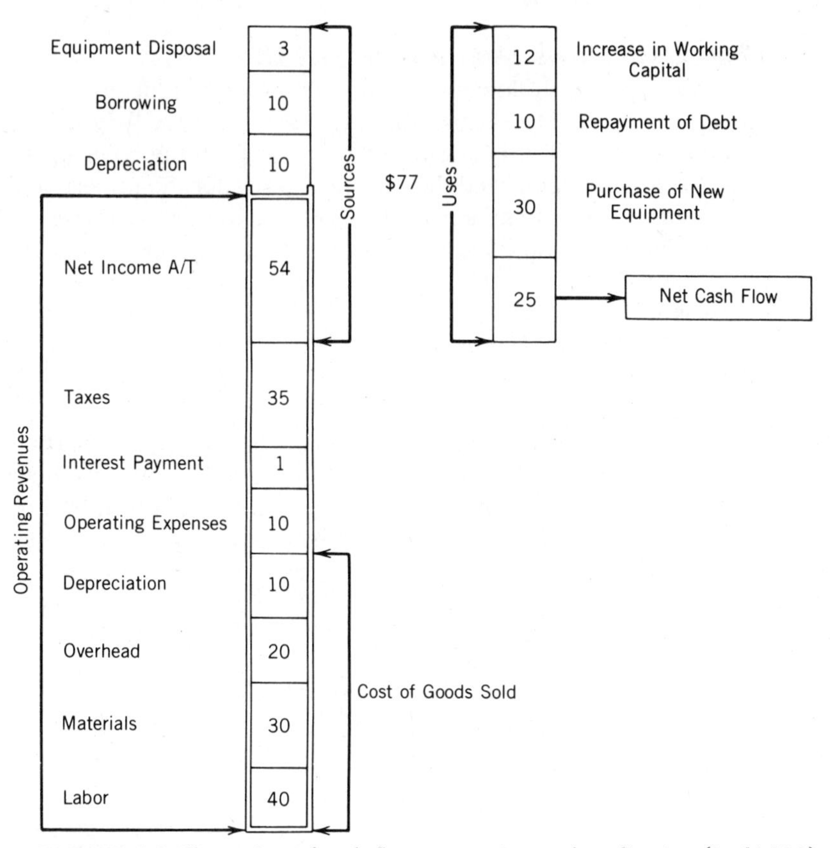


**FIGURE 1.2** Illustration of cash flow generation and application (in $1,000).

| CURRENT ASSETS | | CURRENT LIABILITIES | |
|---|---|---|---|
| Inventories | +\$100,000 | Wage payable | +\$30,000 |
| Accounts receivable | −20,000 | Note payable | −40,000 |
| Net change | +80,000 | Net change | −10,000 |

$$\text{Net change in working capital} = \$80{,}000 - (-10{,}000) = \$90{,}000$$

- The second step is to calculate the total funds provided from operation.

$$\text{Funds provided from operation} = \$408{,}000 + \$20{,}000 = \$428{,}000$$

Since there are neither equipment sales nor borrowing, the total cash generated is \$428,000.

- The third step is to calculate the total uses of cash.

$$\text{Uses of cash} = \text{Equipment purchase} + \text{Net increase in working capital}$$
$$= 400{,}000 + 90{,}000 = \$490{,}000$$

- The fourth step is to determine the net cash flow for the project year.

$$\text{Net cash flow} = 428{,}000 - 490{,}000 = -\$62{,}000 \quad \square$$

### 1.7.2 Cash Flows over the Project Life

When we repeat cash flow estimation over the project life, the result is a series of net cash flows. If the sum of the series is positive, the flow is from the project to the company; if it is negative, the flow is from the company to the project. This series is often called the cash flow series for the project, and the company decides whether to undertake the project on the basis of this estimated cash flow. A typical net cash flow stream for a project might be as follows.

| End of Year | 19x0 | 19x1 | 19x2 | 19x3 | 19x4 |
|---|---|---|---|---|---|
| Project Year | 0 | 1 | 2 | 3 | 4 |
| Net Cash Flow | −\$35,000 | 25,000 | 12,000 | 11,000 | 8,000 |

We commonly use a graphical representation of the foregoing cash transactions, or a cash flow diagram, in which receipts during a period of time are shown by an upward arrow and disbursements during the period are shown by a downward arrow. In fact, such a cash flow diagram will provide a basis for analyzing an investment project.

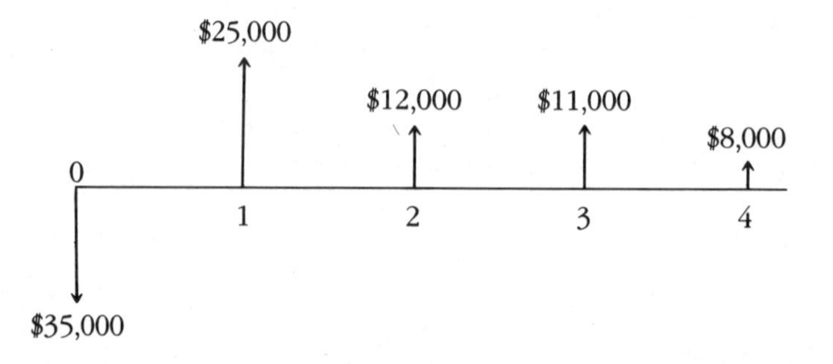

When we have estimated the cash flow series of a project such as that just described, our natural question is, is it worth spending $35,000 now (19x0) to receive a cash inflow stream of $25,000, $12,000, $11,000, and $8,000 over the next four years?

Any proposed investment must be evaluated to ensure, with reasonable confidence, that the benefits of the project will exceed the costs. Whatever methodology is used for the evaluation, certain basic principles remain unchanged. These principles are applicable whether the investment is for a new company, the expansion of an existing facility, or the formation of a new corporate entity. In the chapters ahead we will identify such basic principles, formulate a logical decision framework for economic evaluation, and examine a general analytical methodology for assessing the worth of any proposed capital project. In essence, this is the role of engineering economy.

## 1.8 SUMMARY

The primary purposes of this chapter were (1) to describe the basic capital budgeting function of the company, (2) to describe the basic financial statements and their uses by investors and managers, and (3) to relate the various types of accounting information in defining the cash flows of investment projects.

As shown, financial analysis is designed to determine the relative strengths and weaknesses of a company—whether the company is financially sound and profitable relative to other companies in its industry and whether its position is improving or deteriorating over time. Both investors and managers need such information in order to estimate future cash flows from the company or to detect problems and strengthen weaknesses.

The use of expected earnings (accounting income) to measure the benefits of an investment would require a much more sophisticated accounting system than is currently being used by any corporation. The net income figures resulting from current accounting practices are *not* usable. Moreover, even with improved measures of income, there would remain the question of whether the use of cash flows or net income is more appropriate. The use of cash flows minimizes many of the problems related to measuring income. Therefore, we suggest that the cash flows of the investment be used in the analysis of it.

The concept of funds flow, or generation of cash flow by a project, was considered in detail to demonstrate how to compute the net cash inflows to the company from executed projects. The net cash flow stream, a discrete or continuous function of time, is the basic economic measure of a project used to determine its profitability to the company.

## REFERENCES

1. ANTHONY, R. N., *Management Accounting,* 5th edition, Irwin, Homewood, Ill., 1984.

2. SHARPE, W. F., *Investment,* Prentice-Hall, Englewood Cliffs, N.J., 1978, Chapter 1.

## PROBLEMS

**1.1.** Consider the following balance sheet entries for Tiger Corporation.

a. Compute the firm's working capital.

**BALANCE SHEET**
*As of December 31, 199x*

| | | |
|---|---|---|
| Assets | | |
| Cash | | $ 50,000 |
| Accounts receivable | | 50,000 |
| Inventories | | 150,000 |
| Manufacturing plant at cost | 600,000 | |
| Less: Accumulated depreciation | 300,000 | |
| Net fixed assets | | 300,000 |
| Land | | 150,000 |
| Liabilities and shareholders' equity | | |
| Notes payable | | 50,000 |
| Accounts payable | | 100,000 |
| Long-term mortgage bonds | | 100,000 |
| Preferred stock, 6%, $100 par value (1,000 shares) | | 100,000 |
| Common stock, $10 par value (20,000 shares) | | 200,000 |
| Capital surplus | | 100,000 |
| Retained earnings | | 50,000 |

b. If the firm had the net income after tax of $100,000, what were the earnings per share?

c. When the firm issued its common stock, what was the market price of the stock per share?

**1.2.** The following accounting information was taken from the book of Alpha Manufacturing Company.

| | | | |
|---|---|---|---|
| Beginning inventory | $35,000 | Ending inventory | $23,000 |
| Operating expenses | $22,000 | New purchases | $83,000 |
| Income taxes | $10,000 | Net income | $27,400 |
| Nonoperating revenue | $5,000 | Cash dividend | $12,000 |
| Depreciation | $10,000 | New borrowing | $20,000 |

On the basis of this information, compute the total sales for the period.

**1.3.** The inventory account for item X showed the following information.

| | | |
|---|---|---|
| Beginning balance, Jan. 1 | 100 units @ $1.90 = | $190 |
| Purchases: | | |
| Jan. 15, 1st purchase | 100 units @ $2.00 = | $200 |
| Feb. 12, 2nd purchase | 200 units @ $1.80 = | $360 |
| Feb. 27, 3rd purchase | 250 units @ $2.10 = | $525 |
| Mar. 15, 4th purchase | 150 units @ $2.20 = | $330 |

The ending inventory on March 31 consists of 200 units. Determine the cost of the ending inventory based on two methods of inventory valuation: (a) FIFO and (b) LIFO.

**1.4.** Suppose that a project under consideration is expected to change a firm's financial position for the first year as follows.

| | | |
|---|---|---|
| Sales | | $500,000 |
| Manufacturing costs of sales | | |
| Direct labor | 35,000 | |
| Direct material | 25,000 | |
| Overhead | 10,000 | |
| Depreciation | 30,000 | 100,000 |
| Selling and administrative expenses (directly associated with the project) | | 25,000 |
| Equipment purchase | | 100,000 |
| Decrease in cash revenue of other product | | 10,000 |
| Increase in accounts receivable | | 30,000 |
| Increase in inventory | | 20,000 |
| Increase in current liabilities | | 30,000 |
| Income taxes associated with the project | | 100,000 |
| Repayment of the old loan | | 50,000 |

a. Determine the change in the firm's working capital.
b. Determine the net cash flow for the first project year.

**1.5.** Kendall Manufacturing Company is considering a project that will add a new line of product. The financial data for the first project year are estimated as follows.

| | | |
|---|---|---|
| Sales | | $1,000,000 |
| Cost of goods sold | | |
| Labor | $150,000 | |
| Material | 200,000 | |
| Depreciation | 20,000 | |
| Total | | 370,000 |
| Operating expenses | | 146,000 |
| Interest expenses | | 10,000 |
| Income taxes | | 165,900 |

**ADDITIONAL INFORMATION:**

- The firm will purchase new equipment worth $200,000 at the beginning of the year. The purchase will be financed by paying $100,000 in cash and borrowing the remaining $100,000 from a local bank at 10% interest payable at the end of each year. The $10,000 interest expense shown in the income statement represents this interest payment at the end of the first year.
- The project requires $10,000 in working capital.

Compute the net cash flow from this project during the first year. In doing so, determine the sources of cash and uses of cash, respectively.

**1.6.** Robert Cooper started Wolverine Bakery, Inc. in Ann Arbor, Michigan, on January 1, 1990. The first order of business was to purchase equipment costing $8,000, for which Wolverine Bakery paid $2,000 in cash as a down payment and signed a mortgage note for $6,000 with the vendor, payable in full (plus 12% interest) in January 1991. Cooper intended to depreciate this equipment on a MACRS basis over 5 years. During the first year of operation, Wolverine Bakery had cash sales of

$136,000. Cash expenses included $56,000 for dough ingredients, $44,000 for wages, $1,500 for equipment rentals, $9,600 for store rental, $5,400 for utilities, $2,000 for miscellaneous supplies, and $3,160 in payment of estimated taxes. As of December 31, 1990, there was an unpaid bill of $500 for December utilities and Wolverine Bakery owed $2,300 to its ingredients vendor. Because the vendor delivered frequently, Wolverine Bakery had only $2,200 of inventory on hand at the end of December. There was no beginning inventory on January 1, 1990, because operations did not formally commence until the first delivery of ingredients was made in the middle of January. A customer had paid Wolverine Bakery a $500 advance for a wedding cake to be delivered on January 10, 1991 and a university fraternity owed Wolverine Bakery $200 for a birthday cake delivered in early December. Based on these accounting transactions, Cooper prepared the income statement and the balance sheets shown in Exhibits 1 and 2.

## Exhibit 1

**WOLVERINE BAKERY, INC.**
INCOME STATEMENT FOR THE YEAR 1990

| | | | |
|---|---|---|---|
| Revenues[1] | | | $136,200 |
| Expenses | | | |
| Dough ingredients | | | |
| Beginning inventory | 0 | | |
| Purchase | 56,000 | | |
| Ending inventory | 2,200 | | |
| Cost of ingredients | | $53,800 | |
| Wages | | 44,000 | |
| Rentals | | 11,100 | |
| Depreciation [$8,000 (20%) MACRS] | | 1,600 | |
| Utilities ($5,400 cash + $500 payable) | | 5,900 | |
| Miscellaneous supplies | | 2,000 | |
| Interest ($6,000 for one year at 12%) | | 720 | |
| Total expenses | | | 119,120 |
| Operating profit before tax | | | 17,080 |
| Income tax (18.50%) | | | 3,160 |
| Net income | | | $13,920 |

NOTE 1: The revenues include the $136,000 cash sales plus the $200 owed Wolverine Bakery for the fraternity party it catered.

## Exhibit 2

**WOLVERINE BAKERY, INC.**
COMPARATIVE BALANCE SHEETS

| | December 31, 1989 | | December 31, 1990 | |
|---|---|---|---|---|
| ***Assets*** | | | | |
| Current assets | | | | |
| Cash | | $5,400 | | $20,540 |
| Accounts receivable | | 0 | | 200 |
| Inventories | | 0 | | 2,200 |
| Fixed assets | | | | |
| Equipment at cost | 0 | | 8,000 | |
| Less accumulated depreciation | 0 | | 1,600 | |
| Net fixed assets | | 0 | | 6,400 |
| Total assets | | $5,400 | | $29,340 |
| ***Liabilities and Owners' Equity*** | | | | |
| Current liabilities | | | | |
| Accounts payable | | $ 0 | | $ 2,800 |
| Deferred revenue[1] | | 0 | | 500 |
| Accrued interest | | 0 | | 720 |
| Long-term liabilities | | | | |
| Mortgage note payable | | 0 | | 6,000 |
| Owners' equity | | | | |
| Contributed capital | | 5,400 | | 5,400 |
| Retained earnings | | 0 | | 13,920 |
| Total equities | | $5,400 | | $29,340 |

NOTE 1: The $500 advance payment for the wedding cake for the next year is not revenue of this accounting period; it appears as $500 deferred revenue on the balance sheet.

Put yourself in Cooper's position and prepare the cash flow statement for the operating year 1990.

# 2

# Interest and Equivalence

## 2.1 INTRODUCTION

Engineering economic analysis is primarily concerned with the evaluation of economic investment alternatives. We often describe these investment alternatives by a cash flow diagram showing the amount and timing of estimated future receipts and disbursements that will result from each decision. Because the time value of money is related to the effect of time and interest on monetary amounts, we must consider both the timing and the magnitude of cash flow. When comparing investment alternatives, we must consider the expected receipts and disbursements of these investment alternatives on the same basis. This type of comparison requires understanding of the concepts of equivalence and the proper use of various interest formulas. In this chapter we will examine a number of mathematical operations that are based on the time value of money, with an emphasis on modeling cash flow profiles.

## 2.2. CASH FLOW PROFILE

An investment project can be described by the amount and timing of expected costs and benefits in the planning horizon. (We will use the terms *project* and *proposal* interchangeably throughout this book.) The terms *costs* and *benefits* represent *disbursements* and *receipts,* respectively. We will use the term *payment* (or *net cash flow*) to denote the receipts less the disbursements that occur at the same point in time. The stream of disbursements and receipts for an investment project over the planning horizon is said to be the *cash flow profile* of the project.

To facilitate the description of project cash flows, we classify them in two categories: (1) discrete-time cash flows and (2) continuous-time flows. The discrete-time cash flows are those in which cash flow occurs at the end of, at the start of, or within discrete time periods. The continuous flows are those in which money flows at a given rate and continuously throughout a given time period. The following notation will be adopted:

$F_n$ = discrete payment occurring at period $n$,

$F_t$ = continuous payment occurring at time $t$.

If $F_n < 0$, $F_n$ represents a net disbursement (cash outflow). If $F_n > 0$, $F_n$ represents a net receipt (cash inflow). We can say the same for $F_t$.

## 2.3 TIME PREFERENCE AND INTEREST

### 2.3.1 Time Preference

Cash flows that occur at different points in time have different values and cannot be compared directly with one another. This fact is often stated simply as "money has a time value." There are several reasons why we must assess cash flows in different periods in terms of time preference.

First, money has a potential *earning power,* because having a dollar now gives us an opportunity to invest this dollar in the near future. In other words, equal dollar amounts available at different points in time have different values based on the opportunity to profit from investment activity.

Second, money has a time value because a user may have a different utility of consumption of dollars (i.e., consider them more or less desirable to use) at different times. The preference for consumption in different periods is measured by the rate of time preference. For example, if we have a rate of time preference of $i$ per time period, we are indifferent toward the prospect of either consuming $P$ units now or consuming $P(1 + i)$ units at the end of the period. The rate of time preference is often called the *interest rate* (or *discount rate*) in economic analysis.

Third, money has time value because the *buying power* of a dollar changes through time. When there is inflation, the amount of goods that can be bought for a certain amount of money decreases as the time of purchase is further in the future. Although this change in the buying power of money is important, we limit our concept of time preference to the fact that money has an *earning power,* or utility of consumption. We will treat the effects of inflation explicitly in a later section, and any future reference to the time value of money will be restricted to the first two aspects. Before considering the actual effect of this time value, we will review the types of interest and how they are calculated.

### 2.3.2 Types of Interest

If an amount of money is deposited in a financial institution, interest accrues (accumulates) at regular time intervals. Each time interval represents an *interest period.* Then the interest earned on the original amount is calculated according to a specified interest rate at the end of the interest period. Two approaches are in use in calculating the earned interest: *simple interest* and *compound interest.*

The first approach considers that the interest earned in any present activity is a linear function of time. Consider the situation in which a present amount $P$ is borrowed from the bank, to be repaid $N$ periods hence by a future amount $F$. The difference, $F - P$, is simply the interest payment $I$ owed to the bank for the use of the principal $P$ dollars. Because the interest earned is directly proportional to the principal, the interest $i$ is called *simple interest* and is computed from

$$\begin{aligned} I &= F - P \\ &= (Pi)N \\ F &= P + (Pi)N \\ &= P(1 + iN) \end{aligned} \tag{2.1}$$

The second approach assumes that the earned interest is not withdrawn at the end of an interest period and is automatically redeposited with the original sum in the next interest period. The interest thus accumulated is called *compound interest.* For example, if we deposit \$100 in a bank that pays 5% compounded annually and leave the interest in the account, we will have

| | | |
|---|---|---|
| after 1 year | \$100 (1.05) = | \$105.00 |
| after 2 years | \$105.00 (1.05) = | \$110.25 |
| after 3 years | \$110.25 (1.05) = | \$115.7625 |

The amount \$115.7625 is greater than the original \$100 plus the simple interest of \$100(0.05)(3), which would be \$115.00, because the interest earned during the first and second periods earns additional interest. Symbolically, we can represent a future amount $F$ at time $N$ in terms of a present amount $P$ at time 0, assuming i% interest per period:

$$\begin{aligned} F_1 &= P(1 + i) && \text{after 1 year} \\ F_2 &= F_1(1 + i) = P(1 + i)^2 && \text{after 2 years} \\ &\vdots \\ F_N &= F_{N-1}(1 + i) = P(1 + i)^N && \text{after } N \text{ years} \end{aligned}$$

or

$$F = P[(1 + i)^N] \tag{2.2}$$

From Eq. 2.2, the total interest earned over $N$ periods with the compound interest is

$$I = F - P = P[(1 + i)^N - 1] \tag{2.3}$$

The additional interest earned with the compound interest is

$$\begin{aligned} \Delta I &= P[(1 + i)^N - 1] - PiN \\ &= P[(1 + i)^N - (1 + iN)] \end{aligned} \tag{2.4}$$

As either $i$ or $N$ becomes large, $\Delta I$ also becomes large, so the effect of compounding is further pronounced.

### *Example 2.1*

Compare the interest earned by $1,000 for 10 years at 9% simple interest with that earned by the same amount for 10 years at 9% compounded annually.

$$\Delta I = 1000[(1 + 0.09)^{10} - (1 + 0.09(10))] = \$467.36$$

The difference in interest payments is $467.36. □

Unless stated otherwise, practically all financial transactions are based on compound interest; however, the length of the interest period for compounding and the interest rate per period must be specified for individual transactions. In the next section we discuss the conventions used in describing the interest period and the compounding period in business transactions.

### 2.3.3 Nominal and Effective Interest Rates

In engineering economic analysis, a year is usually used as the interest period, because investments in engineering projects are of long duration and a calendar year is a convenient period for accounting and tax computation. In financial transactions, however, the interest period may be of any duration—a month, a quarter, a year, and so on. For example, the interest charge for the purchase of a car on credit may be compounded monthly, whereas the interest accrued from a savings account in a credit union may be compounded quarterly. Consequently, we must introduce the terms nominal interest rate and effective interest rate to describe more precisely the nature of compounding schemes.

***Nominal Interest.*** If a financial institution uses more than one interest period per year in compounding the interest, it usually quotes the interest on an annual basis. For example, a year's interest at 1.5% compounded each month is typically quoted as "18% (1.5% × 12) compounded monthly." When the interest rate is stated in this fashion, the 18% interest is called a *nominal interest rate* or *annual percentage rate.* The nominal interest rate, while convenient for a financial institution to use in quoting interest rates on its transactions, does not explain the effect of any compounding during the year. We use the term effective interest rate to describe more precisely the compounding effect of any business transaction.

***Effective Interest Rate.*** The effective interest rate represents the actual interest earned or charged for a specified time period. In specifying such a time period, we may use the convention of either a year or a time period identical to the payment period. The effective interest rate based on a year is referred to as the *effective annual interest rate* $i_a$. The effective interest rate based on the payment period is called the *effective interest rate per payment period* $i$.

We will first look at the expression of the effective annual interest rate. Suppose a bank charges an interest rate of 12% compounded quarterly. This means that the interest rate per period is 3% (12%/4) for each of the 3-month periods during the year. Then interest for a sum of $1 accrued at the end of the year (see Eq. 2.3) is

$$\left(1 + \frac{0.12}{4}\right)^4 - 1 = 0.1255$$

Thus, the effective annual interest rate is 12.55%. Similarly, an interest rate of 12% compounded monthly means that the interest rate per period is 1% (12%/12) for each month during the year. Thus, the effective annual interest rate is

$$\left(1 + \frac{0.12}{12}\right)^{12} - 1 = 0.1268 = 12.68\%$$

Now we can generalize the result as

$$i_a = \left(1 + \frac{r}{M}\right)^M - 1 \tag{2.5}$$

where $i_a$ = the effective annual interest rate,
$r$ = the nominal interest rate per year,
$M$ = the number of interest (compounding) periods per year,
$r/M$ = the interest rate per interest period.

For the special case where $M = 1$ (i.e., one interest period per year, or annual compounding) and $r/M = r$, Eq. 2.5 reduces to $i_a = r = i$. This simply means that with annual compounding we do not need to distinguish between the nominal and effective interest rates.

The result of Eq. 2.5 can be further generalized to compute the effective interest rate in *any payment period.* This results in

$$\begin{aligned} i &= \left(1 + \frac{r}{M}\right)^C - 1 \\ &= \left(1 + \frac{r}{CK}\right)^C - 1 \end{aligned} \tag{2.6}$$

where $i$ = the effective interest rate per payment period,
$C$ = the number of interest periods per payment period,
$K$ = the number of payment periods per year,
$r/K$ = the nominal interest rate per payment period.

In deriving Eq. 2.6, we should note the relationships $M \geq C$ and $M = CK$. Obviously, when $K = 1$, $C$ is equal to $M$, and therefore $i = i_a$. Figure 2.1 illustrates the relationship between the nominal and effective interest rates.

Some financial institutions offer a large number of interest periods per year, such as $M = 365$ (daily compounding). As the number of interest periods $M$ becomes very large, the interest rate per interest period, $r/M$, becomes very small. If $M$ approaches infinity and $r/M$ approaches zero as a limit, the limiting

Situation: Interest is calculated on the basis of 12% compounded monthly. Payments are made quarterly.

$K$ = 4, 4 quarterly payment periods per year
$C$ = 3, 3 interest (compounding) periods per quarter
$r$ = 12%
$M$ = 12, 12 monthly interest (compounding) periods per year
$r/M$ = 1%, the interest rate per month
$r/K$ = 3%, the nominal interest rate per quarter

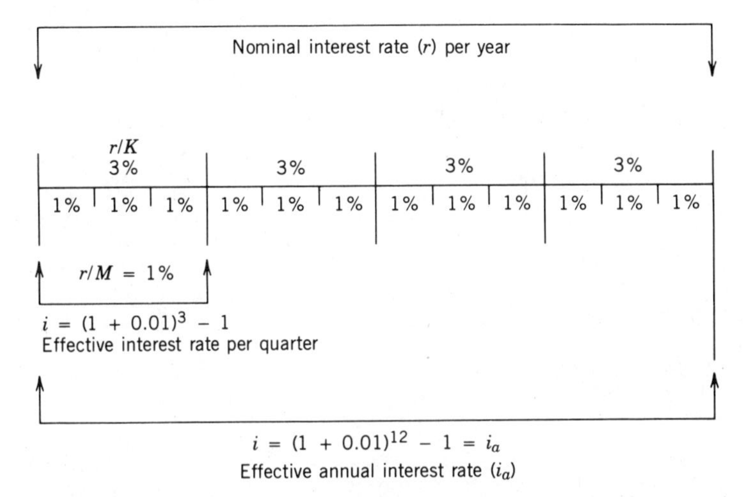


**FIGURE 2.1 Functional relationships of $r$, $i$, and $i_a$ for monthly compounding with quarterly payments.**

condition is equivalent to continuous compounding. By taking limits on both sides of Eq. 2.6, we obtain

$$
\begin{aligned}
i &= \lim_{M\to\infty}\left[\left(1+\frac{r}{M}\right)^{M}-1\right]\\
&= \lim_{CK\to\infty}\left[\left(1+\frac{r}{CK}\right)^{C}-1\right]\\
&= \lim_{CK\to\infty}\left(1+\frac{r}{CK}\right)^{C}-1\\
&= \lim_{CK\to\infty}\left[\left(1+\frac{r}{CK}\right)^{CK}\right]^{1/K}-1\\
&= (e^{r})^{1/K}-1\\
&= e^{r/K}-1 \qquad (2.7)
\end{aligned}
$$

For the effective annual interest rate for continuous compounding, we simply evaluate Eq. 2.7 by setting $K$ to 1. This gives us

$$i_a = e^r - 1 \tag{2.8}$$

### *Example 2.2*

Find the effective interest rate per quarter at a nominal rate of 18% compounded (1) quarterly, (2) monthly, and (3) continuously.

**1.** Quarterly compounding

$$r = 18\%, \quad M = 4, \quad C = 1, \quad K = 4$$

$$i = \left(1 + \frac{0.18}{4}\right)^1 - 1 = 4.5\%$$

**2.** Monthly compounding

$$r = 18\%, \quad M = 12, \quad C = 3, \quad K = 4$$

$$i = \left(1 + \frac{0.18}{12}\right)^3 - 1 = 4.568\%$$

**3.** Continuous compounding

$$r = 18\%, \quad K = 4, \quad (M = \infty, C = \infty)$$

$$i = e^{0.18/4} - 1 = 4.603\%$$

If we deposit $1,000 in a bank for just one quarter at the interest rate and compounding frequencies specified, our balance at the end of the quarter will grow to $1,045, $1,045.68, and $1,046.03, respectively. □

In Example 2.2 we examined how our deposit balance would grow for a time period of one quarter, but these results can be generalized for deposits of any duration. In the sections ahead, we will develop interest formulas that facilitate the interest compounding associated with various types of cash flow and compounding frequencies. For this presentation, we will group the compound interest formulas into four categories by the type of compounding and type of cash flow. We will first consider discrete compounding in which compounding occurs at a discrete point in time: annual compounding, monthly compounding, and so forth.

## 2.4 DISCRETE COMPOUNDING

### 2.4.1 Comparable Payment and Compounding Periods

We first consider the situations for which the payment periods are identical to the compounding periods (annual payments with annual compounding, quarterly payments with quarterly compounding, monthly payments with monthly compounding, and so forth).

***Single Sums.*** In the simplest situation we deposit a single sum of money $P$ in a financial institution for $N$ interest periods. To determine how much can be accumulated by the end of $N$ periods, we may use the result developed in Eq. 2.2,

$$F = P(1 + i)^N \tag{2.9}$$

The factor $(1 + i)^N$ is called the *single-payment compound amount factor* and is available in tables indexed by $i$ and $N$. It is represented symbolically by $(F/P, i, N)$. Note that where payment and compounding periods are identical, the effective interest rate is simply $i = r/M$. This transaction can be portrayed by the cash flow diagram shown in Figure 2.2. (Note the time scale convention: the first period begins at $n = 0$ and ends at $n = 1$.)

For example, consider a deposit of $1,000 for 8 years in an individual retirement account (IRA) that earns an interest rate of 11% compounded annually. The balance of the account at the end of 8 years will be

$$F = \$1000(1 + 0.11)^8 = \$2,304.54$$

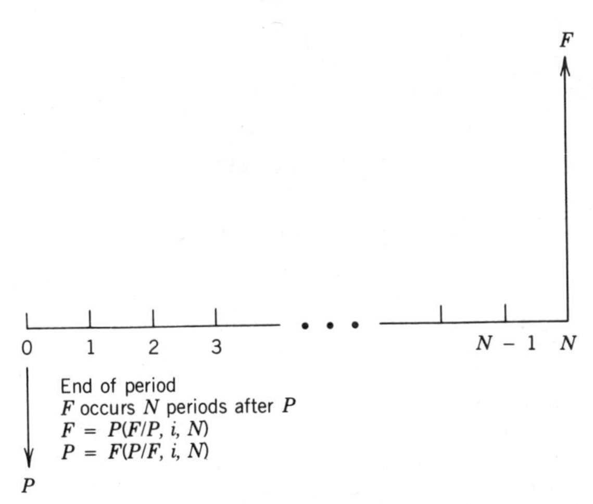


**FIGURE 2.2** Cash flow diagram for a single payment.

If the account earns the interest at the rate of 11% compounded quarterly, the balance becomes

$$F = \$1000\left(1 + \frac{0.11}{4}\right)^{32} = \$2,382.42$$

If we wish to know what sum $P$ we must deposit with a bank now, at $i$% compounded periodically, in order to have a future sum $F$ in $N$ periods, we can solve Eq. 2.9 for $P$.

$$P = F[(1 + i)^{-N}] \tag{2.10}$$

The bracketed term is called the *single-payment present-worth factor,* designated by $(P/F, i, N)$.

For example, we will have \$100 at the end of 3 years if we deposit \$86.38 in a 5% interest-bearing account:

$$P = \$100\overset{P/F,5\%,3}{(0.8638)} = \$86.38$$

***Uniform Series.*** Most transactions with a financial institution involve more than two flows. If we have equal, periodic flows, we can develop formulas for determining beginning and ending balances. For example, an amount $A$ deposited at the *end* of each compounding period in an account paying $i$% will grow to an amount after $N$ periods of

$$A \sum_{n=1}^{N} (1 + i)^{N-n} = A\left[\frac{(1 + i)^N - 1}{i}\right] \tag{2.11}$$

The term in brackets is called the *uniform-series compound amount factor,* or *equal-series compound amount factor,* and is represented by $(F/A, i, N)$. The transaction can be portrayed by the cash flow diagram shown in Figure 2.3. In deriving the summation results in Eq. 2.11, we refer the reader to Table 2.1, which contains closed-form expressions for selected finite summations that are useful in developing interest formulas.

The inverse relationship to Eq. 2.11 yields the *uniform-series sinking-fund factor,* or *sinking-fund factor,*

$$A = F\left[\frac{i}{(1 + i)^N - 1}\right] \tag{2.12}$$

designated by $(A/F, i, N)$. The name derives from a historical practice of depositing a fixed sum at the end of each period into an interest-bearing account (a sinking fund) to provide for replacement moneys for fixed assets.

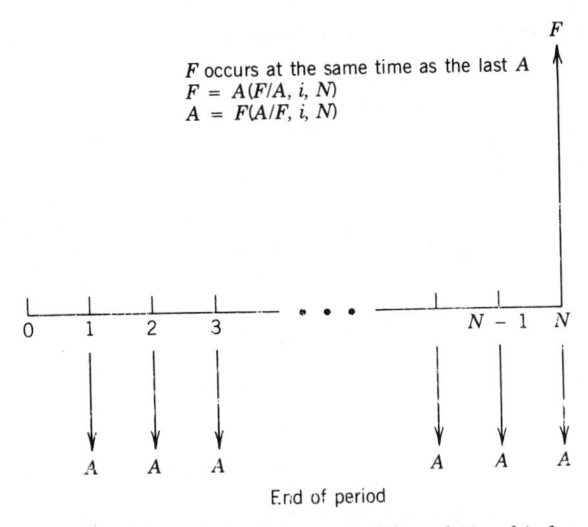


**FIGURE 2.3** **Cash flow diagram of the relationship between *A* and *F*.**

**Table 2.1** ***Summations Useful in Deriving Interest Formulas***

Geometric series

$$\sum_{n=0}^{N} x^n = 1 + x + x^2 + \cdots + x^N = \frac{1 - x^{N+1}}{1 - x}$$

where $x \neq 1$

If $-1 < x < 1$, then

$$\sum_{n=0}^{\infty} x^n = 1 + x + x^2 + x^3 + \cdots = \frac{1}{1 - x}$$

Arithmetic–geometric series

$$\sum_{n=0}^{N} nx^n = 0 + x + 2x^2 + \cdots + Nx^N = \frac{x[1 - (N + 1)x^N + Nx^{N+1}]}{(1 - x)^2}$$

where $x \neq 1$

If $-1 < x < 1$, then

$$\sum_{n=0}^{\infty} nx^n = 0 + x + 2x^2 + 3x^3 + \cdots = \frac{x}{(1 - x)^2}$$

Educational endowment funds can be constructed conveniently by using the sinking-fund factor: to build a \$12,000 fund in 18 years at 5% compounded annually requires

$$A = \$12{,}000\overset{A/F,5\%,18}{(0.0356)} = \$427.20 \quad \text{at the end of each year}$$

The relationships among *P, F,* and *A* can be manipulated to relate a series of equal, periodic flows (defined by *A*) to a present amount *P*. Substituting Eq. 2.11 into Eq. 2.9 yields

$$P = A\left[\frac{(1+i)^N - 1}{i(1+i)^N}\right] \tag{2.13}$$

and its inverse

$$A = P\left[\frac{i(1+i)^N}{(1+i)^N - 1}\right] \tag{2.14}$$

The bracketed term in Eq. 2.13 is the *uniform-series present worth factor,* designated by (*P*/*A, i, N*). The term in Eq. 2.14 is the *uniform-series capital recovery factor,* or simply the *capital recovery factor,* represented by (*A*/*P, i, N*). Figure 2.4 shows the cash flow transactions associated with these factors. The latter factor can be used to determine loan repayment schedules so that principal and interest are repaid over a given time period in equal end-of-period amounts.

To illustrate the use of *A*/*P* and *P*/*A* factors, consider a commercial mortgage at 8% over 20 years, with a loan principal of \$1 million. If equal year-end payments are desired, each annual payment must be

$$\$1{,}000{,}000\overset{A/P,8\%,20}{(0.10185)} = \$101{,}850$$

The loan schedule can then be constructed as in Table 2.2. The interest due at $n = 1$ is 8% of the \$1 million outstanding during the first year. The \$21,850 left over is applied to the principal, reducing the amount outstanding in the second year to \$978,150. The interest due in the second year is 8% of \$978,150, or \$78,252, leaving \$23,598 for repayment of the principal. At $n = 20$, the last \$101,850 payment is just sufficient to pay the interest on the outstanding loan principal and to repay the outstanding principal.

Such an equal-payments scheme is also common for home mortgages and automobile loans. In each period a decreasing amount of interest is paid, leaving a larger amount to reduce the principal. Each reduction of loan principal increases an owner's equity in the item by a corresponding amount.

The series present worth factor can be useful for determining the outstanding balance of a loan at any time, as portrayed in Table 2.2. At the end of the fifth year, for example, we still owe 15 payments of \$101,850. The value of those

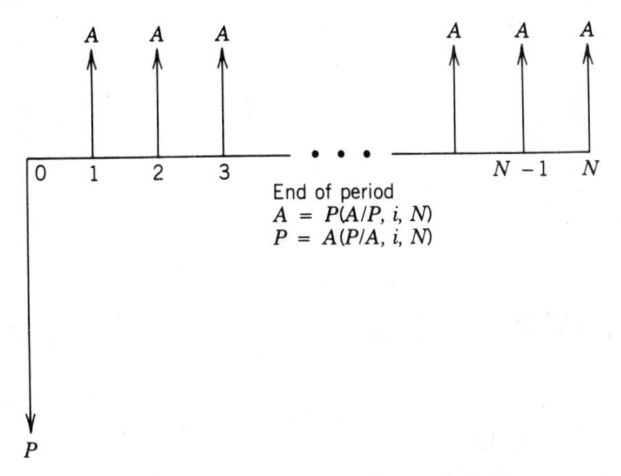


**FIGURE 2.4** Equal-payment series and single present amount.

**Table 2.2** ***A Loan Repayment Schedule Showing Principal and Interest Payments***

| Year | Beginning Loan Balance | Interest Payment | Principal Payment | Total Payment |
|---|---|---|---|---|
| 1 | 1,000,000* | 80,000 | 21,850 | 101,850 |
| 2 | 978,150 | 78,252 | 23,598 | 101,850 |
| 3 | 954,552 | 76,364 | 25,486 | 101,850 |
| 4 | 929,066 | 74,325 | 27,525 | 101,850 |
| 5 | 901,541 | 72,123 | 29,727 | 101,850 |
| 6 | 871,814 | 69,745 | 32,105 | 101,850 |
| 7 | 839,709 | 67,177 | 34,673 | 101,850 |
| 8 | 805,036 | 64,403 | 37,447 | 101,850 |
| 9 | 767,589 | 61,407 | 40,443 | 101,850 |
| 10 | 727,146 | 58,172 | 43,678 | 101,850 |
| 11 | 683,468 | 54,677 | 47,173 | 101,850 |
| 12 | 636,295 | 50,904 | 50,946 | 101,850 |
| 13 | 585,349 | 46,828 | 55,022 | 101,850 |
| 14 | 530,327 | 42,426 | 59,424 | 101,850 |
| 15 | 470,903 | 37,672 | 64,178 | 101,850 |
| 16 | 406,725 | 32,538 | 69,312 | 101,850 |
| 17 | 337,413 | 26,993 | 74,857 | 101,850 |
| 18 | 262,556 | 21,005 | 60,845 | 101,850 |
| 19 | 181,711 | 14,537 | 87,313 | 101,850 |
| 20 | 94,398 | 7,452 | 94,398 | 101,850 |

*All figures are rounded to nearest dollars.

NOTE: Loan Amount = $1,000,000
Loan life = 20 years
Loan interest = 8% compounded annually
Equal annual payment size = $1,000,000(A/P, 8%, 20)
= $1,000,000(0.10185)
= $101,850

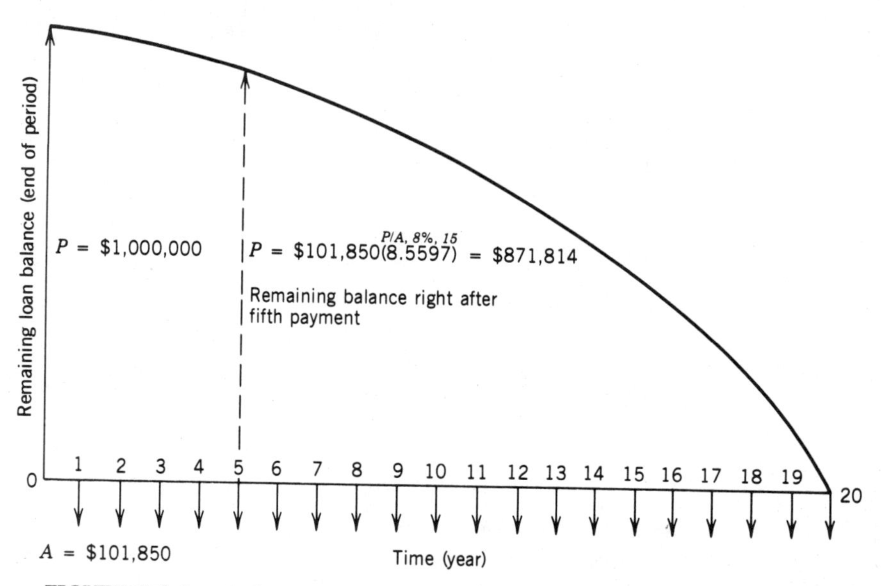


**FIGURE 2.5** loan balance as a function of time ($n$).

payments at time 5 can be represented as in Figure 2.5 with the time scale shifted by 5 and is found from the equation to be

$$P = \$101{,}850\overset{P/A,8\%,15}{(8.5595)} = \$871{,}785$$

which is the same as the $871,814 in Table 2.2.

## *Example 2.3*

Suppose we are in the market for a medium-sized used car. We have surveyed the dealers' advertisements in the newspaper and have found a car that should fulfill our needs. The asking price of the car is $7,500, and the dealer proposes that we make a $500 down payment now and pay the rest of the balance in equal end-of-month payments of $194.82 each over a 48-month period. Consider the following situations.

1. Instead of using the dealer's financing, we decide to make a down payment of $500 and borrow the rest from a bank at 12% compounded monthly. What would be our monthly payment to pay off the loan in 4 years? To find $A$,

$$i = \frac{12\%}{12 \text{ periods}} = 1\% \text{ per month}$$

$$N = (4 \text{ years})(12 \text{ periods per year}) = 48 \text{ periods}$$

$$A = P(A/P, i, N) = (\$7{,}500 - \$500)\overset{A/P,1\%,48}{(0.0263)} = \$184.34$$

**2.** We are going to accept the dealer's offer but we want to know the effective rate of interest per month that the dealer is charging. To find $i$, let $P =$ \$7,000, $A =$ \$194.82, and $N = 48$.

$$\$194.82 = \$7{,}000(A/P, i, 48)$$

The satisfying value $i$ can be found by trial and error from

$$(A/P, i, 48) = \frac{i(1+i)^{48}}{(1+i)^{48}-1} = 0.0278$$

to be $i = 1.25\%$ per month. This value is used to find the nominal annual interest rate used by the dealer.

$$\begin{aligned} r = (i)(M) &= (1.25\% \text{ per month})(12 \text{ months per year}) \\ &= 15\% \text{ per year} \end{aligned}$$

Then the effective annual interest used by the dealer is simply

$$i_a = \left(1 + \frac{0.15}{12}\right)^{12} - 1 = 16.08\% \quad \square$$

***Linear Gradient Series.*** Many engineering economy problems, particularly those related to equipment maintenance, involve cash flows that change by a constant amount ($G$) each period. We can use the gradient factors to convert such gradient series to present amounts and equal annual series. Consider the series

$$F_n = (n-1)G, \qquad n = 1, 2, \ldots, N \tag{2.15}$$

As shown in Figure 2.6, the gradient $G$ can be either positive or negative. If $G > 0$, we call the series an increasing gradient series. If $G < 0$, we have a decreasing gradient series. We can apply the single-payment present-worth factor to each term of the series and obtain the expression

$$P = \sum_{n=1}^{N} (n-1)G(1+i)^{-n} \tag{2.16}$$

Using the finite summation of a linear function in Table 2.1, we obtain

$$P = G\left[\frac{1-(1+Ni)(1+i)^{-N}}{i^2}\right] \tag{2.17}$$

The resulting factor in brackets is called the *gradient series present-worth factor* and is designated $(P/G, i, N)$.

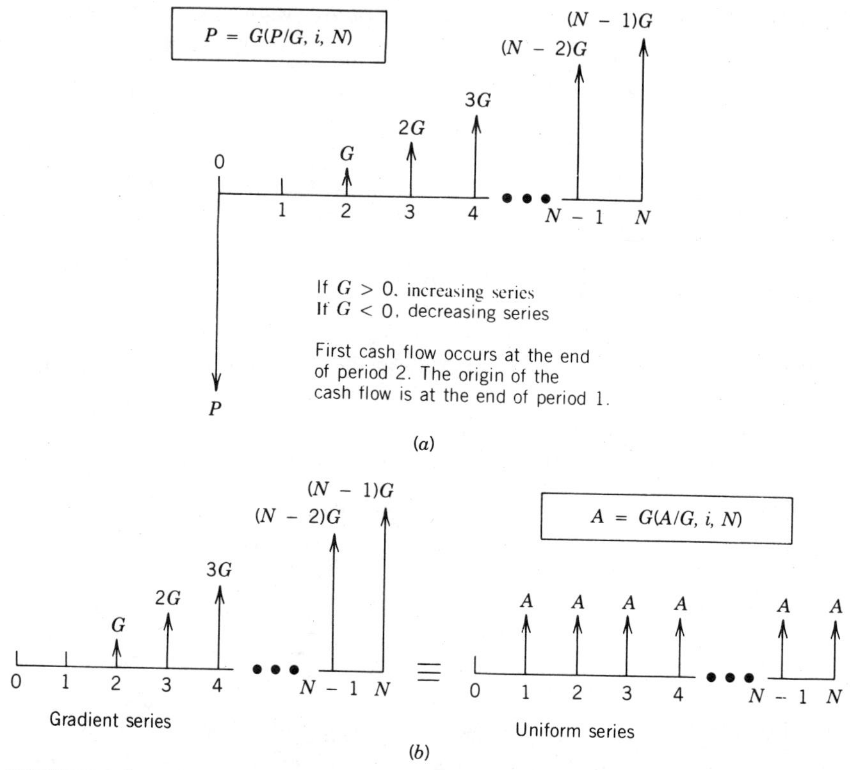


**FIGURE 2.6** Cash flow diagram for a gradient series. (*a*) A strictly gradient series. (*b*) Conversion factor from a gradient series to a uniform series.

A uniform series equivalent to the gradient series can be obtained by substituting Eq. 2.17 into Eq. 2.14 for $P$,

$$A = G\left[\frac{1}{i} - \frac{N}{(1+i)^N - 1}\right] \tag{2.18}$$

where the resulting factor in brackets is referred to as the *gradient-to-uniform-series conversion factor* and is designated $(A/G, i, N)$.

To obtain the future-worth equivalent of a gradient series, we substitute Eq. 2.18 into Eq. 2.14 for $A$.

$$F = \frac{G}{i}\left[\frac{(1+i)^N - 1}{i} - N\right]$$

$$= \frac{G}{i}\left[(F/A, i, N) - N\right] \tag{2.19}$$

### *Example 2.4*

An example of the use of a gradient factor is to find the future amount of the following series with $i = 10\%$ per period.

| $n$ | 0 | 1 | 2 | 3 | 4 | 5 | 6 | 7 | 8 |
|---|---|---|---|---|---|---|---|---|---|
| $F_n$ | 0 | 100 | 106 | 112 | 118 | 124 | 130 | 136 | 142 |

The constant portion of 100 is separated from the gradient series of 0,0,6,12, . . . , 42.

| $n$ | 0 | 1 | 2 | 3 | 4 | 5 | 6 | 7 | 8 |
|---|---|---|---|---|---|---|---|---|---|
| $F_n$ | 0 | 100 | 100 | 100 | 100 | 100 | 100 | 100 | 100 |
| | 0 | 0 | 6 | 12 | 18 | 24 | 30 | 36 | 42 |

We can quickly verify that the portion of the strict gradient series will accumulate to \$206.15.

$$F = 100\overset{F/A,10\%,8}{(11.436)} + \frac{6}{0.1}[\overset{F/A,10\%,8}{(11.436)} - 8]$$

$$= 1{,}143.60 + 206.15$$

$$= 1{,}349.76 \quad \square$$

***Geometric Series.*** In many situations periodic payments increase or decrease over time, not by a constant amount (gradient) but by a constant percentage (geometric growth). If we use $g$ to designate the percentage change in the payment from one period to the next, the magnitude of the $n$th payment, $F_n$, is related to the first payment, $F_1$, by

$$F_n = F_1(1 + g)^{n-1}, \qquad n = 1, 2, \ldots, N \tag{2.20}$$

As illustrated in Figure 2.7, $g$ can be either positive or negative, depending on the type of cash flow. If $g > 0$ the series will increase, and if $g < 0$ the series will decrease.

To find an expression for the present amount $P$, we apply the single-payment present-worth factor to each term of the series

$$P = \sum_{n=1}^{N} F_1(1 + g)^{n-1}(1 + i)^{-n} \tag{2.21}$$

Bringing the term $F_1(1 + g)^{-1}$ outside the summation yields

$$P = \frac{F_1}{1 + g}\sum_{n=1}^{N}\left(\frac{1 + g}{1 + i}\right)^n \tag{2.22}$$

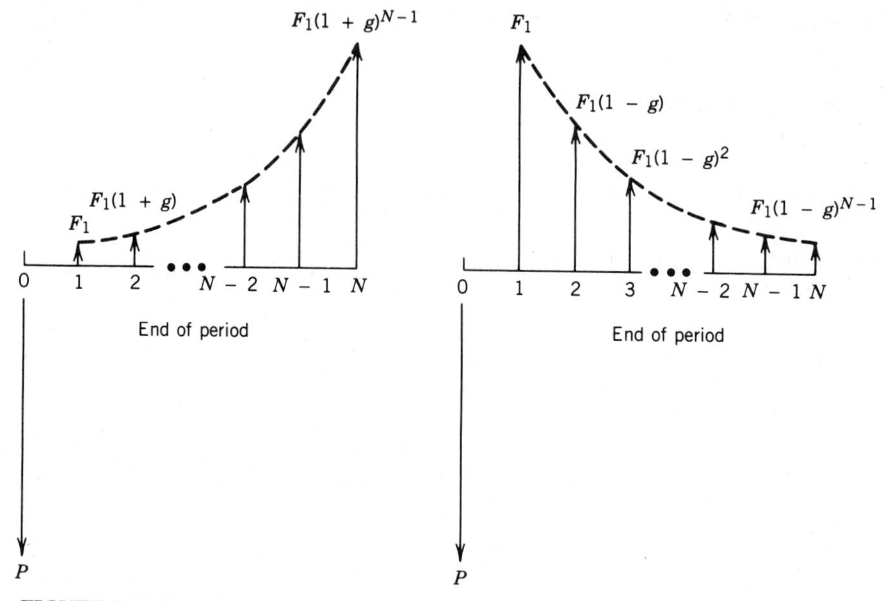


**FIGURE 2.7** Cash flow diagram of the geometric series.

The summation in Eq. 2.22 represents the first $N$ terms of a geometric series, and the closed-form expression for the partial geometric summation yields the following relationship.

$$P = \begin{cases} F_1\left[\dfrac{1-(1+g)^N(1+i)^{-N}}{i-g}\right] & i \neq g \\ \dfrac{NF_1}{1+i} & i = g \end{cases} \tag{2.23}$$

This present-worth factor is designated $(P/A, g, i, N)$.

The future-worth equivalent of the geometric series is obtained by substituting Eq. 2.23 into Eq. 2.5 to find $(F/A, g, i, N)$.

$$F = \begin{cases} F_1\left[\dfrac{(1+i)^N(1+i)^{-N}}{i-g}\right] & i \neq g \\ NF_1(1+i)^{N-1} & i = g \end{cases} \tag{2.24}$$

We may use an alternative expression of Eq. 2.23 as shown in [4]. In Eq. 2.22 we may rewrite the term $(1+g)/(1+i)$ as

$$\frac{1+g}{1+i} = \frac{1}{1+g'} \tag{2.25}$$

or

$$g' = \frac{1+i}{1+g} - 1$$

We then substitute Eq. 2.25 back into Eq. 2.22 to obtain

$$P = \frac{F_1}{1+g} \sum_{n=1}^{N} (1+g')^{-n} \tag{2.26}$$

The summation term constitutes the uniform-series present-worth factor for $N$ periods. Therefore,

$$\begin{aligned} P &= \frac{F_1}{1+g}\left[\frac{g'(1+g')^N}{(1+g')-1}\right] \\ &= \frac{F_1}{1+g}(P/A, g', N), g \neq i \end{aligned} \tag{2.27}$$

If $g < i$, then $g' > 0$, and we can use the $(P/A, g', N)$ factor to find $P$. If $g = i$, then $g' = 0$, and the value of $(P/A, g', N)$ will be $N$. The geometric-series factor thus reduces to $P = F_1N/(1+g)$. If $g > i$, then $g' < 0$. In this case, no table values can be used to evaluate the $P/A$ factor, and it will have to be calculated directly from a formula. Table 2.3 summarizes the interest formulas developed in this section and the cash flow situations in which they should be used.

### *Example 2.5*

A mining company is concerned about the increasing cost of diesel fuel for their mining operation. A special piece of mining equipment, a tractor-mounted ripper, is used to loosen the earth in open-pit mining operations. The company thinks that the diesel fuel consumption will escalate at the rate of 10% per year as the efficiency of the equipment decreases. The company's records indicate that the ripper averages 18 gallons per operational hour in year 1, with 2,000 hours of operation per year. What would the present worth of the cost of fuel for this ripper be for the next five years if the interest rate is 15% compounded annually?

Assuming that all the fuel costs occur at the end of each year, we determine the present equivalent fuel cost by calculating the fuel cost for the first year:

$$F_1 = (\$1.10/\text{gal})(18 \text{ gal/hr})(2{,}000 \text{ hr/year}) = \$39{,}600/\text{year}$$

$$(g = 0.10, N = 5, i = 0.15)$$

Then, using the appropriate factors in Eq. 2.23, we compute

$$P = \$39{,}600\left[\frac{1-(1+0.10)^5(1+0.15)^{-5}}{0.15-0.10}\right] = \$157{,}839.18$$

**Table 2.3** ***Summary of Discrete Compounding Formulas with Discrete Payments***

| Flow Type | Factor Notation | Formula | Cash Flow Diagram |
|---|---|---|---|
| Single | Compound amount (F/P, i, N) | $F = P(1 + i)^N$ | 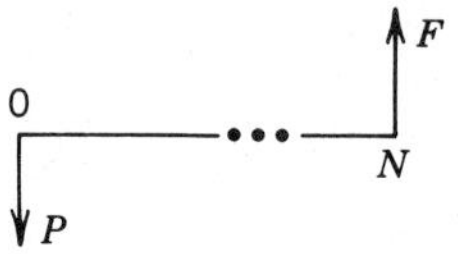 0 … N; P, F |
| | Present worth (P/F, i, N) | $P = F(1 + i)^{-N}$ | |
| Equal payment series | Compound amount (F/A, i, N) | $F = A\left[\frac{(1+i)^N - 1}{i}\right]$ | 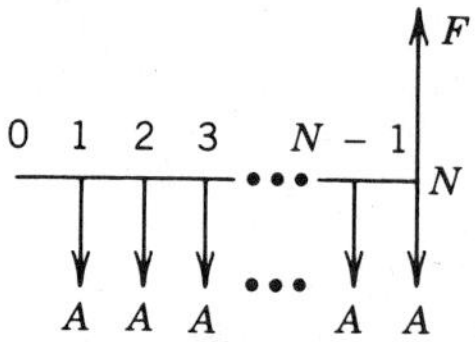 0 1 2 3 … N − 1 N; A A A … A A; F |
| | Sinking fund (A/F, i, N) | $A = F\left[\frac{i}{(1+i)^N - 1}\right]$ | |
| | Present worth (P/A, i, N) | $P = A\left[\frac{(1+i)^N - 1}{i(1+i)^N}\right]$ | 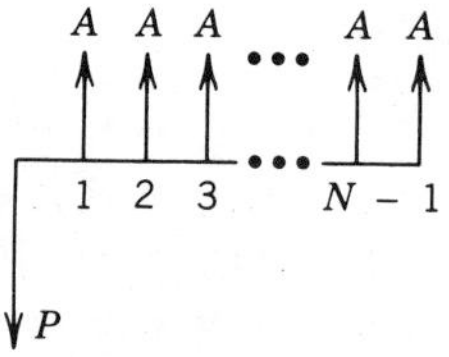 A A A … A A; 1 2 3 … N − 1; P |
| | Capital recovery (A/P, i, N) | $A = P\left[\frac{i(1+i)^N}{(1+i)^N - 1}\right]$ | |
| Gradient series | Uniform gradient Present worth (P/G, i, N) | $P = G\left[\frac{(1+i)^N - iN - 1}{i^2(1+i)^N}\right]$ | 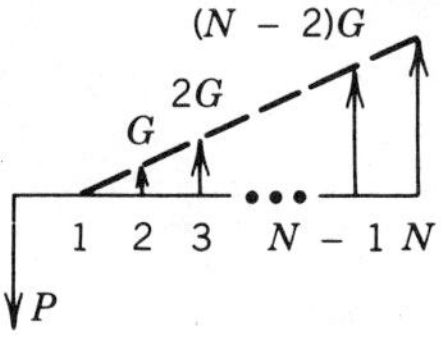 G, 2G, (N − 2)G; 1 2 3 … N − 1 N; P |
| | Geometric gradient Present worth (P/A, g, i, N) | $P = \begin{cases} F_1\left[\frac{1 - (1+g)^N(1+i)^{-N}}{i - g}\right] \\ \frac{NF_1}{1+i} \quad (\text{if } i = g) \end{cases}$ | 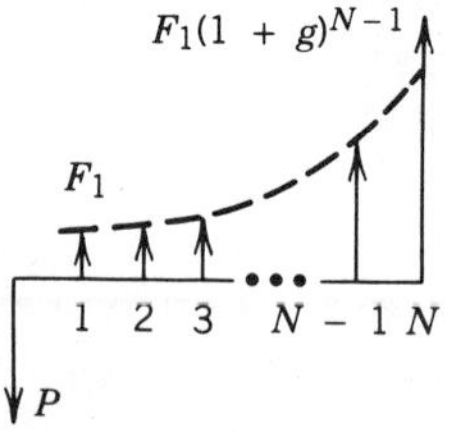 $F_1$, $F_1(1+g)^{N-1}$; 1 2 3 … N − 1 N; P |

Source: Park [3].

Using the alternative formula in Eq. 2.27, we first compute

$$g' = \frac{1.15}{1.10} - 1 = 0.04545$$

We then obtain

$$P = \frac{39{,}600}{1.10} \overset{P/A,\ 4.545\%,5}{(4.38442)} = 157{,}839.20$$

Although Eq. 2.27 looks more compact than Eq. 2.23, it does not provide any computational advantage in this example. □

All the interest formulas developed in Table 2.3 are applicable only to situations in which the compounding period coincides with the payment period. In the next section we discuss situations in which we have noncomparable payment and compounding periods.

### 2.4.2 Noncomparable Payment and Compounding Periods

Whenever the payment period and the compounding period do not correspond, we approach the problem by finding the effective interest rate based on the payment period and then using this rate in the compounding interest formulas in Table 2.3.

The specific computational procedure for noncomparable compounding and payment periods is as follows.

1. Identify the number of compounding periods per year ($M$), the number of payment periods per year ($K$), and the number of interest periods per payment period ($C$).
2. Compute the effective interest rate per payment period, using Eq. 2.6.

$$i = \left(1 + \frac{r}{M}\right)^C - 1$$

3. Find the total number of payment periods.

$$N = K \quad \text{(number of years)}$$

4. Use $i$ and $N$ in the appropriate formula given in Table 2.3.

#### *Example 2.6*

What is the present worth of a series of equal quarterly payments of $1,000 that extends over a period of 5 years if the interest rate is 8% compounded monthly? The variables are

$K = 4$ payment periods per year

$M = 12$ compounding periods per year

$C = 3$ interest periods per payment period (quarter)

$r = 8\%$

$$i = \left(1 + \frac{r}{M}\right)^C - 1 = \left(1 + \frac{0.08}{12}\right)^3 - 1 = 2.0133\% \text{ per quarter}$$

$N = (5)(4) = 20$ payment periods

Then the present amount is

$$P = A(P/A, i, N)$$
$$= \$1000(P/A, 2.0133\%, 20) = \$16{,}330.37 \quad \square$$

In certain situations the compounding periods occur *less* frequently than the payment periods. Depending on the financial institution involved, no interest may be paid for funds deposited during an interest period. The accounting methods used by most firms record cash transactions at the end of the period in which they have occurred, and any cash transactions that occur within a compounding period are assumed to have occurred at the end of that period. Thus, when cash flows occur daily but the compounding period is monthly, we sum the cash flows within each month (ignoring interest) and place them at the end of each month. The modified cash flows become the basis for any calculations involving the interest factors.

In the extreme situation in which payment occurs more frequently than compounding, we might find that the cash flows continuously throughout the planning horizon on a somewhat uniform basis. If this happens, we can also apply the approach discussed earlier (integrating instead of summing all cash flows that occur during the compounding period and placing them at the end of each compounding period) to find the present worth of the cash flow series. In practice, we avoid this cumbersome approach by adopting the funds flow concept, which is discussed in the next section.

### *Example 2.7*

Consider the cash flow diagram shown in Figure 2.8*a*, where the time scale is monthly. If interest is compounded quarterly, the cash flows can be relocated as shown in Figure 2.8*b*. The cash flow shown in Figure 2.8*b* is equivalent to the cash flow in Figure 2.8*a* for quarterly compounding. After the equivalent cash flow is determined, we can proceed as previously discussed for the situation in which the compounding periods and the payment periods coincide.

Let $i = 3\%$ per quarter. Then the present worth of cash flow given in Figure 2.8*a* is equivalent to the present worth of cash flow given in Figure 2.8*b*. Since $G = \$90$,

$$P = \$330\overset{P/A,3\%,8}{(7.0197)} + \$90\overset{P/G,3\%,8}{(23.4806)} = \$4{,}429.76 \quad \square$$

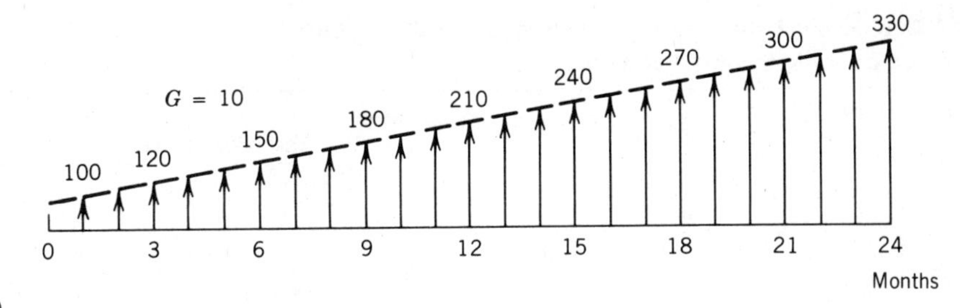


(a)

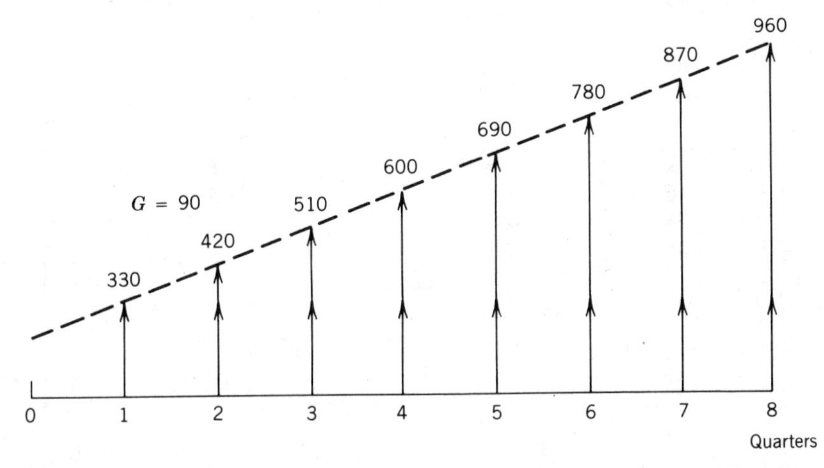


(b)

**FIGURE 2.8** Example of cash flows where compounding is less frequent than payment. (*a*) Original cash flows. (*b*) Equivalent quarterly cash flows.

## 2.5 CONTINUOUS COMPOUNDING

### 2.5.1 Discrete Payments

When payments occur at discrete points in time but interest is permitted to compound an infinite number of times per year (that is, continuously in time), we have the special instance of more frequent compounding than payments discussed in Section 2.4.2. Therefore, we approach the problem in the following way.

1. Identify the payment periods per year ($K$).
2. Compute $i = e^{r/K} - 1$ by using Eq. 2.7.
3. Find the total number of payment periods.

$$N = K \quad \text{(number of years)}$$

4. Use $i$ and $N$ in the appropriate interest formulas given in Table 2.3.

We can derive a new family of interest factors under continuous compounding by substituting $e^r - 1$ for $i$ when payments are annual and $e^{r/K} - 1$

**Table 2.4** ***Summary of Continuous Compounding Formulas with Annual Payments***

| Flow Type | Factor Notation | Formula | Cash Flow Diagram |
|---|---|---|---|
| Single | Compound amount $(F/P, r, N)$ | $F = P(e^{rN})$ |  |
| | Present worth $(P/F, r, N)$ | $P = F(e^{-rN})$ | |
| Equal payment series | Compound amount $(F/A, r, N)$ | $F = A\left(\frac{e^{rN} - 1}{e^{r} - 1}\right)$ | 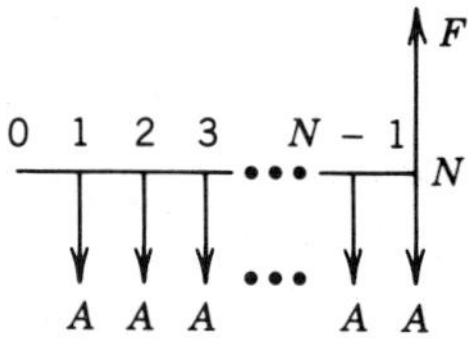 |
| | Sinking fund $(A/F, r, N)$ | $A = F\left(\frac{e^{r} - 1}{e^{rN} - 1}\right)$ | |
| | Present worth $(P/A, r, N)$ | $P = A\left[\frac{e^{rN} - 1}{e^{rN}(e^{r} - 1)}\right]$ | 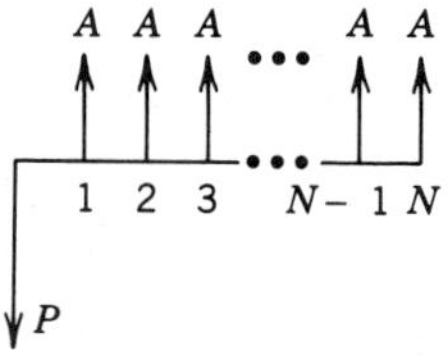 |
| | Capital recovery $(A/P, r, N)$ | $A = P\left[\frac{e^{rN}(e^{r} - 1)}{e^{rN} - 1}\right]$ | |
| Gradient series | Uniform gradient Present worth $(P/G, r, N)$ | $P = G\left[\frac{e^{rN} - 1 - N(e^{r} - 1)}{e^{rN}(e^{r} - 1)^2}\right]$ | 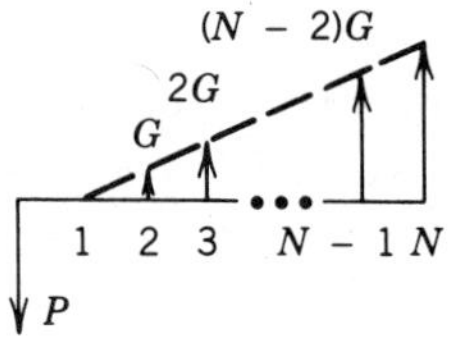 |
| | Geometric gradient Present worth $(P/A, g, r, N)$ | $P = \begin{cases} F_1\left[\frac{1 - e^{(g-r)N}}{e^{r} - e^{g}}\right] \\ \frac{NF_1}{e^{r}} \quad (\text{if } g = e^{r} - 1) \end{cases}$ | 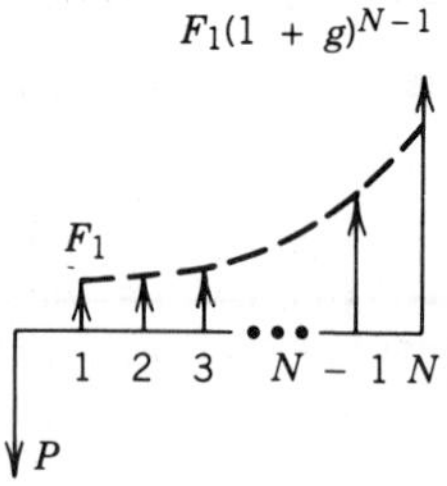 |

Source: Park [3].

for $i$ when payments are more frequent than annual. Table 2.4 summarizes the resulting compound interest factors for annual payments.

### *Example 2.8*

What is the present worth of a uniform series of year-end payments of $500 each for 10 years if the interest rate is 8% compounded continuously?

$$\begin{aligned} \text{Let } r &= 0.08 \\ i &= e^r - 1 = 8.33\% \\ N &= 10 \\ A &= \$500 \end{aligned}$$

Then

$$P = A\left[\frac{e^{rN} - 1}{e^{rN}(e^r - 1)}\right] = \$500(6.6117) = \$3{,}305.85$$

Using the discrete compounding formula with $i = 8.33\%$, we also find that

$$P = A\overset{P/A,8.33\%,10}{(6.6117)} = \$3{,}305.85 \quad \square$$

### *Example 2.9*

A series of equal quarterly payments of $1,000 extends over a period of 5 years. What is the present worth of this quarterly time series at 8% interest compounded continuously?

Since the payments are quarterly, the calculations must be quarterly. The required calculations are

$$\frac{r}{K} = \frac{8\%}{4 \text{ quarters}} = 2\% \text{ per quarter compounded continuously}$$

$$i = e^{r/K} - 1 = e^{0.02} - 1 = 0.0202 = 2.02\% \text{ per quarter}$$

$$N = (4 \text{ payment periods per year})(5 \text{ years}) = 20 \text{ periods}$$

$$P = A\left[\frac{e^{(r/K)N} - 1}{e^{(r/K)N}(e^{r/K} - 1)}\right] = \$1{,}000(15.3197) = \$16{,}319.70$$

Using the discrete compounding formula with $i = 2.02\%$, we also find that

$$P = A\overset{P/A,2.02\%,20}{(16.3197)} = \$16{,}319.70 \quad \square$$

### 2.5.2 Continuous Cash Flows

It is often appropriate to treat cash flows as though they were continuous rather than discrete. An advantage of the continuous flow representation is its

**Table 2.5** ***Summary of Interest Factors for Continuous Cash Flows with Continuous Compounding***

| Type of Cash Flow | Cash Flow Function | Parameters: To Find | Parameters: Given | Algebraic Notation | Factor Notation |
|---|---|---|---|---|---|
| Uniform (step) | 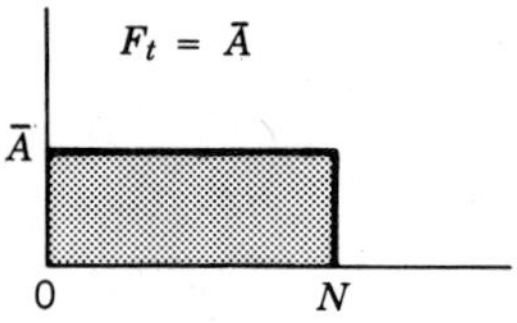 $F_t = \bar{A}$ | $P$ | $\bar{A}$ | $\bar{A}\left(\frac{e^{rN}-1}{re^{rN}}\right)$ | $(P/\bar{A}, r, N)$ |
| | | $\bar{A}$ | $P$ | $P\left(\frac{re^{rN}}{e^{rN}-1}\right)$ | $(\bar{A}/P, r, N)$ |
| | | $F$ | $\bar{A}$ | $\bar{A}\left(\frac{e^{rN}-1}{r}\right)$ | $(F/\bar{A}, r, N)$ |
| | | $\bar{A}$ | $F$ | $F\left(\frac{r}{e^{rN}-1}\right)$ | $(\bar{A}/P, r, N)$ |
| Gradient (ramp) | 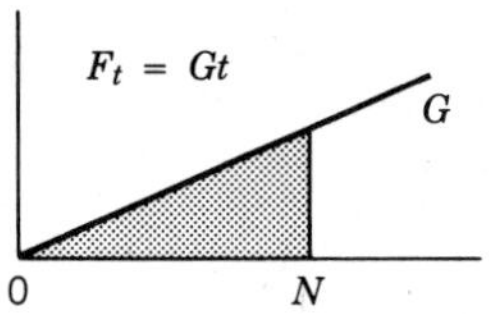 $F_t = Gt$ | $P$ | $G$ | $\frac{G}{r^2}(1 - e^{-rN}) - \frac{G}{r}(Ne^{-rN})$ | |
| Decay | 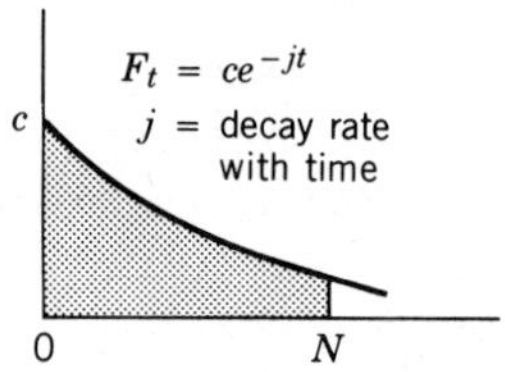 $F_t = ce^{-jt}$, $j$ = decay rate with time | $P$ | $c, j$ | $\frac{c}{r+j}(1 - e^{-(r+j)N})$ | |
| Exponential | 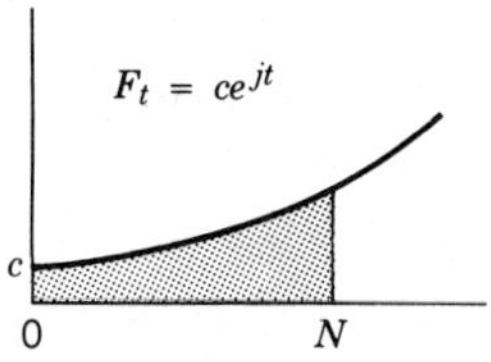 $F_t = ce^{jt}$ | $P$ | $c, j$ | $\frac{c}{r-j}(1 - e^{-(r-j)N})$ | |
| Growth | 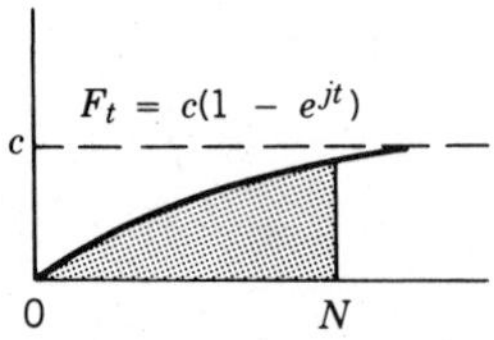 $F_t = c(1 - e^{jt})$ | $P$ | $c, j$ | $\frac{c}{r}(1 - e^{-rN}) - \frac{c}{r+j}(1 - e^{-(r+j)N})$ | |

flexibility for dealing with patterns other than the uniform and gradient ones. Some of the selected continuous cash flow functions are shown in Table 2.5.

To find the present worth of a continuous cash flow function under continuous compounding, we first recognize that the present-worth formula for a discrete series of cash flows with discrete compounding is

$$P = \sum_{n=0}^{N} F_n(1 + i)^{-n}$$

Since $F_n$ becomes a continuous function $F_t$ and the effective annual interest rate $i$ for continuous compounding is $e^r - 1$, integration of the argument instead of summation yields

$$P = \int_0^N (F_t)e^{-rt}\, dt \tag{2.28}$$

[Note that $n \to t$, $F_n \to F_t$, $\sum_{n=0}^{N} \to \int_0^N$, and $(1 + i)^{-n} \to e^{-rt}$.] Then the future value equivalent of $F_t$ over $N$ periods is simply

$$F = \int_0^N F_t e^{rt}\, dt \tag{2.29}$$

To illustrate the continuous flow concept, consider $F_t$ to be a uniform flow function when an amount flows at the rate $\bar{A}$ per period for $N$ periods. (This cash flow function is presented in Table 2.5 and is expressed as $F_t = \bar{A}$, $0 \leq t \leq N$.) Then the present-worth equivalent is

$$P = \int_0^N \bar{A}e^{-rt}\, dt = \bar{A}\left(\frac{e^{rN} - 1}{re^{rN}}\right) = \bar{A}\left(\frac{1 - e^{-rN}}{r}\right) \tag{2.30}$$

The resulting factor in parenthesis in (2.30) is referred to as the *funds flow present-worth factor* and is designated $(P/\bar{A}, r, N)$. The future-worth equivalent is obtained from

$$F = \int_0^N \bar{A}e^{rt}\, dt = \bar{A}\left(\frac{e^{rN} - 1}{r}\right) \tag{2.31}$$

The resulting factor $(e^{rN} - 1)/r$ is called the *funds flow compound amount factor* and is designated $(F/\bar{A}, r, N)$. Since the relationships of $\bar{A}$ to $P$ and $F$ are given by Eqs. 2.30 and 2.31, we can easily solve for $\bar{A}$ if $P$ or $F$ is given. Table 2.5 summarizes all the funds flow factors necessary to find present-worth and future-worth equivalents for a variety of cash flow functions.

As a simple example, we compare the present-worth figures obtained in two situations. We deposit $10 each day for 18 months in a savings account that has an interest rate of 12% compounded daily. Assuming that there are 548 days in the 18-month period, we compute the present worth.

$$P = 10\overset{P/A,0.032877\%,548}{(501.4211)} = \$5{,}014.21$$

Now we approximate this discrete cash flow series by a uniform continuous cash flow profile (assuming continuous compounding). In doing so, we may define $\bar{A}$ as

$$\bar{A} = 10(365) = \$3{,}650/\text{year}$$

Note that our time unit is a year. Thus, an 18-month period is 1.5 years. Substituting these values back into Eq. 2.30 yields

$$P = \int_0^{1.5} 3{,}650e^{-0.12t} = \frac{3{,}650}{0.12}\left(1 - e^{-0.18}\right)$$
$$= \$5{,}010.53$$

The discrepancy between the values obtained by the two methods is only \$3.68.

### *Example 2.10*

A county government is considering building a road from downtown to the airport to relieve congested traffic on the existing two-lane divided highway. Before allowing the sale of a bond to finance the road project, the court has requested an estimate of future toll revenues over the bond life. The toll revenues are directly proportional to the growth of traffic over the years, so the following growth cash flow function (with units in millions of dollars) is assumed to be reasonable.

$$F_t = 5(1 - e^{-0.10t})$$

Find the present worth of toll revenues at 6% interest compounded continuously over a 25-year period.

Expanding $F_t$ gives us

$$F_t = 5 - 5e^{-0.10t}$$

If we let $f(t)_1 = 5$ and $f(t)_2 = -5e^{-0.10t}$, the present-worth equivalent for each function would be

$$P_1 = \int_0^{25} 5e^{-0.06t}\,dt = 5\left[\frac{e^{0.06(25)} - 1}{(0.06)e^{0.06(25)}}\right] = \$64.74$$

$$P_2 = \int_0^{25} -5e^{-(0.10 + 0.06)t}\,dt = -5\left[\frac{e^{0.16(25)} - 1}{(0.16)e^{0.16(25)}}\right] = -\$30.68$$

and

$$P = P_1 + P_2 = \$34.06$$

The present worth of toll revenues over a 25-year period amounts to \$34.06 million. This figure could be used for bond validation. □

## 2.6 EQUIVALENCE OF CASH FLOWS

### 2.6.1 Concept of Equivalence

When we compare two cash flows, we must compare their characteristics on the same basis. By definition, two cash flows are equivalent if they have the same economic effect. More precisely, two cash flows are equivalent at interest $i$ if we can convert one cash flow into the other by using the proper compound interest factors. For example, if we deposit \$100 in a bank for 3 years at 8% interest compounded annually, we will accumulate \$125.97. Here we may say that, at 8% interest, \$100 at time 0 is equivalent to \$125.97 at time 3.

Consider another example in which an individual has to choose between two options. Option I is to receive a lump sum of \$1,000 now. Option II is to receive \$600 at the end of each year for 2 years, which provides \$1,200 over the 2-year period. Our question is what interest rate makes these two options equivalent. To answer the question, we need to establish a common base in time to convert the cash flows. Three common bases are the equivalent future value $F$, the equivalent present value $P$, and the equivalent annual value $A$. Future value is a measure of the cash flow relative to some "future planning horizon," considering the earning opportunities of the intermediate cash receipts. The present value represents a measure of future cash flow relative to the time point "now" with provisions that account for earning opportunities. The annual equivalent value determines the equal payments on an annual basis. The uniform cash flow equivalent might be the more appropriate term to use. The conceptual transformation from one type of cash flow to another is depicted in Figure 2.9.

For our example, we will use $F$ as a base of reference value and set the planning horizon at the end of year 2. To find the future value of option I, we may use an $(F/P, i, A)$ factor.

$$F_{\text{I}} = 1000(F/P, i, 2) = 1000(1 + i)^2$$

For option II, we may use an $(F/A, i, n)$ factor.

$$F_{\text{II}} = 600(F/A, i, 2) = 600(1 + i) + 600$$

If we specify $i$, we can easily evaluate $F_{\text{I}}$ and $F_{\text{II}}$. Table 2.6 summarizes these values at selected interest rates. We observe from the table that $F_{\text{I}} = F_{\text{II}} = \$1{,}278$ at $i = 13\%$. In other words, the two options are equivalent if the individual can earn a 13% interest from the investment activity. We also observe that at $i = 13\%$ $P_{\text{I}} = P_{\text{II}}$ and $A_{\text{I}} = A_{\text{II}}$. This is not surprising, because the present value amount is merely the future amount times a constant. The same can be said for the annual

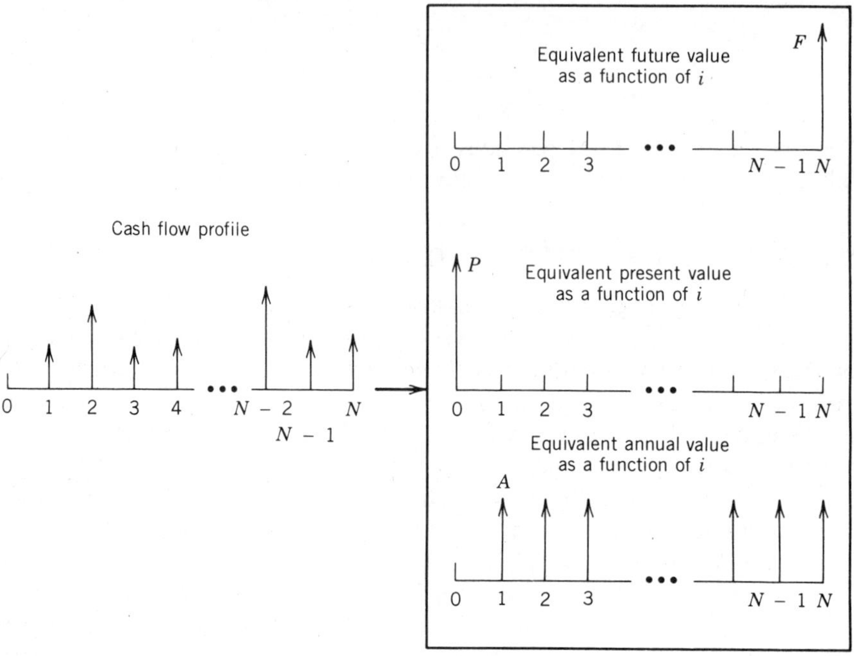


**FIGURE 2.9** Conversion to equivalent bases.

## Table 2.6 *Equivalence Calculations*

| $i$ (%) | Option I $F_I = 1{,}000(1 + i)^2$ | Option II $F_{II} = 600(1 + i) + 600$ | Equivalence |
|---|---|---|---|
| 0 | \$1,000 | \$1,200 | |
| 5 | 1,103 | 1,230 | |
| 7 | 1,145 | 1,242 | $F_I < F_{II}$ |
| 12 | 1,254 | 1,272 | |
| 13 | 1,277 | 1,277 | $F_I = F_{II}$ |
| 15 | 1,323 | 1,290 | $F_I > F_{II}$ |
| 20 | 1,440 | 1,320 | |

equivalent value amount. Therefore, we should expect that any equivalent value that directly compares future value amounts could just as well compare present value amounts or annual equivalent amounts without affecting the selection outcome.

### 2.6.2 Equivalence Calculations with Several Interest Factors

Thus far we have used only single factors to perform equivalence calculations. In many situations, however, we must use several interest factors to obtain an equivalent value. To show this, we will take an example from home financing instruments offered by many banks. The particular financing method to be considered is called the graduated-payment method (GPM). This mortgage financing is designed for young people with low incomes but good earning prospects. (The term mortgage refers to a special loan for buying a piece of property such as a house.) The Department of Housing and Urban Development (HUD) initiated the GPM with a fixed interest rate for 30 years. During the first 5 or 10 years the monthly payments increase in stair-step fashion each year, allowing buyers to make a lower monthly payment in the beginning; the payments then level off at an amount higher than those of a comparable conventional fixed-rate mortgage. The monthly payment is applied to both principal and interest and can carry negative amortization. (The loan balance actually grows instead of decreasing under negative amortization when monthly payments are lower than monthly loan interests.) Our question is how the monthly payments are computed for a certain loan amount, interest rate, and life of the loan.

Let

$P$ = loan amount,

$A$ = monthly payment for the first year,

$i$ = loan interest rate per month,

$g$ = annual rate of increase in the monthly payment,

$K$ = number of years the payment will increase,

$N$ = number of months to maturity of the loan.

Figure 2.10*a* illustrates the cash flow transactions associated with the GPM. From the lender's view, lending the amount $P$ now should be equivalent to a transaction in which the monthly payments are as shown in Figure 2.10*a*. To establish the equivalence relation between $P$ and $A$ with fixed values of $i$, $g$, $K$, and $N$, we convert each group of 12 equal monthly payments to a single present equivalent amount at the beginning of each year. Then the remaining $(N - 12K)$ equal payments are converted to a single present equivalent amount at $n = 12K$. The equivalent cash flow after this transformation should look like Figure 2.10*b*. To find the present equivalent value of this transformed cash flow, we simply calculate

$$P = A(P/A, i, 12) + A(1 + g)(P/A, i, 12)(P/F, i, 12)$$
$$+ A(1 + g)^2(P/A, i, 12)(P/F, i, 24) + \cdots$$

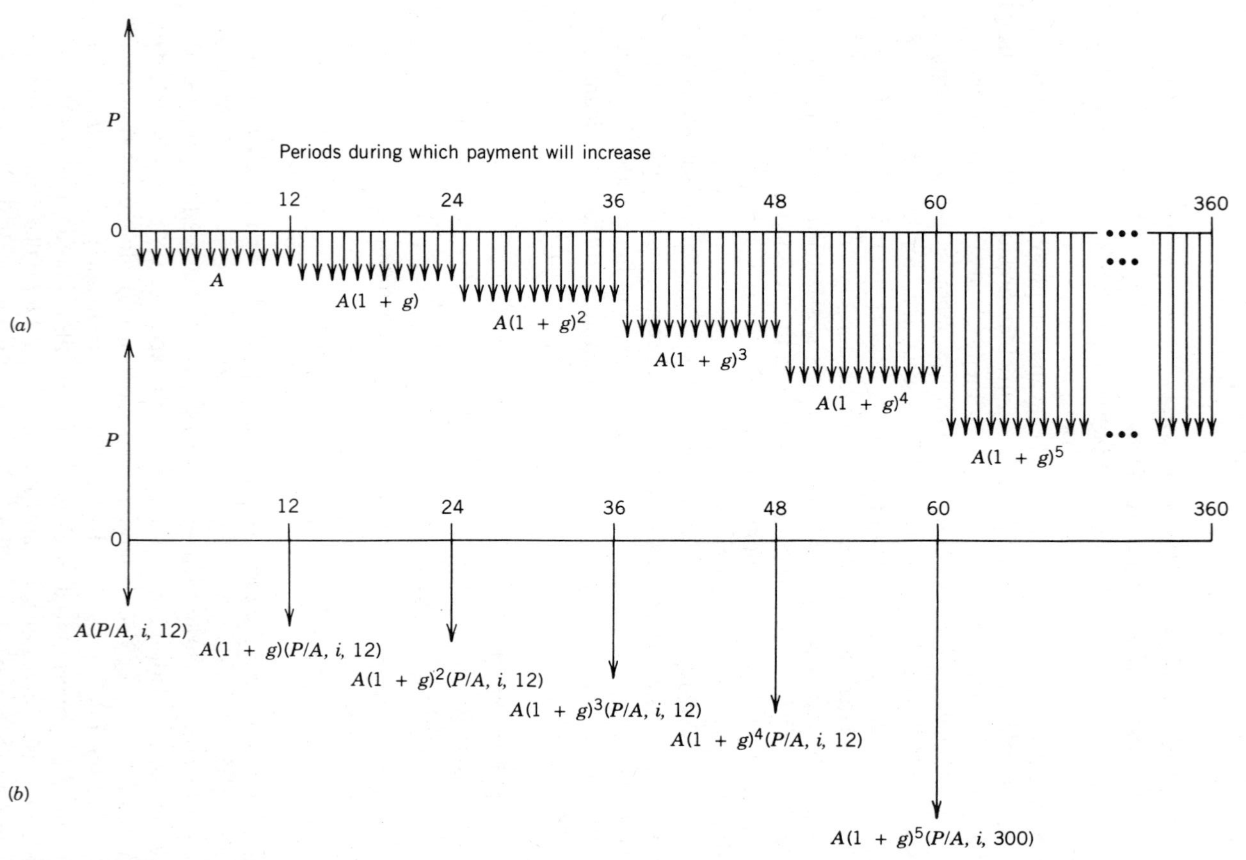


**FIGURE 2.10** Cash flow diagram of a typical GPM loan. (*a*) Loan transactions (monthly). (*b*) Equivalent transactions.

$$+ A\{(1 + g)^{K-1} (P/A, i, 12)[P/F, i, 12(K - 1)]\}$$
$$+ A(1 + g)^{K}(P/A, i, N - 12K)(P/F, i, 12K) \quad (2.32)$$

We multiply each term by $(1 + i)^{12}$ or $(F/P, i, 12)$.

$$\begin{aligned} P(1 + i)^{12} = {} & A(P/A, i, 12)(1 + i)^{12} + A(1 + g)(P/A, i, 12) \\ & + A(1 + g)^{2}(P/A, i, 12)(1 + i)^{-12} + \cdots \\ & + A(1 + g)^{K-1}(P/A, i, 12)(1 + i)^{-12(K-2)} \\ & + A(1 + g)^{K}(P/A, i, N - 12K)(1 + i)^{-12(K+1)} \end{aligned} \quad (2.33)$$

We multiply each term in Eq. 2.32 by $1 + g$.

$$\begin{aligned} P(1+g) = {} & A(1+g)(P/A, i, 12) + A(1+g)^{2}(P/A, i, 12)(1+i)^{-12} \\ & + A(1 + g)^{3}(P/A, i, 12)(1 + i)^{-24} + \cdots \\ & + A(1 + g)^{K}(P/A, i, 12)(1 + i)^{-12(K-1)} \\ & + A(1 + g)^{K+1}(P/A, i, N - 12K)(1 + i)^{-12K} \end{aligned} \quad (2.34)$$

Now we subtract Eq. 2.34 from Eq. 2.33 and solve for $A$ to get

$$\begin{aligned} A = {} & P[(1 + i)^{12} - (1 + g)]\{(1 + g)^{K}(1 + i)^{-12(K+1)}[(P/A, i, N - 12K) \\ & - (P/A, i, 12)] + [(P/A, i, 12)(1 + i)^{12} \\ & - (1 + g)^{K+1}(P/A, i, N - 12K)(1 + i)^{-12K}]\}^{-1} \end{aligned} \quad (2.35)$$

For an example of such an equivalence calculation, consider the following data:

$P$ = \$45,000,
$i = \frac{3}{4}\%$ per month (9% compounded monthly),
$g$ = 5% per year,
$K$ = 5 years (no further increase in monthly payment after the sixth year),
$N$ = 360 months (30 years).

Evaluating Eq. 2.35 with these figures yields

$$\begin{aligned} A &= 45{,}000[0.04387]\{0.8916\,[107.7267] + 12.5076 - 101.9927\}^{-1} \\ &= \$300.18/\text{month} \end{aligned}$$

Then the monthly payment will increase the second year to \$315.19, the third year to \$330.95, the fourth year to \$347.50, the fifth year to \$364.87, and for the remaining years to \$383.11.

***Example 2.11***

The following two cash flow transactions are said to be equivalent in terms of economic desirability at an interest rate of 10% compounded annually. Determine the unknown value *A*.

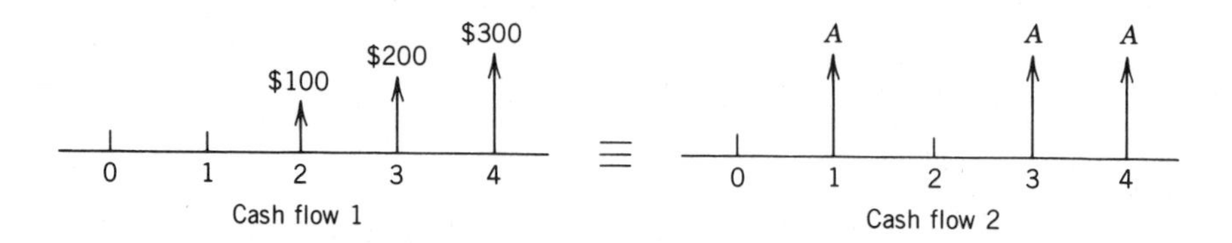


We will first use the present equivalent as the basis of comparison. Cash flow 1 represents a strict gradient series, whereas cash flow 2 can be viewed as an equivalent payment series with the second payment missing. Therefore, the equivalence would be expressed by

$$100\overset{P/G,10\%,4}{(4.3781)} = A\left[\overset{P/F,10\%,1}{(0.9091)} + \overset{P/F,10\%,3}{(0.7513)} + \overset{P/F,10\%,4}{(0.6830)}\right]$$

Solving for $A$ yields

$$A = \$186.83$$

If we use the annual equivalent as the basis of comparison, we compute

$$100\overset{A/G,10\%,4}{(1.3812)} = A - A\overset{P/F,10\%,2}{(0.8264)}\ \overset{A/P,10\%,4}{(0.3155)}$$

Solving for $A$ yields $A = \$186.83$ again. The second approach should be computationally more attractive because it takes advantage of the cash flow pattern and thus requires fewer interest factors in the computation. □

## 2.7 EFFECT OF INFLATION ON CASH FLOW EQUIVALENCE

Up to this point we have shown how we properly account for the time value of money in equivalence calculations in the absence of inflation. In this section we present methods that incorporate the effect of inflation in our equivalence calculations.

### 2.7.1 Measure of Inflation

***Definition.*** Before discussing the effect of inflation on equivalence calculations, we need to discuss how we measure inflation. In simple terms, the results of investment activity are stated in dollars, but the dollar is an imperfect unit of measure because its value changes from time to time. Inflation is the term used to describe a decline in the value of the dollar. For example, if we deposit $1,000 in a one-year savings certificate and withdraw $1,090 a year later, we say that our rate of return has been 9%—and it has, as long as those dollars we withdraw at year's end actually purchase 9% more. If inflation has reduced the value of the dollar by 10%, our 9% positive investment return in dollars is actually about a 1% loss in economic value or purchasing power. Inflation is thus a measure of the decline in the purchasing power of the dollar.

***Measure.*** The decline in purchasing power can be measured in many ways. Consumers may judge inflation in terms of the prices they pay for food and other goods; economists record this measure in the form of the consumer price index (CPI), which is based on sample prices in a "market basket" of purchases. We should note that consumer prices do not always behave like wholesale prices or commodity prices, and as a result, a dollar's worth varies depending on what is bought.

There is another measure of the dollar's value that reflects the average purchasing power of the dollar as it applies to all goods and services in the economy—the gross national product implicit price deflator (GNPIPD). The GNPIPD is computed and published quarterly by the U.S. Department of Commerce, Bureau of Economic Analysis.

Various cost indices are also available to the estimator. A government index listing is given by the *Statistical Abstract of the United States,* a yearly publication that includes material, labor, and construction costs. The Bureau of Labor Statistics publishes the monthly *Producer Price Index* and covers some 3,000 product groupings.

***Average Inflation Rate.*** To account for the effect of inflation, we utilize an annual percentage rate that represents the annual increase in prices over a one-year period. Because the rate each year is based on the previous year's price, this inflation rate has a compounding effect. For example, prices that increase at the rate of 5% per year in the first year and 8% per year in the second year, with a starting base price of $100, will increase at an average inflation rate of 6.49%.

$$\underbrace{\underbrace{100(1 + 0.05)}_{\text{first year}}(1 + 0.08)}_{\text{second year}} = 113.40$$

Let $f$ be the average annual inflation rate. Then we equate

$$100(1 + f)^2 = 113.40$$
$$f = 6.49\%$$

The inflation rate itself may be computed from any of the several available indices. With the CPI value, the annual inflation rate may be calculated from the expression

$$\text{Annual inflation rate for year } n = \frac{\text{CPI}_n - \text{CPI}_{n-1}}{\text{CPI}_{n-1}} \tag{2.36}$$

For example, with $\text{CPI}_{1990} = 270$ and $\text{CPI}_{1989} = 260$, the annual inflation rate for year 1990 is

$$\frac{270 - 260}{260} = 0.0385 \text{ or } 3.85\%$$

As just indicated, we can easily compute the inflation rates for the years with known CPI values. However, most equivalence calculations for projects require the use of cash flow estimates that depend on expectations of *future* inflation rates. The methods used by economists to estimate future inflation rates are many and varied. Important factors to consider may include historical trends in rates, predicted economic conditions, professional judgment, and other elements of economic forecasting. The estimation of future inflation rates is certainly a difficult task; a complete discussion of this subject is beyond the scope of this text but can be found elsewhere [1]. Our interest here is in how we use these rates in equivalence calculations, when they are provided.

### 2.7.2 Explicit and Implicit Treatments of Inflation in Discounting

We will present three basic approaches for calculating equivalence values in an inflationary environment that allow for the simultaneous consideration of changes in earning power and changes in purchasing power. The three approaches are consistent and, if applied properly, should result in identical solutions. The first approach assumes that cash flow is estimated in terms of *actual dollars,* and the second uses the concept of *constant dollars.* The third approach uses a combination of actual and constant dollars and is discussed in Section 2.7.3.

***Definition of Inflation Terminology.*** To develop the relationship between actual-dollar analysis and constant-dollar analysis, we will give precise definitions of several inflation-related terms, borrowed from Thuesen and Fabrycky [4].

*Actual dollars* represent the out-of-pocket dollars received or expended at any point in time. Other names for them are then-current dollars, current dollars, future dollars, inflated dollars, and nominal dollars.

*Constant dollars* represent the hypothetical purchasing power of future receipts and disbursements in terms of the purchasing dollars in some base year. (The base year is normally time zero, the beginning of the investment.) We will assume that the base year is always time zero unless specified otherwise. Other names are real dollars, deflated dollars, and today's dollars.

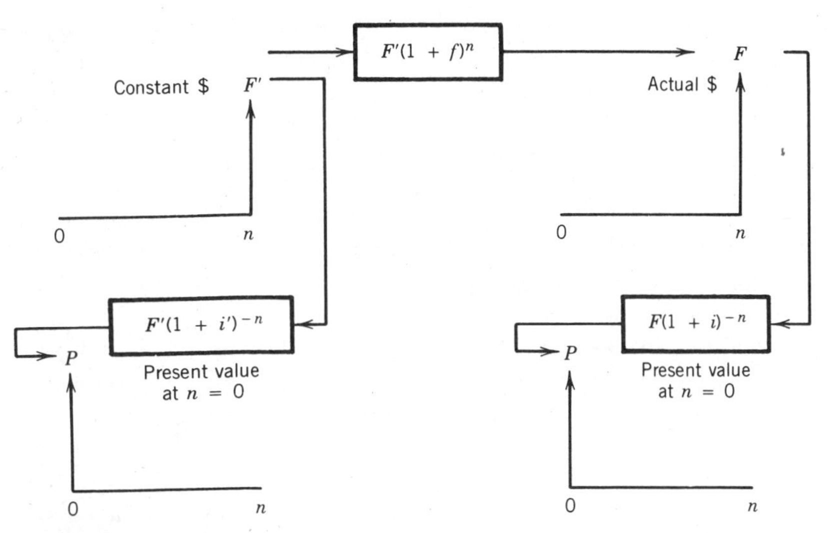


**FIGURE 2.11 Relationships of $i$, $i^1$ and $f$.**

As shown in Figure 2.11, to find the present value equivalent of this actual dollar, we should use the market interest rate $i$ in

$$P = F(1 + i)^{-n} \tag{2.38}$$

If the cash flow is already given in constant dollars with the inflation effect removed, we should use $i'$ to account for only the earning power of the money. To find the present value equivalent of this constant dollar at $i'$, we use

$$P = F'(1 + i')^{-n} \tag{2.39}$$

The $P$ values must be equal at time zero, and equating the results of Eqs. 2.38 and 2.39 yields

$$\begin{aligned} F(1 + i)^{-n} &= F'(1 + i')^{-n} \\ &= F(1 + f)^{-n}(1 + i')^{-n} \\ (1 + i)^{-n} &= (1 + f)^{-n}(1 + i')^{-n} \\ (1 + i) &= (1 + f)(1 + i') \\ &= 1 + f + i' + i'f \end{aligned}$$

or

$$i = i' + f + i'f \tag{2.40}$$

*Market interest rate* ($i$) represents the opportunity to earn as reflecte the actual rates of interest available in the financial market. The inte rates used in previous sections are actually market interest rates. ( designation $i$ is used consistently throughout this book to represent terest rates available in the marketplace.) When the rate of inflation creases, there is a corresponding upward movement in market inte rates. Thus, the market interest rates include the effects of both the earn power and the purchasing power of money. Other names are combin interest rate, nominal interest rate, minimum attractive rate of return, a inflation-adjusted discount rate.

*Inflation-free interest rate* ($i'$) represents the earning power of mon isolated from the effects of inflation. This interest rate is not quoted b financial institutions and other investors and is therefore not general known to the public. This rate can be computed, however, if the mark interest rate and inflation rate are known. Naturally, if there is no inflatio in an economy, $i$ and $i'$ should be identical. Other names are real interes rate, true interest rate, and constant-dollar interest rate.

*General inflation rate* ($f$) represents the average annual percentage of increase in prices of goods and services. The market inflation rate is expected to respond to this general inflation rate. *Escalation rate* ($e$) represents a specific inflation rate applicable to a specific segment of the economy. It is sometimes used in contracts.

It is important to recognize that there is a relationship between inflation and interest rate. For example, the historical rate on AAA bonds is about 2.5% to 3% above the general inflation rate as measured by the CPI [1]. In addition, the rate of return (ROR) required by well-managed companies on their investments must be at some level above the inflation rate. In the next section we will derive the mathematical relationships of $i$, $i'$, and $f$.

***Relationships of i, i′, and f.*** We must first establish the relationship between actual dollars and constant dollars. Suppose we estimate a future single payment $F'$ that occurs at the end of the $n$th period in terms of constant dollars (primes indicate constant dollars). To translate this constant-dollar amount into the actual dollars at the end of the $n$th period, we use

$$F = F'(1 + f)^n$$

Solving for $F'$ yields

$$F' = F(1 + f)^{-n} \tag{2.37}$$

where $F'$ = constant-dollar expression for the cash flow at the end of the $n$th period,

$F$ = actual-dollar expression for the cash flow at the end of the $n$th period.

Solving for $i'$ yields

$$i' = \frac{i - f}{1 + f} \tag{2.41}$$

As an example, say that the inflation rate is 6% per year and the market interest rate is known to be 15% per year. Calculating $i'$ gives us

$$i' = \frac{0.15 - 0.06}{1 + 0.06} = 8.49\%$$

To summarize, the interest rate that is applicable in equivalence calculations depends on the assumptions used in estimating the cash flow. If the cash flow is estimated in terms of actual dollars, the market interest rate ($i$) should be used. If the cash flow is estimated in terms of constant dollars, the inflation-free interest rate ($i$) should be used. In subsequent sections we will give more detailed examples of how the two interest rates are used in equivalence calculations.

***Actual-Dollar versus Constant-Dollar Analysis.*** If cash flow is represented in constant dollars (such as 1990 dollars), an inflation-free discount rate $i$ (say 5% to 15%) may be appropriate for a profitable business. If cash flow is represented in inflated dollars, a market interest rate (say 15% to 25%) may be appropriate. Often, the difficulty lies in determining the nature of the cash flow. In this section, we will consider two cases and explain how the analyses in terms of actual and constant dollars can be used.

**Case 1:** Projections in physical units can often be translated into constant-dollar projections by using a constant-dollar price per unit and then converted to present value by using an inflation-free discount rate.

## *Example 2.12*

SM Manufacturing Company makes electric meters of the type with which utility companies measure electricity consumption by users. SM has projected the sale of its meters by using data on new housing starts and deterioration and replacement of existing units. The price per unit should keep up with the wholesale price index (WPI). In 1990 the price per unit is $25. To achieve the production and sales projected in the following, SM needs to invest $75,000 now (in 1990). Other costs remain unchanged.

| $n$ | 0 | 1 | 2 | 3 | 4 | 5 | 6 | 7 |
|---|---|---|---|---|---|---|---|---|
| Unit Sales | — | 1,000 | 1,100 | 1,200 | 1,200 | 1,300 | 1,300 | 1,200 |
| $ Inflow | — | 25,000 | 27,500 | 30,000 | 30,000 | 32,500 | 32,500 | 30,000 |

SM thinks it should earn a 5% inflation-free rate of return (ROR) on any investment.

This is an easy problem because all figures are in constant (1990) dollars. Just discount the dollar inflows at 5%. For example, present value would be

$$
\begin{aligned}
P &= -75{,}000 + 25{,}000(1/1.05) + 27{,}500(1/1.05)^2 \\
&\quad + 30{,}000(1/1.05)^3 + 30{,}000(1/1.05)^4 + 32{,}500(1/1.05)^5 \\
&\quad + 32{,}500(1/1.05)^6 + 30{,}000(1/1.05)^7 \\
&= 95{,}386 \text{ in 1990 dollars} \quad \square
\end{aligned}
$$

**Case 2:** If projections in dollars are made with numerical and statistical techniques, they will very likely reflect some inflationary trend. If they do, we should use a market interest rate or a two-step approach in which we first convert to constant dollars and then compute present value by using an inflation-free discount rate.

### *Example 2.13*

U.S. Cola Company (USCC) is studying a new marketing scheme in southeast Georgia. By examining a similar project conducted from 1977 to 1989 and using nonlinear statistical regression, the analysts have projected additional dollar profits from this new marketing practice as follows.

| Year | 1 (1991) | 2 | 3 | 4 | 5 | 6 |
|---|---|---|---|---|---|---|
| Additonal Profit | 100,000 | 120,000 | 150,000 | 200,000 | 150,000 | 100,000 |

An investment of $500,000 is required now (1990) to fund the project. USCC is accustomed to obtaining a 20% ROR on its projects during these inflation-ridden times.

Statistical regression on dollar sales inevitably reflects any inflationary trends during the study period (1977 to 1989 in this example), so we may conclude that the dollar profits are represented in inflated, actual dollars. The 20% discount rate was developed for today's inflationary economy, so it can be used to compute a present value:

$$
\begin{aligned}
P &= -500{,}000 + 100{,}000(1/1.2) + 120{,}000(1/1.2)^2 \\
&\quad + 150{,}000(1/1.2)^3 + 200{,}000(1/1.2)^4 \\
&\quad + 150{,}000(1/1.2)^5 + 100{,}000(1/1.2)^6 \\
&= -56{,}306 \text{ in 1990 dollars}
\end{aligned}
$$

(Note that in the sign convention used a minus sign means cash outflow.) □

### *Example 2.14*

The scenario is the same as in example 2.13, but we assume that inflation is projected to be 9% per year, and we do the analysis by first converting to constant dollars. USCC expects at least a 10% inflation-free return on its investments. Noting that

$$(1 + 0.1)(1 + 0.09) = 1.990 \cong (1 + 0.2)$$

we judge this to be a reasonable translation. We first deflate the cash flow at 9%.

| $n$ | 1 | 2 | 3 | 4 | 5 | 6 |
|---|---|---|---|---|---|---|
| $F'_n$ | 91,743 | 101,002 | 115,828 | 141,685 | 97,490 | 59,627 |

Now we compute a present value using the constant-dollar cash flow with appropriate interest rate of 10%:

$$\begin{aligned} P &= -500{,}000 + 91{,}743(1/1.1) + 101{,}002(1/1.1)^2 \\ &\quad + 115{,}828(1/1.1)^3 + 141{,}685(1/1.1)^4 \\ &\quad + 97{,}490(1/1.1)^5 + 59{,}627(1/1.1)^6 \\ &= -55{,}137 \text{ in 1990 dollars} \end{aligned}$$

This value agrees closely with that obtained by using actual dollars and the market interest rate of 20%; the discrepancy comes from the fact that $1.199 \neq 1.200$. □

***Composite Cash Flow Elements with Different Escalation Rates.*** The equivalence calculation examples in the previous sections were all based on the assumption that all cash flows respond to the inflationary trend in a uniform manner. Many project cash flows, however, are composed of several cash flow elements with different degrees of responsiveness to the inflationary trend. For example, the net cash flow elements for a certain project may comprise sales revenue, operating and maintenance costs, and taxes. Each element may respond to the inflationary environment to a varying degree. In computing the tax element alone, we need to isolate the depreciation element. With inflation, sales and operating costs are assumed to increase accordingly. Depreciation would be unchanged, but taxes, profits, and thus the net cash flow usually would be higher. (A complete discussion of the effect of inflation on the after-tax cash flow will be given in Chapter 4.) Now we will discuss briefly how we compute the equivalence value with such cash flows.

In complex situations there may be several inflation rates. For example, an apartment developer might project physical unit sales, building costs in actual dollars using a building cost index, and sales revenue in actual dollars using a real estate price index, and then find the equivalent present value using an interest rate that reflects the consumer price index.

### *Example 2.15*

This more complex example illustrates the apartment building project. Base year cost per unit is \$15,000 and selling price per unit is \$20,000. The building cost index is projected to increase 11% next year and 10% more the following year. The real estate price index is expected to jump 15% next year and then level off at a 13% increase per year. We will use a market interest rate of 15%, hoping that it will yield an inflation-free return of 5% when the general inflation rate is 9% to 10% (to be precise, $f$ = 9.52%).

| Item | $n$: 0 | 1 | 2 | 3 |
|---|---|---|---|---|
| Units built | 200 | 250 | 200 | — |
| Units sold | — | 200 | 250 | 200 |
| Costs (thousands) | 3,000 | 3,750(1.11) | 3,000(1.11)(1.1) | — |
| Revenues (thousands) | — | 4,000(1.15) | 5,000(1.15)(1.13) | $4{,}000(1.15)(1.13)^2$ |
| Net flow (thousands) (actual \$) | −3,000 | +438 | +2,835 | +5,874 |
| ($P/F$, 15%, $n$) | 1 | 0.8696 | 0.7561 | 0.6575 |

$$P = -3{,}000 + 438(0.8696) + 2{,}835(0.7561) + 5{,}874(0.6575)$$

$$P = \$3{,}387{,}000 \text{ in base year (time 0) dollars} \quad \square$$

### 2.7.3 Home Ownership Analysis during Inflation

A personal decision of wide and continuing interest is whether it is more economical to buy a home or to rent during an inflationary environment. In this section we will illustrate how this decision can be made on a rational basis by applying the concepts of actual and constant dollars.

***Renting a House.*** To make a meaningful comparison, let's estimate the current rent of a two-bedroom apartment as \$400 per month plus \$60 per month for basic utilities (heating and cooling but not telephone, water, and sewer). Both costs have a tendency to increase with inflation, so let's project a 10% inflation rate, which gives us the following monthly costs per year.

| $n$ | 1 | 2 | 3 | | 10 |
|---|---|---|---|---|---|
| Rent | 400 | 440 | 484 | . . . | 943 |
| Utilities | 60 | 66 | 73 | . . . | 141 |

We selected a planning period of 10 years because realtors tell us that very few people live in the same house for the period of a home mortgage (typically 25 to 30 years). Of course, when you rent an apartment you are free to switch every year, and we'll assume a fairly uniform market of rents with no rent control (this situation occurs when the vacancy rate is 5% to 10%). Let's use a market interest rate of 15% (annual compounding) to compute the present value of apartment living costs (approximate, since we collapse all monthly flows to the year's end).

$$P = (-460)(12)/1.15 + (-506)(12)/(1.15)^2 + \cdots + (-1{,}084)(12)/(1.15)^{10}$$
$$= -39{,}610 \text{ in time 0 dollars}$$

Alternatively, we can compute an inflation-free discount rate $i'$ to be used with constant dollars by applying Eq. 2.40.

$$0.15 = i' + 0.1 + 0.1i'$$
$$i' = 0.0455$$

We must also convert 460 to 460/1.1 = 418.18. Thus, a present value using the constant-dollar cash flow is

$$P = (-418.18)(12)(P/A,\ 4.55\%,10)$$

and

$$(P/A,\ 4.55\%,\ 10) = \left[\frac{(1.0455)^{10} - 1}{0.0455(1.0455)^{10}}\right] = 7.8933$$

so

$$P = (-418.18)(12)(7.8933)$$
$$= -39{,}610 \text{ in time 0 dollars}$$

***Buying a House.*** Now we must estimate the cash flow for a house or condominium. The purchase cost will be $60,000. "Wait a minute!" you say. "I've seen those $60,000 units and they're too old, too small, or too far away, or built like apartments." Right. It's difficult to compare the space and quality of an apartment with those of a house, but it is not fair to compare a two-bedroom apartment with a new, close-in home or condominium containing 1,500 or more square feet. Therefore, the $60,000 home is a more appropriate comparison. If you finally decide to spend $80,000, you're allocating more money to your residence than when you lived in apartments, but you'll get more space, privacy, convenience, return, and so forth.

We will try for 95% financing, which means that we need a $3,000 down payment plus about another $3,000 for closing costs, for a cash requirement of about $6,000.

The mortgage interest rate might be 14.5% (total 14.5/12 = 1.208% per month) on a fixed-rate 30-year mortgage. So the monthly payment is

$$57{,}000\ (A/P,\ 1.208\%,\ 360) = (57{,}000)\left[\frac{0.01208(1.01208)^{360}}{(1.01208)^{360} - 1}\right]$$

$$= (57{,}000)(0.012242)$$

$$= \$697.815 = \$698/\text{month}$$

The mortgage balance remaining after our 10-year comparison period is

$$697.815(P/A,\ 1.208\%,\ 240) = 697.815\left[\frac{(1.01208)^{240} - 1}{0.01208(1.1208)^{240}}\right]$$

$$= (697.815)(78.143) = \$54{,}529$$

We will have paid off less than 5% of the loan in 10 years, which is not unusual for these mortgages. Approximately 97% of our monthly payments will be interest, which is tax deductible:

| | | | |
|---|---|---|---|
| (698)(12)(10) | = | \$83,760 | total payments |
| 57,000 − 54,529 | = | \$ 2,471 | principal repayments |
| | | \$81,289 | interest payments |

We will assume a 40%[1] marginal income tax rate (federal plus state) and sufficient other deductions to make the interest reduce our tax by

$$(698)(0.97)(0.40) = \$271/\text{month}$$

So the after-tax cost of the mortgage is only \$698 − \$271 = \$427.

Real estate taxes are estimated to be \$600 per year, or \$50/month, and these are also tax deductible, which saves us \$20/month for an after-tax cost of \$30/month. These taxes will increase at about 10% per year.

Basic taxes and utilities will be about \$60/month for a condominium and \$100/month for a house, so let's use \$80/month, with 10% inflation. Homeowner's insurance is slightly higher than renter's insurance, so we allow \$100 per year. Maintenance can be another \$300 per year. The monthly total of these items is \$33/month, inflating at 10%. Our home will appreciate in value at about 7% per year and sell at

$$60{,}000(1.07)^{10} = \$118{,}029$$

After paying a 6% realtor's commission and the mortgage balance, we keep

$$(118{,}029)(0.94) - 54{,}529 = \$56{,}418$$

[1]A 30% tax rate may be more reasonable for many homeowners. We will leave this for the reader to do as an exercise (see Problem 2.23).

(We assume no capital gain tax on this amount.) Now we're ready to compute $P$.

$$
\begin{aligned}
P = &-6{,}000 && \text{constant dollars}\\
&- (427)(12)(P/A,\ 15\%,\ 10) && \text{actual dollars}\\
&- (30/1.1)(12)(P/A,\ 4.55\%,\ 10) && \\
&- (80/1.1)(12)(P/A,\ 4.55\%,\ 10) && \text{constant dollars}\\
&- (33/1.1)(12)(P/A,\ 4.55\%,\ 10) && \\
&+ 56{,}418(P/F,\ 15\%,\ 10) && \text{actual dollars}
\end{aligned}
$$

Note carefully that we use 15% for actual-dollars expenses and 4.55% for constant-dollars expenses. We could convert the real estate taxes, utilities, incremental insurance, and maintenance to actual dollars by using 10% and then using 15% for discounting, but that is too much work. Our method produces the same numerical results.

$$
\begin{aligned}
P = &-6{,}000\\
&- (427)(12)(5.0188)\\
&- (130)(12)(7.8933)\\
&+ (56{,}418)(0.2472)\\
= &-30{,}080 \text{ in constant dollars}
\end{aligned}
$$

This cost is \$9,530 *less* than renting. In this example the present value costs in constant dollars for home ownership are about 76% of the present value costs for renting. The big difference comes from the fact that you are using \$57,000 of someone else's money to buy an asset that resells at two times its purchase price. You pay interest on the loan, but this is partly offset by the rent you would pay in an apartment.

Notice that the house was assumed to appreciate at 7%, compared with a mortgage interest rate of 14.5% nominal (15.5% effective per year). Many people think home ownership makes sense only if the mortgage interest rate is below the real estate appreciation rate. This is not true, as the example demonstrates.

We also used a 15% market interest rate, versus 10% general inflation and 7% real estate inflation. We might question the sensitivity of the results to these factors. In Table 2.7 we show some results of a sensitivity analysis in which we vary the inflation rate, the real estate appreciation rate, and the rent. We can see that there is a wide range of parameter values where buying is better. In fact, many people have benefited financially from home ownership during inflation. The home ownership analysis could be based on the principle of monthly payment and monthly compounding without collapsing all monthly flows to year end. We will leave this for the reader to do as an exercise (see Problem 2.22).

**Table 2.7** ***Sensitivity Analysis: Buy versus Rent Decision***

| Inflation $f$: | 5% | | | 10% | | | 15% | | |
|---|---|---|---|---|---|---|---|---|---|
| Market Interest $i$: | 10% | | | 15% | | | 20% | | |
| Real Estate Appreciation Rate: | 0% | 2.5% | 5% | 5% | 7.5% | 10% | 5% | 10% | 15% |
| Rent | | | | | | | | | |
| 350 | −36.6 | −36.6 | −36.6 | −35.3 | −35.3 | −35.3 | −34.1 | −34.1 | −34.1 |
| | −49.6 | −43.5 | −35.9 | −34.8 | −28.8 | −21.4 | −33.4 | −24.6 | −11.4 |
| | 136 | 119 | 98 | 99 | 82 | 61 | 98 | 72 | 33 |
| 400 | −41.1 | −41.1 | −41.1 | −39.6 | −39.6 | −39.6 | −38.3 | −38.3 | −38.3 |
| | −49.6 | −43.5 | −35.9 | −34.8 | −28.8 | −21.4 | −33.4 | −24.6 | −11.4 |
| | 121 | 106 | 87 | 88 | 73 | 54 | 87 | 64 | 30 |
| 450 | −45.5 | −45.5 | −45.5 | −43.9 | −43.9 | −43.9 | −42.4 | −42.4 | −42.4 |
| | −49.6 | −43.5 | −35.9 | −34.8 | −28.8 | −21.4 | −33.4 | −24.6 | −11.4 |
| | 109 | 96 | 79 | 79 | 66 | 49 | 79 | 58 | 27 |

NOTES: Each triplet of entries consists of present value of rental cash flow in thousands, present value of ownership cash flow in thousands, and percentage ratio of ownership flow to rental flow.

Other parameters:

| | | | |
|---|---|---|---|
| 5% down | $60,000 home cost | $80/month utilities (home) | $300/year maintenance (home) |
| 5% closing costs | 14.5% mortgage rate | $600/year real estate taxes | 6% realtor's commission |
| 30-year mortgage | 40% marginal tax rate | $100/year incremental insurance | 10-year planning period |

## 2.8 SUMMARY

In this chapter we have examined the concept of the time value of money and the equivalence of cash flows. Discrete compound interest formulas have been derived for converting present sums, future sums, uniform series, gradient series, and geometric series to specified points in time. We also discussed the concepts of nominal interest rate and effective interest rate, which led to the idea of continuous compounding. Continuous-compounding formulas were then derived for both discrete and continuous cash flows.

We discussed the measures of inflation and the effects of inflation on equivalence calculations. We presented two basic approaches that may be used in equivalence calculations to offset the effects of changes in purchasing power. In the actual-dollar analysis, we include an inflation component in estimating cash flows so that a market interest rate is used to find the equivalence value. In the constant-dollar approach we express the cash flows in terms of base-year dollars and use an inflation-free interest rate to compute the equivalent value at the specified points in time. We also showed that if these approaches are applied correctly, they should lead to identical results.

## REFERENCES

1. BUCK, J. R., and C. S. PARK, *Inflation and Its Impact on Investment Decisions,* Industrial Engineering and Management Press, Institute of Industrial Engineers, Norcross, Ga., 1984.
2. FLEISCHER, G. A., and T. L. WARD, "Classification of Compound Interest Models in Economic Analysis," *The Engineering Economist,* Vol. 23, No. 1, pp. 13–29, Fall 1977.
3. PARK, C. S., *Modern Engineering Economic Analysis,* Addison–Wesley, Reading, Mass., 1990.
4. THUESEN, G. J., and W. J. FABRYCKY, *Engineering Economy,* 7th edition, Prentice–Hall, Englewood Cliffs, N.J., 1989.
5. WHITE, J. A., M. H. AGEE, and K. E. CASE, *Principles of Engineering Economic Analysis,* 3rd edition, Wiley, New York, 1989.

## PROBLEMS

**2.1.** A typical bank offers you a Visa credit card that charges interest on unpaid balance at a 1.5% per month compounded monthly. This means that the nominal interest (annual percentage) rate for this account is $A$ and the effective annual interest rate is $B$. Suppose your beginning balance was $500 and you make only the required minimum *monthly* payment (payable at the end of each month) of $20 for next 3 months. If you made no new purchases with this card during this period, your unpaid balance will be $C$ at the end of 3 months. What are the values of $A$, $B$, and $C$?

**2.2.** In January 1989, C&S, the largest mutual savings bank in Georgia, published the following information: interest, 7.55%; effective annual yield, 7.842%. The bank did not explain how the 7.55% is connected to the 7.842%, but you can figure out that the compounding scheme used by the bank should be ______.

**2.3.** How many years will it take an investment to double if the interest rate is 12% compounded (a) annually, (b) semiannually, (c) quarterly, (d) monthly, (e) weekly, (f) daily, and (g) continuously?

**2.4.** Suppose that $1,000 is placed in a bank account at the end of each *quarter* over the next 10 years. Determine the total accumulated value (future worth) at the end of 10 years where the interest rate is 8% compounded *quarterly*.

**2.5.** What equal-payment series is required to repay the following present amounts?

a. $10,000 in 4 years at 10% interest compounded annually with 4 annual payments.

b. $5,000 in 3 years at 12% interest compounded semiannually with 6 semiannual payments.

c. $6,000 in 5 years at 8% interest compounded quarterly with 20 quarterly payments.

d. $80,000 in 30 years at 9% interest compounded monthly with 360 monthly payments.

**2.6.** Suppose that $5,000 is placed in a bank account at the end of each quarter over the next 10 years. Determine the total accumulated value (future worth) at the end of 10 years when the interest rate is

a. 12% compounded annually.
b. 12% compounded quarterly.
c. 12% compounded monthly.
d. 12% compounded continuously.

**2.7.** What equal *quarterly* payments will be required to repay a loan of $10,000 over 3 years if the rate of interest is 8% compounded *continuously?*

**2.8.** Compute the present worth of cash flow that has a triangular pattern with 12% interest compounded continuously.

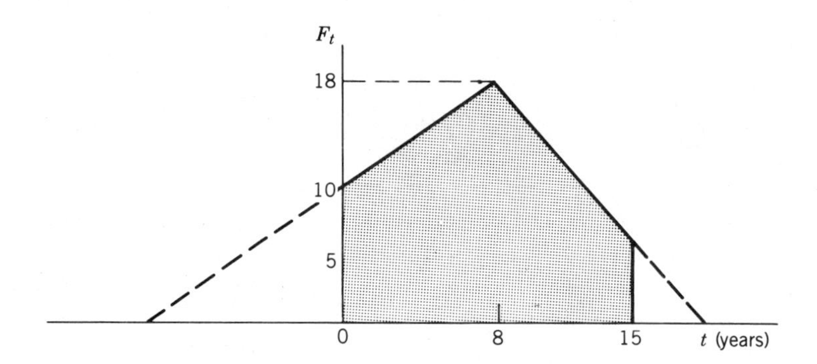


**2.9.** Suppose a uniformly increasing continuous cash flow (a ramp) accumulates $600 over 3 years. Find the present worth of this cash flow under continuous compounding at $r = 12\%$.

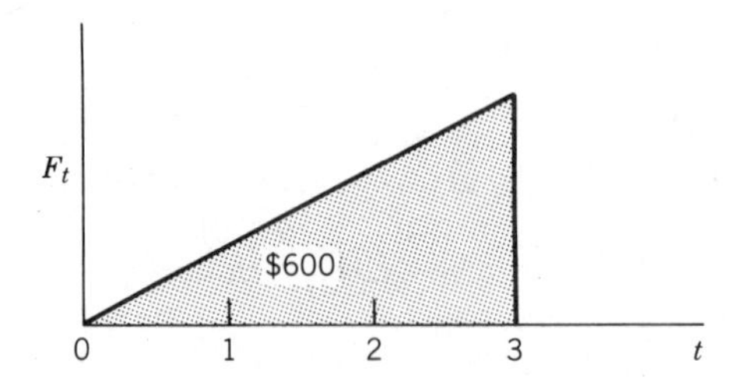

**2.10.** For computing the equivalent equal-payment series ($A$) of the following cash flow with $i = 10\%$, which of the following statements is (are) correct?

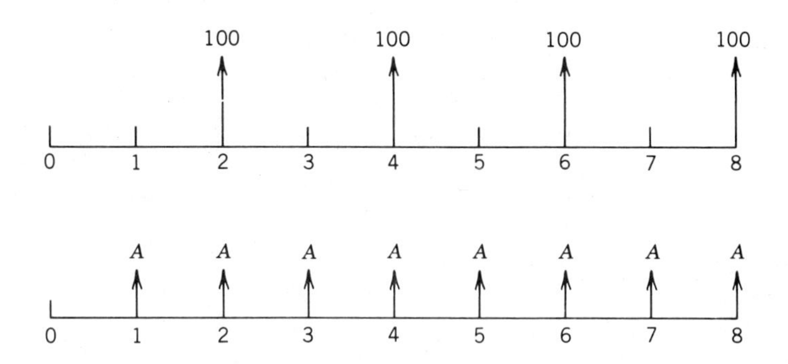


a. $A = 100(P/A, 10\%, 4)(A/P, 10\%, 8)$

$A = [100(P/F, 10\%, 2) + 100(P/F, 10\%, 4) + 100(P/F, 10\%, 6) + 100(P/F, 10\%, 8)](A/P, 10\%, 8)$

c. $A = 100(A/F, 10\%, 2)$

d. $A = 100(P/A, 21\%, 4)(A/P, 10\%, 8)$

e. $A = 100(F/A, 10\%, 4)(A/F, 10\%, 8)$

f. $A = 100(F/A, 21\%, 4)(A/F, 10\%, 8)$

**2.11.** The following equation describes the conversion of a cash flow into an equivalent equal-payment series with $n = 8$. Draw the original cash flow diagram. Assume an interest rate of 10% compounded annually.

$$A = [-1{,}000 - 1{,}000(P/F, 10\%, 1)](A/P, 10\%, 8)$$
$$+ [3{,}000 + 500(A/G, 10\%, 4)](P/A, 10\%, 4)(P/F, 10\%, 1)(A/P, 10\%, 8)$$
$$+ 750(F/A, 10\%, 2)(A/F, 10\%, 8)$$

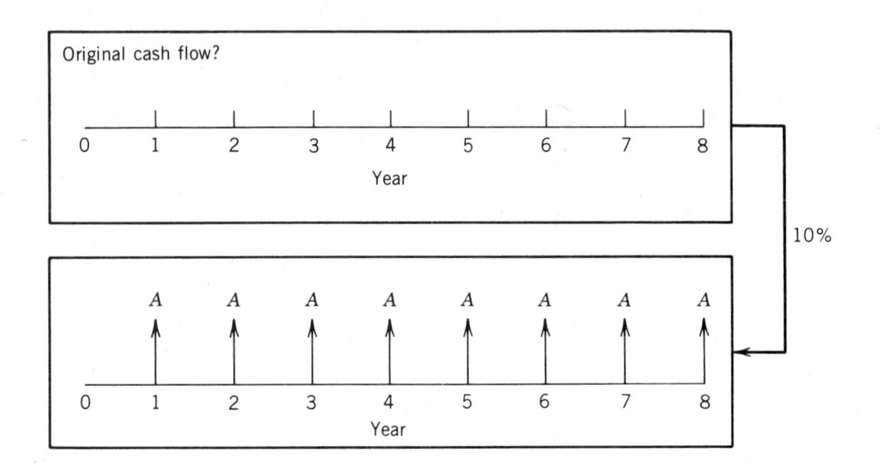

**2.12.** The following two cash flow transactions are said to be equivalent at 10% interest compounded annually. Find the unknown value $X$ that satisfies the equivalence.

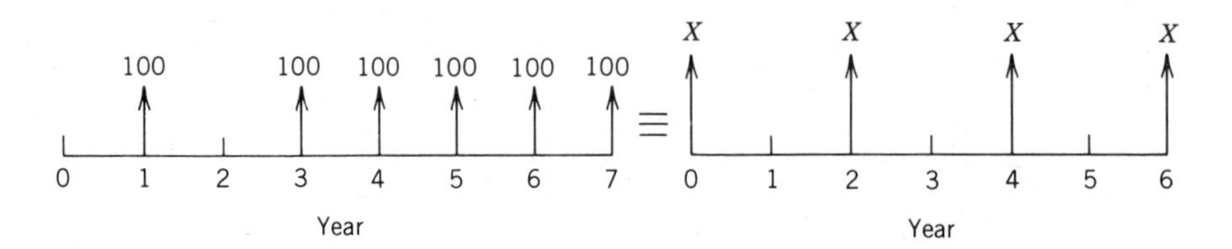


**2.13.** Suppose you have the choice of investing in (1) a zero-coupon bond that costs \$513.60 today, pays nothing during its life, and then pays \$1,000 after 5 years or (2) a municipal bond that costs \$1,000 today, pays \$67 in interest *semiannually,* and matures at the end of 5 years. Which bond would provide the higher *yield to maturity* (or return on your investment)?

**2.14.** You borrow $B$ dollars from your bank, which adds on the total interest before computing the monthly payment (add-on interest). Thus, if the quoted nominal interest rate (annual percentage rate) is $r$% and the loan is for $N$ months, the total amount that you agree to repay is

$$B + B(N/12)(r/100)$$

This is divided by $N$ to give the amount of each payment, $A$.

$$A = B(1/N + r/1200)$$

This is called an add-on loan. But the true rate of interest that you are paying is somewhat more than $r$%, because you do not hold the amount of the loan for the full $N$ months.

a. Find the equation to determine the true rate of interest $i$ per month.
b. Plot the relationship between $r$ and $i$ as a function of $N$.
c. For $B$ = \$10,000, $N$ = 36 months, and $r$ = 8%, find the effective annual borrowing rate per year.
d. Identify the lending situation in which the true interest rate $i$ per month approaches to $r/12$.

**2.15.** John Hamilton is going to buy a car worth \$10,000 from a local dealer. He is told that the add-on interest rate is only 1.25% per month, and his monthly payment is computed as follows:

Installment period = 30 months
Interest = 30(0.0125)(\$10,000) = \$3,750
Credit check, life insurance, and processing fee = \$50
Total amount owed = \$10,000 + \$3,750 + \$50 = \$13,800
Monthly payment size = \$13,800/30 = \$460 per month

What is the effective rate that John is paying for his auto financing?

a. Effective interest rate per month?
b. Effective annual interest rate?
c. Suppose that John bought the car and made 15 such monthly payments (\$460). Now he decides to pay off the remaining debt with one lump sum payment at the time of the sixteenth payment. What should the size of this payment be?

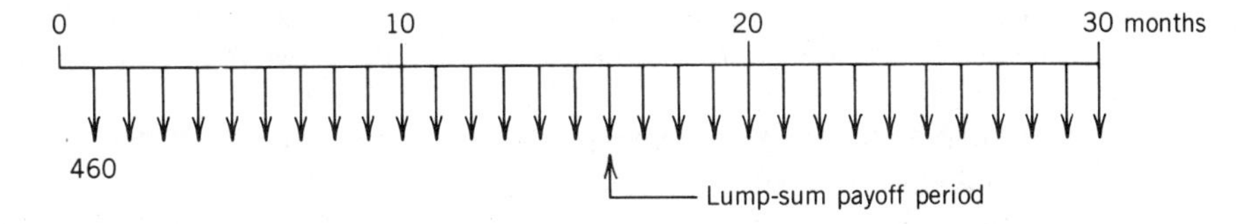


**2.16.** A pipeline was built 3 years ago to last 6 years. It develops leaks according to the relation

$$\log N = 0.07T - 2.42, \qquad T > 30$$

where $N$ is the total number of leaks from installation and $T$ is the time in months from installation. It costs \$500 to repair a leak. If money is worth 8% per year, and without considering any tax effect, how much can be spent now for a cathodic system that will reduce leaks by 75%? (Adapted from F. C. Jelen and J. H. Black, *Cost and Optimization,* McGraw–Hill, New York, 1983.)

**2.17.** A market survey indicates that the price of a 10-oz jar of instant coffee has fluctuated over the last few years as follows:

| Period | −4 | −3 | −2 | −1 | 0 | 1 |
|---|---|---|---|---|---|---|
| Price (\$) | 2.83 | 3.13 | 3.47 | 4.67 | 5.83 | ? |

a. Assuming that the base period (price index = 100) is period −4 (four periods ago), compute the average price index for this instant coffee.
b. Estimate the price at time period 1, if the current price trend is expected to continue.

**2.18.** The annual operating costs of a small electrical generating unit are expected to remain the same (\$200,000) if the effects of inflation are not considered. The best estimates indicate that the annual inflation-free rate of interest ($i'$) will be 5% and the annual inflation rate ($f$) 6%. If the generator is to be used 3 more years, what is the present equivalent of its operating costs using *actual-dollar analysis?*

**2.19.** You want to know how much money to set aside now to pay for 1,000 gallons of home heating oil each year for 10 years. The current price of heating oil is \$1.00 per gallon, and the price is expected to increase at a 10% compound price change each year for the next 10 years. The money to pay for the fuel oil will be set aside now in a bank savings account that pays 6% annual interest. How much money do you have to place in the savings account now, if payment for the fuel is made by end-of-year withdrawals?

**2.20.** An investment of \$100,000 is required to expand a certain production facility in a manufacturing company. The firm estimates that labor costs will be \$150,000 for the first year but will increase at the rate of 8% over the previous year's expenditure. Material costs, on the other hand, will be \$400,000 for the first year but will increase at the rate of 10% per year due to inflation. If the firm's inflation-free interest rate ($i'$) is 10% and the average general inflation rate ($f$) is expected to be 5% over the next 5 years, determine the total present equivalent operating expenses (with no tax consideration) for the project.

**2.21.** A couple with a 7-year-old daughter want to save for their child's college expenses in advance. Assuming that the child enters college at age 18, they estimate that an amount of \$20,000 per year in terms of today's dollars will be required to support the child's college expenses for 4 years. The future inflation rate is estimated to be

6% per year and they can invest their savings at 8% compounded quarterly.

a. Determine the equal quarterly amounts the couple must save until they send their child to college.

b. If the couple has decided to save only $500 each quarter, how much will the child have to borrow each year to support her college education?

**2.22.** Consider the problem of renting versus buying a home given in Section 2.7.3. Recall that the analysis was performed on the basis of annual payments with annual compounding. Repeat the analysis using monthly payments and monthly compounding.

**2.23.** Consider again the problem of renting versus buying a home given in Section 2.7.3. Recall that the tax rate used in the analysis was 40%, which seems too high. Repeat the analysis using a tax rate of 30%. Does a lower tax rate make the buying option more attractive?

# 3

# Transform Techniques in Cash Flow Modeling

## 3.1 INTRODUCTION

In Chapter 2 equivalence calculations were made by the proper use of the various interest formulas. In particular, with the interest rate and the compounding schemes specified, we showed how to convert various cash flow profiles into equivalent present values. In many situations, however, the cash flow patterns may take more complex forms than those discussed in Chapter 2. If they do, transform methods are often used to accomplish the same equivalence calculations with less computational effort and in a more routine manner. These methods are the *Z*-transform and Laplace transform methods. We will show in this chapter how they may be used in the modeling and analysis of economic situations involving either a discrete or a continuous time series of cash flow.

We will first discuss the concept of present value and its relationship to transform theory. Some useful properties of transforms will be presented, and their applications to economic model building will be discussed. Many examples are offered to aid the reader in understanding these powerful techniques. The reader will see that application of these transform formulas eliminates many of the calculations that are required when conventional interest formulas are used in complicated equivalence calculations.

## 3.2 Z-TRANSFORMS AND DISCRETE CASH FLOWS

### 3.2.1 The Z-Transform and Present Value

Consider that the function $f(n)$ describes the cash flow magnitude at the discrete point in time $n$. Then the equivalent present value of this cash flow series over an infinite time horizon at an interest rate $i$, assuming a discrete compounding principle, is

$$PV(i) = \sum_{n=0}^{\infty} f(n)(1 + i)^{-n} \tag{3.1}$$

Hill and Buck [6] recognized that the general form of the summation in (3.1) bears a striking resemblance to the definition of $Z$-transforms, the only difference being a definition of variables. That is, when a general discrete time series is described by a function $f(nT)$, where $T$ is an equidistant time interval and $n$ is an integer, the $Z$-transform of the time series $f(nT)$ is defined as

$$F(z) = \sum_{n=0}^{\infty} f(nT)z^{-n} \tag{3.2a}$$

With $T = 1$,

$$F(z) = \sum_{n=0}^{\infty} f(n)z^{-n} = Z\{f(n)\} \tag{3.2b}$$

where $z$ is a complex variable. If we replace $z$ with the interest rate $1 + i$ and set the constant-length time interval $T$ to unity (that is, the compounding period is the unit of time, monthly or yearly), Eq. 3.2 becomes

$$F(z) = \sum_{n=0}^{\infty} f(n)(1 + i)^{-n} \tag{3.3}$$

where $i$ is the interest rate for a compounding period. Throughout this chapter the value of $T$ will be set to unity so that the compounding period can be assumed to be the unit of time. In the literature of mathematics, we find a transformation essentially the same as our $Z$-transform but expressed in positive powers of $z$:

$$F'(z) = \sum_{n=0}^{\infty} f(n)z^{n} \tag{3.4}$$

In this book we use the definition in (3.2) because the expressions for the corresponding $Z$-transform are analogous to those for present values. It should be obvious, however, that both transformations have the same purpose and application, and that one transform is converted to the other by the relations

$$F'(z) = F\left(\frac{1}{z}\right), \qquad F(z) = F'\left(\frac{1}{z}\right) \tag{3.5}$$

In the construction of $Z$-transforms, the following notation will be used. If $f(n)$ represents the discrete $f$ function, $F(z)$ will represent the transform. In addition, as a shorthand notation, the transform pair will be denoted by $f(n) \leftrightarrow F(z)$. This double arrow is symbolic of the uniqueness of the one-to-one correspondence between $f(n)$ and $F(z)$. Thus, if $Z\{g(n)\} = G(z)$, we write $g(n) \leftrightarrow G(z)$. This lowercase–uppercase correspondence will be adhered to throughout this chapter.

For a cash flow sequence of infinite duration, the resulting $Z$-transform will be an infinite series involving inverse powers of $z$. This series can be expressed as a rational fraction in $z$, provided that the series converges. These so-called closed-form expressions will be especially convenient for our computations. For expressing a $Z$-transform as a ratio of polynomials in $z$, two important identities of infinite series will be needed:

$$\sum_{n=0}^{\infty} a^n = \frac{1}{1-a} \quad \text{provided } |a| < 1 \tag{3.6}$$

and

$$\sum_{n=0}^{\infty} (1+n)a^n = \frac{1}{(1-a)^2} \quad \text{provided } |a| < 1 \tag{3.7}$$

Now consider the sequence of function $f(n) = a^n$. The $Z$-transform is

$$F(z) = \sum_{n=0}^{\infty} f(n)z^{-n} = \sum_{n=0}^{\infty} a^n z^{-n} = \sum_{n=0}^{\infty} \left(\frac{a}{z}\right)^n$$

Using Eq. 3.6, we obtain

$$F(z) = \frac{1}{1-a/z} = \frac{z}{z-a} \quad \text{if } \left|\frac{a}{z}\right| < 1 \tag{3.8}$$

In other words, the infinite geometric series $a^n$ converges to $z/(z-a)$ if $|z| > |a|$. For ease of conversion, the table of transform pairs of $f(n)$ and $F(z)$ is provided (see Table 3.1).

Many cash flow transactions have a finite time duration. Because transforms are defined for series with infinite time horizons, it is necessary to introduce additional techniques to provide a methodology that is applicable to finite time horizons. We will examine some properties of the $Z$-transform in the following section.

### 3.2.2 Properties of the Z-Transform

Many useful properties of the $Z$-transform are discussed in the literature of mathematics, probability theory, and operations research [4,5,7]. We will focus on two important properties that are most relevant in equivalence calculations: linearity and translation.

**Table 3.1** ***A Short Table of Z-Transform Pairs***

| Standard Pattern | Original Function, $f(n)$ | Z-Transform, $F(z)$ | Present Value of $f(n)$ Starting at $n=0$ and Continuous over the Infinite Time Horizon |
|---|---|---|---|
| Step (uniform series) | $C$ | $C\left(\frac{z}{z-1}\right)$ | $C\left(\frac{1+i}{i}\right)$ |
| Ramp (gradient series) | $Cn$ | $C\left[\frac{z}{(z-1)^2}\right]$ | $C\left(\frac{1+i}{i^2}\right)$ |
| Geometric | $Ca^n$ | $C\left(\frac{z}{z-a}\right)$ | $C\left(\frac{1+i}{1+i-a}\right)$ |
| Decay | $Ce^{-jn}$ | $C\left(\frac{z}{z-e^{-j}}\right)$ | $C\left(\frac{1+i}{1+i-e^{-j}}\right)$ |
| Growth | $C(1-e^{-jn})$ | $C\left(\frac{z}{z-1}-\frac{z}{z-e^{-j}}\right)$ | $C\left(\frac{1+i}{i}-\frac{1+i}{1+i-e^{-j}}\right)$ |
| Impulse (single payment) | $C\delta(n-k)$ | $C\left(z^{-k}\right)$ | $C(1+i)^{-k}$ |

NOTE: $C$ = pattern scale factor
$n$ = time index for compounding periods
$i$ = effective interest rate for a compounding period.
$a$ = pattern base factor
$j$ = pattern rate factor
$\delta$ = impulse function
$k$ = number of time periods before the impulse occurs

***Linearity.*** The $Z$-transform is a linear operation. Thus, when a sequence can be expressed as a sum of other sequences, the following result will be useful.

$$f(n) = C_1 f_1(n) + C_2 f_2(n) \leftrightarrow F(z)$$

$$f_1(n) \leftrightarrow F_1(z)$$

$$f_2(n) \leftrightarrow F_2(z)$$

then

$$F(z) = C_1\, \mathrm{F}_1\,(z) + C_2 F_2\,(z) \tag{3.9}$$

This linearity property makes it possible to combine component time forms and to amplify general time patterns by the scale factor $C$ to represent the

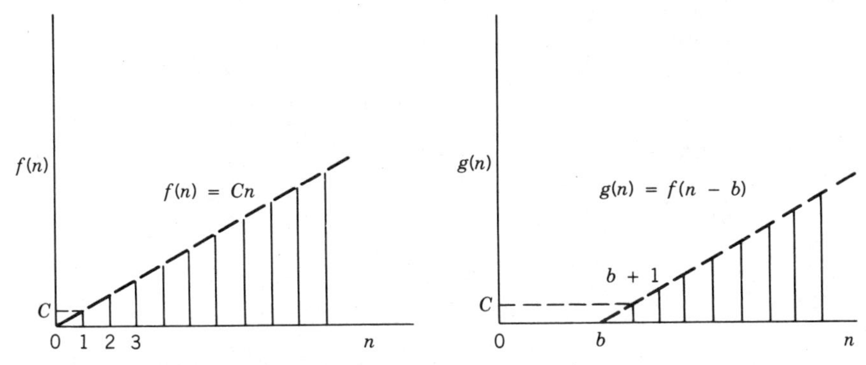


**FIGURE 3.1** Ramp pattern transaction (gradient series).

proportion of the component. By adding it to or subtracting it from the scaled transforms of other components, we are able to describe the composite Z-transform of the entire stream of components.

***Translation with Time Advance.*** To consider a composite of cash flows that start at various points in time, we seek the relation between the Z-transform of sequences and their shifted version. Consider the sequence $g(n)$ obtained from $f(n)$ by shifting $f(n)$ to the right by $b$ units of time. This situation is illustrated in Figure 3.1, in which the function $f(n)$ takes a ramp pattern. Since the sequence $g(n)$ is 0 for $n < b$, we can define the sequence $g(n)$ in terms of $f(n)$ as

$$g(n) = \begin{cases} f(n-b) & \text{for } n \geq b \\ 0 & \text{for } n < b \end{cases} \tag{3.10}$$

To find the transform of this time-shifted function, we use the property of the unit step function. If we take the unit step function and translate it $b$ units to the right to get $u(n - b)$, we obtain the function shown in Figure 3.2*a*. Mathematically, we denote this by

$$u(n-b) = \begin{cases} 1 & \text{for } n \geq b \\ 0 & \text{for } n < b \end{cases} \tag{3.11}$$

Notice that the shifted unit step function in Figure 3.2*a* has no values for $n < b$ but is equal to 1 for $n \geq b$. The product $f(n - b)u(n - b)$ will be zero for $n < b$ and will equal $f(n - b)$ for $n \geq b$. This product form shown in Figure 3.2*c* defines precisely the shifted ramp function we defined in Figure 3.1*b*. More generally, we define such a function as

$$g(n) = f(n-b)u(n-b) = \begin{cases} f(n-b) & \text{for } n \geq b \\ 0 & \text{for } n < b \end{cases} \tag{3.12}$$

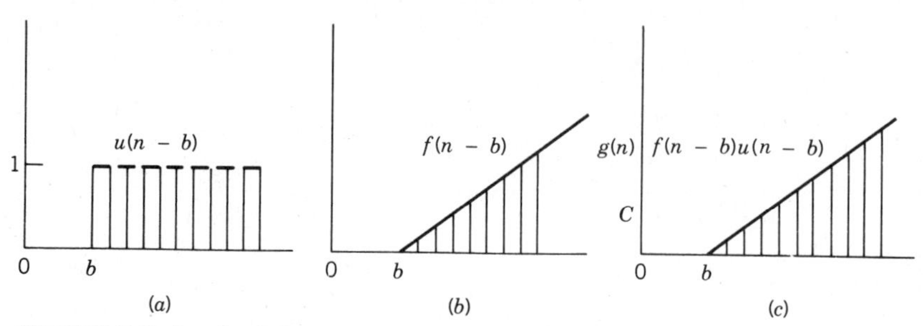


**FIGURE 3.2** Graph of the ramp function with translation and cutoff.

Then the $Z$-transform of the above expression is defined as

$$G(z) = z^{-b} F(z) = (1 + i)^{-b} F(z) \tag{3.13}$$

The quantity $z^{-b}$ in the $Z$-transform simply reflects the fact that the start of the function $f(n)$ has been shifted forward in time by $b$ units. Thus, if $f(n) = Cn$, shifting $f(n)$ to the right by $b$ units and taking its $Z$-transform generates $z^{-b}[Cz/(z - 1)^2]$. Expressing this in terms of the present value and replacing $z$ with $(1 + i)$, we obtain $PV(i) = [C(1 + i)^{1-b}]/i^2$.

***Translation with Cutoff.*** Many realistic cash flow functions extend over finite time horizons. Another scheme of translation property is useful in finding the $Z$-transforms for these translated cash flow functions. Consider the function $g(n)$ shown in Figure 3.3*c*. This function is basically the truncated ramp function $f(n)$ in Figure 3.3*b* with the added feature of a delayed turn-on at time $b$, where $b$ is an integer. By using the translation property discussed in the last section and multiplying the ramp function $f(n)$ by a unit step, we can express the desired truncated function $g(n)$ as

$$g(n) = f(n)u(n - b) = \begin{cases} f(n) & \text{for } b \leq n \\ 0 & \text{otherwise} \end{cases} \tag{3.14}$$

and the $Z$-transform of this product expression is

$$G(z) = z^{-b} Z\{f(n + b)\} \tag{3.15}$$

Unlike the situation in Figure 3.2*c*, the origin of the function $f(n)$ remains unchanged, but the first transaction begins at time $b$. Thus, it is important to recognize the functional distinction between $f(n)u(n - b)$ and $f(n - b)u(n - b)$. That is, an expression $f(n - b)u(n - b)$ similar to the one illustrated by Figure 3.2*c* will appear in shifting the sequence $f(n)$ to the right by $b$ units in time; and its first transaction also starts at time $b$.

Now the $Z$-transform and present value expression for the ramp function

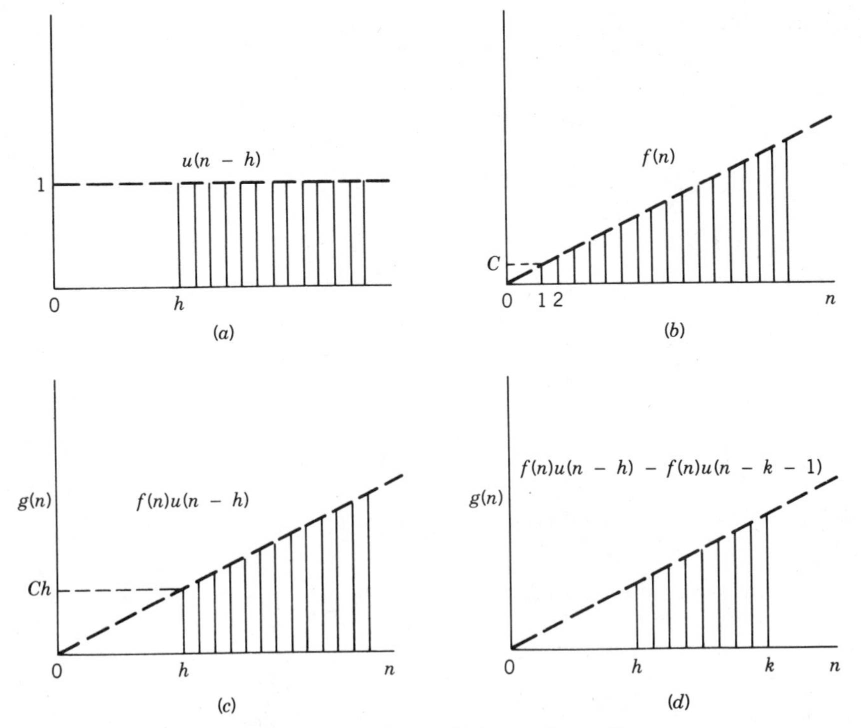


**FIGURE 3.3** Ramp function with translation and cutoff.

with a delayed turn-on at time $h$ shown in Figure 3.3*c* can easily be found. Since $f(n) = Cn$, the transform of $g(n)$ is

$$\begin{aligned} G(z) &= z^{-h}Z\{f(n + h)\} \\ &= z^{-h}Z\{Cn + Ch\} \\ &= Cz^{-h}\left[\frac{z}{(z-1)^2} + \frac{hz}{z-1}\right] \\ &= \frac{Cz^{1-h}}{(z-1)^2}[1 + h(z-1)] \end{aligned} \tag{3.16}$$

By replacing $z$ with $(1 + i)$, we obtain the present value expression

$$PV(i) = \frac{C(1+i)^{1-h}}{i^2}(1 + hi) \tag{3.17}$$

Suppose we want to find the present value of the series shown in Figure 3.3*d*. This function is the same ramp function with the delayed turn-on at time $h$ but also with a turn-off at time $k$, where $h$ and $k$ are integers. By using the property of the unit step function, we can express the desired ramp translation with turn-on and turn-off as follows.

$$g(n) = f(n)u(n - h) - f(n)u(n - k - 1) \tag{3.18}$$

The transform of this function will be

$$\begin{aligned} G(z) &= z^{-h}Z\{f(n + h)\} - z^{-(k+1)}Z\{f(n + k + 1)\} \\ &= z^{-h}Z\{Cn + Ch\} - z^{-(k+1)}Z\{Cn + C(k + 1)\} \\ &= Cz^{-h}\left[\frac{z}{(z - 1)^2} + \frac{hz}{z - 1}\right] - Cz^{-(k+1)}\left[\frac{z}{(z - 1)^2} + \frac{(k + 1)z}{z - 1}\right] \\ &= \frac{C}{(z - 1)^2}(z^{1-h} - z^{-k}) + \frac{C}{z - 1}[hz^{1-h} - (k + 1)z^{-k}] \end{aligned} \tag{3.19}$$

In terms of the present value expression, we have

$$PV(i) = \frac{C}{i^2}[(1 + i)^{1-h} - (1 + i)^{-k}] + \frac{C}{i}[h(1 + i)^{1-h} - (k + 1)(1 + i)^{-k}] \tag{3.20}$$

### *Example 3.1*

As an example of the use of Eq. 3.20, suppose that estimates of certain end-of-year expenses are \$300 for the third year, \$400 for the fourth year, and \$500 for the fifth year. If the effective interest rate is 15%, what is the equivalent present value?

The gradient series can be expressed as

$$f(n) = 100n \quad \text{where } 3 \le n \le 5$$

With $C = 100$, $h = 3$, $k = 5$, and $i = 0.15$, we obtain

$$\begin{aligned} PV(15\%) &= \frac{100}{(0.15)^2}[(1.15)^{-2} - (1.15)^{-5}] + \frac{100}{0.15}[3(1.15)^{-2} - 6(1.15)^{-5}] \\ &= \$674.54 \quad \square \end{aligned}$$

***Translation with Impulses.*** Suppose we want to find the transform of an impulse function $g(n)$ as given in Figure 3.4c. This type of impulse function may represent the salvage value of an item at time $h$, when the salvage value $f(n)$ decreases exponentially over time. To obtain the transforms of such impulse functions, we need to define a Kronecker delta function that corresponds to a unit impulse function as shown in Figure 3.4c. That is,

$$\delta(n - h) = \begin{cases} 1 & \text{for } n = h \\ 0 & \text{otherwise} \end{cases} \tag{3.21}$$

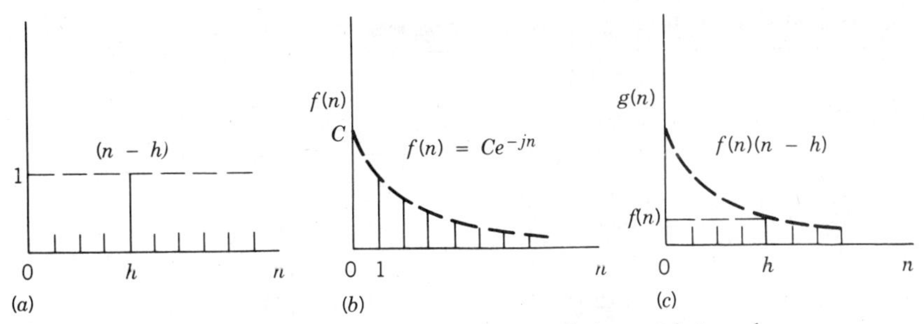


**FIGURE 3.4** Kronecker delta function and translation with impulse.

By multiplying the salvage value function $f(n)$ by the unit impulse function, we obtain an expression in which the salvage value occurs only at time $h$, as desired. Formally, we may write this product expression as

$$g(n) = f(n)\delta(n - h) = \begin{cases} f(h) & \text{for } n = h \\ 0 & \text{otherwise} \end{cases} \tag{3.22}$$

Since $f(h)$ is a constant and the transform of the shifted unit impulse function $\delta(n - h)$ is $z^{-h}$, the transform of the product form yields

$$G(z) = z^{-h}f(h) = f(h)(1 + i)^{-h} \tag{3.23}$$

and this is exactly the present value expression for a single payment. If we define $f(h) = Ce^{-jh}$, we can find the present value expression

$$PV(i) = Ce^{-jh}(1 + i)^{-h} \tag{3.24}$$

where $j$ is the pattern rate factor for a decay function.

The linearity and translation properties just discussed provide many of the necessary analytical tools for finding the Z-transforms of realistic discrete time series encountered in economic analysis. Table 3.2 summarizes some other useful operational rules for the Z-transform. (See [7].)

### 3.2.3 Development of Present Value Models

We develop two types of present value models that correspond to the timing of the start of the original cash flow function. They are the extensive models and the simplified models.

***Extensive Present Value Models.*** The extensive models represent cash flow functions that are shifted forward in time but switched on only at time $h$ ($h \geq b$) and then terminated at time $k$ ($k > h$) (see Figure 3.5). This function is basically the shifted ramp (gradient series) in Figure 3.1$b$ with a delayed turn-on at time $h$ and a turn-off at time $k$, where $h$ and $k$ are integers.

**Table 3.2** ***Some Properties of the Z-Transform***

| Operational Rule | Original Function, $f(n)$ | Z-Transform, $F(z)$ |
|---|---|---|
| Linearity | $C_1 f_1(n) + C_2 f_2(n)$ | $C_1 F_1(z) + C_2 F_2(z)$ |
| Damping | $a^{-n} f(n)$ | $F(az)$ |
| Shifting to the right | $f(n-k)u(n-k),\ k \geq 0$ | $z^{-k}F(z)$ |
| Shifting to the left | $f(n+k),\ k \geq 0$ | $z^k\left[F(z) - \sum_{n=0}^{k-1} f(n)z^{-n}\right]$ |
| Differencing of $f(n)$ | $\Delta f(n) = f(n+1) - f(n)$ | $(z-1)F(z) - zf(0)$ |
| | $\nabla f(n) = f(n) - f(n-1)u(n-1)$ | $\frac{z-1}{z}F(z)$ |
| Summation of $f(n)$ | $\sum_{j=0}^{n} f(j)$ | $\frac{z}{z-1}F(z)$ |
| Periodic sequences | $f(n+k) = f(n)$, period $k$ | $\frac{z^k}{z^k - 1}\sum_{n=0}^{k-1} f(n)z^{-n}$ |
| Convolution | $f(n) * g(n)$ | $F(z) * G(z)$ |

To find the correct transform, we use the translation properties of Eqs. 3.12 and 3.14. The function $g(n)$ can then be expressed by multiplying the shifted gradient series by a unit step function. The resulting functional expression is

$$g(n) = [u(n-b) - u(n-k-1)]f(n-b) \tag{3.25}$$

Since $f(n) = Cn$ (gradient series), we may rewrite $f(n-b)$ as

$$f(n-b) = C(n-b) = Cn - Cb = f(n) - f(b)$$

Thus, we may also rewrite $g(n)$ as

$$\begin{aligned} g(n) &= f(n)[u(n-b) - u(n-k-1)] \\ &\quad + f(b)[u(n-k-1) - u(n-b)] \end{aligned} \tag{3.26}$$

Note that $f(b)$ is a constant $Cb$. Using the transform results of Eq. 3.18, we obtain

$$\begin{aligned} G(z) = {} & \frac{C}{(z-1)^2}(z^{1-b} - z^{-k}) + \frac{C}{z-1}[(b-b)z^{1-b} \\ & - (k+1-b)z^{-k}] \end{aligned} \tag{3.27}$$

By replacing $z$ with $1+i$, we obtain the present value expression of this extensive model.

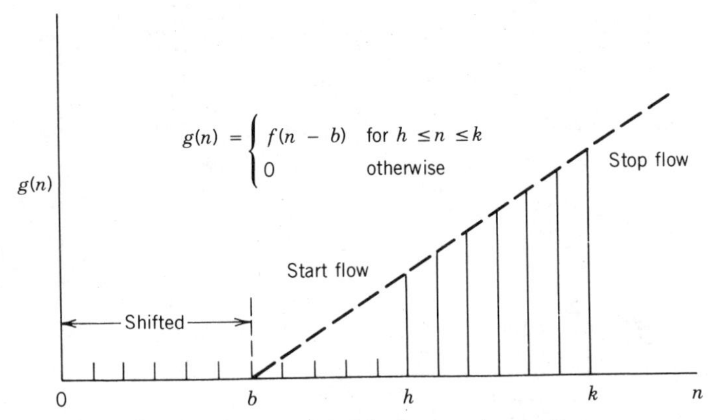


**FIGURE 3.5** Extensive model of ramp time pattern.

$$PV(i) = \frac{C}{i^2}[(1 + i)^{1-h} - (1 + i)^{-k}] + \frac{C}{i}[(h - b)(1 + i)^{1-h}] - (k + 1 - b)(1 + i)^{-k}] \tag{3.28}$$

If we use the conventional engineering economy notation, the present value of this shifted-gradient series is

$$PV(i) = [Cb(P/A, i, k - h + i) + C(P/G, i, k - h + 1)]\ (P/F, i, h - 1) \tag{3.29}$$

If we converted these factor notations to algebraic form, the final form would be as long an expression as Eq. 3.28. Table 3.3 provides the extensive models of other discrete cash flow patterns.

## *Example 3.2*

A 20-MW oil-burning power plant now under construction is expected to be in full commercial operation in 4 years from now. The fuel cost for this new plant is a function of plant size, thermal conversion efficiency (heat rate), and plant utilization factor. Because of inflation, the future price of oil will increase. The annual fuel cost is then represented by the following expression,

$$f(n) = (S)(H)(U)\left(\frac{8{,}760\ \text{hr/year}}{10^6}\right)P_0\ (1 + f)^{n-1}$$

where $f(n)$ = annual fuel cost at the end of the $n$th operating year,
$S$ = plant size in kW (1 MW = 1,000 kW),
$H$ = heat rate (Btu/kW·hr),
$U$ = plant utilization factor,
$f$ = average annual fuel inflation rate,
$P_0$ = starting price of fuel per million Btu during the first year of operation.

**Table 3.3** ***Extensive Discrete Present Value Models***

| Cash Flow Pattern | Function | Typical Cost Example | Present Value, $PV(i)$ |
|---|---|---|---|
|  $C$, $g(n)$; $b$, $h$, $k$ | Step $g(n) = f(n-b) = C$ | Operating costs | $\frac{C}{i}[(1+i)^{1-h} - (1+i)^{-k}]$ |
| 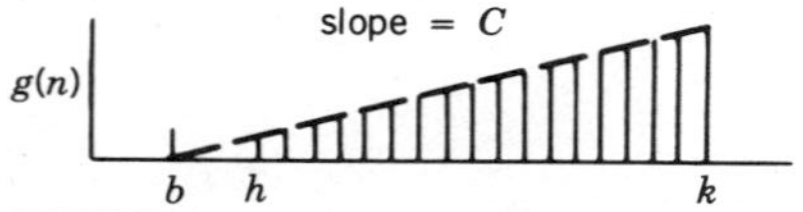 $g(n)$; slope = $C$; $b$, $h$, $k$ | Ramp $g(n) = f(n-b) = C(n-b)$ | Maintenance and deterioration | $\frac{C}{i^2}[(1+i)^{1-h} - (1+i)^{-k}] + \frac{C}{i}[(h-b)(1+i)^{1-h} - (k+1-b)(1+i)^{-k}]$ |
| 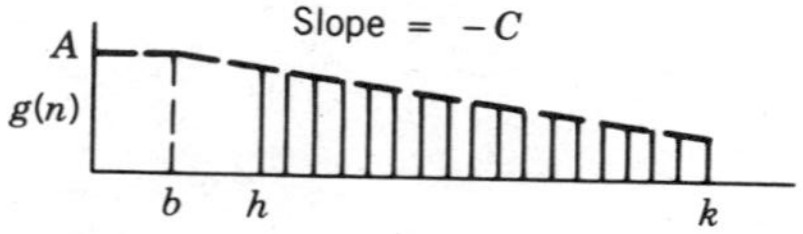 $A$, $g(n)$; Slope = $-C$; $b$, $h$, $k$ | Decreasing Ramp $g(n) = f(n-b) = A - C(n-b)$ | Value depreciation costs | $\frac{(1+i)^{1-h}}{i}\left[A - \frac{C}{i} - C(h-b)\right] - \frac{(1+i)^{-k}}{i}\left[A - \frac{C}{i} - C(k+1-b)\right]$ |
| 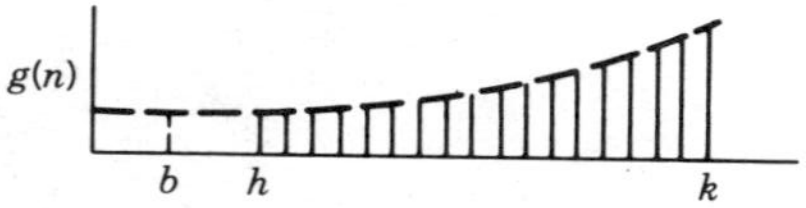 $g(n)$; $b$, $h$, $k$ | Geometric series $g(n) = f(n-b) = Ca^{(n-b)}$ | Inflationary costs | $\frac{Ca^{h-b}}{1+i-a}[(1+i)^{1-h} - a^{k+1-h}(1+i)^{-k}]$ |
| 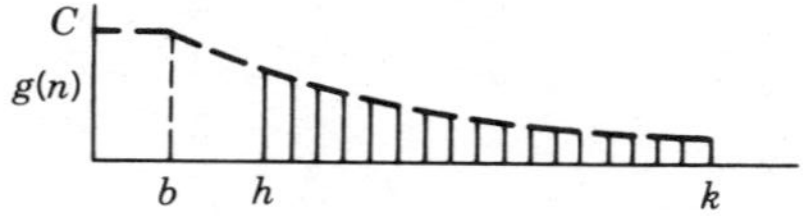 $C$, $g(n)$; $b$, $h$, $k$ | Decay $g(n) = f(n-b) = Ce^{-j(n-b)})$ | Start-up and learning costs | $\frac{C(1+i)}{1+i-e^{-j}}\left[\frac{e^{-j(h-b)}}{(1+i)^h} - \frac{e^{-j(k+1-b)}}{(1+i)^{k+1}}\right]$ |
| 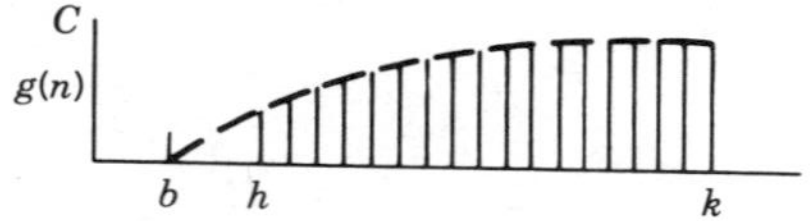 $C$, $g(n)$; $b$, $h$, $k$ | Growth $g(n) = C(1 - e^{-j(n-b)})$ | Wear-in maintenance costs | $\frac{C(1+i)}{i(1+i-e^{-j})}\left[\frac{(1+i-e^{-j}) - ie^{-j(h-b)}}{(1+i)^h} - \frac{(1+i-e^{-j}) - ie^{-j(k+1-b)}}{(1+i)^{k+1}}\right]$ |

Assume that $S = 20{,}000$ kW, $H = 10{,}000$ Btu/kW·hr, $U = 0.20$, $f = 0.07$, and $P_0 =$ \$4.5 per million Btu during year 4. The expected life of the plant is 15 years. What is the present value of the total fuel cost at the beginning of construction (now) if the annual market rate of interest is 18%?

With the parameters as specified, the annual fuel cost function is

$$f(n) = 1{,}576{,}800\ (1 + 0.07)^{n-1}, \qquad 1 \le n \le 15$$

To find the present value of the total fuel cost at the beginning of construction, we rewrite $f(n)$ to obtain $g(n)$.

$$\begin{aligned} g(n) &= f(n - 4) \\ &= 1{,}576{,}800\ (1 + 0.07)^{n-5}, \qquad 5 \le n \le 19 \\ &= 1{,}473{,}645(1.07)^{n-4} \end{aligned}$$

Now we can use the geometric series formula given in Table 3.3. We identify $C = 1{,}473{,}645$, $a = 1.07$, $b = 4$, $h = 5$, $k = 19$, and $i = 0.18$, which yield

$$\begin{aligned} PV(18\%) &= \frac{1{,}473{,}645(1.07)}{0.11}[(1.18)^{-4} - (1.07)^{15}\ (1.18)^{-19}] \\ &= \$5{,}689{,}941 \quad \square \end{aligned}$$

***Simplified Present Value Models.*** The simplified models are defined as those with cash flows that have no delayed turn-on ($b = 0$) and that terminate after $k$ time units. The procedure for finding the $Z$-transform for this type of simplified form was illustrated in the previous section (see Figure 3.3*d*). Table 3.4 summarizes the present value models for some other common cash flow patterns. These simplified present value models correspond, in fact, to the traditional tabulated interest factors found in engineering economy textbooks. They simplify the use of this transform methodology when the modified features of cash flow patterns are not required.

### 3.2.4 Extension to Future and Annual Equivalent Models

The future equivalent values at the end of period $N$ can easily be obtained from the present values shown in Tables 3.3 and 3.4 by multiplying through by $(1 + i)^N$. Similarly, annual equivalent values over period $N$ are determined by multiplying the present values by the factor $i/[1 - (1 + i)^{-N}]$.

$$FV(i) = PV(i)[(1 + i)^N]$$

$$AE(i) = PV(i)\left[\frac{i}{1 - (1 + i)^{-N}}\right]$$

Consequently, all the $Z$-transforms in Tables 3.3 and 3.4 may be directly converted to a future or annual equivalent value by applying these elementary algebraic

**Table 3.4** ***Simplified Discrete Present Value Models***

| Cash Flow Pattern | Function | Typical Cost Example | Present Value, $PV(i)$ |
|---|---|---|---|
| $f(n)$; $0$ $h = 1$ $k$ | $f(n) = C$ | Operating costs | $\frac{C}{i}[1 - (1 + i)^{-k}]$ |
| $f(n)$; $0$ $h = 1$ $k$ | $f(n) = Cn$ | Maintenance and deterioration | $\frac{C}{i^2}[1 - (1 + i)^{-k}]$ $+ \frac{C}{i}[1 - (k + 1)(1 + i)^{-k}]$ |
| $f(n)$; $0$ $h = 1$ $k$ | $f(n) = A - Cn$ | Value depreciation costs | $\frac{1}{i}\left(A - \frac{C}{i} - C\right)$ $- \frac{(1 + i)^{-k}}{i}\left[A - \frac{C}{i} - C(k + 1)\right]$ |
| $f(n)$; $b = h = 1$ $k$ | $f(n) = Ca^n$ | Inflationary costs | $\frac{Ca}{1 + i - a}[1 - a^k(1 + i)^{-k}]$ |
| $f(n)$; $0$ $h = 1$ $k$ | $f(n) = Ce^{-jn}$ | Start-up and learning costs | $\frac{C(1 + i)}{1 + i - e^{-j}}\left[\frac{e^{-j}}{1 + i} - \frac{e^{-j(k+1)}}{(1 + i)^{k+1}}\right]$ |
| $f(n)$; $0$ $h = 1$ $k$ | $f(n) = C(1 - e^{-jn})$ | Wear-in maintenance costs | $\frac{C(1 + i)}{1 + i - e^{-j}}\left[\frac{1 - e^{-j}}{i} - \frac{1 + i - e^{-j} - ie^{-j(k+1)}}{i(1 + i)^{k+1}}\right]$ |

manipulations as needed. All the Z-transforms derived in the previous sections are based on the assumption that the compounding periods and the payment occurrences coincide. In situations in which the compounding periods occur more frequently than the receipt of payments, one can find the effective interest rate for the payment period and use it in the Z-transforms developed in Tables 3.3 and 3.4 (see Section 2.4.2).

### 3.2.5 Applications of Z-Transforms

In this section we will demonstrate the application of the Z-transform to the solution of equivalence problems. Two uses will be illustrated: profit margin analysis and calculation of the present value of interest payments.

***Profit Margin Analysis.*** Consider that a new production facility under construction is expected to be in full commercial operation 2 years from now. The plant is expected to have an initial profit margin of $5 million per year. Find the present value of the total profit margin at 10% interest compounded annually for 20 years of operation if

i. Profit margin and plant performance stay level.

$$g(n) = 5, \qquad 3 \leq n \leq 22$$

ii. Performance traces a learning curve whereby the profit margin grows in each year.

$$g(n) = 5(2 - e^{-0.10(n-3)}), \qquad 3 \leq n \leq 22$$

iii. Performance traces the same growth curve, but the profit margin shrinks at a rate of $e^{-0.03(n-3)}$ so that

$$g(n) = 5e^{-0.03(n-3)}(2 - e^{-0.10(n-3)}), \qquad 3 \leq n \leq 22$$

These three cases are illustrated in Figure 3.6.

For case i, the cash flow diagram is a shifted step function with $C = 5$, $b = 3$, $h = 3$, and $k = 22$. From Table 3.3 the equivalent present value for this shifted step function is

$$PV(i) = \frac{5}{0.1}[(1.1)^{-2} - (1.1)^{-22}] = 50(0.7036) = \underline{\$35.18}$$

For case ii, the growth cash flow function may be regarded as a linear combination of a shifted step function and a shifted decay function. That is,

$$\begin{aligned} \text{Growth function} &= \text{Step function} - \text{Decay function} \\ g(n) &= 5(2) - 5e^{-0.10(n-3)} \end{aligned}$$

Thus, for the step function, the corresponding parameters would be $C = 10$, $b = 3$, $h = 3$, and $k = 22$, and for the decay function they would be $C = 5$, $j = 0.10$,

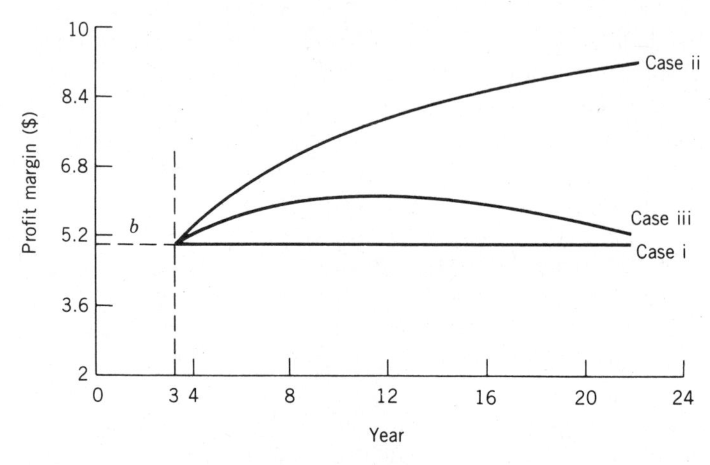


| $n$ | Case i | Case ii | Case iii |
|---|---|---|---|
| 3 | 5 | 5 | 5 |
| 4 | 5 | 5.475813 | 5.313978 |
| 5 | 5 | 5.906346 | 5.562387 |
| 6 | 5 | 6.295909 | 5.754028 |
| 7 | 5 | 6.6484 | 5.896602 |
| 8 | 5 | 6.967347 | 5.996851 |
| 9 | 5 | 7.255942 | 6.060672 |
| 10 | 5 | 7.517073 | 6.093221 |
| 11 | 5 | 7.753355 | 6.099005 |
| 12 | 5 | 7.967152 | 6.08196 |
| 13 | 5 | 8.160603 | 6.045523 |
| 14 | 5 | 8.335645 | 5.992693 |
| 15 | 5 | 8.494029 | 5.926083 |
| 16 | 5 | 8.637341 | 5.847972 |
| 17 | 5 | 8.767016 | 5.76034 |
| 18 | 5 | 8.884349 | 5.664912 |
| 19 | 5 | 8.990518 | 5.563184 |
| 20 | 5 | 9.086582 | 5.456453 |
| 21 | 5 | 9.173506 | 5.345845 |
| 22 | 5 | 9.252158 | 5.232331 |

**FIGURE 3.6** Profit margin analysis.

$b = 3$, $h = 3$, and $k = 22$. From Table 3.3 we obtain the $Z$-transform of this composite function as follows.

$$PV(i) = \frac{10}{0.1}\,[(1.1)^{-2} - (1.1)^{-22}] - \frac{5(1.1)}{1.1 - e^{-0.1}}\left[\frac{1}{(1.1)^3} - \frac{e^{-2.0}}{(1.1)^{23}}\right]$$

$$= 70.36 - 20.74 = \underline{\$49.61}$$

As expected, the total profit margin has increased significantly compared with case i, where no learning effect is appreciable.

For case iii, $g(n)$ is also a linear combination of two similar types of decay function. That is,

$$g(n) = 10e^{-0.03(n-3)} - 5e^{-0.13(n-3)}$$

The first decay function has parameter values of $C = 10$, $b = 3$, $h = 3$, $j = 0.03$, and $k = 22$. The second decay function has $C = 5$, $b = 3$, $h = 3$, $j = 0.13$, and $k = 22$. Thus, from Table 3.3 the $Z$-transform of this combination yields

$$\begin{aligned} PV(i) &= \frac{10(1.1)}{1.1 - e^{-0.03}}\left[\frac{1}{(1.1)^3} - \frac{e^{-0.60}}{(1.1)^{23}}\right] \\ &\quad - \frac{5(1.1)}{1.1 - e^{-0.13}}\left[\frac{1}{(1.1)^3} - \frac{e^{-2.60}}{(1.1)^{23}}\right] \\ &= 58.58 - 18.41 = \underline{\$40.17} \end{aligned}$$

***Analysis of Loan Transactions.*** The repayment schedule for most loans is made up of a portion for the payment of principal and a portion for the payment of interest on the unpaid balance. In economic analysis the interest paid on borrowed capital is considered as a deductible expense for income tax computation. Therefore, it is quite important to know how much of each payment is interest and how much is used to reduce the principal amount borrowed initially. To illustrate this situation, suppose that we want to develop an expression for the present value of the interest components of a uniform repayment plan. Let

$A$ = the equal annual repayment amount,

$B$ = the amount borrowed,

$i_b$ = the borrowing interest rate per period,

$N$ = the maturity of the loan (period).

Then the annual payments will be

$$A = B(A/P, i_b, N) = B\,\frac{i_b(1 + i_b)^N}{(1 + i_b)^N - 1} \tag{3.30}$$

Each payment is divided into an amount that is interest and a remaining amount for reduction of the principal. Let

$I_n$ = portion of payment $A$ at time $n$ that is interest,

$B_n$ = portion of payment $A$ at time $n$ that is used to reduce the remaining balance,

$A = I_n + B_n$, where $n = 1, 2, \ldots, N$,
$U_n$ = unpaid balance at the end of period $n$, with $U_0 = B$.

The relation of these parameters is illustrated in Figure 3.7. Since the interest payment is based on the unpaid principal that remains at the end of each period, the interest accumulation in the first year is simply $i_bB$. Thus, the first payment $A$ consists of an interest payment $i_bB$ and a principal payment of $A - i_bB$. The unpaid balance remaining after the first payment would be $U_1 = B - (A - i_bB) = B(1 + i_b) - A$.. Consequently, the interest charge for the second year would be $i_bU_1$, and the size of the net principal reduction associated with the second payment would be $A - i_bU_1$. In other words, the unpaid balance remaining after the second payment would be

$$\begin{aligned} U_2 &= U_1 - (A - i_bU_1) \\ &= U_1(1 + i_b) - A \end{aligned} \tag{3.31}$$

The amount of principal remaining to be repaid right after making the $n$th payment can be found with the recursive relationship

$$U_n = U_{n-1}(1 + i_b) - A$$

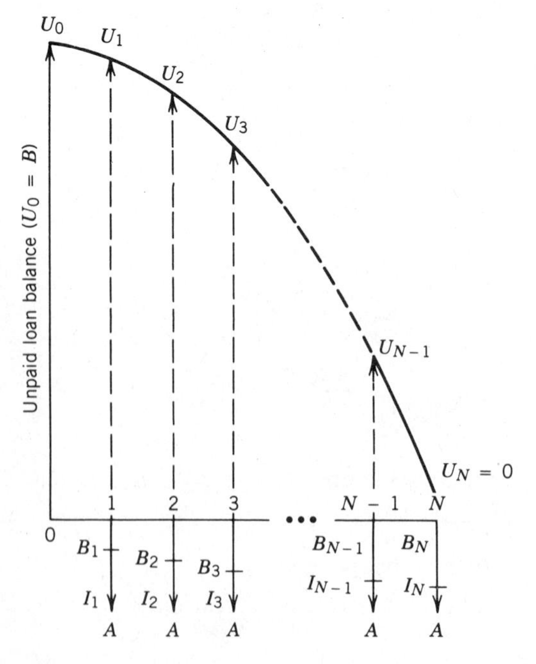


**FIGURE 3.7** Loan transactions—unpaid balance as functions of $B_n$ and $I_n$.

It follows immediately that

$$\begin{aligned} U_n &= B(1 + i_b)^n - A[(1 + i_b)^{n-1} + (1 + i_b)^{n-2} + \cdots + 1] \\ &= B(1 + i_b)^n - \frac{A}{i_b}[(1 + i_b)^n - 1] \\ &= \left(B - \frac{A}{i_b}\right)(1 + i_b)^n + \frac{A}{i_b}, \qquad n = 0, 1, \ldots, N - 1 \end{aligned} \tag{3.32}$$

Now we can express the amount of interest payment required at the end of period $n + 1$.

$$\begin{aligned} I_{n+1} &= i_b U_n \\ &= \underbrace{(Bi_b - A)(1 + i_b)^n}_{\text{geometric series}} + \underbrace{A}_{\text{step function}}, \qquad n, = 0, 1, \ldots, N - 1 \end{aligned} \tag{3.33}$$

Finally, the total present value of these interest payments at an interest rate of $i$ over the loan life of $N$ periods is defined as

$$PV(i) = \sum_{n=0}^{N-1} I_{n+1}(1 + i)^{-(n+1)} \tag{3.34}$$

Let $g(n) = I_{n+1}$, where $g(n)$ is the sum of a geometric and a uniform series. From Table 3.4, the $Z$-transform of the geometric series portion is obtained by letting $a = (1 + i_b)$, $C = Bi_b - A$, $b = h = 1$, and $k = N$.

$$PV_1(i) = \frac{Bi_b - A}{i - i_b}[1 - (1 + i_b)^N(1 + i)^{-N}], \quad \text{where } i \neq i_b \tag{3.35}$$

The transform of the step function portion is found by substituting $C = A$, $b = h = 1$, and $k = N$.

$$PV_2(i) = \frac{A}{i}[(1 - (1 + i)^{-N}] \tag{3.36}$$

Finally, the transform of $g(n)$ *is found to be*

$$\begin{aligned} PV(i) &= PV_1(i) + PV_2(i) \\ &= \frac{Bi_b - A}{i - i_b}\left[1 - \left(\frac{1 + i_b}{1 + i}\right)^N\right] + \frac{A}{i}[1 - (1 + i)^{-N}] \end{aligned} \tag{3.37}$$

To illustrate the use of this formula, suppose that \$50,000 is borrowed at 8% annual interest and is to be repaid in ten equal annual payments. Determine the total present value of these interest payments associated with the loan transaction at a discount rate of 15%. Since we have $B = \$50{,}000$, $i_b = 8\%$, $N = 10$ years, and $i = 15\%$, the payment size $A$ is

$$A = \$50{,}000(A/P,\ 8\%,\ 10) = \$7{,}451.47$$

Then the total present value is

$$\begin{aligned} PV(15\%) &= \frac{\$50{,}000(0.08) - \$7{,}451.57}{0.15 - 0.08}\left[1 - \left(\frac{1.08}{1.15}\right)^{10}\right] \\ &\quad + \frac{\$7{,}451.47}{0.15}\left[1 - \frac{1}{(1.15)^{10}}\right] \\ &= -\$49{,}306.70(1 - 0.53365) + \$49{,}676.47(1 - 0.24718) \\ &= \$14{,}403.26 \end{aligned}$$

It may be of interest to compare the use of Eq. 3.37 with that of the conventional discounting formula developed by Brooking and Burgess [1]. They use the expression

$$\begin{aligned} PV(i) &= B\left\{(A/P,\ i_b,\ N)\left[[(P/A,\ i,\ N) - \frac{(P/F,\ i_b,\ N) - (P/F,\ i,\ N)}{i - i_b}\right]\right\} \\ &= A\left[\frac{1 - (1 + i)^{-N}}{i} - \frac{(1 + i_b)^{-N} - (1 + i)^{-N}}{i - i_b}\right] \end{aligned} \tag{3.38}$$

Our method may be numerically verified with the traditional method as follows.

$$\begin{aligned} PV(15\%) &= \$7{,}451.47(5.0188 - \frac{0.4632 - 0.2472}{0.07}) \\ &= \$7{,}451.47(5.0188 - 3.0857) \\ &= \$14{,}404.47 \end{aligned}$$

The slight difference is due to rounding errors.

## 3.3 LAPLACE TRANSFORMS AND CONTINUOUS CASH FLOWS

Up to this point we have discussed only discrete cash flow functions. In this section we will extend the modeling philosophy to continuous cash flow functions. The Laplace transform method offers a modeling flexibility similar to that of the $Z$-transform for computing present values for many forms of continuous cash flow functions.

### 3.3.1 Laplace Transform and Present Value

As shown in Section 2.5.2, the present value of the infinite continuous cash flow streams, assuming continuous compounding, is given by the expression

$$PV(r) = \int_0^\infty f(t)e^{-rt}\,dt \tag{3.39}$$

where $f(t)$ = continuous cash flow function of the project,
$r$ = nominal interest rate $[r = \ln(1 + i)]$,
$t$ = time expressed in years,
$e^{-rt}$ = discount function.

As Buck and Hill [2] recognized, the general form of this integral bears a close resemblance to the definition of the Laplace transforms. That is, if the function $f(t)$ is considered to be piecewise continuous, then the Laplace transform of $f(t)$, written $L\{f(t)\}$, is defined as a function $F(s)$ of the variable $s$ by the integral

$$L\{f(t)\} = F(s) = \int_0^\infty f(t)e^{-st}\,dt \tag{3.40}$$

over the range of values of $s$ for which the integral exists. Replacing $s$ in Eq. 3.40 with the continuous compound interest rate $r$ simply generates Eq. 3.39; thus, taking a Laplace transform on the cash flow function $f(t)$ is equivalent to computing the present value of the cash flow streams over an infinite horizon time.

In the construction of Laplace transforms, we will use the following notation. If $f(t)$ represents the time domain continuous function, then $F(s)$ will represent its transform. As for the $Z$-transform, this lowercase–uppercase correspondence will be used throughout the text. As a shorthand notation, the transform pair will be denoted by

$$f(t) \leftrightarrow F(s)$$

For example, to find the transform of a linear function $f(t) = t$, $t > 0$, we directly evaluate Eq. 3.40.

$$F(s) = \int_0^\infty te^{-st}\,dt = \frac{1}{s^2} \tag{3.41}$$

and find that the transform pair is

$$t \leftrightarrow \frac{1}{s^2}$$

The transforms of some causal time functions that are typically encountered are shown in Table 3.5. The function $u(t)$ in this table represents the unit

**Table 3.5** ***A Short Table of Laplace Transform Pairs****

| Standard Cash Flow Pattern | Cash Flow Function, $f(t)$ | Laplace Transform, $F(s)$ | Present Value, (Infinite), $PV(r)$ |
|---|---|---|---|
| Unit step<br>$f(t)$ 1<br>$t$ | $f(t) = u(t) = \begin{cases} 1 & t > 0 \\ 0 & \text{otherwise} \end{cases}$ | $1/s$ | $1/r$ |
| Delayed unit step<br>$f(t)$<br>$b$ $t$ | $f(t) = u(t - b) \quad b > 0$ | $e^{-bs}/s$ | $e^{-br}/r$ |
| Ramp<br>$f(t)$<br>$t$ | $f(t) = t$ | $1/s^2$ | $1/r^2$ |

## Table 3.5 (Continued)

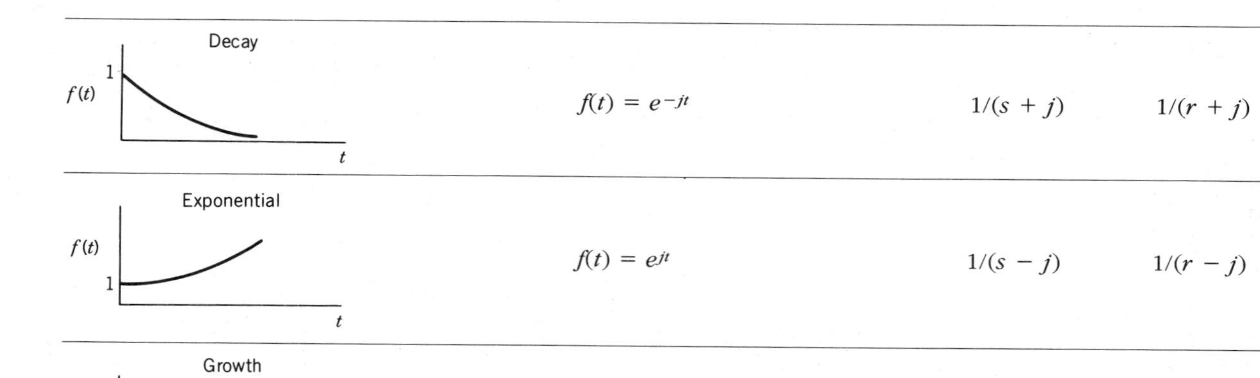


| Standard Cash Flow Pattern | Cash Flow Function, $f(t)$ | Laplace Transform, $F(s)$ | Present Value, (Infinite), $PV(r)$ |
|---|---|---|---|
| Decay | $f(t) = e^{-jt}$ | $1/(s + j)$ | $1/(r + j)$ |
| Exponential | $f(t) = e^{jt}$ | $1/(s - j)$ | $1/(r - j)$ |
| Growth | $f(t) = 1 - e^{-jt}$ | $\frac{1}{s} - \frac{1}{s + j}$ | $j/r(r + j)$ |

*See [7] for a complete Laplace function table.

step function with jump at $t = 0$, and $u(t - a)$ denotes the unit step function with jump at $t = a$. The special property of this function is discussed in the next section.

### 3.3.2 Properties of Laplace Transforms

In this section we will examine some useful operational properties of the Laplace transform. As in the $Z$-transform analysis, the properties most relevant to modeling cash flows are linearity and translation.

***Linearity.*** If we define

$$f_1(t) \leftrightarrow F_1(s) \quad \text{and} \quad f_2(t) \leftrightarrow F_2(s)$$

then

$$c_1 f_1(t) + c_2 f_2(t) \leftrightarrow c_1 F_1(s) + c_2 F_2(s) \tag{3.42}$$

This follows from the linearity property of integrals of Eq. 3.40. Suppose we define $f(t)$ as

$$f(t) = 1 + t + \tfrac{1}{2}t^2$$

The transform is

$$L\{f(t)\} = L\{1\} + L\{t\} + L\{\tfrac{1}{2}t^2\}$$

Using Eqs. 3.42 and 3.40 along with the transform results in Table 3.5, we obtain

$$F(s) = \frac{1}{s} + \frac{1}{s^2} + \frac{1}{s^3}$$

***Translation with Time Delay.*** Consider Figure 3.8, in which the function $g(t)$ is obtained from $f(t)$ by shifting the graph of $f(t)$ $b$ units on the time scale to the right. Mathematically, we define such a function as

$$g(t) = \begin{cases} f(t - b) & \text{for } t \geq b \\ 0 & \text{for } t < b \end{cases} \tag{3.43}$$

To find the transform of this type of cash flow function that starts after a delay of $b$ time units, we utilize the property of the unit step function $u(t)$. If we take the unit step function and translate it $b$ units to the right to get $u(t - b)$, we obtain the function shown in Table 3.5. Mathematically, we denote this by

$$u(t - b) = \begin{cases} 1 & \text{for } t \geq b \\ 0 & \text{for } t < b \end{cases} \tag{3.44}$$

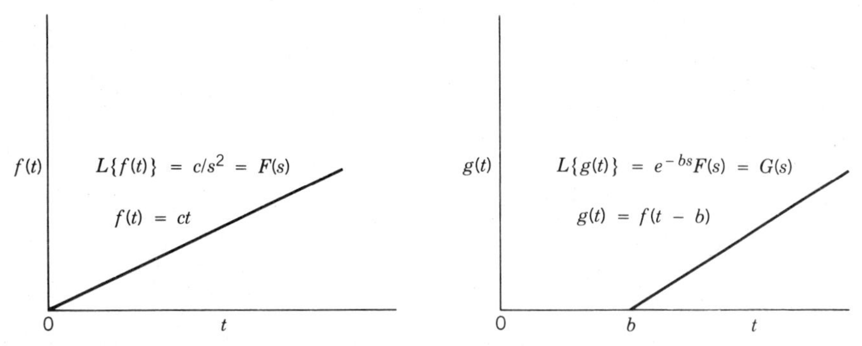


**FIGURE 3.8** Translation of continuous ramp pattern.

Then the product $g(t)u(t - b)$ or $f(t - b)u(t - b)$ will be defined as

$$g(t) = f(t - b)u(t - b) = \begin{cases} f(t - b) & \text{for } t \geq b \\ 0 & \text{for } t < b \end{cases} \tag{3.45}$$

The Laplace transform of $g(t)$ given by Eq. 3.45 is

$$L\{g(t)\} = e^{-bs}F(s) \tag{3.46}$$

Accordingly, a cash flow that starts later than $t = 0$ can be treated as if it started immediately and then a correction for the delayed start can be made with the discount factor $e^{-sb}$ $(= e^{-rb})$. This feature proves to be very useful when developing present value models with a composite of delayed turn-on cash flows.

***Translation with Cutoff.*** Another translation property of interest is turning cash flow streams on and off as desired. To illustrate the concept, suppose we wish to find the Laplace transform of a ramp function with features of a delayed turn-on at time $h$ and a turn-off at time $k$. This function is illustrated in Figure 3.9. Mathematically, we denote such a function by

$$g(t) = f(t)[u(t - h) - u(t - k)] \tag{3.47}$$

The first unit step begins the transactions at $t = h$ and the second stops the transactions at $t = k$. The Laplace transform of this $g(t)$ is defined by

$$G(s) = (e^{-hs} - e^{-ks})\left[F(s) + \frac{f(h)}{s}\right] \tag{3.48}$$

Some care must be exercised in using the time delay theorem. The reader should note the subtle functional difference that $f(t)u(t - h)$ is not a simple time-shifted function $[f(t - h)u(t - h)]$.

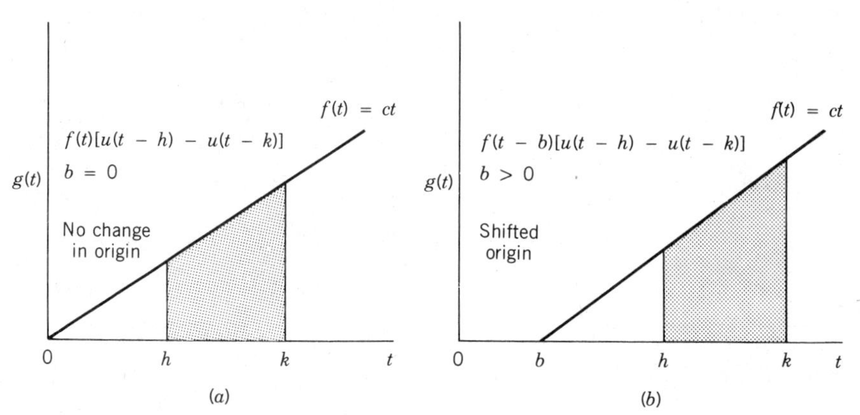


**FIGURE 3.9** A continuous ramp function with translation and cutoff.

$$f(t)u(t-h) \neq f(t-h)u(t-h)$$

This difference is illustrated in Fig. 3.10. We can rewrite the function as

$$f(t)u(t-h) = f(t-h)u(t-h) + f(h)u(t-h) \tag{3.49}$$

Since $f(h)$ is a constant, the Laplace transform of Eq. 3.49 is found by using Eq. 3.46:

$$\begin{aligned} L\{f(t)u(t-h)\} &= e^{-hs}F(s) + \frac{f(h)e^{-hs}}{s} \\ &= e^{-hs}L\{f(t+h)\} \end{aligned} \tag{3.50}$$

Therefore, the transform of Eq. 3.47 can be expressed as

$$L\{g(t)\} = G(s) = e^{-hs}L\{f(t+h\} - e^{-ks}L\{f(t+k)\} \tag{3.51}$$

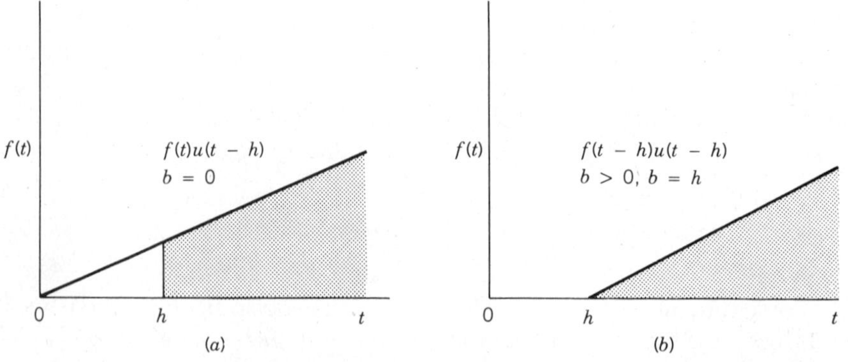


**FIGURE 3.10** Functional difference between $f(t)u(t-h)$ and $f(t-h)u(t-h)$.

### *Example 3.3*

Suppose a cash flow function is given by

$$f(t) = 5t, \qquad 10 \leq t \leq 20$$

Using Eq. 3.48 and Table 3.5, we obtain

$$\begin{aligned} G(s) &= e^{-10s}L\{5(t+10)\} - e^{-20s}L\{5(t+20)\} \\ &= e^{-10s}\left(\frac{5}{s^2} + \frac{50}{s}\right) - e^{-20s}\left(\frac{5}{s^2} + \frac{100}{s}\right) \\ &= \frac{5}{s^2}\left(e^{-10s} - e^{-20s}\right) + \frac{50}{s}\left(e^{-10s} - 2e^{-20s}\right) \end{aligned}$$

With a nominal interest rate of 10% ($r = s = 0.1$), the total present value is

$$\begin{aligned} PV(10\%) &= \frac{5}{(0.1)^2}(e^{-1} - e^{-2}) + \frac{50}{0.1}(e^{-1} - 2e^{-2}) \\ &= 116.27 + 48.60 = \$164.87 \end{aligned}$$

Our method may be numerically verified by direct integration of the cash flow function.

$$PV(10\%) = \int_{10}^{20} 5te^{-0.1t}\,dt = \$165$$

Once again, the slight difference is due to rounding errors. □

***Translations with Impulses.*** Suppose we want to find the transform of an impulse function $f(t)$ shown in Figure 3.11. This type of impulse function may represent the salvage value of an asset at $t = h$ when the salvage value $f(t)$ decreases exponentially over time. To obtain the transform of such an impulse function, we need to define a Kronecker delta function that corresponds to a unit impulse at $t = h$. That is,

$$\delta(t - h) = \begin{cases} 1 & \text{for } t = h \\ 0 & \text{otherwise} \end{cases} \tag{3.52}$$

By multiplying the salvage value function $f(t)$ by the unit impulse function, we obtain

$$g(t) = f(t)\delta(t - h) = \begin{cases} f(h) & \text{for } t = h \\ 0 & \text{otherwise} \end{cases} \tag{3.53}$$

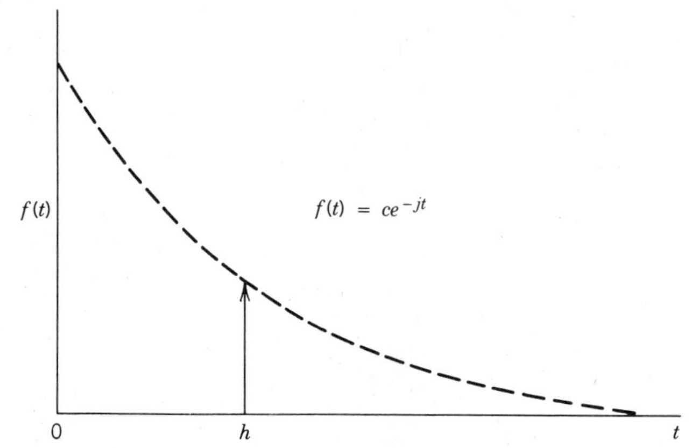


**FIGURE 3.11** Example of an impulse cash flow function—decay.

Since $f(h)$ is a constant, the transform of the product form yields

$$g(s) = e^{-hs}f(h) \tag{3.54}$$

which is the present value expression for a single payment.

Many other useful operational rules can be used in modeling continuous cash flow functions, such as scaling, periodic functions, and convolutions. These are summarized in Table 3.6. (See also Muth [7].)

### 3.3.3 Development of Continuous Present Value Models

Two types of present value models are needed, corresponding to the start of the original cash flow function. These are the extensive models and the simplified models. Figure 3.12 illustrates the modeling concept of both the extensive and the simplified forms of the ramp time form.

***Extensive Present Value Models.*** The computational procedure for finding the correct extensive present value model was discussed in the previous section. Formulas for directly computing the present values of these extensive models of five common cash flow time forms are presented in Table 3.7. To examine the modeling concept again, consider the exponential time forms of cash flow given in Table 3.7.

Let $f(t) = ce^{jt}$, where $c$ is the scale factor and $j$ is the growth rate with time. To obtain a geometric time form shifted to the right by $b$ time units, we define $g(t) = f(t - b)$. To denote the added feature of a delayed turn-on at $t = h$ and a turn-off at $t = k$, we write

$$g(t) = f(t - b)[u(t - h) - u(t - k)]$$

**Table 3.6** ***Summary of Operational Rules of the Laplace Transform***

| Operational Rule | Original Function | Laplace Transform |
|---|---|---|
| Linearity | $C_1f_1(t) + C_2f_2(t)$ | $c_1F_1(s) + c_2F_2(s)$ |
| Change of scale | $f(at),\quad a > 0$ | $\frac{1}{a}F(s)$ |
| Shifting to the right | $f(t - a)u(t - a),\quad a > 0$ | $e^{-as}F(s)$ |
| Shifting to the left | $f(t + a),\quad a < 0$ | $e^{as}\left[F(s) - \int_0^a e^{-st}f(t)\,dt\right]$ |
| Damping | $e^{-at}f(t)$ | $F(s + a)$ |
| Differentiation of $F(s)$ function | $tf(t)$ | $-\frac{d}{ds}F(s)$ |
| Integration of $F(s)$ function | $\frac{f(t)}{t}$ | $\int_s^\infty F(u)\,du$ |
| Differentiation of $f(t)$ | $\frac{d}{dt}f(t)$ | $sF(s) - f(0^+)$ |
| | $\frac{d^n}{dt^n}f(t)$ | $s^nF(s) - s^{n-1}f(0^+) - s^{n-2}fE(0^+) - \cdots - f^{(n-1)}(0^+)$ |
| Integration of f(t) | $\int_0^t f(u)\,du$ | $\frac{1}{s}F(s)$ |
| Periodic function | $f(t) = f(t + T),\quad T = \text{period}$ | $F(s) = \frac{1}{1 - e^{-sT}}\int_0^T e^{-st}f(t)dt$ |
| Convolution | $f_1(t) * f_2(t)$ | $F_1(s)F_2(s)$ |

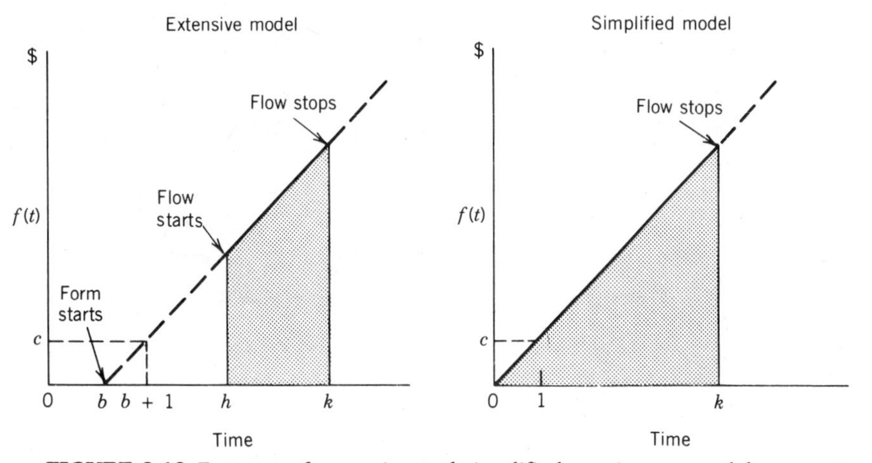


**FIGURE 3.12** Features of extensive and simplified continuous models.

**Table 3.7** ***Extensive Continuous Present Value Models***

| Time Form | $f(t)$ | $PV(r)$ |
|---|---|---|
| Step | $c$ | $\frac{c}{r}(e^{-br} - e^{-kr})$ |
| Ramp | $ct$ | $\frac{c}{r^2}(e^{-br} - e^{-kr}) + \frac{c}{r}[(h - b)e^{-hr} - (k - b)e^{-kr}]$ |
| Decay | $ce^{-jt}$ | $\frac{ce^{+bj}}{r + j}(e^{-h(j+r)} - e^{-k(j+r)})$ |
| Growth | $c(1 - e^{-jt})$ | $\frac{c}{r}(e^{-hr} - e^{-kr}) - \frac{ce^{bj}}{r + j}(e^{-h(j+r)} - e^{-k(j+r)})$ |
| Exponential | $ce^{jt}$ | $\frac{ce^{-bj}}{r - j}(e^{h(j-r)} - e^{k(j-r)}), \quad j \neq r$ |

Since $f(t) = ce^{jt}$, $f(t - b) = ce^{j(t-b)}$. Therefore, we may rewrite $g(t)$ as

$$
\begin{aligned}
g(t) &= ce^{j(t-b)}[u(t - h) - u(t - k)] \\
&= (e^{-bj})(ce^{jt})[u(t - h) - u(t - k)] \\
&= (e^{-bj})f(t)[u(t - h) - u(t - k)]
\end{aligned} \tag{3.55}
$$

From Eq. 3.47, the transform of $g(t)$ yields

$$L\{g(t)\} = e^{-bj}[e^{-hs}L\{f(t + h) - e^{-ks}L\{f(t + k)\}]$$

To evaluate $L\{f(t + h)\}$ and $L\{f(t + k)\}$, we simply expand the original function $f(t) = ce^{jt}$

$$
\begin{aligned}
L\{f(t + h)\} &= \{ce^{j(t+h)}\} = ce^{jh}L\{f(t)\} = ce^{jh}F(s) \\
L\{f(t + k)\} &= \{ce^{j(t+k)}\} = ce^{jk}L\{f(t)\} = ce^{jk}F(s)
\end{aligned}
$$

Since $F(s) = 1/(s - j)$ for $f(t) = e^{jt}$, but with $s = r$, we have

$$L\{g(t)\} = \frac{ce^{-bj}}{r - j}\left(e^{h(j-r)} - e^{k(j-r)}\right) \tag{3.56}$$

### *Example 3.4*

Consider a cash inflow stream that starts at $t = 2$ (years) and increases \$1,000 per year uniformly until $t = 10$. Table 3.7 reveals that the ramp is the proper time form for the cash flow. This time form has the scale parameter of $c = \$1,000$. Assume that the pattern starts at $b = 2$, the cash flow begins immediately after that at $h = 2$, and the flow stops at $k = 10$. The present value at the nominal rate of interest 10% is

$$
\begin{aligned}
PV(10\%) &= \frac{\$1,000}{(0.1)^2}(e^{-0.2} - e^{-1}) + \frac{\$1,000}{0.1}(0 - 8e^{-1}) \\
&= \$45,085.13 - \$29,430.35 = \$15,654.78 \quad \square
\end{aligned}
$$

***Simplified Present Value Models.*** When there is no shift in time form and no delayed turn-on, the extra factors in the extensive model become cumbersome. In other words, if $b = h = 0$, we can further simplify the formulas in Table 3.7. The reader may notice that the simplified models correspond to the traditional tabulated interest factors (funds flow factors) used in most engineering economy textbooks. These are summarized in Table 3.8.

***Present Values of Impulse Cash Flows.*** Single instantaneous cash flows are referred to as "impulses" to distinguish them from the continuous flow streams examined in the previous sections. Frequently, it is necessary to describe a cash

**Table 3.8** ***Simplified Continuous Present Value Models***

| Time Form | $f(t)$ | $PV(r)$ |
|---|---|---|
| 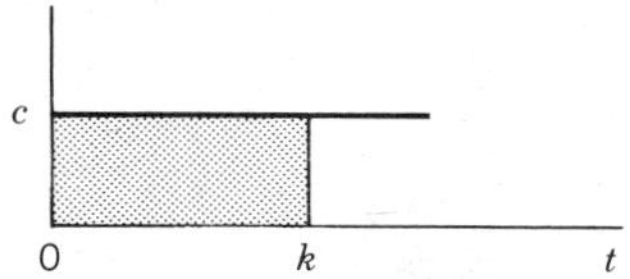 Step (c; 0, k, t) | $c$ | $\frac{c}{r}(1 - e^{-kr})$ |
| Ramp (0, k, t) | 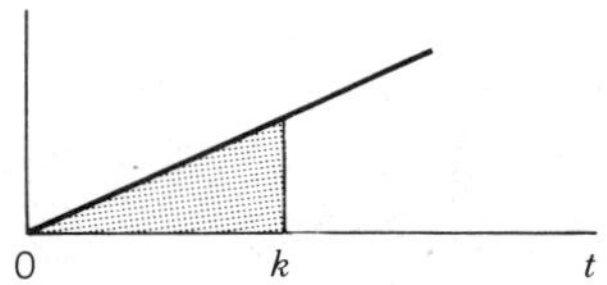 $ct$ | $\frac{c}{r^2}(1 - rke^{-kr})$ |
| Decay (c; 0, k, t) | 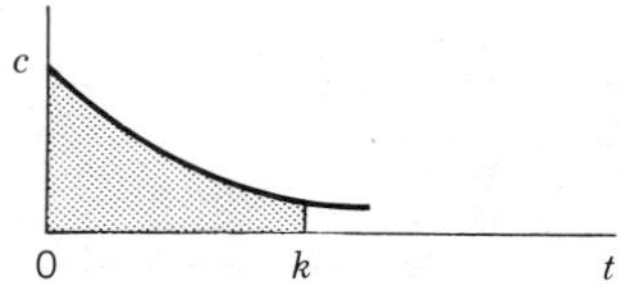 $ce^{-jt}$ | $\frac{c}{r+j}(1 - e^{-k(j+r)})$ |
| Growth (c; 0, k, t) | 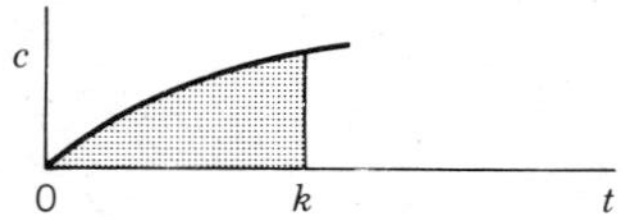 $c(1 - e^{-jt})$ | $\frac{c}{r}(1 - e^{-kr}) - \frac{c}{r+j}(1 - e^{-k(j+r)})$ |
| Exponential (c; 0, k, t) | 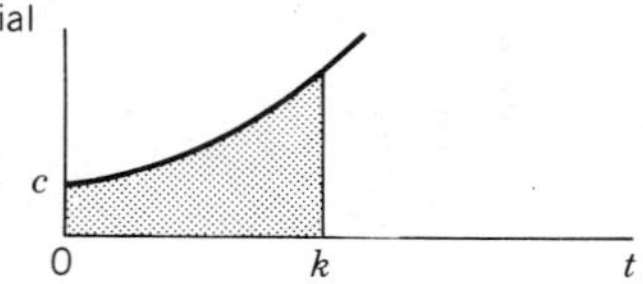 $ce^{jt}$ | $\frac{c}{r-j}(1 - e^{k(j-r)})$ |

impulse that changes in magnitude over time according to some time form. As an example, a salvage value from the sale of a machine decreases gradually with the age of the machine, but the actual value received is a single flow at the time of disposal. Present value formulas corresponding to such a cash impulse, following the four time forms but occurring only at time $T$, are summarized in Table 3.9. These present value formulas are derived from Eq. 3.54.

### *Example 3.5*

Suppose that the salvage value of an automobile can be described by a decay time form with an initial value of \$6,000. The decay rate with time is given as 0.3. Find the present value of the salvage value that occurs at the end of 5 years at a nominal interest rate of 10% compounded continuously. Let $c = 6{,}000$, $r = 0.3$, $j = 0.1$, and $T = 5$. Then

$$PV(10\%) = ce^{-(j+r)T} = \$6{,}000e^{-2.0} = \$812.01 \quad \square$$

**Table 3.9** ***Present Values of Impulse Cash Flows***

| Time Form | $PV(r)$ |
|---|---|
| Step: $f(t) = c$ | 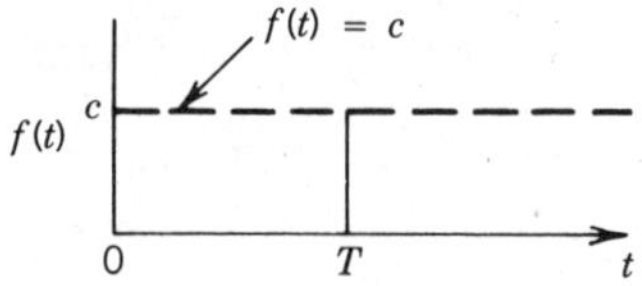 $ce^{-rT}$ |
| Ramp: $f(t) = ct$ | 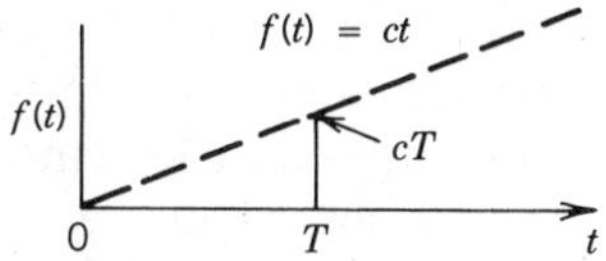 $cTe^{-rT}$ |
| Decay: $f(t) = ce^{-jt}$ | 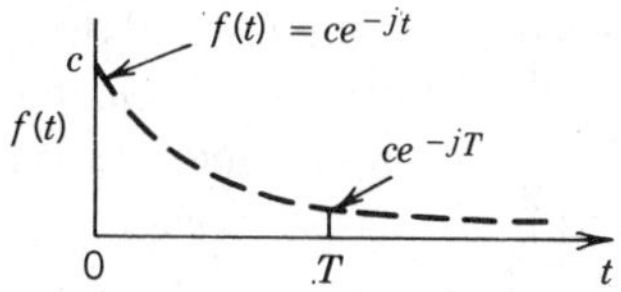 $ce^{-(j+r)T}$ |
| Growth: $c(1 - e^{-jT})$ | 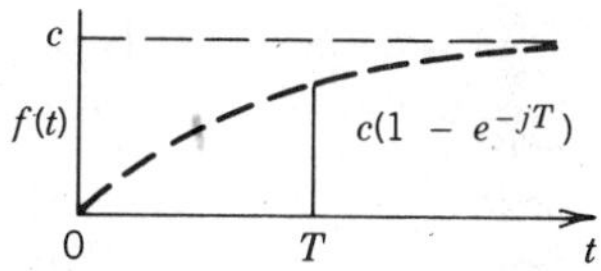 $ce^{-rT}(1 - e^{-jT})$ |

### 3.3.4 Extension to Future and Annual Equivalent Models

The future equivalent values at the end of period $T$ can easily be obtained from the present value formulas shown in Tables 3.7, 3.8, and 3.9 simply by multiplying through by $e^{rT}$. Similarly, annual values of equivalent cash flow streams are defined here as the annual cash flow of a step time form starting immediately, terminating at the same time as the equivalent stream, and possessing equal present value. Accordingly, the present value of a step (uniform) time form with the annual cash flow of $\bar{A}$ dollars may be equated to the present value formulas of the other time forms. Solving for $\bar{A}$ gives us the equivalent annual value.

#### *Example 3.6*

Consider Example 3.5 and find the equivalent annual value at a nominal interest rate of 10% compounded continuously. Since the present value of the ramp time form that extends over a 10-year period is \$15,654.78, the annual equivalent cash flow stream of $\bar{A}$ dollars per year is determined as follows. From Table 3.8, the present value of the step time form with $b = h = 0$, $c = \bar{A}$, $k = 10$, and $r = 0.1$ yields

$$\frac{\bar{A}}{0.1}(1 - e^{-1}) = \$15{,}654.78$$

The satisfying value of $\bar{A}$ is the equivalent annual value, which is $\bar{A} =$ \$2,476.55. □

### 3.3.5 Application of the Laplace Transform

***Description of the Basic Inventory System.*** Consider the simplest imaginable type of inventory system in which there is only a single item. The demand rate for this item is assumed to be deterministic and a constant $\lambda$ units per year. The fixed cost of placing an order in dollars is $A$. The unit cost of the item in dollars is $C$. Let $I_0$ be the inventory carrying charge (measured in the units of dollars per year per dollar of investment in inventory) exclusive of the rate of return (i.e., of the opportunity cost). We will further assume that the procurement lead time is a constant and that the system is not allowed to be out of stock at any point in time. Orders for the item are received in lots of $Q$ units. The problem is to determine the optimal value of $Q$.

Figure 3.13 depicts the inventory behavior of this model with respect to time. Since the order quantity $Q$ and the demand rate $\lambda$ are constant, the inventory level of the first cycle $T$ is

$$I(t) = Q - \lambda t, \qquad 0 \le t \le T \tag{3.57}$$

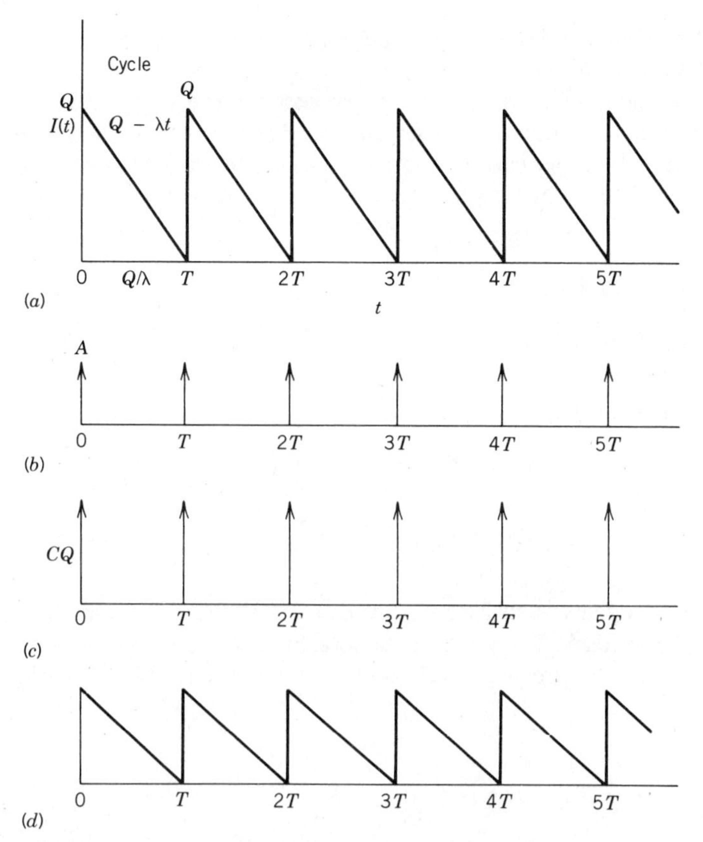


**FIGURE 3.13** Inventory behavior: quantity and const as functions of time. (*a*) Inventory positions. (*b*) Ordering costs. (*c*) Purchase costs. (*d*) Inventory costs.

Note that $I(t) = 0$ at $t = T$ and $T = Q/\lambda$. Let

$$r = \text{the nominal interest rate,}$$
$$f(t)_1 = \text{the ordering cost per cycle,}$$
$$f(t)_2 = \text{the purchase cost per cycle,}$$
$$f(t)_3 = \text{the inventory carrying cost per cycle.}$$

Then the inventory cost for the first cycle is given by

$$f(t) = f(t)_1 + f(t)_2 + f(t)_3 \tag{3.58}$$

where

$$f(t)_1 = A,$$
$$f(t)_2 = CQ,$$
$$f(t)_3 = I_0 C \int_0^T (Q - \lambda t)\, dt.$$

Equation 3.58 represents the inventory cost per cycle *without* considering the effect of the time value of money.

To find the present value of the inventory cost for the first cycle, we assume that the ordering and purchase costs will occur only at the beginning of the cycle, but that the inventory carrying cost will occur continuously over the cycle. With these assumptions, the Laplace transform of the inventory cost function is

$$\begin{aligned} F(s) &= A + CQ + L\{I_0C(Q - \lambda t)\} \\ &= A + CQ + I_0C\left[\frac{Q}{s} - \frac{\lambda}{s^2}(1 - e^{-sT})\right] \end{aligned} \tag{3.59}$$

After substituting $s = r$ and $T = Q/\lambda$ back into Eq. 3.59, we find that the present value expression is

$$PV(r)_{\text{cycle}} = A + CQ + I_0C\left[\frac{Q}{r} - \frac{\lambda}{r^2}\,(1 - e^{-rQ/\lambda})\right] \tag{3.60}$$

Since the cycle repeats itself forever, we can use the Laplace transform property of periodic functions. If we denote the total inventory cost over infinite cycles as $g(t)$, $G(s)$ can be expressed in terms of the transform of the first cycle $F(s)$.

$$G(s) = F(s)\,\frac{1}{1 - e^{-sT}} \tag{3.61}$$

$$\begin{aligned} PV(r)_{\text{total}} &= PV(r)_{\text{cycle}}\left(\frac{1}{1 - e^{-rQ/\lambda}}\right) \\ &= \frac{1}{1 - e^{-rQ/\lambda}}\left(A + CQ + \frac{I_0CQ}{r}\right) - \frac{I_0C\lambda}{r^2} \end{aligned} \tag{3.62}$$

Differentiating $PV(r)_{\text{total}}$ with respect to $Q$ and equating the result to zero gives us

$$(1 - e^{-rQ/\lambda})\left(C + \frac{I_0C}{r}\right) - \left(A + CQ + \frac{I_0CQ}{r}\right)\left(\frac{r}{\lambda}e^{-rQ/\lambda}\right) = 0 \tag{3.63}$$

An exact analytical solution of (3.63) for $Q$ is not normally possible, but a numerical solution may be obtained by using the Newton–Raphson method [8]. An approximate solution (within about 2%) can be obtained more easily, however, by using a second-order approximation for the exponential term.

$$\begin{aligned} e^{-rQ/\lambda} &= 1 - \left(\frac{r}{\lambda}\right)Q + \left(\frac{r}{\lambda}\right)^2 Q^2\,\frac{1}{2!} \\ &\quad - \left(\frac{r}{\lambda}\right)^3 Q^3\,\frac{1}{3!} + \cdots \end{aligned} \tag{3.64}$$

Since $0<r<1$ (in general), we can ignore the terms $(r/\lambda)^3$ and higher. Then, substituting the first three terms into Eq. 3.63 and solving for $Q$, we obtain

$$Q^* \triangleq \left[\frac{2A}{(I_0 + r)C}\right]^{1/2} \tag{3.65}$$

With $A = \$10$, $C = \$5/\text{unit}$, $I_0 = 0.1$, $r = 0.1$, and $\lambda = 100$ units per year, the optimal order quantity is about

$$Q^* \triangleq \left[\frac{2(15)(100)}{(0.1 + 0.2)5}\right]^{1/2} = 36.51$$

The numerical solution obtained by the Newton–Raphson method would be $Q^* = 36.23$.

## 3.4 SUMMARY

The $Z$-transform and the Laplace transform can be used in a wide variety of cash flow models, and in many situations these methodologies are more efficient than the traditional approach. This chapter was intended to (1) introduce the transform methodologies, (2) provide alternative techniques for modeling cash flows that are interrupted or impulses that follow a particular time form, and (3) demonstrate the use of this methodology in equivalence calculations. We do not recommend the transform analysis for modeling simple cash flow transactions because there is not much savings in computation. The transform analysis will provide definite computational advantages, however, for complex cash flow functions.

## REFERENCES


1. Brooking, S. A., and A. R. Burgess, "Present Worth of Interest Tax Credit," *The Engineering Economist,* Vol. 21, No. 2, Winter 1976, pp. 111–117.
2. Buck, J. R., and T. W. Hill, "Laplace Transforms for the Economic Analysis of Deterministic Problems in Engineering," *The Engineering Economist,* Vol. 16, No. 4, 1971, pp. 247–263.
3. Buck, J. R., and T. W. Hill, "Additions to the Laplace Transform Methodology for Economic Analysis," *The Engineering Economist,* Vol. 20, No. 3, 1975, pp. 197–208.
4. Giffin, W. C., *Transform Techniques for Probability Modeling,* Academic Press, New York, 1975.
5. Grubbstrom, R. W., "On the Application of the Laplace Transform to Certain Economic Problems," *Management Science,* Vol. 13, No. 7, 1967, pp. 558–567.
6. Hill, T. W., and J. R. Buck, "Zeta Transforms, Present Value, and Economic Analysis," *AIIE Transactions,* Vol. 6, No. 2, 1974, pp. 120–125.
7. Muth, E. J., *Transform Methods with Applications to Engineering and Operations Research,* Prentice–Hall, Englewood Cliffs, N.J., 1977.

8. PARK, C. S., and Y. K. SON, "The Effect of Discounting on Inventory Lot Sizing Models," *Engineering Costs and Production Economics,* Vol. 16, No. 1, 1989, pp. 35–48.

9. REMER, D. S., J. C. TU, D. E. CARSON, and S. A. GANIY, "The State of the Art of Present Worth Analysis of Cash Flow Distributions," *Engineering Costs and Production Economics,* Vol. 7, No. 4, 1984, pp. 257–278.

## PROBLEMS

**3.1.** Consider a cash flow stream for which the monthly profits are $\$1{,}000e^{-0.1n}$ for $n = 1, 2, 3, \ldots, 12$ months and the nominal interest rate is 12%. Find the present value under

a. 12% compounded annually.
b. 12% compounded quarterly.
c. 12% compounded monthly.
d. 12% compounded continuously.

**3.2.** Consider the discrete cash flow patterns shown in the accompanying illustration.

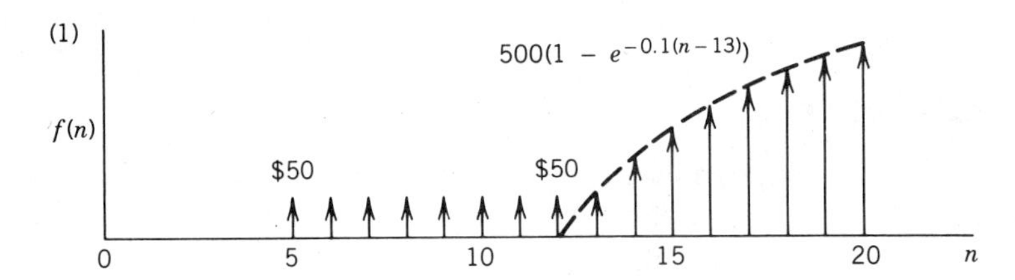


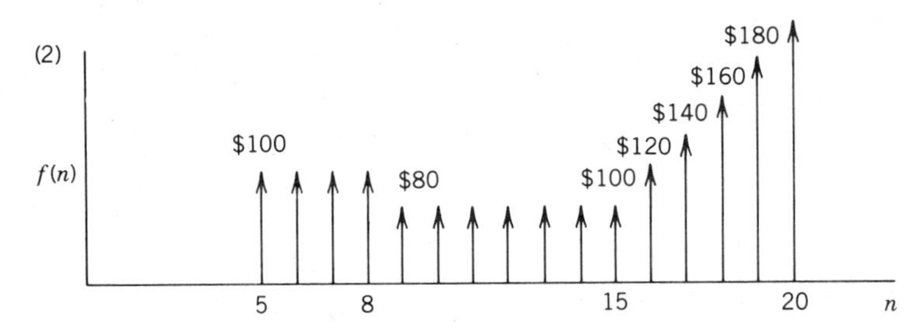


a. Compute the present value of each cash flow series using the conventional interest formulas at $i = 10\%$.
b. Compute the present value of each cash flow series using the discrete transform results at $i = 10\%$.
c. Compute the annual equivalent value of each cash flow series over 20 years.

**3.3.** Consider the retirement schedule for a \$100,000 bond issue by a city, which is to be proportional to the city's anticipated growth. If this anticipated growth tends to follow the general growth pattern of

$$f(n) = C(1 - e^{-0.087n})$$

and the bond interest rate is 5%, find an increasing repayment over 20 years.

**3.4.** Suppose you borrow \$100,000 at an interest rate of 9% compounded monthly over 30 years to finance a home. If your interest rate is 1% per month, compute the present value of the total interest payment of the loan.

**3.5.** Consider the following cost and return components of a machine tool.

a. The initial cost of $8,000.

b. A uniform operating cost of $800 each year.

c. Maintenance costs, which increase at a rate of $400 each year.

d. Annual start-up costs, which decay at the rate of 1.0 from an upper limit of $1,000 initially.

e. A single salvage value return, which decays at the rate of 0.5 with age from the initial cost of $8,000.

Compute the present value of these five cash flow components over 10 years.

**3.6.** Consider a machine that now exists in condition $j$ and generates earnings at the uniform continuous rate of $A_j$ dollars per year. If at some time $T$ the machine's condition changes from $j$ to $k$, its earning rate will instantaneously change from $A_j$ to $A_k$. We will inspect the machine exactly one year from now. You may treat the time value of money in terms of a nominal interest rate of $r$ compounded continuously.

a. If the machine's condition changes to $k$ at time $T$, where $T$ is in time interval between 0 and 1, what is the present value of its earnings for the year?

b. If the machine remains in condition $j$ for the entire year, what is the present value of its earnings for the year?

**3.7.** Suppose a uniformly increasing continuous cash flow (a ramp) accumulates $1,000 over 4 years. The continuous cash flow function is expressed as

$$f(t) = ct, \qquad 0 \leq t \leq 4$$

Assume that $r = 12\%$ compounded continuously.

a. Find the slope $c$.

b. Compute the present equivalent of this continuous series.

c. Compute the future value of this continuous series.

**3.8.** Find the present value of the following quadratic cash flow at 10% interest compounded continuously,

$$f(t) = \$200 + 45t - 3t^2$$

a. if $0 \leq t \leq 10$.

b. if $0 \leq t \leq \infty$.

**3.9.** A chemical process for an industrial solvent generates a continuous after-tax cash flow $f(t)$ of $250,000 per year for a 10-year planning horizon.

a. Find the present value of this cash flow stream over 10 years if money is worth 12% compounded continuously.

b. The profit per year is expected to increase continuously because of increased productivity and can be expressed as

$$f(t) = 250{,}000(2.0 - e^{-0.2t})$$

where $t$ is time in years. Find the present value of this cash flow stream.

c. Productivity increases as in part b, but competition reduces the profit continuously by 8% per year. Find the present value of the cash flow.

**3.10.** Consider the following simple inventory system. A stock of $Q$ units is produced at a rate of $a_p$ units per day for a period $T_p$. It is then necessary to leave the batch in stock for a period of $T_d$, during which sorting, inspection, and painting are carried out. A quantity $Q_1$ is then supplied to the assembly line at the rate of $a$ units per day for a period $T_c$. The supply to the assembly is intermittent, so that after a supply period $T_c$ there is an interval $T_0$ before supply is resumed for another period $T_c$, and so on. Assuming that the relationship between $Q$ and $a$ is defined as $Q = ka$, $k$ is an integer, and $b$ stands for a holding cost of one unit per unit time, answer the following questions.

a. Draw the level of inventory position as a function of time $t$.

b. Assuming continuous compounding at a nominal rate of $r$, find the expression of present value of the total inventory cost over one complete cycle. (One cycle is defined as a time interval in which the entire stock $Q$ is depleted.)

c. With $Q$ = 1,000 units, $a_p$ = 10 units/day, $T_p$ = 100 days, $T_d$ = 50 days, $a$ = 5 units/day, $T_c$ = 80 days, $T_0$ = 55 days, $b$ = \$5 per unit per year, and $r$ = 12% compounded continuously, find the total present value using the formula developed in part b.

**3.11.** Consider an inventory system in which an order is placed every $T$ units of time. It is desired to determine the optimal value of $Q$ by maximizing the average annual profit. This profit is the revenue less the sum of the ordering, purchasing, and inventory carrying costs. All demands will be met from inventory so that there are never any back orders or lost sales. We assume that the demand rate $\lambda$ is known with certainty and does not change with time. If the on-hand inventory does not continually increase or decrease with each period, the quantity ordered each time will be $Q = \lambda T$. To minimize carrying charges, the on-hand inventory when a procurement arrives should be zero. Suppose that $A$ is the fixed cost of placing an order, $C$ is the cost of one unit, $I$ is the inventory carrying charge, and $R$ is the unit sales price. For simplicity, we select the time origin as a point just prior to the arrival of an order so that nothing is on hand at the time origin.

a. If $r$ is the nominal interest compounded continuously, find the optimal $Q$ that maximizes the present value of all future profits.

b. As a specific example, consider a situation in which $A$ = \$15, $C$ = \$35, $I$ = 0.10, $r$ = 10%, $\lambda$ = 1,500 units per year, and $R$ = \$60 per unit.

**3.12.** Develop the $Z$-transform result for the decay function, $g(n) = Ce^{-j(n-b)}$, shown in Table 3.3.

**3.13.** Develop the $Z$-transform result for the growth function, $g(n) = C(1 - e^{-j(n-b)})$, shown in Table 3.3. Knowing that this growth function is the sum of $C$ and $-Ce^{-j(n-b)}$, use the linearlity property.

**3.14.** Develop the Laplace transform result for the growth function shown in Table 3.7.

# 4

# Depreciation and Corporate Taxation

## 4.1 INTRODUCTION

In this chapter we give a brief treatment of the major aspects of the U.S. corporate tax law. The emphasis is on the depreciation and tax treatment of assets used for production and distribution in a trade or business. Depreciation and taxes and their effect on cash flow represent one of the *most important* aspects of investment projects. It is safe to say that no major corporate investment decision is made without a careful analysis of the tax effects involved.

Section 4.2 gives an overview of the tax structure for U.S. corporations and the role played by depreciation in determining taxable income. The major depreciation methods in use today, including the methods for the 1981 and 1986 tax laws, are described in Section 4.3. In Section 4.4 the view shifts from the effect of the total firm to the effect of individual projects on after-tax cash flows. For a more complete treatment of depreciation and tax law, the reader should consult the references at the end of the chapter. The material in this chapter is intended to be a guide for educational purposes. Readers are advised to seek expert counsel before undertaking specific investment projects and policies.

## 4.2 CORPORATE TAX RATES

### 4.2.1 Tax Structure for Corporations

The U.S. government and the respective states, counties, and municipalities levy a variety of taxes on the assets and income of corporations that conduct business within their jurisdictions.

- Ad valorem taxes, levied against tangible and intangible property.
- Licenses, levied against business establishments and certain assets.
- Excise taxes, levied against commodities and products at a certain point of processing.
- Sales taxes, levied against commodities and products at the point of retail sale.

- Transfer taxes, levied against property at the point of sale.
- Income taxes, levied against taxable income, as defined by law.

In this chapter we concentrate on income taxes and their effect on cash flow. In particular, we discuss ordinary income taxes in Sections 4.2 and 4.3. The tax treatment in this chapter is based on the 1986 U.S. Internal Revenue Code [8].

The income tax is applied to taxable income, which is defined as gross income minus allowable deductions. The allowable deductions include

salaries and wages,
rent,
interest,
advertising,
pension plans,
qualified research expenses,
depreciation, amortization, and depletion.

The federal tax is on a graduated structure (see Figure 4.1 and Table 4.1).

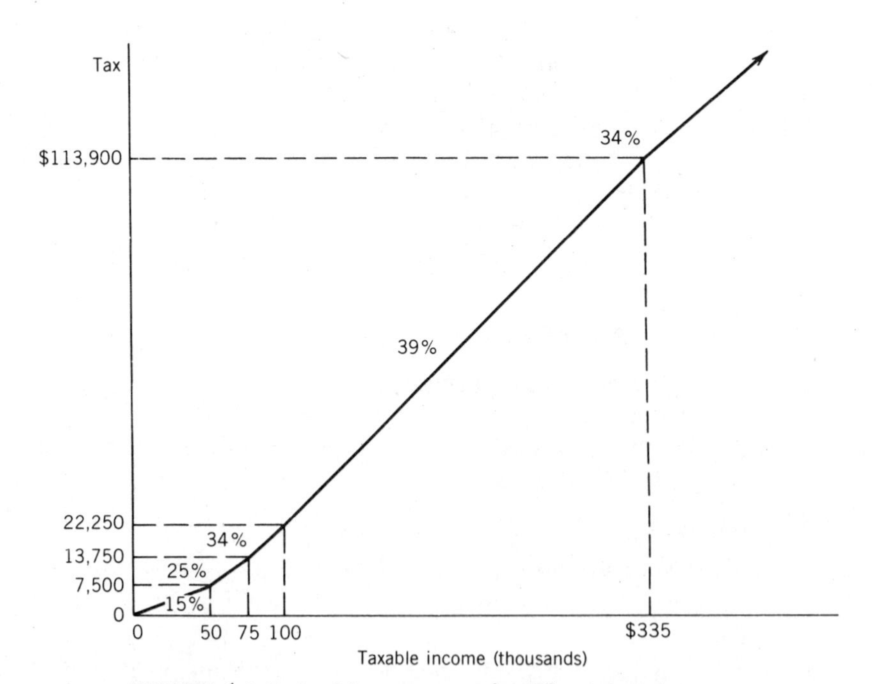


**FIGURE 4.1 Federal income taxes for U.S. corporations.**

**Table 4.1** ***Federal Tax Structure for Corporations, 1988 Tax Year***

| Taxable Income, X | Tax Rate | Tax Computation |
|---|---|---|
| $0–50,000 | 15% | (0.15)(X) |
| $50,001–75,000 | 25% | $7,500 + (0.25)(X − 50,000) |
| $75,001–100,000 | 34% | $13,750 + (0.34)(X − 75,000) |
| $100,001–335,000 | 39% | $22,250 + (0.39)(X − 100,000) |
| Above $335,000 | 34% | $113,900 + (0.34)(X − 335,000) |

*The 39% actually consists of the 34% rate and a 5% surtax. See Fleischer and Smith [4].

### *Example 4.1*

In tax year 1990, UVW Corporation had revenues of $175,000 and expenses of $30,000 for salaries, $40,000 for wages, $10,000 for interest, and $30,000 for depreciation. Its taxable income is

$$\$175{,}000 - (30{,}000 + 40{,}000 + 10{,}000 + 30{,}000) = \$65{,}000$$

Its income tax is

$$\begin{aligned} 15\% \text{ of } \$50{,}000 &= \$7{,}500 \\ 25\% \text{ of } \$15{,}000 &= \underline{\phantom{\$}3{,}750} \\ \text{Income tax} &= \$11{,}250 \end{aligned}$$

Alternatively, the tax is $7,500 + 0.25(65,000 − 50,000) = $11,250. □

### 4.2.2 Depreciation and Its Relation to Income Taxes

Depreciation is of special significance when computing income taxes, for it is an accounting expense that reduces taxable income and hence reduces taxes, but it does not represent a cash flow during this accounting period. Instead, the cash flow may have occurred at time 0 when the asset was purchased, or it may be spread over a loan repayment period that is different from the lifetime used for depreciation. For the time being, we will assume that an asset is always paid for at time 0. In Section 4.4 we will consider asset financing.

This special property of depreciation—of being a noncash item that reduces taxes—can be illustrated by comparing Example 4.1 with Example 4.2.

### *Example 4.2*

In tax year 1990 XYZ Corporation had revenues of $175,000 and expenses of $30,000 for salaries, $70,000 for wages, and $10,000 for interest. The taxable income of the corporation is

$$\$175{,}000 - (30{,}000 + 70{,}000 + 10{,}000) = \$65{,}000$$

Its income tax is $11,250, as in Example 4.1. □

Although the taxable incomes and tax amounts are the same for UVW and XYZ, the after-tax cash flows are different. Obtaining precise after-tax cash flows requires more accounting information (about the funds flow), but we can obtain good estimates by assuming that the salary, wage, and interest expenses correspond to cash outflows during the year and the revenues correspond to cash inflows. For a stable company the boundary effects at January 1 and December 31 tend to cancel. So if UVW and XYZ have no other major cash flows, we can estimate as follows.

| Cash Flows | UVW Corp. | XYZ Corp. |
|---|---|---|
| Revenues | $+175,000 | $+175,000 |
| Salaries | −30,000 | −30,000 |
| Wages | −40,000 | −70,000 |
| Interest | −10,000 | −10,000 |
| Income taxes | −11,250 | −11,250 |
| After-tax cash flow | $+83,750 | $+53,750 |

XYZ has $30,000 less in cash flow because instead of the $30,000 depreciation expenses the company had $30,000 additional wages. Of course, UVW had a large negative cash flow at time 0 when it bought the asset.

People often say that one gets a "cash flow from depreciation." This is incorrect. What they mean is that the after-tax cash flow can be approximated by adding to after-tax profit (accounting income) the amount of depreciation expense. This concept is illustrated for UVW and XYZ.

| Item | UVW Corp. | XYZ Corp. |
|---|---|---|
| Taxable income | $65,000 | $65,000 |
| Income taxes | −11,250 | −11,250 |
| After-tax profit | 53,750 | 53,750 |
| Depreciation | 30,000 | 0 |
| After-tax cash flow | $83,750 | $53,750 |

The arithmetic steps for obtaining after-tax profit are identical to those for obtaining after-tax cash flow from operations, with the exception of depreciation, which is excluded from the after-tax cash flow computation (it is needed only for computing income taxes). Adding depreciation to after-tax profit cancels the operation of subtracting it from revenues, and we thus obtain after-tax cash flow.

It is clear that depreciation, through its influence on taxes, plays an extremely important role in project cash flow analysis. The timing of the depreciation amounts can affect the feasibility of an investment project, so we will investigate some alternative depreciation strategies in Section 4.3.5.

### 4.2.3 Use of Effective and Marginal Income Tax Rates in Project Evaluations

Recall XYZ Corporation in Example 4.2, and assume that in 1990 the company is considering a project with the following characteristics in tax year 1991.

| | |
|---|---|
| Revenues | \$50,000 |
| Salaries | 5,000 |
| Wages | 20,000 |
| Interest | 5,000 |
| Taxable income | \$20,000 |

The XYZ management wishes to evaluate the impact of the project on after-tax cash flow. Because of the graduated income tax rate, the project cannot be considered in a vacuum, so we must assume something about the other operations of XYZ.

#### *Example 4.3*

Assume the base operations of XYZ for 1990 are forecast as shown in Example 4.2. The new project has the following impact.

$$\begin{aligned}\text{Taxable income} &= \$65{,}000 + (50{,}000 - 5{,}000 - 20{,}000 - 5{,}000)\\ &= \$85{,}000\end{aligned}$$

Income tax is \$13,750 + 0.34(85,000 − 75,000) = \$17,150. The additional income tax of \$17,150 − 11,250 = \$5,900 results in an after-tax profit contribution of \$20,000 − 5,900 = \$14,100. □

The \$5,900 tax on \$20,000 income in Example 4.3, a rate of 29.5%, represents a *marginal* rate. It is applied to the new project with the assumption that the base operations of XYZ would remain unchanged, as given in Example 4.2. The 29.5% really represents an average of two separate marginal rates: 25% applied to the first \$10,000 taxable income of the new project and 34% applied to the second \$10,000.

The *effective* tax rates, both before and after considering a new project, would be

| | | |
|---|---|---|
| Before: | \$11,250/65,000 | = 17.3% |
| After: | \$17,150/85,000 | = 20.2% |

It would be improper to apply either of these two rates when considering a project that is in addition to base operations, which are expected to continue anyway. The effective tax rate is useful when evaluating the overall financial performance of a company.

A large, profitable corporation with continuing base operations will then have a marginal federal tax rate of 34% and an effective federal tax rate near 34%. Small corporations and those that fluctuate between losses and profits will have marginal and effective rates that vary. Estimating a prospective marginal tax rate under such circumstances may be difficult, and the only solution may be scenario analysis.

## 4.3 DEPRECIATION METHODS AND REGULATIONS

### 4.3.1 Depreciation Regulations, Notation

Since the beginning of the federal income tax, depreciation methods for tax purposes have been regulated by the U.S. Department of the Treasury. Because there have been numerous changes in the tax code, the complete set of depreciation rules is complex and beyond the scope of this book. We will summarize the major features of the tax law as it applies to assets used for production and distribution in a trade or business. More detailed information can be found in the references at the end of this chapter.

Many other countries, particularly industrialized nations, have depreciation rules that are similar to those in the United States. Often the only changes are in parameter values and depreciation lifetimes. The analyst planning to work in a foreign country will find the description in Section 4.3.2 a useful introduction to the methods used elsewhere.

Throughout the rest of this chapter we will use the following notation.

$\alpha$ — in the declining balance method, the fraction of book balance taken as depreciation the next period

ACRS — accelerated cost recovery system of depreciation

ADR — asset depreciation range

A/T — after taxes

$B$ — loan size

$BB_n$ — book balance or accounting value of asset after period $n$

$D_n$ — depreciation amount at end of period $n$

DB — declining-balance method of depreciation

DDB, 2.0 DB, 200% DB — double-declining-balance method

1.5 DB, 150% DB — 150% declining-balance method

$F$ — salvage value, estimated salvage value

$f$ — general inflation rate

$i$ — interest rate for discounting

$i'$ — real or inflation-free interest rate

$i_b$ — loan interest rate

$IP_n$ — interest payment on loan at end of period $n$

$LB_n$ — loan principal balance at end of period $n$

MACRS — modified ACRS of depreciation

$N$ — lifetime used for depreciation; project lifetime

| | |
|---|---|
| $n$ | time index |
| $P$ | installed cost of asset |
| $PP_n$ | principal repayment on loan at end of period $n$ |
| $PV$ | net present value |
| SL | straight-line method of depreciation |
| SOYD | sum-of-the-years'-digits method of depreciation |
| Sec. 179 | Section 179 expense deduction |
| $t_m$ | marginal corporate income tax rate |
| $TP_n$ | total loan payment at end of period $n$ |

Generally, an asset will be depreciated according to one of three methods. The method depends mainly on when the asset was placed in service but also on some other factors, such as whether the asset was new or used, from whom it was acquired, whether it was domestic or imported, whether it is used within or outside the United States, and the percentage of its use for business. The major divisions by date placed in service are

1. Assets placed in service before 1981—depreciated by one of the conventional methods.
2. Assets placed in service during the years 1981 through 1986—depreciated according to the ACRS methods.
3. Assets placed in service after 1986—depreciated according to the MACRS methods.

As discussed later, the ACRS and MACRS methods are derived from the conventional methods. Further, we may still use the straight-line method as an alternative to ACRS and MACRS.

### 4.3.2 Conventional Methods

For assets placed in service before 1981, the taxpayer must continue to use one of the conventional methods. In addition, some states have not adopted the ACRS and MACRS methods for state income tax computation. The three most popular methods are straight-line, sum-of-the-years'-digits, and declining balance. A preliminary step in the use of any one of these methods was the determination of the depreciation lifetime. In the *absence* of reliable asset data, the taxpayer was obliged to refer to a U.S. Treasury document on asset depreciation range (ADR).

***Straight-Line (SL).*** The straight-line method spreads the amount evenly over the depreciation lifetime of $N$ years.

$$D_n = (P - F)/N, \qquad n = 1, \ldots, N \tag{4.1}$$

The book balance, or accounting book value of the asset at end of year $n$, is given by

$$BB_n = P - n(P - F)/N, \qquad n = 1, \ldots, N \tag{4.2}$$

This method is used on any depreciable property, including intangible property.

### Example 4.4

An asset costs $200,000 and has an estimated salvage value of $50,000. Compute SL depreciation using a lifetime of 5 years.

$$D_n = (200{,}000 - 50{,}000)/5 = \$30{,}000$$

The depreciation each year is $30,000, and the book balance decreases each year by that amount:

$$BB_1 = 200{,}000 - 30{,}000 = \$170{,}000$$
$$BB_2 = 170{,}000 - 30{,}000 = \$140{,}000$$
$$\vdots$$
$$BB_5 = 80{,}000 - 30{,}000 = \$50{,}000$$

At the end of the depreciation lifetime the book balance equals the salvage value that was estimated at the beginning. □

***Sum-of-the-Years'-Digits (SOYD).*** A faster depreciation method is the SOYD method. First, the taxpayer takes the digits $1, \ldots, N$ and obtains their sum.

$$\sum_{n=1}^{N} = \frac{N(N+1)}{2}$$

The depreciation in the first year is based on $N$ divided by this sum, that in the second year is based on $(N - 1)$ divided by this sum, and so forth.

$$D_n = \frac{(N - n + 1)(P - F)}{N(N + 1)/2} \tag{4.3}$$

$$BB_n = \left(\frac{N - n}{N}\right)\left(\frac{N - n + 1}{N + 1}\right)(P - F) + F \tag{4.4}$$

### Example 4.5

The asset in Example 4.4 is to be depreciated using SOYD. The sum of the digits $1 + 2 + 3 + 4 + 5 = 15$. So the fractions are 5/15, 4/15, 3/15, 2/15, and 1/15.

Depreciation amounts are

$$D_1 = (5/15)(200{,}000 - 50{,}000) = \$50{,}000$$
$$D_2 = (4/15)(200{,}000 - 50{,}000) = \$40{,}000$$
$$\vdots$$
$$D_5 = (1/15)(200{,}000 - 50{,}000) = \$10{,}000$$

The book balances are more easily obtained as follows.

$$BB_1 = 200{,}000 - D_1 = 200{,}000 - 50{,}000 = \$150{,}000$$
$$BB_2 = 150{,}000 - D_2 = 150{,}000 - 40{,}000 = \$110{,}000$$
$$\vdots$$
$$BB_5 = 60{,}000 - D_5 = 60{,}000 - 10{,}000 = \$50{,}000 \quad \square$$

***Declining Balance (DB).*** The declining-balance method allocates each year a given fraction of the book balance at the end of the previous year. The fraction is obtained as follows.

$$\alpha = (100\%/N)(\text{multiplier}) \tag{4.5}$$

The multiplier depends on the type of asset, the date it was placed in service, and whether it was old or new when placed in service. In the United States the multiplier is generally 1.5 or 2.0. If 1.5, the method is called 1.5 DB, 150% DB, or one and one-half declining balance. If 2.0, it is referred to as 2.0 DB, 200% DB, or double declining balance (DDB).

Thus, the taxpayer has

$$D_n = \alpha BB_{n-1} = \alpha P(1 - \alpha)^{n-1} \tag{4.6}$$

$$BB_n = P(1 - \alpha)^n \tag{4.7}$$

Notice that these formulas do not contain the term for the salvage value $F$. Equation 4.7 shows that the declining-balance method yields an implied salvage value after $N$ years. If the implied value is lower than $F$, the depreciation amounts are adjusted so that $BB_N = F$.

## *Example 4.6*

The asset in Example 4.4 is to be depreciated by using DDB.

$$\alpha = (100\%/5)(2.0) = 40\%$$
$$D_1 = (0.4)(200{,}000) = \$80{,}000 \qquad BB_1 = 200{,}000 - 80{,}000 = \$120{,}000$$

$$D_2 = (0.4)(120{,}000) = \$48{,}000 \qquad BB_2 = 120{,}000 - 48{,}000 = \$72{,}000$$
$$D_3 = (0.4)(72{,}000) = \$28{,}800 \qquad BB_3 = 72{,}000 - 28{,}800 = \$43{,}200$$
$$D_{33} \rightarrow 22{,}000 \qquad BB_3 \rightarrow \$50{,}000$$

Since $BB_3$ is less than $F = \$50{,}000$, we adjust $D_3$ to be \$22,000, making $BB_3 =$ \$50,000. Both $D_4$ and $D_5$ are zero, and $BB_4$ and $BB_5$ remain at \$50,000. Figure 4.2 shows how the three conventional methods of depreciation reduce book values. □

***Switching.*** If the implied salvage value when a DB method is used is greater than the estimated salvage value $F$, the taxpayer usually switches to the SL method.

## *Example 4.7*

The asset in Example 4.4 is to be depreciated by using DDB, but assuming $F = 0$:

$$D_1 = \$80{,}000 \qquad BB_1 = \$120{,}000$$
$$D_2 = \$48{,}000 \qquad BB_2 = \$72{,}000$$
$$D_3 = \$28{,}800 \qquad BB_3 = \$43{,}200$$
$$D_4 = (0.4)(43{,}200) = \$17{,}280 \qquad BB_4 = 43{,}200 - 17{,}280 = \$25{,}920$$
$$D_5 = (0.4)(25{,}920) = \$10{,}368 \qquad BB_5 = 25{,}920 - 10{,}368 = \$15{,}552$$
$$D_5 \rightarrow \$25{,}920 \qquad BB_5 \rightarrow 0$$

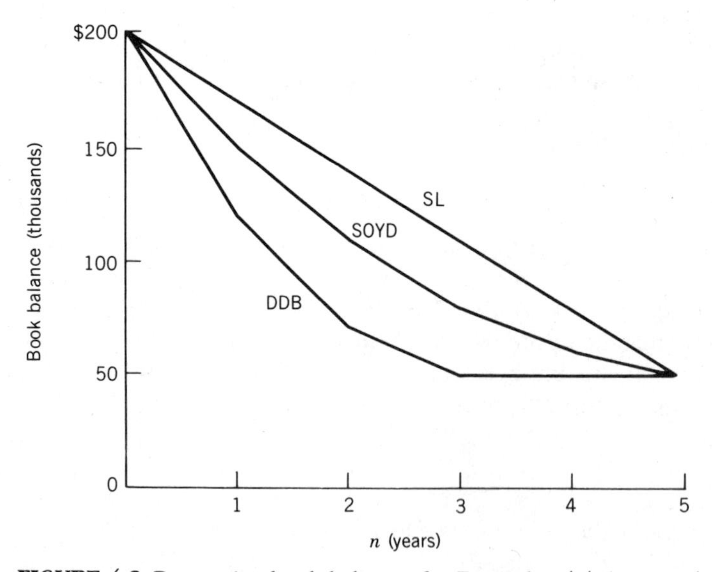


**FIGURE 4.2** Decreasing book balances for Examples 4.4 through 4.6.

Since $BB_5$ exceeds $F = 0$, the taxpayer adjusts $D_5$ to be \$25,920, making $BB_5 = 0$. Here the taxpayer switched after the fourth year. He or she could have switched after the third year.

$$D_4 = \frac{BB_3}{(N - 3)} = \frac{43{,}200}{2} = \$21{,}600$$

Then $BB_4 = 43{,}200 - 21{,}600 = \$21{,}600$, $D_5 = \$21{,}600$, and $BB_5 = 0$. □

In Example 4.7 the question is the optimal time to switch. The question is relevant today for two reasons.

1. Many pre-1981 assets are currently being depreciated by using DB and are therefore eligible for switching.
2. The MACRS *requires* switching from DB to SL.

There are two cases to consider, depending on the implied salvage value of the DB method:

$$BB_N = P(1 - \alpha)^N$$

**Case 1.** $BB_N < F$. Here the implied salvage value is *lower* than the estimated salvage value $F$. (This situation occurs only for pre-1981 assets.) Depreciation terminates when the book balance reaches $F$, as in Example 4.6. For 150% DB and DDB it is *always* preferable to use DB without switching over to SL. (See Problem 4.8 at the end of the chapter.)

**Case 2.** $BB_N > F$. Here it is better to switch, since otherwise we would have less total depreciation. The question is when to switch. It happens that there is a very easy rule for finding the optimal time to switch. If depreciation in any year by DB is less than it would be by a switch to SL, it is time to switch and thereafter use the SL method. (See Problem 4.9 at the end of the chapter.)

### *Example 4.8*

With $P = \$10{,}000$, $F = \$500$, and $N = 8$, find the optimal time to switch from DDB to SL. Find $PV(10\%)$ of switched and unswitched depreciation amounts:

Here $\alpha = (100\%)(2)/8 = 0.25$. The implied salvage value is $BB_8 = 10{,}000(0.75)^8 = \$1{,}001$, which exceeds 500. We tabulate the DDB values and the switched SL values in Table 4.2. The optimal time to switch is after year 5, which gives the following depreciation schedule.

| $n$ | 1 | 2 | 3 | 4 | 5 | 6 | 7 | 8 |
|---|---|---|---|---|---|---|---|---|
| $D_n$ | \$2,500 | 1,875 | 1,406 | 1,055 | 791 | 624 | 624 | 624 |

**Table 4.2** ***Double-Declining-Balance Depreciation Switched to Straight Line, Example 4.8***

| | DDB | | SL | |
|---|---|---|---|---|
| $n$ | Book Balance, Beginning Year $n$ | $D_n$ | Remaining Life | $D_n$ If Switched at Year $n$ |
| 1 | $10,000 | $2,500 | 8 | $1,188 |
| 2 | 7,500 | 1,875 | 7 | 1,000 |
| 3 | 5,625 | 1,406 | 6 | 854 |
| 4 | 4,219 | 1,055 | 5 | 744 |
| 5 | 3,164 | 791 | 4 | 666 |
| 6 | 2,373 | 593 | 3* | 624 |
| 7 | 1,780 | 445 | 2 | 640 |
| 8 | 1,335 | 334 +501† | 1 | 835 |

*Optimal time to switch.
†Adjustment for loss on sale of asset.

The *PV* of the switched policy is $7,054, based on $9,500 total depreciation, compared with a *PV* of the unswitched DB policy of $7,043, based on $8,999 total depreciation plus $501 loss on the sale of the asset. □

***Half-Year Convention.*** In some cases the taxpayer is allowed or required to use the half-year convention; only a half-year's depreciation can be taken the first year, and the remaining half-year's depreciation is taken after period *N*.

### *Example 4.9*

The asset in Example 4.4 is to be deprecated by using SL, a lifetime of 5 years, the half-year convention, and zero salvage value.

$$D_1 = (200{,}000/5)(0.5) = \$20{,}000$$
$$D_n = 200{,}000/5 = \$40{,}000, \quad n = 2,3,4,5$$
$$D_6 = (200{,}000/5)(0.5) = \$20{,}000 \quad \square$$

***Methods Based on Asset Usage.*** For some assets it is desirable to compute depreciation on the basis of asset usage. Typically, hours of usage or production units are estimated for the lifetime of the asset. From the total amount to be depreciated, one computes an hourly or unit depreciation charge, which is then applied to actual hours or units. These methods are still allowable under the current tax law.

### 4.3.3 Multiple-Asset Depreciation

The depreciation methods are not difficult to apply to a single asset. Even for 20 assets, for example, they do not represent an accounting burden. Consider, however, the situation of a power company with thousands of utility poles, a national auto rental company with thousands of vehicles, or an office organization with hundreds of desks, chairs, typewriters, calculators, computers, and so forth. If each $50 chair had to be depreciated separately by item accounting, the burden on the accountants would be heavy. The adjustments on asset disposal would be even more onerous, requiring knowledge of the salvage value of each individual asset. Since some assets are disposed of before the end of their depreciation lifetimes and others afterward, numerous adjustments would occur.

To simplify matters in such cases, many organizations use multiple-asset depreciation. Assets are typically grouped into vintage accounts. For example, an organization might have the following accounts.

3-year property, installed in 1983
3-year property, installed in 1984
3-year property, installed in 1985
5-year property, installed in 1983
5-year property, installed in 1984
etc.

For each vintage account there must be established a depreciation reserve account, which is an accounting device used to record all the depreciation expenses claimed over the years for the assets in the account.

One of the more popular ways to treat asset disposal is simply to add the retirement proceeds to the depreciation reserve. Thus, if 50 assets from a vintage account are sold for a lump sum, that sum is added to the depreciation reserve account. The method of computing depreciation depends on the type of account and is similar to item accounting [4]. When the last asset of the account is retired, an adjustment occurs: if the original cost of the assets in the account exceeds the depreciation reserve, insufficient depreciation has been claimed, and this difference is then claimed. The depreciation reserve account is not allowed to exceed the original cost of the asset group (this might happen if retirement proceeds exceeded estimates). If the depreciation reserve exceeds the original cost, any excess is treated as a gain and taxed that year.

### 4.3.4 1981 Accelerated Cost Recovery System (ACRS)

The 1981 Economic Recovery Tax Act made major changes in the depreciation charges allowed for tax accounting, and these charges apply to most assets placed in service during the period 1981 through 1986. The major changes from the conventional methods are the following.

1. The taxpayer must use either the ACRS table percentages or the alternative SL method. The SOYD and DB methods are generally not allowed. This restriction is not so bad, since ACRS is derived from DB with switching to SL or SOYD, as shown in Blank and Smith [2].

**Table 4.3** ***ACRS Property Types, for Assets Placed in Service during Period 1981–1986***

| Property Type | ADR* Class Life | Typical Examples of Property |
|---|---|---|
| 3-year | ADR ≤ 4 | Personal manufacturing-oriented assets, autos, light trucks, research equipment |
| 5-year | 5 ≤ ADR | Personal, manufacturing-oriented assets |
| | 5 ≤ ADR ≤ 18 | Public utility assets |
| 10-year | ADR ≤ 12.5 | Buildings and structural components |
| | 18 ≤ ADR ≤ 25 | Public utility assets |
| 15-, 18-, and 19-year real property† | 12.5 ≤ ADR | Buildings and structural components |
| 15-year public utility property | 25 ≤ ADR | Public utility property |

*ADR, asset depreciation range—guideline published by the IRS.
†Depends on date placed in service.

2. The taxpayer is restricted in the lifetimes used for depreciation. Again, this isn't so bad, since lifetimes are short for ACRS percentages and up to 45 years for SL.
3. Salvage value is ignored for computing depreciation charges.

Even though the old system has been replaced, the ADR life tables must be used to obtain the property types or classes. The property types or classes are shown in Table 4.3.

Next, the taxpayer must decide whether to use *ACRS percentages* or the *alternative SL method.* If ACRS percentages are selected, the taxpayer will use either Table 4.4 (for personal property) or a table like Table 4.5 (for real property). Table 4.4 was derived from 150% DB with switching to SL, with the half-year convention built in. Table 4.5 is similar to the 15-year column in Table 4.4, except that the depreciation is based on the month the asset is placed in service.

If the alternative SL method is selected, the taxpayer may use longer periods depending on the asset's property class, as follows.

3-year property: 3, 5, or 12 years
5-year property: 5, 12, or 25 years
10-year property: 10, 25, or 35 years
15-year property: 15, 35, or 45 years

For personal property, the *half-year convention* must be used. For example, using a 3-year life would result in depreciation amounts of 16.67%, 33.33%, 33.33%, and 16.67% for years 1 through 4. Real property depreciation must be based on the month the asset is placed in service.

The *Section 179* expense deduction replaces the old additional 20% first-year depreciation [3]. It could be applied to all personal depreciable assets.

**Table 4.4** ***ACRS Percentages for Personal Property Placed in Service 1981–1986***

| | Depreciation Percentages | | | |
|---|---|---|---|---|
| | Recovery Period, years | | | |
| Year $n$ | 3 | 5 | 10 | 15 (public utility) |
| 1 | 25 | 15 | 8 | 5 |
| 2 | 38 | 22 | 14 | 10 |
| 3 | 37 | 21 | 12 | 9 |
| 4 | | 21 | 10 | 8 |
| 5 | | 21 | 10 | 7 |
| 6 | | | 10 | 7 |
| 7 | | | 9 | 6 |
| 8 | | | 9 | 6 |
| 9 | | | 9 | 6 |
| 10 | | | 9 | 6 |
| 11 | | | | 6 |
| 12 | | | | 6 |
| 13 | | | | 6 |
| 14 | | | | 6 |
| 15 | | | | 6 |

**Table 4.5** ***ACRS Percentages for Real Property Placed in Service January 1, 1981 Through March 15, 1984***

| | Depreciation Percentages* | | | | | | | | | | | |
|---|---|---|---|---|---|---|---|---|---|---|---|---|
| | Month Property Placed in Service | | | | | | | | | | | |
| Recovery Year, $n$ | 1 | 2 | 3 | 4 | 5 | 6 | 7 | 8 | 9 | 10 | 11 | 12 |
| 1 | 12 | 11 | 10 | 9 | 8 | 7 | 6 | 5 | 4 | 3 | 2 | 1 |
| 2 | 10 | 10 | 11 | 11 | 11 | 11 | 11 | 11 | 11 | 11 | 11 | 12 |
| 3 | 9 | 9 | 9 | 9 | 10 | 10 | 10 | 10 | 10 | 10 | 10 | 10 |
| 4 | 8 | 8 | 8 | 8 | 8 | 8 | 9 | 9 | 9 | 9 | 9 | 9 |
| 5 | 7 | 7 | 7 | 7 | 7 | 7 | 8 | 8 | 8 | 8 | 8 | 8 |
| 6 | 6 | 6 | 6 | 6 | 7 | 7 | 7 | 7 | 7 | 7 | 7 | 7 |
| 7 | 6 | 6 | 6 | 6 | 6 | 6 | 6 | 6 | 6 | 6 | 6 | 6 |
| 8 | 6 | 6 | 6 | 6 | 6 | 6 | 6 | 6 | 6 | 6 | 6 | 6 |
| 9 | 6 | 6 | 6 | 6 | 5 | 6 | 5 | 5 | 5 | 6 | 6 | 6 |
| 10 | 5 | 6 | 5 | 6 | 5 | 5 | 5 | 5 | 5 | 5 | 6 | 5 |
| 11 | 5 | 5 | 5 | 5 | 5 | 5 | 5 | 5 | 5 | 5 | 5 | 5 |
| 12 | 5 | 5 | 5 | 5 | 5 | 5 | 5 | 5 | 5 | 5 | 5 | 5 |
| 13 | 5 | 5 | 5 | 5 | 5 | 5 | 5 | 5 | 5 | 5 | 5 | 5 |
| 14 | 5 | 5 | 5 | 5 | 5 | 5 | 5 | 5 | 5 | 5 | 5 | 5 |
| 15 | 5 | 5 | 5 | 5 | 5 | 5 | 5 | 5 | 5 | 5 | 5 | 5 |
| 16 | — | — | 1 | 1 | 2 | 2 | 3 | 3 | 4 | 4 | 4 | 5 |

*Percentages do not apply to low-income housing.

NOTE: Similar tables apply to real property placed in service March 16, 1984 through May 8, 1984 (18-year property) and May 9, 1984 through December 31, 1986 (19-year property).

However, there were aggregate annual dollar limits per tax return, with a limit of $5,000 per year for tax years 1982 through 1986. Any Section 179 deduction must be subtracted from the value of the asset before computing depreciation.

### *Example 4.10*

Apply ACRS percentages to the asset in Example 4.4, treating it as a 5-year property and claiming $5,000 of Sec. 179 expenses.

Sec. 179 expense deduction = $5,000 (year 1)

Adjusted basis is 200,000 − 5,000 = $195,000

$$D_1 = (0.15)(195{,}000) = \$29{,}250 \qquad BB_1 = 195{,}000 - 29{,}500 = \$165{,}750$$

$$D_2 = (0.22)(195{,}000) = \$42{,}900 \qquad BB_2 = 165{,}750 - 42{,}900 = \$122{,}850$$

$$\vdots$$

$$D_3 = D_4 = D_5 = (0.21)(195{,}000) = \$40{,}950 \qquad BB_5 = 0$$

□

It should be noted that the major effect of the ACRS system was to accelerate the depreciation of assets. This was achieved by use of *shorter lifetimes* and by *ignoring salvage values* in computing depreciation.

### 4.3.5 Tax Reform Act of 1986 (MACRS)

The Tax Reform Act of 1986 made numerous changes in both individual and corporate tax law. In addition to reduced overall tax rates and fewer brackets, the major corporate tax changes are as follows.

1. The investment tax credit (ITC) was eliminated effective January 1, 1986. Originally introduced in 1962 and modified, repealed, and reinstated over the years, the ITC represents a credit against taxes for qualified property [9]. For some types of property the ITC was 10%; its repeal thus represents a significant change.
2. Property class definitions were changed and some new classes were introduced. The overall effect was to increase somewhat the depreciation lifetimes for personal property.
3. Depreciation write-offs for a more narrowly defined 3-year class are faster.
4. Depreciation write-offs for real property were substantially stretched out, with only the SL method allowed.
5. Statutory percentage tables were replaced by specifications of depreciation methods.
6. The limit on Sec. 179 expense deduction was raised to $10,000, but a phaseout provision for large companies was introduced.
7. The new methods rely heavily on the half-year, mid-quarter, and mid-month conventions.

**Table 4.6** ***MACRS Property Classifications***

| Recovery Period | ADR Midpoint Class | Typical Property |
|---|---|---|
| 3-year | ADR* $\leq 4$ | Tools, horses |
| 5-year | $4 < \text{ADR} \leq 10$ | Automobiles, light trucks, high-tech equipment, computerized telephone-switching systems, energy (solar) property |
| 7-year | $10 < \text{ADR} \leq 16$ | Manufacturing equipment, agricultural structures |
| 10-year | $16 < \text{ADR} \leq 20$ | Vessels, barges, tugs |
| 15-year | $20 < \text{ADR} \leq 25$ | Wastewater plants, telephone distribution plant or similar utility property |
| 20-year | $25 \leq \text{ADR}$ | Municipal sewers, electric power plant |
| 27.5-year | | Residential rental property |
| 31.5-year | | Nonresidential real property, including elevators and escalators |

*ADR, asset depreciation range—guideline published by the IRS.

Table 4.6 shows the MACRS property classifications. To classify personal property one must determine the ADR class life by using a U.S. Treasury Department Publication [8].

The method specified for depreciating items in the 3-, 5-, 7-, and 10-year classes is to use the DDB method over 3, 5, 7, and 10 years, respectively, with the half-year convention. In addition, optimal switching to SL is *required.* For property in the 15- and 20-year classes, use 150% DB with optimal switching to SL, with the half-year convention. Only the SL method is allowed for property in the 27.5- and 31.5-year classes, with a mid-month convention. Salvage values are ignored in the MACRS.

### *Example 4.11*

Apply the MACRS to the asset in Example 4.4, treating it as 5-year property and claiming no Sec. 179 deduction. The result is shown in Table 4.7. The optimal time to switch to SL is during year 5 (or year 4). Note the resulting percentages in the last column of the table, which apply to any asset in the 5-year class. □

Since there are only six cases to consider for switching to SL, it is convenient to find the optimal time to switch and the resulting percentages for each case. These results are shown in Table 4.8. (Table 4.8 was constructed by using conventional round-off methods; it might be superseded by U.S. Treasury Department tables using different round-off methods.) Table 4.8 thus plays the same role for the MACRS as Tables 4.4 and 4.5 do for the ACRS.

The phaseout provision for the *Sec. 179 expense deduction* has the effect of reducing the $10,000 limit by the amount by which eligible property placed in service that year exceeds $200,000. Thus, if more than $210,000 of eligible property is placed in service, no Sec. 179 deduction is allowed. The deduction is

**Table 4.7** ***Example 4.11 Showing MACRS Application***

| | DDB | | SL | | | |
|---|---|---|---|---|---|---|
| $n$ | $BB_{n-1}$ | $D_n$ | Remaining Life | $D_n$ If Switched at Year $n$ | Resulting $D_n$ | Resulting Percentage |
| 1 | $200,000 | $40,000 | 5 | $20,000 | $40,000 | 20.00% |
| 2 | 160,000 | 64,000 | 4½ | 35,556 | 64,000 | 32.00 |
| 3 | 96,000 | 38,400 | 3½ | 27,429 | 38,400 | 19.20 |
| 4 | 57,600 | 23,040 | 2½ | 23,040 | 23,040 | 11.52 |
| 5* | 34,560 | 13,824 | 1½ | 23,040 | 23,040 | 11.52 |
| 6 | N/A | N/A | ½ | | 11,520 | 5.76 |

*Optimal time to switch.
NOTES: $\alpha = (100\%/5)(2.0) = 40\%$. Half-year convention applies.

also limited by the taxable income derived from the active conduct of any trade or business during the tax year, as calculated without the deduction. Thus, the deduction cannot have the effect of generating an after-tax loss.

The *mid-quarter convention* replaces the half-year convention for personal property in some instances. If the aggregate amount of all personal property that is placed in service during the last 3 months of any tax year exceeds 40% for the year, then the mid-quarter convention applies. All personal property is then treated as being installed (or disposed of) at the midpoint of the quarter during which actual installation (or disposal) occurs. The annual depreciation amounts that would normally be taken are then adjusted as follows.

| Quarter of Tax Year | Percentage for Installed Items | Percentage for Disposed Items |
|---|---|---|
| First | 87.5% | 12.5% |
| Second | 62.5 | 37.5 |
| Third | 37.5 | 62.5 |
| Fourth | 12.5 | 87.5 |

Unfortunately, once an asset is subject to the mid-quarter convention, Table 4.8 no longer applies; depreciation for the asset must be tracked individually during its lifetime [4,8].

For taxpayers who do not wish to obtain the fast write-offs from the DDB-switch-to-SL and the 150%-DB-switch-to-SL methods, the *alternative SL* method allows the use of SL with zero salvage value. Whereas the alternative SL method for ACRS allowed some choice in the depreciation lifetime, the new law generally requires use of the class life from the ADR table. As in the 1981–1986 tax law, the alternative SL method is required for certain types of property: (1) tangible property used mainly outside the United States, (2) property financed by tax-exempt methods, (3) some imported property, and (4) intangible property. In addition, certain "listed property" that is not used 50% or more for business must be depreciated by the alternative SL method. Such listed property includes

## Table 4.8 *MACRS Percentages*

| Year | Property: | Personal* | | | | | | Real† | |
|---|---|---|---|---|---|---|---|---|---|
| | Class: | 3 | 5 | 7 | 10 | 15 | 20 | 27.5 | 31.5 |
| | Depreciation Rate: | DDB | DDB | DDB | DDB | 150% DB | 150% DB | 3.64% SL | 3.17% SL |
| 1 | | 33.33% | 20% | 14.29% | 10% | 5% | 3.75% | *A* | *B* |
| 2 | | 44.44 | 32 | 24.49 | 18 | 9.5 | 7.22 | 3.64% | 3.17% |
| 3 | | 14.81 | 19.2 | 17.49 | 14.4 | 8.5 | 6.68 | 3.64 | 3.17 |
| 4 | | 7.41 | 11.52 | 12.49 | 11.52 | 7.69 | 6.18 | | |
| 5 | | | 11.52 | 8.92 | 9.22 | 6.93 | 5.71 | | |
| 6 | | | 5.76 | 8.92 | 7.37 | 6.23 | 5.28 | | |
| 7 | | | | 8.92 | 6.55 | 5.90 | 4.89 | | |
| 8 | | | | 4.46 | 6.55 | 5.90 | 4.52 | | |
| 9 | | | | | 6.55 | 5.90 | 4.46 | | |
| 10 | | | | | 6.55 | 5.90 | 4.46 | | |
| 11 | | | | | 3.28 | 5.90 | 4.46 | | |
| 12 | | | | | | 5.90 | 4.46 | | |
| 13 | | | | | | 5.90 | 4.46 | | |
| 14 | | | | | | 5.90 | 4.46 | | |
| 15 | | | | | | 5.90 | 4.46 | | |
| 16 | | | | | | 2.45 | 4.46 | | |
| 17 | | | | | | | 4.46 | | |
| 18 | | | | | | | 4.46 | | |
| 19 | | | | | | | 4.46 | | |
| 20 | | | | | | | 4.46 | | |
| 21 | | | | | | | 2.23 | | |
| 22 | | | | | | | | | |
| 23 | | | | | | | | | |
| 24 | | | | | | | | | |
| 25 | | | | | | | | | |
| 26 | | | | | | | | 3.64 | |
| 27 | | | | | | | | *C* | |
| 28 | | | | | | | | *E* | |
| 29 | | | | | | | | | |
| 30 | | | | | | | | | |
| 31 | | | | | | | | | 3.17 |
| 32 | | | | | | | | | *D* |
| 33 | | | | | | | | | *F* |

*Includes optimal switching to SL.

†Let $j$ = month asset placed in service (mid-month convention);
$A = (3.64\%)(12.5 - j)/12$ and $B = (3.17\%)(12.5 - j)/12$.
For $j \leq 6$: $C = (3.64\%)(5.5 + j)/12$, $D = (3.17\%)(5.5 + j)/12$, $E = 0$, and $F = 0$.
For $6 < j$: $C = 3.64\%$, $D = 3.17\%$, $E = (3.64\%)(j - 6.5)/12$, and $F = (3.17\%)(j - 6.5)/12$.

**NOTE**: Table reflects conventional round-off methods and may differ slightly from official IRS tables.

passenger automobiles, photographic equipment, phonographic and video equipment, and most home computers.

Most of the provisions of the MACRS became effective on January 1, 1987. The repeal of the ITC was effective on January 1, 1986. In addition, the taxpayer had the option of using the MACRS for any property beginning on August 1, 1986.

#### 4.3.6. Adjustments to Income Taxes

The elimination of the 60% long-term capital gains deduction means that any profit on the sales of an asset is subject to ordinary income tax. Effectively, the distinction between short term and long term has been eliminated, which leads to a simpler treatment. Some features are held over from the old tax law, however [8]. The following example illustrates the calculation of gain for a single asset sold before the end of its recovery period.

***Example 4.12***

Assume that the asset in example 4.11 is sold during the fifth year for $50,000. Calculate the gain.

Following the half-year convention, it is assumed that the asset is retired at the midpoint of the year, and thus the depreciation for year 5 is (0.5)(23,040) = $11,520. This results in total depreciation of $176,960 (refer to Table 4.7), and a remaining book value of $200,000 − 176,960 = $23,040. The gain is thus $50,000 − 23,040 = $26,960, which is taxed at ordinary corporate income tax rates. □

Note that if the asset had been sold during the sixth year, all the depreciation in Table 4.7 would have been allowed, resulting in a gain of $50,000. Any Sec. 179 expense deduction that was taken for the asset must be included in the calculation for remaining book value.

The disposal within the first three years of use of an asset for which Sec. 179 deduction was claimed requires an extra calculation to compute the *recapture* of Sec. 179 expense deduction. Conceptually, the depreciation amounts are recalculated as though no Sec. 179 deduction had been claimed. Computationally, the method involves use of the depreciation percentages on the amount of Sec. 179 deduction, as shown in the following example.

***Example 4.13***

Apply the MACRS to the asset in Example 4.4, treating the asset as a 5-year property and claiming $8,000 of Sec. 179 deduction. Assume the asset is sold during the third year for $80,000. Calculate the gain.

$\alpha = 40\%$

Sec. 179 expense deduction = $8,000 (year 1)

Adjusted basis is $200,000 − 8,000 = $192,000

| ASSET | | SEC. 179 | |
|---|---|---|---|
| $D_1 = (192{,}000)(0.2) =$ | \$38,000 | $(8{,}000)(0.2) =$ | \$1,600 |
| $D_2 = (192{,}000)(0.32) =$ | 61,440 | $(8{,}000)(0.32) =$ | 2,560 |
| $D_3 = (192{,}000)(0.5)(0.192) =$ | 18,432 | $(8{,}000)(0.5)(0.192) =$ | 768 |
| Total = | \$118,272 | Total = | \$4,928 |

The gain for the asset is (selling price $-$ $BB_n$).

$$\$80{,}000 - (192{,}000 - 118{,}272) = \$6{,}272 \quad \square$$

The \$6,272 gain in Example 4.13 is treated as ordinary income. The *recapture* of Sec. 179 deduction is

$$\$8{,}000 - 4{,}928 = \$3{,}072$$

which amount is *also* treated as ordinary income. The after-tax cash flow from the sale (assuming $t_m$ doesn't change) is

$$\$80{,}000 - (6{,}272 + 3{,}072)t_m$$

Note that the recaptured \$3,072 corresponds to the \$8,000 multiplied by the remaining depreciation percentages that would have been applied to the asset had it not been sold early.

$$\$8{,}000[(0.5)(0.192) + 0.1152 + 0.1152 + 0.0576] = \$3{,}072$$

Observe also that if the Sec. 179 deduction had not been claimed, the gain on the sale would have been

$$\$6{,}272 - 3{,}072 = \$3{,}200$$

(See Problem 4.15 at the end of the chapter.)

### 4.3.7 Depreciation Strategies

The changes specified in the Tax Reform Act of 1986 greatly reduced the choices that had previously existed for the taxpayer. The bulk of the work done on optimal depreciation strategies [3,6] is no longer applicable. Optimal switching from DB to SL is reflected in the tables for most situations; otherwise (when the mid-quarter convention applies) the time is found by comparing DB and SL amounts as in Table 4.7. The only significant choices that remain concern the selection of the MACRS percentages or the alternative SL method. The other choices possible represent mainly fine tuning for particular situations.

Generally, a stable corporation in the top (34%) tax bracket should select the MACRS percentages over the alternative SL method to obtain the benefit of the faster depreciation. The total depreciation will be the same, provided the asset is retained for the depreciation period(s). If the marginal tax rate remains unchanged, either method then provides the same total shield of revenues from income taxes.

### *Example 4.14*

An asset in the 5-year property class cost $600,000 and has zero estimated salvage value after 6 years of use. The asset will generate annual revenues of $1,000,000 and require $200,000 annual labor and $100,000 annual material expenses. There are no other revenues or expenses. Compare after-tax cash flows based on MACRS percentages and the alternative SL method with a 5-year class life. No Sec. 179 expense deduction is to be taken.

Table 4.9 shows the results of the calculations. We have included in the table the cash flow for purchase at time 0. Several observations can be made.

- The totals in the last column for depreciation, total expenses, taxable income, tax, profit A/T, which span time 0 through time 6, are the *same* for both methods.
- The different depreciation methods result in different *time patterns* for taxable income, tax, profit A/T, and cash flow A/T.
- Compared with the alternative SL method, the MACRS percentages result in lower early profit A/T but higher early cash flow A/T. The reverse occurs during the later years.
- The total cash flow A/T from time 0 through time 6, including purchase flow, is the *same* for both methods. Also, it is *equal* to the total profit A/T, because the tax structure and the company's marginal tax rate remain constant during the 6 years and there are no side effects from other cash flows. □

Which of the two methods in Example 4.14 would be preferred? We can obtain the present value of each after-tax cash flow to help us choose. Using $i$ = 10%, we have

| Depreciation Method | *PV* of Cash Flow A/T | *PV* of Depreciation |
|---|---|---|
| MACRS percentages | $1,569,876 | $463,956 |
| Alternative SL method | 1,559,764 | 434,217 |

In this example we would select the MACRS percentages method, since it gives the larger *PV* by about $10,000. If we compare the *PV*s of depreciation amounts, we obtain the same ranking, as shown. The accelerated method gives a greater *PV* of depreciation and, through the tax shield, a greater *PV* of after-tax cash flow.

Notice that the difference in *PV*s of depreciation, multiplied by $t_m$, equals the difference in *PV*s of A/T cash flows. For example,

**Table 4.9** *Comparison of MACRS Percentages and the Alternative SL Method, Example 4.14*

| Depreciation Method | *n*: 0 | 1 | 2 | 3 | 4 | 5 | 6 | Totals |
|---|---|---|---|---|---|---|---|---|
| *MACRS %s* | | | | | | | | |
| Revenues | | $1,000,000 | 1,000,000 | 1,000,000 | 1,000,000 | 1,000,000 | 1,000,000 | 6,000,000 |
| Expenses | | | | | | | | |
| Labor | | $200,000 | 200,000 | 200,000 | 200,000 | 200,000 | 200,000 | 1,200,000 |
| Material | | $100,000 | 100,000 | 100,000 | 100,000 | 100,000 | 100,000 | 600,000 |
| Depreciation | | $120,000 | 192,000 | 115,200 | 69,120 | 69,120 | 34,560 | 600,000 |
| Total | | $420,000 | 492,000 | 415,200 | 369,120 | 369,120 | 334,560 | 2,400,000 |
| Taxable income | | $580,000 | 508,000 | 584,800 | 630,880 | 630,880 | 665,440 | 3,600,000 |
| Income tax* | | $197,200 | 172,720 | 198,832 | 214,499 | 214,499 | 226,250 | 1,224,000 |
| Profit A/T | | $382,800 | 335,280 | 385,968 | 416,381 | 416,381 | 439,190 | 2,376,000 |
| Cash flow | −$600,000 | 502,800 | 527,280 | 501,168 | 485,501 | 485,501 | 473,750 | 2,376,000 |
| ***Alternative SL*** | | | | | | | | |
| Revenues | | $1,000,000 | 1,000,000 | 1,000,000 | 1,000,000 | 1,000,000 | 1,000,000 | 6,000,000 |
| Expenses | | | | | | | | |
| Labor | | $200,000 | 200,000 | 200,000 | 200,000 | 200,000 | 200,000 | 1,200,000 |
| Material | | $100,000 | 100,000 | 100,000 | 100,000 | 100,000 | 100,000 | 600,000 |
| Depreciation | | $60,000 | 120,000 | 120,000 | 120,000 | 120,000 | 60,000 | 600,000 |
| Total | | $360,000 | 420,000 | 420,000 | 420,000 | 420,000 | 360,000 | 2,400,000 |
| Taxable income | | $640,000 | 580,000 | 580,000 | 580,000 | 580,000 | 640,000 | 3,600,000 |
| Income tax* | | $217,600 | 197,200 | 197,200 | 197,200 | 197,200 | 217,600 | 1,224,000 |
| Profit A/T | | $422,400 | 382,800 | 382,800 | 382,800 | 382,800 | 422,400 | 2,376,000 |
| Cash flow | −$600,000 | 482,400 | 502,800 | 502,800 | 502,800 | 502,800 | 482,400 | 2,376,000 |

*Income tax = $113,900 + 34% of taxable income above $335,000.

| *PV* of Depreciation | | | *PV* of A/T Cash Flow | |
|---|---|---|---|---|
| MACRS %s | Alt. SL | | MACRS %s | Alt. SL |
| ($463,956 | − 434,217)(0.34) | = | $1,569,876 | − 1,559,764 |
| | 29,739(0.34) | = | $10,112 | |

We will discuss this relationship in more detail in Section 4.4.

The *PV* advantage of the MACRS percentages method depends on the depreciation lifetimes and the interest rate used for discounting. As long as the corporation's marginal tax rate is *constant,* it is better to use the MACRS percentages. However, when major changes in the marginal tax rate are expected to occur, the alternative SL method might be advantageous; a case-by-case analysis is necessary. (See Problem 4.13.)

Fine-tuning strategies that yield smaller benefits time asset acquisitions to *avoid the mid-quarter convention.* Clearly, if a small change in the date on which an asset is placed in service can avoid the mid-quarter rule, it is usually to the company's benefit.

## 4.4 AFTER-TAX CASH FLOW ANALYSIS

In this chapter we have shown many examples in which the final numbers obtained are the after-tax cash flows. That is as it should be. It is the after-tax cash flow that can be distributed to the owners of the corporation without liquidating assets or can be reinvested to yield future profits. In this section we provide a framework for analysis of cash flows deriving from revenues, current expenses, depreciation, financing costs, capital gains, and inflation.

### 4.4.1 Generalized Cash Flows

The acceptance of a major project by a small company, or by one in the lower tax brackets, is likely to change the company's marginal tax rate, as in Example 4.3. For such situations we typically use the income statement approach to estimate after-tax cash flows with and without the project. Table 4.9 illustrates this approach.

If we are performing the analysis for a large, profitable corporation, and the acceptance or rejection of the project will not change the marginal tax rate (likely the rate of 34%), we can use a more convenient approach. This approach involves *generalized cash flows.* If we have identified all the financial elements related to a project, we obtain the contribution of the project to after-tax cash flow as follows.

After-tax (A/T) cash flow from project at time $n$

= − (investment at time $n$)

+ (after-tax proceeds from sale of investment at time $n$)

+ (bank loan at time $n$)

− (loan principal repayment at time $n$)

+ ($t_m$) (depreciation at time $n$)

$+ (1 - t_m)$ (revenues at time $n$)
$- (1 - t_m)$ (expenses at time $n$ for such items as labor, materials, and interest)

where $t_m$ is the marginal tax rate. We have seen most of these elements in previous examples, except for loans. Loan proceeds and principal repayments do not affect taxes, but loan interest is a tax-deductible expense.

### *Example 4.15*

XYZ Corporation is considering the installation of a new computer system for its marketing department. The system costs \$200,000 installed, will generate additional revenues of \$50,000/year, and will save \$40,000/year in labor costs. The system will be financed by a \$150,000 bank loan repayable in three equal annual principal installments plus 15% interest on the outstanding loan balance. It will be depreciated by using the alternative SL method and 5-year property type. No Sec. 179 deduction will apply. The useful life of this system is 10 years, at which time it will be sold for \$20,000. The marginal tax rate is 40%, made up of 34% federal tax plus state and local income taxes (state and local tax laws are assumed similar to federal tax laws). Find the year-by-year after-tax cash flow for the project.

Table 4.10 shows the calculations. The following notes explain some of the items; the numbers refer to lines in the table.

2. Since book value at time 10 is zero, gain on sale is \$20,000 − 0 = \$20,000. This is taxed at the ordinary rate of 40%, leaving proceeds of \$12,000.
5. Depreciation by the alternative SL method writes off 20% a year except for years 1 and 6, when the half-year convention gives 10%.
7. Since the figure for labor represents a savings, it is a positive cash flow.
8. Interest on the loan is

$$n = 1, \quad (0.15)(150) = \$22.5$$
$$n = 2, \quad (0.15)(100) = \$15$$
$$n = 3, \quad (0.15)(\ 50) = \$7.5$$

The net cash flow can now be evaluated by some standard technique. For example, if $i = 20\%$,

$$\begin{aligned} PV(20\%) = -50 & \\ &- 1.5(0.8333) + 62(0.3349) \\ &+ 11(0.6944) + 54(0.2791) \\ &+ 14.5(0.5787) + 54(0.2326) \\ &+ 70(0.4823) + 54(0.1938) \\ &+ 70(0.4019) + 66(0.1615) = \$96.2 \quad \square \end{aligned}$$

**Table 4.10** ***After-Tax Cash Flow Computations Using Generalized Cash Flows, Example 4.15***

| Item | *n*: 0 | 1 | 2 | 3 | 4–5 | 6 | 7–9 | 10 |
|---|---|---|---|---|---|---|---|---|
| 1. Investment | −$200* | | | | | | | |
| 2. Proceeds from sale | | | | | | | | +12 |
| 3. Bank loan | +$150 | | | | | | | |
| 4. Principal repayment | | −$50 | −50 | −50 | | | | |
| 5. (0.4) (depreciation) | | +$8 | +16 | +16 | +16 | +8 | | |
| 6. (0.6) (revenues) | | +$30 | +30 | +30 | +30 | +30 | +30 | +30 |
| 7. (0.6) (labor and material) | | +$24 | +24 | +24 | +24 | +24 | +24 | +24 |
| 8. (0.6) (interest) | | −$13.5 | −9 | −4.5 | | | | |
| 10. Net cash flow after taxes | −$50 | −1.5 | +11 | +14.5 | +70 | +62 | +54 | +66 |

*All figures are in thousands.

This method of computing net cash flow after taxes has two advantages. (1) It is particularly suited to situations in which revenues from the project are zero or are unknown. (2) It simplifies sensitivity analysis for individual items, such as labor savings, without having to recalculate taxable income, taxes, profit A/T, and so forth.

### 4.4.2 Effects of Depreciation Methods

The advantage of one depreciation method over another can be determined by evaluating the *PV*s of the $+t_m D_n$ cash flow streams, as shown in Example 4.14. Note that, when using the alternative SL method, the depreciation lifetime is often longer than for the MACRS percentages.

### 4.4.3 Effects of Financing Costs

Since interest is tax-deductible, companies in high tax brackets incur lower after-tax financing costs. In addition to the interest rate and tax bracket effects, the loan repayment method has an effect.

***Equal Principal Repayments.*** With a loan of size $B$ to be repaid at interest $i_b$ in $N$ periods, we have

$$PP_n = B/N, \qquad n = 1, \ldots, N \tag{4.8}$$

the principal repayment at the end of each period, and

$$IP_n = i_b B[1 - (n - 1)/N], \qquad n = 1, \ldots, N \tag{4.9}$$

the interest payment at the end of each period.

Using a periodic discount rate of $i$, we have the *PV* of the net after-tax cash flow for the loan.

$$PV(i) = B - \frac{B}{N}(P/A, i, N) - \sum_{n=1}^{N} \frac{(1 - t_m) i_b B[1 - (n - 1)/N]}{(1 + i)^n} \tag{4.10}$$

The last part of Eq. 4.10 is a gradient series, and

$$\begin{aligned} PV(i) = B &- [B/N + (1 - t_m) i_b B](P/A, i, N) \\ &+ (1 - t_m)(i_b B/N)(P/G, i, N) \end{aligned} \tag{4.11}$$

When $i = (1 - t_m)i_b$, the *PV* of the loan cash flow is zero (see Problem 4.18).

***Equal Principal plus Interest Payments.*** Let

$TP_n = IP_n + PP_n$, the total loan payment at the end of each period

$LB_n$ = loan principal balance at the end of each period

Then

$$TP_n = B(A/P, i_b, N), \qquad n = 1, \ldots, N \tag{4.12}$$

$$IP_n = i_b LB_{n-1}, \qquad n = 1, \ldots, N \tag{4.13}$$

$$\begin{aligned} LB_n &= LB_{n-1} - PP_n \\ &= (LB_{n-1}) - B(A/P, i_b, N) + i_b (LB_{n-1}) \\ &= (1 + i_b)LB_{n-1} - B(A/P, i_b, N) \end{aligned} \tag{4.14}$$

We can use Eq. 3.37 to develop an expression for the $PV$ of the loan cash flow. We have three components: loan proceeds of $B$ at time 0, equal principal plus interest payments of $A = B(A/P, i_b, N)$, and the tax shield of the interest payments. Thus, we have

$$PV(i) = B - A(P/A,i,N) + t_m\left\{\frac{Bi_b - A}{i - i_b}\left[1 - \left(\frac{1 + i_b}{1 + i}\right)^N\right] + \frac{A}{i}[1 - (1 + i)^{-N}]\right\} \tag{4.15}$$

### *Example 4.16*

A loan of $10,000 at 10% annual interest is to be repaid in 5 years. Evaluate the effects on after-tax cash flow of the marginal tax rate, the discount rate, and equal principal payments versus equal total payments.

Table 4.11 shows the loan schedules and after-tax cash flows for the two repayment methods with $t_m = 34\%$. Table 4.12 gives the $PV$s for various tax rates and discount rates. To illustrate, using $t_m = 34\%$, $i = 15\%$, and equal principal repayments, we have

$$PV(15\%) = 10{,}000 - 2{,}660\overset{P/A,15\%,5}{(3.35216)} + 132\overset{P/G,15\%,8}{(5.775)} = \$1{,}846$$

Several observations can be made.

1. Higher tax rates increase the $PV$ of the A/T loan cash flow. This is caused by the tax deductibility of interest.
2. Higher discount rates increase the $PV$. This effect is natural for a loan.
3. There is little difference in $PV$s between the two repayment methods, 6% or less in this example. The method of equal total payments involves borrowing larger amounts after time 1, and the $PV$s change to a greater extent.
4. The $PV$'s in the table are proportional to $B$ = $10,000. The table can be used for other $B$ values, with $N = 5$. □

**Table 4.11** *After-Tax Cash Flows for Loan Using 34% Tax Rate, Example 4.16*

| $n$ | Loan Balance $LB_{n-1}$ | Interest Payment $IP_n$ | Principal Payment $PP_n$ | Total Payment $TP_n$ | Interest Tax Shield | A/T Cash Flow |
|---|---|---|---|---|---|---|
| *1. Equal principal repayments* | | | | | | |
| 1 | $10,000 | $1,000 | $2,000 | $3,000 | $340 | −$2,660 |
| 2 | 8,000 | 800 | 2,000 | 2,800 | 272 | −2,528 |
| 3 | 6,000 | 600 | 2,000 | 2,600 | 204 | −2,396 |
| 4 | 4,000 | 400 | 2,000 | 2,400 | 136 | −2,264 |
| 5 | 2,000 | 200 | 2,000 | 2,200 | 68 | −2,132 |
| *2. Equal principal plus interest payment* | | | | | | |
| 1 | 10,000.00 | 1,000.00 | 1,637.97 | 2,637.97 | 340.00 | −2,297.97 |
| 2 | 8,362.03 | 836.20 | 1,801.77 | 2,637.97 | 284.31 | −2,353.66 |
| 3 | 6,560.26 | 656.03 | 1,981.94 | 2,637.97 | 223.05 | −2,414.92 |
| 4 | 4,578.32 | 457.83 | 2,180.14 | 2,637.97 | 155.66 | −2,482.31 |
| 5 | 2,398.18 | 239.82 | 2,398.18 | 2,638.00* | 81.54 | −2,556.46 |

*Amount is adjusted to compensate for cumulative rounding to nearest cent.

NOTES: $IP_n = (10\%)LB_{n-1}$

$TP_n = IP_n + PP_n$

Interest tax shield = $(34\%)IP_n$

A/T cash flow = $-TP_n$ + interest tax shield

**Table 4.12** *Present Value of A/T Cash Flow for Loan in Example 4.16*

| $i$ | $t_m$: 0.50 | 0.40 | 0.34 | 0.30 | 0.25 | 0.15 | 0 |
|---|---|---|---|---|---|---|---|
| 0 | −$1,500 | −1,800 | −1,980 | −2,100 | −2,250 | −2,550 | −3,000 |
| | −$1,595 | −1,914 | −2,105 | −2,233 | −2,392 | −2,711 | −3,190 |
| 0.025 | −$708 | −992 | −1,162 | −1,275 | −1,417 | −1,700 | −2,125 |
| | −$752 | −1,053 | −1,233 | −1,353 | −1,504 | −1,804 | −2,256 |
| 0.05 | 0 | −268 | −429 | −536 | −671 | −939 | −1,341 |
| | 0 | −284 | −455 | −568 | −711 | −995 | −1,421 |
| 0.075 | $636 | 382 | 229 | 127 | 0 | −254 | −636 |
| | $673 | 404 | 242 | 135 | 0 | −269 | −673 |
| 0.10 | $1,209 | 967 | 822 | 726 | 605 | 363 | 0 |
| | $1,277 | 1,022 | 869 | 766 | 639 | 383 | 0 |
| 0.125 | $1,727 | 1,497 | 1,359 | 1,267 | 1,152 | 921 | 576 |
| | $1,822 | 1,579 | 1,433 | 1,336 | 1,215 | 972 | 607 |
| 0.15 | $2,197 | 1,977 | 1,846 | 1,758 | 1,648 | 1,428 | 1,099 |
| | $2,314 | 2,083 | 1,944 | 1,851 | 1,736 | 1,504 | 1,157 |

NOTES: First entry is *PV* for equal principal repayments.
Second entry is *PV* for equal total payments.
$t_m$ = marginal tax rate.
$i$ = discount rate for *PV*.
Loan amount $B$ = $10,000, $N = 5$, $i_b = 10\%$.
*PV* includes original loan amount, principal repayments, and interest payments.

### *Example 4.17*

Rework Example 4.14 with the inclusion of a \$500,000 loan at $i_b = 10\%$ for 5 years, to be repaid in equal principal installments. Use MACRS percentages but do not take Sec. 179. Compare *PV*s at $i = 10\%$ with and without the loan.

Table 4.13 shows the computations. We can make the following observations.

1. *PV* at $i = 10\%$ of the A/T cash flow is \$1,610,989. The difference between this amount and the \$1,569,876 without financing, or \$41,113, is attributed directly to the financing. The entry in Table 4.12 for 10% and 34% is + 822 for a \$10,000 loan. For a \$500,000 loan the *PV* is $(50)(822) = \$41,100 \approx \$41,113$. The financing at 10% costs only $(10\%)(1 - 0.34) = 6.6\%$ after taxes and increases *PV* at $i = 10\%$.
2. Cash flow A/T for the 6 years, including purchase, salvage, and loan, is equal to profit A/T. Again, this is because $t_m$ remains unchanged, and there are no side effects. □

Before we leave this section, we might ask whether it is appropriate to use for discounting a value $i$ different from the loan interest rate $i_b$. The value $i$ represents the corporation's time value trade-offs and reflects partially the investment opportunities available. Thus, there is nothing illogical in borrowing money at $i_b$, with an after-tax rate of $(1 - t_m)i_b$, and evaluating cash flows with a different rate $i$. Presumably, the money will be invested to earn a rate $i$ or greater. Table 4.12 shows the advantage of borrowing at a rate $i_b$ where $(1 - t_m)i_b$ is less than $i$: the *PV* is positive. This subject is treated in more detail in Chapter 5.

### 4.4.4 Effects of Inflation

In Chapter 2 it was shown that if a cash flow series increases with inflation, we have a choice of discounting actual dollars by using a market interest rate or discounting constant dollars by using an inflation-free, or real, interest rate. The *PV* is the same in either case. In an inflationary environment a company is usually able to compensate for increasing material and labor prices by raising its selling prices. However, the depreciation tax shield is based on historical cost, and it loses value as inflation drives up the general price level. Inflationary increases in salvage value also lead to increased taxes. Finally, inflation causes "bracket creep," whereby real profits are shifted into higher tax brackets. In the next example we use the income statement approach for measuring the *PV* loss caused by inflation.

### *Example 4.18*

An asset costing \$100,000 generates revenues of \$90,000/year for 6 years. Labor and material expenses are \$30,000 and \$5,000 per year, respectively. Depreciation is the alternative SL method with a 5-year lifetime and no Sec. 179 deduction. Salvage value is expected to be zero. There are no other revenues or

**Table 4.13** ***After-Tax Cash Flows with Financing, Example 4.17***

| | $n$: 0 | 1 | 2 | 3 | 4 | 5 | 6 | Totals |
|---|---|---|---|---|---|---|---|---|
| Revenues | | $1,000,000 | 1,000,000 | 1,000,000 | 1,000,000 | 1,000,000 | 1,000,000 | 6,000,000 |
| Expenses | | | | | | | | |
| Labor | | $200,000 | 200,000 | 200,000 | 200,000 | 200,000 | 200,000 | 1,200,000 |
| Material | | $100,000 | 100,000 | 100,000 | 100,000 | 100,000 | 100,000 | 600,000 |
| Depreciation | | $120,000 | 192,000 | 115,200 | 69,120 | 69,120 | 34,560 | 600,000 |
| Interest | | $50,000 | 40,000 | 30,000 | 20,000 | 10,000 | 0 | 150,000 |
| Total | | $470,000 | 532,000 | 445,200 | 389,120 | 379,120 | 334,560 | 2,550,000 |
| Taxable income | | $530,000 | 468,000 | 554,800 | 610,880 | 620,880 | 665,440 | 3,450,000 |
| Income tax* | | $180,200 | 159,120 | 188,632 | 207,699 | 211,099 | 226,250 | 1,173,000 |
| Profit A/T | | $349,800 | 308,880 | 366,168 | 403,181 | 409,781 | 439,190 | 2,277,000 |
| Loan principal cash flow | +$500,000 | −100,000 | −100,000 | −100,000 | −100,000 | −100,000 | 0 | 0 |
| Purchase | −$600,000 | | | | | | | |
| Cash flow A/T | −$100,000 | 369,800 | 400,880 | 381,368 | 372,301 | 378,901 | 473,750 | 2,277,000 |

*Income tax = $113,900 + 34% of taxable income above $335,000.

expenses. Calculate the *PV* of A/T cash flows with $i = 10\%$ and zero inflation. Then rework the example with 6% annual inflation.

Table 4.14 shows the calculations for both cases in the income statement format. For simplicity, all cash flows and inflation effects occur at year end, and we follow the convention given in Chapter 2 that a full year of inflation occurs by time 1. For the zero-inflation case we have for the A/T cash flow a $PV(10\%) =$ \$114,464. For the 6% inflation case the cash flows are expressed in actual dollars; to evaluate the *PV* we will adjust the original 10% discount rate. The appropriate rate for the inflation case then is

$$\begin{aligned} i &= i' + f + i'f \\ &= 0.1 + 0.06 + (0.1)(0.06) = 0.166, \\ &\quad \text{or } 16.6\% \end{aligned}$$

Thus, the *PV* of the A/T cash flow with 6% inflation is \$111,721. This is \$2,743 less than for the zero-inflation case. □

We can attribute this \$2,743 loss in Example 4.18 to two factors: (1) inflation-caused loss of value of the tax shield from depreciation and (2) tax bracket creep. The first factor can be quantified by using generalized cash flows. Observe that in the zero-inflation case the marginal tax rate remains constant at 15%. For zero inflation,

$$\begin{aligned} PV(10\%) &= -\$100{,}000 && \text{investment} \\ &\quad + (0.15)(10{,}000)\overset{P/A,10\%,6}{(4.35526)} && \text{depreciation} \\ &\quad + (0.15)(10{,}000)\overset{P/A,10\%,4}{(3.16987)}\overset{P/F,10\%,1}{(0.90909)} && \text{depreciation} \\ &\quad + (0.85)(90{,}000)\overset{P/A,10\%,6}{(4.35526)} && \text{revenues} \\ &\quad - (0.85)(35{,}000)\overset{P/A,10\%,6}{(4.35526)} && \text{labor and material} \\ &= -100{,}000 + 6{,}533 + 4{,}323 + 333{,}177 - 129{,}569 \\ &= \$114{,}464 \end{aligned}$$

For 6% inflation and $t_m = 0.15$, we obtain

$$\begin{aligned} PV(16.6\%) &= -\$100{,}000 && \text{investment} \\ &\quad + (0.15)(10{,}000)\overset{P/A,16.6\%,6}{(3.62692)} && \text{depreciation} \\ &\quad + (0.15)(10{,}000)\overset{P/A,16.6\%,4}{(2.76500)}\overset{P/F,16.6\%,1}{(0.85763)} && \text{depreciation} \\ &\quad + (0.85)(90{,}000)\overset{P/A,10\%,6}{(4.35526)} && \text{revenues} \end{aligned}$$

**Table 4.14** *Effects of Inflation on Cash Flows, Example 4.18*

| Cash Flow | $n$: 0 | 1 | 2 | 3 | 4 | 5 | 6 | Totals |
|---|---|---|---|---|---|---|---|---|
| ***Zero Inflation*** | | | | | | | | |
| Revenues | | $90,000 | 90,000 | 90,000 | 90,000 | 90,000 | 90,000 | 540,000 |
| Expenses | | | | | | | | |
| Labor | | $30,000 | 30,000 | 30,000 | 30,000 | 30,000 | 30,000 | 180,000 |
| Material | | $5,000 | 5,000 | 5,000 | 5,000 | 5,000 | 5,000 | 30,000 |
| Depreciation | | $10,000 | 20,000 | 20,000 | 20,000 | 20,000 | 10,000 | 100,000 |
| Total | | $45,000 | 55,000 | 55,000 | 55,000 | 55,000 | 45,000 | 310,000 |
| Taxable income | | $45,000 | 35,000 | 35,000 | 35,000 | 35,000 | 45,000 | 310,000 |
| Income tax* | | $6,750 | 5,250 | 5,250 | 5,250 | 5,250 | 6,750 | 34,500 |
| Profit A/T | | $38,250 | 29,750 | 29,750 | 29,750 | 29,750 | 38,250 | 195,500 |
| Cash flow A/T | −$100,000 | $48,250 | 49,750 | 49,750 | 49,750 | 49,750 | 48,250 | 195,500 |
| ***6% Inflation*** | | | | | | | | |
| Revenues | | $95,400 | 101,124 | 107,191 | 113,623 | 120,440 | 127,667 | 665,445 |
| Expenses | | | | | | | | |
| Labor | | $31,800 | 33,708 | 35,730 | 37,874 | 40,147 | 42,556 | 221,815 |
| Material | | $5,300 | 5,618 | 5,955 | 6,312 | 6,691 | 7,093 | 36,969 |
| Depreciation | | $10,000 | 20,000 | 20,000 | 20,000 | 20,000 | 10,000 | 100,000 |
| Total | | $47,100 | 59,326 | 61,685 | 64,186 | 66,838 | 59,649 | 358,784 |
| Taxable income | | $48,300 | 41,798 | 45,506 | 49,437 | 53,602 | 68,018 | 306,661 |
| Income tax* | | $7,245 | 6,270 | 6,826 | 7,416 | 8,401 | 12,005 | 48,163 |
| Profit A/T | | $41,055 | 35,528 | 38,680 | 42,021 | 45,201 | 56,013 | 258,498 |
| Cash flow A/T | −$100,000 | $51,055 | 55,528 | 58,680 | 62,021 | 65,201 | 66,013 | 258,498 |

*Income tax calculated using schedule in Table 4.1.

$$- (0.85)(35{,}000)\overset{P/A,10\%,6}{(4.35526)} \qquad \text{labor and material}$$

$$= -100{,}000 + 5{,}440 + 3{,}557 + 333{,}177 - 129{,}569$$

$$= \$112{,}605$$

The difference between these two values is the reduction in value of the tax shield from depreciation.

$$\$114{,}464 - 112{,}605 = \$1{,}859 \text{ loss}$$

or

$$(6{,}533 + 4{,}323) - (5{,}440 + 3{,}557) = 1{,}859$$

The remainder of the loss, \$2,743 − 1,859 = \$884, is caused by bracket creep. In years 5 and 6 some of the taxable income falls in the 25% bracket.

The example in this section reflects a 6% inflation rate. The effects are more pronounced, of course, with higher inflation rates and longer depreciation and asset lifetimes.

### 4.4.5 Cash Flow Analysis for Tax-Exempt Corporations

If an organization pays no income tax or capital gains tax, there is no tax shield from depreciation. Therefore, the depreciation method is of interest primarily for cost analysis and management control. Such organizations typically use SL and depreciation lifetimes that agree closely with productive lifetimes of assets.

## 4.5 SUMMARY

In this chapter we have covered the major aspects of income taxes for corporations engaged in manufacturing and trade in the United States. A central theme is how depreciation acts as a tax shield to increase after-tax cash flow. After introducing conventional depreciation methods, we presented the ACRS methods of the 1981 tax law, followed by the MACRS methods of the 1986 tax law.

The last section presented the generalized cash flow approach, which is one of the more powerful analysis techniques for situations in which the marginal tax rate is constant. It is especially useful for sensitivity analysis, and it enables us conveniently to make evaluations of the loss in *PV* caused by inflation.

We clearly recognize the *ever-changing* nature of the tax law. We believe that the student who understands the analysis procedures given in this chapter will be able to make suitable adjustments to reflect future changes in tax law. Thus, although we are presenting the 1986 U.S. tax law and relevant analysis procedures, in a larger context we are presenting an *approach* to the analysis of *any* tax and depreciation law.

## REFERENCES

1. AMERICAN TELEPHONE & TELEGRAPH COMPANY, *Engineering Economy,* 3rd edition, McGraw–Hill, New York, 1977.
2. BLANK, L. T., and D. R. SMITH, "A Comparative Analysis of the Accelerated Cost Recovery System as Enacted by the 1981 Economic Recovery Tax Act," *The Engineering Economist,* Vol. 28, No. 1, pp. 1–30, 1982.
3. BUSSEY, L. E., *The Economic Analysis of Industrial Projects,* Prentice–Hall, Englewood Cliffs, N.J., 1978, Chs. 4 and 5.
4. FLEISCHER, G. E., and D. R. SMITH, "Tax Reform Act of 1986," *Industrial Engineering,* February 1987, pp. 42–50 (Part 1), and March 1987, pp. 22–28 (Part 2).
5. GRANT, E. L., and P. T. NORTON, JR., *Depreciation,* Ronald Press, New York, 1955.
6. MARSTON, A., R. WINFREY, and J. C. HEMPSTEAD, *Engineering Valuation and Depreciation,* Ames, Iowa State University Press, 1963.
7. SCHOOMER, B. A., JR., "Optimal Depreciation Strategies for Income Tax Purposes," *Management Science,* Vol. 12, No. 12, pp. B552–B579, 1966.
8. U.S. DEPARTMENT OF THE TREASURY, Publication 534, *Depreciation,* 1989, and Publication 544, *Sales and Other Dispositions of Assets,* 1989.
9. WHITE, J. A., M. H. AGEE, and K. E. CASE, *Principles of Engineering Economic Analysis,* 3rd edition, Wiley, New York, 1989, Ch. 6.

## PROBLEMS

**4.1.** An asset is purchased and installed in 1972 at a cost of \$36,000. The asset has an 8-year depreciation life and an estimated salvage value of zero. Determine the depreciation amount $D_n$ for each of the first 3 years and the book balance (book value) $BB_n$ at the end of each of those years, using the following depreciation methods: SL, SOYD, 150% DB, and DDB. Ignore half-year conventions, ITC, and Sec. 179.

**4.2.** A large, stable, profitable corporation borrowed \$15,000, with the principal to be repaid in five annual installments of \$1,000 (end of year 1), \$2,000 (end of year 2), . . . , \$5,000 (end of year 5). Interest is 12% per year on the outstanding loan balance. With a marginal tax rate of 40% and an interest rate of 15%, find the net present value of the loan. Discuss the algebraic sign of the net present value you obtained.

**4.3.** In 1990 your corporation had revenues of \$1,000,000, payroll expenses of \$150,000, material expenses of \$200,000, and depreciation expenses of \$80,000. Your corporation purchased and installed new assets worth \$200,000, financed entirely by company funds (depreciation on these assets is included in the previous figures). In addition, the company made the second annual payment on a \$2,000,000, 15-year, 10% mortgage loan. The mortgage is repayable in 15 equal annual payments (principal plus interest). Estimate the cash flow after taxes for 1990. Use the income statement approach.

**4.4.** Rework Problem 4.3 for the situation of a large, stable, profitable corporation with a marginal tax rate of 40%. Use the generalized cash flow method.

**4.5.** An asset costing \$40,000 was installed in 1989 and depreciated using MACRS percentages for the 5-year property class. In 1991 the asset was sold. Calculate the net cash flow after taxes resulting from the sale of the asset. The company is a large, stable, profitable corporation with a marginal tax rate of 34%. The selling price of the asset was (a) \$25,000; (b) \$10,000.

**4.6.** A proposed project requires an investment of $120,000 for an asset in the 5-year property class. Annual revenues of $200,000 and operating costs of $40,000 are anticipated. It is expected that the project will be terminated after 4 years and the asset sold for $50,000 on the last day of the fourth year. The project will be financed partly by a loan for $80,000 at 20% interest, with principal repaid in four equal annual installments. The marginal tax rate is 45%, and alternative SL depreciation is to be used. Obtain the after-tax cash flow year by year.

**4.7.** Your company has just purchased a new machine that costs $10,000. The machine has an expected life of 10 years and salvage value of $700. The depreciation life is 10 years. Compare the present values of the depreciation that would be obtained, using SL, SOYD, DDB, or a combination of two methods. Ignore ITC and Sec. 179.

**4.8.** For a pre-1981 asset depreciated by either 150% DB or DDB, show that if at the end of the depreciation life the implied book balance is *less* than the estimated salvage value, it is better *not* to switch to SL.

**4.9.** For an asset depreciated by a DB method, show that the optimal time to switch to SL is when the depreciation by DB in any year is less than it would be by switching to SL.

**4.10.** A heavy-lift crane costing $160,000 is to be depreciated by the production units (hours of usage) method. The crane has an estimated life of 12,000 hours and a zero estimated salvage value. Calculate the depreciation for the first 3 years if the usage is 1,400, 2,000, and 1,900 hours in years 1, 2, and 3, respectively.

**4.11.** Derive the MACRS percentages for the 3-, 7-, and 10-year property classes.

**4.12.** Derive the MACRS percentages for the 15- and 20-year property classes.

**4.13.** Construct a situation for a company in which the alternative SL method might be preferred over the MACRS percentages method. Define carefully the criteria and the conditions of the situation.

**4.14.** Consider the asset in Examples 4.11 and 4.12. Would there be any advantage in postponing the sale of the asset until the sixth year so that all the year 5 and year 6 depreciation could be claimed as depreciation expenses?

**4.15.** Rework Example 4.13 assuming that the Sec. 179 deduction was *not* claimed.

**4.16.** An asset in the 7-year property class cost $800,000 and has an estimated salvage value of $100,000 after 8 years of use. The asset will generate annual revenues of $1,300,000 and require annual labor expenses of $250,000 and material expenses of $120,000. There are no other revenues or expenses for the company. Compare after-tax cash flows by using MACRS percentages and the alternative SL method. The life of the project is 8 years. Take advantage of Sec. 179 if possible.

**4.17.** A company is planning to add a new section to its manufacturing plant at a cost of $400,000, of which $100,000 is in a building (31.5-year property) and $300,000 in equipment (7-year property). It is estimated that labor and materials will cost $200,000 per year. The building will be financed by a 10-year loan at 12% interest, with principal repaid in ten equal installments. In addition, a $150,000, 5-year loan will be used to help finance the equipment; this loan is at 14%, with principal to be repaid in five equal installments. Equipment depreciation will be by the alternative SL method. The useful life of the equipment is 10 years, at which time it will have zero salvage value. The useful life of the building is nearly 30 years, but for purposes of analysis assume a salvage value equal to book value after 10 years. The marginal tax rate is 40% and the company's interest rate is 10%. Determine an equal annual revenue stream that will result in a present value of zero for this project.

**4.18.** Prove that when the interest rate $i$ used for discounting cash flows is related to the borrowing interest rate by

$$i = (1 - t_m)i_b$$

the $PV$ of the loan cash flow is zero.

**4.19.** A loan of $1,000,000 at 15% is to be repaid in 30 years. Evaluate the $PV$ of the loan cash flow for (a) equal principal repayments and (b) equal total payments. The marginal tax rate is 40% and the interest rate is 12%.

**4.20.** Rework Example 4.18 with 12% annual inflation. In addition, assume that the salvage value is 10% of the initial cost when there is no inflation and that the salvage value increases with inflation. Assume a marginal tax rate of 40%.

**4.21.** For a company with an interest rate of 10% when there is zero inflation, calculate the net present value of each of the following cash flow streams. The company is a large, stable, profitable firm with a marginal tax rate of 40%.

a. A revenue stream of $1,000 per year, for 20 years, with zero inflation.
b. A revenue stream of $1,000 per year, for 20 years, obtained from a long-term, fixed-price contract, when inflation is 4.5% per year.
c. A revenue stream of $1,000 per year, for 20 years, under conditions of flexible pricing, when inflation is 9% per year.
d. The tax shield of depreciation of $600 per year, for 20 years, with zero inflation.
e. The tax shield of depreciation of $600 per year, for 20 years, with 9% inflation per year.

**4.22.** Consider the following investment, in an inflation-free economy. Gamma High-Tech Company is considering the purchase of a $10,000 asset that will be used for only 2 years (project life). The salvage value of this asset at the end of 2 years is expected to be $4,000. The asset will generate annual revenue of $20,000 but is expected to have an annual expense of $5,000. The investment will be classified as a 3-MACRS property (tax life) with annual depreciation allowances of $3,333, $4,444, $1,481, and $741. The marginal income tax rate for the firm is 40%. The firm's inflation-free interest rate ($i'$) is 10%.

a. Determine the after-tax cash flow (ATC) for this investment project.
b. The firm expects an average inflation rate ($f$) of 5% during the project period, but it also expects an 8% annual increase in revenue and a 6% annual increase in expense. No increase in salvage value is expected. Compute the present value of this investment.

**4.23.** Mobile Electric Company expects to have taxable incomes of $320,000 from its regular business over the next 2 years. You have been asked by the president of the company to evaluate the residential wiring project for a proposed apartment complex during year 0. This 2-year project requires purchase of new equipment for $50,000. The equipment falls into the MACRS 5-year class with regular depreciation allowances of 20, 32, 19.2, 11.52, 11.52, and 5.76%. The machine will be sold after 2 years for $30,000. The project will bring in an additional annual revenue of $100,000, but it is expected to incur an additional annual operating cost of $40,000. The project requires an investment in working capital in the amount of $10,000. The investment in working capital will be recovered when the machine is sold.

a. What is the marginal tax rate to use in evaluating the acquisition of the equipment during years 1 and 2, respectively?
b. With the project, what is the effective tax rate of the firm during project year 1?
c. If the firm elects to use the alternative MACRS straight-line depreciation over the

property class life, what would be the depreciation allowances for the first 2 years?

d. What are the *project*'s net cash flows (A/T) in years 0, 1, and 2?
e. If the firm expects to borrow the initial investment at 10% over a 2-year period (equal annual payments of $28,810), what are the *project*'s net cash flows?
f. In part e, if the firm expects an annual inflation of 5% ($f$) during the project period, compute the after-tax cash flows in actual dollars. What is the equivalent present value of this amount at time 0, if the firm's inflation-free interest rate ($i'$) is 10%?
g. With the result in part f compute the firm's inflation loss in present value. How much additional annual revenue (in actual dollars) is required to make up this inflation loss?
h. In part g, determine how much additional before-tax annual revenue the project has to generate in *actual dollars* (equal amount) to make up the inflation loss.

# 5

# Selecting a Minimum Attractive Rate of Return

## 5.1 INTRODUCTION

The interest rate used for discounting cash flows has a direct effect on the outcome of project evaluation and comparison. A change in the interest rate can change the accept–reject decision for an individual project, and it can alter the choice of the "best" from among several projects. An extremely important aspect of engineering economic analysis is, therefore, the selection of such a rate. The rate is usually designated simply as the *interest rate,* or the *discount rate.* Because of its use in accept–reject decisions, it is also known as the *minimum attractive rate of return, MARR.*

In this chapter we address three major issues. The first is the computation of costs of capital from individual sources. The second is the proper use of weighted cost of capital, given a fixed capital structure of the firm. The third is the method of computing a weighted cost of capital under different conditions of capital structure, growth, and risk. The ordering here is based on complexity; the third issue is much discussed in the literature, and we attempt only to outline the major lines of thought. The three issues are presented after a brief section on financing sources.

## 5.2 INVESTMENT AND BORROWING OPPORTUNITIES

Investment and financing decisions are made in a dynamic environment. The decision maker may have available one or more investment projects, with only limited time for making a decision, and some knowledge about different financing sources and their respective costs. The same or different investment projects may be available some months or several years later, along with similar or different financing sources. The decision maker's problem is to select now the projects that will add to the firm's revenues and at the same time maintain the firm's financial strength so that it will be able to undertake profitable investments in the future.

It is helpful to have information about future potential investments, even if it is only general information of a statistical nature. Similarly, knowing that certain financing sources behave in reasonably consistent and predictable ways aids the decision maker.

### 5.2.1 Future Investment Opportunities

If all the investment opportunities available to a firm during a period of a year were ranked, they might form a distribution such as that shown in Figure 5.1*a*, the *investment opportunity curve.* Assuming that we can specify an investment efficiency criterion based on the interest rate concept, the distribution shows the declining interest rate that each successive investment would earn. The curve eventually flattens when it reaches risk-free liquid investments, such as U.S. Treasury securities and money market funds. The actual investments could take a variety of forms, including development of new products, expansion of market effort, modernization of equipment, energy conservation measures, and buying shares in other companies.

The investment opportunity curve will change yearly. If the changes are not drastic, we can use the information in the curve to develop the *MARR.* If the annual changes are extensive, we may be forced to use one of the techniques discussed in later chapters (Chapters 8, 10, and 11).

### 5.2.2 Financing Sources

The most common sources of financing for companies are the following:

cash advances (typical in the construction industry),

bank loans (short and long term),

loans against inventory,

loans against accounts receivable,

delaying payment on accounts payable,

bond issues,

preferred stock issues,

common stock issues, and

retained earnings.

Some of these funding sources are tied to assets, such as mortgages secured by buildings, loans secured by machinery, and loans secured by inventory. In addition, bond interest and principal repayment have precedence over dividend payments on stock, and preferred stock dividends have precedence over common stock dividends. Thus, bonds are considered safer than preferred stock, and preferred stock safer than common stock, for the same corporation.

Let us defer the subject of asset-based financing until Section 5.5.5. Then we can rank all the borrowing opportunities by annual interest cost and construct a *borrowing opportunity curve* as in Figure 5.1*b*. The intersection of the investment opportunity curve with the borrowing opportunity curve, shown in Figure 5.1*c*, represents the point at which the firm should operate. It should

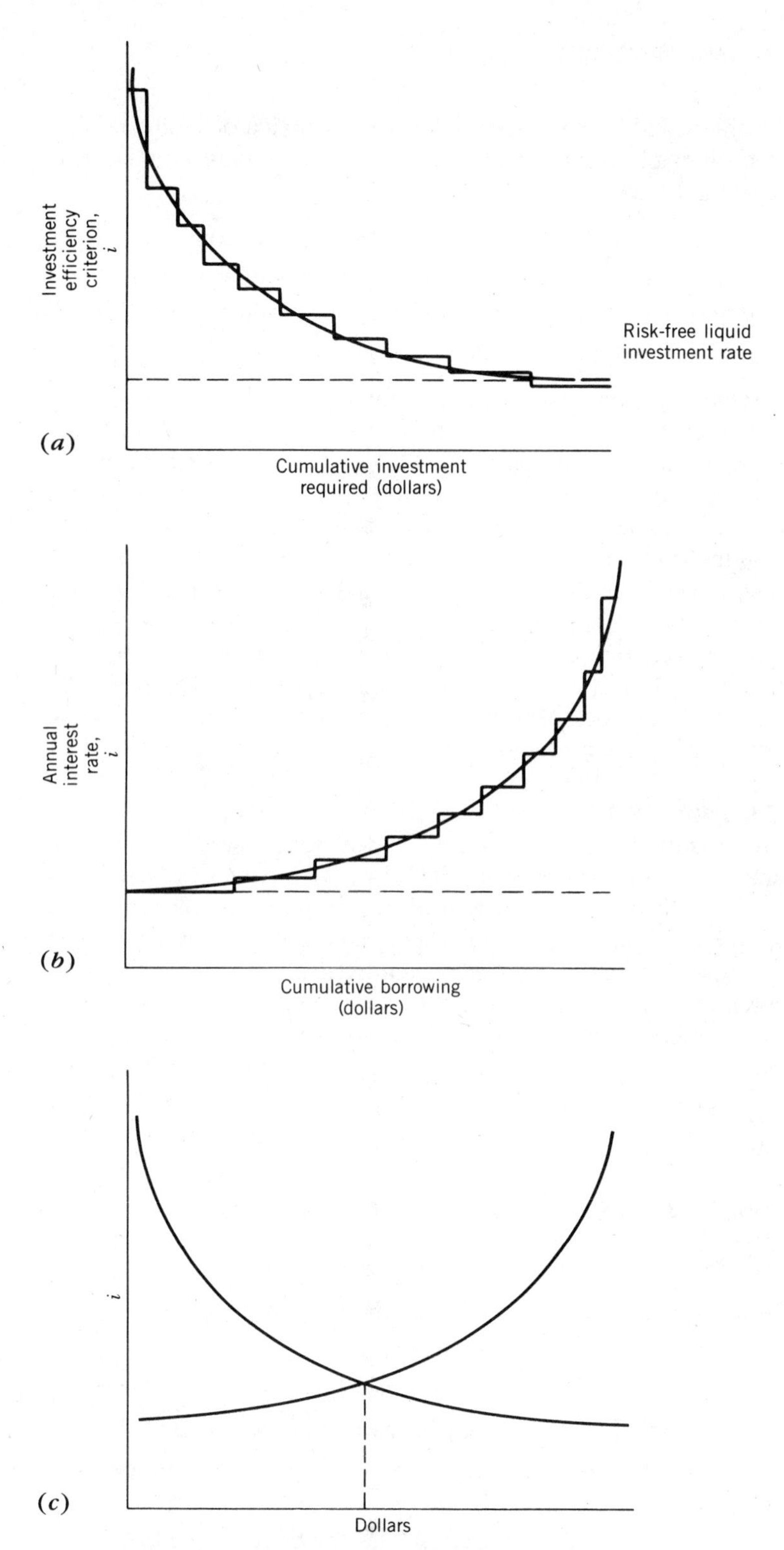


**FIGURE 5.1** (*a*) Investment opportunity curve. (*b*) Borrowing opportunity curve. (*c*) Intersection of the two curves.

accept all investment and borrowing opportunities to the left of the intersection point and reject all opportunities to the right. Any other solution would require one or more of the following.

1. Giving up a profitable investment–borrowing combination.
2. Accepting an unprofitable investment–borrowing combination.
3. Replacing an investment opportunity with a less profitable one.
4. Replacing a borrowing opportunity with a costlier one.

Hence, the intersection point is the optimal solution.

The foregoing analysis is oversimplified in that we have ignored the difference between short-term and long-term effects. In any one year we may be making investment decisions about 3% to 10% of the total investment of a company. We have also ignored the effects of making investments and borrowing money, and the resulting strength of the firm, on the *costs* of borrowing. We will treat these issues later in the chapter. The issue of project divisibility is covered in later chapters.

For large corporations the bulk of the financing is achieved with bonds and common stock, referred to as *debt* and *equity,* respectively. To simplify the presentation in other parts of the chapter, we will treat the borrowing as consisting of these two types of opportunities.

### 5.2.3 Capital Rationing

In many situations the firm is not willing or not allowed to invest and borrow all the way to the intersection point of the two curves in Figure 5.1*c*. Uncertainties in profit projections may lead management to "play safe" and operate somewhere to the left of the intersection. Bankers usually apply financial ratios to compute the maximum credit extended to a customer. Expectations of better investment opportunities in future years, with corresponding cash needs, may be another valid reason for not operating at the intersection point. In such situations any idle funds are usually invested in relatively risk-free, liquid investments. In Chapter 8 we examine some special cases of such capital rationing.

## 5.3 COSTS OF CAPITAL FROM INDIVIDUAL SOURCES

We define here some terms used in the chapter.

| | |
|---|---|
| $A_e$ | annual expense associated with interest payments |
| $B$ | total debt capital (bonds), in dollars |
| $DPS_n$ | dividend payment per share of stock paid at end of year $n$ |
| $DPS$ | Annual dividend payment per share of stock, assumed constant |
| $EPS_0$ | Earnings per share, after interest and taxes for the period just ended |
| $F$ | redemption value, or face value, of the bond |
| $g$ | geometric rate of growth in dividend |

| | |
|---|---|
| $i_b$ | nominal borrowing interest rate, per year |
| $i_{bp}$ | nominal borrowing interest rate, per period |
| $i_e$ | equity interest rate per year |
| $IRR$ | internal rate of return |
| $k$ | tax-adjusted weighted-average cost of capital |
| $k_b$ | effective annual after-tax rate on debt (bond) |
| $k_{bp}$ | effective periodic (quarterly) after-tax interest rate on debt |
| $M$ | number of bond interest periods per year |
| $MARR$ | minimum attractive rate of return |
| $N$ | maturity of the bond, in years |
| $P$ | selling price of the bond, between the firm issuing the bond and the underwriter |
| $P_0$ | current price per share of common (preferred) stock |
| $r$ | nominal (coupon) interest rate per year |
| $S$ | total equity capital (shares) in dollars |
| $S_e$ | selling expenses associated with the bond issue incurred by the issuing firm |
| $t_m$ | marginal corporate income tax (combined state and federal) |
| $V$ | $B + S$, total capital; also general symbol for value of the firm |

### 5.3.1 Debt Capital

We saw in Chapter 4 that interest paid on loans is a tax-deductible expense. So the after-tax cost of loans and bonds is less than the before-tax loan interest rate and bond yield rate. We now examine some specific types of debt instruments. We will express all costs in terms of effective annual rates.

***Bank Loans.*** There are several cases to consider, depending on the number of interest payments and income tax payments per year. In the simplest case, interest and taxes are payable annually. Then the effective annual after-tax rate of the loan is

$$k_b = i_b(1 - t_m) \tag{5.1}$$

We implicitly assume that the marginal tax rate remains the same.

### *Example 5.1*

A 12% bank loan, with interest payable annually, for a company with a marginal tax rate of 35% and annual tax payments has an after-tax cost of

$$(0.12)(1 - 0.35) = 0.0780 = 7.80\% \quad \square$$

Most corporations, however, pay income taxes more frequently. In the United States, for example, tax payments are made quarterly, based on estimated income statements. So the tax benefit of the interest expense is taken quarterly. This has the effect of reducing the effective annual after-tax rate. Unfortunately, the method for computing the precise cost is more complicated. It requires solving the following equation for $k_{bp}$, the effective debt cost per period (quarter).

$$1 + (i_b t_m/4)(P/A, k_{bp}, 4) - (1 + i_b)(P/F, k_{bp}, 4) = 0 \tag{5.2}$$

Here we are solving an equivalence problem for an unknown interest rate, per quarter. We then convert to the cost per year.

$$k_b = (1 + k_{bp})^4 - 1 \tag{5.3}$$

### *Example 5.2*

Use the data in Example 5.1, but assume quarterly tax payments, and find the effective annual after-tax cost of the loan. Here

$$\frac{i_b t_m}{4} = \frac{(0.12)(0.35)}{4} = 0.0105$$

$$1 + (0.0105)(P/A, k_{bp}, 4) - (1.12)(P/F, k_{bp}, 4) = 0$$

The solution, by trial and error, is $k_{bp} = 0.018673$. Then

$$k_b = (1 + 0.018673)^4 - 1 = 0.07681 = 7.681\% \quad \square$$

Another common case is that in which interest and tax payments are made quarterly. Here we may use the expression

$$k_b = [1 + (i_{bp}/4)(1 - t_m)]^4 - 1 \tag{5.4}$$

This expression assumes that quarterly tax payments reflect quarterly interest payments, not one-fourth of annual interest payments. (In any case, the difference is slight for $i_b$ below 20%.) Note, however, that the popular expression [2] for $k_b$, namely $[(1 + i_b/4)^4 - 1](1 - t_m)$, is not applicable to U.S. corporations, since it assumes quarterly compounding but annual tax payments.

### *Example 5.3*

Use the data in Example 5.1, but assume quarterly interest and tax payments and compute the effective annual after-tax cost of the loan. This problem can be solved two ways, depending on what the 12% means. If the 12% represents a *nominal* annual rate, the borrowing interest rate per period is

$$i_{bp} = 0.12/4 = 0.03$$

Then $k_b = [1 + (0.03)(0.65)]^4 - 1 = 0.0803 = 8.03\%$. If the 12% represents an *effective* annual rate, then

$$i_{bp} = (1 + 0.12)^{0.25} - 1 = 0.02874$$

and

$$k_b = [1 + (0.02874)(0.65)]^4 - 1 = 0.07684 = 7.684\% \qquad \square$$

The three preceding examples show that there are slight differences in the effective annual after-tax cost, depending on compounding and tax payment timing. For higher loan rates, say above 20%, these differences become larger in absolute terms, but they are still relatively small compared with the inherent variability of loan rates. For this reason use of either Eq. 5.1 or Eq. 5.4 is recommended. (The problems at the end of the chapter explore this issue in more detail.)

When there are discount points on a loan, the borrower does not receive the full amount of the loan, but payments are calculated on the full amount. The effective loan rate is thereby increased. For tax purposes the discount is prorated over the life of the loan. The exact method for finding $k_b$ requires setting up a loan schedule, similar to the examples in Section 4.3, and obtaining the complete after-tax cash flow.

## *Example 5.4*

A loan for $1,000 is to be repaid with annual payments over 3 years at 20% interest. The loan is discounted 24 points, which means that the time 0 proceeds are $1,000(1 − 0.24) = $760. With $t_m = 30\%$, find the effective annual after-tax rate.

The annual payments are $(1{,}000)\overset{A/P,20\%,3}{(0.47473)} = \$474.73$. The loan schedule is as follows.

| Item | *n*: 1 | 2 | 3 |
|---|---|---|---|
| 1. Loan balance, BOP* | $1,000.00 | 725.27 | 395.59 |
| 2. Interest payment | 200.00 | 145.05 | 79.12 |
| 3. Principal payment | 274.73 | 329.68 | 395.59 |
| 4. Total payment, 2 + 3 | 474.73 | 474.73 | 474.71† |
| 5. Interest tax benefit (2)(0.3) | 60.00 | 43.52 | 23.74 |
| 6. Discount tax benefit ($240/3)(0.3) | 24.00 | 24.00 | 24.00 |
| 7. A/T cash outflow, 4 − 5 − 6 | 390.73 | 407.21 | 426.97 |

*BOP means beginning of period.
†Amount is adjusted to compensate for cumulative rounding to the nearest cent.

We now solve the equivalence problem.

$$\$760 = \frac{390.73}{1 + k_b} + \frac{407.21}{(1 + k_b)^2} + \frac{426.97}{(1 + k_b)^3}$$

By trial and error we obtain $k_b = 0.2776 = 27.76\%$. □

The loan schedule method is tedious. There are two approximation methods that tend to give good results.

*Method 1:* Find the interest rate that makes the *PV* of the borrower's before-tax cash flow equal to zero. Multiply this rate by $(1 - t_m)$ to obtain an estimate of $k_b$.

*Method 2:* Multiply the loan interest rate by $(1 - t_m)$. Use this rate to compute a hypothetical annual payment by multiplying the full loan amount by a factor, $(A/P)$. Adjust the hypothetical annual payment by the discount tax benefit. Using the adjusted hypothetical annual payments and the net loan proceeds, find an interest rate that makes the *PV* of this cash flow equal to zero. This rate is an estimate of $k_b$.

### *Example 5.5*

Apply approximation methods 1 and 2 to the data in Example 5.4.

*Method 1:* The borrower's before-tax cash flow is

| $n$ | 0 | 1 | 2 | 3 |
|---|---|---|---|---|
| Flow | +\$760.00 | −474.73 | −474.73 | −474.71 |

Using $\$760 = 474.73(A/P, i, 3)$, we find $i = 0.3941$. Thus, $k_b \cong (0.3941)(1.0 - 0.3) = 0.2759 = 27.59\%$. The error here is 0.17% absolute and 0.62% relative.

*Method 2:* We obtain $(20\%)(1.0 - 0.3) = 14\%$ and find the hypothetical annual payment.

$$(\$1{,}000)(\overset{A/P,14\%,3}{0.43073}) = \$430.73$$

The discount tax benefit is $(\$240/3)(0.3) = \$24.00$, so the adjusted hypothetical annual payment is $\$430.73 - 24.00 = \$406.73$. We now set up the following cash flow and find the interest rate so that $PV = 0$.

| $n$ | 0 | 1 | 2 | 3 |
|---|---|---|---|---|
| Flow | +\$760.00 | −474.73 | −474.73 | −474.71 |

Using $\$760 = 406.73(A/P, i, 3)$, we find $i = 0.2800$. Thus, $k_b \cong 0.2800 = 28.00\%$. The error here is 0.24% absolute and 0.86% relative. □

Either method certainly seems good enough for practical use, but method 1 is easier to use. (See the problems at the end of the chapter.) The example shows that it is important to treat properly the discount points. The $k_b$ in Example 5.4 is higher than the before-tax $i_b$ because of the 24 points discount.

***Bonds.*** The after-tax cost of bonds depends, in the same way as bank loans, on the timing of interest and tax payments. In addition, there are issuing expenses and discounts (or premiums) to be considered. We will assume quarterly tax payments. Many analysts ignore the selling expenses $S_e$ and annual expenses $A_e$, but we will include them for completeness. The selling expenses $S_e$ must be prorated over the maturity period $N$ for tax purposes. The annual expenses $A_e$ are tax-deductible each year. The discount, or difference between redemption value $F$ and selling price $P$, must be spread over the maturity period by one of two methods.

**Case 1:** For bonds issued before July 1, 1982, by prorating,

**Case 2:** For bonds issued after July 1, 1982, by the yield method.

For case 1, to find $k_b$ we must find $k_{bp}$ by solving an equation similar to Eq. 5.2.

$$(P - S_e) - \frac{Fr}{M}\sum_{n=1}^{MN}(1 + k_{bp})^{-4n/M} - A_e\sum_{n=1}^{N}(1 + k_{bp})^{-4n}$$

$$+ \left(\frac{Fr}{4} + \frac{F-P+S_e}{4N}\right)t_m\sum_{n=1}^{4N}(1 + k_{bp})^{-n} - F(1 + k_{bP})^{-4N} = 0 \quad (5.5)$$

The first term represents the net proceeds from the bond issue; the second term, the series of interest payments; the third term, the annual expenses; the fourth term, the tax benefits from the interest payments, discount, and selling expenses; and the last term, the redemption of the bond. After solving for $k_{bp}$ by trial and error, we find $k_b$ by using Eq. 5.3.

For practical application Eq. 5.5 is needlessly complex. Most corporate bond issues in the United States have semiannual interest payments. We can combine the interest payment with its tax benefit on a quarterly basis and get a very good approximation. The relatively small annual expenses can also be treated quarterly. These changes give the following equation to be solved for $k_{bp}$.

$$(P - S) + \left[\frac{(F - P + S_e)t_m}{4N} - \frac{(Fr + A_e)(1 - t_m)}{4}\right](P/A, k_{bp}, 4N)$$

$$- F(P/F, k_{bp}, 4N) = 0 \quad (5.6)$$

### *Example 5.6*

A $10,000 bond is sold at a price of $9,000, with associated selling expenses of $50. The maturity is 4 years, with a coupon interest rate of 7%. The marginal tax rate is 40%; annual expenses are negligible. Find the effective annual after-tax interest rate, for case 1 (bond issued before July 1, 1982). We will assume quarterly tax payments and use the approximate form of Eq. 5.6. Here $P = 9{,}000$, $F = 10{,}000$, $N = 4$, $S_e = 50$, $A_e = 0$, $r = 0.07$, and $t_m = 0.40$. Applying (5.6), we have

$$(9{,}000 - 50) + \left[\frac{(10{,}000 - 9{,}000 + 50)(0.40)}{16} - \frac{(700)(0.60)}{4}\right](P/A, k_{bp}, 16)$$
$$- 10{,}000(P/F, k_{bp}, 16) = 0$$

$$(8{,}950) + (26.25 - 105.00)(P/A, k_{bp}, 16) - 10{,}000(P/F, k_{bp}, 16) = 0$$

By trial and error, we obtain $k_{bp} = 0.0153$, which gives, using Eq. 5.3, $k_b = 0.0627 = 6.27\%$. □

For case 2 we must first find the yield to maturity of the bond. Using a method similar to Eq. 5.5 but ignoring tax effects, we solve for $i_b$ using

$$P - (Fr)(P/A, i_b, N) - F(P/F, i_b, N) = 0 \tag{5.7}$$

Then, for tax purposes, we treat the bond as a loan at interest rate $i_b$, with negative amortization and a balloon payment. If the bond sells at a premium, the amortization is positive, but there is still the balloon payment. Ignoring the effects of selling expenses $S_e$ and annual expenses $A_e$, we could find $k_b$ by using the same procedure as in Example 5.3.

### *Example 5.7*

Using the data in Example 5.6, find the effective annual after-tax interest rate for case 2 (bond issued after July 1, 1982). Also find $k_b$, ignoring selling expenses $S_e$ and annual expenses $A_e$.

We use Eq. 5.7 to find the yield to maturity, assuming annual compounding, according to the tax law ($M = 1$) [10]:

$$9{,}000 - (10{,}000)(0.07)(P/A, i_{bp}, 4) - 10{,}000(P/F, i_{bp}, 4) = 0$$

This gives $i_b = 10.17\%$.

At this point we can estimate $k_b$, ignoring $S_e$ and $A_e$. Assuming quarterly tax payments, we find

$$i_{bp} = (1.1017)^{0.25} - 1 = 0.0245$$

**Table 5.1** ***After-Tax Cash Flow Calculations for Bond Issued after July 1, 1982, Example 5.7***

| Item | *n*: 1 | 2 | 3 | 4 |
|---|---|---|---|---|
| 1. Adjusted issue price ("loan balance," BOP)* | $9,000.00 | 9,214.94 | 9,451.74 | 9,712.61 |
| 2. "Interest" payment | $914.94 | 936.80 | 960.87 | 987.39 |
| 3. Change in issue price ("principal" payment) | −$214.94 | −236.80 | −260.87 | −287.39 |
| 4. Actual interest paid ("total" payment) | $700.00 | 700.00 | 700.00 | 700.00 |
| 5. Interest tax benefit, (2)(0.40) | $365.98 | 374.72 | 384.35 | 394.96 |
| 6. Selling expense tax benefit, ($50/4)(0.40) | $5.00 | 5.00 | 5.00 | 5.00 |
| 7. A/T cash outflow, 4 − 5 − 6 | $329.02 | 320.28 | 310.65 | 300.04 |

*"Loan balance" at $n = 5$ is $9,712.61 + 287.39 = $10,000.00.

Then

$$k_b \cong [1 + (0.0245)(0.60)]^4 - 1 = 0.0601 = 6.01\%$$

Returning to the general situation, we now construct a loan schedule based on 10.17% and annual repayments, shown in Table 5.1. Using the last line of the table, we construct the entire after-tax bond cash flow (assuming annual interest and tax payments).

| *n* | 0 | 1 | 2 | 3 | 4 |
|---|---|---|---|---|---|
| Flow | +$8,950 | −329.02 | −320.28 | −310.65 | −10,300.04 |

The interest rate that makes the *PV* of this cash flow equal to 0, by trial and error, is $k_b = 0.0620 = 6.20\%$. (This interest rate is known as the internal rate of return.)

If we wish to be more precise, we must convert the annual flows into quarterly flows.

| *n* | 0 | 1–4 | 5–8 | 9–12 | 13–15 | 16 |
|---|---|---|---|---|---|---|
| Flow | +$8,950 | −82.26 | −80.07 | −77.66 | −75.01 | −10,075.01 |

This yields $k_{bp} = 0.0153$ and $k_b = (1.0153)^4 - 1 = 0.0628 = 6.28\%$. Still more precision could be obtained by treating the actual interest payments semiannually, but it is not worth the effort. □

The approximate methods for discounted loans could be used to treat the bond discount and selling expenses. Applying method 1 to Example 5.6, we proceed as follows.

Before-tax cash flow is given by

| $n$ | 0 | 1–7 | 8 |
|---|---|---|---|
| Flow | +\$8,950 | −350 | −10,350 |

Solving for an interest rate to give a $PV$ of 0, we obtain $i = 0.0513$. This converts to

$$k_b \cong [1 + (0.0513)(0.60)]^2 - 1 = 0.0625 = 6.25\%$$

The approximation clearly seems preferable in this example.

Before we leave this section, we present one more approximation formula for $k_b$. This one is simpler and less precise, but it has the advantage of containing no compound interest factors.

$$k_b \cong \frac{[Fr + (F - P + S)/N + A](1 - t_m)}{(F + P - S)/2} \tag{5.8}$$

### *Example 5.8*

Using the data in Example 5.6, and the approximation in Eq. 5.8, find $k_b$.

$$k_b = \frac{[700 + (10{,}000 - 9{,}000 + 50)/4 + 0](0.60)}{(10{,}000 + 9{,}000 - 50)/2}$$

$$= 0.0609 = 6.09\% \quad \square$$

We have solved for $k_b$ with the data in Example 5.6 in six ways.

| | |
|---|---|
| Case 1, Eq. 5.6 | 6.27% |
| Case 2, ignoring $S_e$, $A_e$ | 6.01% |
| Case 2, loan schedule, annual flows | 6.20% |
| Case 2, loan schedule, quarterly flows | 6.28% |
| Case 2, approximation method 1 | 6.25% |
| Cases 1 and 2, Eq. 5.8 | 6.09% |

*All* these are *approximations*. Exact methods would require

**Case 1:** Use of Eq. 5.5.

**Case 2:** Separation of flows in Table 5.1 into semiannual interest payments, quarterly tax benefits, and an analog of Eq. 5.5.

These are left as exercises at the end of the chapter (see Problem 5.3).

It is important to note that the value for $k_b$ obtained from $i_b(1 - t_m)$ is valid *only* if the bond is a perpetuity or if there is no discount and $S_e$ and $A_e$ are zero [6].

***Accounts Payable.*** Without realizing it, many companies use accounts payable as a financing source. It is not uncommon for vendors to give 2% discount if payment on an invoice is received within 10 days. By paying after the 10-day period, the buyer is borrowing money and paying (giving up) the 2% discount. The effective annual cost of this source of financing depends on when the buyer pays. Let us assume that, on the average, the buyer pays 45 days after the invoice date. We will also assume that the buyer incurs no interest expense for delayed payments. In order to be sure of paying within 10 days, the buyer might pay on the seventh day. Thus, the buyer is paying 100%/98% − 1.0 = 0.0204 = 2.04% for use of the money for 45 − 7 = 38 days. There are 9.6 38-day compounding periods per year, so we have, with $t_m = 0.30$, for example,

$$k_b = [1 + (0.0204)(0.70)]^{9.6} - 1 = 0.149 = 14.9\%$$

In this section on debt capital we have presented a variety of methods for computing the effective annual after-tax interest rate. Some of the complexity is due entirely to the tax law. In other instances we have attempted to be precise in order to show the principles involved. In practice, most analysts use one of the approximate methods. We hope this section contributes to an understanding of why and when the approximate methods perform adequately.

### 5.3.2. Equity Capital

A detailed explanation of the cost of equity capital would require several chapters, or perhaps an entire book. In this section we provide some valuation methods with only brief explanations and references. This will enable students and analysts to proceed more quickly to Section 5.4, where we discuss the use of weighted cost of capital. In Section 5.5 we present a brief discussion of equity valuation and specification of a weighted cost of capital.

***Preferred Stock.*** The computation here is straightforward: we need only know, or estimate, the share price of a new offering of preferred stock. Then we obtain the equity interest rate.

$$i_e = DPS/P_0 \tag{5.9}$$

The notation implies that we receive the proceeds (number of shares × $P_0$) on issuing the stock at time 0. The dividends are paid from after-tax corporate earnings.

***Common Stock.*** The simplest valuation model for common stock considers only dividend payments. The value of a share to a stockholder is

$$P_0 = \sum_{n=0}^{\infty} DPS_n(1 + i_e)^{-n} \tag{5.10}$$

If dividends are *constant*, this results in

$$P_0 = \frac{DPS}{i_e} \quad \text{or} \quad i_e = \frac{DPS}{P_0} \tag{5.9}$$

The share price in Eq. 5.10 is before (ex ante) the current dividend $DPS_0$. If the current dividend was just paid, the $DPS_0$ term in the summation is eliminated and $P_0$ is the price after (ex post) the current dividend. As for preferred stock, dividends are paid from after-tax earnings.

In actuality, dividend payments are not constant. Many investors actively seek *growth* companies, and their valuation includes share appreciation besides dividends, which are also expected to grow. One such model results in the following expression [4]:

$$i_e = DPS_0/P_0 + g \tag{5.11}$$

The growth rate $g$ can be estimated by using the ratio of retained earnings to current share price [8].

$$g = (EPS_0 - DPS_0)/P_0 \tag{5.12}$$

Equation 5.11 then becomes

$$i_e = EPS_0/P_0 \tag{5.13}$$

This expression is deceptive in its simplicity, for it is the end result of extensive derivation. Implicit in the use of Eq. 5.13 is the assumption that any expected future change in earnings is already reflected in the current share price. Equation 5.13 implies that in order to attract a $P_0$ share price on new issues, the company must assure potential investors of earnings per share at the level of $EPS_0$. We will use Eq. 5.13 in calculating the weighted cost of capital.

***Retained Earnings.*** Most modern authorities in financial theory agree that retained earnings provide the means for a company to increase its future dividends. Therefore, a form such as Eq. 5.13 already reflects the effect of retained earnings through the $P_0$ term. Hence, we do not need to consider retained earnings explicitly in a weighted-average cost of capital.

## 5.4 USE OF A WEIGHTED-AVERAGE COST OF CAPITAL

In this section we present two related ways of specifying and using a cost of capital. As in Section 5.3.2, we defer most of the discussion to Section 5.5. We assume that the corporation is operating in a stable situation: The ratio of debt financing to equity financing remains constant; the costs of individual financing sources remain the same; the ratio of dividend growth, if any, remains constant; and the marginal tax rate remains the same. For simplicity, we assume annual cash flows.

The general formula for computing a weighted-average cost of capital is

$$k = \frac{k_b B}{V} + \frac{i_e S}{V} \tag{5.14}$$

If there are several types of debt, of equity, or of both, the formula is expanded to

$$k = \sum_j \frac{k_{bj} B_j}{V} + \sum_l \frac{i_{el} S_l}{V} \tag{5.15}$$

where the subscripts $j$ and $l$ refer to the different sources and $V$ is redefined to include all sources.

When the debt consists of bonds issued at par (no discount or premium) and there are no selling expenses or annual expenses, we use Eq. 5.1 and rewrite Eq. 5.14 as

$$k = \frac{i_b(1 - t_m)B}{V} + \frac{i_e S}{V} \tag{5.16}$$

This form shows that the effect of income taxes is to reduce not only $i_b$ but also $k$.

### *Example 5.9*

A medium-size firm has the following financing sources.

| | |
|---|---|
| Bank loans | \$10,000,000 outstanding, at 16% |
| Mortgage | \$30,000,000 outstanding, at 11% |
| Bonds | \$90,000,000 outstanding, at 13% |
| Preferred stock | 1,000,000 shares, market price \$20/share |
| Common stock | 5,000,000 shares, market price \$18/share |

The preferred dividend is fixed at $3.00. The common dividend fluctuates but has averaged $4.00 for the last several years. Obtain the tax-adjusted weighted-average cost of capital when $t_m = 35\%$.

Without additional detailed information, we assume that the loans, mortgage, and bond rates are annual before-tax rates, and that the effects of points, discounts, selling expenses, and annual expenses are negligible. This allows us to use Eq. 5.1. For bank loans, $k_b = (0.16)(0.65) = 0.104$, etc. For preferred stock, $i_e = \$3.00/20.00 = 0.15$, etc. The amount for preferred equity capital is (\$20)(1,000,000 shares) $= \$20 \times 10^6$, etc. We then obtain a weighted average.

| 1<br>Source | 2<br>Amount, $10^6$ | 3<br>% | 4<br>B/T cost | 5<br>A/T cost | 6<br>3×5 |
|---|---|---|---|---|---|
| Bank loans | 10 | 4.2 | 0.16 | 0.1040 | 0.4368 |
| Mortgage | 30 | 12.5 | 0.11 | 0.0715 | 0.8938 |
| Bonds | 90 | 37.5 | 0.13 | 0.0845 | 3.1688 |
| Preferred stock | 20 | 8.3 | — | 0.1500 | 1.2450 |
| Common stock | 90 | 37.5 | — | 0.2222 | 8.3325 |
| Total | $240 | 100.0% | | | 14.0769%<br>≈ 14.08% |

We find $k = 14.08\%$. □

The value of 14.08% in Example 5.9 represents a *historic* value. For decision-making purposes we would require information about future capital costs.

The purpose of computing a cost of capital is to provide a *MARR* against which we can compare investment efficiency. Investment criteria are presented in the next chapter, but we have already used one criterion, the *internal rate of return,* designated *IRR* or $i^*$. The *IRR* is the (not necessarily unique) interest rate that makes the *PV* of a cash flow equal to zero (we have also used the *PV* criterion since Chapter 2). One way to evaluate an investment is to compare its *IRR* against *MARR;* if the *IRR* exceeds *MARR,* the investment is favorable. Another way is to compute *PV*(*MARR*); a positive value implies a favorable investment. The cost of capital to use for *MARR* depends on the type of cash flow.

### 5.4.1 Net Equity Flows

In Chapter 4 we devoted considerable attention to obtaining after-tax cash flows, including situations involving debt financing. When the cash flow computations reflect interest, taxes, and debt repayment, what is left is called *net equity flow. The proper rate for discounting net equity flows, when using* PV, *is* $i_e$, *the equity interest rate.*

### *Example 5.10*

A corporation wishes to establish a new marketing venture that is expected to increase revenues over the next 5 years. The venture requires asset investment

**Table 5.2** ***Net Equity Flows for Example 5.10***

| Item | *n*: 0 | 1 | 2 | 3 | 4 | 5 |
|---|---|---|---|---|---|---|
| 1. Revenues | — | $57,500 | 63,000 | 68,500 | 74,000 | 79,500 |
| 2. Expenses | | | | | | |
| 3. Operating | — | $10,000 | 10,000 | 10,000 | 10,000 | 10,000 |
| 4. Interest | — | $10,000 | 8,000 | 6,000 | 4,000 | 2,000 |
| 5. Depreciation | — | $50,000 | 40,000 | 30,000 | 20,000 | 10,000 |
| 6. Total expenses | — | $70,000 | 58,000 | 46,000 | 34,000 | 22,000 |
| 7. Taxable income | — | −$12,500 | 5,000 | 22,500 | 40,000 | 57,500 |
| 8. Income tax, 60% | — | −$7,500 | 3,000 | 13,500 | 24,000 | 34,500 |
| 9. Profit A/T | — | −$5,000 | 2,000 | 9,000 | 16,000 | 23,000 |
| 10. Loan principal cash flow | +$100,000 | −20,000 | −20,000 | −20,000 | −20,000 | −20,000 |
| 11. Asset purchase | −$150,000 | — | — | — | — | — |
| 12. Net equity flow | −$50,000 | +25,000 | +22,000 | +19,000 | +16,000 | +13,000 |
| 13. Return *on* equity | — | $15,000 | 12,000 | 9,000 | 6,000 | 3,000 |
| 14. Return *of* equity | — | $10,000 | 10,000 | 10,000 | 10,000 | 10,000 |

NOTES: Line 4 is based on loan principal declining $20,000/year.
Line 5 is based on SOYD with $N$ = 5. (For tax purposes we should use MACRS.)
Line 8: negative tax at time 1 is a tax credit.
Line 12 = lines 5 + 9 + 10 + 11.
Line 13 is based on 30% of unrecovered equity.
*PV*(30%) of line 12 = 0.

of $150,000, to be financed with 33.3% equity and 66.7% debt. The equity interest rate is 30%. The loan interest rate is 10%, with the loan principal to be repaid in equal annual installments. Depreciation is SOYD and zero salvage value is expected; no investment tax credit or Sec. 179 expense deduction is to be used. Increased revenues are expected to be

| *n* | 1 | 2 | 3 | 4 | 5 |
|---|---|---|---|---|---|
| Increased Revenues | $57,500 | 63,000 | 68,500 | 74,000 | 79,500 |

Annual operating expenses will increase $10,000. The marginal tax rate is 60% (combined federal, state, and city rate). Evaluate this venture by using net equity flows. The calculations are shown in Table 5.2, following the principles given in Chapter 4. Let us evaluate the net equity flows in line 12 of the table with *PV*(30%).

$$PV(30\%) = -\$50{,}000 + 25{,}000\overset{P/A,30\%,5}{(2.4356)} - 3{,}000\overset{P/G,30\%,5}{(3.6297)}$$

$$= 1 \cong 0$$

The *IRR* for the same cash flow is 30%. Both criteria indicate that the investment breaks even financially. □

### 5.4.2 After-Tax Composite Flows

If net equity flows are to be evaluated by using $i_e$, why has so much been written about the weighted-average cost of capital? The answer is that we can evaluate investments without *explicitly* treating the debt flows (both interest and principal) by using $k$ from Eq. 5.14. The debt financing is treated *implicitly*. This method is particularly appropriate when debt financing, such as bonds, is not identified with individual investments but rather enables the company to engage in a set of investments. It is also suitable for decentralized decision making.

We will apply the method to Example 5.10, with the calculations shown in Table 5.3. We obtain

$$\text{Taxable income} = \text{Revenues} - \text{Operating expenses} - \text{Depreciation} \quad (5.17)$$

Equation 5.17 contains no interest expense term! Then for $n > 0$

$$\text{After-tax debt plus equity flow} = (\text{Taxable income})(1 - t_m) + \text{Depreciation} \quad (5.18)$$

The time 0 flow is simply the total investment, or $150,000 in the example.

Using Eq. 5.16, we obtain the tax-adjusted weighted-average cost of capital.

$$k = (0.10)(1 - 0.6)(66.7\%) + (0.3)(33.3\%) = 12.67\%$$

**Table 5.3** ***After-Tax Composite Flows for Example 5.10***

| Item | *n*: 0 | 1 | 2 | 3 | 4 | 5 |
|---|---|---|---|---|---|---|
| 1. Revenues | — | $57,500 | 63,000 | 68,500 | 74,000 | 79,500 |
| 2. Expenses | | | | | | |
| 3. Operating | — | $10,000 | 10,000 | 10,000 | 10,000 | 10,000 |
| 4. Depreciation | — | $50,000 | 40,000 | 30,000 | 20,000 | 10,000 |
| 5. Total expenses | — | $60,000 | 50,000 | 40,000 | 30,000 | 20,000 |
| 6. Taxable income | — | −$2,500 | 13,000 | 28,500 | 44,000 | 59,500 |
| 7. Income tax, 60% | — | −$1,500 | 7,800 | 17,100 | 26,400 | 35,700 |
| 8. Profit A/T | — | −$1,000 | 5,200 | 11,400 | 17,600 | 23,800 |
| 9. After-tax composite flow | −$150,000 | 49,000 | 45,200 | 41,400 | 37,600 | 33,800 |

NOTES: Line 4 is based on SOYD with $N = 5$.
Line 7: negative tax at time 1 is a tax credit.
Line 9 = lines 4 + 8 for $t > 0$.
*PV*(12.67%) of line 9 = 0.

Evaluating the cash flow in the last line of the table gives

$$\begin{aligned} PV(12.67\%) &= -\$150{,}000 + 49{,}000(\overset{P/A,12.67\%,5}{3.5450}) - 3{,}800(\overset{P/G,12.67\%,5}{6.2514}) \\ &= -1 \cong 0 \end{aligned}$$

Since $PV(12.67\%) = 0$, the $IRR = 12.67\%$.

Again, the venture is expected to break even financially. *Here the after-tax composite flow was evaluated by using the tax-adjusted weighted-average cost of capital.* The conclusion we have reached in this example about the desirability of the investment is the same. The fact that this is a general result has been proved [9]. Notice in Example 5.10 that even though both debt and equity are reduced, as shown in lines 10 and 14 of Table 5.2, the *debt–equity ratio remains constant!* This is necessary to obtain the agreement between decisions made by using $i_e$ and $k$ [1].

## 5.5 SPECIFYING THE WEIGHTED-AVERAGE COST OF CAPITAL

In this section we outline some of the major arguments that lead to specific conclusions regarding the weighted-average cost of capital. Much of the presentation is based on Haley and Schall [5]. We make no attempt to be complete, nor do we try to present all views. We will use the following notation in addition to that previously defined.

| | |
|---|---|
| $D_n$ | depreciation amount at end of year $n$ |
| $DP$ | constant total dividend payments per year |
| $DP_n$ | total divident payments to time 0 stockholders at end of year $n$ |
| $E_{an}$ | adjusted earnings, earnings adjusted for net investment, at end of year $n$ |
| $E_n$ | equity earnings, after interest and taxes, at end of year $n$ |
| $f$ | annual inflation rate |
| $F_n$ | net cash flow at end of year $n$ |
| $FIN_n$ | financing obtained from owners at end of year $n$ |
| $i_{ef}$ | equity interest rate, adjusted to account for inflation |
| $INV_n$ | investment in company assets at end of year $n$ |
| $IP_n$ | interest payment on debt (bonds) at end of year $n$ |
| $k_{bf}$ | after-tax cost of debt capital, adjusted to account for inflation |
| $k_f$ | tax-adjusted weighted-average cost of capital, adjusted to account for inflation |
| $k_v$ | firm's overall capitalization rate |
| $NINV_n$ | net investment, investment adjusted for asset depreciation, at end of year $n$ |

$S_n$ total value of equity capital at end of year $n$

$V_u$ unlevered firm value (with no debt)

$Y_n$ total cash flow, total payments to capital suppliers, at end of year $n$

$Y_{nj}$ payment to capital supplier $j$ at end of year $n$

### 5.5.1 Basic Valuation Forms

We begin with the cost of stock financing under certainty and perfect capital markets. Perfect capital markets imply no transaction and information costs, no restraints on transactions, divisibility of assets, and competitive markets where no party can affect prices. We wish to examine the effect of dividend payments on the value of the firm.

The fundamental valuation equation given by Eq. 5.10 is expressed here in aggregate form.

$$V = \sum_{n=0}^{\infty} DP_n(1 + i_e)^{-n} \tag{5.19}$$

Dividend payments are part of the outflows of a firm. The complete cash flows of a firm at time 0 are summarized by the inflow–outflow equation:

$$F_0 + FIN_0 = DP_0 + INV_0 \tag{5.20}$$

Net operating revenues plus new financing equals dividend payments plus asset investment. For $n > 0$, we value the firm from the view of the time 0 stockholder, and

$$F_n + FIN_n = DP_n + INV_n + (1 + i_e)FIN_{n-1} \tag{5.21}$$

Here the $DP_n$ is to the original time 0 stockholders. Equations 5.20 and 5.21 are referred to as "money pump" equations. From (5.21) we obtain

$$DP_n = F_n + FIN_n - INV_n - (1 + i_e)FIN_{n-1} \tag{5.22}$$

Substituting into (5.19) and simplifying repeatedly gives

$$V = \sum_{n=0}^{\infty} (F_n - INV_n)(1 + i_e)^{-n} \tag{5.23}$$

But

$$F_n - INV_n = DP_n - FIN_n \tag{5.24}$$

The right side of (5.24) represents the cash payments to all suppliers of capital. Equation 5.23 indicates that the *value of the firm is independent of dividend policy*. It depends only on net operating revenues and asset investment.

Another form of the valuation uses

$$E_n = F_n - D_n \tag{5.25}$$

Equity earnings equal net operating revenues less decline in value of assets. It doesn't matter if $D_n$ accurately reflects value decline or not, for $E_n$ will be adjusted accordingly. We cannot simply discount the earnings stream $E_n$, but we must deduct investment to cover asset value decline, to avoid reducing future earnings. We define net investment as

$$NINV_n = INV_n - D_n \tag{5.26}$$

and adjusted earnings

$$\begin{aligned} E_{an} &= E_n - NINV_n \\ &= (F_n - D_n) - (INV_n - D_n) \\ &= F_n - INV_n \end{aligned} \tag{5.27}$$

The last right-hand term in (5.27) is exactly equal to cash flow. Thus,

$$V = \sum_{n=0}^{\infty} E_{an}(1 + i_e)^{-n} \tag{5.28}$$

or

$$V = \sum_{n=0}^{\infty} (E_n - NINV_n)(1 + i_e)^{-n} \tag{5.29}$$

If *EPS* represents equity earnings adjusted for net investment, we obtain Eq. 5.13, presented earlier in the chapter. Equations 5.19, 5.23, and 5.28 represent three equivalent forms of the firm's value.

### 5.5.2 Valuation with Debt and Taxes

We begin this analysis by considering the case of no taxes. As before, we operate in perfect capital markets. In addition, we assume perpetual debt, no retirement of stock, and no new stock or debt. The question is whether investment and financing decisions will raise bond values at the expense of stockholders or vice versa.

We recall the definition of $V$ as the sum of debt and equity capital.

$$V = B + S \tag{5.30}$$

We recognize that $k_b$ and $i_e$ depend on the debt ratio $B/V$. Too high a ratio leads to a risk of default, causing $B$ to decline and $k_b$ to increase. On the other hand, for "safe" values of $B/V$ an increase in $B$ can reduce cash payments to capital suppliers and increase $V$.

Total cash flow is given by

$$Y_n = F_n - INV_n = i_b B + i_e S \tag{5.31}$$

We now employ the *value additivity principle,* which says that the value of a sum of income streams received by a set of investors equals the sum of the values of the individual streams, no matter how the streams are apportioned. In other words,

$$V = \text{Value of} \sum_j \sum_n Y_{nj} = \sum_j \text{Value of} \sum_n Y_{nj} \tag{5.32}$$

without regard to the split of each $Y_n$ and the investors indexed by $j$. Thus, the choice of financing does not affect the firm's value $V$.

When we add taxes to the model, the cash flow becomes

$$Y_n = [(1 - t_m)F_n + t_m D_n - INV_n] + t_m IP_n \tag{5.33}$$

By shifting to more debt financing, we can increase the cash flow and firm value, up to a point. In (5.33) the summation and discounting over time of the term in brackets represents value to stockholders in the absence of debt. This is called the unlevered value $V_u$. If $B = 0$, all of $Y_n$ goes to stockholders. The summation and discounting of the $t_m IP_n$ terms result in a value of $t_m B$. This is so because the value of the $IP_n$ stream is $B$. Since

$$V = S + B = \sum_{n=0}^{\infty} Y_n(1 + k_v) \tag{5.34}$$

where $k_v$ is the firm's overall capitalization rate, we obtain

$$V = V_u + t_m B = S + B \tag{5.35}$$

Figure 5.2 shows how $V$ varies with $B/V$ according to Eq. 5.35.

### 5.5.3 The Firm's Capitalization Rate

We now proceed to examine the effect of debt ratio on the firm's capitalization rate $k_v$. We start with no taxes, no asset investment, and no new debt

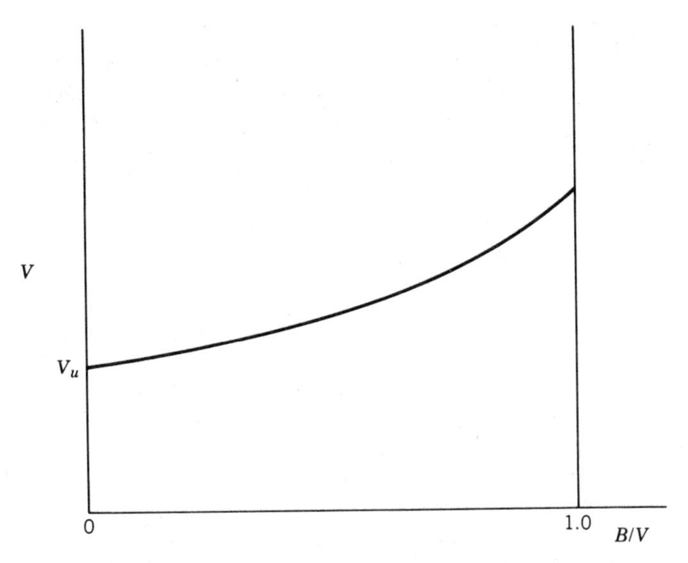


**FIGURE 5.2** Relation of firm value to debt ratio *B/V*.

or equity issues. We expect the firm to generate a constant $F_c = F_n$ forever. This revenue will be paid as dividends and interest.

$$F_c = DP_n + IP_n, \qquad n = 0, 1, 2, \ldots \tag{5.36}$$

We know that

$$k_v = F_c/V \tag{5.37}$$

and

$$i_e = DP/S, \qquad DP = i_e S \tag{5.38}$$

$$i_b = IP/B, \qquad IP = i_b B \tag{5.39}$$

so we obtain

$$k_v = \frac{i_b B}{V} + \frac{i_e S}{V} \tag{5.40}$$

which is consistent with (5.14) when $t_m = 0$.

For a given stream of $Y_n$, the value of the firm does not depend on the debt ratio. Neither does $k_v$. The market capitalizes the streams to stockholders and bondholders to keep $V$ constant. Therefore, other variables must compensate so that $k_v$ in (5.40) remains the same. We have

$$S = V - B$$

$$\frac{S}{V} = 1 - \frac{B}{V} \tag{5.41}$$

Substituting into (5.40) gives

$$k_v = \frac{i_b B}{V} + \left(1 - \frac{B}{V}\right) i_e \tag{5.42}$$

and

$$i_e = \frac{k_v - i_b B/V}{1 - B/V} \tag{5.43}$$

Figure 5.3 shows the results of this derivation. At zero debt $i_e = k_v$. At 100% debt the bondholders are essentially acting as stockholders, so we expect $i_b = k_v$.

When we add taxes we have

$$Y_n = (1 - t_m)F_n + t_m IP_n \tag{5.44}$$

Recall that $INV_n$ and $D_n$ are both zero in this derivation. If $Y_n$ is negative, we assume tax losses can be sold. This is an ideal case, but more realism would lengthen the presentation [5]. Recall the valuation

$$V = V_u + t_m B \tag{5.35}$$

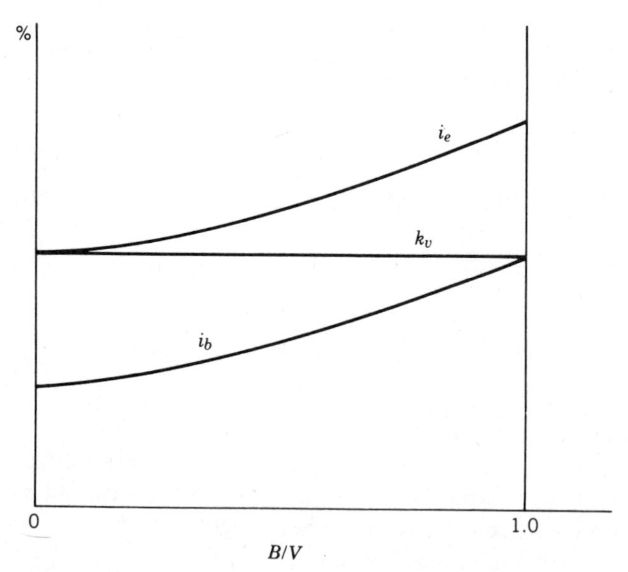


**FIGURE 5.3** Relation of capital costs to debt ratio $B/V$, no taxes.

$$1 = \frac{V_u}{V} + \frac{t_m B}{V} \tag{5.45}$$

or

$$V = \frac{V_u}{1 - t_m B/V} \tag{5.46}$$

From (5.37) and (5.44) we obtain

$$k_v = \frac{(1 - t_m)F_n}{V} + \frac{t_m IP_n}{V} \tag{5.47}$$

Since $i_b = IP_n/B$ we have

$$k_v = \frac{(1 - t_m)F_n}{V} + \frac{t_m i_b B}{V} \tag{5.48}$$

Using (5.46) gives

$$k_v = \frac{(1 - t_m)F_n(1 - t_m B/V)}{V_u} + \frac{t_m i_b B}{V} \tag{5.49}$$

The ratio $(1 - t_m)F_n/V_u$ is the equity interest rate for unlevered firms $i_u$. Thus,

$$k_v = i_u - t_m(i_u - i_b)(B/V) \tag{5.50}$$

This capitalization rate depends on the debt ratio $B/V$! Since $i_u > i_b$ we would generally minimize $k_v$ at a point different from that at which we maximize $V$. As before, we argue that $V$ is maximized at $B/V = 1.0$, as shown in Fig. 5.2.

### 5.5.4 Obtaining a Cutoff Rate

We now examine the situation in which the firm can raise debt and equity capital, in a ratio of existing $B/V$, to finance new investments. We assume that the overall risk of the firm is unchanged and that any new earnings are constant. We wish to maximize the increase in the firm's value.

$$\max(\Delta V - INV_0) \tag{5.51}$$

Letting the symbol ′ represent the situation before undertaking the investment, we have

$$\begin{aligned} V' &= V'_u + t_m B' \\ V &= V_u + t_m B \end{aligned} \tag{5.52}$$

We also assume

$$B = B' + B'' \tag{5.53}$$

The old bonds do not change value as a result of new ones, where $B''$ represents the new bonds issued. For a fixed $INV_0$ we are interested in

$$\Delta V = (V_u - V'_u) + t_m B'' \tag{5.54}$$

We have from (5.44)

$$V'_u = (1 - t_m)F'_c/i'_u$$

and

$$V_u = (1 - t_m)F_c/i_u \tag{5.55}$$

Since $F_c = F'_c + \Delta F_c$, we have

$$\Delta V_u = (1 - t_m)\,\Delta F_c/i'_u \tag{5.56}$$

Using (5.54) gives

$$\Delta V = \frac{(1 - t_m)\,\Delta F_c}{i'_u} + t_m B'' \tag{5.57}$$

Since $B'/V' = B/V$, $B'' = (B/V)\Delta V$, and

$$\Delta V = \frac{(1 - t_m)\,\Delta F_c}{i'_u} + t_m(B/V)\,\Delta V \tag{5.58}$$

Simplifying, we obtain

$$\Delta V = \frac{(1 - t_m)\,\Delta F_c}{i'_u(1 - t_m B/V)} \tag{5.59}$$

We wish to express

$$\Delta V = (1 - t_m)\,\Delta F_c/k \tag{5.60}$$

where $k$ is the cutoff rate for deciding on investments. Combining (5.59) and (5.60) yields, after lengthy manipulation,

$$k = i'_u(1 - t_m B/V)$$

$$k = \frac{i_e S}{V} + \frac{(1 - t_m)i_b B}{V} \tag{5.61}$$

This is the tax-adjusted weighted-average cost of capital! The analysis implies that to increase the firm's value, we should evaluate the net operating cash flows from the investments (the after-tax composite flows), as in Eq. 5.60, and decide on the basis of $\Delta V$, or $PV$ of the investment cash flow.

### 5.5.5 Other Issues

***Market Imperfections.*** The argument for the imperfect market is similar, but the criterion for investment acceptance is the same: use $k$ as defined by Eq. 5.14 or 5.61, as appropriate. The significance of this and the previous derivations is that *we do not need to know* $i_u$! *We simply calculate* $k$ on the basis of the observed or predicted values of $i_e$ and $i_b$.

The "optimal" $B/V$ ratio must consider bankruptcy risks, takeover risks, restricted credit lines, and so forth. For these reasons a firm may wish to take a long-range view of the preferred debt ratio. Recognizing the variability of $F_n$ and $INV_n$, a firm might operate below the optimal debt ratio in order to provide an equity cushion for emergencies.

***Asset-Based Financing.*** Opportunities often come up for special debt financing related to assets, such as buildings and vehicles. We must be extremely careful in this situation. Will the financing change the *functional relationship* between $B/V$ and $k_b$ and $i_e$? If there is no good evidence for a "yes" answer, the special financing should be ignored and the investment simply evaluated by $k$. If there is strong evidence that the financing will allow the firm to operate at a higher debt ratio with the same $k_b$ and $i_e$, a more detailed analysis is needed.

One way to approach such a situation is to split the related investment into an ordinary debt–equity combination and a special debt so that the sum equals the investment. For example, suppose $B/V$ is historically 0.5 and a project is considered for which 75% of the investment would be asset-based debt. One might estimate (but how?) that up to one-third of the debt, or 25% of the investment, does not contribute to raising the $k_b$ and $i_e$ functions. Then the investment can be treated as being financed by 25% special debt and 75% ordinary (50%, 50%) debt–equity combination. To evaluate the project we

Obtain net operating revenue less depreciation.

Subtract interest on one-third of the debt.

Compute taxes and subtract taxes.

Subtract principal repayment on one-third of the debt.

Treat the resulting flow at $k$.

In essence, this is a hybrid of the net equity flows method and the after-tax composite flows method.

***Marginal Cost of Capital.*** The issue of the marginal cost of capital is often clouded by semantics. Three types of situations are often discussed.

1. The difference between historical and future costs of capital.

2. Shifts *along* the $k_b$ and $i_e$ functions caused by changes in debt ratio.
3. Shifts (raising or lowering) of the $k_b$ and $i_e$ functions brought about by a change in the risk class of a firm.

The first situation does not require any discussion. Most analysts recognize the need to predict future cost of capital. The third is somewhat difficult to handle because of the data required, although investment firms have specialists who perform this work.

That leaves us with the second situation, which gets to the question of optimal debt ratio, or optimal capital structure. If a firm is at the optimal debt ratio, there is no incentive to move from this point, except possibly because of transaction costs and minimum issue sizes. If that is the case, the firm should still aim, in the long run, for its optimal debt ratio, and the cost of capital should be based on it. If a firm is *not* at its optimal debt ratio, it should move toward it, all the while using a cost of capital based on the optimal ratio. We should mention that determining the optimal debt ratio is a rather challenging task.

### 5.5.6 Effect of Inflation

Normally the market corrects for expected inflation in the observed values of $i_e$ and $k_b$. Nevertheless, it is of interest to show the relationship between the rates with and without inflation [3].

For the all-equity situation, the previously specified dividend payments $DP_n$ must grow at the inflation rate $f$. Thus,

$$DP_n = (DP_0)(1 + f)^n \tag{5.62}$$

In addition,

$$E_n = S_n - S_{n-1} + (DP_0)(1 + f)^n \tag{5.63}$$

Equation 5.63 says that equity earnings must be sufficient to cover the inflated dividends but also to cover the increased value of the equity. The required equity value at any point in time is

$$S_n = S_0(1 + f)^n \tag{5.64}$$

Using $i_e = DP_0/S_0$ and $i_{ef} = DP_n/S_{n-1}$, we obtain

$$i_{ef} = (1 + i_e)(1 + f) - 1 \tag{5.65}$$

This equation is similar to one in Chapter 2 for the market (nominal) interest rate in terms of the inflation-free (real) interest rate.

A similar derivation for debt gives

$$k_{bf} = (1 + k_b)(1 + f) - 1 \tag{5.66}$$

The tax-adjusted weighted-average cost of capital then becomes

$$k_f = (1 + k)(1 + f) - 1 \tag{5.67}$$

In Eq. 5.67 we assume that the firm acquires sufficient new debt each year to maintain a constant debt ratio.

## 5.6 SUMMARY

Selecting a minimum attractive rate of return, *MARR*, is a complex and controversial task. Most of the theoretical work has been based on too many nice assumptions, which has led practitioners to be skeptical. However, a growing body of theoretical and empirical work has provided support for the *tax-adjusted weighted-average cost of capital, to be applied to after-tax composite flows.* Alternatively, the *equity interest* rate can be applied to *net equity flows,* provided the debt ratio of the investment is the same as the firm's optimal debt ratio. In practice, composite flows with $k$ are easier to apply and less prone to misinterpretation.

Much of the chapter is concerned with details of calculating the effective annual after-tax interest rate on debt. We have provided some approximation methods that we hope will simplify such computations.

Project viability, as evaluated by the investment criteria in the next chapter, is extremely sensitive to *MARR.* The concepts and procedures in this chapter, therefore, should be applied carefully.

A word about notation is in order before we proceed to the next chapter. In this chapter we have used $k$ and other symbols for discounting cash flows. Because of the popularity of the symbol $i$, however, in subsequent chapters we will use it almost exclusively to represent the cost of capital for discounting. One should not forget that $i$ represents either the equity interest rate when used with new equity flows or the tax-adjusted weight-average cost of capital when used with after-tax composite flows.

## REFERENCES


1. BERANEK, W., "The Cost of Capital, Capital Budgeting, and the Maximization of Shareholder Wealth," *Journal of Financial and Quantitative Analysis,* Vol. 10, No. 1, pp. 1–18, March 1975.
2. BUSSEY, L. E., *The Economic Analysis of Industrial Projects,* Prentice–Hall, Englewood Cliffs, N.J., 1978 (see Ch. 6).
3. FLEISCHER, G. A., *Engineering Economy, Capital Allocation Theory,* Brooks/Cole, Monterey, Calif., 1984 (see Ch. 11).
4. GORDON, M. J., and L. I. GOULD, "The Cost of Equity Capital: A Reconsideration," *Journal of Finance,* Vol. 33, No. 3, pp. 849–861, 1978.
5. HALEY, C. W., and L. D. SCHALL, *The Theory of Financial Decisions,* 2nd edition, McGraw–Hill, New York, 1979 (see Chs. 2, 11, 13).
6. MOORE, W. T., and M. L. JOSE, "An Alternative Approach to Computing the After-Tax

Cost of Debt: A Note," *The Engineering Economist,* Vol. 29, No. 2, pp. 150–155, Winter 1984.

7. Oso, J. B., "The Proper Role of the Tax-Adjusted Cost of Capital in Present Value Studies," *The Engineering Economist,* Vol. 24, No. 1, pp. 1–12, 1978.
8. Solomon, E., *The Theory of Financial Management,* Columbia University Press, New York, 1963.
9. Ward, T. L., and W. G. Sullivan, "Equivalence of the Present Worth and Revenue Requirements Methods of Capital Investment Analysis," *AIIE Transactions,* Vol. 13, No. 1, pp. 29–40, 1981.
10. *1989 Federal Tax Course,* Commerce Clearing House, Chicago, Ill., 1988.

## PROBLEMS

**5.1.** Find the effective annual after-tax cost of a 15% bank loan for a company with a marginal tax rate of 40%, if

a. Interest and taxes are payable annually.
b. Interest is payable annually and taxes are payable quarterly.
c. Interest and taxes are payable quarterly, and the 15% represents a *nominal* annual rate.
d. Interest and taxes are payable quarterly, and the 15% represents an *effective* annual rate.

**5.2.** A loan for $20,000 is to be repaid with annual payments over 5 years at 14% interest. The loan is discounted 8 points. With a marginal tax rate of 40%, find the effective annual after-tax interest cost

a. Using the exact method.
b. Using approximation method 1.
c. Using approximation method 2.

**5.3.** Find the *exact* effective annual after-tax interest cost for Example 5.6 for (a) case 1 and (b) case 2.

**5.4.** Find the effective annual after-tax interest cost from accounts payable for the situation in which a company pays its bills 30 days, on average, after the invoice date. By not paying within 10 days after the invoice date, the company gives up a 1.5% discount.

**5.5.** A company has the following liabilities.

| | |
|---|---|
| Bank loans, @ 10% | $300,000 |
| Accounts payable | 700,000 |
| Bonds, @ 8% | 1,000,000 |
| Total | $2,000,000 |

Accounts payable carry no interest charge and are not eligible for any discount. In addition, there are 12,500 shares of common stock outstanding. The current market price is $100 each, and the current earnings per share, after taxes, are $16. With a tax rate of 50%, what is the company's after-tax cost of capital?

**5.6.** Slick Enterprises has the following capital structure.

| | |
|---|---|
| Loans, 10% | $100,000 |
| Loans, 15% | 100,000 |
| Accounts payable | 300,000 |
| Mortgage, 8% | 400,000 |
| Total liabilities | 900,000 |

| | |
|---|---|
| Common stock | 100,000 |
| 10,000 × $10 | 100,000 |
| Total liability and equity | $1,000,000 |

Accounts payable over 30 days old incur a cost of 1½% per month. About half the accounts are older than 30 days. Common stock has a market price of $15 and earnings per share of $3.50 after taxes, of which $1.50 is paid as dividends.

a. Obtain a weighted average cost of capital, assuming a marginal tax rate of 40%.
b. Is your numerical answer in part a a good value to use for minimum attractive rate of return? Explain.

**5.7.** As vice-president of finance of the DEF Corporation, you must determine a cost of capital to be used by the project planning staff for evaluating your projects. Your balance sheet is as follows.

BALANCE SHEET
*March 31, 1989*

| Assets | | Liabilities | |
|---|---|---|---|
| Cash | $250,000 | Bank loan | $200,000 |
| Inventory | 700,000 | Accounts payable | 400,000 |
| Equipment | 200,000 | | |
| Building | 350,000 | | |
| | | Equity | |
| | | Common stock | $600,000 |
| | | Retained earnings | 300,000 |

The bank loan is at 12% per year, compounded quarterly. The large value for accounts payable reflects the fact that DEF Corporation is very slow in paying its bills: on the average, you pay 3 months after the invoice date. In your industry there are no discounts for early payment, so you simply pay as late as possible. (Some of your suppliers have threatened to stop doing business with you.) The DEF Corporation has 30,000 shares of common stock outstanding. The most recent market price was $42 per share. After-tax earnings per share are $4.25, and dividends are $2.00 per share. The corporation is in the 40% tax bracket. What after-tax cost of capital should you use?

**5.8.** Suppose a company had the following capital structure:

| | |
|---|---|
| Bonds @ 8% | $13,000,000 |
| Loans @ 10% | 2,000,000 |
| Loans @ 15% | 2,000,000 |
| Loans @ 20% | 3,000,000 |

Common stock: 1,000,000 shares; market price, $20/share; current earnings per share, $2

a. Determine an average cost of capital. Assume an income tax rate of 40%.
b. Determine the marginal cost of capital.
c. Which do you think would be more appropriate for this company in evaluating a project that required an investment of $1,000,000 and had an after-tax *IRR* of 8½%?

**5.9.** What is the difference between average cost of capital and marginal cost of capital? Which is better to use for establishing *MARR?*

**5.10.** A division of a company expects the revenues from a project to be $1,000,000/year for a period of 6 years. Labor expenses will be \$200,000/year and material expenses \$100,000/year. In addition, a \$600,000 asset must be purchased for the project. This asset will be depreciated by using the alternate MACRS with a 5-year life; the salvage value is expected to be zero. A \$500,000 loan at 10% for 5 years will be used to finance the asset. The loan principal will be repaid in five equal annual installments. The marginal tax rate is 40%. Evaluate the after-tax cash flow from this project by using

a. Net equity flows, with $i_e = 20\%$.

b. Equity plus interest flows, with $k = 12\%$.

***PART TWO***

# DETERMINISTIC ANALYSIS

# 6 Measures of Investment Worth —Single Project

## 6.1 INTRODUCTION

In this chapter we focus primarily on evaluating individual projects by the application of various numerical criteria. In our analysis we treat investment projects as almost the same as securities (stocks, bonds, and so on). Both investment projects and securities normally require initial outlays in orger to provide a later sequence of cash receipts. The major difference is that investment projects are not marketable and securities are. When it is necessary to distinguish between projects and securities in our discussion, it will be done. Otherwise, the assumption can be made that the analyses are identical.

Ten different criteria are discussed in this chapter. The net present value (*PV*) criterion is considered the standard measure of investment, and the other measures are discussed and compared with it. The *PV* criterion and its economic interpretation by means of the project balance concept are discussed in Section 6.2. The internal rate of return (*IRR*) criterion, Solomon's average rate of return (*ARR*) criterion, and modified internal rate of return (*MIRR*) criterion are defined in Section 6.3 and are compared with the *PV* criterion. In Section 6.4 alternative measures, benefit–cost ratios, are presented, and again they are compared with the *PV* criterion. The payback period of an investment is discussed in Section 6.5. Finally, the time-dependent measure of investment worth is developed in Section 6.6. In discussing the various measures, we need to make certain assumptions about the investment settings.

### 6.1.1 Initial Assumptions

In the following investment worth analysis, we assume that the *MARR* (or cost of capital) is known to the decision maker. We also assume a stable, perfect capital market and complete certainty about investment outcomes. In a perfect capital market a firm can raise as much cash as it wants at the going rate of interest, or the firm has sufficient funds to accept all profitable investments. A perfect capital market makes it possible for a firm to invest as much cash as it wants at the market rate of interest. Since the firm may already have undertaken all profitable investments, the market rate of interest is assumed to measure the return on the firm's marginal investment opportunities. Having complete certainty about an investment means that the firm has perfect knowledge of the present and future cash flows associated with the project. Because of this knowledge, the firm finds it unnecessary to make any allowance for uncertainty in project evaluation.

These assumptions describe what might be called the ideal investment situation, quite different from the real-world situation. By setting aside certain complications, however, these assumptions will allow us to introduce the topic of investment analysis at a much simpler level than we otherwise could. In later chapters these assumptions will be removed and the analysis extended to more realistic situations, in which none of these assumptions is fully satisfied.

### 6.1.2 Notation

To discuss the various evaluation criteria, we will use the following common notation for cash flow representation.

- $n$ time, measured in discrete compounding periods
- $i$ opportunity interest rate (*MARR*), or market interest rate
- $C_0$ initial investment at time 0, a positive amount
- $b_n$ revenue at end of period $n$, $b_n \geq 0$
- $c_n$ expense at the end of period $n$, $c_n \geq 0$
- $N$ project life
- $F_n$ net cash flow at the end of period $n$ ($F_n = b_n - c_n$; if $b_n \geq c_n$, then $F_n \geq 0$; if $b_n < c_n$, then $F_n < 0$)

Figure 6.1 illustrates this notation with a cash flow diagram. Additional notation pertaining to a specific criterion will be defined later as necessary. It must be emphasized that all cash flows represent the *cash flows after taxes.*

## 6.2 THE NET PRESENT VALUE CRITERION

We will use the concept of equivalence to develop the net present value (*PV*) criterion for evaluating investment worth. The future value and annual equivalent criteria are variations of the *PV* criterion found by converting the *PV* into either the future value or the annual equivalent by using the same interest rate. In this section we define and discuss the interpretation of these three criteria.

### 6.2.1 Mathematical Definition

***The* PV *Criterion.*** Consider a project that will generate cash receipts of $b_n$ at the end of each period $n$. The present value of cash receipts over the project life, $B$, is expressed by

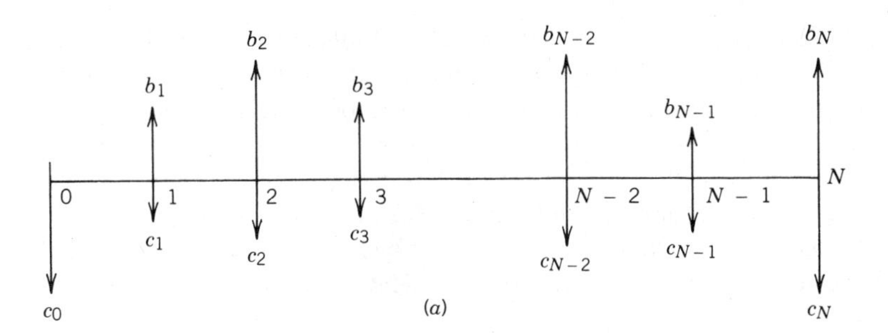


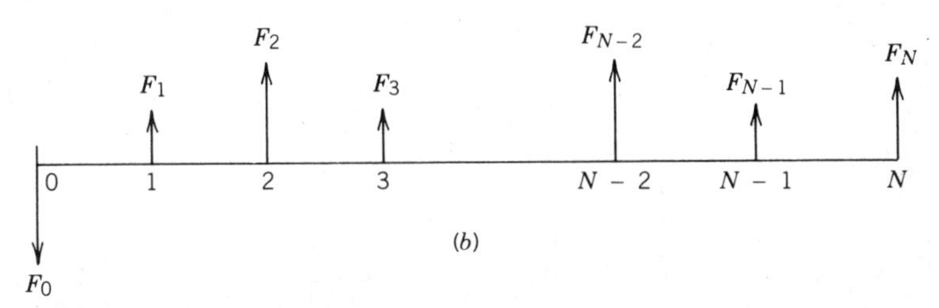


**FIGURE 6.1.** Notation conventions. (a) Gross cash flow. (b) Net cash flow.

$$B = \sum_{n=0}^{N} \frac{b_n}{(1 + i)^n} \tag{6.1}$$

Assume that the cash expenses (including the initial outlay associated with the project) at the end of each period are $c_n$. The present value expression of cash expenses, $C$, is

$$C = \sum_{n=0}^{N} \frac{c_n}{(1 + i)^n} \tag{6.2}$$

Then the *PV* of the project [denoted by $PV(i)$] is defined by the difference between $B$ and $C$; that is,

$$PV(i) = \sum_{n=0}^{N} \frac{b_n - c_n}{(1 + i)^n} = \sum_{n=0}^{N} \frac{F_n}{(1 + i)^n} \tag{6.3a}$$

The $F_n$ will be positive if the corresponding period has a net cash inflow and negative if there is a net cash outflow. The foregoing computation of the *PV* is based on a rate of interest that remains constant over time. The *PV* could be computed with different rates of interest over time, in which case we would label the $n$th period's rate of interest as $i_n$. The *PV* expression is then

$$PV(i_n, n) = F_0 + \frac{F_1}{1 + i_1} + \frac{F_2}{(1 + i_1)(1 + i_2)} + \cdots \tag{6.3b}$$

For simplicity, we assume here a single rate of interest in computing the *PV*. We further assume compounding at discrete points in time. A continuous compounding process or continuous cash flows can be handled according to the procedures outlined in Chapter 2.

A positive *PV* for a project represents a positive surplus, and we should accept the project if sufficient funds are available for it. A project with a negative *PV* should be rejected, because we could do better by investing in other projects at the opportunity rate or outside the market. The decision rule expressed simply is

If $PV(i) > 0$, accept.
If $PV(i) = 0$, remain indifferent.
If $PV(i) < 0$, reject.

***Future Value Criterion.*** As a variation of the *PV* criterion, the future value (*FV*) criterion measures the economic value of a project at the end of the project's life, *N*. Converting the project cash flows into a single payment concentrated at period *N* produces a cash flow equal to *FV*.

$$FV(i) = \sum_{n=0}^{N} F_n(1 + i)^{N-n}$$

$$= PV(i)(1 + i)^N \qquad (6.4)$$

From another view, if we borrowed and lent at *i*, operated the project, and left all extra funds to accumulate at *i*, we would have a value equal to $FV(i)$ at the end of period *N*. If this value is positive, the project is acceptable. If it is negative, the project should be rejected. As expected, the decision rule for the *FV* criterion is the same as that for the *PV* criterion.

If $FV(i) > 0$, accept.
If $FV(i) = 0$, remain indifferent.
If $FV(i) < 0$, reject.

***Annual Equivalent Criterion.*** The annual equivalent (*AE*) criterion is another basis for measuring investment worth that has characteristics similar to those of the *PV* criterion. This similarity is evident when we consider that any cash flow can be converted into a series of equal annual payments by first finding the *PV* for the original series and then multiplying the *PV* by the capital recovery factor.

$$AE(i) = PV(i)\left[\frac{i(1 + i)^N}{(1 + i)^N - 1}\right] = PV(i)(A/P, i, N) \qquad (6.5)$$

Because the factor $(A/P, i, N)$ is positive for $-1 < i < \infty$, the *AE* criterion should provide a consistent basis for evaluating an investment project as the previous criteria have done.

> If $AE(i) > 0$, accept.
> If $AE(i) = 0$, remain indifferent.
> If $AE(i) < 0$, reject.

## *Example 6.1*

This example will serve to illustrate the use of the *PV* criterion. Consider a project that requires a \$1,000 initial investment with the following patterns of cash flow.

| | End of Period $n$ | | | | | |
|---|---|---|---|---|---|---|
| Cash Flow | 0 | 1 | 2 | 3 | 4 | 5 |
| Receipt ($b_n$) | \$0 | 500 | 500 | 500 | 500 | 500 |
| Expense ($c_n$) | \$1,000 | 100 | 140 | 180 | 220 | 260 |
| Net Flow ($F_n$) | −\$1,000 | 400 | 360 | 320 | 280 | 240 |

The cash flow diagram is shown in Figure 6.2. Assume the firm's *MARR* is 10%. Substituting $F_n$ values into Eq. 6.3 and varying $i$ values ($0 \le i \le 40\%$), we obtain Table 6.1 and Figure 6.3. We then find that the project's *PV* decreases monoto-

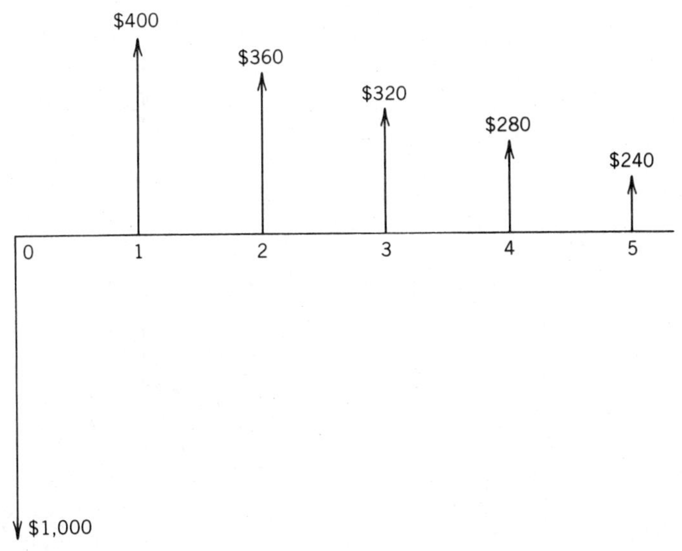


**FIGURE 6.2. Cash flow diagram for Example 6.1.**

**Table 6.1** ***Net Present Values PV(i) at Varying Interest Rate i, Example 6.1***

| $i$ (%) | $PV(i)$ | $i$ (%) | $PV(i)$ |
|---|---|---|---|
| 0 | $600.00 | 21% | −$19.5 |
| 1 | 556.96 | 22 | −38.84 |
| 2 | 515.77 | 23 | −57.30 |
| 3 | 476.33 | 24 | −75.15 |
| 4 | 438.54 | 25 | −92.43 |
| 5 | 402.31 | 26 | −109.15 |
| 6 | 367.56 | 27 | −125.34 |
| 7 | 334.21 | 28 | −141.03 |
| 8 | 302.19 | 29 | −156.23 |
| 9 | 271.42 | 30 | −170.96 |
| 10 | 241.84 | 31 | −185.25 |
| 11 | 213.40 | 32 | −199.11 |
| 12 | 186.03 | 33 | −212.56 |
| 13 | 159.68 | 34 | −225.61 |
| 14 | 134.31 | 35 | −238.29 |
| 15 | 109.86 | 36 | −250.60 |
| 16 | 86.29 | 37 | −262.56 |
| 17 | 63.55 | 38 | −274.19 |
| 18 | 41.62 | 39 | −285.49 |
| 19 | 20.45 | 40 | −296.48 |
| 20 | 0 | | |

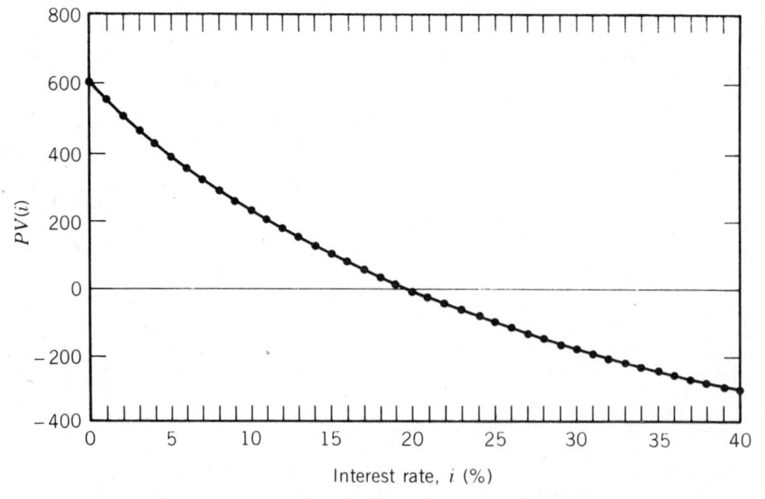


**FIGURE 6.3.** Plot of $PV(i)$ as a function of $i$, Example 6.1.

nically with the firm's $i$. The project has a positive $PV$ if the firm's interest rate ($MARR$) is below 20% and a negative $PV$ if the $MARR$ is above 20%. At $i = 10\%$, the $PV$ (the equivalent present value to the firm of the total surplus) is \$241.84.

Using Eqs. 6.4 and 6.5, we find

$$FV(10\%) = \$241.84(F/P, 10\%, 5) = \$389.49$$

$$AE(10\%) = \$241.84(A/P, 10\%, 5) = \$58.0$$

Since both $FV(10\%)$ and $AE(10\%)$ are positive, the project is considered viable under these criteria. □

### 6.2.2 Economic Interpretation Through Project Balance

An alternative way to interpret the economic significance of these criteria is through the project balance concept. In this section we define the project balance concept and then explain how these criteria are related to the terminal project balance.

***Project Balance Concept.*** The *project balance* describes the net equivalent amount of dollars tied up in or committed to the project at each point in time over the life of the project. We will use $PB(i)_n$ to denote the project balance at the end of period $n$ computed at the opportunity cost rate ($MARR$) of $i$. We will assume that the cost of having money tied up in the project is not incurred unless it is committed for the entire period. To show how the $PB(i)_n$ are computed, we consider the project described in Example 6.1. (See Figure 6.2.)

The project balance at the present time ($n = 0$) is just the investment itself.

$$PB(10\%)_0 = -\$1{,}000$$

At $n = 1$, the firm has an accumulated commitment of \$1,100, which consists of the initial investment and the associated cost of having the initial investment tied up in the project for one period. However, the project returns \$400 at $n = 1$. This reduces the firm's investment commitment to \$700, so the project balance at $n = 1$ is

$$PB(10\%)_1 = -\$1{,}000(1 + 0.1) + \$400 = -\$700$$

This amount becomes the net amount committed to that project at the beginning of period 2. The project balance at the end of period 2 is

$$PB(10\%)_2 = -\$700(1 + 0.1) + \$360 = -\$410$$

This represents the cost of having \$700 committed at the beginning of the second year along with the receipt of \$360 at the end of that year.

We compute the remaining project balances similarly.

$$PB(10\%)_3 = -\$410(1.1) + \$320 = -\$131.00$$
$$PB(10\%)_4 = -\$131(1.1) + \$280 = \$135.90$$
$$PB(10\%)_5 = \$135.90(1.1) + \$240 = \$389.49$$

Notice that the firm fully recovers its initial investment and opportunity cost at the end of period 4 and has a profit of \$135.90. Assuming that the firm can reinvest this amount at the same interest rate ($i = 10\%$) in other projects or outside the market, the project balance grows to \$389.49 with the receipt of \$240 at the end of period 5. The project is then terminated with a net profit of \$389.49.

If we compute the present value equivalent of this net profit at time 0, we obtain

$$PV(10\%) = \$389.49(P/F, 10\%, 5) = \$241.84$$

The result is the same as that obtained when we directly compute the present value of the project at $i = 10\%$. Table 6.2 summarizes these computational results.

***Mathematical Derivation.*** Defining the project balance mathematically based on the previous example yields the recursive relationship

$$PB(i)_n = (1 + i)PB(i)_{n-1} + F_n \tag{6.6}$$

where $PB(i)_0 = F_0$ and $n = 0, 1, 2, \ldots, N$.

We can develop an alternative expression for the project balance from Eq. 6.6 by making substitutions as follows.

$$\begin{aligned} PB(i)_0 &= F_0 \\ PB(i)_1 &= (1 + i)F_0 + F_1 \\ PB(i)_2 &= (1 + i)[(1 + i)F_0 + F_1] + F_2 \\ &= F_0(1 + i)^2 + F_1(1 + i) + F_2 \end{aligned}$$

so that at any period $n$

$$PB(i)_n = F_0(1 + i)^n + F_1(1 + i)^{n-1} + \cdots + F_n \tag{6.7}$$

The terminal project balance is then expressed by

$$\begin{aligned} PB(i)_N &= F_0(1 + i)^N + F_1(1 + i)^{N-1} + \cdots + F_N \\ &= \sum_{n=0}^{N} F_n(1 + i)^{N-n} \\ &= FV(i) \end{aligned} \tag{6.8}$$

Note that $PB(i)_N$ is the future value of the project.

**Table 6.2** ***Project Balance Computations for the Project in Example 6.1***

| Item | $n$: 0 | 1 | 2 | 3 | 4 | 5 |
|---|---|---|---|---|---|---|
| Beginning project balance, $PB(i)_{n-1}$ | \$0 | −1,000 | −700 | −410 | −131 | +135.90 |
| Interest owed, $i[PB(i)_{n-1}]$ | \$0 | −100 | −70 | −41 | −13.10 | 13.59 |
| Cash receipt, $F_n$ | −1,000 | 400 | 360 | 320 | 280 | 240 |
| Ending project balance, $PB(i)_n$ | −\$1,000 | −\$700 | −\$410 | −\$131 | \$135.90 | \$389.49 $PB(i)_N$ |

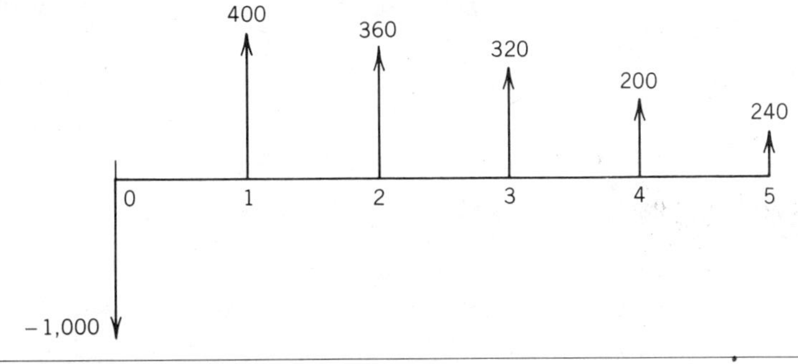


$$PV(10\%) = PB(10\%)_5(1 + 0.1)^{-5} = \$389.49(P/F, 10\%, 5) = \$241.84$$

$$PV(10\%) = -\$1{,}000 + 400(1.1)^{-1} + 360(1.1)^{-2} + 320(1.1)^{-3} + 280(1.1)^{-4} + 240(1.1)^{-5} = \$241.84$$

***Economic Interpretation.*** If $PB(i)_N > 0$, we can say that the firm recovers the initial investment plus any interest owed, with a profit at the end of the project. If $PB(i)_N = 0$, the firm recovers only the initial investment plus interest owed and breaks even. If $PB(i)_N < 0$, the firm ends up with a loss by not being able to recover even the initial investment and interest owed. Naturally, the firm should accept a project only if $PB(i)_N > 0$. The present equivalent amount of this terminal profit is

$$PV(i) = \frac{PB(i)_N}{(1 + i)^N} = \frac{FV(i)_N}{(1 + i)^N} \tag{6.9}$$

The factor $1/(1 + i)^N$ is always positive for $-1 < i < \infty$. This implies that the $PV(i)$ will be positive if and only if $PB(i)_N > 0$ [14].

Now the meaning of the $PV$ criterion should be clear; accepting a project with $PV(i) > 0$ is equivalent to accepting a project with $PB(i)_N > 0$. Because the $PV$ and the future value are measures of equivalence that differ only in the times at which they are stated, they should provide identical results. The analysis and discussion should also make clear why we consider $PV$ as the baseline, or correct, criterion to use in a stable, perfect capital market with complete certainty.

## 6.3 INTERNAL RATE-OF-RETURN CRITERION

### 6.3.1 Definition of IRR

***Mathematical Definition.*** The internal rate of return (*IRR*) is another time-discounted measure of investment worth similar to the *PV* criterion. The *IRR* of a project is defined as the rate of interest that equates the *PV* of the entire series of cash flows to zero. The project's *IRR*, $i^*$, is defined mathematically by

$$PV(i^*) = \sum_{n=0}^{N} \frac{F_n}{(1 + i^*)^n} = 0 \tag{6.10}$$

Multiplying both sides of Eq. 6.10 by $(1 + i^*)^N$, we obtain

$$\begin{aligned} PV(i^*)(1 + i^*)^N &= \sum_{n=0}^{N} F_n(1 + i^*)^{N-n} \\ &= FV(i^*) = 0 \end{aligned} \tag{6.11}$$

The left-hand side of Eq. 6.11 is, by definition, the future value (terminal project balance) of the project.

If we multiply both sides of Eq. 6.10 by the capital recovery factor, we obtain the relationship $AE(i^*) = 0$ (see Eq. 6.9). Alternatively, the *IRR* of a project may be defined as the rate of interest that equates the future value, terminal project balance, and annual equivalent value of the entire series of cash flows to zero.

$$PV(i^*) = FV(i^*) = PB(i^*)_N = AE(i^*) = 0 \tag{6.12}$$

***Computational Methods.*** Note that Eq. 6.11 is a polynomial function of $i^*$. A direct solution for such a function is not generally possible except for projects with a life of four periods or fewer. Instead, two approximation techniques are in general use, one using iterative procedures (a trial-and-error approach) and the other using Newton's approximation to the solution of a polynomial.

An iterative procedure requires an initial guess. To approximate the *IRR,* we calculate the *PV* for a certain interest rate (initial guess). If this *PV* is not zero, another interest rate is tried. A negative *PV* usually indicates that the choice is too high. We continue approximating until we reach the two bounds that contain the answer. We then interpolate to find the closest approximation to the *IRR*(*s*).

The Newton approximation to a polynomial $f(X) = 0$ is made by starting with an arbitrary approximation of $X$ and forming successive approximations by the formula

$$X_{j+1} = X_j - \frac{f(X_j)}{f'(X_j)} \qquad (6.13)$$

where $f'(X_j)$ is the first derivative of the polynomial evaluated at $X_j$. *The process is continued until we observe* $X_j \cong X_{j-1}$.

### Example 6.2

Consider a project with cash flows −\$100, 50, and 84 at the end of periods 0, 1, and 2, respectively. The present value expression for this project is

$$PV(i) = -\$100 + \frac{50}{1+i} + \frac{84}{(1+i)^2}$$

Let $X = 1/(1+i)$. Our polynomial, the present value function, is then

$$f(X) = -100 + 50X + 84X^2$$

The derivative of this polynomial is

$$f'(X) = 50 + 168X$$

Suppose the first approximation we make is

$$X_1 = 0.8696 \qquad (i = 0.15)$$

The second approximation is

$$X_2 = 0.8696 - \frac{-100 + 50(0.8696) + 84(0.8696)^2}{50 + 168(0.8696)}$$
$$= 0.8339$$

The third approximation is

$$X_3 = 0.8339 - \frac{-100 + 50(0.8339) + 84(0.8339)^2}{50 + 168(0.8339)}$$
$$= 0.8333$$

Further iterations indicate that $X = 0.8333$ or $i^* = 20\%$. (With any approximation we are limited by rounding, so when we get the same answer twice in the sequence of approximations, we stop. □

Although the calculations in Newton's method are relatively simple, they are time-consuming if many iterations are required. The use of a computer is eventually necessary. (When we program the computer, it is wise to set tolerance limits on the degree of accuracy required to avoid unnecessary iterations.)

***Uniqueness of* i*.** The existence of a unique *IRR* is of special interest in applying the *IRR* investment worth criterion. Consider a project with cash flows of −$10, $47, −$72, and $36 at the end of periods 0, 1, 2, and 3, respectively. Applying Eq. 6.10 and solving for $i$ gives us three roots: 20%, 50%, and 100%. This really should not surprise us, since Eq. 6.10 is a third-degree polynomial for the project. Here the plot of *PV* as a function of interest rate crosses the $i$ axis several times, as illustrated in Figure 6.4. As we will see in later sections, multiple *IRR*s hinder the application of the *IRR* criterion, and we do not recommend the *IRR* criterion in such cases. In this section we will focus on the problem of whether a unique *IRR* for a project can be predicted by the cash flow stream.

One way to predict an upper limit on the number of positive roots of a polynomial is to apply Descartes' rule of signs.

> ***Descartes' Rule.*** The number of real positive roots of an $n$th-degree polynominal with real coefficients is never greater than the number of sign changes in the sequence of the coefficients.

Letting $X = 1/(1 + i)$, we can write Eq. 6.10 as

$$F_0 + F_1X + F_2X^2 + \cdots + F_NX^N = 0 \tag{6.14}$$

Thus, we need examine only the sign changes in $F_n$ to apply the rule. For example, if the project has outflows followed by inflows, there is only one sign change and hence at most one real positive root.

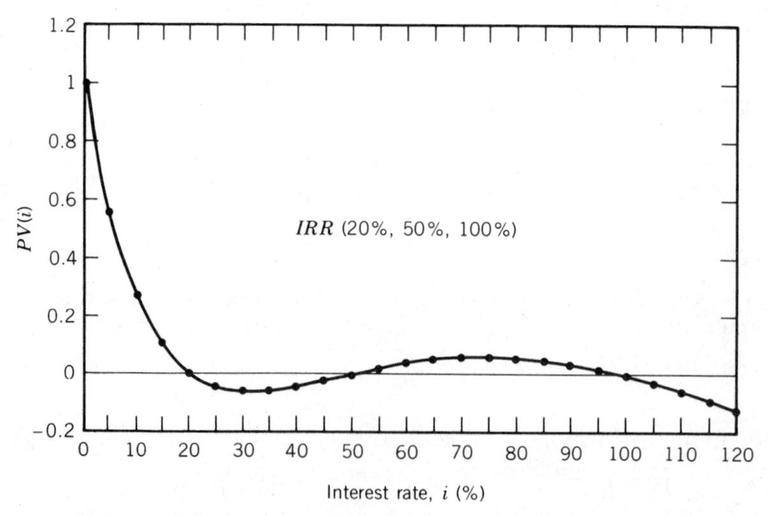


**FIGURE 6.4. Multiple internal rates of return.**

The Norstrom criterion [5] provides a more discriminating condition for the uniqueness of the root in the interval ($0 < i^* < \infty$).

> ***Norstrom Criterion.*** Consider a cash flow series $F_0, F_1, F_2, ..., F_N$. Form the auxiliary series $S_n = \sum_{j=0}^{n} F_j$, $n = 0, 1, ..., N$. If the series $S_n$ starts negative and changes sign only once, there exists a unique positive real root. (6.15)

Additional criteria for the uniqueness of roots do exist, but they are rather tedious to apply and will not be discussed here. Bernhard [5] discusses these additional criteria and provides another general method for detecting the uniqueness of *IRR*.

### *Example 6.3*

To illustrate the use of both Descartes' rule and the Norstrom criterion, consider the following pattern of cash flows.

| $n$ | 0 | 1 | 2 |
|---|---|---|---|
| $F_n$ | −\$100 | \$140 | −\$10 |

Descartes' rule implies that the maximum number of positive real roots is less than or equal to two, which indicates that there may be multiple roots. There are two sign changes in $F_n(-, +, -)$.

To apply the Norstrom criterion, we first compute the cumulative cash flow stream, $S_n$.

$$S_0 = F_0 = -\$100$$

$$S_1 = F_0 + F_1 = -\$100 + \$140 = \$40$$

$$S_2 = F_0 + F_1 + F_2 = \$40 - \$10 = \$30$$

The criterion indicates a unique positive, real root for the problem because there is only one sign change in the $S_n$ series $(-, +, +)$. In fact, the project has a unique *IRR* at $i^* = 32.45\%$. □

### 6.3.2 Classification of Investment Projects

In discussing the *IRR* criterion, we need to distinguish between simple and nonsimple investments. Investment projects are further classified as pure or mixed investments.

***Simple versus Nonsimple.*** A *simple* investment is defined as one in which there is only one sign change in the net cash flow ($F_n$). A *nonsimple* investment is one whose net cash outflows are not restricted to the initial period but are interspersed with net cash inflows throughout the life of the project. In other

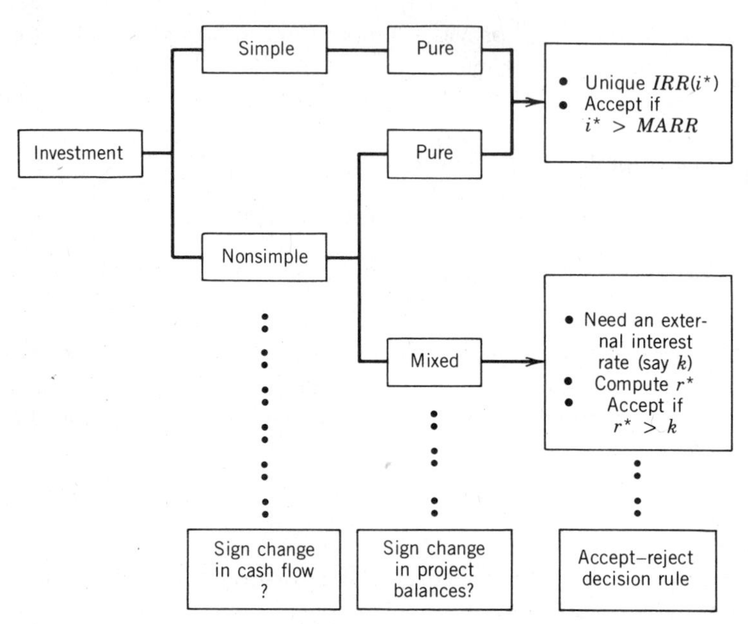


**FIGURE 6.5.** **Classification of investment projects.**

words, when there is more than one sign change in the net cash flow, the project is called a nonsimple project.

***Pure versus Mixed.*** A *pure* investment is defined as an investment whose project balances computed at the project's *IRR*, $PB(i^*)_n$, are either zero or negative throughout the life of the project (with $F_0 < 0$). The implication of nonpositivity of $PB(i^*)_n$ for all values of $n$ is that the firm has committed (or "lent") funds in the amount of $PB(i^*)_n$ dollars to the project for time $n$ to time $n + 1$. In other words, the firm does not "borrow" from the project at any time during the life of the project.

A *mixed* investment, in contrast, is defined as any investment for which $PB(i^*)_n > 0$ for some values of $n$ and $PB(i^*)_n \leq 0$ for the remaining values of $n$. These sign changes in $PB(i^*)_n$ indicate that at some times during the project's life $[PB(i^*)_n < 0]$ the firm acts as an "investor" in the project and at other times $[PB(i^*)_n > 0]$ the firm acts as a "borrower" from the project.

***Classification by $i_{min}$.*** An alternative way of distinguishing between pure and mixed investments is to compute the value of $i_{\min}$, the smallest interest rate that makes $PB(i)_n \leq 0$ for $n = 0, 1, 2, \ldots, N - 1$. Then we evaluate the sign of $PB(i_{\min})_N$, the terminal project balance. If $PB(i_{\min})_N \geq 0$, the project is a pure investment. If $PB(i_{\min})_N < 0$, the project is a mixed investment.

If $PB(i_{\min})_N > 0$, we can find some *IRR*, $i^* > i_{\min}$, that will set $PB(i^*)_N$ to zero. Then use of a higher interest rate will simply magnify the negativity of $PB(i)_n$. Thus, the condition of $i^* \geq i_{\min}$ will ensure the nonpositivity of $PB(i^*)_n$ for $0 \leq n \leq N - 1$. This is the definition of a pure investment.

If $PB(i_{\min})_N < 0$, we can expect that $i^* < i_{\min}$, which will set $PB(i^*)_N$ to zero. Because $i_{min}$ is the minimum rate at which the nonpositivity condition $[PB(i_{min}) \leq 0]$ satisfies $0 \leq n \leq N - 1$, we know that $PB(i^*)_n$ is not always zero or negative for $0 \leq n \leq N - 1$. This implies that the project is a mixed investment.

Figure 6.5 illustrates the final classification scheme that provides the basis for the analysis of investments under the *IRR* criterion. Note that simple investments are always classified as pure investments. (See the proof in Bussey [6].) As we will see, the phenomenon of multiple *IRR*s occurs only in the situation of a mixed investment. Although a simple investment is always a pure investment, a pure investment is not necessarily a simple investment, as we will see in Example 6.4.

### Example 6.4

We will illustrate the distinction between pure and mixed investments with numerical examples. Consider the following four projects with known $i^*$ values.

| End of Period | Project | | | |
|---|---|---|---|---|
| $n$ | A | B | C | D |
| 0 | −\$100 | −\$100 | −\$100 | −\$100 |
| 1 | −100 | 140 | 50 | 470 |
| 2 | 200 | −10 | −50 | −720 |
| 3 | 200 | | 200 | 360 |
| *IRR* | $i^* = 41.42\%$ | $i^* = 32.45\%$ | $i^* = 29.95\%$ | $i^* = 20\%, 50\%, 100\%$ |

Table 6.3 summarizes the project balances from these projects at their respective *IRR*s. Project A is the only simple project; the rest are nonsimple. Projects A and C are pure investments, whereas projects B and D are mixed investments. As seen in project B, the existence of a unique *IRR* is a necessary but not a sufficient condition for a pure investment.

**Table 6.3** ***Project Balances, Example 6.4***

| | | | End of Period $n$ | | | |
|---|---|---|---|---|---|---|
| Project | *IRR* | | 0 | 1 | 2 | 3 |
| | | $F_n$ | −\$100 | −100 | 200 | 200 |
| A | 41.42% | $PB(i^*)_n$ | −\$100 | −241.42 | −141.42 | 0 |
| | | $F_n$ | −\$100 | 140 | −10 | |
| B | 32.45% | $PB(i^*)_n$ | −\$100 | 7.55 | 0 | |
| | | $F_n$ | −\$100 | 50 | −50 | 200 |
| C | 29.95% | $PB(i^*)_n$ | −\$100 | −79.95 | −153.90 | 0 |
| | | $F_n$ | −\$100 | 470 | −720 | 360 |
| | 20% | $PB(20\%)$ | −\$100 | 350 | −300 | 0 |
| | 50% | $PB(50\%)$ | −\$100 | 320 | −240 | 0 |
| D | 100% | $PB(100\%)$ | −\$100 | 270 | −180 | 0 |

In distinguishing pure and mixed investments, we could use the $i_{min}$ test. We will show how this is done for project D. Since $N = 3$, we need to consider $PB(i)_0$, $PB(i)_1$, and $PB(i)_2$.

$$PB(i)_0 = -100$$
$$PB(i)_1 = PB(i)_0(1 + i) + 470 = -100i + 370$$
$$PB(i)_2 = PB(i)_1(1 + i) - 720 = -100i^2 + 270i - 350$$

Since $PB(i)_0 < 0$, we find the smallest value of $i$ that makes both $PB(i)_1$ and $PB(i)_2$ nonpositive. The minimum value is 370%. Now we evaluate $PB(i_{min})_3$ to find

$$PB(370\%)_3 = -720(4.70) + 360 = -\$3024 < 0$$

Since $PB(i_{min})_3 < 0$, project D is a mixed investment. □

### 6.3.3 IRR and Pure Investments

According to the *IRR* criterion, a pure investment should be accepted if its *IRR* is above the *MARR* (or cost of capital) to the firm. We will show why this decision rule can produce an accept–reject decision consistent with the *PV* criterion.

Recall that pure investments have the following characteristics.

1. Net investment throughout the life of the project.
2. Existence of unique $i^*$.
3. $PB(i^*)_n \leq 0$ for $0 \leq n \leq N - 1$, and $PB(i^*)_N = 0$.
4. $PB(i)_N\left[\frac{1}{(1 + i)^N}\right] = PV(i)$

and if $i = i^*$,

$$PB(i)_N = 0 \rightarrow PV(i) = 0$$

We will first consider computing $PB(i)_N$ with $i > i^*$. Here $i$ is the *MARR* (or cost of capital) to the firm. Since $PB(i^*)_n \leq 0$ for $0 \leq n \leq N - 1$ and $PB(i^*)_N = 0$, the effect of a higher compounding rate is to magnify the negativity of these project balances. This implies that $PB(i)_N < PB(i^*)_N = 0$. From Eq. 6.9, this also implies that $PV(i) < 0$. If $i = i^*$, then $PB(i)_N = PB(i^*)_N = 0$ so that $PV(i) = 0$. If $i < i^*$, then $PB(i)_N > 0$, indicating that $PV(i) > 0$. Hence we accept the investment. This proves the equivalence of the *PV* and *IRR* criteria for accept–reject decisions concerning simple investments. These relationships are illustrated in Figure 6.6.

| |
|---|
| If $i < i^*$, accept. |
| If $i = i^*$, remain indifferent. |
| If $i > i^*$, reject. |

When a firm makes a pure investment, it has funds committed to the project over the life of the project and at no time takes a loan from the project. Only in such a

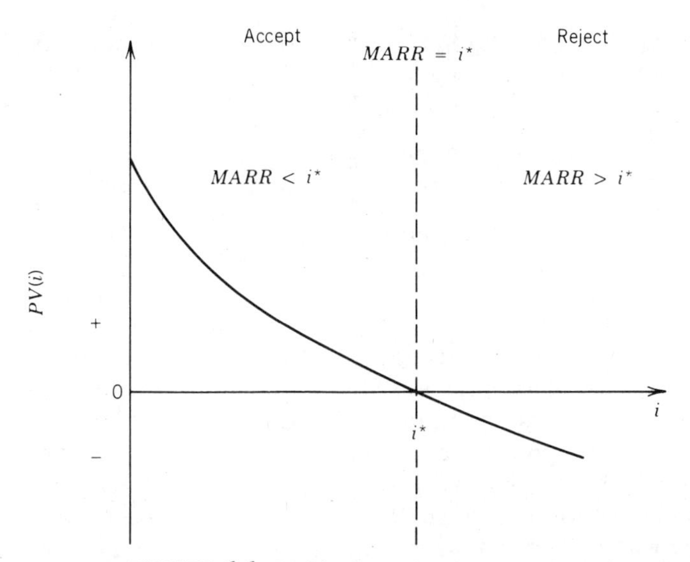


**FIGURE 6.6.** ***PV(i)*** **of simple investment as a function of** ***i*.**

situation is a rate of return concept *internal* to the project. Then the *IRR* can be viewed as the interest rate *earned* on the committed project balance (unrecovered balance, or negative project balance) of an investment, *not* the interest earned on the initial investment. The reader should keep this in mind, since it is a point not generally understood by many practitioners.

### *Example 6.5*

Consider the project described in Example 6.1. (Note that the project is a simple and pure investment.) The project was acceptable at $i = 10\%$ by the *PV* criterion. We find that the *IRR* of this project is 20% by solving for $i^*$ in Eq. 6.10.

$$PV(i^*) = -\$1,000 + \frac{\$400}{1 + i^*} + \frac{\$360}{(1 + i^*)^2} + \frac{\$320}{(1 + i^*)^3} + \frac{\$280}{(1 + i^*)^4} + \frac{\$240}{(1 + i^*)^5} = 0$$

Since $i^* > 10\%$, the project should be acceptable. The economic interpretation of the 20% is that the investment under consideration brings in enough cash to pay for itself in 5 years and also to provide the firm with a return of 20% on its invested capital over the project life.

Expressed another way, suppose that a firm obtains all its capital by borrowing from a bank at the interest rate of exactly 20%. If the firm invests in the project and uses the cash flow generated by the investment to pay off the principal and interest on the bank loan, the firm should come out exactly even on the transaction. If the firm can borrow the funds at a rate lower than 20%, the project should be profitable. If the borrowing interest rate is greater than 20%, acceptance of the project would result in losses. This break-even characteristic makes the *IRR* a popular criterion among many practitioners. □

### 6.3.4 *IRR* and Mixed Investments

Recall that the mixed investments have the following characteristics.

1. More than one sign change in cash flow.
2. Possibility of multiple rates of return.
3. Mixed signs in $PB(i^*)_n$.

The difficulty in mixed investments is determining which rate to use for the acceptance test, if any. The mixed signs in $PB(i^*)_n$ indicate that the firm has funds committed to the project part of the time [$PB(i^*)_n < 0$ for some values of $n$] and takes a "loan" from the project the rest of the time [$PB(i^*)_n > 0$ for some value of $n$]. Because of this lending and borrowing activity, there is no rate of return concept internal to the project. The return on such mixed investments tends to vary with the external interest rate (i.e., cost of capital) to the firm.

To circumvent this conceptual difficulty, we may modify the procedure for computation by compounding positive project balances at the cost of borrowing capital, $k$, and negative project balances at the return on invested capital (*RIC*), $r$. (We use the symbol $r$ because the return on invested capital of a mixed project is generally not equal to the *IRR*, $i^*$, of the project.) Since the firm is never indebted to the project for pure investment, it is clear that $k$ does not enter into the compounding process; hence this *RIC* is independent of $k$, the cost of capital to the firm. Two approaches may be used in computing $r$: the trial-and-error approach and the analytical approach.

***Trial-and-Error Approach.*** The trial-and-error approach is similar to finding a project's internal rate of return. For a given cost of capital, $k$, we first compute the project balances from an investment with a somewhat arbitrarily selected $r$ value. Since it is hoped that projects will promise a return of at least the cost of capital, a value of $r$ close to $k$ is a good starting point for most problems. For a given pair of $(k, r)$, we calculate the last project balance and see whether it is positive, negative, or zero. Suppose the last project balance $PB(r, k)_N$ is negative—what do we do then? A nonzero terminal project balance indicates that the guessed $r$ value is not the true $r$ value. We must lower the $r$ value and go through the process again. Conversely, if the $PB(r, k)_N > 0$, we raise the $r$ value and repeat the process.

### *Example 6.6*

To illustrate the method described, consider the following cash flow of a project.

| $n$ | 0 | 1 | 2 |
|---|---|---|---|
| $F_n$ | −$1,000 | 2,900 | −2,080 |

Suppose that the cost of capital, $k$, is known to be 15%. For $k = 15\%$, we must compute $r^*$ by trial and error.

For $k = 15\%$ and trial $r = 16\%$,

$$PB(16, 15)_0 = -1{,}000 \qquad = -\$1{,}000$$
$$PB(16, 15)_1 = -1{,}000(1 + 0.16) + 2{,}900 = \$1{,}300 \quad [\text{use } r, \text{ since } PB(16, 15)_0 < 0]$$
$$PB(16, 15)_2 = 1{,}300(1 + 0.15) - 2{,}080 = -\$585 \quad [\text{use } k, \text{ since } PB(16, 15)_1 > 0]$$

The terminal project balance is not zero, indicating that $r^*$ is not equal to our 16% trial $r$. The next trial value should be smaller than 16% because the terminal balance is negative (−585). After several trials, we conclude that for $k = 15\%$, $r^*$ is approximately at 9.13%. To verify the results,

$$PB(9.13, 15)_0 = -1{,}000 \qquad = -\$1{,}000$$
$$PB(9.13, 15)_1 = -1{,}000(1 + 0.0913) + 2{,}900 = \$1{,}808.70$$
$$PB(9.13, 15)_2 = 1{,}808.70(1 + 0.15) - 2{,}080 = 0$$

Since $r^* < k$, the investment is not profitable. Note that the project would also be rejected under the $PV$ analysis at $MARR = i = k = 15\%$.

$$PV(15\%) = -1{,}000 + 2{,}900(P/F, 15\%, 1) - 2{,}080(P/F, 15\%, 2)$$
$$= -\$51.04 < 0 \quad \square$$

***Analytical Approach.*** The most direct procedure for determining the functional relationship between $r$ and $k$ of a mixed investment is to write out the expression for the future value of the project. Since the project balance of a mixed investment is compounded at either $r$ or $k$, depending on the sign of the project balance, the terminal (future) balance of the project, denoted by $PB(r, k)_N$, is a function of two variables. The following steps can be used to determine the $RIC$, $r$.

**Step 1:** Find $i_{\min}$ by solving for the smallest real rate for which all $PB(i_{\min})_n \le 0$, for $n = 1, \ldots, N - 1$. This is usually done by a trial-and-error method.

**Step 2:** Find $PB(i_{\min})_N$.

a. If $PB(i_{\min})_N \ge 0$, the project is a pure investment.

(1) Find the $IRR$, $i^*$, for which $PB(i^*)_N = 0$; $i^* = r^*$ for a pure investment.

(2) Apply the decision rules given in step 5.

b. If $PB(i_{\min})_N < 0$, the project is a mixed investment and it is necessary to proceed with step 3.

**Step 3:** Calculate $PB(r, k)_n$ according to the following.

$$PB(r, k)_0 = F_0$$
$$PB(r, k)_1 = \begin{cases} PB(r, k)_0(1 + r) + F_1 & \text{if } PB(r, k)_0 \le 0 \\ PB(r, k)_0(1 + k) + F_1 & \text{if } PB(r, k)_0 > 0 \end{cases}$$
$$\vdots$$
$$PB(r, k)_n = \begin{cases} PB(r, k)_{n-1}(1 + r) + F_n & \text{if } PB(r, k)_{n-1} \le 0 \\ PB(r, k)_{n-1}(1 + k) + F_n & \text{if } PB(r, k)_{n-1} > 0 \end{cases}$$

To determine the positivity or negativity of $PB(r, k)_n$ at each period, set $r = i_{min}$, knowing that $r \le i_{min}$. (See Problem 6.10.)

**Step 4:** Determine the value of $r^*$ by solving the equation $PB(r, k)_N = 0$.

**Step 5:** Apply the following set of decision rules to accept or reject the project.

> If $r^* > k$, accept.
> If $r^* = k$, remain indifferent.
> If $r^* < k$, reject.

### *Example 6.7*

Consider the project cash flows given in Example 6.6.

| End of Period $n$ | 0 | 1 | 2 |
|---|---|---|---|
| Cash Flow $F_n$ | −$1,000 | 2,900 | −2,080 |

There are two sign changes in the ordered sequence of cash flows (−, +, −). The project has two *IRR*s, corresponding to $i^*_1 = 30\%$ and $i^*_2 = 60\%$. To derive the functional relationship between the return on invested capital, $r$, and the cost of capital, $k$, we apply the algorithm described in the preceding section.

**Step 1:** Find the $i_{min}$ that satisfies the following two equations ($N = 2, N - 1 = 1$).

$$PB(i)_0 = -1{,}000 < 0$$

$$PB(i)_1 = -1{,}000(1 + i) + 2{,}900$$
$$= -1{,}000i + 1{,}900 \le 0$$

Since $PB(i)_0 < 0$, we need only find the smallest $i$ that satisfies $PB(i)_1 \le 0$. The value of $i_{min}$ is 190%.

**Step 2:** Calculate $PB(i_{min})_N$.

$$PB(i_{min})_2 = (-1{,}000i_{min} + 1900)(1 + i_{min}) - 2{,}080$$
$$= -2{,}080$$

Since $PB(i_{min})_2 < 0$, the project is a mixed investment.

**Step 3:** Calculate $PB(r, k)_n$.

$$PB(r, k)_0 = -1{,}000$$

Since $PB(r, k)_0 < 0$, we use $r$.

$$PB(r, k)_1 = -1{,}000(1 + r) + 2{,}900$$
$$= -1{,}000r + 1{,}900$$

Since $r$ cannot exceed $i_{\min}$, $PB(r, k)_1 \geq 0$. Then we use $k$.

$$PB(r, k)_2 = (-1{,}000r + 1{,}900)(1 + k) - 2{,}080$$

**Step 4:** Find the solution of $PB(r, k)_2 = 0$.

$$r = 1.9 - \frac{2.08}{1 + k} \tag{6.16}$$

The graph of Eq. 6.16 is shown in Figure 6.7. We observe the following characteristics.

1. First, since $\frac{dr}{dk} = \frac{2.08}{(1 + k)^2} > 0$, $r$ is a monotonically increasing function of $k$. This means that the higher the cost that the firm places on borrowing funds from the project, the higher the return it will require on the invested capital.
2. Second, if we set $r = k$ in Eq. 6.16, we have $r = k = i^*$. Equation 6.16 intersects the 45° line $r = k$ twice, once at $k = 30\%$ and again at $k = 60\%$. With $r = k = i^*$, the terminal project balance $PB(r, k)_2$ decreases to $PB(i^*)_2 = 0$. Solving $PB(i^*)_2 = 0$ for $i^*$ yields the *IRR* of the project. In other words, this mixed investment has multiple rates of return ($i_1^* = 30\%$, $i_2^* = 60\%$). Therefore, the roots $i^*$ for mixed investment are the values of the return on invested capital, $r$, when the cost of borrowed money, $k$, is assumed to be equal to $r$.
3. Third, applying the decision rule, we have

| | |
|---|---|
| If $30\% < k < 60\% \rightarrow r^* > k$, | accept the project. |
| If $k = 30\%$ or $k = 60\% \rightarrow r^* = k$, | remain indifferent. |
| If $k < 30\%$ or $k > 60\% \rightarrow r^* < k$, | reject the project. |

4. Fourth, the decision we make will be consistent with the decision derived from applying the *PV* criterion when $i = MARR = k$. The *PV* of the project at an interest rate of $k$ can be expressed as

$$PV(k) = -\$1{,}000 + \frac{\$2{,}900}{1 + k} - \frac{\$2{,}080}{(1 + k)^2} \tag{6.17}$$

which is also depicted in Figure 6.7.

The following comments about the *PV* function are in order. First, the *IRR* is by definition the solution to the equation $PV(k) = 0$. Therefore, we observe that $PV(k)$ intersects the horizontal axis at $k = 30\%$ and at $k = 60\%$. Second, since $PV(k)$ is positive only in the range $30\% < k < 60\%$, the *PV* criterion gives the same accept–reject signal as the *IRR* criterion. □

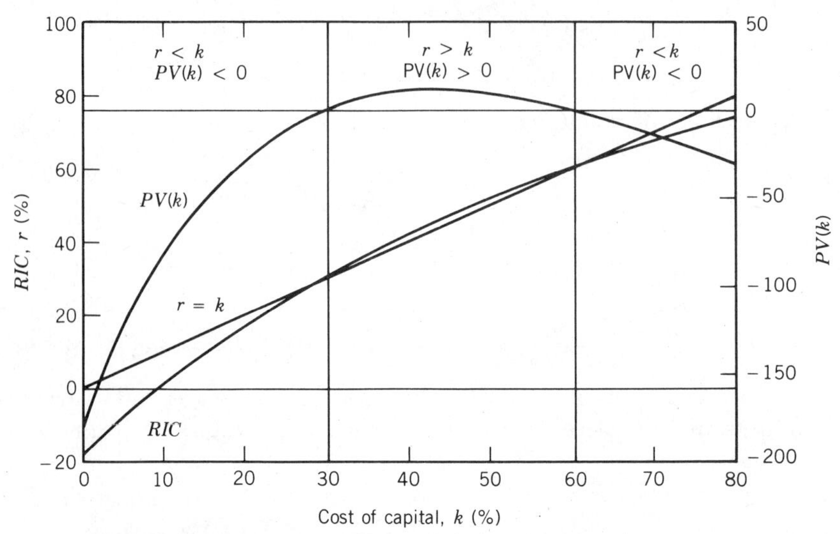


**FIGURE 6.7.** ***RIC*** **and** ***PV*** **as functions of** ***k*****, Example 6.7.**

### 6.3.5 Modified Rate of Return

An alternative way of approaching mixed investments is to modify the procedure for computing the rate of return by making explicit and consistent assumptions about the interest rate at which intermediate receipts from projects may be reinvested. This reinvestment could be either in other projects or in the outside market. This procedure is similar to the previous use of two different rates $(r, k)$ in the computation of the *project balance.* This section reviews some of the methods for applying the procedure.

***Solomon's Average Rate of Return (ARR).*** A different way of looking at a project is to ask the following question. Suppose we take the net revenues $F_n (F_n > 0)$ and reinvest them each year at $i$, letting them accumulate until time $N$. What rate of interest does investment $C_0$ have to earn to reach the same accumulated value in $N$ periods [15]?

Mathematically, we wish to find $s$ to solve the equation

$$\underbrace{C_0(1+s)^N}_{\text{alternative investment}} = \underbrace{\sum_{n=1}^{N} F_n(1+i)^{N-n}}_{\text{current investment}} \tag{6.18}$$

With known $s$, the acceptance rule is

| |
|---|
| If $s > i = MARR$, accept. |
| If $s = i$, remain indifferent. |
| If $s < i$, reject. |

We can easily show that the *ARR* criterion is completely consistent with the *PV* criterion [2]. Recall that for a given project with $F_0 < 0$ but $F_n > 0$ for $1 \leq n \leq N$, the *PV* acceptance rule is

$$\sum_{n=0}^{N} F_n(1 + i)^{-n} > 0 \tag{6.19}$$

Substituting $C_0$ for $F_0$ (note that $F_0 = -C_0$) gives us

$$C_0 < \sum_{n=1}^{N} F_n(1 + i)^{-n} \tag{6.20}$$

Multiplying both sides of Eq. 6.18 by $(1 + i)^{-N}$ yields

$$C_0(1 + s)^N(1 + i)^{-N} = \sum_{n=1}^{N} F_n(1 + i)^{-n} \tag{6.21}$$

By comparing Eqs. 6.20 and 6.21, we can deduce that

$$C_0(1 + s)^N(1 + i)^{-N} > C_0$$

or

$$(1 + s)^N > (1 + i)^N \tag{6.22}$$

This implies that $s > i$, which is the *ARR* acceptance condition.

### *Example 6.8*

Consider the cash flows shown in Figure 6.2, where $C_0 = \$1{,}000$, $F_1 = \$400$, $F_2 = \$360$, $F_3 = \$320$, $F_4 = \$280$, and $F_5 = \$240$. Substituting these values into Eq. 6.18, we obtain

$$\begin{aligned} 1{,}000(1 + s)^5 &= 400(1.1)^4 + 360(1.1)^3 + 320(1.1)^2 + 280(1.1) + 240 \\ &= \$2{,}000 \end{aligned}$$

Solving for $s$ yields 15%. This tells us that we can invest \$1,000 in the project, reinvest the proceeds at our opportunity rate (*MARR*) of 10%, and have \$2,000 at time 5. If we do not wish to invest in the project but still wish to earn \$2,000, the original \$1,000 would have to earn 15% per period. Since the *MARR* is 10%, we are clearly better off accepting the project. If $s$ had been less than $i = 10\%$, we would have rejected the project. □

***Modified Internal Rate of Return (MIRR).*** As a variation of the *ARR* procedure, we may make explicit the expected reinvestment rate of intermediate incomes

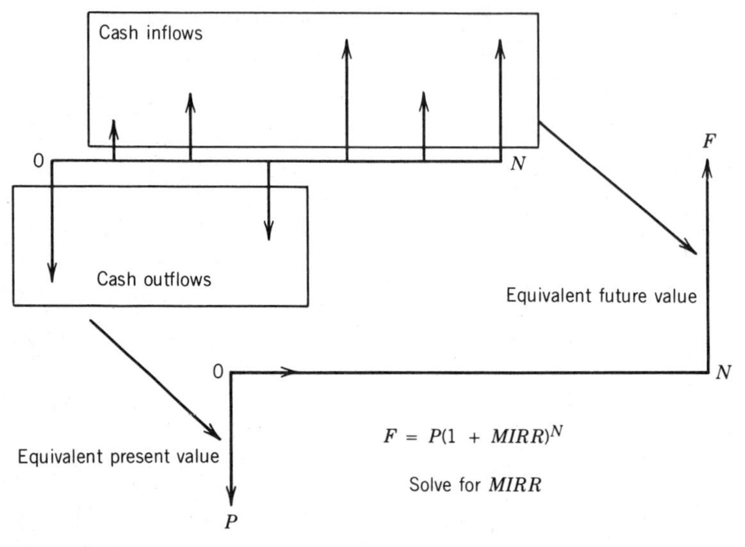


**FIGURE 6.8.** Illustration of the *MIRR* concept.

and costs and reduce them to an equivalent initial cost and a terminal project balance, a procedure known as the modified internal rate of return (*MIRR*) [12] or the external rate of return. In this way a unique *IRR* can be computed. This *MIRR* is defined by

$$\frac{\text{Future value of net cash inflow}}{\text{Present value of net cash outflow}}$$

$$= \frac{\sum_{n=0}^{N} \max(F_n, 0)(1+i)^{N-n}}{-\sum_{n=0}^{N} \min(F_n, 0)(1+i)^{-n}} = (1 + MIRR)^N \tag{6.23}$$

where $\max(F_n, 0) = F_n$ if $F_n > 0$, otherwise $F_n = 0$; $\min(F_n, 0) = F_n$ if $F_n < 0$, otherwise $F_n = 0$; and $i$ is the *MARR* to the firm. The meaning of the *MIRR* is illustrated in Figure 6.8.

By rearranging terms in Eq. 6.23, we can rewrite it as

$$\sum_{n=0}^{N} \max(F_n, 0)(1+i)^{N-n} = \left[-\sum_{n=0}^{N} \min(F_n, 0)(1+i)^{-n}\right](1 + MIRR)^N \tag{6.24}$$

If the cash outflow is restricted to the first period, $n = 0$, the *MIRR* is exactly the same as the *ARR, s.* The acceptance rule is then

| |
|---|
| If *MIRR* $> i$, accept. |
| If *MIRR* $= i$, remain indifferent. |
| If *MIRR* $< i$, reject. |

The *MIRR* will always give a unique solution and is also consistent with the *PV* criterion. The *MIRR* will always exceed the alternative rate whenever the investment sequence has a positive *PV* at *i.* This can be visualized from the following equations.

The project acceptance condition by the *PV* criterion is

$$\sum_{n=0}^{N} \max(F_n, 0)(1+i)^{-n} \quad > \quad -\sum_{n=0}^{N} \min(F_n, 0)(1+i)^{-n} \tag{6.25}$$

$$\begin{array}{ccc}\text{Present value of} & & \text{Present value of}\\ \text{net cash inflow} & > & \text{net cash outflow}\end{array}$$

Multiplying both sides of Eq. 6.24 by $(1+i)^{-N}$ yields

$$\sum_{n=0}^{N} \max(F_n, 0)(1+i)^{-n}$$

$$= \left[-\sum_{n=0}^{N} \min(F_n, 0)(1+i)^{-n}\right](1+MIRR)^{N}(1+i)^{-N} \tag{6.26}$$

From the relation given in Eq. 6.25, we can say

$$\left[-\sum_{n=0}^{N} \min(F_n, 0)(1+i)^{-n}\right](1+MIRR)^{N}(1+i)^{-N}$$

$$> -\sum_{n=0}^{N} \min(F_n, 0)(1+i)^{-n} \tag{6.27}$$

Simplifying the terms above gives

$$(1+MIRR)^{N} > (1+i)^{N} \tag{6.28}$$

which indicates that $MIRR > i$.

There are three other variations of the *MIRR* [4], but these indices (including *ARR* and *MIRR*) have numerical values distinctly different from one another, and without additional information provided, these rates are considerably more complex to use than the simple *PV* criterion.

### *Example 6.9*

Using an example from [4], we will illustrate the method of computing the *MIRR*. Assume that $i = 6\%$, and the cash flow components are

| Cash Flow | $n$: 0 | 1 | 2 | 3 |
|---|---|---|---|---|
| $b_n$ | \$0 | 3 | 2 | 25 |
| $c_n$ | \$10 | 1 | 5 | 2 |
| $F_n$ | −\$10 | 2 | −3 | 23 |

$$\text{Present value of net cash outflow} = +10 + 3(1 + 0.06)^{-2} = \$12.67$$

$$\text{Future value of net cash inflow} = 2(1 + 0.06)^{2} + 23 = \$25.25$$

Using Eq. 6.24, we find

$$25.25 = 12.67(1 + MIRR)^3$$

$$MIRR = 25.84\%$$

Since $MIRR > 6\%$, the project should be acceptable. Note that $PV(6\%) = \$8.53 > 0$, so the *MIRR* result is consistent with the *PV* criterion. □

## 6.4 BENEFIT–COST RATIOS

Another way to express the worthiness of a project is to compare the inflows with the investment. This leads to three types of benefit–cost ratios: the aggregate benefit–cost ratio (Eckstein $B/C$), the netted benefit–cost ratio (simple $B/C$), and the Lorie–Savage ratio.

Let $B$ and $C$ be the present values of cash inflows and outflows defined by Eqs. 6.1 and 6.2. We will split the equivalent cost $C$ into two components, the initial capital expenditure and the annual costs accrued in each successive period. Assuming that an initial investment is required during the first $m$ periods, while annual costs accrue in each period following, the components are defined as

$$I = \sum_{n=0}^{m} c_n(1 + i)^{-n} \tag{6.29}$$

$$C' = \sum_{n=m+1}^{N} c_n(1 + i)^{-n} \tag{6.30}$$

and $C = I + C'$.

The following example will be used to demonstrate the application of different $B/C$ ratio criteria.

| Cash Flow | $n$: 0 | 1 | 2 | 3 | 4 | 5 |
|---|---|---|---|---|---|---|
| $b_n$ | \$0 | 0 | 10 | 10 | 20 | 20 |
| $c_n$ | \$10 | 5 | 5 | 5 | 5 | 10 |
| $F_n$ | −\$10 | −5 | 5 | 5 | 15 | 10 |

With $i = 10\%$, we define

$$N = 5$$

$$m = 1$$

$$B = 10(1.1)^{-2} + 10(1.1)^{-3} + 20(1.1)^{-4} + 20(1.1)^{-5} = \$41.86$$

$$C = 10 + 5(1.1)^{-1} + 5(1.1)^{-2} + 5(1.1)^{-3} + 5(1.1)^{-4} + 10(1.1)^{-5} = \$32.06$$

$$I = 10 + 5(1.1)^{-1} = \$14.55$$

$$C' = 5(1.1)^{-2} + 5(1.1)^{-3} + 5(1.1)^{-4} + 10(1.1)^{-5} = \$17.51$$

$$PV(10\%) = B - C = \$9.80$$

### 6.4.1 Benefit–Cost Ratios Defined

***Aggregate B/C Ratio.*** The aggregate $B/C$ ratio introduced by Eckstein [7] is defined as

$$R_A = \frac{B}{C} = \frac{B}{I + C'}, \qquad I + C' > 0 \tag{6.31}$$

To accept a project, the $R_A$ must be greater than 1. Historically, this ratio was developed in the 1930s in response to the fact that in public projects the user is generally not the same as the sponsor. To have a better perspective on the user's benefits, we need to separate them from the sponsor's costs. If we assume that for a project $b_n$ represents the user's benefits and $c_n$ the sponsor's costs, the ratio is

$$R_A = \frac{41.86}{14.55 + 17.51} = 1.306$$

The ratio exceeds 1, which implies that the user's benefits exceed the sponsor's costs. Public projects usually also have benefits that are difficult to measure, whereas costs are more easily quantified. In this respect, the Eckstein $B/C$ ratio lends itself readily to sensitivity analysis with respect to the value of benefits. We will discuss this measure in greater detail in Chapter 14.

***Netted B/C Ratio.*** As an alternative expression in defining their terms, some analysts consider only the initial capital expenditure as a cash outlay, and equivalent benefits become net benefits (i.e., revenues minus annual outlays). This alternative measure is referred to as the *netted benefit–cost ratio,* $R_N$, and is expressed by

$$R_N = \frac{B - C'}{I}, \qquad I > 0 \tag{6.32}$$

The advantage of having the benefit–cost ratio defined in this manner is that it provides an index indicating the net benefit expected per dollar invested, sometimes called a *profitability index*. Again, for a project to remain under consideration, the ratio must be greater than 1. For our example, the $R_N$ is

$$R_N = \frac{41.86 - 17.51}{14.55} = 1.674$$

Note that this is just a comparison of the present value of net revenues ($F_n$) with the present value of investment. Since $R_N > 1$, there is a surplus at time 0 and the project is favorable. The use of this criterion also had its origin in the evaluation of public projects in the 1930s.

***Lorie–Savage Ratio.*** As a variation on $R_N$, the Lorie–Savage ($L$–$S$) ratio is defined as

$$L\text{–}S = \frac{B - C}{I} = \frac{B - C'}{I} - 1 = R_N - 1 > 0 \tag{6.33}$$

Here the comparison is between the surplus at time 0 and the investment itself. If the ratio is greater than 0, the project is favorable. Clearly, the $R_N$ $B/C$ and the $L$–$S$ $B/C$ ratios will always yield the same decision for a project, since both the ratios and their respective cutoff points differ by 1.0. For our example, $L$–$S$ = $1.674 - 1 = 0.674 > 0$. Thus the $L$–$S$ ratio also indicates acceptance of the project.

### 6.4.2 Equivalence of B/C Ratios and PV

Using the notation in Section 6.4.1, we can state the *PV* criterion for project acceptance as

$$\begin{aligned} PV(i) &= B - C \\ &= B - (I + C') > 0. \end{aligned} \tag{6.34}$$

By transposing the term $(I + C')$ to the right-hand side and dividing both sides by $(I + C')$, we have

$$\frac{B}{I + C'} > 1 \qquad (I + C' > 0)$$

which is exactly the decision rule for accepting a project with the $R_A$ criterion. On the other hand, by transposing the term $I$ to the right-hand side and dividing both sides of the equation by $I$, we obtain

$$\frac{B - C'}{I} > 1 \qquad (I > 0)$$

which is exactly the decision rule for accepting a project with the $R_N$ criterion. In other words, use of $R_A$ or use of $R_N$ will lead to the same conclusion about the initial acceptability of a single project, as long as $I > 0$ and $I + C' > 0$. Notice that these $B/C$ ratios will always agree with each other for an individual project, since $I$ and $C'$ are nonnegative.

$$\frac{B}{I + C'} > 1 \longleftrightarrow B > I + C' \longleftrightarrow B - C' > I$$

$$B > I + C' \;\updownarrow\; PV(i) = B - (I + C') > 0$$

$$B - C' > I \;\updownarrow\; \frac{B - C'}{I} > 1 \;\updownarrow\; \frac{B - C'}{I} - 1 > 0$$

Although *ARR* does not appear to be related to the benefit–cost ratios, it does, in fact, yield the same decisions for a project. From Eq. 6.18 we have

$$C_0(1 + s)^N = \sum_{n=1}^{N} F_n(1 + i)^{N-n}$$

Expressed differently,

$$\begin{aligned} I(1 + s)^N(1 + i)^{-N} &= \sum_{n=1}^{N} F_n(1 + i)^{-n} \\ &= B - C' \\ \frac{B - C'}{I} &= \left(\frac{1 + s}{1 + i}\right)^N \end{aligned} \qquad (6.35)$$

We require $s > i$ for project acceptance, so we must have

$$\frac{B - C'}{I} > 1$$

## 6.5 PAYBACK PERIOD

A popular rule-of-thumb method for evaluating projects is to determine the number of periods needed to recover the original investment. In this section we present two procedures for assessing the payback period of an investment.

### 6.5.1 Payback Period Defined

***Conventional Payback Period.*** The payback period (*PP*) is defined as the number of periods it will take to recover the initial investment outlay. Mathematically, the payback period is computed as the smallest value of $n$ that satisfies the equation

$$\sum_{n=0}^{n_p} F_n \geqslant 0 \tag{6.36}$$

This payback period ($n_p$) is then compared with the maximum acceptable payback period ($n_{max}$) to determine whether the project should be accepted. If $n_{max} > n_p$, the proposed project will be accepted. Otherwise, the project will be rejected.

Obviously the most serious deficiencies of the payback period are that it fails to consider the time value of money and that it fails to consider the consequences of the investment after the payback period.

***Discounted Payback Period.*** As a modification of the conventional payback period, one may incorporate the time value of money. The method is to determine the length of time required for the project's equivalent receipts to exceed the equivalent capital outlays.

Mathematically, the discounted payback period $Q$ is the smallest $n$ that satisfies the expression

$$\sum_{n=0}^{Q} F_n(1 + i)^{-n} \geqslant 0 \tag{6.37}$$

where $i$ is the *MARR*.

If we multiply both sides of Eq. 6.37 by $(1 + i)^Q$, we should obtain

$$\sum_{n=0}^{Q} F_n(1 + i)^{Q-n} \geqslant 0 \tag{6.38}$$

Notice that Eq. 6.38 is the definition of project balance $PB(i)_n$. Thus, the discounted payback period is alternatively defined as the smallest $n$ that makes $PB(i)_n \geq 0$.

### 6.5.2 Popularity of the Payback Period

Clearly, the payback period analysis is simple to apply and, in some cases, may give answers approximately equivalent to those provided by more sophisticated methods. A number of authors have tried to show an equivalence between the payback period and other criteria, such as *IRR*, under special circumstances [11]. For example, Gordon [8] interpreted the payback period as an indirect, though quick, measure of return. With a uniform stream of receipts, the reciprocal of the payback period is the *IRR* for a project of infinite life and is a good approximation to this rate for a long-lived project.

Weingartner [19] analyzed the basic reasons why the payback period measure is so popular in business. One reason is that the payback period can function like many other rules of thumb to shortcut the process of generating information and then evaluating it. Payback reduces the information search by focusing on the time when the firm expects to "be made whole again." Hence, it allows the decision maker to judge whether the life of the project past the break-even (bench mark) point is sufficient to make the undertaking worthwhile.

In summary, the payback period gives some measure of the rate at which a project will recover its initial outlay. This piece of information is not available from either the *PV* or the *IRR*. The payback period may not be used as a direct figure of merit, but it may be used as a constraint: no project may be accepted unless its payback period is shorter than some specified period of time.

### *Example 6.10*

Suppose that a firm is considering a project costing $10,000, the life of the project is 5 years, and the expected net annual cash flows at the end of the year are as follows (assume *MARR* = 10%).

| Cash Flow | *n*: 0 | 1 | 2 | 3 | 4 | 5 |
|---|---|---|---|---|---|---|
| $F_n$ | −$10,000 | $3,000 | $3,000 | $4,000 | $3,000 | $3,000 |
| Cumulative $F_n$ | −$10,000 | −7,000 | −4,000 | 0 | 3,000 | 6,000 |
| *PV*(10%) | −$10,000 | 2,727 | 2,479 | 3,005 | 2,049 | 1,862 |
| Cumulative present value | −$10,000 | −7,273 | −4,794 | −1,789 | 260 | 2,122 |

The conventional payback period is 3 years, whereas the discounted payback period is 3.87 years. This example demonstrates how consideration of the time value of money in payback analysis can produce different results. Clearly, this discounted measure is conceptually better than the conventional one, but both measures fail to indicate the overall profitability of the project. □

## 6.6 TIME-DEPENDENT MEASURE OF INVESTMENT WORTH

The project balance, which measures the equivalent loss or profit of an investment project as a function of time, is a recent development that provides additional insight into investment decisions. In Section 6.2.2 we defined the project balance and demonstrated its calculation both mathematically and through examples. This section presents particular characteristics of project balance profiles and their economic interpretation. We also discuss some possible measures of investment desirability based on these profiles. (The material presented in this section is based on the analysis given by Park and Thuesen [14].)

### 6.6.1 Areas of Negative and Positive Balances

Recall that the project balance is defined by

$$PB(i)_n = \sum_{j=0}^{n} F_j(1 + i)^{n-j}, \qquad n = 0, 1, 2, \ldots, N$$

By plotting $PB(i)_n$ as a function of time $n$, we can trace the time path of project balance as shown in Figure 6.9. This time path is referred to as the *project balance pattern,* and it provides the basic information about the attractiveness of a particular investment proposal as a function of its life. The shaded area represents the period of time during which the project balance has negative values, that is, the time during which the initial investment plus interest is not fully recovered. This area is referred to as the *area of negative balance* (*ANB*). Mathematically, the area is represented by

$$ANB = \sum_{n=0}^{Q-1} PB(i)_n \tag{6.39}$$

where $Q$ is the discounted payback period [the first period in which $PB(i)_n \geq 0$]. Since the value $PB(i)_n$ for $n < Q$ represents the magnitude of negative balance

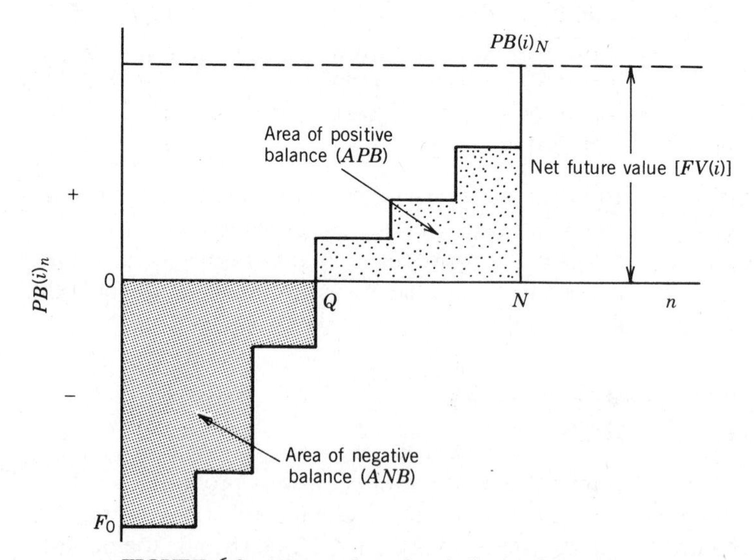


**FIGURE 6.9.** A general project balance diagram.

of the project at the end of period $n$, it is equivalent to the amount of possible loss if the project is terminated at this time. With certainty, the *ANB* can be interpreted as the total amount of dollars to be tied up for the particular investment option. The smaller the *ANB*, the more flexible the firm's future investment options. Therefore, the smaller the *ANB* for a project, the more attractive the project is considered, assuming that the expected terminal profits for other projects are the same.

Point $Q$ on the horizontal axis in Figure 6.9 represents the discounted payback period, which indicates how long it will be before the project breaks even. Therefore, the smaller the $Q$ for a project, the more desirable the project is considered, if other things are equal. (See the mathematical definition in Eq. 6.38.)

The stippled area in Figure 6.9 represents the period of time during which the $PB(i)_n$ maintains a positive project balance. This area is referred to as the *area of positive balance* (*APB*). The initial investment of the project has been fully recovered, so receipts during this time period contribute directly to the final profitability of the project. Symbolically, the area is represented by

$$APB = \sum_{n=Q}^{N-1} PB(i)_n \tag{6.40}$$

The project balance diagram during these periods can be interpreted as the rate at which the project is expected to accumulate profits. This is certainly an important parameter which affects project desirability when decisions are made about the retirement of projects. Since the values $PB(i)_n$ for $n > Q$ represent the magnitude of positive project balance, there is no possible loss even though the project is terminated in a period before the end of its life or no additional receipts are received. Thus, $PB(i)_n$ becomes the net equivalent dollars earned.

Finally, the last project balance $PB(i)_N$ represents the net future value of the project (or terminal profit) at the end of its life. The *PV* of the project can be found easily by a simple transformation, shown in Eq. 6.9.

### 6.6.2. Investment Flexibility

To illustrate the basic concept of investment flexibility and its discriminating ability compared with the traditional measures of investment worth (e.g., *PV*), we consider the hypothetical investment situation shown in Figure 6.10*a*. Projects 1 and 2 have single-payment and uniform-series cash flows, respectively. Projects 3 and 4 are gradient series, one being an increasing gradient series and the other a decreasing gradient series. All the projects require the same initial investment and have a service life of 3 years. All the projects would have an equivalent future value of $63.40 at a *MARR* of 10% [or *PV*(10%) = $47.63]. This implies that no project is preferable to the others when they are compared on the basis of present value.

Plotting the project balance pattern for each project provides additional information that is not revealed by computing only present value equivalents

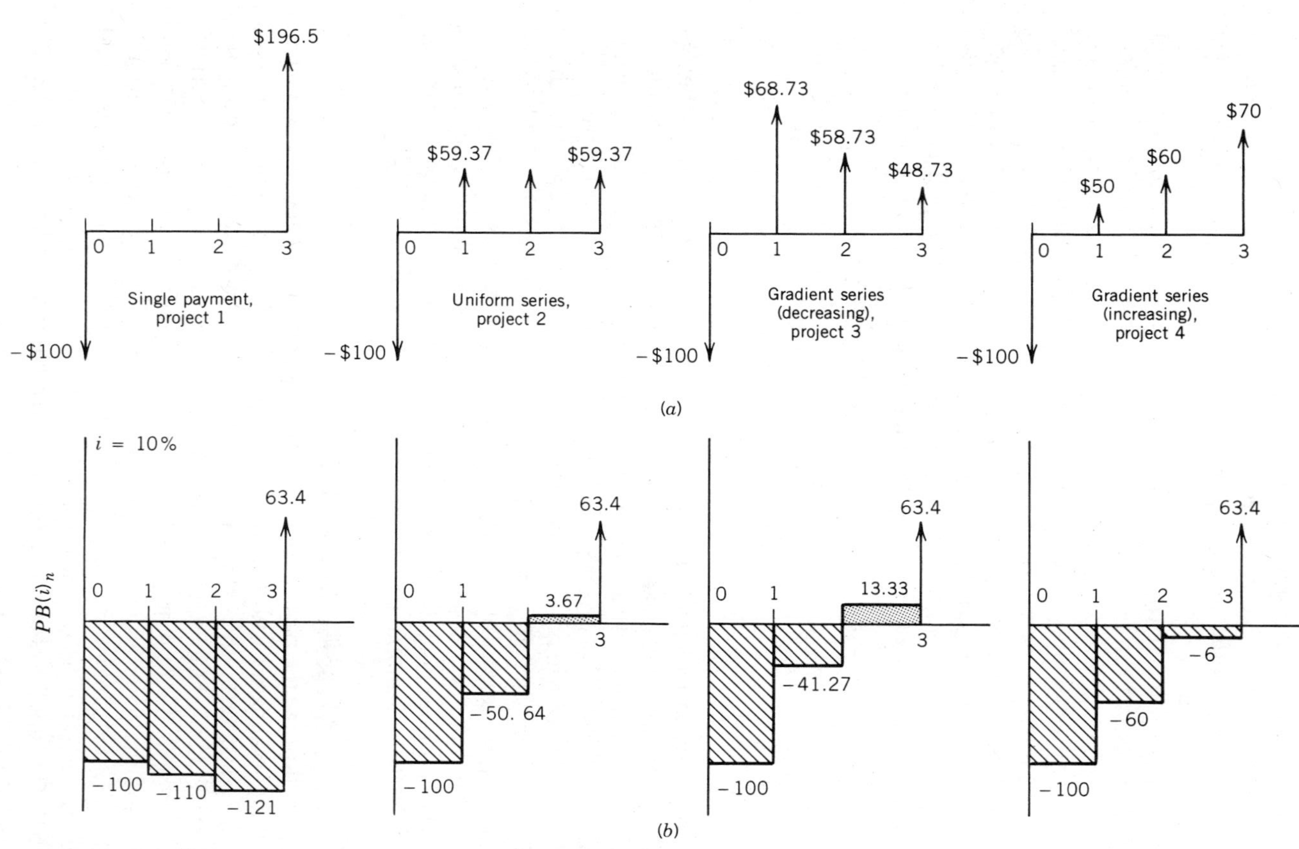


**FIGURE 6.10.** Project balance for four cash flow patterns.

**Table 6.4** ***Statistics of Project Balance Patterns for Projects 1, 2, 3, and 4***

| Project Number | Cash Flow Pattern | Future Value *FV*(10%) | *ANB* | *APB* | *Q* |
|---|---|---|---|---|---|
| 1 | Single payment | $63.4 | 331.00 | 0 | 3 |
| 2 | Uniform series | $63.4 | 150.63 | 3.67 | 2 |
| 3 | Gradient series (decreasing) | $63.4 | 141.27 | 13.33 | 2 |
| 4 | Gradient series (increasing) | $63.4 | 166.00 | 0 | 3 |

(see Figure 6.10*b*). For example, a comparison of project 1 with project 3 in terms of the shape of the project balance pattern shows that project 3 recovers its initial investment within 2 years, whereas project 1 takes 3 years to recover the same initial investment. This, in turn, indicates that project 3 would provide more flexibility in future investment activity to the firm than project 1. By selecting project 3, the investor can be sure of being restored to his or her initial position within a short span of time. Similar one-to-one comparisons can be made among all four projects. Table 6.4 summarizes the statistics obtained from the balance patterns for each project shown in Figure 6.10*b*.

Table 6.4 shows that project 3 appears to be most desirable, even though its terminal profitability is equal to those of the other projects, because its *ANB* is the smallest and its *APB* is the largest among the projects. As discussed in Section 6.6.1, the small value of *ANB* implies more flexibility in the firm's future investment activity. In other words, an early resolution of the negative project balance would make funds available for attractive investment opportunities that become

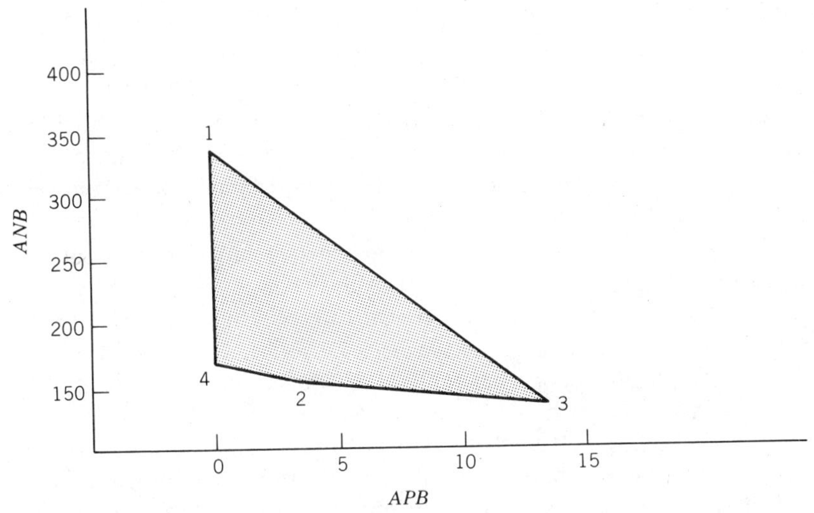


**FIGURE 6.11.** Plot of *APB* against *ANB* for different cash flow patterns.

available in the subsequent decision periods. One-to-one comparisons of the projects in terms of *ANB* and *APB* can be depicted graphically (see Figure 6.11). From Figure 6.9, it becomes evident that the project balance parameters such as *ANB* and *APB* reflect the changes in the cash flow patterns over time. Since project 3 represents the highest *APB* with the smallest *ANB*, project 3 appears to be the most desirable. Of course, the environment in which the decision is made and individual preferences will dictate which of these parameters should be used so that the economic implications of an investment project are fully understood.

## 6.7 SUMMARY

In this chapter we showed the following.

1. The *PV*, *FV*, and *AE* will always yield the same decision for a project. We consider *PV* as the baseline, or "correct," criterion to use in a stable, perfect capital market with complete certainty about investment outcomes.
2. The distinction between pure and mixed investments is needed to determine whether the return on invested capital is independent of the cost of capital.
3. Only for a pure investment is there a rate of return concept internal to the project. For pure investments, the *IRR* and *PV* criteria result in identical acceptance and rejection decisions.
4. The return on invested capital for a mixed project varies directly with the cost of capital. The phenomenon of multiple *IRR*s, which occurs only in the situation of a mixed investment, is actually a manifestation of the existence of this basic functional relationship. The *RIC* is consistent with the *PV* criterion.
5. *ARR* and *MIRR* will also always yield the same decision for a project, consistent with the *PV* criterion.
6. $R_A$, $R_N$, and $L$–$S$ ratios will give the same accept–reject decisions for an individual project. The *PV* and these $B/C$ ratios will always agree.
7. Neither the payback period nor the discounted payback period should be considered as a criterion, since they may not agree with *PV*. They may be used as additional constraints in the decision-making process, but they should be used with caution.
8. The project balance diagram provides quantitative information about four important characteristics associated with the economic desirability of an investment project. Two of these characteristics, net future value (terminal project balance) and discounted payback period, have generally been a part of conventional economic analyses. However, the other two characteristics, *ANB* and *APB*, have not been considered. Possible applications of the project balance indicate that a variety of measurements can be devised that reflect particular characteristics of the investment project under consideration. The project balance at the end of a project, $PB(i)_N$, is identical to the *FV* criterion.

In conclusion, we can say that the *PV* criterion is superior among the traditional measures of investment worth because of its ease of use, robustness, and consistency.

## REFERENCES


1. BALDWIN, R. H., "How to Assess Investment Proposals," *Harvard Business Review,* Vol. 27, No. 3, pp. 98–104, May–June 1959.
2. BERNHARD, R. H., "Discount Methods for Expenditure Evaluation—A Clarification of Their Assumptions," *Journal of Industrial Engineering,* Vol. 18, No. 1, pp. 19–27, January–February 1962.
3. BERNHARD, R. H., "A Comprehensive Comparison and Critique of Discounting Indices Proposed for Capital Investment Evaluation," *The Engineering Economist,* Vol. 16, No. 3, pp. 157–186, Spring 1971.
4. BERNHARD, R. H., "Modified Rates of Return for Investment Project Evaluation—A Comparison and Critique," *The Engineering Economist,* Vol. 24, No. 3, pp. 161–167, Spring 1979.
5. BERNHARD, R. H., "Unrecovered Investment, Uniqueness of the Internal Rate and the Question of Project Acceptability," *Journal of Financial and Quantitative Analysis,* Vol. 12, No. 1, pp. 33–38, March 1977.
6. BUSSEY, L. E., *The Economic Analysis of Industrial Projects,* Prentice–Hall, Englewood Cliffs, N.J., 1978.
7. ECKSTEIN, O., *Water Resource Development: The Economics of Project Evaluation,* Harvard University Press, Cambridge, Mass., 1958.
8. GORDON, M., "The Payoff Period and the Rate of Profit," *Journal of Business,* Vol. 28, No. 4, pp. 253–260, October 1955.
9. KAPLAN, S., "A Note on a Method for Precisely Determining the Uniqueness or Nonuniqueness of the Internal Rate of Return for a Proposed Investment," *Journal of Industrial Engineering,* Vol. 26, No. 1, pp. 70–71, January–February 1965.
10. KAPLAN, S., "Computer Algorithms for Finding Exact Rates of Return," *Journal of Business,* Vol. 40, No. 4, pp. 389–392, October 1967.
11. LEVY, H., and M. SARNAT, *Capital Investment and Financial Decisions,* 2nd edition, Prentice–Hall, Englewood Cliffs, N.J., 1983.
12. LIN, S., "The Modified Internal Rate of Return and Investment Criterion," *The Engineering Economist,* Vol. 21, No. 4, pp. 237–248, Summer 1976.
13. MAO, J. C. T., *Quantitative Analysis of Financial Decisions,* Macmillan, Toronto, 1969.
14. PARK, C. S., and G. J. THUESEN, "Combining Concepts of Uncertainty Resolution and Project Balance for Capital Allocation Decisions," *The Engineering Economist,* Vol. 24, No. 2, pp. 109–127, Winter 1979.
15. SOLOMON, E., "The Arithmetic of Capital-Budgeting Decision," *Journal of Business,* Vol. 29, No. 2, pp. 124–129, April 1956.
16. TEICHROEW, D., A. A. ROBICHEK, and M. MONTALBANO, "Mathematical Analysis of Rates of Return under Certainty," *Management Science,* Vol. 11, No. 3, pp. 395–403, January 1965.
17. TEICHROEW, D., A. A. ROBICHEK, and M. MONTALBANO, "An Analysis of Criteria for Investment and Financing Decisions under Certainty," *Management Science,* Vol. 12, No. 3, pp. 151–179, November 1965.

18. WEINGARTNER, H. M., "The Excess Present Value Index: A Theoretical Basis and Critique," *Journal of Accounting Research,* Vol. 1, No. 2, pp. 213–224, Autumn 1963.
19. WEINGARTNER, H. M., "Some New Views on the Payback Period and Capital Budgeting Decision," *Management Science,* Vol. 15, No. 12, pp. B594–B607, August 1969.

## PROBLEMS

All cash flows given in this problem set represent the cash flows after taxes, unless otherwise mentioned.

**6.1.** Consider the following sets of investment projects.

| Project | After-Tax Cash Flows | | | |
|---|---|---|---|---|
| | $n$: 0 | 1 | 2 | 3 |
| A | −$10,000 | 0 | 0 | 19,650 |
| B | −$10,000 | 5,937 | 5,937 | 5,937 |
| C | −$10,000 | 6,873 | 5,873 | 4,873 |
| D | −$10,000 | 5,000 | 6,000 | 7,000 |

a. Compute the net present value of each project at $i = 10\%$.
b. Compute the project balance of each project as a function of the project year.
c. Compute the future value of each project at $i = 10\%$.
d. Compute the annual equivalent of each project at $i = 10\%$.

**6.2.** In Problem 6.1

a. Graph the net present value of each project as a function of $i$.
b. Graph the project balances (at $i = 10\%$) of each project as a function of $n$.
c. From the graphical results in part b of Problem 6.1, which project appears to be the safest to undertake if there is some possibility of premature termination of the projects at the end of year 2?

**6.3.** Consider the following set of independent investment projects.

| Project | $F_n$ | | | | | | |
|---|---|---|---|---|---|---|---|
| | $n$: 0 | 1 | 2 | 3 | 4 | 5 | 6–20 |
| 1 | −100 | 50 | 50 | 50 | 50 | −750 | 100 |
| 2 | −100 | 30 | 30 | 30 | 10 | 10 | |
| 3 | −16 | 92 | −170 | 100 | | | |

Assume $MARR(i) = 10\%$ for the following questions.

a. Compute the present value for each project and determine the acceptability of each project.
b. Compute the future value of each project at the end of each project period and determine its acceptability.
c. Compute the annual equivalent of each project and determine the acceptability of each project.
d. Compute the project value of each project at the end of 20 years with variable *MARR*s: 10% for $n = 0$ to $n = 10$ and 15% for $n = 11$ to $n = 20$.
e. Compute the project balance as a function of $n$ for project 2.
f. Compute Solomon's average rate of return (*ARR*) for project 2, and determine the acceptability of the project.

g. Compute the modified internal rate of return (*MIRR*) for project 3 and determine the project's acceptability.

**6.4.** Consider the following project balance profiles for proposed investment projects.

| | | Project Balance (End of Year) | | | | | |
|---|---|---|---|---|---|---|---|
| Project | $i$ | $PB(i)_0$ | $PB(i)_1$ | $PB(i)_2$ | $PB(i)_3$ | $PB(i)_4$ | $PB(i)_5$ |
| A | 10% | −$1,000 | −1,000 | −900 | −690 | −359 | 105 |
| B | 0 | −$1,000 | −800 | −600 | −400 | −200 | 0 |
| C | 15 | −$1,000 | −650 | −348 | −100 | 85 | 198 |
| D | 18 | −$1,000 | −680 | −302 | −57 | 233 | 575 |
| E | 20 | −$1,000 | −1,200 | −1,440 | −1,328 | −1,194 | −1,000 |
| F | 12.9 | −$1,000 | −530 | −99 | −211 | −89 | 0 |

Project balance figures are rounded to dollars.

a. Compute the present value of each investment.
b. Determine the cash flows for each project.
c. Identify the future value of each project.
d. What would the internal rates of return be for projects B and F?

**6.5.** Consider the following sequence of cash flows.

| Project | $n$: 0 | 1 | 2 | 3 |
|---|---|---|---|---|
| A | −10 | 5 | −5 | 20 |
| B | 100 | −216 | 116 | |

a. Descartes' rule of sign indicates ______ possible rates of return for project A, but the Norstrom rule indicates ______ real root(s) because there are ______ sign change(s) in the $S_n$ series. The rates of return is are ______.
b. For project B, determine the range of *MARR* for which the project would be acceptable.
c. Compute the *MIRR* for both projects. ($i$ = 6%)
d. Compute $i_{\min}$ for both projects and compute the return on invested capital at $k$ = 6%.

**6.6.** Consider the following set of investment projects.

| | After-Tax Cash Flow | | | | | |
|---|---|---|---|---|---|---|
| Project | $n$: 0 | 1 | 2 | 3 | 4 | 5 |
| 1 | −$10 | 60 | −120 | 80 | | |
| 2 | −$225 | 100 | 100 | 100 | 100 | |
| 3 | | 100 | 50 | 0 | −230 | |
| 4 | −$100 | 50 | 50 | 50 | −100 | 600 |
| 5 | −$100 | 300 | −100 | 500 | | |

a. Classify each project as either simple or nonsimple.
b. Compute the internal rate(s) of return for each project.
c. Classify each project as either a pure or a mixed investment.
d. Assuming that *MARR* = $i$ = $k$ = 10%, determine the acceptability of each project based on the rate-of-return principle.
e. For all mixed projects, compute the *MIRR*s.

**6.7.** Consider the following set of investment projects.

| Project | After-Tax Cash Flow | | | | | |
|---|---|---|---|---|---|---|
| | $n$: 0 | 1 | 2 | 3 | 4 | 5 |
| 1 | −60 | 70 | −20 | 240 | | |
| 2 | −100 | 50 | 100 | | | |
| 3 | −800 | 400 | −100 | 400 | 400 | −100 |
| 4 | −160 | 920 | −1,700 | 1,000 | | |
| 5 | −450 | −200 | 700 | −60 | 2,000 | −500 |

a. Compute the *PV* for each project. ($i$ = 12%).
b. Classify each project as either simple or nonsimple.
c. Compute the internal rate(s) of return for each project.
d. Classify each project as either a pure or a mixed investment.
e. Assuming that $MARR = i = k = 12\%$, determine the acceptability of each project based on the rate-of-return principle.

**6.8.** Consider the following set of investment projects.

| Project | After-Tax Cash Flow | | | | |
|---|---|---|---|---|---|
| | $n$: 0 | 1 | 2 | 3 | 4 |
| 1 | −$10 | −30 | 80 | −30 | |
| 2 | −$70 | 50 | 23 | 11 | |
| 3 | −$50 | 25 | 102 | −100 | 392 |
| 4 | −$100 | 500 | −600 | | |
| 5 | −$110 | 10 | 100 | 50 | |
| 6 | −$10 | 60 | −110 | 60 | |

a. Classify each project as either simple or nonsimple.
b. Compute the internal rate(s) of return for each project.
c. Classify each project as either a pure or a mixed investment.
d. Assuming that $MARR = i = k = 10\%$, determine the acceptability of each project based on the rate-of-return principle.

**6.9.** Consider the following series of cash flows for an investment project.

| $n$ | 0 | 1 | 2 | 3 |
|---|---|---|---|---|
| $F_n$ | −$500 | 1,000 | 3,000 | −4,000 |

a. Find $i_{min}$ for this investment.
b. Determine whether this is a mixed investment.
c. If this is a mixed investment, derive the functional relationship between the *RIC*, $r^*$, and the cost of capital, $k$.
d. Assume $k$ = 10%. Determine the value of $r^*$ and the acceptability of this investment.

**6.10.** Prove that $r \leq i_{min}$, in relation to the project balance.

**6.11.** Consider the following set of investment projects.

| Project | After-Tax Cash Flow | | |
|---|---|---|---|
| | $n$: 0 | 1 | 2 |
| 1 | −$1,000 | 500 | 840 |
| 2 | −$2,000 | 1,560 | 944 |
| 3 | −$1,000 | 1,400 | −100 |

Assume $MARR = i = 12\%$ in the following questions.

a. Compute the internal rate of return for each project. If there is more than one rate of return, identify all the rates.

b. Determine the acceptability of each project based on the rate-of-return principle.

**6.12.** Consider the projects described in Problem 6.3.

a. Compute the rate of return (internal rate of return) for each project.

b. Plot the present value as a function of interest rate ($i$) for each project.

c. Classify each project as either simple or nonsimple. Then reclassify each project as either a pure or a mixed investment.

d. Now determine the acceptability of each project by using the rate-of-return principle. Use $MARR(i) = 10\%$.

**6.13.** Consider the following investment project at $MARR = 10\%$.

| Cash Flow | $n$: 0 | 1 | 2 | 3 | 4 | 5 | 6 | 7 | 8 | 9 | 10 |
|---|---|---|---|---|---|---|---|---|---|---|---|
| $b_n$ | | | | 100 | 100 | 200 | 300 | 300 | 200 | 100 | 50 |
| $c_n$ | \$200 | 100 | 50 | 20 | 20 | 100 | 100 | 100 | 50 | 50 | 30 |
| $F_n$ | −\$200 | −100 | −50 | 80 | 80 | 100 | 200 | 200 | 150 | 50 | 20 |

a. Identify the values of $N$, $m$, $B$, $C$, $I$, and $C'$.

b. Compute $R_A$, $R_N$, and the $L$–$S$ ratio.

c. Compute the $PV(10\%)$.

**6.14.** Consider the investment situation in which an investment of $P$ dollars at $n = 0$ is followed by a series of equal annual positive payments $A$ over $N$ periods. If it is assumed that $A$ dollars are recovered each year, with $A$ being a percentage of $P$, the number of years required for payback can be found as a function of the rate of return of the investment. That is, knowing the relationship $A = P(A/P,i^*,N)$, or

$$A = P\left[\frac{i^*(1 + i^*)^N}{(1 + i^*)^N - 1}\right]$$

we can rewrite the relationship as

$$i^* = \frac{A}{P} - \frac{A}{P}\left(\frac{1}{1 + i^*}\right)^N$$

Note that $A/P$ is the payback reciprocal, $R_p = 1/n_p$. Rearranging terms yields

$$R_p = \frac{i^*}{1 - (1 + i^*)^{-N}}$$

This relationship provides a convenient equation for carrying out a numerical analysis of the general relation between the payback reciprocal and the internal rate of return.

a. Develop a chart that estimates the internal rate of return of a project as a function of payback reciprocal.

b. Consider a project that requires an initial investment of \$1,000 and has annual receipts of \$500 for 5 years. This project has a payback period of 2 years, giving $R_p = 0.5$. Verify that the project has the internal rate of return of 41.04% from the chart developed in part a.

**6.15.** Johnson Chemical Company is considering investing in a new composite material processing project after a 3-year period of research and process development.

*R&D cost:* \$3 million over a 3-year period, with an annual R&D growth rate of 50%/year (\$0.63 million at the beginning of year 1, \$0.95 million at the beginning of year 2, and \$1.42 million at the beginning of year 3). These R&D expenditures will be expensed rather than amortized for tax purposes.

*Capital investment:* \$5 million at the beginning of year 4, depreciated over a 7-year period using MACRS percentages.

*Process life:* 10 years.

*Salvage value:* 10% of initial capital investment at the end of year 10.

*Total sales:* \$100 million (at the end of year 4) with a sales growth rate of 10%/year (compound growth) during the first 6 years and −10% (negative compound growth)/year for the remaining process life.

*Out-of-pocket expenditures:* 80% of annual sales.

*Working capital:* 10% of annual sales (considered as an investment at the beginning of each year and recovered fully at the end of year 10)

*Marginal tax rate:* 40%.

*Minimum attractive rate of return (MARR):* 18%.

a. Compute the net present value of this investment and determine whether the project should be pursued.
b. Compute the rate of return on this investment.
c. Compute the benefit–cost ratio for this investment.
d. Compute the annual equivalent for this project.

# 7

# Decision Rules for Selecting among Multiple Alternatives

## 7.1. INTRODUCTION

In the previous chapter we presented ten different criteria for measuring the investment worth of an individual project. For an individual project all ten criteria yield consistent answers for the accept–reject decision. Which one to use is therefore a question of convenience and habit. When we *compare* projects, however, the situation is quite different. Naive or improper application of various criteria can lead to conflicting results. Fortunately, the *proper use of any of the ten criteria will always result in decisions consistent with PV analysis, which we consider the baseline, or "correct," criterion.*

In Section 7.2 we present some preliminary steps that must be taken before analysis can begin: formulating mutually exclusive alternatives and ordering them. Section 7.3 is the main part of the chapter, and here we present the criteria and decision rules for comparing alternatives. In Section 7.4 we examine some of the more detailed aspects of the "assumptions" behind the decision criteria and consider other writings on the subject. Section 7.5 treats the subject of unequal lives, which becomes important in service projects; benefits of service projects are unknown or not measured. Finally, there is a brief discussion of investment timing in Section 7.6.

As in Chapter 6, we assume that the *MARR* is known and that we operate in a stable, perfect capital market with complete certainty about the outcome of investments. The firm can therefore borrow funds at the *MARR* and invest any excess funds at the same rate. The firm's ability to borrow may be *limited,* however, which differs from the situation assumed in Chapter 6.

## 7.2. FORMULATING MUTUALLY EXCLUSIVE ALTERNATIVES

We need to distinguish between projects that are independent of one another and those that are dependent. We say that two or more projects are *independent* if the accept–reject decision of one has no influence, except for a possible budgetary reason, on the accept–reject decision of any of the others. We call this a *set of independent projects.* Typical examples are projects that derive revenues from different markets and require different technical resources.

Two or more projects are *mutually exclusive* if the acceptance of any one precludes the acceptance of any of the others. We call this a *set of mutually exclusive projects.* An example is a set of projects, each of which requires full-time use of a single, special-purpose machine. If we select a particular project, the machine becomes unavailable for any other use.

Two projects are *dependent* if the acceptance of one requires the acceptance of another. For example, the decision to add container ship dock facilities in an existing harbor may require a decision to increase the depth of the harbor channel. The container ship dock project is dependent on the channel project. Notice that the channel project does *not* depend on the dock project, however, since an increase in channel depth can benefit the conventional docks. If the channel project also depended on the container dock project, we would combine the two into one project.

Before applying any investment criterion to selecting among projects, we follow this procedure.

1. Reject any individual project that fails to meet the criterion acceptance test, *unless* some other project that passes the test depends on it. This step is not absolutely necessary, but it speeds later computations.
2. Form all possible, feasible *combinations* with the remaining projects. We call this step formulating mutually exclusive alternatives.
3. *Order* the alternatives formed in step 2, usually, but not always, by the investment required at time 0, $c_0$. If there is an overall budget limit, we may at this step eliminate any alternatives that exceed the limit.

### *Example 7.1*

A chemical company is considering the manufacture of two products, A and B. The market demand for each of these products is independent of the demand for the other. Product A may be produced by either process x or process y, and product B by either process y or process z. It is inefficient to use more than one process to manufacture a particular product, and no process may be used to manufacture more than one product. *Formulate* all mutually exclusive investment alternatives. Table 7.1 presents the eight alternatives. The first one is the *do-nothing* alternative, which should always be included. We then list all alternatives that consist of a single product for manufacture, followed by all feasible combinations of two products. Since Ay and By are inherently mutually exclusive, we do not consider the combination. Nor do we consider combinations such as Ax, Ay, since they are mutually exclusive according to the problem statement. □

**Table 7.1** ***Mutually Exclusive Investment Alternatives, Example 7.1***

| Alternative | Product–Process Combinations Included |
|---|---|
| 1 | None |
| 2 | Ax |
| 3 | Ay |
| 4 | By |
| 5 | Bz |
| 6 | Ax, By |
| 7 | Ax, Bz |
| 8 | Ay, Bz |

## *Example 7.2*

A marketing manager is evaluating strategies for three market areas, A, B, and C. The strategy selected in any one area is independent of that in any other area. Only one strategy is to be selected for each area. There are two strategies for A, 1 and 2; three for B, 1, 2, and 3; and three for C, 1, 2, and 3. Strategy 1 for any area is a do-nothing strategy. *Formulate* all mutually exclusive investment alternatives. Table 7.2 presents the (2)(3)(3) = 18 alternatives. □

After formulating all possible, feasible combinations, we *treat* them as a set of mutually exclusive alternatives. The cash flow for any alternative is simply the sum of the cash flows of the included projects. Since we consider all possible, feasible combinations, we must obtain the optimal combination. The reason for defining mutually exclusive alternatives is related to the properties of an invest-

**Table 7.2** ***Mutually Exclusive Investment Alternatives, Example 7.2***

| | Strategy Selected for Each Market Area | | | | Strategy Selected for Each Market Area | | |
|---|---|---|---|---|---|---|---|
| Alternative | A | B | C | Alternative | A | B | C |
| 1 | 1 | 1 | 1 | 10 | 2 | 1 | 1 |
| 2 | 1 | 1 | 2 | 11 | 2 | 1 | 2 |
| 3 | 1 | 1 | 3 | 12 | 2 | 1 | 3 |
| 4 | 1 | 2 | 1 | 13 | 2 | 2 | 1 |
| 5 | 1 | 2 | 2 | 14 | 2 | 2 | 2 |
| 6 | 1 | 2 | 3 | 15 | 2 | 2 | 3 |
| 7 | 1 | 3 | 1 | 16 | 2 | 3 | 1 |
| 8 | 1 | 3 | 2 | 17 | 2 | 3 | 2 |
| 9 | 1 | 3 | 3 | 18 | 2 | 3 | 3 |

ment worth criterion. If we are considering the projects as wholly or partially independent, can we be sure our criterion will always lead to the best combination, no matter which project we examine first? With mutually exclusive alternatives we avoid this type of problem, because we have specific rules for ordering the alternatives before applying the investment worth criterion.

The *ordering* of the alternatives depends on which criterion is to be applied. There are four classifications.

1. Time 0 investment, $c_0$: order the alternatives by increasing $c_0$. Applies to *PV, FV, AE, PB,* and *ARR*.
2. *I*, the *PV*($i$) of initial investments $c_0, c_1, ..., c_m$: order by increasing *I*. Here $i$ is the *MARR*. Applies to $R_N$ and *L–S*.
3. *C*, the *PV*($i$) of all expenditures, consisting of initial investment plus annual expenses: order by increasing *C*. Again, $i$ is the *MARR*. Applies to $R_A$ and *MIRR*.
4. *PV*(0%) of all cash flows: order by increasing *PV*(0%). When there are ties, order by increasing first derivative of *PV*(0%). Applies to *IRR* and *RIC*.

These ordering rules are designed to facilitate the application of the criteria, as shown in the next section. They are not the only rules. For example, *any* ordering rule will work with *PV, FV, AE,* and *PB*. In addition, we can sometimes use ordering rule 1, based on $c_0$, with the other criteria, provided we modify the decision rules. These modifications often result in cumbersome variations and thus are usually avoided.

## 7.3 APPLICATION OF INVESTMENT WORTH CRITERIA

### 7.3.1 Total Investment Approach

This approach applies the investment criterion separately to each mutually exclusive alternative. Example 7.3 illustrates the approach.

### *Example 7.3*

Two mutually exclusive alternatives, *j* and *k*, are being considered as shown in Table 7.3. Apply the various criteria to each alternative, using *MARR* = 10%. The results are shown in the lower part of Table 7.3. (The derivation of the results in the table is left as an exercise; see Problem 7.3.) □

***Opposite Ranking Phenomenon.*** Four of the criteria seem to indicate that alternative *k* is the better choice, whereas the other six give numerically higher ratings for *j*. We have here an example of the *opposite ranking phenomenon*. The cause of the discrepancy is that some of the criteria are *relative* measures of investment worth and others are *absolute* measures. The resolution of this conflict, for the situation of perfect capital markets and complete certainty, is given by the *incremental approach* in Section 7.3.2.

**Table 7.3** ***Total Investment Approach, Example 7.3***

| | Alternative $j$ | | | Alternative $k$ | | |
|---|---|---|---|---|---|---|
| Time | Outflow | Inflow | Net Flow | Outflow | Inflow | Net Flow |
| 0 | $1,000 | 0 | −$1,000 | $2,000 | 0 | −$2,000 |
| 1 | 2,000 | 2,475 | 475 | 5,000 | 5,915 | 915 |
| 2 | 1,000 | 1,475 | 475 | 6,000 | 6,915 | 915 |
| 3 | 500 | 975 | 475 | 7,000 | 7,915 | 915 |

| Criterion* | Value for $j$ | Value for $k$ | Alternative with Larger Value |
|---|---|---|---|
| $PV$ | $181 | $275 | $k$ |
| $FV$ | $241 | $367 | $k$ |
| $AE$ | $ 73 | $111 | $k$ |
| $PB_N$ | $242 | $367 | $k$ |
| $IRR$ | 20% | 18% | $j$ |
| $ARR$ | 16% | 15% | $j$ |
| $MIRR$ | 12% | 11% | $j$ |
| $R_A$ | 1.045 | 1.016 | $j$ |
| $R_N$ | 1.182 | 1.138 | $j$ |
| $L\text{–}S$ | 0.182 | 0.138 | $j$ |

*$i$ = 10% for all criteria.

At this point, we argue that *when we apply the total investment approach*, the $PV$, $FV$, $AE$, and $PB_N$ give the *correct answer*. This is so because maximizing these criteria maximizes the future wealth of the firm. This point is proved in detail in Section 7.4.1. Before we resolve the discrepancies between $PV$ and the other criteria, some special cases are considered.

***Consistency Within Groups.*** The consistency within groups of the criteria is not coincidence but rather a fundamental characteristic. If the lifetimes of all alternatives are the same and $-100\% < i$, it is easy to show that the following groups will always show internal consistency in ranking mutually exclusive alternatives.

*PV, FV, AE and* $PB_N$. The four criteria, $PV$, $FV$, $AE$, and $PB$, will always agree among themselves.
If

$$PV(i)_j < PV(i)_k$$

then

$$(F/P, i, N)PV(i)_j < (F/P, i, N)PV(i)_k$$

and

$$FV(i)_j < FV(i)_k$$

In addition,

$$(A/P, i, N)PV(i)_j < (A/P, i, N)PV(i)_k$$

and

$$AE(i)_j < AE(i)_k \tag{7.1}$$

The $PB_N(i)$ is the same as $FV(i)$, so we complete the proof.

These criteria measure the surplus in an investment alternative over and above investment of $i = MARR$. It does not matter when we measure the surplus in comparing alternatives—at time 0, at time $N$, or spread equally over the life of the alternative. If one alternative has a greater time 0 surplus than another, its time $N$ surplus will also be greater, and so forth. The surplus is measured in dollars (or other currency unit), and hence these criteria are *absolute* measures of investment worth. This argument again reinforces the *correctness of using PV, FV, AE, and* $PB_N$ *with the total investment approach.*

For example, the addition of alternative $m$ to $j$, where $PV(10\%)_m$ equals 0, does not change the $PV$ measure of $j$:

| Net Cash Flow | n: 0 | 1 | 2 | 3 | PV(10%) |
|---|---|---|---|---|---|
| Alternative $m$ | −$5,000 | 0 | 0 | 6,655 | 0 |
| Alternative $j$ | −$1,000 | 475 | 475 | 475 | 181 |
| Alternative $j + m$ | −$6,000 | 475 | 475 | 7,130 | 181 |

*$R_N$, L–S, and ARR.* The Lorrie–Savage ratio $L–S$ is simply the netted benefit–cost ratio minus one, or $L–S = R_N - 1$, so we need only compare Solomon's average rate of return, $ARR$, with $R_N$. In addition to equal lifetimes and $-100\% < i$, we assume the initial investment occurs only at time 0 (other outlays are annual operating expenses). Then $I = c_0$, and $R_N = (B - C')/c_0$.
Assume

$$R_{Nj} > R_{Nk}$$

Then

$$\frac{B_j - C'_j}{c_{0j}} > \frac{B_k - C'_k}{c_{0k}}$$

or

$$\sum_{n=1}^{N} \frac{F_{nj}(1 + i)^{-n}}{c_{0j}} > \sum_{n=1}^{N} \frac{F_{nj}(1 + i)^{-n}}{c_{0k}}$$

where $F_{nj}$ is the net cash flow for alternative $j$ at the end of period $n$. In addition,

$$\sum_{n=1}^{N} \frac{F_{nj}(1 + i)^{N-n}}{c_{0j}} > \sum_{n=1}^{N} \frac{F_{nk}(1 + i)^{N-n}}{c_{0k}}$$

Substituting from Eq. 6.18, we have

$$(1 + s_j)^N > (1 + s_k)^N$$

and

$$s_j > s_k \tag{7.2}$$

If the initial investment extends beyond time 0, the result need not hold (see Problem 7.16 at the end of the chapter).

*$R_A$, MIRR.* The aggregate benefit–cost ratio, $R_A$, and the modified internal rate of return as defined by Eq. 6.23, *MIRR,* will always agree. Assume

$$R_{Aj} > R_{Ak}$$

or

$$\frac{B_j}{I_j + C'_j} > \frac{B_k}{I_k + C'_k}$$

From the definitions of $B$, $I$, and $C'$, Eqs. 6.1, 6.29, and 6.30, we substitute and obtain

$$\frac{\sum_{n=0}^{N} b_{nj}(1+i)^{-n}}{\sum_{n=0}^{N} c_{nj}(1+i)^{-n}} > \frac{\sum_{n=0}^{N} b_{nk}(1+i)^{-n}}{\sum_{n=0}^{N} c_{nk}(1+i)^{-n}}$$

Then

$$\frac{\sum_{n=0}^{N} b_{nj}(1+i)^{N-n}}{\sum_{n=0}^{N} c_{nj}(1+i)^{N-n}} > \frac{\sum_{n=0}^{N} b_{nk}(1+i)^{N-n}}{\sum_{n=0}^{N} c_{nk}(1+i)^{N-n}}$$

Using Eq. 6.23, we obtain

$$(1 + MIRR_j)^N > (1 + MIRR_k)^N$$

$$MIRR_j > MIRR_k \tag{7.3}$$

*IRR.* The internal rate of return, *IRR,* or return on invested capital, *RIC,* for mixed investments does not necessarily agree with any of the other criteria.

***Special Cases.*** In some special cases there will be agreement across some of the groups [2]. If each alternative has the same initial investment, the *PV* and $R_N$ groups will give consistent rankings. If each alternative has a constant net cash flow during its lifetime, *IRR* (or *RIC*) will agree with the $R_N$ group.

***Modification of Criteria To Include Unspent Budget Amounts.*** Some authors advocate a modification of the investment criteria to include the effects of left-over funds [8,12]. Applying this concept to alternatives $j$ and $k$ with *IRR*, we would add to alternative $j$ an additional investment of \$1,000 earning interest at $MARR = 10\%$ and returning $1{,}000(A/P, 10\%, 3) = \$402$ each year. The argument is that we have \$2,000 to invest at time 0; otherwise we would not consider alternative $k$. The augmented cash flow, designated by some as a *total cash flow*, becomes for alternative $j$

| Time | Original Net Flow, Alternative $j$ | Unspent 1,000 Earning 10% | Total Flow |
|---|---|---|---|
| 0 | −\$1,000 | −\$1,000 | −\$2,000 |
| 1 | 475 | 402 | 877 |
| 2 | 475 | 402 | 877 |
| 3 | 475 | 402 | 877 |

The *IRR* for the total flow is 15%, which is less than the 18% for $k$. Thus, *IRR*, applied to the total cash flow, agrees with *PV*. Notice that this agreement between the criteria can be derived from the special cases just mentioned.

A similar approach has been proposed for benefit–cost ratios [8]. The extent to which the total cash flow approach ensures consistent ranking by the various criteria does not appear to have been fully examined. (See Problem 7.24 at the end of the chapter.) The following example illustrates opposite ranking with the same initial investment.

## *Example 7.4*

Relevant summary data for alternatives $p$ and $q$ are given below. Evaluate the alternatives by using $R_A$ and $R_N$. Here $I$ is assumed to be $c_0$.

| Item | Alternative $p$ | Alternative $q$ |
|---|---|---|
| Time 0 investment, $c_0$ | \$100 | \$100 |
| *PV* of annual expenses, $C'$ | 10 | 0 |
| *PV* of annual receipts, $B$ | 220 | 205 |

Computing $R_A$ and $R_N$, we have

| Ratio | Alternative $p$ | Alternative $q$ |
|---|---|---|
| $R_A$ | 2.00 | 2.05 |
| $R_N$ | 2.10 | 2.05 |

□

In Example 7.4 alternative $p$ has a smaller $R_A$ value but a larger $R_N$ value. Since the classification of a cash flow element as a user benefit or a sponsor cost is often arbitrary, the use of a total cash flow approach is questionable.

### 7.3.2 Incremental Analysis

Investment alternatives can have opposite ranking because some criteria are *relative* measures of investment worth. The resolution of the discrepancy requires incremental analysis. The general approach is as follows.

1. *Order* the investment alternatives by the ordering rule specified for the criterion in Section 7.2.
2. Apply the criterion to the cash flow of the first alternative.
3. a. If the criterion value is favorable, go to step 4.
   b. If the criterion value is unfavorable, select the next alternative in order. Continue until an alternative with a favorable criterion value is obtained. (If none is obtained, reject all alternatives.) Go to step 4.
4. Apply the criterion to the cash flow *difference* between the next alternative in order and the one most recently evaluated favorably.
5. Repeat step 4 until no more alternatives exist. Accept the last alternative for which the cash flow difference was evaluated favorably.

See Example 7.6 at the end of Section 7.3 for a comprehensive application of these rules.

***Irrelevance of Ordering for PV, FV, AE, and $PB_N$.*** The ordering rule for criteria *PV, FV, AE,* and $PB_N$ is by increasing time 0 investment, $c_0$. This rule is based on convention but is not required. For these four *absolute* measures of investment worth, the *ordering is irrelevant; furthermore, the incremental analysis always agrees with the total investment approach, which is optimal for perfect capital markets and complete certainty.*

If

$$PV(i)_j < PV(i)_k$$

then

$$PV(i)_{k-j} > 0 \quad (7.4)$$

by the definition of *PV* and the distributive rule of multiplication. In addition,

$$PV(i)_{k-j} = -PV(i)_{j-k} \quad (7.5)$$

and since the other three criteria always agree with *PV,* the ordering of alternatives is irrelevant for this group.

Applying these rules to Example 7.3, we obtain the ordering based on $c_0$: $j$, $k$. We have $PV(10\%)_j = \$181$, which is favorable.

We than examine the cash flow difference between $k$ and $j$.

| | Net Flow | | |
|---|---|---|---|
| $n$ | Alt. $j$ | Alt. $k$ | $k - j$ |
| 0 | −$1,000 | −$2,000 | −$1,000 |
| 1 | 475 | 915 | 440 |
| 2 | 475 | 915 | 440 |
| 3 | 475 | 915 | 440 |

We have

$$PV(10\%)_{k-j} = -1{,}000 + 440\overset{P/A,\ 10\%,\ 3}{(2.4869)} = \$94$$

There are no more alternatives, and we accept $k$, the last one for which the cash flow difference was evaluated favorably.

*If we had started with the larger* time 0 investment, we would have evaluated $k$ and found it favorable with a $PV$ of $275. The cash flow difference between $j$ and $k$ is

| $n$ | 0 | 1 | 2 | 3 |
|---|---|---|---|---|
| $j - k$, Cash Flow | $1,000 | −440 | −440 | −440 |

The $PV(10\%)_{j-k} = -\$94$, and we again accept $k$.

***Agreement on Increments Between PV and Other Criteria.*** Let us compare with $PV$ any *one* of the other criteria, from the set $IRR$ (or $RIC$), $ARR$, $MIRR$, $R_A$, $R_N$, $L\text{–}S$. We will use the ordering rule for the other criterion, since we just showed that for $PV$ the ordering is irrelevant. The ordering rules are designed so that each increment appears to be an investment when evaluated by the criterion, as

**Table 7.4** ***Incremental Analysis, Example 7.3***

| Criterion | Value for $j$ | Favorable? | Next Increment | Value for Increment | Favorable? | Final Choice |
|---|---|---|---|---|---|---|
| $PV$ | $181 | Yes | $k-j$ | $94 | Yes | $k$ |
| $FV$ | $241 | Yes | $k-j$ | $125* | Yes | $k$ |
| $AE$ | $73 | Yes | $k-j$ | $38 | Yes | $k$ |
| $PB_N$ | $242 | Yes | $k-j$ | $125 | Yes | $k$ |
| $IRR$ | 20% | Yes | $k-j$ | 15% | Yes | $k$ |
| $ARR$ | 16% | Yes | $k-j$ | 14% | Yes | $k$ |
| $MIRR$ | 12% | Yes | $k-j$ | 10.3% | Yes | $k$ |
| $R_A$ | 1.045 | Yes | $k-j$ | 1.007 | Yes | $k$ |
| $R_N$ | 1.182 | Yes | $k-j$ | 1.093 | Yes | $k$ |
| $L\text{–}S$ | 0.182 | Yes | $k-j$ | 0.093 | Yes | $k$ |

*Values do not add to 367 because of rounding.

opposed to a loan, for example. Examining each increment by both criteria will yield the *identical sequence of accept–reject decisions,* because the other criterion *always agrees with PV for an individual project, or cash flow,* as shown in Chapter 6. Therefore, using *incremental analysis with any of the ten criteria will result in optimal decisions.* Some of the ordering rules may give different *sequences* of increments, but since each criterion agrees step by step with *PV,* and since *PV* is indifferent to ordering, the *final decisions will be the same.*

Table 7.4 contains the relevant data for all ten criteria as applied to Example 7.3. For each the ordering is *j, k.* Again, the derivation of table entries is left as an exercise; see Problem 7.4.

***Alternative Derivations.*** In this section we provide some alternative algebraic derivations to show the correctness of incremental analysis. Space limits prevent us from presenting all of them, and some are left as chapter problems. These proofs also illustrate the logic behind the ordering rules.

$R_N$. If

$$R_{N,k-j} > 1$$

then

$$(B_{k-j} - C'_{k-j})/I_{k-j} > 1$$

and

$$B_{k-j} - C'_{k-j} > I_{k-j}$$

Since

$$I_{k-j} > 0 \quad \text{by the ordering rule,}$$

$$B_k - C'_k - I_k > B_j - C'_j - I_j$$

or

$$PV_k > PV_j \tag{7.6}$$

We can also reverse the step sequence. Thus, the netted benefit–cost ratio, when used with incremental analysis, always agrees with *PV.*

$R_A$. If

$$R_{A,\,k-j} > 1$$

then

$$B_{k-j}/C_{k-j} > 1$$

and

$$B_{k-j} > C_{k-j}$$

Since

$$C_{k-j} > 0 \quad \text{by the ordering rule}$$

$$B_k - C_k > B_j - C_j$$

or

$$PV_k > PV_j \tag{7.7}$$

***Detailed Rules for IRR.*** Many practitioners apply *IRR* by using incremental analysis and ordering based on time 0 investment, $c_0$. Most of the time this presents no difficulties. Figure 7.1 shows the *PV* functions of the two alternatives *j* and *k* in Example 7.3. The ordering of alternatives by $c_0$ is first *j*, then *k*. If $MARR = i_1$, then $i_j^* > MARR$, so alternative *j* is favorable. The difference cash flow ($k - j$) has an *IRR* of $i_F > MARR$, so we accept *k*. (We call the *IRR* $i_F$ for Fisher's intersection, described in Section 7.4.) From the graph in Figure 7.1, it is clear that $PV(i_1)_k > PV(i_1)_j$, so we are consistent with *PV*. If $MARR = i_2$, then again $i_j^* > MARR$. But the *IRR* for ($k - j$) is less than *MARR*, so our final choice is alternative *j*. Again, we have consistency with *PV*, since $PV(i_2)_k < PV(i_2)_j$.

But what if the ordering is first *k*, then *j*, and the *PV* functions are similar to those in Figure 7.1? The next example illustrates this situation.

## *Example 7.5*

Compare the following two alternatives by using *IRR* with incremental analysis based on ordering by $c_0$, with $MARR = 10\%$.

| | Net Flow | | |
|---|---|---|---|
| $n$ | Alt. $w$ | Alt. $j$ | $j - w$ |
| 0 | −$900 | −$1,000 | −$100 |
| 1 | −350 | 475 | 825 |
| 2 | 915 | 475 | −440 |
| 3 | 915 | 475 | −440 |

The ordering by $c_0$ is first *w*, then *j*.

Examining *w*, we have $i_w^* = 19\% > 10\%$, so it is favorable. The difference cash flow has multiple sign changes, so multiple roots are possible. However, in the range 0 to 100% there is only one *IRR*: 16%. This value exceeds *MARR*, so we would accept *j*. However, $PV(10\%)_j = 181$, which is *less* than $PV(10\%)_w = 225$!

The explanation is that the cash flow difference ($j - w$) represents a borrowing activity, despite the negative time 0 flow. The *PV* function for ($j - w$) begins negative (which is characteristic of borrowing activities), crosses the horizontal axis near 16%, and then continues upward. The other root is at $i = 660\%$. (Applying *RIC*, we obtain $r_{j-w}^* = -38.64\%$ at $k = i = 10\%$, so we accept *w*.)

| $i$ | 0 | 10 | 16 | 20 | 50 | 100 | 200 | 500 | 660 | 700 |
|---|---|---|---|---|---|---|---|---|---|---|
| $PV(i)_{j-w}$ | −155 | −44 | 2 | 27 | 124 | 148 | 110 | 23 | 0 | −5 |

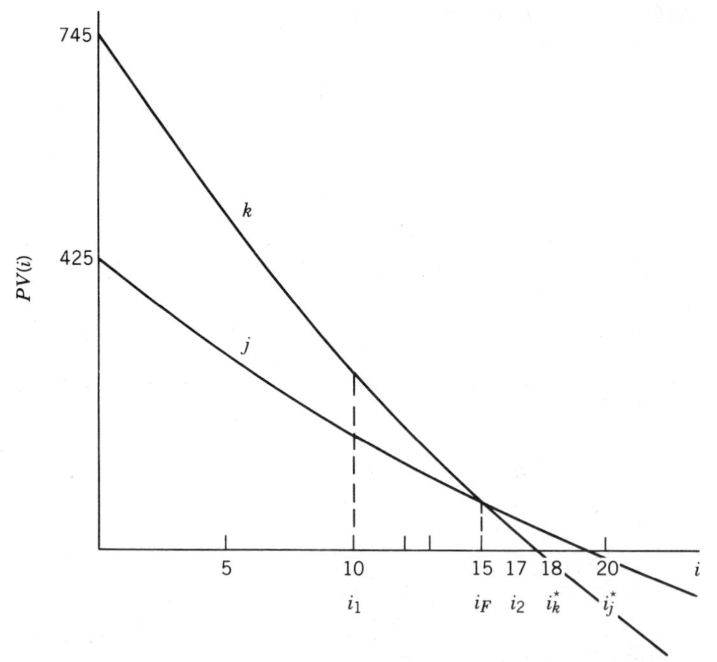


**FIGURE 7.1.** *IRR* **for cash flow difference (*k* - *j*), Example 7.3.**

From this example, it is clear that *ordering by* $c_0$ for *IRR* can lead to *incorrect* results.

Various other circumstances can cause problems for the practitioner accustomed to ordering by $c_0$. These include the situations in which the time 0 investment is the same for two alternatives. To remedy these difficulties, Wohl has recently developed a set of strict rules for applying *IRR* with incremental analysis [1, 13]. These rules result in complete consistency with *PV.*

For many applications the ordering rule based on *PV*(0%) clears up any inconsistencies between *IRR* and *PV.* When there are multiple roots, the *IRR* criterion can be replaced by return on invested capital, *RIC.* Multiple roots inevitably cause more computational work, whether we use *RIC,* the strict rules for *IRR* that require obtaining all roots, or plotting of the *PV* function. Plotting the *PV* function need be neither difficult nor time-consuming, and the plot contains at least as much information as is obtained by the other methods. In essence, we have argued for use of the *PV* criterion.

## *Example 7.6*

We end this section with a comprehensive example that demonstrates the incremental analysis technique for several of the criteria. Three independent projects, A, B, and C, are to be evaluated by using *PV,* $R_N$, $R_A$, and *IRR,* with *MARR* = 10%. (Note that each of the four groups is represented.) There is a time 0 expenditure budget of $4,000. The cash flows for the three projects are given in the upper left portion of Table 7.5. Select the best project or projects.

**Table 7.5** ***Preliminary Data for Example 7.6***

| Alternative: | 1 | 2 | 3 | 4 | 5 | 6 |
|---|---|---|---|---|---|---|
| Project: | A | B | C | A + B | A + C | B + C |
| Outflows, $n$ = 0 | \$1,000 | 900 | 3,000 | 1,900 | 4,000 | 3,900 |
| 1 | \$2,000 | 1,265 | 5,000 | 3,265 | 7,000 | 6,265 |
| 2 | \$1,000 | 6,000 | 5,000 | 7,000 | 6,000 | 11,000 |
| 3 | \$500 | 7,000 | 5,000 | 7,500 | 5,500 | 12,000 |
| Inflows, $n$ = 0 | \$0 | 0 | 0 | 0 | 0 | 0 |
| 1 | \$2,475 | 915 | 6,336 | 3,390 | 8,811 | 7,251 |
| 2 | \$1,475 | 6,915 | 6,336 | 8,390 | 7,811 | 13,251 |
| 3 | \$975 | 7,915 | 6,336 | 8,890 | 7,311 | 14,251 |
| Net flows, $n$ = 0 | −\$1,000 | −900 | −3,000 | −1,900 | −4,000 | −3,900 |
| 1 | \$475 | −350 | 1,336 | 125 | 1,811 | 986 |
| 2 | \$475 | 915 | 1,336 | 1,390 | 1,811 | 2,251 |
| 3 | \$475 | 915 | 1,336 | 1,390 | 1,811 | 2,251 |
| $c_0$ | \$1,000 | 900 | 3,000 | 1,900 | 4,000 | 3,900 |
| $m$ | 0 | 2 | 0 | 0,2 | 0 | 2,0 |
| $I(10\%)$ | \$1,000 | 7,009 | 3,000 | 8,009 | 4,000 | 10,009 |
| $C'(10\%)$ | \$3,020 | 5,259 | 12,434 | 8,279 | 15,455 | 17,693 |
| $C(10\%)$ | \$4,020 | 12,268 | 15,434 | 16,288 | 19,455 | 27,702 |
| $PV(0\%)$ | \$425 | 580 | 1,008 | 1,005 | 1,433 | 1,588 |
| $B(10\%)$ | \$4,202 | 12,493 | 15,757 | 16,695 | 19,958 | 28,250 |
| $PV(10\%)$ | \$181 | 225 | 322 | 407 | 504 | 548 |
| $R_N$ | 1.182 | 1.032 | 1.108 | NA | NA | NA |
| $R_A$ | 1.045 | 1.018 | 1.021 | NA | NA | NA |
| $IRR$, % | 20.0 | 18.8 | 16.0 | NA | NA | NA |

NOTE: $m$ is the period of the initial investments $c_0, c_1, ..., c_m$.

**Preliminary screening.** The lower left portion of Table 7.5 shows the relevant data for screening the projects individually. Each of the three projects, A, B, and C, is acceptable by each of the four criteria, $PV$, $R_N$, $R_A$, and $IRR$, using $MARR$ = 10%. (We expected the agreement on the individual projects by the criteria.)

**Form investment alternatives.** Since we know that the budget is \$4,000 and that A, B, and C are independent, we can form three combinations: (A + B), (A + C), and (B + C). We thus have six investment alternatives (in addition to the do-nothing alternative). For the three alternatives composed of combinations of projects, the cash flows are shown in the upper right portion of Table 7.5, and the data needed for applying the criteria are shown in the lower right.

**Order the alternatives.** Here we apply the ordering rules specified in Section 7.2.

For $PV$, order by $c_o$: alternatives 2, 1, 4, 3, 6, 5.
For $R_N$, order by $I$: alternatives 1, 3, 5, 2, 4, 6.

For $R_A$, order by $C$: alternatives 1, 2, 3, 4, 5, 6.
For *IRR*, order by $PV(0\%)$: alternatives 1, 2, 4, 3, 5, 6.

We are now ready to apply the incremental method with the four criteria.

***PV*(10%)**

Alt. 2 vs. do nothing, or B vs. do nothing: $PV(10\%)_{2-0} = \$225 > 0$, so alt. 2 is *favorable.*

Alt. 1 vs. alt. 2, or A vs. B: $PV(10\%)_{1-2} = -\$44 < 0$, so alt. 1 is *not* favored over alt. 2. Note that we are using the relation $PV(i)_{x-y} = PV(i)_x - PV(i)_y$ to save ourselves some work. The cash flow difference (1 − 2) is the same as (A − B).

Alt. 4 vs. alt. 2, or (A + B) vs. A: $PV(10\%)_{4-2} = \$182 > 0$, so alt. 4 is *favored* over alt. 2. Note that the difference (4 − 2) is just the cash flow for A.

Alt. 3 vs. alt. 4, or C vs. (A + B): $PV(10\%)_{3-4} = -\$85 < 0$, so alt. 3 is *not* favored over 4.

Alt. 6 vs. alt. 4, or (B + C) vs. (A + B): $PV(10\%)_{6-4} = \$141 > 0$, so alt. 6 is *favored* over alt. 4. The difference (6 − 4) is the same as (C − B).

Alt. 5 vs. alt. 6, or (A + C) vs. (B + C): $PV(10\%)_{5-6} = -\$44 < 0$, so alt. 5 is *not* favored over alt. 6. The difference (5 − 6) is the same as (A − B), which was evaluated earlier.

The last alternative favorably evaluated is 6, so we accept projects B and C with a total *PV* of $548.

**$R_N$, the netted benefit–cost ratio**

Alt. 1 vs. do nothing, or A vs. do nothing: $R_{N,1-0} = 1.182 > 1$, so alt. 1 is *favorable.*

Alt. 3 vs. alt. 1, or C vs. A:

$$R_{N,3-1} = \frac{(\$15{,}757 - 4{,}202) - (\$12{,}434 - 3{,}020)}{\$3{,}000 - 1{,}000} = 1.071 > 1$$

so alt. 3 is *favored* over alt. 1.

Alt. 5 vs. alt. 3, or (A + C) vs. C: The difference (5 − 3) is the same as the cash flow for A. So $R_{N,5-3} = 1.182 > 1$, and alt. 5 is *favored* over alt. 3.

Alt. 2 vs. alt. 5, or B vs. (A + C):

$$R_{N,2-5} = \frac{(\$12{,}493 - 19{,}958) - (\$5{,}259 - 15{,}455)}{\$7{,}009 - 4{,}000} = 0.908 < 1$$

so alt. 2 is *not* favored over alt. 5.

Alt. 4 vs. alt. 5, or (A + B) vs. (A + C):

$$R_{N,4-5} = \frac{(\$16{,}695 - 19{,}958) - (\$8{,}279 - 15{,}455)}{\$8{,}009 - 4{,}000} = 0.976 < 1$$

so alt. 4 is *not* favored over 5. The difference (4 − 5) is the same as (B − C).

Alt. 6 vs. alt. 5, or (B + C) vs. (A + C):

$$R_{N,6-5} = \frac{(\$28{,}250 - 19{,}958) - (\$17{,}693 - 15{,}455)}{\$10{,}009 - 4{,}000} = 1.007 > 1$$

so alt. 6 is *favored* over alt. 5. The difference (6 − 5) is the same as (B − A).

We accept projects B and C, which constitute alternative 6, the last one favorably accepted. This decision agrees with *PV* analysis, as expected.

**$R_A$, The aggregate benefit–cost ratio**

Alt. 1 vs. do nothing, or A vs. do nothing: $R_A$,1−0 = 1.045 > 1, so alt. 1 is *favorable.*

Alt. 2 vs. alt. 1, or B vs. A:

$$R_{A,2-1} = \frac{\$12{,}493 - 4{,}202}{\$12{,}268 - 4{,}020} = 1.005 > 1$$

so alt. 2 is *favored* over alt. 1.

Alt. 3 vs. alt. 2, or C vs. B:

$$R_{A,3-2} = \frac{\$15{,}757 - 12{,}493}{\$15{,}434 - 12{,}268} = 1.031 > 1$$

so alt. 3 is *favored* over alt. 2.

Alt. 4 vs. alt. 3, or (A + B) vs. C:

$$R_{A,4-3} = \frac{\$16{,}695 - 15{,}757}{\$16{,}288 - 15{,}434} = 1.098 > 1$$

so alt. 4 is *favored* over alt. 3.

Alt. 5 vs. alt. 4, or (A + C) vs. (A + B): The difference (5 − 4) is the same as (C − B), which was evaluated in the comparison of alt. 3 vs. alt. 2. So $R_{A,5-4}$ = 1.031 > 1, and alt. 5 is *favored* over alt. 4.

Alt. 6 vs. alt. 5, or (B + C) vs. (A + C): The difference (6 − 5) is the same as (B − A), which was evaluated in the comparison of alt. 2 vs. alt. 1. So $R_A$ $_{,6-5}$ = 1.005 > 1, and alt. 6 is *favored* over alt. 5.

Again, our final selection is alternative 6, or projects B and C.

***IRR*, internal rate of return, and *RIC*, return on invested capital**

Alt. 1 vs. do nothing, or A vs. do nothing: $i^*_{1-0}$ = 20.0% > 10%, so alt. 1 is *favorable.*

Alt. 2 vs. alt. 1, or B vs. A: The difference cash flow is +$100, −825, +440, +440. The multiple sign changes suggest two roots, and if we refer to

Example 7.5, we see that the roots are 16% and 660%. Applying *RIC,* we obtain $r^* = 15.03\%$, so alt. 2 is *favored* over alt. 1.

As an alternative, consider a more fundamental approach to the analysis of the cash flow for alt. 2 vs. alt. 1. In Example 7.5 the opposite cash flow, that is, −$100, +825, −440, −440, was determined to be a borrowing activity. The cash flow +100, −825, +440, +440 is an investment activity, despite the initial inflow.

| $i$ | 0 | 10 | 16 | 20 | 50 | 100 |
|---|---|---|---|---|---|---|
| $PV(i)_{2-1}$ | $155 | 44 | −2 | −27 | −124 | −148 |

Notice that our decision here to accept the cash flow +$100, −825, +440, +440 using $i = 10\%$ is consistent with the decision in Example 7.5 to reject the opposite cash flow using $i = 10\%$. With $i^*_{2-1}$ near 16% > 10%, alt. 2 is *favored* over alt. 1.

Alt. 4 vs. alt. 2, or (A + B) vs. B: The difference cash flow is just that for A. So $i^*_{2-1} = 20.0\% > 10\%$, and alt. 4 is *favored* over alt. 2.

Alt. 3 vs. alt. 4, or C vs. (A + B): The difference cash flow is −$1,100, +1,211, −54, −54. Again, we have multiple sign changes, but Norstrom's auxiliary series $S_n$ is −$1,100, +111, +57, +3. This guarantees a unique, positive, real root (see Section 6.3.1). With $i^*_{3-4} = 0.3\% < 10\%$, alt. 3 is *not* favored over alt. 4.

Alt. 5 vs. alt. 4, or (A + C) vs. (A + B): The difference cash flow is the same as (C − B), or −$2,100, +1,686, +421, +421. This is a pure investment with a unique root of $i^*_{5-4} = 13.5\% > 10\%$, so alt. 5 is *favored* over alt. 4.

Alt. 6 vs. alt. 5, or (B + C) vs. (A + C): We have a repeat of the cash flow for alt. 2 vs. alt. 1, and $i^*_{6-5} = 16\% > 10\%$, so alt. 6 is *favored over alt. 5.*

Again, but after considerable work, our final selection is alternative 6, or projects B and C.

We thus arrive at the same final selection by using incremental analysis with each of the four criteria. □

Several conclusions are drawn from Example 7.6.

1. Correct ordering for evaluation is essential for all criteria except *PV* and the related *FV, AE,* and $PB_N$.
2. Although the ordering is different for each of the criteria used here, the final results are consistent with the fundamental criterion of *PV.*
3. The *IRR* criterion is particularly troublesome to apply, especially when we take differences between combination alternatives. The *RIC* concept is difficult to apply, and sometimes it is easier to obtain the *PV(i)* function.
4. *PV* with the *total investment approach* is by far the *easiest method* to apply. In this example we need to compute the *PV* of each of the three projects, add the appropriate *PVs* to obtain the *PV* of each combination alternative, and simply select the alternative with the largest *PV.*

## 7.4 REINVESTMENT ISSUES

We will begin this section with a simple example that puzzles most students when they encounter it for the first time.

### *Example* 7.7.

Given projects A and B, which is preferred, project A with *MARR* = 5% or project B with *MARR* = 10%?

| | Cash Flows | | | |
|---|---|---|---|---|
| Project | $n$: 0 | 1 | 2 | 3 |
| A | −\$1,000 | 600 | 500 | 300 |
| B | −\$1,000 | 300 | 200 | 1,000 |

If we compute *PV*s, we obtain $PV(5\%)_A = \$284$, $PV(10\%)_B = \$189$. Most students (and practitioners, too) select project A because of its higher *PV*.

But how can we compare a *PV* computed at 5% with one computed at 10%? The interest rate used for discounting certainly implies something about reinvestment opportunities, as discussed in Chapter 5. Trying to compare the two projects as stated in the example is tantamount to trying to compare projects in different economic environments. Projects A and B might represent investment opportunities in two different countries, with different reinvestment rates and restrictions on repatriating cash flows. Or perhaps projects A and B occur in different regulatory environments, and the decision maker assumes that after project selection the firm will reinvest its cash flows in the chosen environment.

If we are eventually to recover the reinvested cash, by repatriating it in the one situation or by returning it to the firm's treasury in the second, it does not make sense to compare A and B by using *PV*. *PV* measures the surplus of funds a project generates over and above a minimum rate, and in this example the minimum rates differ. Instead, let's compute the total cash available at time 3 for each option.

**Direct computation**

Project A: $600(1.05)^2 + 500(1.05)^1 + 300 = \$1{,}487$

Project B: $300(1.1)^2 + 200(1.1)^1 + 1{,}000 = \$1{,}583$

**Computation from *PV***

Project A: $(284 + 1{,}000)(1.05)^3 = \$1{,}486 \approx \$1{,}487$

Project B: $(189 + 1{,}000)(1.1)^3 = \$1{,}583$

We see that project B produces more cash at time 3, which is a direct result of the higher reinvestment rate, 10% for project B versus 5% for project A. It is clear that the reinvestment rate plays a crucial role in the analysis. □

### 7.4.1 Net Present Value

Virtually all writers on engineering economics agree that the *PV* criterion is based on the assumption of reinvestment at the interest rate used for calculating *PV*. In Section 6.2 we assumed that positive cash flows would be reinvested at the outside, or market, interest rate, the same rate used for obtaining *PV*. In Chapter 5 we explained that the equity interest rate is the outside rate from the view of the equity holder. Whichever assumptions we make, we represent the rate by $i$.

In a perfect capital market we can borrow and lend unlimited amounts at the market interest rate. In this chapter we have modified that assumption to reflect a limited borrowing ability. But we still assume that we can *lend unlimited amounts* by investing at a market interest rate. This is the same as assuming reinvestment at the market interest rate. In this situation, maximizing *PV* is the same as maximizing the future cash of the firm.

Assume that we have two mutually exclusive alternatives, $j$ and $k$, with cash flows $F_{nj}$, $n = 0, \ldots, N_j$, and $F_{nk}$, $n = 0, \ldots, N_k$. Further, assume that outlays occur only at time 0 and that we have a budget of $M$, which is greater than either time 0 outlay. The $MARR = i$. Select a horizon time $N$ as the greater of $N_j$ and $N_k$.

We have by definition

$$PV(i)_j = \sum_{n=0}^{N_j} F_{nj}(1 + i)^{-n}, \qquad PV(i)_k = \sum_{n=0}^{N_k} F_{nk}(1 + i)^{-n} \tag{7.8}$$

Now let's obtain the future cash at time $N$ for the three possible decisions. Say $N = N_j$, for example.

*Decision 1, do nothing*

$$\text{Future cash at time } N = M(1 + i)^N \tag{7.9}$$

Unspent amounts are invested at $i$, which is consistent with the reinvestment assumption.

*Decision 2, select j*

Future cash at time $N$

$$= (M + F_{0j})(1 + i)^N + \sum_{n=1}^{N} F_{nj}(1 + i)^{N-n}$$

$$= M(1 + i)^N + \sum_{n=0}^{N} F_{nj}(1 + i)^{N-n}$$

$$= M(1 + i)^N + (1 + i)^N \sum_{n=0}^{N} F_{nj}(1 + i)^{-n}$$

$$= M(1 + i)^N + PV(i)_j(1 + i)^N \tag{7.10}$$

The future cash is the same as for do nothing plus the $PV(i)_j$ shifted to time $N$. For $j$ the $N_j = N$, so the shifted *PV* is the *FV*.

*Decision 3, select k*

Future cash at time $N$

$$= (M + F_{0k})(1 + i)^N + \sum_{n=1}^{N_k} F_{nk}(1 + i)^{N_k - n}(1 + i)^{N - N_k}$$

$$= M(1 + i)^N + \sum_{n=0}^{N} F_{nk}(1 + i)^{N-n}$$

$$= M(1 + i)^N + PV(i)_k(1 + i)^N \tag{7.11}$$

At the end of the project life the accumulated cash from reinvesting project inflows is left to earn interest until time $N$.

The $PV$ of the do-nothing alternative is zero. In each case the future cash at time $N$ is equal to the initial amount $M$ times $(F/P, i, N)$ plus $PV(i)(F/P, i, N)$. Thus, by selecting the alternative with maximum $PV(i)$, we maximize future cash, assuming reinvestment at $i$.

Let us return to Example 7.7 and evaluate A and B with $i = 5\%$.

$$PV(5\%)_A = \$284, \qquad PV(5\%)_B = \$331$$

Here project B is preferred. Computing future cash amounts, we have

Project A: $(1{,}000 + 284)(1.05)^3 = \$1{,}486$

Project B: $(1{,}000 + 331)(1.05)^3 = \$1{,}541$

Again, project B is preferred, in agreement with our theoretical analysis.

### 7.4.2 Internal Rate of Return

Some authors have argued that implicit in the use of the *IRR* is an assumption of reinvestment at the project *IRR* [1, 3, 4]. It is difficult to prove or disprove what someone had in mind in stating the *IRR* criterion or using it. Instead, we show in this section the results of selecting alternatives with *IRR* under some special circumstances.

Let us first compute *IRR* for the projects in Example 7.7.

$$\text{Project A: } -\$1{,}000 + \frac{600}{1+i} + \frac{500}{(1+i)^2} + \frac{300}{(1+i)^3} = 0$$

$$i_A^* = 21.48\%$$

$$\text{Project B: } -\$1{,}000 + \frac{300}{1+i} + \frac{200}{(1+i)^2} + \frac{1{,}000}{(1+i)^3} = 0$$

$$i_B^* = 18.33\%$$

If we simply select A over B on the basis of its higher *IRR,* we would be in conflict with the *PV*s calculated at 5%: \$1,486 for A and \$1,541 for B.

We might ask whether there is a value $i$ for which *PV* favors project A. The

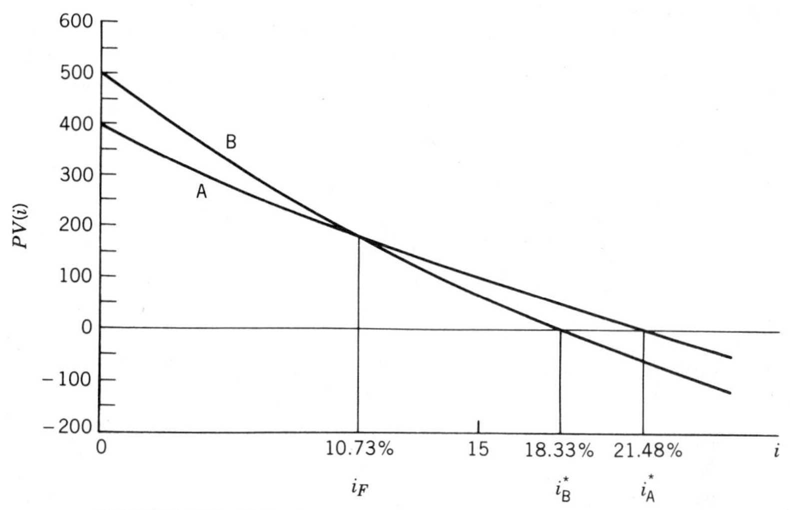


**FIGURE 7.2.** **Fisher's intersection, Example** 7.7.

*PV* curves for A and B are similar to those in Figure 7.1, with B starting higher than A but crossing the horizontal axis sooner. Figure 7.2 shows the curves for A and B. Clearly, *PV*(21.48%) is greater for A than for B. The point of intersection is at 10.73%. This point is also called *Fisher's intersection* or the *rate of return over cost* [5]. For any value of $i$ equal to or greater than 10.73%, the *PV* criterion prefers project A. Fisher's intersection is also the *IRR* of the difference cash flow between A and B: \$0, 300, 300, −700.

If we are uncertain about the reinvestment rate in selecting one of two alternatives, calculating Fisher's intersection can be useful. In our example any reinvestment rate greater than 10.73% would lead us to prefer A over B, and B would be favored at rates lower than 10.73%. This approach becomes cumbersome when more than two alternatives are compared.

Let us now examine the consequences of *assuming reinvestment at IRR.* If each investment alternative has its excess cash reinvested at its *IRR,* we could select among alternatives by choosing the one with the highest *IRR.* Say that alternative $j$ has the highest *IRR,* $i_j^*$. Then for all others $PV(i_j^*)$ must be less than zero. (If we also have borrowing alternatives with upward-sloping *PV* curves, we must modify the acceptance rules.)

But how sensible is the assumption of reinvestment at *IRR*? Not very sensible at all. Do different *IRR*s imply different reinvestment rates? We don't think so. And with what should we compare the *IRR* if reinvestment is at that rate? The entire discussion is rather fruitless and provides little help for decision making.

The *IRR,* along with *RIC,* is a useful criterion when it is used with correct ordering and the incremental method. Capital that remains invested in a project grows at the *IRR* of the project, and cash released would be invested to grow at the *MARR* (or the cost of capital when this rate is used in *PV* calculation) [9]. When two alternatives are compared, Fisher's intersection is useful if the reinvestment rate is not known with certainty.

### 7.4.3 Benefit–Cost Ratio

A similar argument can be presented for the aggregate $B/C$ ratio in relation to Fisher's intersection when we are comparing two alternatives [3]. We can demonstrate the logic by applying it to Example 7.7. We assume for simplicity that all investment and operating expenditures occur at time 0 and that the flows from time 1 to time 3 are benefits. Thus, we have

$$B_A = \frac{\$600}{1+i} + \frac{500}{(1+i)^2} + \frac{300}{(1+i)^3}$$

$$I_A + C'_A = 1{,}000$$

$$B_B = \frac{\$300}{1+i} + \frac{200}{(1+i)^2} + \frac{1{,}000}{(1+i)^3}$$

$$I_B + C'_B = 1{,}000$$

The $i$ value for which the $B/(I + C')$ ratios are equal must satisfy the following expression.

$$\frac{\dfrac{\$600}{1+i} + \dfrac{500}{(1+i)^2} + \dfrac{300}{(1+i)^3}}{1{,}000} = \frac{\dfrac{\$300}{1+i} + \dfrac{200}{(1+i)^2} + \dfrac{1{,}000}{(1+i)^3}}{1{,}000}$$

or

$$\frac{\$300}{1+i} + \frac{300}{(1+i)^2} - \frac{700}{(1+i)^3} = 0$$

But this last expression simply yields the *IRR* of the difference cash flow between A and B.

We conclude this section on reinvestment issues by observing that much has been written on the subject, but not all is of use in decision making. The reinvestment rate assumed is critical for alternative selection, and the assumed value should be based on the concepts in Chapter 5. Use of the *PV* criterion implies reinvestment at the rate used for *PV* calculations. Fisher's intersection is useful when comparing two alternatives, but it becomes cumbersome with more than two.

In the real world the reinvestment rates may depend on the time period and on which investments have been accepted. In Chapter 8 we present some mathematical programming approaches that can be used to model such problems.

## 7.5 COMPARISON OF PROJECTS WITH UNEQUAL LIVES

Comparing projects with unequal lives can be particularly troublesome, for a number of different situations must be considered. Furthermore, many of the methods presented in textbooks have underlying assumptions that are not always clearly stated. Unfortunately, competing projects often have unequal lives, especially in engineering studies for which only costs (not benefits) are known. Problems in this class are more difficult than those for which all benefits are known, and they require more assumptions to be made. Another aspect of the unequal-lives situation is that of repeatability. Decisions involving projects that are likely to be repeated can often be made conveniently by easier methods.

We thus have the following classifications of cases:

1. *Service projects,* for which no revenues or benefits are estimated, or the revenues or benefits do not depend on the project. Here we must select a *study period* common to all alternatives. There are two general cases.
   a. Repeatability is likely.
   b. Repeatability is unlikely.
2. *Revenue projects,* for which all benefits and costs are known. Here the *study period* may be different for each alternative, provided we have a well-specified reinvestment rate.
   a. Repeatability is likely.
   b. Repeatability is unlikely.

These four cases will lead to (and in some instances force us into making) various assumptions concerning reinvestment, salvage values, and characteristics of the repeated projects.

Notice that in this section we are not trying to determine the best life of any individual project that is likely to be repeated. This type of decision is covered in detail in Chapter 16. We now present some of the more common ways of treating unequal lives.

### 7.5.1 Common Service Period Approach

If the benefit from a project is needed for a much longer period than the individual life of the project, it may be convenient to assume repeatability of identical projects.

## *Example* 7.8.

The Historical Society of New England must repaint its showcase headquarters building. The choice is between a latex paint that costs \$12.00/gallon and an oil paint that costs \$26.00/gallon. Each gallon would cover 500 square feet; labor is the same for both, 1 hour per 100 square feet at \$18.00/hour. The latex paint has an estimated life of 5 years, compared with 8 years for the oil paint. With $i = 8\%$, which paint should be selected?

Let us assume that after either the 5- or the 8-year period the building would be repainted repeatedly with the same paint and that the same costs would apply, as shown in Figure 7.3. The lowest common multiple of 5 and 8 is 40, so we will use 40 as the *common service period.* This becomes the *study period.*

For latex paint, we have the initial painting and seven repaintings.

$$PV(8\%) = \left(\frac{\$12.00}{500} + \frac{\$18.00}{100}\right)[1 + (P/F, 8\%, 5) + (P/F, 8\%, 10) + \cdots + (P/F, 8\%, 35)]$$

$$= (\$0.204)[1 + \overset{P/A,\ 46.9\%,\ 7}{(1.9866)}] = \$0.609 \text{ per square foot}$$

Note: $1.469 = (1.08)^5$.

For oil paint, there are four repaintings plus the initial painting.

$$PV(8\%) = \left(\frac{\$26.00}{500} + \frac{\$18.00}{100}\right)[1 + (P/F, 8\%, 8) + (P/F, 8\%, 16) + \cdots + (P/F, 8\%, 32)]$$

$$= (\$0.232)[1 + \overset{P/A,\ 85.1\%,\ 4}{(1.0751)}] = \$0.481 \text{ per square foot}$$

Note: $1.851 = (1.08)^8$.

The *PV* of the oil paint per square foot is considerably less, so the oil paint should be the choice. □

In Example 7.8 a service period of 40 years seem reasonable. The number of repaintings needed with each type of paint will depend on the technology of paint, so we may or may not need exactly seven (latex) or four (oil) repaintings. The validity of the analysis also depends on the costs of paint and labor remaining constant. If we assume constant-dollar prices, this may be a reasonable assumption. But then our interest rate of 8% must represent an inflation-free rate $i'$. Thus, many assumptions are necessary to make the approach valid.

An easier way to solve Example 7.8 is to use annual equivalents. The *AE* of each 40-year cash flow is the same as that of the corresponding 5- or 8-year cash flow.

For latex paint, computing from a 5-year life, we have

$$AE(8\%) = \left(\frac{\$12.00}{500} + \frac{\$18.00}{100}\right)\overset{A/P,\ 8\%,\ 5}{(0.2505)} = \$0.0511 \text{ per square foot}$$

Computing from a 40-year period, we have

$$AE(8\%) = \$0.609\overset{A/P,\ 8\%,\ 40}{(0.0839)} = \$0.0511 \text{ per square foot}$$

For oil paint, computing from an 8-year life, we have

$$AE(8\%) = \left(\frac{\$26.00}{500} + \frac{\$18.00}{100}\right)\overset{A/P,\ 8\%,\ 8}{(0.1740)} = \$0.0403 \text{ per square foot}$$

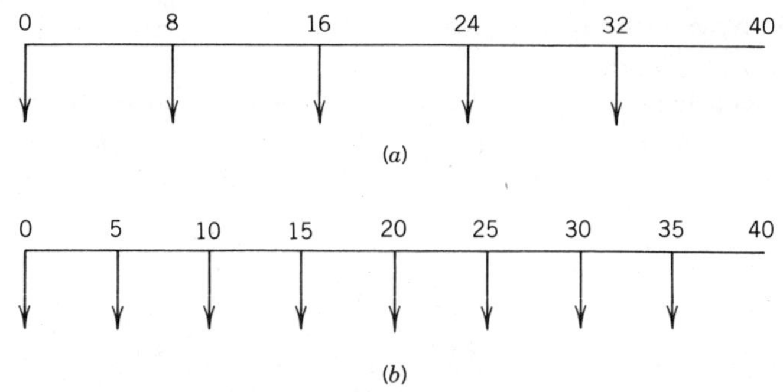


**FIGURE 7.3.** Common service period approach, Example 7.8: (*a*) oil, five paintings, (*b*) latex, eight paintings.

Computing from a 40-year period, we have

$$AE(8\%) = \$0.481\overset{A/P,\ 8\%,\ 40}{(0.0839)} = \$0.0403 \text{ per square foot}$$

With annual equivalents there is another possible interpretation regarding a common service period. We could assume that after the initial period, either 5 or 8 years, the building would be repainted with the type of paint that has the lower *AE* cost. Thus, if oil paint had the lower *AE* cost, the sequences would be

Latex paint (0 → 5), oil paint (5 → )
Oil paint (0 → 8), oil paint (8 → )

After time 5 the *AE* costs are the same. If latex paint had the lower *AE* cost, the sequences would be

Latex paint (0 → 5), latex paint (5 → )
Oil paint (0 → 8), latex paint (8 → )

After time 8 the *AE* costs are the same.

What circumstances would allow us to ignore the costs beyond time 5 when oil paint has the lower *AE* cost or time 8 when latex had the lower *AE* cost? An infinite service period with unchanging costs! Then we could simply look at the first 8 years; thereafter, costs would be identical. Actually, we do not need all these assumptions for reasonable accuracy in decision making. A long service period, say 30 years, and gradual changes in costs and technology will usually lead to the same decision about the initial choice of paint. Thus, we can minimize the *PV* of a long service period by selecting the alternative with the lower *AE* cost for an initial life.

The common service period approach is often used for analyzing *service projects,* for which no revenues or benefits are estimated or whose revenues or benefits are independent. The approach also can be applied to *revenue projects,* whose costs *and* benefits are known. In this situation we must be even more careful about our assumptions, especially regarding benefits.

### 7.5.2 Estimating Salvage Value of Longer-Lived Projects

If repeatability of projects is not likely for service projects, we must assume something about the salvage value of the longer-lived project. The next example shows how we can *explicitly* incorporate salvage values for assets with value remaining beyond the *required service period.*

#### *Example 7.9*

A highway contractor requires a ripper–bulldozer for breaking loose rock without the use of explosives, for a period of 3 years at about 2,000 hours/year. The smaller model, A, costs \$300,000, has a life of 8,000 hours, and costs \$40,000/year to operate. The larger model, B, costs \$450,000, has a life of 12,000 hours, and costs \$50,000/year to operate. Model B will perform adequately under all circumstances, whereas for model A some extra drilling is expected at an annual cost of \$35,000. With a marginal tax rate of 40%, units of production depreciation, and $i$ = 15%, which model should be purchased?

Since either model's lifetime exceeds the required service period (also the *study period*) of 3 years, we must assume something about the used equipment at that time. Let us assume that after 3 years model A would be sold for \$60,000 and model B for \$190,000. The after-tax cash flows for each alternative are given in Table 7.6 and shown in Figure 7.4. Model A has the lower *PV* of costs and would be preferred. □

**Table 7.6** ***Explicit Salvage Values, Example 7.9***

| | After-Tax Cash Flows (thousands) | | | |
|---|---|---|---|---|
| Model | *n*: 0 | 1 | 2 | 3 |
| Model A | | | | |
| Investment | −\$300 | | | |
| Depreciation, (300/8,000) (2,000) (0.4) | | +\$30 | +30 | +30 |
| Operating costs, (40) (0.6) | | −\$24 | −24 | −24 |
| Drilling costs, (35) (0.6) | | −\$21 | −21 | −21 |
| Salvage value | | | | +60 |
| Tax credit on salvage, (75 − 60) (0.4) | | | | +6 |
| Totals | −\$300 | −15 | −15 | +51 |
| Model B | | | | |
| Investment | −\$450 | | | |
| Depreciation, (450/12,000) (2,000) (0.4) | | +30 | +30 | +30 |
| Operating costs, (50) (0.6) | | −30 | −30 | −30 |
| Salvage value | | | | +190 |
| Tax credit on salvage, (225 − 190) (0.4) | | | | +14 |
| Totals | −\$450 | 0 | 0 | +\$204 |

$PV(15\%)_A = -\$291$, $PV(15)_B = -\$316$

The outcome of Example 7.9 depends very much on the salvage values received for the used equipment. We estimated these values by using 1 − (hours used/lifetime in hours)$^{0.8}$. What effect would higher salvage values have, say with an exponent of 1.5 instead of 0.8?

*Model A:*

$$[1 - (0.75)^{1.5}](300{,}000) = \$105{,}000 \text{ salvage value}$$

$$\text{Change in cash flow} = (105{,}000 - 60{,}000)(0.6) = +\$27{,}000$$

$$\text{New } PV = -291{,}000 + 27{,}000/(1.15)^3 = -\$273{,}000$$

*Model B:*

$$[1 - (0.5)^{1.5}](450{,}000) = \$291{,}000 \text{ salvage value}$$

$$\text{Change in cash flow} = (291{,}000 - 190{,}000)(0.6) = +\$61{,}000$$

$$\text{New } PV = -316{,}000 + 61{,}000/(1.15)^3 = -\$276{,}000$$

The numbers have changed to the point that intangible factors are likely to determine the selection.

What would happen if we evaluate models A and B by using *AE*s for their respective lives? First, we need some terminal salvage values; assume 10%, which gives \$30,000 and \$45,000, respectively. Second, we need to make some assumption about the extra drilling costs for model A; assume they would continue during the fourth year. The annual cash flows would be, as in Table 7.6, −\$15,000 for model A and and \$0 for model B. The positive salvage values

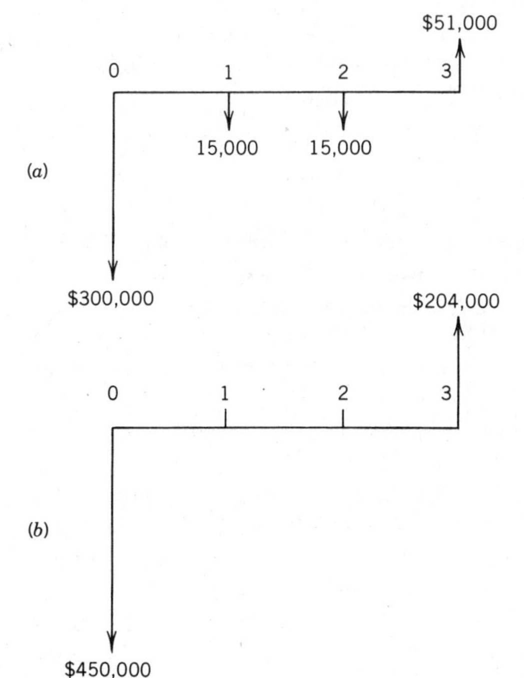


**FIGURE 7.4.** After-tax cash flows using explicit salvage value estimates, Example 7.9: (*a*) model A, (*b*) model B.

would result in depreciation recapture, so the net salvage proceeds would be (30,000)(0.6) = \$18,000 for model A and (45,000)(0.6) = \$27,000 for model B. Thus

$$AE(15\%)_A = -300{,}000\overset{A/P,\ 15\%,\ 4}{(0.3503)} - 15{,}000 + 18{,}000\overset{A/F,\ 15\%,\ 4}{(0.2003)}$$
$$= -\$116{,}485$$

$$AE(15\%)_B = -450{,}000\overset{A/P,\ 15\%,\ 6}{(0.2642)} + 0 + 27{,}000\overset{A/F,\ 15\%,\ 6}{(0.1142)}$$
$$= -\$115{,}807$$

The question at this point is not whether the foregoing analysis is valid, for the problem statement in Example 7.9 implies that it is not. (Many analysts use this method, nevertheless). Rather, we pose this question: Are there 3-year salvage values for models A and B that, when used in a 3-year analysis, yield these *AE* costs?

The answer is yes, and we can calculate the values as follows [7].

$$AE(15\%)_A = -300{,}000(A/P,\ 15\%,\ 3) - 15{,}000 + F_A(A/F,\ 15\%,\ 3)$$
$$= -116{,}485$$

or

$$AE(15\%)_A = -(300{,}000 - F_A)(0.4380) - 15{,}000 - F_A(0.15)$$
$$= -116{,}485$$

This gives $F_A$ = \$103,872, net proceeds after taxes, which implies a selling price of [103,872 − (0.4)(75,000)]/(0.6) = <u>\$123,120</u>. Similarly,

$$AE(15\%)_B = -(450{,}000 - F_B)\overset{A/P,\ 15\%,\ 3}{(0.4380)} - F_B(0.15) = -\$115{,}807$$

and $F_B$ = \$282,267, net proceeds after taxes, giving a selling price of [282,267 − (0.4)(225,000)]/(0.6) = <u>\$320,445</u>. (The derivation of the expression for selling price before depreciation recapture is left as an exercise; see Problem 7.24.)

These 3-year salvage values of \$123,120 and \$320,445 for models A and B, respectively, will result in *AE* costs of −\$116,485 and −\$115,807. From another point of view, if we make the selection decision between A and B by using *AE*s over 4 and 6 years, respectively, we are *implicitly* assuming these 3-year salvage values. Figure 7.5 shows these conversions. In Example 7.9 the contractor needs the equipment for only 3 years and will either sell the selected equipment after the 3-year period or use it elsewhere. If the equipment is sold, the contractor will receive a salvage value. If it is used elsewhere, the contractor considers the ***unused value*** to be equivalent to the implied salvage value [7, 10]. In this example we might be skeptical of 3-year salvage value ratios of 41% and 71% for bulldozer models A and B, respectively.

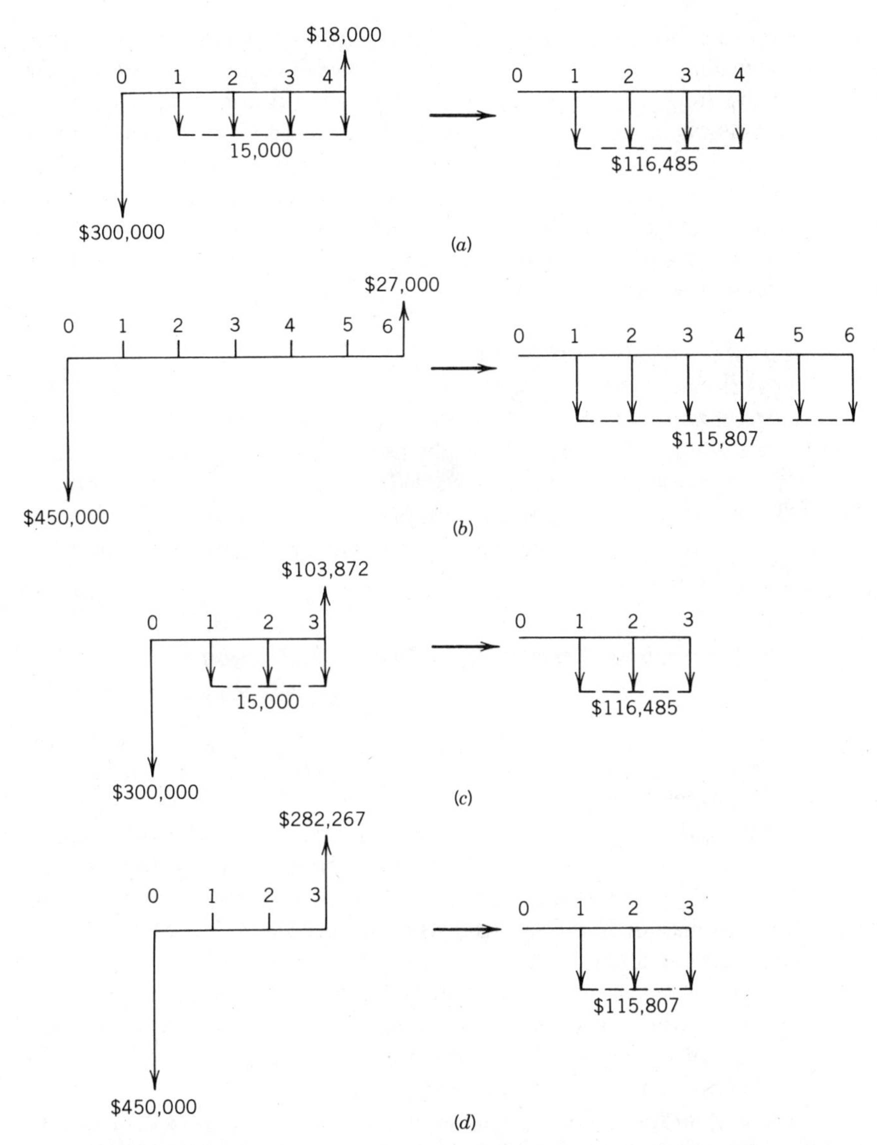


**FIGURE 7.5.** Conversions to annual equivalents, Example 7.9: (*a*) model A, original life; (*b*) model B, original life; (*c*) model A, 3-year life with implied salvage value; (*d*) model B, 3-year life with implied salvage value.

Sometimes one or more of the projects will have a life shorter than the required service period. One way to analyze such a situation is to *assume explicitly* how the requirement would be satisfied, for instance, by leasing an asset or by subcontracting. The *study period* then coincides with the *required service period,* which is desirable. If we use *AE* over the short life, we are *assuming explicitly* that we can lease an asset or subcontract for the remainder of the required service period at an annual cost equal to the *AE.*

In summary, when using *AE*s over the original, unequal lifetimes of projects, we are making *implicit* assumptions about salvage values or leasing costs at the end of the study period. Because most analysts do not understand these assumptions clearly and calculation of the implied values is not straightforward (especially for an after-tax analysis), we recommend that any salvage values and leasing costs used be *estimated explicitly.* The study period should equal the required service period. We do not recommend using *AE*s over unequal lifetimes for comparing service projects or using a study period different from the required service period.

### 7.5.3 Reinvestment Issues When Revenues Are Known

The presentation in Section 7.4.1 proved that, when cash inflows are reinvested at $i$, we will maximize future cash by using $PV(i)$ as a selection criterion. That proof applies for comparing projects with unequal lives as well as of those with equal lives. We thus have a way to compare revenue projects with unequal lives: use $PV(i)$. (If the reinvestment rate is not known, we must use the techniques presented in later chapters.)

### 7.5.4 Summary, Treatment of Unequal Lives

We summarize this section in terms of the classification given at the beginning.

1. Service projects
   a. Repeatability is likely.
      i. Use *AE* for each project's life. This is the easy method and is applicable in a greater variety of circumstances than the following method.
      ii. Use *PV* with a common service period. This is a tedious method and it requires or implies stricter assumptions than those in part i.
   b. Repeatability is not likely.
      i. *Explicitly* estimate salvage values for any assets with a remaining value at the end of the required service period. If an asset life falls short of the required service period, explicitly estimate the cost of leasing an asset or subcontracting.
      ii. Using *AE* for each project's life involves *implicit* estimates of salvage value or of the value of productive use after the required service period or both. This method should be used *only* if these implicit values are calculated and judged realistic.
2. Revenue projects
   a. Repeatability is likely.
      i. Use *AE* for each project's life.
      ii. Alternatively, use *PV* with a common service period.
   b. Repeatability is not likely. Use *PV* for each project's life.

It is particularly important to understand the assumptions underlying each of these methods.

## 7.6 DECISIONS ON THE TIMING OF INVESTMENTS

Sometimes it is possible to change the implementation timing of an investment. There are various reasons why this could occur, related to technology, marketing, production costs, financing costs, and so forth. We discuss briefly some of these situations and indicate how they can be treated analytically. For each situation it is understood that the same investment project with different implementation times should be treated as a set of mutually exclusive projects (a project would be implemented only once, if at all).

A rapidly *changing technology* may be a good reason to consider a timing change. Computer equipment, electronic instrumentation, aircraft, and the like change fast enough that a delay of one or two years in acquiring assets may result in significant differences in operating costs and performance capabilities. Such situations must be evaluated individually.

When the investment involves producing and marketing a product, the *product life cycle* should be considered [6].

| PERIOD | CHARACTERISTICS |
|---|---|
| Early years | Product still being developed. |
| | High unit costs. |
| | Relatively small market. |
| Middle years | Product design is stable. |
| | Production economies have been achieved. |
| | Peak annual sales for product. |
| Late years | Product is being replaced by new ones. |
| | Annual sales are declining. |

Companies with technological strengths try to be leaders and hope to get a marketing advantage by producing an item during its early years. Companies with production and marketing strengths avoid the high development costs and wait until the product design is stable; they will then attempt to produce and market the product at a lower price. During the late years of the product life cycle, the advantage rests with low-cost producers who have widespread marketing organizations. By evaluating its own capabilities, a company can decide how best to utilize its strengths.

*Differential inflation* rates for first cost have been used to argue for earlier construction of civil works and power plants. For methods for dealing with inflation, see Chapter 2. In the case of nuclear power plants, a positive differential inflation rate combined with more complex technology (related to safety measures) has brought about the cancellation of many planned facilities. Had they been constructed five or ten years earlier, they might have been successful investments.

*Changing financing costs* are often cited for delaying planned investments. Here we must be careful to separate the effects of raising more capital, perhaps by borrowing, from those of investment in a project. If the financing is not tied directly to the proposed project, a high borrowing cost should be viewed in the context of the company's overall cost of capital and capital structure; see Chapter 5. Viewed in this way, a high current borrowing cost may or may not raise the weighted-average cost of capital sufficiently to make a project undesirable. When borrowing costs are high, a company with financial strength may gain a significant market advantage by investing in new products and services. Delaying investments because of high rates may be shortsighted. Again, each situation must be evaluated by itself.

When we compare different timing decisions for the same project, a common point in time should be selected for the comparison. For example, *PV* at time 0 (a specific date) can be used. Here it is particularly important to have a good estimate of *MARR*, because different lateral time shifts of cash flows for two or more projects may distort the comparison if the *MARR* does not accurately reflect reinvestment opportunities.

## 7.7 SUMMARY

In this chapter we have shown how the proper use of any of the ten decision criteria presented in Chapter 6 will lead to correct decisions when we select among competing projects. The final selection will be consistent with *PV*, which is the correct, or baseline, criterion to use in a stable, perfect capital market with complete certainty. The necessary steps for proper use of a criterion are

1. Preliminary screening to eliminate unfavorable projects.
2. Forming mutually exclusive alternatives.
3. Ordering the alternatives (not always by the time 0 investment).
4. Applying the incremental procedure.

The total investment approach is guaranteed to work only with *PV, FV, AE,* and *PB*. Moreover, for these four criteria one can use arbitrary ordering with the incremental procedure. Detailed rules apply to *IRR* and make it particularly difficult to use properly.

Use of *PV* implies reinvestment at the rate used for computing *PV*. Since the other criteria, when used properly, give the same project selection, it can be argued that their use also implies reinvestment at the same rate. It is clear that the discount rate, designated *MARR*, must be selected carefully; see Chapter 5. Much has been written about the reinvestment rate implied by use of other criteria, especially *IRR*, but this criterion is of relatively little use for decision making, with the exception of Fisher's intersection.

When comparing projects with unequal lives, one must distinguish between service projects and revenue projects. The likelihood of repeatability affects the analysis techniques to be used. Finally, any salvage value assumptions should be stated clearly and treated explicitly.

This chapter is the last one dealing with "traditional" engineering economic analysis techniques. The next chapter considers more complex decision environments, still assuming certainty. Later chapters deal with variable cash flows and other uncertainties.

## REFERENCES


1. AU, T., and T. P. AU, *Engineering Economics for Capital Investment Analysis,* Allyn and Bacon, Boston, 1983.
2. BERNHARD, R. H., "A Comprehensive Comparison and Critique of Discounting Indices Proposed for Capital Investment Evaluation," *The Engineering Economist,* Vol. 16, No. 3, pp. 157–186, Spring 1971.
3. BUSSEY, L. E., *The Economic Analysis of Industrial Projects,* Prentice–Hall, Englewood Cliffs, N.J., 1978 (see Ch. 8).
4. DEGARMO, E. P., W. G. SULLIVAN, and J. R. CANADA, *Engineering Economy,* 7th edition, Macmillan, New York, 1984 (see Chs. 5 and 6).
5. FISHER, I., *The Theory of Interest,* Macmillan, New York, 1930.
6. KAMIEN, M. I., and N. L. SCHWARTZ, "Timing of Innovations under Rivalry," *Econometrica,* Vol. 40, No. 1, pp. 43–59, 1972.
7. KULONDA, D. J., "Replacement Analysis with Unequal Lives," *The Engineering Economist,* Vol. 23, No. 3, pp. 171–179, Spring 1978.
8. LEVY, N. S., "On the Ranking of Economic Alternatives by the Total Opportunity ROR and B/C Ratios—A Note," *The Engineering Economist,* Vol. 26, No. 2, pp. 166–171, Winter 1981.
9. LOHMANN, J. R., "The IRR, NPV and the Fallacy of the Reinvestment Rate Assumptions," *The Engineering Economist,* Vol. 33, No. 4, pp. 303–330, Summer 1988.
10. SAXENA, U., and A. GARG, "On Comparing Alternatives with Different Lives," *The Engineering Economist,* Vol. 29, No. 1, pp. 59–70, Fall 1983.
11. THEUSEN, G. J., and W. J. FABRYCKY, *Engineering Economy,* 7th edition, Prentice–Hall, Englewood Cliffs, N.J., 1989 (see Ch. 7, Sec. 8.3).
12. WHITE, J. A., M. H. AGEE, and K. E. CASE, *Principles of Engineering Economic Analysis,* 3rd edition, Wiley, New York, 1989 (see Ch. 5).
13. WOHL, M., "A New Ordering Procedure and Set of Decision Rules for the Internal Rate of Return Method," *The Engineering Economist,* Vol. 30, No. 4, pp. 363–386, Summer 1985.


## PROBLEMS

**7.1.** A company has the capability of manufacturing four products. There are three plants, with product capabilities as follows.

| | |
|---|---|
| Plant A | Products 1, 2, 4 |
| Plant B | Products 2, 3 |
| Plant C | Products 1, 3, 4 |

For various reasons, the company does not produce the *same* product in more than *two* plants. In addition, any particular plant is used to produce only *one* product. Form all possible combinations of plants and products that the company should consider.

**7.2.** If there are four independent investment proposals A, B, C, and D, form all possible investment alternatives with them.

**7.3.** Apply the ten investment criteria to projects $j$ and $k$ in Example 7.3 to derive the results in Table 7.3.

**7.4.** Apply the incremental procedure to projects $j$ and $k$ in Example 7.3 to derive the results in Table 7.4.

**7.5.** Consider the four projects with cash flows as shown.

| Project | $n$: 0 | 1 | 2 | 3 |
|---|---|---|---|---|
| A | −1,000 | 900 | 500 | 100 |
| B | −1,000 | 600 | 500 | 500 |
| C | −2,000 | 900 | 900 | 800 |
| D | +1,000 | −402 | −402 | −402 |

Before proceeding to the questions, we will need to obtain *FV* for each project by using *MARR* = 10%, 20%.

a. Explain why the *FV* criterion prefers A over B at 20% when it prefers B over A at 10%.

b. With *MARR* = 10%, how much money would you have at time 3 if you invested $1,000 of your own money in A? In B?

c. Which of the following situations would you prefer?
   i. *MARR* = 10%; you invest $1000 in B.
   ii. *MARR* = 20%; you invest $1000 in A.
   Explain your answer.

d. With *MARR* = 10%, how much money would you have at time 3 if you invested $2,000 of your own money in C?

e. Explain why the *FV* criterion prefers A over C at 10%, even though in situation d the cash at time 3 is greater than that in situation b (for project A).

f. What is the *IRR* for D? Would you accept D with *MARR* = 20%? How would you modify the *IRR* acceptance rule when examining project D?

g. Suppose A and B are mutually exclusive projects. Which project would you select using *MARR* of 10% and the *IRR* criterion?

**7.6.** Your company is faced with three independent proposals:

| Project | $n$: 0 | 1 | 2 | 3 |
|---|---|---|---|---|
| A | −1,000 | 500 | 500 | 500 |
| B | −1,500 | 1,000 | 200 | 1,000 |
| C | −3,000 | 1,300 | 1,300 | 1,300 |

a. With a budget of $3,000 at time 0 and *MARR* = 8%, which project or projects should you choose? Use *FV*.

b. How much cash would you have at time 3? Answer this part by performing a minimum of computations.

c. Could you use *IRR* to obtain the answer to part a? Do you foresee any potential difficulties?

**7.7.** Consider the following three mutually exclusive projects. Each has a lifetime of 20 years and *MARR* = 15%.

| Project | Investment | Annual User Benefits | Annual Sponsor Costs |
|---|---|---|---|
| A | 1,000 | 400 | 160 |
| B | 800 | 300 | 110 |
| C | 1,500 | 360 | 50 |

a. Select the best project, using the *PV* criterion.
b. Select the best project, using the aggregate benefit–cost ratio.

**7.8.** Consider the following four mutually exclusive projects. Use the incremental method with *PV* and the aggregate cost–benefit ratio to select the best project. Each has a lifetime of 20 years, and *MARR* = 8%.

| Project | Investment | Annual User Benefits | Annual Sponsor Costs |
|---|---|---|---|
| A | 978 | 500 | 100 |
| B | 1,180 | 492 | 60 |
| C | 1,390 | 550 | 120 |
| D | 1,600 | 630 | 140 |

**7.9.** Use *IRR* to select the best of the following three mutually exclusive projects. Each has a lifetime of 10 years, and *MARR* = 15%.

| Project | Investment | Annual Net Cash Flow |
|---|---|---|
| A | 5,000 | 1,400 |
| B | 10,000 | 2,500 |
| C | 8,000 | 1,900 |

**7.10.** Use *IRR* to select the best of the following three independent projects. Each has a lifetime of 5 years, and *MARR* = 8%. The investment budget is $13,000.

| Project | Investment | Annual Net Cash Flow |
|---|---|---|
| A | 5,000 | 1,319 |
| B | 7,000 | 1,942 |
| C | 8,500 | 2,300 |

**7.11.** Use the netted benefit–cost ratio to select the best of the following four mutually exclusive projects. Each has a lifetime of 5 years, and *MARR* = 12%.

| Project | Investment | Annual Net Cash Flow |
|---|---|---|
| A | 10,000 | 4,438 |
| B | 14,000 | 5,548 |
| C | 12,000 | 5,048 |
| D | 5,000 | 2,774 |

**7.12.** Rework Problem 7.11 with the assumption that the projects are independent and the investment budget is $16,000.

**7.13.** Listed are cash flows for three independent proposals. Use the netted benefit–cost ratio to select the best proposal or proposals with *MARR* = 12% and an investment budget of $34,000.

| Project | *n*: 0 | 1 | 2 | 3 | 4 |
|---|---|---|---|---|---|
| A | −10,000 | 4,175 | 4,175 | 4,175 | 4,175 |
| B | −17,500 | 10,025 | 3,025 | 7,025 | 7,025 |
| C | −15,000 | 6,025 | 6,025 | 6,025 | 6,025 |

**7.14.** Listed are data for three mutually exclusive proposals. Use the aggregate benefit–cost ratio to select the best proposal with *MARR* = 10%.

| | Proposal A | | Proposal B | | Proposal C | |
|---|---|---|---|---|---|---|
| *n* | Costs | Benefits | Costs | Benefits | Costs | Benefits |
| 0 | 10,000 | — | 14,000 | — | 17,000 | — |
| 1 | 1,000 | 5,500 | 4,000 | 10,000 | 1,000 | 10,000 |
| 2 | 1,000 | 5,500 | 4,000 | 10,000 | 1,000 | 3,000 |
| 3 | 1,000 | 5,500 | 4,000 | 10,000 | 1,000 | 10,000 |
| 4 | 1,000 | 5,500 | 4,000 | 10,000 | 1,000 | 10,000 |

**7.15.** Apply *IRR* to the selection in problem 7.14.

**7.16.** Construct an example in which Solomon's average rate of return yields an answer inconsistent with the netted benefit–cost ratio. Use the total investment approach.

**7.17.** Prove that if each investment alternative has the same initial investment, then *PV* agrees with the netted benefit–cost ratio. Use the total investment approach.

**7.18.** Prove that if each investment alternative has a constant net cash flow during its lifetime, then *IRR* agrees with the netted benefit–cost ratio. Use the total investment approach.

**7.19.** What modifications are needed in the accept–reject rules for the aggregate benefit–cost ratio if the ordering for the incremental procedure is by the time 0 investment?

**7.20.** Prove, or disprove by counterexample, that consistency is obtained across all four groups of investment criteria. Use the total investment approach:

a. When the total invested in each alternative is the same.

b. When the total invested in each alternative is the same, and the lifetimes of all alternatives are the same.

**7.21.** Prove that incremental analysis with Solomon's average rate of return yields the same answer as *PV* analysis.

**7.22.** Use the *common service period* approach to compare the following two options. *MARR* = 12%; ignore taxes.

i. Initial cost of $1,000, annual costs of $300, salvage value of $100, 10-year lifetime.

ii. Initial cost of $1,300, annual costs of $270, salvage value of $200, 12-year lifetime.

Is the length of the common service period plausible?

**7.23.** A manufacturer requires a chemical finishing process for a product produced under contract for a period of 4 years. Three options are available.

i. Process device A, which costs $100,000, has annual operating and labor costs of $60,000 and an estimated salvage value of $10,000 after 4 years.

ii. Process device B, which costs $150,000, has annual operating and labor costs of $50,000 and an estimated salvage value of $30,000 after 6 years.

iii. Subcontracting at $100,000 per year.

a. Which option would you recommend? *MARR* = 10%.

b. What is the salvage value of process device B after 4 years that would cause the manufacturer to be indifferent in choosing between it and process device A?

c. What options should the manufacturer consider if the required service period is 5 years? 7 years?

**7.24.** Derive the selling price before depreciation recapture for the assets in Example 7.9.

# 8

# Deterministic Capital Budgeting Models

## 8.1 INTRODUCTION

In the previous chapter we determined that we should select from among multiple alternatives by choosing the one with the maximum net present value (*PV*) or by using the incremental approach with one of several criteria. There are two important characteristics of the problems solved in the previous chapter.

1. We could easily formulate and list all mutually exclusive alternatives of interest.
2. There is an underlying assumption of ability to borrow and lend unlimited amounts at a single, fixed interest rate. When budget limits are imposed, the borrowing ability at time 0 is restricted, and we are left with a single, fixed interest rate for future lending, or reinvestment.

In this chapter we relax these assumptions. We consider problems in which budget limits are imposed during several time periods, the projects have interdependencies, and there are different, but known, borrowing and lending opportunities. In short, we examine problems for which it would be exceedingly difficult to specify all mutually exclusive alternatives. This type of analysis is called capital budgeting. In keeping with the sense of Part Two of this book, we assume certainty with respect to all information. Linear programming (LP) is a convenient tool for analyzing such situations, and we give a brief introduction to its use in Section 8.2.

We will also see that *PV* maximization is not necessarily our best objective, for different reinvestment rates are possible. In the pure capital rationing model (Section 8.3), which allows no external borrowing and lending, this situation has been the focus of much academic controversy during the last twenty years. We include a brief review of the major arguments, not from the view of favoring any one of them but rather to give the reader an important historical perspective on the subject.

The inclusion of borrowing and lending opportunities (Section 8.4) leads to more realistic operational models. In some situations the previously mentioned academic controversy disappears, and in others it reappears. Weingartner's horizon model (Section 8.5) provides the analyst with a convenient way of avoiding these issues, while yielding solutions consistent with *PV* analysis of situations allowing unlimited borrowing and lending. Bernhard's general model (Section 8.6) allows for the use of dividends and other terms in the objective function, and for a variety of linear and nonlinear constraints.

In Section 8.7 we finally consider the situation of integer restrictions, which we have avoided until now because it requires more difficult mathematical analysis. Multiple objectives are discussed in Section 8.8. Following the summary, the chapter ends with a case study illustrating the application of Bernhard's general model to a dividend-terminal-wealth problem.

## 8.2 THE USE OF LINEAR PROGRAMMING MODELS

Because linear programming models are so widely used in capital budgeting, we present a brief introduction here. In this section we illustrate the application of LP in a typical example. A word of caution is in order here: the example given is *not* intended to represent the best principles of capital budgeting but *rather to illustrate* the use of LP. The various methods of capital budgeting (for the deterministic case) are given in the following sections.

### *Example 8.1*

Table 8.1 presents data for Example 8.1, which concerns five investment projects. There are budget limits of $4,400 and $4,000 at time 0 and time 1, respectively; these limits do not apply to any funds generated by the projects themselves. We note the sign convention that inflows are positive and outflows negative. Most of the projects require investment during the first two years before they return any funds. All the projects are simple investments with unique, positive, real *IRR*s, and for a sufficiently low *MARR*, say 20%, all have positive *PV*s. But from the budget limits it is clear that we cannot accept all of them; hence the capital rationing problem. Moreover, project 5 starts to provide cash inflow at time 1, when all the others require outflows, so we would like to consider this advantage of project 5. (The solution to Example 8.1 follows in the text). □

### 8.2.1 Criterion Function To Be Optimized

A variety of criterion functions could be optimized.

- Maximize the *PV* of the cash flows of the selected projects.
- Maximize the *IRR* of the total cash flow of the selected projects.

**Table 8.1** ***Data for Example 8.1***

| Cash Flow at Time | Project 1 | 2 | 3 | 4 | 5 |
|---|---|---|---|---|---|
| 0 | −$1,000 | −$1,200 | −$2,000 | −$2,500 | −$3,000 |
| 1 | −2,000 | −2,400 | −2,100 | −1,300 | 900 |
| 2 | 2,000 | 2,500 | 3,000 | 2,000 | 1,400 |
| 3 | 2,900 | 3,567 | 3,000 | 2,000 | 1,600 |
| 4 | 0 | 0 | 1,308 | 2,000 | 1,800 |
| 5 | 0 | 0 | 0 | 2,296 | 955 |
| *PV*(20%) | $400 | 600 | 700 | 850 | 900 |
| *IRR*, % | 29.1 | 31.3 | 29.7 | 28.9 | 32.2 |

Budgets for external sources of funds: $n$ = 0, $4,400; $n$ = 1, $4,000

- Maximize the "utility" of the dividends that can be paid from the cash flows of the selected projects.
- Maximize the cash that can be accumulated at the end of the planning period.

Other functions could be used. The important thing is that the function is clearly expressed in terms of the decision variables for project acceptance or rejection and that it is (we hope) linear.

Let us see, for *illustration* purposes, the *PV* of the cash flows of the selected projects,

$$\underset{j}{\text{Max}} \sum_{j} p_j x_j \tag{8.1}$$

where $p_j$ is the *PV* of project $j$, using $i$ = *MARR*, and
$x_j$ is a project selection variable, with $0 \le x_j \le 1$.

Using a *MARR* value of $i$ = 20%, we obtain

| $j$ | 1 | 2 | 3 | 4 | 5 |
|---|---|---|---|---|---|
| $p_j$ | $400 | 600 | 700 | 850 | 900 |

Thus Eq. 8.1 becomes, for Example 8.1,

$$\max \$400x_1 + 600x_2 + 700x_3 + 850x_4 + 900x_5 \tag{8.2}$$

The project selection variables are continuous in this linear formulation. A value of $x_j = 0$ means that the project is not selected, a value of $x_j = 1$ implies complete acceptance, and a fractional values implies partial acceptance. We will leave aside the question of the practicality of fractional acceptance. In some industries, such as oil and gas exploration, fractional acceptance is common practice; generally, though, it is not possible to accept fractional projects without changing the nature of their cash flows. (We will consider integer restrictions in Section 8.7.)

### 8.2.2 Multiple Budget Periods

The budget limits for Example 8.1 can be expressed by linear constraints on the selection variables,

$$-\sum_j a_{nj} x_j \leq M_n, \qquad n = 0, 1, ..., N \tag{8.3}$$

where $a_{nj}$ = cash flow for project $j$ at time $n$, inflows having a plus sign, and outflows a minus sign,
$M_n$ = budget limit on externally supplied funds at time $n$, and
$N$ = end of the planning period.

Notice that $M_n$ represents only the funds from sources other than the projects. The equation states that project outflows minus project inflows at time $n$ must be less than the budget limit on funds from other sources at time $n$. A negative value for $M_n$ implies that the set of selected projects must *generate* funds. (Equation 8.3 states that cash outflows $\leqslant$ cash inflows + $M_n$). Note that the absence of a budget limit is not equivalent to $M_n$ being zero; the former implies a positive, unbounded $M_n$ value. Equations of the type (8.3) are usually called budget constraints or cash balance equations. Inflows and outflows for borrowing, lending, and dividend payments may also be included; these are discussed in later sections.

Applying the equation to Example 8.1, we obtain two constraints,

$$n = 0: \qquad \$1{,}000x_1 + 1{,}200x_2 + 2{,}000x_3 + 2{,}500x_4 + 3{,}000x_5 \leqslant 4{,}400$$

$$n = 1: \qquad \$2{,}000x_1 + 2{,}400x_2 + 2{,}100x_3 + 1{,}300x_4 - 900x_5 \leqslant 4{,}000 \tag{8.4}$$

The advantage of project 5 at time 1 is clearly apparent here; setting $x_5 = 1$ increases the amount available for other projects by \$900. There are no stated limits for times 2, 3, 4, and 5, so we need not write constraints for these times.

### 8.2.3 Project Limits and Interdependencies

The limits on the selection variables given following Eq. 8.1 are presented here again.

$$x_j \leqslant 1, \qquad j = 1, \ldots, J \tag{8.5}$$

The nonnegativity constraints are expressed separately:

$$x_j \geqslant 0, \qquad j = 1, \ldots, J \tag{8.6}$$

It is also possible to have interdependencies among project selection variables. Some common types are the following.

1. Mutual exclusivity—when a subset of projects form a mutually exclusive set.

$$x_j + x_k + x_m \leqslant 1 \tag{8.7}$$

The selection of one project precludes the selection of either of the other two in Eq. 8.7. Note that a complete interpretation is possible only if the $x_j$ are restricted to integers.

2. Contingency—when execution of one project depends on execution of another.

$$x_j - x_k \leqslant 0 \tag{8.8}$$

Here $x_j$ cannot be selected unless $x_k$ is also selected.

3. Complementary and competitive projects—when the selection of two projects changes the cash flows involved. For complementary projects inflows are greater than the sum of the individual project inflows; the opposite is true for competitive projects. Such situations can be handled by defining a new project for the combination and then establishing mutual exclusivity,

$$x_j + x_k + x_m \leqslant 1$$

where $x_m$ is a combination of $j$ and $k$. (If there are many such situations, the method becomes cumbersome.)

We will not impose interdependencies in Example 8.1, in order to keep the duality analysis simple at this point. That type of treatment is given in Section 8.5.

### 8.2.4 LP Formulation of Lorie–Savage Problem

In LP terminology, the *primal problem* formulation of the capital budgeting problem is given symbolically and numerically by Table 8.2. This version summarizes the relationships that have been presented so far in this chapter. This version of the problem is also designated as the LP formulation of the Lorie–Savage problem [15], after the two economists who stated the original form of the project selection problem. Their concern with the problem came from the inadequacies of the *IRR* method to deal with budget limitations and project interdependencies. Our analysis in the next section follows closely the work of Weingartner [20], who applied LP to the Lorie–Savage problem.

### 8.2.5 Duality Analysis

For every *primal problem* in linear programming, there is a related *dual problem* formulation. Table 8.3 presents both the symbolic and numeric versions of the dual problem for Example 8.1. The dual formulation is a minimization problem stated in terms of the $\rho_n$ and $\mu_n$. By making appropriate conversions from minimization to maximization and from $\geqslant$ to $\leqslant$, we can easily show that the dual formulation of the problem in Table 8.3 is the same as the formulation given in Table 8.2. In other words, the dual of the dual is the primal, and our specific designations are based on habit and convenience.

**Table 8.2** ***Primal Problem Formulation for Maximizing PV for Example 8.1 (Lorie–Savage Formulation)***

*Symbolic*

$$\underset{j}{\text{Max}} \sum_j p_j x_j \tag{8.1}$$

s.t.*

$[\rho_n]$ $$-\sum_j a_{nj} x_j \le M_n, \quad n = 0, 1, \ldots, N \tag{8.3}$$

$[\mu_j]$ $$x_j \le 1, \quad j = 1, \ldots, J \tag{8.5}$$

$$x_j \ge 0, \quad j = 1, \ldots, J \tag{8.6}$$

where $p_j$ = *PV* of project $j$ using $i$ = *MARR*,
$x_j$ = project selection variable,
$a_{nj}$ = cash flow for project $j$ at time $n$; inflows have a plus sign, outflows have a minus sign,
$M_n$ = budget limit on externally supplied funds at time $n$,
$N$ = end of the planning period,
$\rho_n, \mu_j$ = dual variables for the primal constraints.

*Numeric*

$$\text{Max } \$400x_1 + 600x_2 + 700x_3 + 850x_4 + 900x_5 \tag{8.2}$$

s.t.

$[\rho_0]$ $$\$1{,}000x_1 + 1{,}200x_2 + 2{,}000x_3 + 2{,}500x_4 + 3{,}000x_5 \le \$4{,}400$$

$[\rho_1]$ $$\$2{,}000x_1 + 2{,}400x_2 + 2{,}100x_3 + 1{,}300x_4 - 900x_5 \le \$4{,}000 \tag{8.4}$$

$[\mu_1]$ $x_1 \le 1$
$[\mu_2]$ $x_2 \le 1$
$[\mu_3]$ $x_3 \le 1$ (8.5)
$[\mu_4]$ $x_4 \le 1$
$[\mu_5]$ $x_5 \le 1$

All $x_j \ge 0, j = 1, \ldots, 5$ (8.6)

*The abbreviation s.t. stands for subject to.

The economic interpretation of the dual problem is to establish prices for each of the scarce resources so that the minimum total possible would be paid for the consumption of the resources, while ensuring that the resources used for any project cost as much as or more than the value of the project, the project *PV* in this case [5]. We have two categories of resources here. The first category is cash, represented by cash at time 0 and by cash at time 1; the dual variables $\rho_0$ and $\rho_1$ represent the prices, respectively. The second category consists of the projects themselves: a project is considered a scarce resource in the sense that we have the opportunity to execute only one of each. The dual variables $\mu_1, \ldots, \mu_5$ correspond to the upper-bound constraints of the projects and represent the respective prices for the project opportunities.

**Table 8.3** ***Dual Problem Formulation for Example 8.1, Solution to Primal and Dual***

*Symbolic*

$$\text{Min} \sum_n \rho_n M_n + \sum_j \mu_j \tag{8.9}$$

s.t.

$$[x_j] \quad -\sum_n a_{nj}\rho_n + \mu_j \geq p_j, \quad j = 1, \ldots, J \tag{8.10}$$

$$\rho_n \geq 0, \quad n = 0, 1, \ldots, N \tag{8.11}$$

$$\mu_j \geq 0, \quad j = 1, \ldots, J \tag{8.12}$$

where $\rho_n$ = dual variable for budget constraint,
$\mu_j$ = dual variable for project upper bound.

*Numeric*

$$\text{Min} \quad 4{,}400\rho_0 + 4{,}000\rho_1 + \mu_1 + \mu_2 + \mu_3 + \mu_4 + \mu_5 \tag{8.13}$$

s.t.

$$\begin{array}{lllllllll}
[x_1] & +1{,}000\rho_0 & +\ 2{,}000\rho_1 & +\ \mu_1 & & & & & \geq 400 \\
[x_2] & +1{,}200\rho_0 & +\ 2{,}400\rho_1 & & +\ \mu_2 & & & & \geq 600 \\
[x_3] & +2{,}000\rho_0 & +\ 2{,}100\rho_1 & & & +\ \mu_3 & & & \geq 700 \\
[x_4] & +2{,}500\rho_0 & +\ 1{,}300\rho_1 & & & & +\ \mu_4 & & \geq 850 \\
[x_5] & +3{,}000\rho_0 & -\ \ 900\rho_1 & & & & & +\ \mu_5 & \geq 900
\end{array} \tag{8.14}$$

$$\rho_n, \mu_j \geq 0$$

*Solution*

Primal variables: $x_1 = 0.22, x_2 = 1.00, x_3 = 0.0, x_4 = 1.0, x_5 = 0.16$
Dual variables: $\rho_0 = 0.3130, \rho_1 = 0.0435$
$\mu_1 = 0.0, \mu_2 = 120.0, \mu_3 = 0.0, \mu_4 = 10.9, \mu_5 = 0.0$
Objective function value: \$1,682

If the primal problem is feasible and bounded, there is an optimal solution to both problems. At such an optimum we have, from the dual constraint,

$$\mu_j^* \geq p_j + \sum_n a_{nj}\rho_n^* \tag{8.15}$$

where the asterisk refers to values of the primal and dual variables at the optimum. We know from complementary slackness [5] that if $x_j^* > 0$, the dual constraint is met exactly, and since all dual variables are nonnegative, we have

$$0 \leq \mu_j^* = p_j + \sum_n a_{nj}\rho_n^* \tag{8.16}$$

The $\mu_j^*$ represents the opportunity value of project *j*, and it is equal to the *PV* plus the cash inflows less any cash outflows evaluated by the $\rho_n^*$. Hence, for all projects that are accepted fractionally or completely,

$$-\sum_n a_{nj}\rho_n^* \leq p_j \tag{8.17}$$

Equation 8.17 states that in order for a project to be accepted, its *PV* must be equal to or greater than the cash outflows minus cash inflows evaluated by the $\rho_n^*$.

Again from complementary slackness, if $x_j^* < 1$, then $\mu_j^* = 0$. So for fractionally accepted projects (8.17) becomes

$$-\sum_n a_{nj}\rho_n^* = p_j \tag{8.18}$$

For rejected projects we also have $\mu_j^* = 0$ and, using Eq. 8.15,

$$-\sum_n a_{nj}\rho_n^* \geq p_j \tag{8.19}$$

In other words, the cash outflows minus cash inflows, evaluated by the $\rho_n^*$, exceed (or equal) the *PV* of the project.

We can demonstrate these conditions by using the optimal values of the LP problem given in Table 8.3. For project 1, fractionally accepted, applying (8.18) gives

$$(\$1{,}000)(0.313) + (2{,}000)(0.0435) = \$400 = PV$$

The value of cash inflows minus outflows equals the *PV*.

For project 2, completely accepted, applying (8.16) gives

$$\mu_2^* = \$600 - (1{,}200)(0.313) - (2{,}400)(0.0435) = \$120 > 0$$

The opportunity cost of the project is \$120, the difference between the *PV* and the cash outflows minus the cash inflows.

For project 3, rejected, applying (8.19) gives

$$(\$2{,}000)(0.313) + (2{,}100)(0.0435) = \$717 \geqslant \$700$$

Here the cash outflows minus inflows are worth more than the *PV*, which explains the rejection.

These types of project evaluation, or project pricing, with the dual variables, are fundamental to the LP modeling and analysis of capital budgeting problems. We will see more of this type of analysis in the following sections.

## 8.3 PURE CAPITAL RATIONING MODELS

The type of model given in Table 8.2 has been extensively analyzed, criticized, and modified during the last twenty years. In this section we attempt to summarize the major arguments so that the reader will obtain a historical perspective on the situation. We do not go into great detail, because the arguments are presented better elsewhere [21] and because the major conclusion to be drawn is that the pure capital rationing (PCR) model is of extremely limited applicability. This fact reinforces the fundamental notion that one must fully understand the assumptions embedded in any mathematical model before attempting to use it.

### 8.3.1 Criticisms of the PV Model

Among the first to criticize the *PV* model (as in Table 8.2) were Baumol and Quandt [3]. They identified three major flaws.

1. There is no provision in the model for investment outside the firm or for dividend payments.
2. The model does not provide for carryover of unused funds from one period to the next.
3. Assuming that we have an appropriate discount rate $i$ for computing the *PV* of each project, this rate is valid in general only for the situation of unlimited borrowing and lending at that rate. Since we have borrowing limits implicitly stated in the budget constraints, an externally determined discount rate is inappropriate.

The first two objections can easily be overcome. For example, investment outside the firm, including lending activities, can easily be represented by new projects. Define project 6 to be lending from time 0 to time 1 at 15%. Then we set $a_{06} = -1$ and $a_{16} = 1.15$ and place no upper bound (or a very large bound) on $x_6$. Similarly, variables can be defined for divided payments and included in the budget constraints. We would also need to include dividends in the objective function, which implies knowledge of the discount rate appropriate for the owner(s) or shareholders of the firm in order to discount correctly the future dividends. Later, we will see some different methods for including dividends in the objective function.

The third objection is a serious one and requires more attention. To illustrate the difficulties arising from it, let us analyze Example 8.2.

#### *Example 8.2*

Table 8.4 presents the data for Example 8.2 along with the optimal LP solution. Example 8.2 is somewhat similar to Example 8.1: projects 1, 2, and 4 are the same; projects 3 and 5 are slightly changed so their *PV*s are negative; project 6 is added to the set; and the budget limits are changed.

Notice that projects 2 and 4 are completely accepted, as they were in

**Table 8.4** ***Data and Solution for Example 8.2***

| Cash Flow at Time | Project | | | | | |
|---|---|---|---|---|---|---|
| | 1 | 2 | 3 | 4 | 5 | 6 |
| 0 | −\$1,000 | −\$1,200 | −\$2,000 | −2,500 | −\$3,000 | \$1,000 |
| 1 | −2,000 | −2,400 | −2,100 | −1,300 | 900 | −700 |
| 2 | 2,000 | 2,500 | 3,000 | 2,000 | 1,400 | −700 |
| 3 | 2,900 | 3,567 | 2,621 | 2,000 | 1,600 | 0 |
| 4 | 0 | 0 | 0 | 2,000 | 211 | 0 |
| 5 | 0 | 0 | 0 | 2,296 | 0 | 0 |
| *PV* (20%) | \$400 | 600 | −150 | 850 | −250 | −70 |

Budgets for external sources of funds: $n = 0$, \$3,000; $n = 1$, \$5,000

*Solution*

Primal variables: $x_1 = 0.30, x_2 = 1.0, x_3 = 0.0, x_4 = 1.0, x_5 = 0.0, x_6 = 1.0$
Dual variables: $\rho_0 = 0.3189, \rho_1 = 0.0405$
$\mu_1 = 0.0, \mu_2 = 120.0, \mu_3 = 0.0, \mu_4 = 0.0, \mu_5 = 0.0, \mu_6 = 220.5$
Objective function value: \$1,500

Example 8.1. The dual variables for the budget constraints, $\rho_0$ and $\rho_1$, do not have their optimal values changed much, so the pricing of projects 2 and 4 is similar.

$$\text{Project 2:} \quad \mu_2 = \$120 = \$600 - (1{,}200)(0.3189) - (2{,}400)(0.0405)$$
$$\text{Project 4:} \quad \mu_4 = 0 = \$850 - (2{,}500)(0.3189) - (1{,}300)(0.0405)$$

Here we have an example of a completely accepted project with $\mu_j = 0$. Project 1 is again accepted fractionally, and it prices out at zero.

$$\text{Project 1:} \quad \mu_1 = 0 = \$400 - (1{,}000)(0.3189) - (2{,}000)(0.0405)$$

We can demonstrate Eq. 8.19 for a rejected project with negative *PV*.

$$\text{Project 3:} \quad (\$2{,}000)(0.3189) + (2{,}100)(0.0405) = 723 > -150$$

This result is hardly surprising since project 3 has only outflows during the critical times and has a negative *PV*.

The real surprise is that project 6, with a negative *PV* of −\$70, is accepted. Pricing out by using Eq. 8.16 yields

$$\text{Project 6:} \quad \mu_6 = \$221 = -70 + (1{,}000)(0.3189) - (700)(0.0405)$$

The value of the \$1,000 inflow at time 0, less the value of the \$700 outflow at time 1, more than overcomes the negative *PV* and makes project 6 desirable. (The \$700 outflow at time 2 is worth zero since there is no constraint on money at this time.) The extra \$1,000 when it is needed most enables us to select more of the other projects and thereby increase the overall *PV* of the projects selected. □

Project 6 in Example 8.2 has the cash flow pattern of a borrowing activity. Since its $IRR = 26\%$, we are effectively borrowing at a periodic rate of 26% in order to maximize overall *PV* at 20%! This example clearly demonstrates the philosophical conflict in using an interest rate for *PV* maximization when we are faced with a budget limitation. If we have available a borrowing opportunity at a different, higher interest rate, we could be induced to borrow at a rate higher than that used for computing *PV*. The budget limits, in effect, invalidate the use of an externally determined discount rate. The inclusion of lending opportunities and dividend payments does not solve the difficulty, so various authors have attempted other approaches, some of which are discussed in the following.

### 8.3.2 Consistent Discount Factors

In reformulating the *PV* model to eliminate the incompatibility presented, Baumol and Quandt defined a model in which the discount rates between periods are determined by the model itself [3]. On the basis of our previous notation, their revised model is

$$\max_{x_j, \rho_n} \sum_n \sum_j a_{nj} \frac{\rho_n}{\rho_0} x_j \tag{8.20}$$

s.t.[1]
$[\rho_n]$

$$-\sum_j a_{nj} x_j \le M_n, \qquad n = 0, 1, \ldots, N \tag{8.3}$$

$$x_j \ge 0, \qquad j = 1, \ldots, J \tag{8.6}$$

The terms $\rho_n/\rho_0$ represent the discount factors from time 0 to time $n$. Whenever the discount factors are so defined, we will designate them as *consistent discount factors.* Notice the absence of project upper-bound constraints (8.5).

A typical dual constraint has the form

$$-\sum_n a_{nj}\rho_n \ge \sum_n a_{nj} \frac{\rho_n}{\rho_0} \tag{8.21}$$

or

$$\left(-1 - \frac{1}{\rho_0}\right) \sum_n a_{nj}\rho_n \ge 0$$

But the dual variables are nonnegative, so

$$\sum_n a_{nj}\rho_n \le 0$$

[1]The abbreviation s.t. stands for subject to.

In the primal objective function the term $\rho_0$ can be placed before the summation signs; thus each coefficient of $x_j$ is nonpositive. The objective function must therefore have an optimal solution of zero with all $x_j = 0$. In addition, the solution to the dual objective function

$$\min_{\rho_n} \sum_n M_n \rho_n \tag{8.22}$$

with $M_n > 0$ will be zero, with all $\rho_n = 0$. The zero value of $\rho_0$ in the denominator of the primal objective function (8.20) renders that function indeterminate.

With this line of reasoning, Baumol and Quandt rejected *PV* models. They then formulated a model with an objective function that is linear in dividend payments. We will not present this model here but instead examine the PCR line that was pursued by others.

Atkins and Ashton [1] criticized the approach of Baumol and Quandt because there were no upper-bound constraints on the projects and the consequent interpretation of dual variables was absent. The discount factors $\rho_n/\rho_0$ are determined by the marginal productivities of capital in the various time periods. In the absence of upper bounds on projects, any project that is accepted is also partially rejected. Hence, the discounted cash flow of that project *must* be zero.

The implication of this reasoning is that projects must have upper bounds placed on them to avoid the phenomenon of each accepted (and, at the same time, rejected) project having a *PV*, based on consistent discount factors $d_n = \rho_n/\rho_0$, equal to zero. In addition, the Atkins and Ashton model allows for funds to be carried forward at a lending rate of interest. The final modification is the interpretation of the discount factors when one of the $\rho_n$ becomes zero: the equivalent form $\rho_n = d_n\rho_0$ avoids these difficulties.

The method for finding a *consistent optimal solution* (an optimal set of $x_j$ and $d_n = \rho_n/\rho_0$) consists of identifying and evaluating the Kuhn–Tucker stationary points [17] of the problem. In the PCR model there are potentially many consistent solutions, whereas in the situation with lending there is only one solution. In general, this is a rather unsatisfactory procedure because of the large number of such points.

Freeland and Rosenblatt [8] pursued the PCR model (with project upper-bound constraints) further and obtained several interesting results.

- The value of the objective function at a consistent optimal solution equals

$$\frac{1}{2}\sum_j \mu_j^* \text{ (property 2).}$$

- For the PCR case (no lending or borrowing allowed) an objective function value different from zero can be obtained only if some of the $M_n$ values have opposite signs.
- If the objective function value for a consistent optimal solution is not zero, there are alternative optimal discount factors $d_n$.

A more recent article by Hayes [12] on the same topic has further clarified the issue for the situation in which *all budgets are fully expended.* Hayes's analysis assumes upper bounds on projects and lending from one period to the next, but his major result does not depend on the lending activities. If the budgets are fully utilized in all periods except the last (the horizon), the optimal set of projects is independent of discount factors and may be obtained by maximizing the cash at the end of the last period (at the horizon). To see why this result is true, let us reexamine the *PV* model.

$$\underset{x_j}{\text{Max}} \sum_n \sum_j a_{nj} d_n x_j \tag{8.23}$$

s.t.

$$[\rho_n] \qquad -\sum_j a_{nj} x_j = M_n, \qquad n = 0, 1, \ldots, N-1 \tag{8.24}$$

$$[\rho_N] \qquad -\sum_j a_{Nj} x_j + l_N = M_N \tag{8.25}$$

$$x_j \leq 1, \qquad j = 1, \ldots, J \tag{8.5}$$

$$x_j \geq 0, \qquad j = 1, \ldots, J \tag{8.6}$$

where $d_n$ = discount factor for time $n$,
$l_n$ = cash left over at time $N$, the horizon,

and the other terms are as defined previously. Note that the budget constraints 8.24 and 8.25 are equalities, reflecting the assumption about cash being used up each period. The $l_N$ term measures any leftover cash at time $N$, the horizon, the only time we are allowed to have excess cash in this model.

To obtain the desired result, let us split the objective function.

$$\underset{x_j}{\text{Max}} \sum_j a_{Nj} d_N x_j + \sum_{n=0}^{N-1} \sum_j a_{nj} d_n x_j \tag{8.26}$$

Now we can substitute the constraints into the objective function.

$$\underset{x_j}{\text{Max}}\ d_N(l_N - M_N) - \sum_{n=0}^{N-1} d_n M_n = \underset{x_j}{\text{Max}}\ d_N l_N - \sum_{n=0}^{N} d_n M_n \tag{8.27}$$

Since the summation in Eq. 8.27 is a constant for fixed values of $d_n$, it may be dropped without affecting the solution, and by dividing out the constant $d_N$ we are left with

$$\underset{x_j}{\text{Max}}\ l_N \tag{8.28}$$

subject to constraints 8.24, 8.25, 8.5, and 8.6.

Appropriate discount factors may be obtained by the usual form $d_n = \rho_n/\rho_0$ with $\rho_0 = 1$ or any positive constant. With all budget constraints at equality, it is easy to show that $\rho_{n-1} \geq \rho_n$; therefore, no possibility exists of zero dual variables. In summary, the discount factors are irrelevant for project selection!

The foregoing result is important because it emphasizes the fact that the dual variables for the budget constraints reflect the marginal productivities of capital in the respective time periods. Since all cash flows are automatically reinvested in this closed system, any consumption choices by the owner or owners of the firm have been expressed by the values set for the $M_n$.

In this section we appear to have presented numerous models, summarized extensive analyses, and arrived at very little in terms of a useful *PV* model. That is precisely true. All the arguments and discussion repeatedly point to the following types of conclusions and statements.

- In any *PV* model the budget constraint dual variables must reflect the marginal productivities of capital.
- Project upper-bound constraints and lending activities must be included in order to have a meaningful formulation.
- For certain types of closed systems, in which all budgets are fully expended, the projects are selected by maximizing cash at the horizon.

We will return to this last point in Section 8.5. In the meantime, we will discuss in more detail the inclusion of lending and borrowing opportunities in the *PV* model.

## 8.4 NET PRESENT VALUE MAXIMIZATION WITH LENDING AND BORROWING

### 8.4.1 Inclusion of Lending Opportunities

We can define a lending project as an outflow of cash in one period followed by an inflow, with interest, at a later period. Define $v_n$ to be the amount lent at time $n$, to be repaid at time $n + 1$ with interest $r_n$. We then have coefficients $a_{nj} = -1$ and $a_{n+1,j} = 1 + r_n$. We will define as many lending variables as there are time periods with budget constraints. There are no limits on lending.

Notice that we have defined only one-period loans, which is the common practice. Multiple-period lending could easily be included. The following are some typical examples, all with a constant lending rate.

| | $a_{nj}$ | | |
|---|---|---|---|
| Period | Case 1: Lump Sum Payment | Case 2: Interest Only During Period, Principal at End of Last Period | Case 3: Equal Payments |
| $n$ | $-1$ | $-1$ | $-1$ |
| $n + 1$ | 0 | $r$ | $(A/P, r, 2)$ |
| $n + 2$ | $(1 + r)^2$ | $1 + r$ | $(A/P, r, 2)$ |

The number of variables tends to become somewhat unwieldy with this approach, however, compared with the benefits derived from distinguishing between short-term and long-term lending rates. If the projects consist mainly of financial instruments, it is important to work at this level of detail [11]. Otherwise, the approximation of multiple-period lending with successive one-period lending usually suffices.

The objective function coefficients for the lending activities can be obtained by straightforward discounting at $i = MARR$. Applying this concept to case 2, we have

$$PV = -1 + \frac{r}{1 + i} + \frac{1 + r}{(1 + i)^2}$$

The resulting $PV$ may be positive or negative, depending on whether $r$ is greater than or smaller than $i$, respectively.

If one-period lending opportunities are included in the $PV$ model, and lending is always preferred to doing nothing, we can solve an equivalent problem by simply maximizing the amount of cash at the horizon [1,8]. This result is similar to that obtained by Hayes, as described in Section 8.3.2. With attractive lending opportunities present, all budgets will be fully expended, and Hayes's result can be applied directly.

### 8.4.2 Inclusion of Borrowing Opportunities

We should note that the inclusion of unlimited lending opportunities still has not resolved the philosophical conflict between using an interest rate for $PV$ maximization and having budget constraints. If borrowing opportunities are also unlimited, it appears that we have eliminated the conflict. But in that case we really do not need budget constraints and LP to solve our project selection problem. Investment projects with an *IRR* less than the lending rate would always be rejected, and those with an *IRR* greater than the borrowing rate would be accepted. The selection problem would concern only projects with an *IRR* between the lending and borrowing rates.

There remain two difficulties with such an approach. The first is that we rarely have unlimited borrowing opportunities at one interest rate. (In a practical sense, only an agency of the U.S. government can borrow unlimited amounts). Typically, we can borrow, but only up to a limit. Some of the models presented in the next section have this feature. The second difficulty is related to the interpretation of the interest rate used for $PV$ calculations. If the (different) lending and borrowing rates are specified, any other discount rate must presumably reflect the time preferences of the owners of the firm. There is no philosophical conflict in having three distinct rates for lending by the firm, borrowing by the firm, and discounting to reflect the owners' time preferences. But in this case we need to include dividends in the objective function, as shown in Section 8.6 and Appendix 8.A.

We have again failed to develop a rational *PV* model. The reasons here are similar to those for the PCR case. In the face of limits on borrowing, the decisions about selecting projects must be related, through the interrelationships among project combinations and budget amounts, to the decisions for dividend payments [21]. The marginal productivities of capital from one period to the next determine the dual variables $\rho_n$. Any attempt to ignore these two realities in constructing a project selection model or procedure is bound to have major conceptual flaws.

## 8.5. WEINGARTNER'S HORIZON MODEL

Many of the conceptual issues discussed in the previous two sections can be avoided by ignoring *PV* and concentrating on accumulated cash as the objective. Such models are called horizon models. They typically include borrowing and lending activities and may have other constraints added. These models represent an empirical approach to capital budgeting and should therefore be judged mainly on this basis. The presentation in this section, which is based largely on Weingartner [20], presumes the use of LP and hence allows fractional projects.

### 8.5.1 Equal Lending and Borrowing Rates

The simplest type of horizon model contains budget constraints, project upper bounds, and lending and borrowing opportunities at a common, fixed rate.

#### *Example 8.3*

Table 8.5 presents the data for Example 8.3, and Table 8.6 presents both the symbolic and numeric primal formulations. The projects, 1 through 6, are the same as for Example 8.2. The budget amounts are slightly different from those in Example 8.2, being \$3,000, \$5,000, and \$4,800 at times 0, 1, and 2, respectively. We are using 20% for both borrowing and lending, with no limits on either. The horizon is at time 2, so we are trying to maximize the accumulated cash at this time. Most of the projects, however, have cash flows after time 2, so we discount at 20% these flows back to time 2. For example, $\hat{a}_1$ represents the \$2,900 inflow at time 3 for project 1, discounted at 20% for one period, or \$2,900/1.2 = \$2,417.

The objective function in this horizon model is $v_2 - w_2$, the accumulated cash available (for lending) at time 2, plus the value of posthorizon flows, represented by the $\hat{a}_j$. The cash balance equations 8.30 and 8.31 are similar to those for the *PV* model. A typical constraint says that cash outflows from projects, plus current lending, plus repayment with interest of previous-period borrowing, minus repayment with interest of previous-period lending, minus current borrowing must be less than or equal to the amount of externally supplied funds. The project upper bounds and nonnegativity restrictions complete the model. For simplicity, we have not included any project dependency or exclusivity constraints. (The solution to Example 8.3 follows in the text.) □

**Table 8.5** ***Data for Examples 8.3 and 8.5 (Horizon Is Time 2)***

| Variable Type | | Project 1 | 2 | 3 | 4 | 5 | 6 |
|---|---|---|---|---|---|---|---|
| Cash Flow at Time | 0 | −$1,000 | −$1,200 | −$2,000 | −$2,500 | −$3,000 | $1,000 |
| Same for | 1 | −2,000 | −2,400 | −2,100 | −1,300 | 900 | −700 |
| Examples 8.3 | 2 | 2,000 | 2,500 | 3,000 | 2,000 | 1,400 | −700 |
| and 8.5 | 3 | 2,900 | 3,567 | 2,621 | 2,000 | 1,600 | 0 |
| | 4 | 0 | 0 | 0 | 2,000 | 211 | 0 |
| | 5 | 0 | 0 | 0 | 2,296 | 0 | 0 |
| Example 8.3 $\hat{a}_j$ | | $2,417 | 2,973 | 2,184 | 4,384 | 1,480 | 0 |
| Example 8.5 $\hat{a}_j$ | | $2,230 | 2,744 | 2,016 | 3,767 | 1,356 | 0 |

| | | | |
|---|---|---|---|
| Budgets ($M_n$) | $n = 0$ | $n = 1$ | $n = 2$ |
| Example 8.3 | $3,000 | 5,000 | 4,800 |
| Example 8.5 | $1,000 | 2,000 | 4,800 |
| Lending rates | $n = 0 \to 1$ | $n = 1 \to 2$ | $n = 2 \to 3$ |
| Example 8.3 | 20% | 20% | 20% |
| Example 8.5 | 15% | 15% | 15% |
| Borrowing rates | $n = 0 \to 1$ | $n = 1 \to 2$ | $n = 2 \to 3$ |
| Example 8.3 | 20% | 20% | 20% |
| Example 8.5 | 30% | 30% | 30% |
| Borrowing limits | $n = 0 \to 1$ | $n = 1 \to 2$ | $n = 2 \to 3$ |
| Example 8.5 only | None | 1,000 | None |

The LP solution of the horizon model in Example 8.3 is straightforward and quick, requiring less than one second of time for both processing and input–output on a mainframe computer. Table 8.7 contains the solution for Example 8.3. Projects 1, 2, and 4 were accepted completely, and projects 3, 5, and 6 were rejected. In addition, there was borrowing of $1,700 at time 0 and $2,740 at time 1. At time 2 a total cash accumulation of $8,012 was available for lending.

There is more to the solution of this horizon problem than the numerical results, however. We note several features of the solution.

- The dual variables $\rho_n$ are powers of 1.2.
- The dual variables $\mu_j$ for accepted projects are equal to the *FV*(20%) of these projects.
- Projects with positive *PV*(20%) were accepted and those with negative *PV* (20%) were rejected.

These features are not coincidental but rather are characteristic of the horizon model as presented in Table 8.6. We can verify this by examining the dual formulation, given in Table 8.8.

**Table 8.6** ***Primal Problem Formulation For Horizon Model for Example 8.3***

*Symbolic*

$$\max_{x_j, v_n, w_n} \sum_j \hat{a}_j x_j + v_N - w_N \quad (8.29)$$

s.t.

$[\rho_0]$
$$-\sum_j a_{0j} x_j + v_0 - w_0 \le M_0 \quad (8.30)$$

$[\rho_n]$
$$-\sum_j a_{nj} x_j - (1+r)v_{n-1} + v_n + (1+r)w_{n-1} - w_n \le M_n, \quad n = 1, 2, \ldots, N \quad (8.31)$$

$[\mu_j]$
$$x_j \le 1, \quad j = 1, \ldots, J \quad (8.5)$$

$$x_j \ge 0, \quad j = 1, \ldots, J \quad (8.6)$$

$$v_n, w_n \ge 0, \quad j = 0, \ldots, J \quad (8.32)$$

where $\hat{a}_j$ = horizon time value of cash flows beyond horizon
$x_j$ = project selection variable
$a_{nj}$ = cash flow for project $j$ at time $n$; inflows have a plus sign, outflows have a minus sign
$v_n$ = lending amount from time $n$ to time $n + 1$
$w_n$ = amount borrowed from time $n$ to time $n + 1$
$r$ = interest rate for borrowing and lending
$M_n$ = budget limit on externally supplied funds at time $n$
$N$ = horizon, end of the planning period
$\rho_n, \mu_j$ = dual variables

*Numeric*

$$\text{Max } \$2{,}417x_1 + 2{,}973x_2 + 2{,}184x_3 + 4{,}384x_4 + 1{,}480x_5 + v_2 - w_2$$

s.t.

$[\rho_0]$
$$\$1{,}000x_1 + 1{,}200x_2 + 2{,}000x_3 + 2{,}500x_4 + 3{,}000x_5 - 1{,}000x_6 + v_0 - w_0 \le 3{,}000$$

$[\rho_1]$
$$\$2{,}000x_1 + 2{,}400x_2 + 2{,}100x_3 + 1{,}300x_4 - 900x_5 + 700x_6 - 1.2v_0 + v_1 + 1.2w_0 - w_1 \le 5{,}000$$

$[\rho_2]$
$$-\$2{,}000x_1 - 2{,}500x_2 - 3{,}000x_3 - 2{,}000x_4 - 1{,}400x_5 + 700x_6 - 1.2v_1 + v_2 + 1.2w_1 - w_2 \le 4{,}800$$

$[\mu_1]$ $x_1 \le 1$

$[\mu_2]$ $x_2 \le 1$

$[\mu_3]$ $x_3 \le 1$

$[\mu_4]$ $x_4 \le 1$

$[\mu_5]$ $x_5 \le 1$

$[\mu_6]$ $x_6 \le 1$

all $x_j \ge 0, \quad j = 1, \ldots, 6$

all $v_n, w_n \ge 0, \quad n = 0, 1, 2$

**Table 8.7** ***Solution to Examples 8.3 and 8.5***

| Variable Type | Objective Function | Example 8.3, \$17,786 | Example 8.5, \$9,210 |
|---|---|---|---|
| Project selection | $x_1$ | 1.0 | 0 |
| | $x_2$ | 1.0 | 1.0 |
| | $x_3$ | 0 | 0 |
| | $x_4$ | 1.0 | 0.151 |
| | $x_5$ | 0 | 0 |
| | $x_6$ | 0 | 0.577 |
| Lending | $v_0$ | 0 | 0 |
| | $v_1$ | 0 | 0 |
| | $v_2$ | \$8,012 | \$5,898 |
| Borrowing | $w_0$ | \$1,700 | 0 |
| | $w_1$ | 2,740 | \$1,000 |
| | $w_2$ | 0 | 0 |
| Budget constraint dual variable | $\rho_0$ | 1.44 | 1.622 |
| | $\rho_1$ | 1.20 | 1.317 |
| | $\rho_2$ | 1.00 | 1.0 |
| Project upper-bound dual variable | $\mu_1$ | 577 | 0 |
| | $\mu_2$ | 865 | 137 |
| | $\mu_3$ | 0 | 0 |
| | $\mu_4$ | 1.224 | 0 |
| | $\mu_5$ | 0 | 0 |
| | $\mu_6$ | 0 | 0 |
| Borrowing limit dual variable | $\beta_1$ | — | 0.017 |

From (8.36) and (8.37) we have $\rho_n^* = 1$. The value of \$1 at time $N$ is \$1 in the optimal solution because we do not have time to do anything with it. Similarly, from (8.34) and (8.35) we obtain

$$1 + r \leq \frac{\rho_n^*}{\rho_{n+1}^*} \leq 1 + r, \qquad n = 0, \ldots, N - 1 \tag{8.38}$$

or

$$\frac{\rho_n^*}{\rho_{n+1}^*} = 1 + r \tag{8.39}$$

and

$$\begin{aligned} \rho_n^* &= \rho_{n+1}^*(1 + r) = \rho_{n+2}^*(1 + r)^2 = \cdots = \rho_{n+N-n}^*(1 + r)^{N-n} \\ &= \rho_N^*(1 + r)^{N-n} = (1 + r)^{N-n} \end{aligned} \tag{8.40}$$

The interpretation of the $\rho_n^*$ now becomes clear: they are compound interest factors that reflect the value at the horizon, time $N$, of an additional dollar at time $n$.

**Table 8.8** ***Dual Problem Formulation for Horizon Model with Common, Fixed Rate for Borrowing and Lending***

| | | | |
|---|---|---|---|
| | $\text{Min} \sum_n \rho_n M_n + \sum_j \mu_j$ | | (8.9) |
| s.t. | | | |
| $[x_j]$ | $-\sum_j a_{nj}\rho_n + \mu_j \geq \hat{a}_j,$ | $j = 1, \ldots, J$ | (8.33) |
| $[v_n]$ | $\rho_n - (1 + r)\rho_{n+1} \geq 0,$ | $n = 0, \ldots, N - 1$ | (8.34) |
| $[w_n]$ | $-\rho_n + (1 + r)\rho_{n+1} \geq 0,$ | $n = 0, \ldots, N - 1$ | (8.35) |
| $[v_N]$ | $\rho_N \geq 1$ | | (8.36) |
| $[w_N]$ | $-\rho_N \geq -1$ | | (8.37) |
| | $\rho_n \geq 0,$ | $n = 0, 1, \ldots, N - 1$ | (8.11) |
| | $\mu_j \geq 0,$ | $j = 1, \ldots, J$ | (8.12) |

The analysis of (8.33) is similar to that of (8.10) in Section 8.2.5. If $x_j^* > 0$, the dual constraint is met exactly, and

$$0 \leq \mu_j^* = \hat{a}_j + \sum_n a_{nj}\rho_n^* \tag{8.41}$$

Substituting for $\rho_n^*$ from (8.40), we have

$$\mu_j^* = \hat{a}_j + \sum_n a_{nj}(1 + r)^{N-n} \tag{8.42}$$

The right side of (8.42) is simply the net future value ($FV$) at time $N$ of the cash flows for project $j$. The $\hat{a}_j$ term is the value of posthorizon flows discounted back to time $N$, and the summation is the forward compounding of the other cash flows. For fractionally accepted projects $\mu_j^* = 0$, and thus (8.42) equals zero. For rejected projects $x_j^*$ and $\mu_j^*$ are both equal to zero, so

$$0 \geq \hat{a}_j + \sum_n a_{nj}(1 + r)^{N-n} \tag{8.43}$$

The horizon model with a common, fixed rate for borrowing and lending thus will accept only projects with nonnegative $FV(r)$, or $PV(r)$, and the $FV(r)$ of any rejected project is nonpositive. This agreement between the LP model and the $PV$ criterion reassures us that the model performs as intended. Actually, we would expect the model to perform precisely in this manner, since the unlimited borrowing and lending opportunities at interest rate $r$ are equivalent to the assumptions underlying the *PV* criterion.

### 8.5.2. Lending Rates Less Than Borrowing Rates

Clearly, an LP model that yields the same answers as *PV* analysis is of little use. The true power of the LP model is the ability to represent a great variety of investment opportunities and restrictions, lending and borrowing opportunities, scarce resource restrictions, and so forth. In this section we modify the horizon model of the previous section by having a borrowing rate higher than the lending rate. The only modification needed in the model is in the cash balance equation 8.31, which becomes

$$[\rho_n] \qquad -\sum_j a_{nj}x_j - (1 + r_l)v_{n-1} + v_n + (1 + r_b)w_{n-1}$$

$$- w_n \le M_n, \qquad n = 1, 2, ..., N \tag{8.44}$$

where $r_l$ = lending interest rate,
$r_b$ = borrowing interest rate.

The corresponding changes in the dual problem affect constraints 8.34 and 8.35, which become, respectively,

$$[v_n] \qquad \rho_n - (1 + r_l)\rho_{n+1} \geqslant 0, \qquad n = 0, \ldots, N-1 \tag{8.45}$$

$$[w_n] \qquad -\rho_n + (1 + r_b)\rho_{n+1} \geqslant 0, \qquad n = 0, \ldots, N-1 \tag{8.46}$$

Instead of (8.38) and (8.39), we obtain

$$1 + r_l \le \frac{\rho_n^*}{\rho_{n+1}^*} \le 1 + r_b, \qquad n = 0, ..., N - 1 \tag{8.47}$$

The ratio of the dual variables for the cash balance equations is now restricted to the range (including end points) between the lending and borrowing interest factors. From complementary slackness [5] we can deduce that if we are lending money at time $n$ ($v_n > 0$), then (8.45) and the left part of (8.47) are satisfied as equality. If we are borrowing at time $n$ ($w_n > 0$), then (8.46) and the right side of (8.47) are satisfied as equality. This makes sense, because the lending activity implies that extra dollars at time $n$ would also be lent, leading to 1.2 times the extra dollars at the horizon as extra dollars at time $n$ + 1, and so forth.

***Example 8.4***

We can demonstrate these results with Example 8.4, for which both data and solution are shown in Table 8.9. There are four projects, six budget limits, a

**Table 8.9** ***Data and Solution for Example 8.4 (Lending Rate Less Than Borrowing Rate)***

| Cash Flow at Time | Project 1 | 2 | 3 | 4 | Budget |
|---|---|---|---|---|---|
| 0 | −600 | −$1,200 | −$900 | −$1,500 | $270 |
| 1 | 360 | 480 | 360 | 420 | 150 |
| 2 | 330 | 360 | 330 | 480 | 30 |
| 3 | 60 | 0 | 300 | 510 | 0 |
| 4 | −150 | 660 | 270 | 540 | −60 |
| 5 | 330 | 510 | 240 | 540 | 0 |
| $\hat{a}_j$ | 150 | 300 | 150 | 330 | — |

$r_l = 0.2$, $r_b = 0.3$, $N = 5$

*Solution*

| | | | |
|---|---|---|---|
| $x_1 = 1.0$ | $v_0 = 0$ | $w_0 = 420$ | $\rho_0 = 2.754$ |
| $x_2 = 0$ | $v_1 = 0$ | $w_1 = 0$ | $\rho_1 = 2.119$ |
| $x_3 = 0.1$ | $v_2 = 393$ | $w_2 = 0$ | $\rho_2 = 1.728$ |
| $x_4 = 0$ | $v_3 = 561$ | $w_3 = 0$ | $\rho_3 = 1.440$ |
| $\mu_1 = 67$ | $v_4 = 490$ | $w_4 = 0$ | $\rho_4 = 1.200$ |
| $\mu_2 = 0$ | $v_5 = 943$ | $w_5 = 0$ | $\rho_5 = 1.000$ |
| $\mu_3 = 0$ | | | |
| $\mu_4 = 0$ | | | |

lending rate of 20%, a borrowing rate of 30%, and a horizon at time 5. The negative budget at time 4 means we must generate $60 to be used elsewhere in the firm. Only project 1 is accepted completely; project 3 is accepted fractionally, and the other two are rejected. There are borrowing at time 0 and lending at times 2 through 5. We can demonstrate how Eq. 8.47 indicates borrowing or lending by taking the ratios of the dual variables.

$\rho_0^*/\rho_1^* = 1.3$ borrowing at time 0
$\rho_1^*/\rho_2^* = 1.23$ neither at time 1
$\rho_2^*/\rho_3^* = 1.2$ lending at time 2
$\rho_3^*/\rho_4^* = 1.2$ lending at time 3
$\rho_4^*/\rho_5^* = 1.2$ lending at time 4 □

Everything seems to work according to theory in Example 8.4, but how do we explain the ratio $\rho_1^*/\rho_2^*$ of 1.23, which is strictly between the limits? And if there are no restrictions on borrowing, why is project 3 accepted only fractionally? To answer these questions, let us assume two hypothetical situations in Example 8.3. First, assume everything is as before except that we borrow at time

1, forcing the ratio to be 1.3. The new dual variables can then be obtained as follows.

$$\rho_5 = 1.0$$
$$\rho_4 = \rho_5(1.2) = 1.2$$
$$\rho_3 = \rho_4(1.2) = 1.44$$
$$\rho_2 = \rho_3(1.2) = 1.728$$
$$\rho_1 = \rho_2(1.3) = 2.246$$
$$\rho_0 = \rho_1(1.3) = 2.920$$

Now let us find the corresponding value of $\mu_3$ from Eq. 8.41. In LP terminology, we are pricing out the activity vector for project 3.

$$\$150 + (-900)(2.920) + (360)(2.246) + (330)(1.728) + (300)(1.44) + (270)(1.2) + (240)(1.0) = -\$103$$

The negative value means that we would not introduce project 3 into the LP solution, given the values for the $\rho_n$. In other words, given the other borrowing and lending activities, we are not justified in borrowing at 30% at time 1 in order to accept more of project 3.

Now assume everything is as in the original solution in Table 8.9, except that we lend at time 1, forcing the ratio to be 1.2. The new dual variables are obtained as before, and the pricing of the activity vector yields

$$\rho_5 = 1.0$$
$$\rho_4 = \rho_5(1.2) = 1.2$$
$$\rho_3 = \rho_4(1.2) = 1.44$$
$$\rho_2 = \rho_3(1.2) = 1.728$$
$$\rho_1 = \rho_2(1.2) = 2.074$$
$$\rho_0 = \rho_1(1.3) = 2.696$$

$$\$150 + (-900)(2.696) + (360)(2.074) + (330)(1.728) + (300)(1.44) + (270)(1.2) + (240)(1.0) = \$37$$

The positive value means that if we were lending money at time 1, given the other borrowing and lending activities, we could improve our situation by accepting project 3. The best action is to accept as much as possible without borrowing at time 1. This turns out to be 10%. What has happened is that the marginal productivity of cash at time 1 is determined by project 3.

### 8.5.3 Inclusion of Borrowing Limits, Supply Schedule of Funds

Another typical restriction in the horizon model is a limit on the amount borrowed at a particular time. Example 8.5 is a slight variation on Example 8.3.

***Example 8.5***

Table 8.5 presents the relevant data. The project cash flows are the same, the budgets at times 0 and 1 are reduced, the lending rate is 15%, the borrowing rate is 30%, and a $1,000 limit on borrowing is imposed at time 1. In anticipation of future borrowing at 30%, the $\hat{a}_j$ have been computed by using a discount rate of 30%. The solution for Example 8.5 is given in Table 8.7. Only one project, number 2, is accepted completely, and 4 and 6 are accepted fractionally. The only borrowing activity is at time 1, at the limit of $1,000. □

To analyze the results of Example 8.5, we need to add one more constraint to the primal problem.

$$[\beta_1] \qquad w_1 \leq 1{,}000 \tag{8.48}$$

The changes in the dual formulation are in the objective function and in Eq. 8.46.

$$\text{Min} \sum_n \rho_n M_n + \sum_j \mu_j + 1{,}000\beta_1 \tag{8.49}$$

$$[w_1] \qquad -\rho_1 + (1 + r_b)\rho_2 + \beta_1 \geqslant 0 \tag{8.50}$$

Instead of (8.47) we have

$$(1 + r_l)\rho_2^* \leqslant \rho_1^* \leqslant (1 + r_b)\rho_2^* + \beta_1^* \tag{8.51}$$

The borrowing restriction at time 1 places a premium on funds at time 1 beyond that of the normal borrowing interest factor of 1.3. The nonzero value of $\beta_1^*$ implies that we are borrowing the full amount and would like to borrow more. We can verify the right side of (8.51).

$$1.317 = (1.3)(1.0) + 0.017$$

What has happened in this example is as follows.

- Project 2, with the highest *IRR* of 31.26%, was accepted completely, exhausting the time 0 budget of $1,000. The cheapest method of borrowing was with partial acceptance of project 6, which is equivalent to borrowing at 26%.
- Projects 3 and 5 had negative *PV*(20%) and would not justify borrowing at 26 or 30%.
- Projects 1 and 4 have similar *IRR*s, 29.1% and 28.9%, respectively, for the original cash flows. The *IRR*s are 28.7% and 28.1%, respectively, for the $a_{0j}$, $a_{1j}$, $a_{2j}$, $\hat{a}_j$ flows, which is what the LP program sees. However, project 1 requires twice as much investment at time 1 as at time 0, whereas the opposite is true for project 4. Since time 1 borrowing costs 30% and the time 0 borrowing costs 26% (via project 6), preference is given to project

4. Neither project justifies borrowing at 30% in both periods, so just enough of project 4 is accepted to reach the borrowing limit of $1,000 at time 1.

It should be noted that the pricing operation with Eq. 8.41 is still valid and yields results consistent with the solution in Table 8.7.

The concept of borrowing limit can be generalized to a series of limits, each applicable to a source of loan funds at a designated rate. For example, a firm may be able to borrow an amount, say 2,000, at 22%, an additional 1,000 at 25%, and a final 1,000 at 30%, as shown in Figure 8.1. This representation is called a sloping supply schedule for funds [20]. If we let $w_{kn}$ represent the amount borrowed at the $k$th step at time $n$, the modification to the horizon model is straightforward, as shown in Table 8.10. By convention, we order the borrowing steps in increasing order of cost $r_k$; the LP algorithm will naturally start borrowing at the lowest cost and move to the next step as each limit is reached.

Analysis of the dual formulation is similar to that in Example 8.5. For each $w_{kn}$ in the primal we have a dual constraint.

$$[w_{kn}] \qquad -\rho_n + (1 + r_{kn})\rho_{n+1} + \beta_{kn} \geq 0 \tag{8.56}$$

The dual constraint of interest is the one corresponding to the last step $k$ at time $n$. If we are at the limit on the last step, we can not say anything beyond Eq. 8.56 without the actual value of $\beta^*_{kn}$ from the LP solution. If we are borrowing an amount below the limit on the last step, however, we have

$$\frac{\rho^*_n}{\rho^*_{n+1}} = 1 + r_{kn} \tag{8.57}$$

Equation 8.57 illustrates the nature of the $\rho^*_n$ as indicators of the marginal cost of funds.

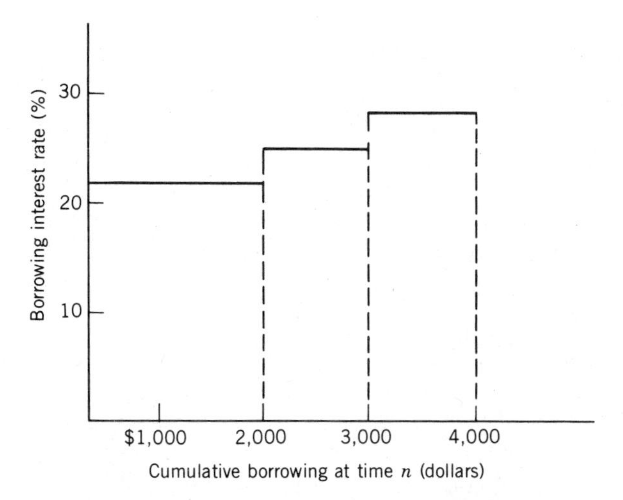


**FIGURE 8.1. Sloping supply schedule of funds.**

**Table 8.10** ***Primal Problem Formulation for Horizon Model with Sloping Supply Schedule for Funds***

$$\text{Max} \sum_j \hat{a}_j x_j + v_N - \sum_k w_{kN} \tag{8.52}$$

s.t.

$$[\rho_0] \quad -\sum_j a_{0j} x_j + v_0 - \sum_k w_{k0} \le M_0 \tag{8.53}$$

$$[\rho_n] \quad -\sum_j a_{nj} x_j - (1 + r_\ell) v_{n-1} + v_n + \sum_k (1 + r_k) w_{k,n-1}$$

$$-\sum_k w_{kn} \le M_n, \qquad n = 1, 2, \ldots, N \tag{8.54}$$

$$[\beta_{kn}] \quad w_{kn} \le B_{kn}, \quad k = 1, \ldots, m; \quad n = 0, \ldots, N \tag{8.55}$$

$$[\mu_j] \quad x_j \le 1, \qquad j = 1, \ldots, n \qquad (8.5)$$

all variables $\ge 0$

where $w_{kn}$ = amount borrowed at $k$th step at time $n$,
$r_k$ = interest rate at $k$th step of borrowing,
$B_{kn}$ = limit on $k$th step at time $n$.
Other terms are as previously described

### *8.5.4 Dual Analysis with Project Interdependencies*

The presence of project interdependencies will affect the use of (8.33) in pricing out project activity vectors in a manner consistent with the LP solution.

### *Example 8.6*

We add a contingency relationship that project 4 cannot be performed without project 5 in Example 8.3. Then we add to the primal the form 8.8, or

$$[\nu] \qquad x_4 - x_5 \leqslant 0$$

and the dual constraints 8.33 would become

$$-\sum_n a_{n4}\rho_n + \mu_4 + \upsilon \ge \hat{a}_4$$

$$-\sum_n a_{n5}\rho_n + \mu_5 - \upsilon \ge \hat{a}_5$$

The pricing operation would reflect a penalty being applied to project 4 and a subsidy being applied to project 5. Since project 4 is so highly favorable, both 4 and 5 would be accepted. Using (8.41), we have

$$\text{Project 4:}\quad \mu_4^* = \$4{,}384 - (2{,}500)(1.44) - (1{,}300)(1.2) + (2{,}000)(1.0) - 360 = \$864$$

$$\text{Project 5:}\quad \mu_5^* = \$1{,}480 - (3{,}000)(1.44) + (900)(1.2) + (1{,}400)(1.0) + 360 = 0$$

The value of $\nu$ is 360, just enough for project 5 to price out at zero, where it can be accepted. □

The mutual exclusivity constraints 8.7 would be handled in a similar manner. In practice, such constraints allow more projects to be fractionally accepted, and the typical end result is use of integer programming (Section 8.7).

## 8.6 BERNHARD'S GENERAL MODEL

All the features of the horizon model with borrowing constraints (Section 8.5.3) are retained in Bernhard's general model for capital budgeting [4]. Bernhard also includes dividends in a nonlinear objective function, with dividends constrained by a horizon posture restriction. In the following sections we will present the model and some general results. With few exceptions, the notation will follow Bernhard's, which is largely consistent with what we have been using. Appendix 8.A presents an application of the model to a dividend–terminal-wealth problem.

### 8.6.1 Model Formulation

The objective function is an unspecified function of dividends and terminal wealth.

$$\text{Max } f(D_1, D_2, \ldots, D_N, G) \tag{8.58}$$

where $D_n$ = dividend paid at time $n$,
$G$ = time $N$ terminal wealth, to be specified in more detail later.

It is assumed that $\partial f/\partial D_n \geq 0$ and $\partial f/\partial G \geq 0$, which imply that more dividends and terminal wealth, respectively, lead to greater utility values. Typically, $f$ is defined to be concave.

The cash balance equations, or budget constraints, contain a liquidity requirement that reflects certain banking practices. The firm is required to maintain $C_n + c_n w_n$ in a bank account. The $C_n$ is a constant representing basic liquidity at time $n$, and the $c_n$ ($0 \leqslant c_n < 1$) is a compensating balance fraction. The amount $C_n + c_n w_n$ earns interest at rate $r_{ln}$. The typical constraint is

$$[\rho_n] \qquad -\sum_j a_{nj}x_j - l_{n-1}(v_{n-1} + c_{n-1}w_{n-1} + C_{n-1}) + (v_n + c_n w_n + C_n) + b_{n-1}w_{n-1} - w_n + D_n \leq M'_n, \qquad n = 0, 1, \ldots, N \tag{8.59}$$

where $M'_n$ = budget limit on externally supplied funds at time $n$,
$l_n$ = lending interest rate factor at time $n$, $1 + r_{ln}$, and
$b_n$ = borrowing interest rate factor at time $n$, $1 + r_{bn}$.

Equation 8.59 states that project outlays, minus previous-period lending, plus current lending, plus previous-period borrowing, minus current borrowing, plus current dividend cannot exceed the budget limit on externally supplied funds at time $n$. Regrouping terms gives

$$[\rho_n] \qquad -\sum_j a_{nj}x_j - l_{n-1}v_{n-1} + v_n + (b_{n-1} - l_{n-1}c_{n-1})w_{n-1} - (1 - c_n)w_n + D_n \leq M_n, \qquad n = 0, 1, \ldots, N \tag{8.60}$$

where

$$M_n = M'_n + (l_{n-1}C_{n-1}) - C_n$$

Group payback restrictions state that at time $n'$ the net outflows on the set of selected projects are recovered.

$$[\psi] \qquad -\sum_j \sum_{n=0}^{n'} a_{nj}x_j \leq 0 \tag{8.61}$$

Scarce material restrictions are defined for a nonmonetary resource, which could be skilled personnel, special equipment, and so forth.

$$[\nu] \qquad \sum_j d_j x_j \leq d \tag{8.62}$$

where $d_n$ = amount of scarce resource consumed by project $j$,
$d$ = total amount of scarce resource available.

The firm is prevented from paying excessive dividends and thus jeopardizing earning capability past the horizon. This is accomplished by a terminal-wealth horizon posture restriction. First it is necessary to define the terminal wealth. After the last dividend $D_N$ at time $N$, the terminal wealth is

$$G = M' + \sum_j \hat{a}_j x_j + v_N + c_N w_N + C_N - w_N$$

where $M'$ is the value at time $N$ of posthorizon cash flows from other sources.

With the inclusion of $M'$ and the liquidity requirement, the definition is the same as the objective function 8.29 of the horizon model. The definition is rewritten as

$$[\phi] \qquad -\sum_j \hat{a}_j x_j - v_N + (1 - c_N) w_N + G = M \qquad (8.63)$$

where $M$ is $M' + C_N$.

The horizon posture restriction states that the terminal wealth must exceed some functional value of the dividends,

$$G \geqslant K + g(D_1, D_2, \ldots, D_N)$$

where $K$ = a nonnegative constant,
$g$ = a function, typically a convex one.

Rewriting, we have

$$[\theta] \qquad -G + g(D_1, D_2, \ldots, D_N) \leqslant -K \qquad (8.64)$$

Borrowing limits for $n = 0, 1, \ldots, N-1$, project upper bounds, and nonnegativity restrictions complete the model. Table 8.11 summarizes the objective function and constraints.

### 8.6.2 Major Results

With a concave objective function 8.58 and a convex constraint 8.64, the Kuhn–Tucker conditions are necessary and sufficient for optimality, and they enable us to make a number of statements about optimal solutions to the general model [17]. Table 8.12 presents the Kuhn–Tucker conditions. We present only the major results that can be obtained from them; derivations are in Bernard [4].

The pricing out of a project activity vector, analogous to (8.33), gives us

$$\mu_j^* \geq A_j^* = \sum_n a_{nj} \rho_n^* + \sum_{n=0}^{n'} a_{nj} \psi - d_j v^* + \hat{a}_j \rho_N^* \qquad (8.74)$$

where we have used the substitution $\phi^* = \rho_N^*$. The role of $A_j^*$ in (8.74) is similar to that of $PV$ in the horizon model.

$$\begin{aligned} &\textbf{Case 1: } \text{If} & x_j^* &= 1, & \mu_j^* &= A_j^* \geq 0 \\ &\textbf{Case 2: } \text{If} & 0 < x_j^* &< 1, & \mu_j^* &= A_j^* = 0 \\ &\textbf{Case 3: } \text{If} & x_j^* &= 0, & \mu_j^* &= 0 \geq A_j^* \end{aligned} \qquad (8.75)$$

We should oberve that absent or nonbinding group payback and scarce material contraints imply $\psi$ and $\upsilon$ values of zero, and $A_j^*$ reduces to $\sum_n a_{nj} \rho_n^* + \hat{a}_j \rho_N^*$.

**Table 8.11** ***Bernhard's General Model***

$$\text{Max } f(D_1, D_2, \ldots, D_N, G) \tag{8.58}$$

s.t.

$$[\rho_n] \quad -\sum_j a_{nj}x_j - l_{n-1}v_{n-1} + v_n + (b_{n-1} - l_{n-1}c_{n-1})w_{n-1} - (1 - c_n)w_n + D_n \le M_n, \quad n = 0, 1, \ldots, N \tag{8.60}$$

$$[\psi] \quad -\sum_j \sum_{n=0}^{n'} a_{nj}x_j \le 0 \tag{8.61}$$

$$[\upsilon] \quad \sum_j d_j x_j \le d \tag{8.62}$$

$$[\phi] \quad -\sum_j \hat{a}_j x_j - v_N + (1 - c_N)w_N + G = M \tag{8.63}$$

$$[\theta] \quad -G + g(D_1, D_2, \ldots, D_N) \le -K \tag{8.64}$$

$$[\beta_n] \quad w_n \le B_n, \quad n = 0, 1, \ldots, N-1 \tag{8.65}$$

$$[\mu_j] \quad x_j \le 1, \quad j = 1, \ldots, J \tag{8.5}$$

$$x_j, v_n, w_n, D_n \ge 0 \tag{8.66}$$

where $x_j$ = project selection variable
$v_n$ = lending amount from time $n$ to $n + 1$,
$w_n$ = borrowing amount from time $n$ to $n + 1$
$D_n$ = dividend paid at time $n$
$a_{nj}$ = cash flow for project $j$ at time $n$ (inflows +)
$\hat{a}_j$ = horizon time value of cash flows beyond horizon
$l_n$ = lending interest rate factor at time $n$, $1 + r_{ln}$
$b_n$ = borrowing interest rate factor at time $n$, $1 + r_{bn}$
$B_n$ = borrowing limit at time $n$
$c_n$ = compensating balance fraction
$M_n$ = budget limit on externally supplied funds at time $n$, adjusted for basic liquidity requirement
$d_j$ = amount of scarce resource consumed by project $j$
$d$ = total amount of scarce resource available
$G$ = terminal wealth at time $N$, after paying $w_N$
$M$ = value at time $N$ of posthorizon cash flows from other sources, adjusted by basic liquidity requirement
$K$ = nonnegative constant representing the minimum acceptable terminal wealth

$\rho_n, \psi, \upsilon, \phi, \theta, \beta_n, \mu_j$ are dual variables

**Table 8.12** ***Kuhn–Tucker Conditions for Bernhard's General Model***

| | | |
|---|---|---|
| $[v_n]$ | $-\rho_n + l_n\rho_{n+1} \leq 0, \qquad n = 0,1, \ldots, N-1$ | (8.67) |
| $[w_n]$ | $(1 - c_n)\rho_n - (b_n - l_nc_n)\rho_{n+1} - \beta_n \leq 0, \qquad n = 0, 1, \ldots, N-1$ | (8.68) |
| $[v_N]$ | $-\rho_N + \phi \leq 0$ | (8.69) |
| $[w_N]$ | $(1 - c_N)\rho_N - (1 - c_N)\,\phi \leq 0$ | (8.70) |
| $[x_j]$ | $\sum\limits_n a_{nj}\rho_n + a_j\phi - d_j\upsilon + \sum\limits_{n=0}^{n'} a_{nj}\psi - \mu_j \leq 0, \qquad j = 1, 2, \ldots, J$ | (8.71) |
| $[D_n]$ | $\left.\dfrac{\partial f}{\partial D_n}\right\vert_{D_n} - \rho_n - \theta \left.\dfrac{\partial g}{\partial D_n}\right\vert_{D_n} \leq 0, \qquad n = 0, 1, \ldots, N$ | (8.72) |
| $[G]$ | $\left.\dfrac{\partial f}{\partial G}\right\vert_G - \phi + \theta \leq 0$ | (8.73) |

SOURCE: Bernard [4].

Turning to the $\rho_n^*$, we let

$$\hat{b}_n = \frac{b_n - l_nc_n}{1 - c_n} \tag{8.76}$$

This $\hat{b}_n$ is the effective borrowing rate. For example, if $b_n = 1.3$, $l_n = 1.2$, and $c_n = 0.2$, in order to borrow a usable \$100, we have to borrow \$125 at 30% and put (0.2)(125) = 25 back in the bank at 20%. Our true borrowing cost is

$$(\$125)(0.3) - (25)(0.2) = 32.5, \text{ or } 32.5\%$$

Equation 8.76 yields the equivalent factor of 1.325. In addition, let

$$\hat{\beta}_n^* = \beta_n^*/(1 - c_n) \tag{8.77}$$

Then we can manipulate (8.67) and (8.68) to yield

$$l_n\rho_{n+1}^* \leqslant \rho_n^* \leqslant \hat{b}_n\rho_{n+1}^* + \hat{\beta}_n^*, \qquad n = 0, 1, \ldots, N-1 \tag{8.78}$$

This equation is similar to (8.51), showing that compensating balance fractions do not necessarily complicate the model once we interpret them as higher effective borrowing rates.

If $v_n^* > 0$, complementary slackness indicates that the left side of (8.77) is satisfied as equality. In this case the ratio $\rho_n^*/\rho_{n+1}^*$ equals the lending rate factor. If the company borrows, $w_n^* > 0$, and the right side is equality. Note that the ratio of dual variables is affected by the value of $\hat{\beta}_n^*$, the dual variable of the borrowing limit. If the borrowing constraint is absent or nonbinding, the $\hat{\beta}_n^*$ drops out and

(8.78) reduces to the analogous result 8.47 for the linear horizon model with time-varying rates.

The general model is a rather flexible framework for capital budgeting. Most of the results have been extended to the cases of linear mixed-integer programming and quadratic mixed-integer programming, respectively [18, 19]. A natural consequence of using any of these models is the need for a complete programming solution; simple acceptance criteria are possible only under very restrictive and simplistic assumptions.

## 8.7 DISCRETE CAPITAL BUDGETING

We have carefully avoided the issue of integer solutions until now, in order to present the concepts and theory of capital budgeting in the simpler LP framework. As we turn to discrete models, two issues face us. The first is practicality. Can we solve efficiently problems with integer restrictions? The second is the question of economic interpretation. Will the dual variables, particularly the $\rho_n$, play the same role in pricing out project opportunities?

### 8.7.1 Number of Fractional Projects in LP Solution

Before we delve into these issues, we briefly review the nature of the solutions to our example problems heretofore. Recall that in Example 8.1 we had a solution vector $\mathbf{x}^* = (0.22, 1, 0, 1, 0.16)$. Two of the five project selection variables had fractional values in the optimal LP solution. There are also two budget constraints in Example 8.1, and there are no project interdependencies. Weingartner [20] proved that in the LP formulation of the Lorie–Savage problem (the *PV* maximization in Table 8.2) the number of fractional projects in the optimal solution cannot exceed the number of budget constraints. An explanation of this fact is based on the following reasoning. If there is only one budget constraint, there need be at most one fractional project. All others would be either more preferable than the fractional one and accepted fully or less preferable and rejected completely. If there are two equally preferable fractional projects, we could adjust the investment amounts until one was completely accepted or rejected. If there are two budget constraints, it may be possible that one fractional project will exhaust the monies remaining after all fully accepted projects are funded, but more than likely two fractional projects will be needed. If there are three fractional projects in the presence of two budget constraints, one will be more (or equally) preferable, and its funding can be increased until it is accepted fully or one of the remaining two is rejected completely. The LP algorithm by nature seeks extreme points and avoids alternative optima with more variables than necessary. This type of inductive reasoning can be applied to three budget constraints and so forth.

Another way to regard the problem in Table 8.2 is as an upper-bounded LP problem [14]. A basic variable is then one whose value is allowed to be between its lower (0) and upper (1) bounds at some particular iteration. The projects' upper-bound constraints are deleted from the constraint matrix in the upper-bounded LP algorithm, and the rank of the constraint matrix is two for Example 8.1. Hence, there are at most two fractional projects in the optimal solution.

In the basic horizon model, in which the lending rate is equal to the borrowing rate for each time period and there are no borrowing limits and project interdependencies, as shown in Example 8.3, there is always an integer optimum solution. This fact is related to the equivalence between this model and the *PV* criterion. When the borrowing rate is greater than the lending rate, we can have fractional projects, as demonstrated by Example 8.4. The maximum number of fractional projects that are possible because the borrowing rate is greater is equal to the number of time periods with $r_l < r_b$, minus one. Moreover, the number of fractional projects may be increased by one for each project interdependency constraint and for each time period with a borrowing limit. The reasoning behind these last results is similar to that given for the *PV* maximization problem.

### 8.7.2 Branch-and-Bound Solution Procedure

Various algorithms have been developed for solving the mixed-integer linear programming problem [9]. It is beyond the scope of this text to deal with them, since many algorithms are designed for special problem structures and require a high level of mathematical sophistication on the part of the user. Instead, we will demonstrate the solution of a small problem with a branch-and-bound solution procedure which can be used by anyone with access to an LP code [17].

### *Example 8.7*

Use Example 8.5 (Table 8.5) as a starting point and obtain an optimal integer solution. Table 8.7 shows the optimal LP solution vector $\mathbf{x}^* = (0, 1, 0, 0.15, 0, 0.58)$ and objective function value $z^* = \$9{,}210$. We will designate this as problem 1. The presence of two fractional project selection variables, $x_4$ and $x_6$, gives us a choice in the procedure. We will arbitrarily select $x_4$ and create two new problems.

Problem 2: Problem 1 with $x_4 = 0$ added as a constraint

Problem 3: Problem 1 with $x_4 = 1$ added as a constraint

We then proceed to solve problems 2 and 3 by using an LP algorithm.

Problem 2: $\mathbf{x}^* = (0.17, 1, 0, 0, 0, 0.37)$, $z^* = \$9{,}205$

Problem 3: $\mathbf{x}^* = (0, 0.89, 0, 1, 0, 1)$, $z^* = \$9{,}031$

The procedure so far has not eliminated all fractional $x_i$ values but has, in fact, created others that did not appear in problem 1. The problem 2 solution has $x_1$ fractional, whereas $x_1$ was an integer in problem 1; the problem 3 solution has a fractional value for $x_2$, which was an integer in problem 1.

Undeterred, we proceed by taking problem 2 and creating from it two new problems.

Problem 4: Problem 2 with $x_1 = 0$ added

Problem 5: Problem 2 with $x_1 = 1$ added

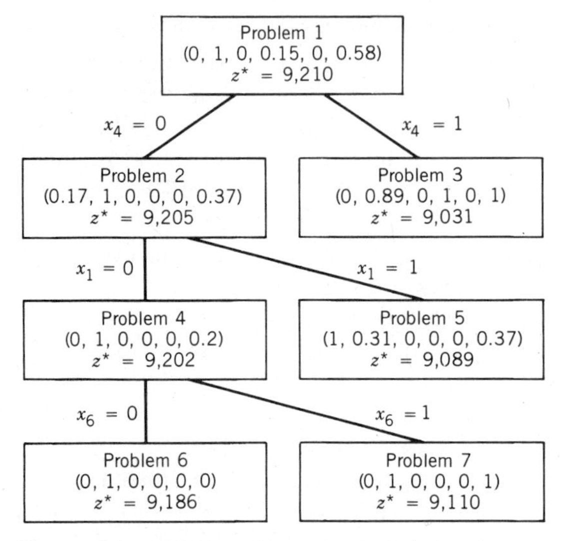


**Figure 8.2. Branch-and-bound solution tree for Example 8.7. Numbers in parentheses are values of x*, the vector of project selection variables.**

Figure 8.2 shows how the problems are derived from one another. The choice of problem 2 over problem 3 is based on the LP solution values, \$9,205 versus 9,031. We know that any integer solution derived from problem 3 (with added constraints such as $x_2 = 0$ or $x_2 = 1$) cannot exceed \$9,031, since the LP solution is always an upper bound on the integer solution. We think that a better integer solution is likely to be derived from problem 2. The LP solutions are

Problem 4: $\mathbf{x}^* = (0, 1, 0, 0, 0, 0.2)$, $z^* = \$9{,}202$

Problem 5: $\mathbf{x}^* = (1, 0.31, 0, 0, 0, 0.37)$, $z^* = \$9{,}089$

Because problem 4 has a better objective function value, we create from it two new problems.

Problem 6: Problem 4 with $x_6 = 0$ added

Problem 7: Problem 4 with $x_6 = 1$ added

The LP solutions are

Problem 6: $\mathbf{x}^* = (0, 1, 0, 0, 0, 0)$, $z^* = \$9{,}186$

Problem 7: $\mathbf{x}^* = (0, 1, 0, 0, 0, 1)$, $z^* = \$9{,}110$

At this point we have two integer solutions, and we can avoid further analysis of problems 6 and 7. We select the better of the two, that from problem 6 with $z^* = \$9{,}186$, and designate it as the incumbent (integer) solution.

Before we decide which of problems 3 and 5 to examine further, we check to see whether either can be ruled out by comparing its upper bound, or LP objective function value, with that of the incumbent. It happens that both have upper bounds less than \$9,186, and we do not examine them further. Problems

3 and 5 have been fathomed. There are no other candidate problems to examine, so we have finished and obtained the optimal solution $\mathbf{x}^* = (0, 1, 0, 0, 0, 0)$ with $z^* = \$9{,}186$. Figure 8.2 depicts the entire search process in tree form. □

In Example 8.7 the optimal integer objective function value is not much below the LP optimum, about 0.3%. However, we cannot generalize such characteristics, because so much depends on the projects, interest rates, and so forth. Note that a typical rounding process applied to the LP solution would give $\mathbf{x} = (0, 1, 0, 0, 0, 1)$, as in problem 7, which is suboptimal. We could conjecture that if we had branched first on $x_6$ instead of $x_4$, we might have reached the integer optimum sooner. Again, it is beyond our scope here to deal with such issues [9]. Our purpose has been to demonstrate an easily available integer solution procedure on a small capital budgeting problem.

### 8.7.3 Duality Analysis for Integer Solutions

Two basic approaches to duality analysis for mixed-integer linear programming have been presented in the literature. We will briefly discuss the first, more difficult method and then concentrate on the second, more straightforward method and its variations.

***Recomputed Dual Variables.*** One method for solving mixed-integer linear programs is to use the cutting-plane procedure [9]. This approach begins with the LP optimum and successively adds constraints that delete portions of the feasible LP space but do not delete any integer solutions. Each time a constraint is added, the LP is solved again. The added constraints, called cutting planes, are derived from the current LP solution. When the current LP solution is an integer in the required variables, the procedure stops. At this point we have dual variables for both the original constraint set and the added constraints.

Gomory and Baumol [10] derived a technique for taking the dual variables for the added constraints and reapportioning them among the original constraint set. The purpose is to obtain a set of dual variables for the original problem only. (Dual variables for the cutting planes would be difficult to interpret in terms of the resources expressed by the original constraint set.) The disadvantage of this approach, apart from its complexity and the need to use the cutting-plane procedure, is that the recomputed duals are not always unique. Furthermore, the interpretation of the dual variables as measuring changes in the objective function value resulting from small changes in resource limits does not always apply in the integer case. Small changes in resource limits can cause jumps in the objective function value [20].

***Penalties and Subsidies.*** Let us assume we have reached the LP problem corresponding to the optimal integer solution in a branch-and-bound integer procedure. (In Example 8.7 this would be problem 6, with $x_1, x_4$, and $x_6$ constrained to be zero). The LP form of the problem will contain a number of constraints that force certain project selection variables to zero and other constraints that force some project selection variables to their upper bounds. To the primal formulation of the horizon model—whether it be the basic model in Table 8.6, the model with a sloping supply schedule of funds in Table 8.10, or a model with time-dependent interest rates and project interdependencies—we would thus add

$$\begin{aligned} x_j &= 0, \quad j \text{ in } J_1 \\ x_j &= 1, \quad j \text{ in } J_2 \end{aligned} \tag{8.79}$$

where $J_1$ = set of projects constrained to be zero in the optimal integer solution,

$J_2$ = set of projects constrained to be at upper bound in the optimal integer solution.

These additional constraints will have corresponding dual variables. At first glance, the dual variables appear to be unconstrained, since the primal constraints 8.79 are equalities. We can, however, reinterpret the constraints as

$$\begin{aligned} &x_j \leqslant \epsilon, \quad j \text{ in } J_1 \\ &x_j \geqslant 1 - \epsilon \quad \text{or} \quad -x_j \leqslant -1 + \epsilon, \quad j \text{ in } J_2 \end{aligned} \tag{8.80}$$

where $\epsilon$ is a very small positive number. (Some LP codes check for variables set at fixed values and delete the corresponding constraints during a preprocessing stage. If this is done, it is necessary to use (8.80) instead of (8.79) to obtain information about the dual variables. An $\epsilon$ value of 0.001 or 0.0001 usually does the trick.)

In the pricing operation of a project constrained to be zero in the optimal integer solution, we then modify (8.33) as follows (assuming no project interdependencies),

$$-\sum_n a_{nj}\rho_n + \mu_j + \gamma_j \geq \hat{a}_j \tag{8.81}$$

where $\gamma_j$ is a dual variable, nonnegative. Rewriting, we have

$$\mu_j^* \geq \sum_n a_{nj}\rho_n^* + \hat{a}_j - \gamma_j^* \tag{8.82}$$

But since $\mu_j^*$ for a rejected project is zero, the $\gamma_j^*$ acts as a penalty (without Eq. 8.79 the $\mu_j^*$ was positive) to force rejection of project $j$.

If the project was constrained to be at its upper bound, Eq. 8.41 becomes

$$0 \leq \mu_j^* = \hat{a}_j + \sum_n a_{nj}\rho_n^* + \gamma_j^* \tag{8.83}$$

In this instance the $\gamma_j^*$ acts as a subsidy to enable acceptance of project $j$. These penalties and subsidies are a natural consequence of forcing the solution to satisfy the integrality requirements. They are known only through solving the mixed-integer programming problem. If we had solved the problem by some other method, such as the cutting-plane or enumeration method, we would still have to set up an LP model with the appropriate constraints of type 8.79 in order to extract the values of the penalties and subsidies.

We can apply these concepts to our integer solution for Example 8.5. The optimal dual variables for problem 6 are

$$\begin{array}{lllll} \rho_0^* = 1.69 & \mu_1^* = 0 & \mu_4^* = 0 & \gamma_1^* = 0 & \beta_1^* = 0 \\ \rho_1^* = 1.3 & \mu_2^* = 96 & \mu_5^* = 0 & \gamma_4^* = 0 & \\ \rho_2^* = 1.0 & \mu_3^* = 0 & \mu_6^* = 0 & \gamma_6^* = 80 & \end{array}$$

The ratio analysis of $\rho_n$ indicates borrowing at times 0 and 1, so projects 1 and 4 do not require penalties. Recall that project 4 was fractionally accepted in the LP optimum; project 1 was zero in the LP optimum but was introduced at an intermediate stage in the branch-and-bound procedure. We may verify the value of $\mu_6^* = 0$, which justifies rejection of project 6 (it was fractionally accepted in the LP optimum).

$$(\$1{,}000)(1.69) - (700)(1.3) - (700)(1.0) - 80 = 0$$

It is possible to reformulate the dual of the horizon model for an integer solution so that

- Projects forced into acceptance receive a subsidy, and projects forced into rejection receive no penalty, *or*
- Projects forced into rejection receive a penalty, but those forced into acceptance receive no subsidy.

The interested reader is referred to Weingartner [20] for further details of this method.

## 8.8 CAPITAL BUDGETING WITH MULTIPLE OBJECTIVES

In many situations it is not possible or desirable to evaluate different investment alternatives by one criterion, such as *PV* or terminal wealth. There are various techniques for dealing with multiple objectives, but most fall into one of three classes: goal programming, interactive multiple-criteria optimization, and nonlinear programming. In this section we provide an example of a goal-programming formulation, and we discuss the interactive approach.

Nonlinear programming can be applied if the decision maker can specify a utility function of the criteria, for example, dividends and terminal wealth. The major disadvantage of the approach seems to be the difficulty of specifying the utility function. Because each application depends so much on the utility function and on subsequent refinements in the solution algorithm, we will not discuss the approach in this section. Appendix 8.A provides an example of applying a quadratic programming algorithm to a problem involving dividends and terminal wealth.

### 8.8.1 Goal Programming

Goal programming is a technique that enables a decision maker to strive toward a number of objectives simultaneously. The first step consists of establishing a goal for each criterion. Next, an objective function is specified for each criterion with respect to this goal. Third, weighting factors are placed on deviations of the objective functions from their goals. Fourth, the separate objective functions are combined into one overall function to be optimized [13].

## *Example 8.8*

Table 8.13 presents data for Example 8.8. The first-year after-tax profit and employment of specialized personnel are considered to be primary goals, and terminal wealth at time 2 is considered a secondary goal. Let us assume the goals are established, respectively, as

Goal 1, first-year after-tax profits: \$2,500

Goal 2, specialized personnel needed: 700 person-hours

Goal 3, terminal wealth at time 2: \$4,000

These imply

$$\begin{aligned} \$2{,}000x_1 + 3{,}000x_2 + 1{,}700x_3 - 500x_4 &\geq \$2{,}500 \\ 100x_1 + 100x_2 + 300x_3 + 400x_4 &= 700 \text{ person-hours} \\ v_2 - w_2 + \$800x_1 + 600x_2 + 3{,}000x_3 + 1{,}000x_4 &\geq \$4{,}000 \end{aligned} \tag{8.84}$$

**Table 8.13** *Goal Programming, Example 8.8*

| | Project | | | |
|---|---|---|---|---|
| Coefficient Type | 1 | 2 | 3 | 4 |
| First-year after-tax profit | \$2,000 | \$3,000 | \$1,700 | −\$500 |
| Specialized personnel needed, person-hours | 100 | 100 | 300 | 400 |
| Cash flow at time | | | | |
| 0 | −\$1,000 | −\$800 | −\$2,000 | −\$200 |
| 1 | 300 | 200 | 1,000 | 100 |
| 2 | 400 | 200 | 1,000 | 200 |
| $\hat{a}_j$ | \$800 | \$600 | \$3,000 | \$1,000 |

Budgets for external sources of funds: $n$ = 0, \$2,000; $n$ = 1, −\$500; $n$ = 2, −\$500

The inequalities for profits and terminal wealth are typical of goals that can be exceeded without penalty. We now define auxiliary variables as follows.

$$\begin{aligned} y_1 &= \$2{,}000x_1 + 3{,}000x_2 + 1{,}700x_3 - 500x_4 - 2{,}500 \\ y_2 &= 100x_1 + 100x_2 + 300x_3 + 400x_4 - 700 \text{ person-hours} \\ y_3 &= v_2 - w_2 + \$800x_1 + 600x_2 + 3{,}000x_3 + 1{,}000x_4 - 4{,}000 \end{aligned} \quad (8.85)$$

We are concerned with measuring positive and negative deviations, so we define components

$$\begin{aligned} y_k^+ &= y_k \quad \text{if } y_k \geq 0 \\ y_k^- &= |y_k| \quad \text{if } y_k < 0 \end{aligned} \qquad k = 1, 2, 3 \quad (8.86)$$

Our overall objective is to minimize some weighted sum of deviations,

$$\text{Min} \sum_k (c_k^+ y_k^+ + c_k^- y_k^-) \quad (8.87)$$

where $c_k^+$ and $c_k^-$ are weighting factors for the deviations. If a \$100 profit deviation is deemed equivalent to a deviation of one specialized employee, we might set $c_1^- = 100$, $c_2^+ = 1$, and $c_2^- = 1$. Since terminal wealth is a secondary goal, set $c_3^- = 10$, an order of magnitude lower. The positive deviations for goals 1 and 3 have no adverse consequences, so $c_1^+ = 0$ and $c_3^+ = 0$. Thus, the objective function becomes

$$\text{Min } 100y_1^- + y_2^+ + y_2^- + 10y_3^- \quad (8.88)$$

The constraints of the problem are of two types. The first type consists of goal constraints, obtained from (8.84).

$$\begin{aligned} \$2{,}000x_1 + 3{,}000x_2 + 1{,}700x_3 - 500x_4 - (y_1^+ - y_1^-) &= \$2{,}500 \\ 100x_1 + 100x_2 + 300x_3 + 400x_4 - (y_2^+ - y_2^-) &= 700 \text{ person-hours} \\ v_2 - w_2 + 800x_1 + 600x_2 + 3{,}000x_3 + 1{,}000x_4 - (y_3^+ - y_3^-) &= \$4{,}000 \end{aligned} \quad (8.89)$$

The second type consists of the original set of constraints. In this example they would be cash balance equations for $n = 0$, 1, and 2, respectively; project upper bounds; and nonnegativity constraints. An LP solution of (8.88) subject to the two types of constraints yields values for the $x_j$, $v_n$, and $w_n$ that result in the "best" set of deviations from the goals. Assuming that lending and borrowing occur at 10%, the solution is:

$$\begin{array}{lllll} x_1 = 0.483 & v_0 = 0 & w_0 = 494.2 & y_1^+ = 1{,}825 & y_1^- = 0 \\ x_2 = 1.0 & v_1 = 0 & w_1 = 93.0 & y_2^+ = 0 & y_2^- = 0 \\ x_3 = 0.51 & v_2 = 496.5 & w_2 = 0 & y_3^+ = 0 & y_3^- = 0 \\ x_4 = 1.0 & & & & \end{array}$$

objective function value = 0

The solution indicates that the first-year after-tax profit will be $4,325, or $1,825 above the goal of $2,500. Goal 2, specialized personnel needed, and goal 3, terminal wealth at time 2, are met exactly. Since there is no penalty for exceeding goal 1, the objective function value is 0. □

A number of variations of the goal-programming technique are suitable for particular circumstances [13]. In all of them care must be taken in formulating the goals and relative weights for deviations.

### 8.8.2 Interactive Multiple-Criteria Optimization

Another approach is to assume the operational setting of optimizing a nonlinear function of the decision variables, *without* knowing the explicit form of the trade-off (utility) function. Instead, we assume the decision maker is able to provide information about the gradient of the function. This information is then used to guide a search process over the domain of the function [6].

To illustrate the concept, let us take the three goals in Example 8.8, described in the previous section, and convert them into three criteria. We assume that we can measure each criterion by a function $f_j$ and that we wish to maximize an overall utility function,

$$\underset{\mathbf{x}}{Max}\ U(f_1, f_2, f_3) \tag{8.90}$$

where $f_1(\mathbf{x})$ = criterion function for first-year after-tax profits,
$f_2(\mathbf{x})$ = criterion function for specialized personnel needed, and
$f_3(\mathbf{x})$ = criterion function for terminal wealth at time 2 (let $\mathbf{v}$ and $\mathbf{w}$ be included in an extended $\mathbf{x}$ vector).

We will be careful to specify the $f_j$ as concave, differentiable functions and assume $U$ is increasing in each $f_j$. Maximization of (8.90) by a steepest-ascent procedure will then lead to a global optimum [23].

The procedure begins with an initial feasible solution $\mathbf{x}^1$. At any iteration $k$ the direction of the search is obtained from

$$\underset{\mathbf{y}^k}{\text{Max}}\ \nabla_{\mathbf{x}^k} U(f_1(\mathbf{x}^k), f_2(\mathbf{x}^k), f_3(\mathbf{x}^k)) \cdot \mathbf{y}^k \tag{8.91}$$

by letting the search direction be $\mathbf{d}^k = \mathbf{y}^k - \mathbf{x}^k$. But (8.91) can be replaced by

$$\underset{\mathbf{y}^k}{\text{Max}} \sum_j c_j^k \nabla_{\mathbf{x}^k} f_j(\mathbf{x}^k) \cdot \mathbf{y}^k \tag{8.92}$$

where

$$c_j^k = \frac{(\partial U/\partial f_j)^k}{(\partial U/\partial f_1)^k} \tag{8.93}$$

In many situations (8.92) is linear and can be solved by LP. In any case, if we can express $f_j$, we can express (8.92). What has happened is that the ratios of the partial derivatives of $U$ with respect to $f_j$ (which result from the breakdown of $\nabla U$) have been replaced by trade-offs $c_j^k$. Each $c_j^k$ measures the reduction in value of criterion function $j$ that the decision maker would tolerate for one unit of increase in the value of criterion function 1, which is taken as a reference point. The trade-offs depend on the current solution and thus are indexed by the iteration counter $k$. They are obtained from the decision maker by an interactive procedure.

After the direction $\mathbf{d}^k$ is determined, the interactive procedure presents a number of solutions in the form of

$$f_1(\mathbf{x}^k + a\mathbf{y}^k), \qquad f_2(\mathbf{x}^k + a\mathbf{y}^k), \qquad f_3(\mathbf{x}^k + a\mathbf{y}^k)$$

where $a$ is the step size, which is typically incremented by one-tenth of $\mathbf{d}^k$. The decision maker provides input again by selecting the preferred combination of $f_1, f_2, f_3$ values, without reference to the utility function $U$.

Given appropriate conditions on the $f_j$ and $U$, the procedure will converge to a global maximum. The great advantage of the procedure is that no explicit form of the function $U$ is required. The decision maker instead is required to provide information about trade-offs among $f_j$ values and to indicate preferences for $f_1, \ldots, f_n$ combinations.

## 8.9 SUMMARY

In this chapter we have presented a number of techniques for capital budgeting under deterministic conditions. The methods are generally designed for selecting among many different investment alternatives (too many to enumerate explicitly) in the presence of budget limits, project interdependencies, and lending and borrowing opportunities. Linear programming is a major tool in the formulation, solution, and interpretation of many of the methods, either as the primary modeling technique or as a subroutine. The pricing of activity vectors is an important concept with direct economic interpretation, and we have devoted considerable space to illustrating the concept.

The models in this chapter may be grouped into three broad classifications. The first is the class of *PV* objective functions. This type suffers from some serious conceptual problems in the reconciliation of the discount rate used and the presence of budget constraints. The second class consists of horizon models; their objective is to maximize the end cash value or the terminal wealth at the end of some planning period. A number of desirable economic interpretations can be derived from such models. Moreover, models of this type are readily extended to include borrowing limits, a sloping supply schedule of funds, and integer restrictions.

The third class is characterized by objective functions containing different types of criterion variables. Bernhard's general model is the first of this type; it includes dividends and terminal wealth in the objective function. Other types discussed are the goal-programming approach and interactive multiple-criteria optimization. Appendix 8.A presents an application of Bernhard's approach to a problem which has dividends and terminal wealth to consider.

## REFERENCES


1. ATKINS, D. R., and D. J. ASHTON, "Discount Rates in Capital Budgeting: A Re-examination of the Baumol & Quandt Paradox," *The Engineering Economist,* Vol. 21, No. 3, pp. 159–171, Spring 1976.
2. BALAS, E., *Duality in Discrete Programming,* Graduate School of Industrial Administration, Carnegie-Mellon University, Pittsburgh, December 1967.
3. BAUMOL, W. J., and R. E. QUANDT, "Investment and Discount Rates under Capital Rationing—A Programming Approach," *Economic Journal,* Vol. 75, No. 298, pp. 317–329, June 1965.
4. BERNHARD, R. H., "Mathematical Programming Models for Capital Budgeting—A Survey, Generalization, and Critique," *Journal of Financial and Quantitative Analysis,* Vol. 4, No. 2, pp. 111–158, 1969.
5. DANTZIG, G. B., *Linear Programming and Extensions,* Princeton University Press, Princeton, N.J., 1963. (See Chapter 12 for a discussion of economic interpretation of dual problem.)
6. DYER, J. S., "A Time-Sharing Computer Program for the Solution of the Multiple Criteria Problem," *Management Science,* Vol. 19, No. 12, pp. 1379–1383, August 1973.
7. FISHER, I., *The Theory of Interest,* Macmillan, New York, 1930 (reprinted by A. M. Kelley, New York, 1961).
8. FREELAND, J. R., and M. J. ROSENBLATT, "An Analysis of Linear Programming Formulations for the Capital Rationing Problem," *The Engineering Economist,* Vol. 24, No. 1, pp. 49–61, Fall 1978.
9. GARFINKEL, R. S., and G. L. NEMHAUSER, *Integer Programming,* Wiley, New York, 1972.
10. GOMORY, R. E., and W. J. BAUMOL, "Integer Programming and Pricing," *Econometrica,* Vol. 28, No. 3, pp. 551–560, 1960.
11. HAMILTON, W. F., and M. A. MOSES, "An Optimization Model for Corporate Financial Planning," *Operations Research,* Vol. 21, No. 3, pp. 677–691, 1973.
12. HAYES, J. W., "Discount Rates in Linear Programming Formulations of the Capital Budgeting Problem," *The Engineering Economist,* Vol. 29, No. 2, pp. 113–126, Winter 1984.
13. IGNIZIO, J. P., *Linear Programming in Single and Multiple Objective Systems,* Prentice–Hall, Englewood Cliffs, N.J., 1982.
14. LASDON, L., *Optimization Theory for Large Systems,* Macmillan, New York, 1970. (See Chapter 6 for upper-bounded algorithm.)
15. LORIE, J. H., and L. J. SAVAGE, "Three Problems in Rationing Capital," *Journal of Business,* Vol. 28, No. 4, pp. 229–239, October 1955; also reprinted in Solomon, E. (ed.), *The Management of Corporate Capital,* Free Press, New York, 1959.
16. MURGA, P., *Capital Budgeting Objective Functions That Consider Dividends and Terminal Wealth,* M.S. thesis, School of Industrial and Systems Engineering, Georgia Institute of Technology, Atlanta, 1978.
17. RAVINDRAN, A., D. T. PHILLIPS, AND J. J. SOLBERG *Operations Research: Principles and Practice,* Wiley, New York, 1987. (See Chapter 4 for branch-and-bound technique. See Chapter 11 for Kuhn–Tucker conditions.)

18. SHARP, G. P., *Extension of Bernhard's Capital Budgeting Model to the Quadratic and Nonlinear Case,* School of Industrial and Systems Engineering, Georgia Institute of Technology, Atlanta, 1983.
19. UNGER, V. E., "Duality Results for Discrete Capital Budgeting Models," *The Engineering Economist,* Vol. 19, No. 4, pp. 237–252, Summer 1974.
20. WEINGARTNER, H. M., *Mathematical Programming and the Analysis of Capital Budgeting Problems,* Prentice–Hall, Englewood Cliffs, N.J., 1963.
21. WEINGARTNER, H. M., "Capital Rationing: *n* Authors in Search of a Plot," *Journal of Finance,* Vol. 32, No. 5, pp. 1403–1431. December 1977.
22. WILKES, F. M., *Capital Budgeting Techniques,* John Wiley & Sons, New York, 1983.
23. ZANGWILL, W. I., *Nonlinear Programming: A Unified Approach,* Prentice–Hall, Englewood Cliffs, N.J., 1969.

## PROBLEMS

**8.1.** You wish to include lending activities at 8% and borrowing activities at 12% in a *PV* LP model. The interest rate used for *PV* calculations is 10%. Define the activity vectors for lending and borrowing opportunities, and write a model formulation for a time horizon of 2 years and three budget constraints.

**8.2.** One of the criticisms of the typical capital budgeting LP model is that only short-term (one-year) lending and borrowing is represented. Can long-term lending and borrowing be included? If so, show how by defining variables and specifying coefficients in the objective function and constraints. Would long-term lending and borrowing be more appropriate in a *PV* LP model or a horizon LP model?

**8.3.** In many decision environments the total number of major projects to be considered is ten or fewer. Thus, enumeration of all combinations would be feasible, since there would be $2^{10} = 1{,}024$ or fewer combinations. In such a case, would it make sense to use a mathematical programming approach? What information would the mathematical programming approach give that is not available from enumeration?

**8.4.** Formulate a *PV* LP model for selecting among the three projects described below. *MARR* = 8%. There is a budget of $13,000 at time 0, and the projects are required to generate $3,500 at time 1 and $1,200 at time 2. The life of each project is 5 years. The projects are independent except that C cannot be selected unless A is also selected. What is the value of extra budget money at time 2?

| Project | Investment | Annual Cash Flow |
|---|---|---|
| A | $5,000 | $1,319 |
| B | 7,000 | 1,942 |
| C | 8,500 | 2,300 |

**8.5.** Formulate a *PV* LP model for selecting among the three projects described below. *MARR* = 15%. There is a budget of $16,000 at time 0, and the projects are required to generate $4,000 at time 1 and $1,300 at time 2. The life of each project is 10 years. The projects are independent except that A cannot be selected unless B is also selected. What is the value of extra budget money at time 2?

| Project | Investment | Annual Cash Flow |
|---|---|---|
| A | $8,000 | $1,900 |
| B | 5,000 | 1,400 |
| C | 10,000 | 2,500 |

**8.6.** Fromulate a horizon LP model for selecting among the three projects described below, with time 2 as the horizon. There is a budget of \$2,000 at time 0, and the projects are required to generate \$500 at time 1 and \$500 at time 2. The life of each project is 20 years. The projects are independent except that C cannot be selected unless A is also selected. The lending rate is 15% and the borrowing rate is 20%, per year. Do you see any obvious difficulty with the application of the horizon model to this particular example?

| Project | Investment | Annual Cash Flow |
|---|---|---|
| A | \$1,000 | \$240 |
| B | 800 | 190 |
| C | 1,500 | 310 |

**8.7.** A horizon LP model was formulated and solved for five independent projects and four budget constraints. The lending rate is 18% and the borrowing rate 25%, per year. There are no posthorizon cash flows. The solution is:

Project selection variables = (0.0, 1.0, 1.0, 0.5, 1.0)

Budget dual variables = (1.7995, 1.475, 1.25, 1.0)

Project dual variables = (10, 240, 310, 0, 110)

a. Indicate whether borrowing or lending occurs in each period.
b. Do you see any difficulty in interpreting the solution of this example?
c. Suppose you wish to evaluate a new independent project.

| Time | 0 | 1 | 2 | 3 |
|---|---|---|---|---|
| Cash flow | −\$1,000 | −1,000 | 2,000 | 1,000 |

What would be your recommendation regarding acceptance?

**8.8.** A horizon LP model was formulated and solved for five independent projects and four budget constraints. The lending rate is 20% and the borrowing rate 25%, per year. There are no posthorizon cash flows. The solution is:

Project selection variables = (1.0, 0.0, 0.4, 1.0, 0.0)

Budget dual variables = (1.8, 1.44, 1.2, 1.0)

Project dual variables = (560, 0, 0, 320, 0)

a. Indicate whether borrowing or lending occurs in each period.
b. Suppose you wished to evaluate a new independent project.

| Time | 0 | 1 | 2 | 3 |
|---|---|---|---|---|
| Cash flow | −\$1,000 | −1,000 | 2,000 | 1,000 |

What would be your recommendation regarding acceptance?

**8.9.** A horizon LP model was formulated and solved for five independent projects and four budget constraints. The lending and borrowing rates are

Time 0 to 1: lend at 15%, borrow at 20%

Time 1 to 2: lend at 15%, borrow at 20%

Time 2 to 3: lend at 18%, borrow at 25%

There are no posthorizon cash flows. The solution is

Project selection variables = (1.0, 0.0, 0.4, 1.0, 0.0)

Budget dual variables = (1.628, 1.357, 1.18, 1.0)

Project dual variables = (560, 0, 0, 320, 0)

a. Indicate whether borrowing or lending occurs in each period.
b. Suppose you wish to evaluate a new independent project.

| Time | 0 | 1 | 2 | 3 |
|---|---|---|---|---|
| Cash flow | −$1,500 | −1,500 | 3,500 | 800 |

What would be your recommendation regarding acceptance?
c. What does the first project contribute to the objective function?

**8.10.** Formulate and solve a horizon LP model for selecting among the five following projects. The lending rate is 13% and the borrowing rate 17%, per year. There are no posthorizon cash flows.

| | Project | | | | | |
|---|---|---|---|---|---|---|
| $n$ | A | B | C | D | E | Budget |
| 0 | −$10,000 | −$5,000 | −$7,500 | −$15,000 | −$20,000 | $30,000 |
| 1 | −5,000 | −12,000 | −8,500 | −3,000 | −5,000 | 25,000 |
| 2 | 2,000 | 15,000 | 11,000 | 2,176 | 14,000 | 30,000 |
| 3 | 4,072 | 3,761 | 1,541 | 2,176 | 16,005 | 35,000 |
| 4 | 16,000 | 1,700 | 4,000 | 2,176 | 8,000 | 10,000 |
| 5 | 18,000 | 1,700 | 12,000 | 2,176 | 10,000 | 20,000 |

Verify that projects with positive future worth are accepted and those with negative future worth rejected. Verify that the ratios of the budget dual variables indicate lending or borrowing.

**8.11.** Rework Problem 8.10 with the inclusion of borrowing limits of $2,000 at time 0 and $2,000 at time 1, at the 17% rate. Unlimited borrowing at 20% is available at times 0 and 1.

**8.12.** Formulate and solve a horizon LP model for selecting among the five projects below. The lending rate is 15% and the borrowing rate 20%, per year. There are no posthorizon cash flows.

| | Project | | | | | |
|---|---|---|---|---|---|---|
| $n$ | A | B | C | D | E | Budget |
| 0 | −$10,000 | −$20,000 | −$15,000 | +$5,000 | −$15,000 | $25,000 |
| 1 | 4,000 | 8,000 | −2,000 | −1,000 | 1,300 | 2,000 |
| 2 | 5,000 | 10,000 | 5,000 | −1,000 | 1,700 | 0 |
| 3 | 4,400 | 3,000 | 7,300 | −3,200 | 6,000 | 2,000 |
| 4 | 2,800 | 7,000 | 6,000 | −1,150 | 4,000 | 1,000 |
| 5 | 1,000 | 6,000 | 7,100 | −800 | 2,700 | 0 |

Verify that projects with positive future worth are accepted and those with negative future worth rejected. Verify that the ratios of the budget dual variables indicate lending or borrowing.

**8.13.** Rework Problem 8.12 with the inclusion of another activity, project F, with cash flow

| n | 0 | 1 | 2 | 3 | 4 | 5 |
|---|---|---|---|---|---|---|
| Cash Flow | −$25,000 | 10,000 | 8,000 | 8,000 | 8,000 | 7,000 |

**8.14.** Rework Problem 8.13 (six projects) with the inclusion of borrowing restrictions of $10,000 per year over the planning period (5 years).

**8.15.** Rework Problem 8.13 (six projects) with the inclusion of a sloping supply schedule of funds. During each year the first $5,000 of borrowing is at 20% and the next $2,500 at 25%, and unlimited borrowing is available at 30%.

**8.16.** A horizon LP model was formulated and solved for selecting among the five projects below. The lending rate is 10% and the borrowing rate 20%, per year. There are no posthorizon cash flows.

| | Project | | | | | |
|---|---|---|---|---|---|---|
| *n* | A | B | C | D | E | Budget |
| 0 | −$1,000 | −$1,200 | −$900 | −$1,000 | $1,000 | $1,000 |
| 1 | 400 | 900 | 300 | 500 | −400 | 1,300 |
| 2 | 800 | 800 | 500 | 700 | −400 | 1,500 |
| 3 | 135 | 700 | 250 | 700 | −400 | 200 |

The optimal solution contains:

Project selection variables = (1.0, 1.0, 0, 1.0, 1.0)

Budget dual variables = (1.452, 1.21, 1.1, 1.0)

a. Indicate whether lending or borrowing occurs during each time period.
b. Determine the project dual variable for project C and for Project E.
c. With a new set of budget amounts, the solution changes.

Budget amounts = ($1,000, −1,000, 1,500, 200)

Project selection variables = (0, 1.0, 0, 1.0, 1.0)

Budget dual variables = (1.584, 1.32, 1.1, 1.0)

Explain why project A is rejected here although it was accepted previously. Express your answer in LP terms, using specific numbers.

d. If you know nothing about the budget amounts, can you say *anything* specific about the acceptance or rejection of projects in this example?
e. Describe a method for determining the optimal value of the objective function for the original problem formulation, if you know the optimal values of the project selection variables and the budget dual variables.

**8.17.** Construct a horizon LP example with lending rate(s) less than borrowing rate(s). Demonstrate the relationships between lending or borrowing activities and the budget dual variables and the relationships between project dual variables and project future values.

**8.18.** Construct a horizon LP example where at least one of the projects is accepted fractionally. Explain why it is accepted fractionally by pricing out the project activity vector.

**8.19.** Construct a horizon LP example with a sloping supply schedule of funds. Validate the relationships between the budget dual variables and tightness of the borrowing constraints.

**8.20.** Construct a horizon integer programming example. Solve it by using a branch-and-bound algorithm. Determine the subsidies and penalties attached to projects that are fractionally accepted in the LP solution, in order to force them to be integer.

**8.21.** Solve the goal-programming example in Section 8.8.1.

**8.22.** Construct an example of Bernhard's general model, using a linear objective function and a linear terminal-wealth posture restriction. Use at least four time periods and different lending or borrowing rates. Try to construct the problem so that at least one project is accepted, at least one is rejected, and at least one borrowing constraint is tight.

**8.23.** The ABC Company has to determine its capital budget for the coming 3 years, for which data (in thousands of dollars) are given in the following table.

| End of Year | Available Investment Capital | Investment Projects | | | | | |
|---|---|---|---|---|---|---|---|
| | | 1 | 2 | 3 | 4 | 5 | 6 |
| 0 | 300 | −50 | −100 | −60 | −50 | −170 | −16 |
| 1 | 100 | −80 | −50 | −60 | −100 | −40 | −25 |
| 2 | 200 | 20 | −20 | −60 | −150 | 50 | −40 |
| Discounted future revenues | | 150 | 210 | 220 | 350 | 200 | 100 |

At the start of year 1 the company has $300,000 available for investment; in year 2 another $100,000 becomes available, and at the start of year 3 an additional $200,000 becomes available. Project 1 requires $50,000 at the start of year 1 and another $80,000 at the start of year 2; at the start of year 3, the project yields $20,000. The yield at the start of year 4 and the discounted yields for later years amount to $150,000. The company can borrow at most $50,000 plus 20% of the money invested so far in the various investment projects at an interest rate of 12% per year. If the company deposits money at the bank, the interest rate is 8%. The company has a bank debt of $10,000, on which it pays 11% interest and which may be repaid at the start of any year. Assume that the company may undertake 100% of each project or take a participation in each project of less than 100%.

a. Formulate the capital budgeting problem by using the horizon model.
b. Find the optimal capital allocations by using a linear programming package.
c. Find the optimal capital budget, assuming that no project can be undertaken partially.

**8.24.** The Micromegabyte Company is a small American manufacturer of microprocessors, which are vital components in many pieces of electronic equipment, including personal computers. In a recent meeting of the board of directors, the company was instructed to engage in the development of various types of software that would go with their microprocessor products. An ad hoc committee has been formed to come up with various proposals that can be initiated in the new fiscal period. The following six proposals were considered to be competitive in the market and to have good profit potentials. Because technology in this field advances rapidly, most software products will have a market life of about 3 years.

| End of Period | Software Projects and Their Cash Flows* 1 | 2 | 3 | 4 | 5 | 6 |
|---|---|---|---|---|---|---|
| 0 | −50 | −100 | −70 | −130 | −250 | −300 |
| 1 | −100 | −50 | −100 | −50 | −100 | −60 |
| 2 | 50 | 100 | 90 | −100 | −60 | 150 |
| 3 | 100 | 50 | 150 | 260 | 300 | 200 |
| 4 | 50 | 30 | 100 | 250 | 150 | 100 |
| 5 | 30 | 30 | 30 | | 100 | |

*All units in $1,000.

The company will have at the start of year 1 (end of period 0) $500,000 available for investment; in year 2 another $200,000 becomes available, and at the start of year 3 an additional $50,000 becomes available.

The company can borrow at most $200,000 over the planning horizon at an interest rate of 12% per year. The company has to repay an old loan of $50,000 over the planning horizon. No partial payment of this loan is allowed, but the company has to pay 13% interest at the end of each year until the loan is paid in full.

Projects 1 and 3 are considered to be mutually exclusive because both projects lead to the same software development but with application on different machines. The company does not have enough resources to support more than one type of operating system. Project 2 is contingent on project 1, and project 4 is contingent on project 3. Projects 2 and 4 are graphics softwares designed to run on the specific operating system.

The company has a total of 10,000 programming hours per year for the first 2 years that can be put into the development of these software projects. Annual programming hour requirements for the projects are estimated to be as follows.

| Year | Project 1 | 2 | 3 | 4 | 5 | 6 |
|---|---|---|---|---|---|---|
| First | 2,000 | 3,000 | 3,000 | 5,000 | 4,000 | 4,000 |
| Second | 3,000 | 4,000 | 3,000 | 4,000 | 5,000 | 4,000 |

The company can always lend any unspent funds at an interest rate of 9%. Determine the firm's best course of action with a horizon time of 4 years.

**8.25.** The National Bank of Maine has $1 billion in total assets, which are offset on the balance sheet by demand deposits, time deposits, and capital accounts of $650 million, $250 million, and $100 million, respectively. The bank seeks your advice on how best to allocate its resources among the following list of assets.

***Bank Assets and Their Expected Rates of Return***

| Asset | Expected Net Return (%) |
|---|---|
| Cash and cash equivalents | 0 |
| Loans | |
| Commercial loans | 5.5 |
| FHA and VA mortgages | 5.0 |
| Conventional mortgages | 6.2 |
| Other loans | 6.9 |
| Investments | |
| Short-term U.S. government securities | 3.0 |
| Long-term U.S. government securities | 4.2 |

In allocating its resources, the bank is now constrained by the following legal and policy considerations.

**Legal restrictions**

- Cash items must equal or exceed 30% of demand deposits.
- Within the loan portfolio, conventional mortgages must not exceed 20% of time deposits.

**Policy guidelines (goals)**

- The management does not wish its total loans to exceed 65% of total assets. Each dollar of deviation from this target will carry a penalty of 3.5 cents per period.
- Within the loan portfolio "commercial loans" are not to exceed 45% or fall below 30% of total loans, and "other loans" are not to exceed the amount of total mortgages. Deviations of 1$ from these targets will carry uniform penalties of 0.8 cent per period.
- To ensure solvency, the management desires to limit its holdings of risk assets, defined as total assets less cash items less short-term U.S. government securities, to seven times the bank's capital accounts. Each dollar of deviation will carry a penalty of 4 cents per period.
- The management wishes to earn a target profit of $50 million. It places no premium on overattainment of this profit objective, but it places a penalty of $1 on each dollar of underattainment. Set up and solve a goal programming formulation for this problem.

## APPENDIX 8.A

# A Dividend-Terminal-Wealth Capital Budgeting Model: A Case Study

Bernhard's general model presented in Section 8.6 is a powerful, flexible tool that enables a decision maker to consider both dividends and terminal wealth as objectives. In this section we present a specific form of this model and demonstrate its application to an example problem. A number of conclusions are drawn concerning the selection of parameters for the objective function and concerning the applicability of the model.

One of the more important characteristics of dividends from the view of the stockholder is the steadiness of the dividend stream. All other factors being equal, small fluctuations about an average level are preferred by many investors. For such investors this steadiness represents a nearly certain income, and they pay for obtaining it. The tools proposed so far for representing that steadiness are of three types: a nondecreasing dividend policy, a policy of required growth in dividends, and a minimum dividend policy. Although helpful, they are somewhat inflexible and do not adequately represent the concept of steadiness.

### MODEL FORMULATION

The objective function of our model, expression 8a.1, is constructed of four types of terms.

- Terminal wealth.
- Dividends paid during the planning period.
- Variability of dividends.
- Deviation from the horizon wealth posture goal.

$$\max_{x_j, v_n, w_n, D_n} \frac{G}{c} + \sum_{n=0}^{N} \frac{D_n}{(1+i)^n} - c_1 \sum_{n=1}^{N} (D_n - D_{n-1})^2$$

$$- c_2\left(\frac{G}{c_3 c} - D_{\text{ave}}\right)^2 \qquad \text{(8a.1)}$$

where $G = \sum_{j=1}^{J} \hat{a}_j x_j + v_N - w_N$

$c = (1 + i)^N$

$$D_{\text{ave}} = \frac{1}{N+1} \sum_{n=0}^{N} \frac{D_n}{(1+i)^n}$$

$i$ = externally determined discount rate reflecting time preferences of owners of the firm,

$c_1, c_2, c_3$ = nonnegative constant parameters,

and other terms are as defined in Table 8.11.

Let us examine this expression part by part. The first term is simply the terminal wealth of the firm (as defined in Section 8.5) discounted back to time 0. Note that we explicitly allow for a discount rate that reflects the time preferences of the stockholders of the firm. The second term consists of the discounted stream of dividends. The third is a penalty term that has the effect of reducing the variability of dividends. The sum of the squared differences for consecutive dividends is minimized according to the $c_1$ parameter assigned. The fourth term represents the difference between the perpetual, annual value of the terminal wealth and the average dividend payment. The parameter $c_3$ divides the terminal wealth $G$ and can be thought of as the inverse of a discount rate for converting $G$ to a perpetual, annual dividend stream. This value is compared to the average time 0 value of dividends paid within the planning period. The difference is squared, weighted by parameter $c_2$, and assigned a minus sign to represent a penalty term. The general effect of the fourth term is to equalize the dividends paid with the potential dividends after the horizon; this is called the horizon wealth posture goal. In many cases $c_3$ will be assigned a value close to $1/i$. In summary, then, the objective function attempts to maximize the present value, using the external discount rate $i$, of the terminal wealth plus dividends, while minimizing variability of dividends and deviation from the horizon wealth posture goal.

The constraints in this model are kept to a minimum, in order to simplify the duality analysis. There are cash balance equations (budget constraints),

$$[\rho_n] \qquad -\sum_j a_{nj} x_j - l_{n-1} v_{n-1} + v_n$$

$$+ b_{n-1} w_{n-1} - w_n + D_n \le M_n, \qquad n = 0, 1, \ldots, N \qquad \text{(8a.2)}$$

where all terms are as defined in Table 8.11.

In addition to project upper bounds and nonnegativity restrictions, there are minimum dividend constraints for each period,

$$[\eta_n] \qquad D_n \geqslant D_{\min}, \qquad n = 0, 1, \ldots, N \qquad (8a.3)$$

$$[\mu_j] \qquad x_j \leqslant 1, \qquad j = 1, \ldots, J \qquad (8a.4)$$

$$x_j, v_n, w_n, D_n \geqslant 0 \qquad (8a.5)$$

## MODEL ANALYSIS

When the terminal wealth $G$ is replaced by its definition in the objective function (8a.1), a quadratic objective function results. Since all the constraints are linear, we can apply a standard duality theory and subsequently a standard solution algorithm [23].

### Transformation to Standard Quadratic Form

To apply the duality theory for quadratic objective function with linear constraints, we need first to transform our problem into the standard form. This will be shown with the aid of tables, so that readers not interested in the duality results may skip this section. Table 8a.1 presents the standard form of the quadratic model in both primal and dual forms. The model defined by (8a.1) through (8a.5) can be redefined in the terms of the general quadratic problem, $q$, $z$, $Q$, $A$, and $b$. Table 8a.2 presents the necessary definitions and transformations for this tedious, but not difficult, task. Notice that each of the vectors and arrays in Table 8a.1 is partitioned to accomplish this. Not all the details of the transformations are given; complete details can be found in Murga [16].

### Duality Analysis for Optimal Solutions

According to Table 8a.1, there are four types of dual constraints, one for each of $q_x$, $q_v$, $q_w$, and $q_D$. These are given in Table 8a.3. We can obtain from these constraints several interesting results. All decision variables will be assumed to be at their optimal values in the following analysis; asterisks have been omitted for clarity.

Let us start with (8a.18). After substituting for corresponding terms, we obtain $J$ inequalities of the following general form.

$$-\sum_{n=0}^{N} a_{nj}\rho_n + \mu_j + \frac{2c_2\hat{a}_j}{c_3c}\left(\frac{G}{c_3c} - D_{\text{ave}}\right) \geq \frac{\hat{a}_j}{c}, \qquad j = 1, 2, \ldots, J \qquad (8a.22)$$

The primal variable associated with this constraint is $x_j$. Equation 8a.22 is analogous to the *pricing operation* of activity vectors in LP, except that in this form we have not eliminated the other primal decision variables from the expression (the $x_j$, $v_n$, and $w_n$ variables are present in the definition of G, and the $D_n$

**Table 8a.1** ***Standard Form of Quadratic Objective Function Subject to Linear Constraints: Primal and Dual Forms***

*Primal*

$$\text{Max} \quad q'z + z'Qz \tag{8a.6}$$

$$\text{s.t.} \quad Az \leq b \tag{8a.7}$$

$$z \geq 0 \tag{8a.8}$$

where $q'$ = $r$-component row vector containing the coefficients of the linear terms $z_k$ of the objective function

$z$ = $r$-component column vector of the primal variables

$Q$ = $(r, r)$ matrix containing elements $c_{kj}$

$$c_{kj} = \begin{cases} \text{coefficient of } z_k z_j \text{ in the objective function if } k = j \\ \frac{1}{2} \text{ times the coefficient of } z_k z_j \text{ in the objective function if } k \neq j \end{cases}$$

$A$ = $(p, r)$ matrix containing the coefficients of the $z_k$ in the set of constraints

$b$ = $p$-component column vector

*Dual*

$$\text{Min} \quad u'b - z'Qz \tag{8a.9}$$

$$\text{s.t.} \quad A'u - 2Qz \geq q \tag{8a.10}$$

$$u \geq 0 \tag{8a.11}$$

where $u'$ is a $p$-component row vector of the dual variables.

variables are present in the definition of $D_{\text{ave}}$). If $x_j$ is accepted, then (8a.22) is satisfied as an equality.

From (8a.19) and (8a.20) we obtain the classical relation among the lending and borrowing rates and the budget constraint dual variables (*ratios of budget dual variables*).

$$l_{n+1} \leqslant \rho_n/\rho_{n+1} \leqslant b_{n+1} \tag{8a.23}$$

We also obtain an expression for $\rho_N$.

$$\rho_N = \frac{2c_2}{c_3 c}\left(\frac{c_3}{2c_2} - \frac{G}{c_3 c} + D_{\text{ave}}\right) \tag{8a.24}$$

Thus, the optimal value of $\rho_N$ is a function of parameters $c_2$, $c_3$, and $i$, the terminal wealth, and the stream of dividends. In the basic horizon model (Section 8.5) it is equal to one.

**Table 8a.2** ***Transformation to Standard Quadratic Form***

Let $z' = (x', v', w', D')$, a row vector of size $[J \times 3(N + 1)]$ where

$$\left.\begin{aligned} x' &= (x_1, x_2, ..., x_J) \\ v' &= (v_0, v_1, ..., v_N) \\ w' &= (w_0, w_1, ..., w_N) \\ D' &= (D_0, D_1, ..., D_N) \end{aligned}\right\} \tag{8a.12}$$

Let $q' = (q'_x, q'_v, q'_w, q'_D)$ = a row vector of size $[J \times 3(N + 1)]$ where

$$\left.\begin{aligned} q'_x &= [\hat{a}_1/c, \hat{a}_2/c, ..., \hat{a}_J/c] \\ q'_v &= [0, 0, ..., 1/c] \\ q'_w &= [0, 0, ..., -1/c] \\ q'_D &= [1, 1/(1 + i), ..., 1/(1 + i)^N] \end{aligned}\right\} \tag{8a.13}$$

Let

$$Q = \begin{bmatrix} Q_{xx} & Q'_{vx} & Q'_{wx} & Q'_{Dx} \\ Q_{vx} & Q_{vv} & Q_{wv} & Q'_{Dv} \\ Q_{wx} & Q_{wv} & Q_{ww} & Q'_{Dw} \\ Q_{Dx} & Q_{Dv} & Q_{Dw} & Q_{DD} \end{bmatrix} \tag{8a.14}$$

where

1. $Q_{xx} = (J, J)$ matrix with entries

$$(Q_{xx})_{kj} = -\frac{c_2 \hat{a}_k \hat{a}_j}{c_3^2 c^2}, \qquad k = 1, 2, ..., N \quad j = 1, 2, ..., N$$

2. $Q_{vx} = [(N + 1), J]$ matrix with nonzero entries just in row $N + 1$,

$$\begin{aligned}(Q_{vx})_{kj} &= -\frac{c_2 \hat{a}_j}{c_3^2 c^2}, \qquad k = N + 1 \quad j = 1, 2, ..., J \\ &= 0 \qquad \text{otherwise}\end{aligned}$$

3. $Q_{wx}$ is defined exactly as $Q_{vx}$ but with opposite sign.
4. $Q_{Dx} = [(N + 1), J]$ matrix with entries

$$(Q_{Dx})_{kj} = \frac{c_2 \hat{a}_j}{c_3(N + 1)(1 + i)^{N+k-1}}, \qquad k = 1, 2, ..., N + 1 \quad j = 1, 2, ..., J$$

5. $Q_{vv} = [(N + 1), (N + 1)]$ matrix with zero entries, except for $k = N + 1, j = N + 1$,

$$(Q_{vv})_{kj} = \frac{c_2}{c_3^2 c^2}, \qquad k = N + 1, \quad j = N + 1$$

*(continued)*

**Table 8a.2** ***Transformation to Standard Quadratic Form (Continued)***

6. $Q_{wv}$ is defined exactly like $Q_{vv}$.
7. $O_{Dv}$ = [(N + 1), (N + 1)] matrix with nonzero entries only in column N + 1,

$$(Q_{Dv})_{kj} = \frac{c_2}{c_3(N+1)(1+i)^{N+k-1}}, \qquad k = 1, 2, \ldots, N+1; \quad j = N+1$$

$$= 0 \qquad \text{otherwise}$$

8. $Q_{ww}$ is defined exactly as $Q_{vv}$.
9. $Q_{Dw}$ is defined exactly as $Q_{Dv}$ but with opposite sign.
10. $Q_{DD}$ = [(N + 1), (N + 1)] symmetric matrix with

$$(Q_{DD})_{kj} = -c_1 - \frac{c_2}{(N+1)^2}, \qquad k = 1, \; j = 1$$

$$= -c_1 - \frac{c_2}{(N+1)c^2}, \qquad k = N+1, \; j = N+1$$

$$= -2c_1 - \frac{c_2}{(N+1)(1+i)^{k+j-2}}, \qquad k = j; \; k \neq 1, \; N+1; \; j \neq 1, \; N+1$$

$$= -c_1 - \frac{c_2}{(N+1)^2(1+i)^{k+j-2}}, \qquad k = 1, 2, \ldots, J, \; j = k-1$$

$$= \frac{c_2}{(N+1)^2(1+i)^{k+j-2}} \qquad \text{otherwise}$$

Let

$$A = \begin{bmatrix} A_{nx} & A_{nv} & A_{nw} & A_{nD} \\ 0 & 0 & 0 & A_D \\ A_x & 0 & 0 & 0 \end{bmatrix} \tag{8a.15}$$

where

1. $A_{nx}$ = [(N + 1), J] matrix,

$$(A_{nx})_{kj} = -a_{kj} \qquad n = k-1; \; k = 1, \ldots, N+1; \; j = 1, \ldots, J$$

2. $A_{nv}$ = [(N + 1), J] matrix whose entries are all zeros except those of the main diagonal and the diagonal just below it.

*(continued)*

**Table 8a.2** ***(Continued)***

$$(A_{nu'})_{kj} = 1, \quad k = 1, 2, ..., N + 1; \quad j = 1, 2, ..., N + 1; \quad k = j$$

$$= -l_{n'}, \quad k = 2, 3, ..., N + 1; \quad j = 1, 2, ..., N; \quad j = k - 1; \quad n = k - 1$$

3. $A_{nw'}$ has a structure similar to that of $A_{nu'}$,

$$(A_{nw'})_{kj} = 1, \quad k = 1, 2, ..., N + 1; \quad j = 1, 2, ..., N + 1; \quad k = j$$

$$= b_{n'}, \quad k = 2, 3, ..., N + 1; \quad j = 1, 2, ..., N; \quad j = k - 1; \quad n = k - 1$$

4. $A_{nD}$ = [(N + 1), (N + 1)] identity matrix.
5. $A_D$ = [(N + 1), (N + 1)] diagonal matrix with −1 in all the entries of the diagonal.
6. $A_x$ = (J, J) identity matrix.
7. All the other entries in A are zeros.

Note that the first row of A corresponds to the cash balance equations, the second row to the minimum dividend policy, and the third row to the project upper bounds.

Let

$$b' = (b'_M, b'_D, b'_x) \tag{8a.16}$$

where

$$b'_M = (M_0, M_1, ..., M_n)$$
$$b'_D = (-D_{min}, -D_{min}, ..., -D_{min})$$
$$b_x = \text{a column of ones}$$

Let

$$u' = (\rho', \eta', \mu') \tag{8a.17}$$

where $\rho$ = vector of dual variables for cash balance equations,

$\eta$ = vector of dual variables for minimum dividend policy,

$\mu$ = vector of dual variables for project upper bounds.

**Table 8a.3** ***Dual Constraints for Dividend–Terminal Wealth Model***

| | |
|---|---|
| $A'_{nx}\rho + A'_x\mu - 2[Q_{xx}x + Q'_{vw}v + Q'_{ux}w + Q'_{Dx}D] \geq q_x$ | (8a.18) |
| $A'_{nv}\rho - 2[Q_{vx}x + Q_{vv}v + Q'_{wu}w + Q'_{Dv}D] \geq q_v$ | (8a.19) |
| $A'_{nw}\rho - 2[Q_{ux}x + Q_{wv}v + Q_{ww}w + Q'_{Dw}D] \geq q_w$ | (8a.20) |
| $A'_{nD}\rho + A'_D\eta - 2[Q_{Dx}x + Q_{DD}v + Q_{Dw}w + Q_{DD}D] \geq q_D$ | (8a.21) |

By substituting (8a.24) into (8a.22), we obtain, for accepted projects, a relationship giving the *opportunity cost* of a project.

$$\mu_j = \sum_{n=0}^{N} a_{nj}\rho_n + \hat{a}_j\rho_N, \qquad x_j > 0 \tag{8a.25}$$

For partially accepted projects $\mu_j$ equals zero in (8a.25). For rejected projects $\mu_j = 0$ and

$$\sum_{n=0}^{N} a_{nj}\rho_n + \hat{a}_j\rho_N \leq 0, \qquad x_j = 0 \tag{8a.26}$$

These results are identical to those for the linear model.

If $D_n > D_{\min}$ for all $n$, we can obtain *expressions for* $\rho_n$ for values of $n$ besides $N$. By manipulating (8a.21), we obtain

$$\rho_0 - \frac{2c_2}{N+1}\left(\frac{G}{c_3c} - D_{\text{ave}}\right) + 2c_1(D_0 + D_1) = 1, \qquad D_0 > D_{\min} \tag{8.27}$$

$$\rho_n - \frac{2c_2}{(N+1)(1+i)^n}\left(\frac{G}{c_3c} - D_{\text{ave}}\right) + 2c_1(D_{n-1} + 2D_n + D_{n+1})$$
$$= \frac{1}{(1+i)^n}, \qquad n = 1, 2, \ldots, N-1; D_n > D_{\min} \tag{8a.28}$$

$$\rho_N\left(1 + \frac{c_3}{N+1}\right) + 2c_1(D_{N-1} + D_N)$$
$$= \frac{1 + c_3/(N+1)}{c}, \qquad D_N > D_{\min} \tag{8a.29}$$

These expressions will be useful in the next section, on ranges of parameter values.

### *Ranges of Parameter Values*

We have seen that the duality analysis of the quadratic model, although tedious, yields some useful results. In this section we analyze further some of these results to determine ranges of parameter values for the constants $c_1$, $c_2$, and $c_3$, in order to have a meaningful model.

The $c_1$ term weights the sum of the squared differences for consecutive dividends. From (8a.27)–(8a.29) we obtain

$$c_1 \leq \frac{1 + [2c_2/(N+1)][G/(c_3c) - D_{\text{ave}}]}{2(D_0 + D_1)}, \qquad D_0 > D_{\min} \tag{8a.30}$$

$$c_1 \leq \frac{1/(1+i)^n + [2c_2/(N+1)(1+i)^n][G/(c_3c) - D_{\text{ave}}]}{2(D_{n-1} + 2D_n + D_{n+1})},$$
$$n = 1, 2, \ldots, N-1; D_n > D_{\min} \tag{8a.31}$$

$$c_1 \leq \frac{1 + c_3/(N+1)}{2c(D_{n-1} + D_n)}, \qquad D_N > D_{\min} \tag{8a.32}$$

The search to find the range of $c_1$ may start with (8a.32), because it is the least complex inequality of those shown. By setting in advance a value for the parameter $c_3$, the decision maker can assume reasonable values for $D_{N-1}$ and $D_N$. These assumptions should be based on the decision maker's subjective judgment about the company's potential earnings within the planning period as well as about the stockholders' income desires. With these assumptions, obtaining an upper bound for $c_1$ represents no problem. The lower bound for $c_1$ basically is zero, an extreme case, in which the model lets the dividends fluctuate freely.

Useful information may also be obtained from (8a.30) and (8a.31). Because the objective of the company is to have the difference between terminal wealth and dividends as low as possible, this difference might be assumed to be zero. In this case

$$c_1 \leq \frac{1}{2(D_0 + D_1)}, \qquad D_0 > D_{\min} \tag{8a.33}$$

$$c_1 \leq \frac{1}{2(D_{n-1} + 2D_n + D_{n-1})(1+i)^n},$$
$$n = 1, 2, \ldots, N-1; D_n > D_{\min} \tag{8a.34}$$

Again, assumptions should be made about the dividends in question. Even though (8a.33) and (8a.34) are helpful, (8a.32) is more useful because this inequality requires fewer assumptions than the other two.

The range for parameter $c_2$ can be obtained from analysis of (8a.24). Since $\rho_N$ should always be nonnegative, we can say that

$$\frac{c_3}{c_2} \geq 2\left(\frac{G}{c_3 c} - D_{\text{ave}}\right) \tag{8a.35}$$

An initial idea of the value of the term in parentheses is needed. Ideally, it would be zero, but we conjecture that the value zero will seldom be obtained in this nonlinear model. The difference value depends on the investments and dividends under analysis.

The range of $c_3$ is set with the idea of equalizing dividends within the planning period with those past the horizon, as indicated by the third term in the objective function (8a.1). Its value will typically be near $1/i$.

We thus have a general procedure for setting the ranges of the parameters.

- Set $c_3$ for the third term in (8a.1), typically near $1/i$.
- Set $c_2$ from (8a.35).
- Set $c_1$ from (8a.32) or alternatively from (8a.33) or (8a.34).

The significance of these results is that they make the dividend–terminal–wealth model much more useful and practical than it would appear at first glance. A model with three parameters of unknown values might require a large number of runs, each representing a different $c_1$, $c_2$, $c_3$ combination, to provide meaningful results. Instead, we have considerably narrowed the range of the parameters and vastly reduced the number of runs required.

## MODEL DEMONSTRATION

To study the behavior of the model, we will demonstrate it on the set of projects represented in Figure 8a.1. The *IRR*s of these projects range from 8.7% to 10%, and a lending rate of 9% and a borrowing rate of 10.1% are used. A horizon at time 5 and an external discount rate of 9.5% are assumed. The behavior of the model is analyzed by using four predetermined values of the parameters $c_1$, $c_2$, and $c_3$ to make 64 runs. Beale's quadratic programming method within the MPOS package was used for the computations [16].

The parameter values were assigned as follows. For $c_3$ the values were set at 1/0.03 to 1/0.06, or 10 to 20. Thus four values were established: 10, 13, 16, and 20. Equation (8a.32) was used with $D_{\min}$ = \$5,000 to establish values for $c_1$ that were less than or equal to $8.45 \times 10^{-5}$. For practical purposes the upper bound was set at $1.0 \times 10^{-4}$ and the lower bound at $1.0 \times 10^{-7}$. The four values selected for $c_1$ were thus $1.0 \times 10^{-7}$, $1.0 \times 10^{-6}$, $1.0 \times 10^{-5}$, and $1.0 \times 10^{-4}$. With $c_1$ and $c_3$ set, Eq. 8a.35 was used to determine $c_2$. With $D_{\min}$ = \$5,000, the difference between average dividends within and past the horizon was set at \$1,000 to \$5,000. The resulting range for $c_2$ was 0.001 to 0.01, yielding the four values 0.001, 0.003, 0.006, and 0.01.

The performance criteria for judging the model results are of three types.

- The terminal wealth $G = \sum_{j=1}^{J} \hat{a}_j x_j + v_N - w_N$

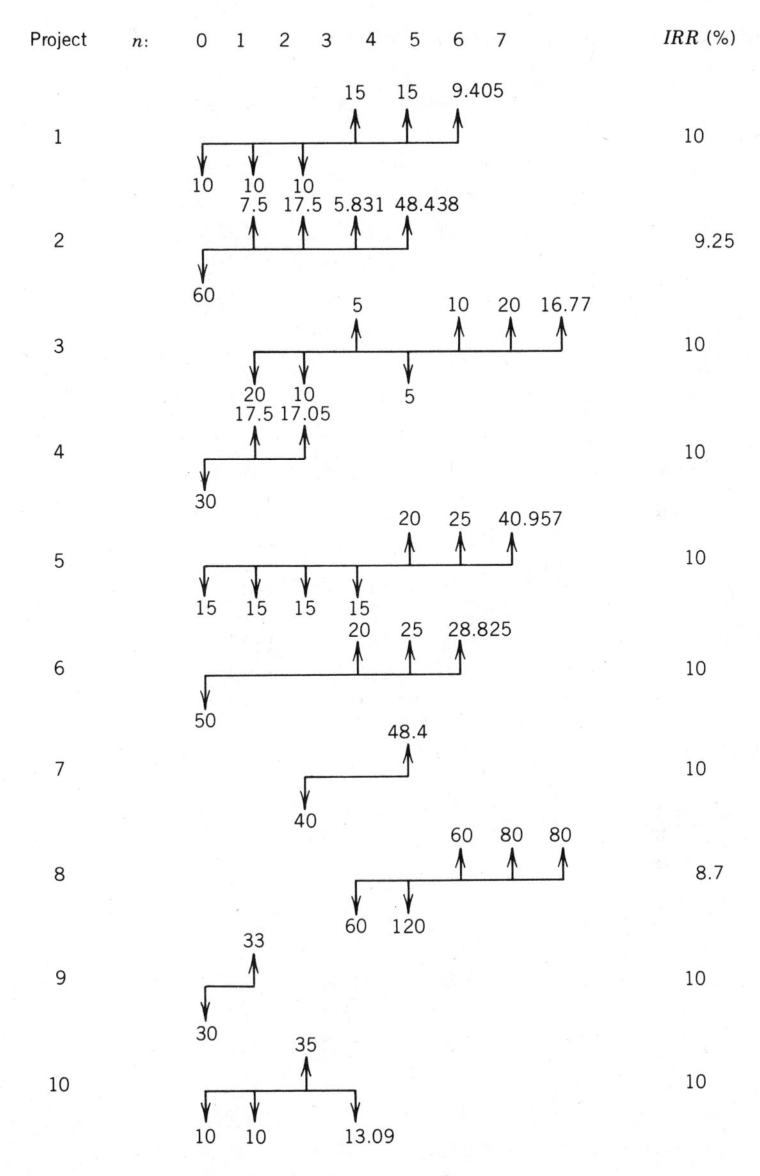


**FIGURE 8a.1.** Cash flows of project set. The budget constraint for $n=0$ is \$70,000 and for other years is \$15,000. All the cash flows are in thousands of dollars.

- The average dividend payment $\overline{D} = \dfrac{1}{N+1} \sum_{n=0}^{N} D_n$
- The standard deviation of the stream of dividends within the planning period $s = \left[ \sum_{n=0}^{N} D_n^2 - (N+1)\overline{D}^2 \right] / N$

**Table 8a.4** ***Terminal Wealth, in Dollars***

| $c_1$ | $c_2$ | $c_3$ = 10 | 13 | 16 | 20 |
|---|---|---|---|---|---|
| $10^{-7}$ | 0.001 | 142,667 | 142,667 | 147,324 | 152,449 |
| | 0.003 | 128,097 | 139,331 | 156,651 | 162,199 |
| | 0.006 | 96,672 | 142,426 | 149,995 | 157,455 |
| | 0.01 | 63,804 | 100,363 | 128,500 | 157,929 |
| $10^{-6}$ | 0.001 | 147,697 | 147,697 | 147,324 | 151,579 |
| | 0.003 | 127,706 | 138,992 | 156,651 | 161,844 |
| | 0.006 | 99,159 | 142,107 | 149,693 | 157,108 |
| | 0.01 | 63,863 | 98,896 | 127,366 | 158,051 |
| $10^{-5}$ | 0.001 | 147,486 | 147,486 | 148,548 | 154,624 |
| | 0.003 | 128,243 | 139,528 | 157,226 | 161,844 |
| | 0.006 | 96,109 | 142,485 | 150,060 | 157,487 |
| | 0.01 | 66,770 | 99,548 | 128,003 | 157,995 |
| $10^{-4}$ | 0.001 | 150,793 | 150,793 | 151,136 | 160,687 |
| | 0.003 | 129,050 | 140,267 | 159,430 | 163,307 |
| | 0.006 | 93,738 | 142,828 | 150,374 | 157,646 |
| | 0.01 | 63,670 | 97,271 | 126,257 | 158,127 |

**Table 8a.5** ***Average Dividend, in Dollars***

| $c_1$ | $c_2$ | $c_3$ = 10 | 13 | 16 | 20 |
|---|---|---|---|---|---|
| $10^{-7}$ | 0.001 | 8,324 | 8,325 | 7,329 | 6,524 |
| | 0.003 | 9,664 | 8,412 | 6,100 | 5,246 |
| | 0.006 | 7,271 | 7,891 | 6,946 | 5,860 |
| | 0.01 | 5,022 | 5,914 | 6,133 | 5,936 |
| $10^{-6}$ | 0.001 | 7,990 | 7,475 | 7,329 | 6,360 |
| | 0.003 | 9,677 | 8,449 | 6,100 | 5,284 |
| | 0.006 | 7,500 | 7,931 | 6,981 | 5,890 |
| | 0.01 | 5,028 | 5,787 | 6,038 | 5,896 |
| $10^{-5}$ | 0.001 | 7,499 | 7,499 | 7,089 | 6,043 |
| | 0.003 | 9,582 | 8,285 | 5,999 | 5,226 |
| | 0.006 | 7,195 | 7,864 | 6,896 | 5,839 |
| | 0.01 | 5,017 | 5,842 | 6,087 | 5,912 |
| $10^{-4}$ | 0.001 | 6,958 | 6,958 | 6,626 | 5,405 |
| | 0.003 | 9,347 | 7,985 | 5,672 | 5,130 |
| | 0.006 | 6,907 | 7,731 | 6,767 | 5,780 |
| | 0.01 | 5,006 | 5,639 | 5,923 | 5,827 |

**Table 8a.6** ***Standard Deviation***

| $c_1$ | $c_2$ | $c_3$ = 10 | 13 | 16 | 20 |
|---|---|---|---|---|---|
| $10^{-7}$ | 0.001 | 7,499 | 7,499 | 3,911 | 3,407 |
| | 0.003 | 2,840 | 3,109 | 2,494 | 604 |
| | 0.006 | 1,486 | 2,095 | 1,847 | 896 |
| | 0.01 | 53.48 | 1,277 | 1,196 | 1,370 |
| $10^{-6}$ | 0.001 | 3,662 | 3,857 | 3,911 | 3,332 |
| | 0.003 | 2,735 | 3,102 | 2,494 | 696 |
| | 0.006 | 1,640 | 2,250 | 1,630 | 866 |
| | 0.01 | 68.17 | 940 | 867 | 1,101 |
| $10^{-5}$ | 0.001 | 3,727 | 3,727 | 2,661 | 2,555 |
| | 0.003 | 2,509 | 2,311 | 1,786 | 554 |
| | 0.006 | 1,288 | 1,809 | 1,353 | 670 |
| | 0.01 | 42.45 | 1,067 | 996 | 1,149 |
| $10^{-4}$ | 0.001 | 1,363 | 1,363 | 1,303 | 994 |
| | 0.003 | 2,464 | 1,539 | 881 | 319 |
| | 0.006 | 992 | 1,442 | 882 | 425 |
| | 0.01 | 14.7 | 565 | 567 | 612 |

The $\overline{D}$ is undiscounted for consistency in the calculation of $s$. Table 8a.4, 8a.5, and 8a.6 present the numerical results for $G$, $\overline{D}$, and $s$, respectively. We will consider each of the parameters in turn.

### Effects of Parameters

Parameter $c_1$, which weights the dividend variability, has a predictable effect on $s$, as shown in Figure 8a.2, although it is not as pronounced for some $c_2$, $c_3$ combinations. On the other hand, its effect on $G$ is negligible, as shown in Figure 8a.3. The effect on $\overline{D}$ is slight and generally restricted to the cases where $c_2$ is at its minimum value of 0.001, shown in Figure 8a.4. In this example, then, the dividend variability is controlled without sacrificing much in terminal wealth. Average dividends are also maintained at the same relative level except where the horizon wealth posture restriction is relatively weak (small values of $c_2$); in these cases $\overline{D}$ is decreased by as much as 17%.

After the effect of $c_1$ is found to be the least among the parameter effects, the plots of $G$ and $s$ against $c_2$ and $c_3$ may be presented without the effect of $c_1$. The values are thus aggregated across $c_1$, and the number of lines is reduced from 16 to 4, which makes the plots much clearer.

Let us now consider parameter $c_2$, which weights the deviation from the horizon wealth posture goal. Figure 8a.5 shows that there is a strong tendency to reduce $G$ at low values of $c_3$, while keeping $G$ at almost the same level when $c_3$ is

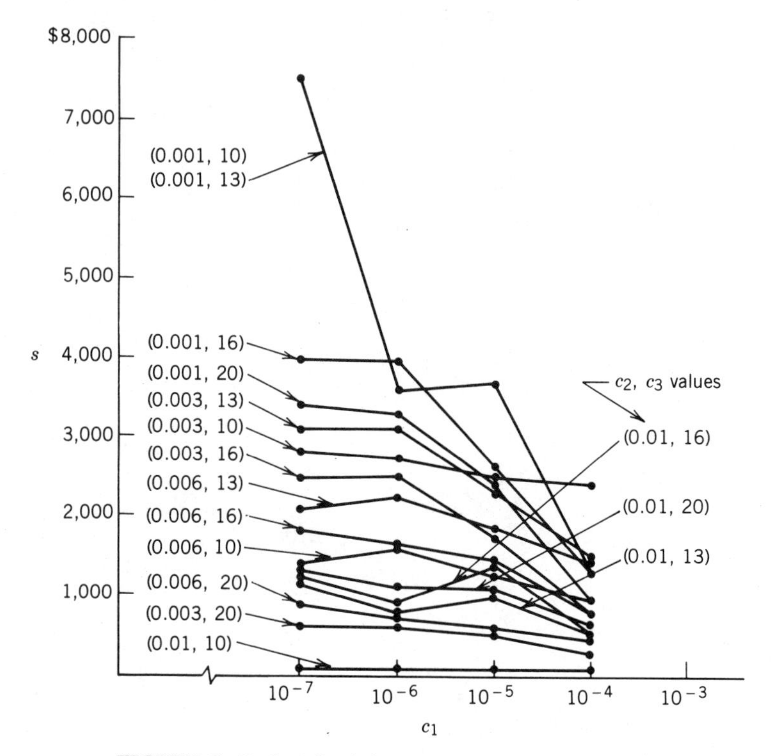


**FIGURE 8a.2.** Standard deviation versus $c_1$.

high. This means that the model loses sensitivity to $c_2$ when there are low posthorizon dividend requirements.

Parameter $c_2$ also reduces the standard deviation of the dividend payments, as may be seen in Figure 8a.6. Note that for $c_2 = 0.01$ and $c_3 = 10$ the standard deviation nearly reaches zero. All would be fine for the stockholders except for the fact that under those conditions the dividends are almost $D_{min}$, and the terminal wealth reaches its lowest value, as may be seen in Tables 8a.4 and 8a.5.

Regarding $\overline{D}$, the effect of $c_2$ is somewhat mixed, as shown in Figure 8a.7. The values for $\overline{D}$ tend to decrease with increasing values of $c_2$. An alternative explanation is that $\overline{D}$ tends to be more controlled with respect to the horizon wealth posture restriction as $c_2$ approaches 0.01, but for smaller $c_2$ values the $c_3$ parameter, interacting with project opportunities, is more important.

Parameter $c_3$ is the one that causes the greatest changes. Recall that $c_3$ is divided into $G$ to yield an estimated perpetual, annual dividend stream. Higher $c_3$ values will tend to increase the terminal wealth, an effect clearly evident in Figure 8a.8. $G$ increases substantially for high values of $c_2$ and remains more stable at low values of the same parameter. This is explained by the fact that $c_2$ is

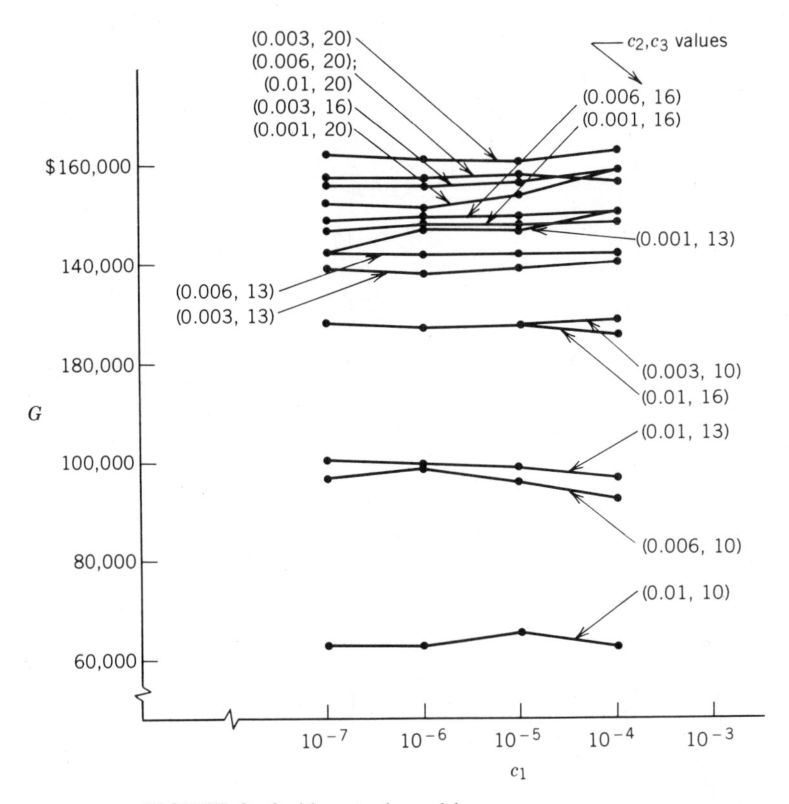


**FIGURE 8a.3.** Terminal wealth versus $c_1$.

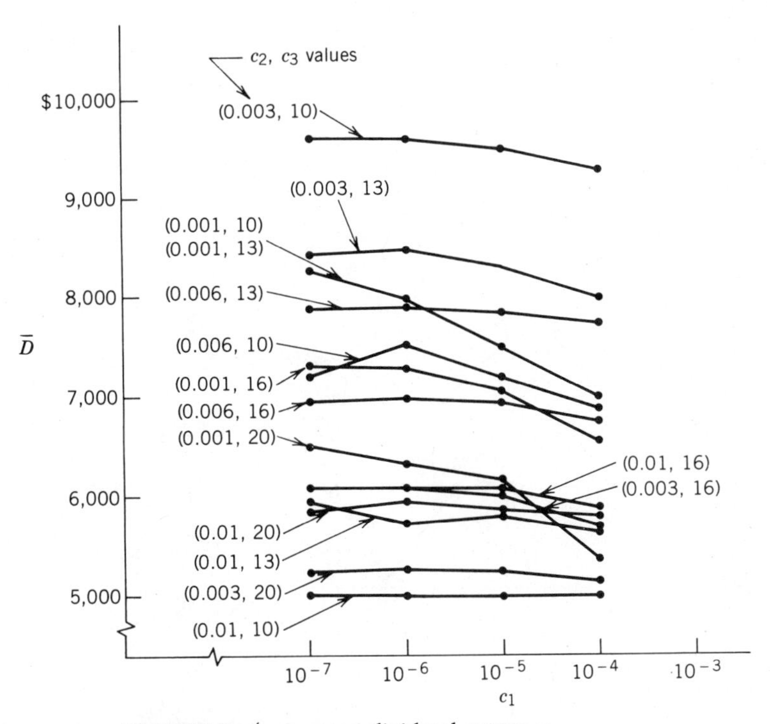


**FIGURE 8a.4.** Average dividend versus $c_1$.

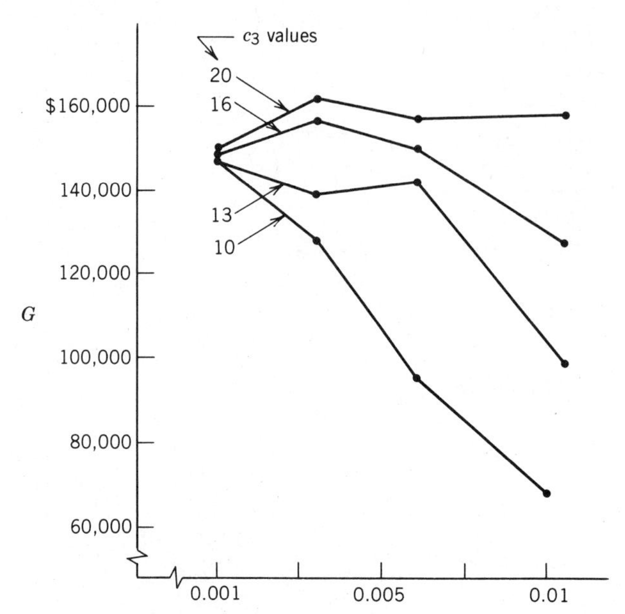


**FIGURE 8a.5.** Terminal wealth versus $c_2$. (Results are aggregated across $c_1$ values.)

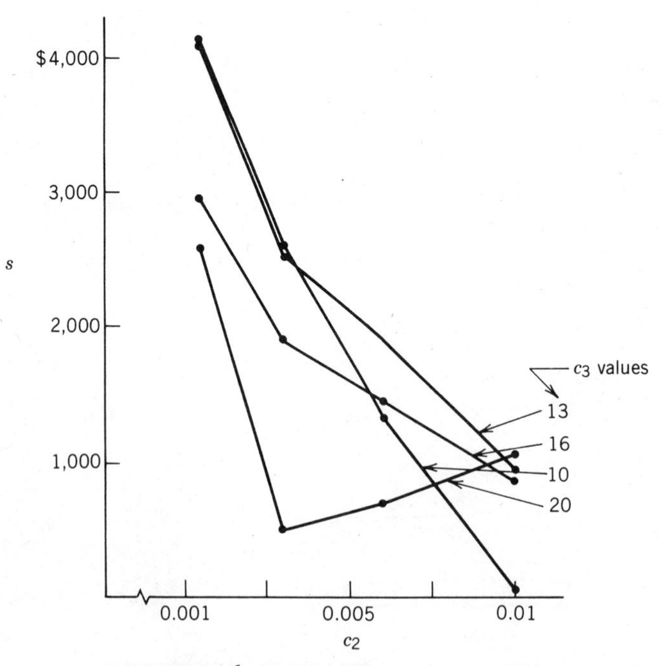


**FIGURE 8a.6.** Standard deviation versus $c_2$. (Results are aggregated across $c_1$ values.)

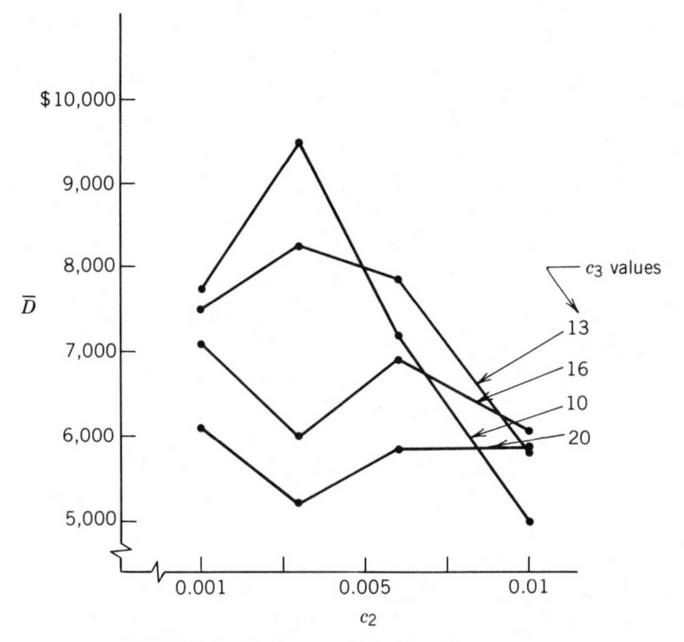


**FIGURE 8a.7.** Average dividend versus $c_2$. (Results are aggregated across $c_1$ values.)

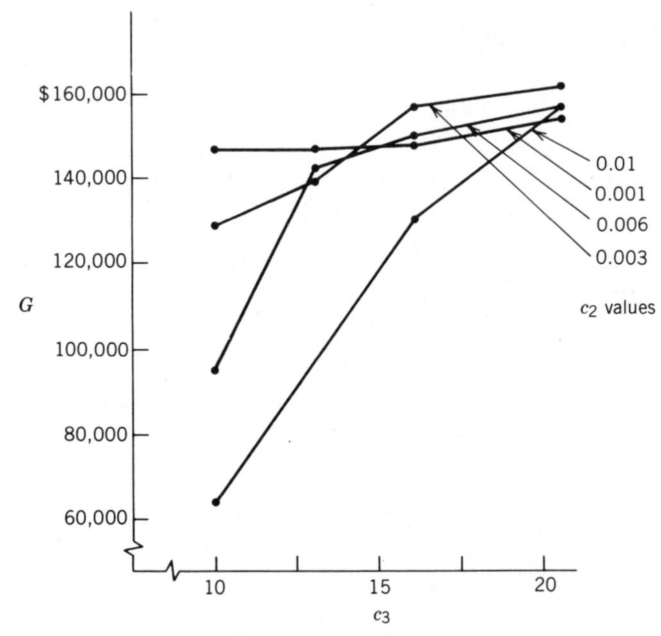


**FIGURE 8a.8.** Terminal wealth versus $c_3$. (Results are aggregated across $c_1$ values.)

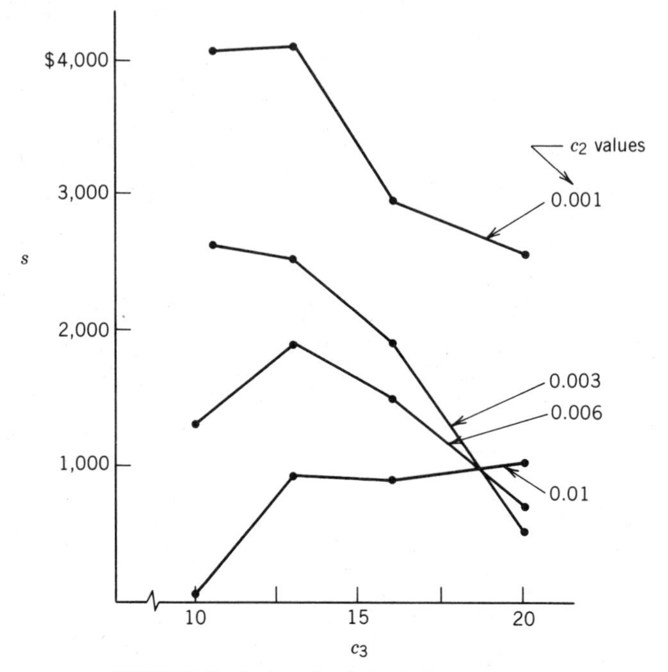


**FIGURE 8a.9.** Standard deviation versus $c_3$. (Results are aggregated across $c_1$ values.)

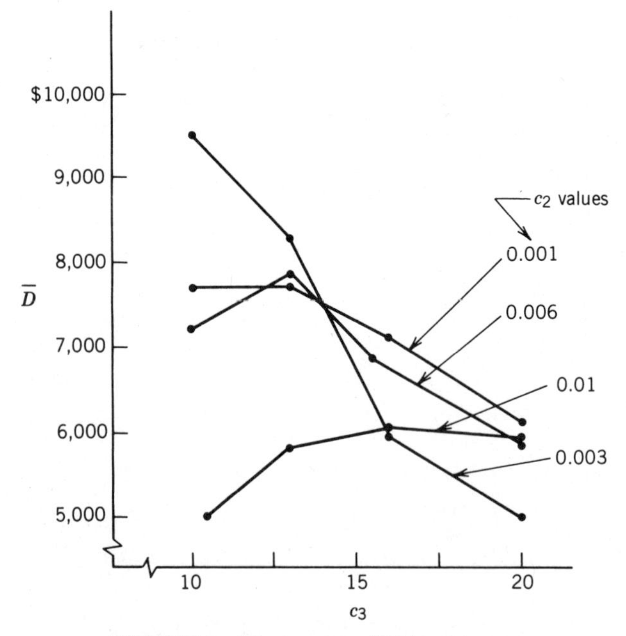


**FIGURE 8a.10.** Average dividend versus $c_3$. Results are aggregated across $c_1$ values.)

the parameter that assigns the weight to the changes made by $c_3$ in the objective function.

Regarding $s$, two different types of behavior attributable to $c_3$ are seen in Figure 8a.9. At high values of $c_2$ the standard deviation increases, whereas at low values of $c_2$ it decreases. The same is true for $\overline{D}$, shown in Figure 8a.10. Evidently, when $c_3 = 20$ the $c_3$ parameter tends to dominate the other two, so that the solutions tend to be similar.

Regarding the acceptance of projects, the most interesting characteristic of the model is that for low values of $c_3$, the projects that have a life beyond the horizon are selected in such a way that the terminal wealth consists basically of residual values of these projects. The contrary is true for high values of $c_3$. This may be seen in Table 8a.7. With $c_3 = 10$, projects 3, 5, and 8 are accepted (only these projects have posthorizon cash flows); the money lent at $n = 5$ has an average of \$41,000; and the terminal wealth is \$128,000. With $c_3 = 20$ none of those projects is selected, and all the terminal wealth consists of the money lent at $n = 5$.

### *Optimality Analysis*

The algorithm used to solve our model (Beale's method) has some difficulties in solving problems with large coefficients. Apparently, the primal solution is not as strongly affected as the dual solution. This was confirmed when the optimality conditions were checked; most of them turned out to be incorrect because of numerical problems. In Murga [16] the optimality conditions are demonstrated for the set of projects with scaled cash flows (divided by 1,000) and unscaled parameter values ($1.0 \times 10^{-7}$, 0.006, 20). Of course, the results are completely different from those presented earlier, and they will not be shown here. The following types of optimality conditions were demonstrated with good numerical accuracy.

- Complementary slackness for $x_j$ and $\mu_j$.
- Complementary slackness for $D_n$ and $\eta_n$.
- Ratio of $\rho_n/\rho_{n+1}$ and lending and borrowing activities.
- Value of $\rho_N$ by Eq. 8a.24.
- Value of $\mu_j$ by Eq. 8a.25.
- Value of $\eta_n$ by Eq. 8a.21.

### *Objectives of the Firm*

The results obtained with the quadratic model were obtained for combinations of three parameters plus specification of $i$, the externally determined discount rate, and $D_{\min}$. The major advantage of the nonlinear programming formulation is the great variety of solutions we may choose from according to our objectives and desires. However, there are sets of solutions that would normally not be considered, such as those for $c_2 = 0.01$ and $c_3 = 10$, in which the company earns little and the stockholders receive the minimum. There are other solutions that apply to classical situations, such as the following.

**Table 8a.7** ***Decision Variables for Selected Runs***

| Parameters: | $c_2$: 0.003 | | | | | | | |
|---|---|---|---|---|---|---|---|---|
| | $c_3$: 10 | | | | 20 | | | |
| | $c_1$: $10^{-7}$ | $10^{-6}$ | $10^{-5}$ | $10^{-4}$ | $10^{-7}$ | $10^{-6}$ | $10^{-5}$ | $10^{-4}$ |
| *Variable type: project selection* | | | | | | | | |
| $x_1$ | 0.49 | 0.46 | 0.45 | 0.38 | 1 | 1 | 1 | 1 |
| $x_2$ | 0 | 0 | 0 | 0 | 0 | 0 | 0 | 0 |
| $x_3$ | 0.90 | 0.90 | 0.91 | 0.91 | 0 | 0 | 0 | 0 |
| $x_4$ | 1 | 1 | 1 | 1 | 0 | 0 | 0 | 0 |
| $x_5$ | 0.78 | 0.78 | 0.78 | 0.79 | 0 | 0 | 0 | 0 |
| $x_6$ | 0 | 0 | 0 | 0 | 1 | 1 | 1 | 1 |
| $x_7$ | 0 | 0 | 0 | 0 | 1 | 1 | 1 | 1 |
| $x_8$ | 0.20 | 0.20 | 0.20 | 0.21 | 0 | 0 | 0 | 0 |
| $x_9$ | 0.44 | 0.44 | 0.43 | 0.41 | 0.06 | 0.05 | 0.06 | 0.07 |
| $x_{10}$ | 0.20 | 0.17 | 0.16 | 0.11 | 0.18 | 0.17 | 0.19 | 0.22 |
| *Lending (thousands)* | | | | | | | | |
| $v_0$ | 0 | 0 | 0 | 0 | 0 | 0 | 0 | 0 |
| $v_1$ | 0 | 0 | 0 | 0 | 0 | 0 | 0 | 0 |
| $v_2$ | 0 | 0 | 0 | 0 | 0 | 0 | 0 | 0 |
| $v_3$ | 0 | 0 | 0 | 0 | 5.7 | 5.4 | 5.8 | 6.6 |
| $v_4$ | 0 | 0 | 0 | 0 | 104.6 | 104.2 | 104.7 | 105.6 |
| $v_5$ | 41.7 | 41.5 | 41.4 | 41.5 | 162.2 | 161.8 | 162.4 | 163.3 |

**Table 8a.7** ***Decision Variables for Selected Runs (Continued)***

| Parameters: | | $c_2$: 0.003<br>$c_3$: 10<br>$c_1$: $10^{-7}$ | $10^{-6}$ | $10^{-5}$ | $10^{-4}$ | 20<br>$10^{-7}$ | $10^{-6}$ | $10^{-5}$ | $10^{-4}$ |
|---|---|---|---|---|---|---|---|---|---|
| *Borrowing (thousands)* | | | | | | | | | |
| | $w_0$ | 0 | 0 | 0 | 0 | 0 | 0 | 0 | 0 |
| | $w_1$ | 0 | 0 | 0 | 0 | 0 | 0 | 0 | 0 |
| | $w_2$ | 0 | 0 | 0 | 0 | 33.5 | 34.0 | 33.3 | 32.3 |
| | $w_3$ | 9.0 | 8.2 | 10.1 | 10.1 | 0 | 0 | 0 | 0 |
| | $w_4$ | 12.2 | 12.0 | 12.4 | 11.9 | 0 | 0 | 0 | 0 |
| | $w_5$ | 0 | 0 | 0 | 0 | 0 | 0 | 0 | 0 |
| *Dividend (thousands)* | | | | | | | | | |
| | $D_0$ | 8.4 | 9.1 | 9.2 | 11.0 | 6.5 | 6.7 | 6.4 | 5.8 |
| | $D_1$ | 10.5 | 11.0 | 10.7 | 11.2 | 5.0 | 5.0 | 5.0 | 5.0 |
| | $D_2$ | 13.4 | 12.6 | 12.4 | 11.2 | 5.0 | 5.0 | 5.0 | 5.0 |
| | $D_3$ | 9.5 | 8.7 | 10.3 | 9.8 | 5.0 | 5.0 | 5.0 | 5.0 |
| | $D_4$ | 11.3 | 11.6 | 9.7 | 8.0 | 5.0 | 5.0 | 5.0 | 5.0 |
| | $D_5$ | 5.0 | 5.0 | 5.0 | 5.0 | 5.0 | 5.0 | 5.0 | 5.0 |
| *Other (thousands)* | | | | | | | | | |
| Terminal wealth, | $G$ | 128.1 | 127.7 | 128.2 | 129.1 | 162.2 | 161.8 | 162.4 | 163.3 |
| Average dividend, | $\overline{D}$ | 9.7 | 9.7 | 9.6 | 9.3 | 5.2 | 5.3 | 5.2 | 5.1 |
| Standard deviation, | $s$ | 2.8 | 2.7 | 2.5 | 2.5 | 0.6 | 0.7 | 0.6 | 0.3 |

1. A company that is new in business should keep most of the money it earns and pay minimum dividends. With $c_1 = 1.0 \times 10^{-4}$, $c_2 = 0.003$, and $c_3 = 20$, the company may obtain that type of solution.
2. A company that wants earnings as high as possible and a fair return for its stockholders with the maximum steadiness should operate near $c_1 = 1.0 \times 10^{-4}$, $c_2 = 0.003$, $c_3 = 13$.
3. A company that prefers to pay the maximum possible dividends with little regard for dividend steadiness or terminal wealth should operate near $c_1 = 1.0 \times 10^{-7}$, $c_2 = 0.003$, and $c_3 = 10$.

A similar analysis may be made for each of the solutions obtained in the tables. More extensive considerations may be used to distinguish among the various solutions. For example, we have not mentioned financial ratios, which are usually constrained to specific ranges by company policies. This means that there may be situations in which borrowing activities cannot be undertaken without violating such policies, and another solution with less borrowing or with a different set of selected projects should be selected.

## SUMMARY

In this section we have presented a dividend–terminal-wealth capital budgeting model and analyzed and solved it with the aid of quadratic programming. A number of parameters allow the user to control dividend variability, terminal wealth, and deviation from the horizon wealth posture goal. The duality analysis of the model yields not only the usual relationships about primal and dual variables and pricing of activity vectors but also ranges for the parameter values. These bounds make the model much more useful in practice than it would appear otherwise. The demonstration of the model shows the effect of the various parameter combinations on terminal wealth, average dividend, and dividend variability.

**PART THREE**

# STOCHASTIC ANALYSIS

# 9
# Utility Theory

## 9.1 INTRODUCTION

In the first two parts of this book we have assumed that decisions are made in a context of complete certainty. The decision makers are characterized as persons wishing to maximize cash flow, the present value of cash flow, or perhaps terminal cash wealth. Cash amounts at different points in time are converted to some common point in time, often time 0, by using an interest rate, and are then added to obtain *PV, FV,* and so forth.

In this third part of the book we relax these ideal assumptions in two important ways.

1. Project cash flow will no longer be regarded as certain. Instead, we will use probability concepts to describe project flows.
2. Decision makers will no longer be assumed to add (linearly) different cash flows at the same point in time or after conversion to the same point in time by use of an interest rate. Instead, small cash flows will usually be given more consideration per dollar than large cash flows.

In this chapter we give a brief introduction of the first concept, the probabilistic description of cash flows. We assume the reader is familiar with the fundamental concepts of probability theory. Probabilistic approaches to investment decisions are given extensive coverage in Chapters 10 to 13.

The principal emphasis in this chapter is on the second concept, namely the utility theory approach to combining and evaluating cash flows. Following the introduction of the concept in this section, the formal statement of utility theory is presented in Section 9.2. In Section 9.3 we discuss the properties of utility functions, followed by the procedures for assessing a utility function by empirical means in Section 9.4. An important operational method, mean–variance analysis, is shown in Section 9.5 to be based on the utility theory concept; mean–variance analysis is presented in depth in Chapters 10 and 11.

### 9.1.1 The Concept of Risk

We may introduce the concept of risk by asking why individual homeowners (with no outstanding mortgage or loan against their homes) would buy fire insurance. The possibility of damage from fire in a particular year is quite low, say 0.01. If the amount of damage caused by a fire is $60,000, we would say the *risk* of fire damage is a 0.01 chance (1% chance) of a $60,000 loss. If the fire insurance premium is $700 per year and the deductible amount on a loss (the amount the individual pays) is $250, then on an *expected monetary value* (*EMV*) basis an individual who buys insurance has the following yearly cost.

| Event | Cost | Probability | Product |
|---|---|---|---|
| No fire occurs | $700 | 0.99 | $693.00 |
| Fire occurs | $700 + 250 = $950 | 0.01 | 9.50 |
| | | Expected annual cost | $702.50 |

Contrast this with the situation of an individual who decides *not* to buy fire insurance.

| Event | Cost | Probability | Product |
|---|---|---|---|
| No fire occurs | 0 | 0.99 | 0 |
| Fire occurs | $60,000 | 0.01 | $600.00 |
| | Expected annual cost | | $600.00 |

Most individual homeowners would clearly prefer to buy fire insurance in order to avoid the risk of a 0.01 chance of a $60,000 loss, even though the expected annual cost of $702.50 is greater than the expected annual cost of

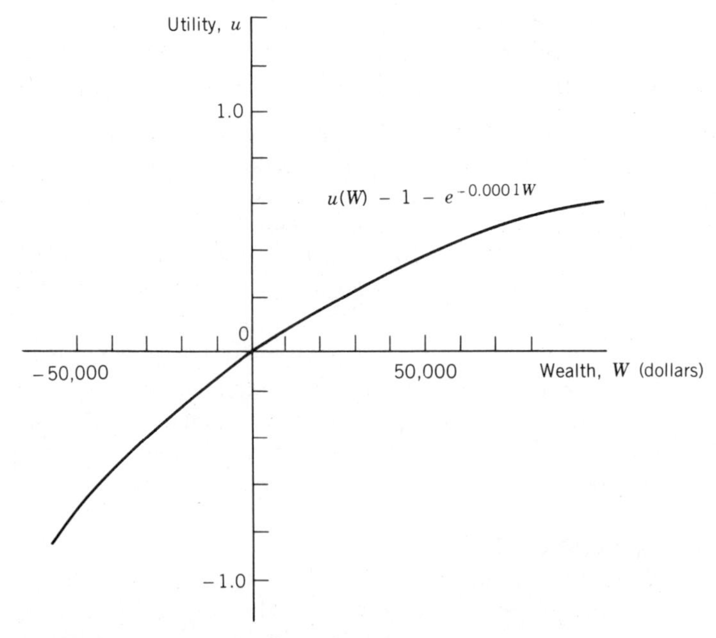


**FIGURE 9.1.** Example of a utility curve.

\$600.00 without insurance. On the other hand, a large corporation with hundreds of retail outlets, facing similar risks and premiums at each outlet, might decide not to buy fire insurance. Such a corporation would become a *self-insurer*. The individual homeowner's way of evaluating the possible \$60,000 loss is different from that of the large corporation, which can presumably make decisions regarding such amounts on an *EMV* basis. A \$60,000 loss could be disastrous for an individual, whereas the large corporation would expect only one such loss per hundred retail outlets.

The individual's behavior, which is *not* based on an *EMV*, can be explained by the concept of *utility*. An example of a *utility function* for an individual, shown in Figure 9.1, has the following selected values. The function is

$$u(W) = 1 - e^{-0.00001W} \tag{9.1}$$

where $W$ is the dollar amount of wealth.

| Wealth, $W$ | Utility Value |
|---|---|
| \$100,000 | 0.63212 |
| 50,000 | 0.39347 |
| 10,000 | 0.09516 |
| 1,000 | 0.00995 |
| 0 | 0 |
| −1,000 | −0.01005 |
| −10,000 | −0.10517 |

The utility function in Figure 9.1 reflects a decreasing incremental value from each incremental dollar of wealth.

Following this line of argument, we can calculate an *expected utility* (*EU*) for our individual homeowner for the two decisions available: buy fire insurance or do not buy it. Let us assume the individual's total wealth, including the home, is \$80,000. If the individual *buys* insurance, the *EU* of this decision for the next year is

| Event | Resulting Wealth, $W$ | Utility | Probability | Product |
|---|---|---|---|---|
| No fire occurs | \$79,300 | 0.54751 | 0.99 | 0.54204 |
| Fire occurs | \$79,050 | 0.54638 | 0.01 | 0.00546 |
| | | Expected utility = $E[u(W)]$ = 0.54750 | | |

If the individual *does not buy* insurance, the *EU* of this decision for the next year is

| Event | Resulting Wealth, $W$ | Utility | Probability | Product |
|---|---|---|---|---|
| No fire occurs | \$80,000 | 0.55067 | 0.99 | 0.54516 |
| Fire occurs | \$20,000 | 0.18127 | 0.01 | 0.00181 |
| | | Expected utility $E[u(W)]$ = 0.54697 | | |

Thus, on the basis of the *EU,* we can explain the decision of an individual homeowner to buy fire insurance even though the *expected annual cost* is higher. Large corporations also make decisions that do not result in the lowest expected annual costs, especially when potential losses are high. Such decisions can also be explained on the basis of *EU.* The difference is the scale of the cash flows; the corporation that is a self-insurer when losses are $60,000 per retail outlet might obtain insurance from an outside source when a single loss could be $25 million.

It is important to distinguish between *risk* and *uncertainty. Risk* applies to situations for which the outcomes are not known with certainty but about which we do have good probability information. Subsequent analysis could then be based on *EMV* or on *EU. Uncertainty* applies to situations about which we do not even have good probability information. In such situations other analysis techniques are appropriate, and the reader is referred to Luca and Raiffa [16].

### 9.1.2 Role of Utility Theory

In the preceding section we used utility theory to reconcile actual behavior with *EMV* decision making. This role of utility theory can be expanded to include behavior that is seemingly irrational because information is incomplete, because individuals have difficulties in establishing ordinal measurement scales, and because multiple-objective functions have been maximized [6]. Empirical behavior of individuals has prompted economists to construct some unusual utility functions. For example, an individual may buy insurance, normally an expected loss in a situation the individual feels offers no other choice. The same individual may buy lottery tickets, virtually always an expected loss in a situation in which the individual *does* have a choice. This type of observed behavior has led economists to hypothesize a compound-shaped utility function [7].

Utility theory can be used to justify the time value of money, as applied in Parts One and Two of this book. Furthermore, by including the effects of uncertainty in the future, we can argue for a discount rate *greater* than the equity rate or the weighted-average cost of capital presented in Chapter 5.

A very important role of utility theory is in the justification of the mean–variance method for analyzing risky cash flows. This is presented in Section 9.5.

It is important to remember that utility theory is both a *prescriptive* and a *descriptive* approach to decision making. The theory tells us how individuals and corporations *should* make decisions, as well as predicting how they *do* make decisions. The *hypothesis* aspect of utility theory should not be forgotten.

### 9.1.3 Alternative Approaches to Decision Making

Two related approaches other than utility theory, have been presented as constructs for decision making. They are based on principles other than expected value with a discount rate based on cost of capital. The first approach uses a risk-adjusted discount rate [2, 10]. Investment projects are assigned to risk classes, based on the uncertainty of the component cash flows. Investments in a "safe" risk class are evaluated by using an interest rate based on cost of capital, whereas investments with more uncertain cash flows are evaluated by using a higher interest rate.

The second approach is based on the concept of general states of wealth at different points in time and the implicit trade-offs an individual or corporation might make among these states [8, 14]. Although conceptually appealing, this choice–theoretic approach is difficult to implement.

## 9.2 PREFERENCE AND ORDERING RULES

In this section we present the formal definition of utility theory as it is commonly interpreted by economists. The theory consists of two parts: the hypothesis about maximizing expected utility and the axioms of behavior.

### 9.2.1 Bernoulli Hypothesis

The basic hypothesis of utility theory is that individuals make decisions with respect to investments in order to *maximize expected utility* [3]. This concept is demonstrated by the following example.

### *Example 9.1*

An individual with a utility function $u(W) = 1 - e^{-0.0001W}$ is faced with a choice between two alternatives. Alternative 1 is represented by the following probability distribution.

| Cash Amount | \$−10,000 | 0 | 10,000 | 20,000 | 30,000 |
|---|---|---|---|---|---|
| Probability | 0.2 | 0.2 | 0.2 | 0.2 | 0.2 |

Alternative 2 is a certain cash amount of \$5,000. The individual has an initial wealth of zero, and no investment is required for either alternative. Which alternative would the individual prefer?

For alternative 1 the expected utility is computed as follows.

| Wealth, $W$ | Utility | Probability | Product |
|---|---|---|---|
| −\$10,000 | −1.7183 | 0.2 | −0.3437 |
| 0 | 0 | 0.2 | 0 |
| 10,000 | 0.6321 | 0.2 | 0.1264 |
| 20,000 | 0.8647 | 0.2 | 0.1729 |
| 30,000 | 0.9502 | 0.2 | 0.1900 |

$$\text{Expected utility} = E[u(W)] = 0.1456$$

For alternative 2 the utility is 0.3935. As this amount is greater than that for alternative 1, the certain cash amount of \$5,000 is preferred to the risky alternative 1, which has a higher expected value of \$10,000.

We can begin with the utility value of 0.1456 and determine a certain cash amount that is exactly equivalent to alternative 1.

$$0.1456 = 1 - e^{-0.0001W}$$

$$e^{-0.0001W} = 0.8544$$

Taking natural logarithms of both sides, we obtain

$$-0.0001W = -0.1574$$

$$W = \$1{,}574$$

The amount \$1,574 is called the *certainty equivalent* (*CE*) of alternative 1. Our individual would prefer any larger certain cash amount to alternative 1, would prefer alternative 1 to any smaller certain cash amount, and would be indifferent about a certain cash amount of \$1,574 and alternative 1. □

***Definition.*** A *certainty equivalent* (*CE*) is a certain cash amount that an individual values as being as desirable as a particular risky option.

### 9.2.2 Axioms of Utility Theory

Individuals are assumed to obey the following rules of behavior in decision making [13, 17, 20].

*Orderability.* We can establish distinct preferences between any two alternatives. For example, given alternatives $A$ and $B$, an individual prefers A to B, shown by $A > B$; prefers $B$ to $A$, shown by $A < B$—we read the symbol $<$ as "is less preferred than"; or is indifferent about choosing between the two, shown by $A \sim B$.

*Transitivity.* The preferences established by ordering are transitive. If $A$ is preferred to $B$, and $B$ is preferred to $C$, then $A$ is preferred to $C$.

$$A > B \quad \text{and} \quad B > C \quad \text{imply } A > C$$

In addition,

$$A \sim B \quad \text{and} \quad B \sim C \quad \text{imply } A \sim C$$

*Continuity.* If $A$ is preferred to $B$ and $B$ is preferred to $C$, there exists a probability $p$ so that the individual is indifferent between receiving $B$ for certain and obtaining $A$ with chance $p$ and $C$ with chance $(1 - p)$. The second alternative is called a lottery involving $A$ and $C$.

$$A > B > C$$

implies that there exists a $p$ so that

$$B \sim \{(p, A), (1 - p, C)\}$$

### *Example 9.2*

Consider the individual with utility function $u(W) = 1 - e^{-0.0001W}$. Find the probability $p$ so that the individual is indifferent between receiving \$20,000 for certain and entering a lottery with chance $p$ of \$30,000 and $(1 - p)$ of \$10,000. The individual's wealth is \$10,000, and there is no cost for either alternative. The comparison is between \$30,000 (the initial \$10,000 plus \$20,000) for certain, a utility of 0.9502, and a chance $p$ of \$40,000 and chance $(1 - p)$ of \$20,000.

$$u(\$40{,}000) = 0.9817$$

$$u(\$20{,}000) = 0.8647$$

$$0.9502 = (p)(0.9817) + (1 - p)(0.8647)$$

Solving for $p$ gives 0.731. □

*Monotonicity.* If two lotteries involve the same two alternatives $A$ and $B$, the individual prefers the lottery in which the preferred alternative has the greater probability of occurring.

$$A > B \quad \text{implies}$$

$$\{(p, A), (1 - p, B)\} > \{(p', A), (1 - p', B)\}$$

if and only if $p > p'$.

*Decomposability.* A risky option containing another risky option may be reduced to its more fundamental components. This axiom, often called the "no fun in gambling" axiom, is best explained by an example.

### *Example 9.3*

Consider a two-stage lottery. In stage 1 there is a 0.5 chance of stopping and receiving nothing and a 0.5 chance of advancing to stage 2. In stage 2 there is a 0.5 chance of receiving \$5,000 and a 0.5 chance of receiving nothing. This lottery may be reduced to its one-stage equivalent of

$$\$0: \quad (0.5) + (0.5)(0.5) = 0.75 \text{ chance}$$

$$\$5{,}000: \quad (0.5)(0.5) = 0.25 \text{ chance} \quad \square$$

*Independence.* A risky option $A$ is preferred to a risky option $B$ if and only if a $[p, (1 - p)]$ chance of $A$ or $C$, respectively, is preferred to a $[p, (1 - p)]$ chance of $B$ or $C$, for arbitrary chance $p$ and risky options $A$, $B$, and $C$.

$$A > B$$

if and only if

$$\{(p, A), (1 - p, C)\} > \{(p, B), (1 - p, C)\}$$

for any $p$, $A$, $B$, and $C$.

The foregoing axioms have been used to derive the Bernoulli hypothesis [17, 20]. There are several different versions of the axioms. Some authors define additional ones or declare that some are embodied in others and thus superfluous.

Psychologists and behaviorally oriented economists each year write numerous papers describing experiments in which individuals systematically violate one or more of these axioms. It is not uncommon for such authors to propose a modification or elaboration of the theory [1, 4, 5, 11, 18]. This point brings us back to the *hypothesis* aspect of utility theory. The theory is an elegant mathematical way to describe real behavior, but it will always be at variance, more or less, with observed behavior.

## 9.3 PROPERTIES OF UTILITY FUNCTIONS

Most economists agree that an individual prefers more wealth to less. Hence, a utility function should be an *increasing,* or at the very least a *nondecreasing,* function of wealth. Other desirable properties are continuity (actually guaranteed by the axioms) and differentiability. The major question is about *risk avoidance* versus *risk seeking*.

### 9.3.1 Risk Attitudes

In all the examples presented so far in this chapter, the individual has been willing to accept a certain cash amount that is *less* than the *EMV* of a risky option. This type of behavior is described as *risk-averse,* or *risk-avoiding* behavior. Risk-averse utility functions, such as the one in Figure 9.1, are *concave* functions of wealth.

It has been suggested that some individuals exhibit *risk-seeking* behavior, as demonstrated by the following example.

### *Example 9.4*

An individual is observed to buy a \$5.00 lottery ticket each week. The possible prizes are represented by random variable $X$, and the chances of winning them are represented by probability $p$ as follows.

| $X =$ | | $p =$ |
|---|---|---|
| | No prize | 0.98889 |
| | \$100 prize | 0.01000 |
| | \$1,000 prize | 0.00100 |
| | \$10,000 prize | 0.00010 |
| | \$100,000 prize | 0.00001 |

Explain the behavior of the individual.

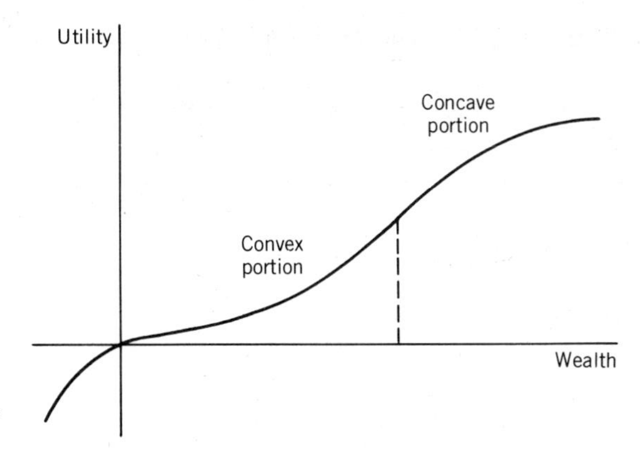


**FIGURE 9.2.** Utility function with a convex portion.

The *EMV* of such a lottery ticket is

$$\begin{aligned} E(X) &= -5 + (0.98889)(0) + (0.01)(100) + (0.001)(1{,}000) \\ &\quad + (0.0001)(10{,}000) + (0.00001)(100{,}000) \\ &= -5 + 0 + 1 + 1 + 1 + 1 \\ &= -1 \end{aligned}$$

We may suggest two possible reasons for the individual to suffer the $1 expected loss each week. The first is that the purchase of a lottery ticket is a form of entertainment, similar to buying tickets to a sports event or a musical performance. The second, and more intriguing possibility, is suggested by the fact that poor people buy disproportionately more lottery tickets than middle-class and wealthy people, especially compared with other expenditures for entertainment. This fact has led many economists to suggest that the utility function for some persons may be *convex,* over a certain range of wealth, as shown in Figure 9.2. The rationale is that a poor person, in order to get out of his or her environment, is willing to take risks that a middle-class or wealthy person would not take [7]. □

We thus have a classification scheme for persons and their respective utility functions.

1. Risk-averse person: concave utility function.
2. Risk-neutral person: linear utility function.
3. Risk-seeking person: convex utility function.

Now let us reconsider the individual in Example 9.1 with utility function $u(W) = 1 - e^{-0.0001W}$. Assume that the individual has a starting wealth of $W_0 = \$20{,}000$ and is presented with the following lottery at no cost.

$$\{(0.5, \$10{,}000), (0.5, \$20{,}000)\}$$

The *CE* for the individual facing this lottery is obtained as follows.

| Event, $X$ | Resulting Wealth, $W$ | Utility | Probability | Product |
|---|---|---|---|---|
| \$10,000 | \$30,000 | 0.95021 | 0.5 | 0.47511 |
| \$20,000 | \$40,000 | 0.98168 | 0.5 | 0.49084 |
| | | | Expected utility $= E[u(W)] =$ | 0.96595 |

$$0.96595 = 1 - e^{-0.0001W}$$
$$e^{-0.0001W} = 0.03405$$
$$-0.0001W = -3.3798$$
$$CE = W = \$33{,}798$$

The difference between the *EMV* of (0.5)(\$30,000) + (0.5)(\$40,000) = \$35,000 and the *CE* of \$33,798 is the *risk premium* (*RP*) the individual is willing to give up to avoid the risky option.

$$\text{Risk premium, } RP = \$35{,}000 - 33{,}798 = \$1{,}202$$

***Definition*** [15]. A *risk premium* is an amount *RP* that solves Eq. 9.2.

$$E\,[u(W_0 + X)] = u[W_0 + E(X) - RP] \tag{9.2}$$

where $W_0$ = the individual's wealth, a constant,
$X$ = random variable representing the cash flow from a risky option,
$RP$ = risk premium.

Here $W = W_0 + X$ is a random variable.

Let us recompute the *CE* for the previous lottery for an individual with the utility function of

$$u(W) = W - (0.00001)(W^2), \qquad 0 \leq W \leq 50{,}000 \tag{9.3}$$

| Event, $X$ | Resulting Wealth, $W$ | Utility | Probability | Product |
|---|---|---|---|---|
| \$10,000 | \$30,000 | 21,000 | 0.5 | 10,500 |
| \$20,000 | \$40,000 | 24,000 | 0.5 | 12,000 |
| | | | Expected utility $= E[u(W)] =$ | 22,500 |

This corresponds to a *CE* of \$34,190 (see Problem 9.7) and a corresponding *RP* of

$$RP = \$35{,}000 - 34{,}190 = \$810$$

The fact that the risk premium is different should not cause us much concern, since the utility functions for the two individuals are different. Let us recompute, however, the risk premiums for *both* individuals assuming an initial wealth of $W_0 = \$30{,}000$. For the individual with $u(W) = 1 - e^{-0.0001W}$, we have

| Event, $X$ | Resulting Wealth, $W$ | Utility | Probability | Product |
|---|---|---|---|---|
| \$10,000 | \$40,000 | 0.98168 | 0.5 | 0.49084 |
| \$20,000 | \$50,000 | 0.99326 | 0.5 | 0.49663 |

Expected utility $= E[u(W)] = 0.98747$

The *CE* is \$43,796, which implies $RP = \$45{,}000 - \$43{,}796 = \$1{,}204$. This amount is not much different from the previous \$1,202. (It actually is the same.)

For the individual with the quadratic utility function, Eq. 9.3, and an initial wealth of $W_0 = \$30{,}000$, we obtain

| Event, $X$ | Resulting Wealth, $W$ | Utility | Probability | Product |
|---|---|---|---|---|
| \$10,000 | \$40,000 | 24,000 | 0.5 | 12,000 |
| \$20,000 | \$50,000 | 25,000 | 0.5 | 12,500 |

Expected utility $= E[u(W)] = 24{,}500$

The *CE* is \$42,930, with a corresponding $RP = \$45{,}000 - \$42{,}930 = \$2{,}070$.

The risk premium *increases* as the individual's wealth increases! In other words, the individual with the quadratic utility function is willing to give up *more* certain cash when faced with a risky option, as his or her wealth increases. Many economists argue that such behavior is not characteristic of intelligent investors. Instead, as their wealth increases, people should be willing to give up a *smaller* risk premium when faced with the same risky option.

### 9.3.2 Types of Utility Functions

Changes in the risk premium as a function of wealth are related to the behavior of the *risk aversion function* [21].

***Definition.*** For a utility function $u$ with first and second derivatives $u'$ and $u''$, respectively, the *risk aversion function* is given by

$$r(W) = -u''(W)/u'(W) \tag{9.4}$$

where $W$ is wealth.

Specifically, if $r(W)$ is *decreasing* as a function of wealth, the risk premium (for a given risky option) decreases as a function of wealth. Similarly, an increasing $r(W)$ implies an increasing *RP*, and a constant $r(W)$ implies a constant *RP*.

A negative exponential function such as

$$u(W) = 1 - e^{-cW}, \qquad c > 0 \tag{9.5}$$

has a constant risk aversion function, since

$$u'(W) = ce^{-cW} \tag{9.5a}$$

$$u''(W) = -c^2e^{-cW} \tag{9.5b}$$

$$r(W) = c^2e^{-cW}/(ce^{-cW}) \tag{9.5c}$$
$$= c$$

This property makes the function appealing to analysts. One does not have to know the wealth of the decision maker to perform analysis regarding *CE*s and *RP*s.

A quadratic function such as

$$u(W) = W - aW^2, \qquad a > 0, \quad W \leq 1/(2a) \tag{9.6}$$

has an increasing risk aversion function, since

$$u'(W) = 1 - 2aW \tag{9.6a}$$

$$u''(W) = -2a \tag{9.6b}$$

$$r(W) = \frac{2a}{1 - 2aW} \tag{9.6c}$$

and the denominator of Eq. 9.6c is less than 1.0.

In Section 9.3.1 we presented the classification of utility functions as follows.

**1.** Risk-averse person: concave utility function,

$$u''(W) < 0 \tag{9.7a}$$

**2.** Risk-neutral person: linear utility function,

$$u''(W) = 0 \tag{9.7b}$$

**3.** Risk-seeking person: convex utility function,

$$u''(W) > 0 \tag{9.7c}$$

We can now add the subclassifications based on the risk aversion function, Eq. 9.4.

a. Decreasing risk aversion,

$$r'(W) < 0 \tag{9.8a}$$

b. Constant risk aversion,

$$r'(W) = 0 \tag{9.8b}$$

c. Increasing risk aversion,

$$r'(W) > 0 \tag{9.8c}$$

An example of a risk-averse utility function with constant risk aversion is the negative exponential function given by Eq. 9.5. An example of a risk-averse utility function with increasing risk aversion is the quadratic function of Eq. 9.6. An example of a risk-averse function with decreasing risk aversion is the logarithmic function.

$$u(W) = \ln(W + d), \qquad d \geq 0 \tag{9.9}$$

In addition, some utility functions have bounded functional values, and others are meaningful only over a bounded domain (range of wealth). Other characteristics are related to the *proportion* of wealth an individual would invest in a risky option [21].

Linear combinations of utility functions, where the weights are positive and all component utility functions have the same subclassification based on Eqs. 9.8a, b, and c, maintain that subclassification [21]. This property is useful when defining a utility function of present value. For example, we can define a utility function for cash $F_n$ received in period $n$, when the utility is measured at time $n$.

$$u_n(F_n) = (F_n)^a, \qquad 0 < a < 1 \tag{9.10}$$

A composite utility function for the vector of cash flows $(F_1, F_2, \ldots, F_n)$ can be expressed as

$$u(F_1, F_2, \ldots, F_n) = \sum_{n=1}^{N} \frac{(F_n)^a}{(1 + i)^n} \tag{9.11}$$

Other functional forms are possible.

## 9.4 EMPIRICAL DETERMINATION OF UTILITY FUNCTIONS

### 9.4.1. General Procedure

The most popular way to determine a utility function is by the certainty equivalent method, whereby information from an individual is elicited by asking questions about lotteries [12]. Either a *numerical* or a *functional* approach can be followed. The numerical approach is presented first, for an individual with zero wealth.

The *numerical* approach requires two reference values for starting. Pick one as \$0 with zero utility and one as \$1,000 with utility 1.0.

$$u(0) = 0 \tag{9.12a}$$

$$u(\$1{,}000) = 1.0 \tag{9.12b}$$

Now present the individual with a lottery involving the nonzero reference point (there is no cost to play).

$$\{(p, \$1{,}000), (1 - p, -\$1{,}000)\} \tag{9.13}$$

The value $p$ that makes the individual indifferent to the lottery results in the following relation.

$$(p)u(\$1{,}000) + (1 - p)u(-\$1{,}000) = u(0) = 0 \tag{9.14}$$

This is so because the individual values the lottery with $p$, the same as not playing, which is equivalent to the individual's current state of zero wealth. If, for example, a value of $p = 0.60$ makes the individual indifferent about playing, then substituting from Eq. 9.12, we have

$$(0.6)(1.0) + (0.4)u(-\$1{,}000) = 0$$
$$u(-\$1{,}000) = -1.5 \tag{9.15}$$

This gives us three value points, and we continue in a similar manner.

For example, we can present the individual with a choice between a certain cash amount of \$1,000 and the following lottery.

$$\{(p, \$10{,}000), (1 - p, \$0)\} \tag{9.16}$$

The value $p$ that causes the individual to be indifferent results in

$$(p)u(\$10{,}000) + (1 - p)u(0) = u(\$1{,}000) \tag{9.17}$$

If $p = 0.35$, for example, substituting and solving gives

$$(0.35)u(\$10{,}000) + (0.65)(0) = 1.0$$
$$u(\$10{,}000) = 2.86 \tag{9.18}$$

Continuing in this manner, we can develop a table as shown here and graphed in Figure 9.3.

| Wealth (dollars) | Utility Value |
|---|---|
| \$20,000 | 3.40 |
| 10,000 | 2.86 |
| 1,000 | 1.00 |
| 0 | 0 |
| −1,000 | −1.50 |
| −2,000 | −4.00 |

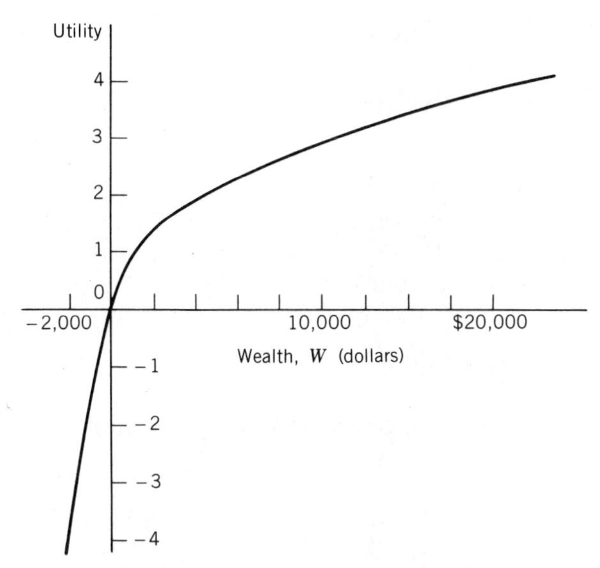


**FIGURE 9.3.** Typical empirically derived utility function.

The *functional* approach requires only one reference value for starting, most often $0 with zero utility. We also hypothesize the *functional* form. For example, assume the individual's utility function is Eq. 9.5,

$$u(W) = 1 - e^{-cW}, \qquad c > 0 \tag{9.5}$$

Next, we present a lottery, such as Eq. 9.13, with no cost to play, and elicit the value $p$ that makes the individual indifferent about playing. If the same value $p = 0.6$ is obtained, we have an equation with one unknown.

$$(0.6)[1 - e^{(-c)(1,000)}] + (0.4)[1 - e^{(-c)(-1,000)}] = 0$$

$$(0.6)(e^{-1,000c}) + (0.4)(e^{1,000c}) = 1 \tag{9.19}$$

This can be solved by trial and error for $c = 0.0004$. Thus, the specific form of Eq. 9.5 is

$$u(W) = 1 - e^{-0.0004W} \tag{9.20}$$

Determining utility functions must be done with extreme care, despite the apparent simplicity of these examples. Inconsistencies and irregular-shaped functions often result. Alternative forms of lotteries are recommended by some to reduce bias in the information-gathering process [19].

### 9.4.2 Sample Results

In this section we present empirical results for the bids in two lottery games.

*Game 1:* A number of individuals (more than 10) submit sealed bids for the right to play a lottery.

$$\{(0.5, \$50), (0.5, -\text{bid})\} \tag{9.21}$$

The highest bidder *must* play the lottery.

*Game 2:* A number of individuals (more than 10) submit sealed bids for the right to play the St. Petersburg game [3]. The highest bidder *must* play. In the St. Petersburg game a coin is tossed repeatedly until it turns up "heads." The payoff is

$$\$(2)^{n-1} \tag{9.22}$$

where $n$ is the first time heads appears. This compound lottery is equivalent to the simple lottery of

$$\{(0.5, \$1), (0.25, \$2), (0.125, \$4), \ldots, [(0.5)^n, (2)^{n-1}], \ldots\} \tag{9.23}$$

The lottery 9.23 has an infinite number of outcomes, and its *EMV* is infinity.

$$\begin{aligned} EMV = E(X) &= (0.5)(1) + (0.25)(2) + (0.125)(4) + \cdots \\ &= 0.5 + 0.5 + 0.5 + \cdots \end{aligned}$$

Table 9.1 shows the results of the bids made by graduate engineering students during the early 1980s. The bidders are ordered by ascending game 1 bids and, for equal game 1 bids, by ascending game 2 bids. Some of the low bids clearly reflect the artificiality of a classroom situation, or perhaps the cash amount in the pocket of a student. Similar artificial distortions can exist in a corporate environment, however, where one may be trying to calibrate a utility function by posing lottery games.

Except for the very low bids of reluctant players, the game 1 bids jump in increments of $5 or more. The lack of bids in amounts of $17 and $22, for example, might lead us to question the continuity axiom. It is apparent that some game 1 bidders—those whose bids were at least $20 (bidders 25 to 31)—thought seriously about the possibility of playing the lottery. With the exception of the highest bidder (who was willing to accept an *EMV* of zero), all showed fairly strong risk aversion. This type of result was expected.

The game 2 bids are more interesting but not so much for the degree of risk aversion shown, which was also expected. Rather, it is interesting to compare the two bids made by the same individual. For example, bidder 27 bid $25 for game 1 and $0.5 for game 2. The $0.5 bid for game 2 is equal to the first payoff in *EMV* terms, so the individual either reflects an unusual utility function or has difficulties assessing probabilities and *EMV* and *EU.* Similar low bids for game 2 were made by bidders 25 and 26. Bidders 18 and 22 offered unusually large sums to play game 2—$12 and $20, respectively.

Such difficulties in assessing *EMV* and *EU,* with resulting inconsistencies, are likely to be experienced by most individuals in society. Recall that the bids

**Table 9.1** ***Results of Bids for Two Lottery Games***

| Bidder | Game 1 Bid | Game 2 Bid |
|---|---|---|
| 1–5 | $1 | $1 |
| 6 | 1 | 2 |
| 7 | 1.5 | 1 |
| 8 | 2 | 1 |
| 9, 10 | 2 | 2 |
| 11 | 5 | 1 |
| 12, 13 | 5 | 2 |
| 14–17 | 5 | 5 |
| 18 | 5 | 12 |
| 19 | 10 | 2 |
| 20 | 10 | 4 |
| 21 | 10 | 5 |
| 22 | 10 | 20 |
| 23 | 15 | 2 |
| 24 | 15 | 4 |
| 25, 26 | 20 | 1 |
| 27 | 25 | 0.5 |
| 28 | 25 | 2.5 |
| 29 | 25 | 4 |
| 30 | 40 | 3 |
| 31 | 50 | 4 |

were made by engineering students with some formal training in probability and statistics. Experiments conducted elsewhere show similar inconsistencies [11, 19]. Thus, the application of utility theory must be performed with great care and caution.

## 9.5 MEAN–VARIANCE ANALYSIS

The *EMV* and *EU* approaches are based on probabilistic expectation over the range of possible outcomes of a risky option. In this section we present arguments for methods that are operationally different but are still based on utility concepts. These operational methods are, in general, more popular and easier to use. Therefore, a theoretical justification is attractive from a modeling point of view. We outline the main arguments and refer the interested reader to detailed sources.

### 9.5.1 Indifference Curves

Take the view of an investor with a quadratic utility function, as in Eq. 9.3, facing a set of alternative lotteries,

$$\{(p, 0), (1 - p, \$X)\}$$

**Table 9.2** ***Lotteries Toward Which an Individual Might Be Indifferent***

| $p$ | $1-p$ | $X$ | $E(X)$ | $Var(X)$, $10^6$ |
|---|---|---|---|---|
| 0 | 1.0 | \$10,000 | \$10,000 | 0 |
| 0.4375 | 0.5625 | 20,000 | 11,250 | 98.4 |
| 0.5714 | 0.4286 | 30,000 | 12,857 | 220.4 |
| 0.6250 | 0.3750 | 40,000 | 15,000 | 375.0 |
| 0.6400 | 0.3600 | 50,000 | 18,000 | 576.0 |

NOTES: Lotteries are of type

$$\{(p, 0), (1-p, \$X)\}$$

2. Utility function is

$$u(W) = W - (0.00001)W^2, \qquad W \leq 50{,}000$$

3. All lotteries have the same $CE$ = \$10,000.

ith $X$ in the range \$10,000 to \$50,000. Table 9.2 shows the lotteries, along with $E(X)$ and $Var(X)$. The $Var(X)$ is the second moment about the mean. It is equal to $E(X^2) - [E(X)]^2$. (See Section 10.2.1 for a more detailed explanation.)

These $E(X)$ and $Var(X)$ values are plotted as curve $U_1$ in Figure 9.4. The lotteries in Table 9.2 have been constructed so that all have a $CE$ of \$10,000; each lottery has the same utility value, and the individual with utility $W - (0.00001)W^2$ would view them indifferently. Curve $U_1$ in Figure 9.4 can thus be interpreted as an indifference function relating $E(X)$ and $Var(X)$. Each combination of $E(X)$, $Var(X)$ on curve $U_1$ has the same utility value.

We could construct other sets of lotteries in which all in a set would have the same utility value. The result would be a family of curves $U_1, U_2, U_3, U_4, \ldots$, one curve corresponding to each set of lotteries. Higher curves represent higher utility values.

Points $A$ and $B$ on curve $U_3$ are valued the same by the individual. A point like $C$ or $D$ that is not on the same curve does not have the same utility as point $A$. Point $C$ is considered less desirable than point $A$ because it has the same $E(X)$ but a higher $Var(X)$. On the other hand, point $D$ is preferred to point $A$ because for the same $Var(X)$ it has a higher $E(X)$. Point $B$ is preferred to point $C$ because of higher $E(X)$ *and* lower $Var(X)$, but by the same reasoning point $D$ is preferred to point $B$. These preference rules are specified in greater detail in Chapter 11. A formal analysis [22] along these lines shows that the mean–variance approach is justified when the investor's utility function is quadratic and the probability distributions of $X$ can be characterized by only two parameters (e.g., normal, lognormal).

### 9.5.2 Coefficient of Risk Aversion

We may observe some characteristics of the utility curves in Figure 9.4. First, the intersection point of a curve with the vertical $E(X)$ axis represents the

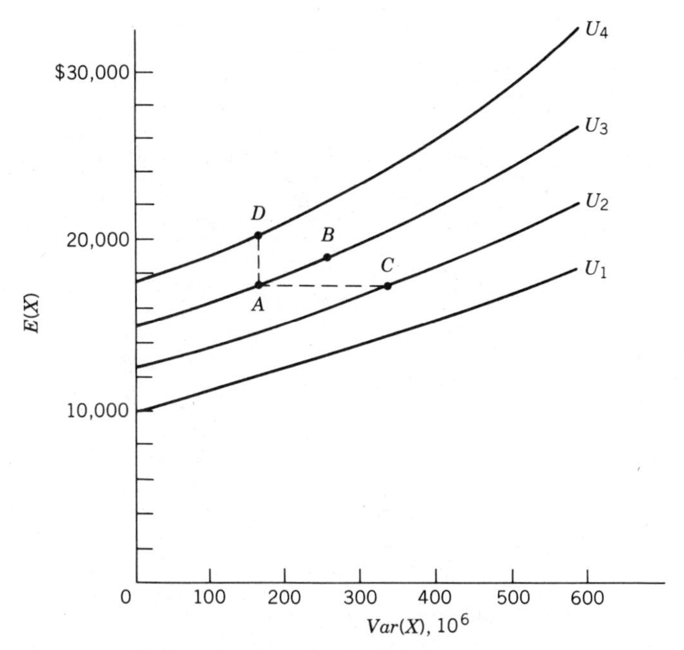


**FIGURE 9.4. Utility indifference curves relating $E(X)$ and $Var(X)$.**

*certainty equivalent* for all the points on that curve. Since such an intersection point has zero $Var(X)$, the cash outcome is certain. Second, the curves have positive slope. This reflects the fact that utility is an *increasing* function of $E(X)$ and a *decreasing* function of $Var(X)$. Third, the curves are concave. One way to explain the concavity of the indifference curves is that as risk increases, much larger increases in $E(X)$ are necessary to maintain the same level of utility for risk-averse individuals.

An approximation to the set of curves in Figure 9.4 might appear as in Figure 9.5. Here all the utility curves are linear and parallel. In Figure 9.5 we can obtain the $CE$ of any point, such as Point $D$, as follows.

$$CE_D = E(D) - \lambda \, Var(D) \tag{9.24}$$

The value $\lambda$ is called the *coefficient of risk aversion* (or sometimes the *risk aversion factor*). It measures the trade-off between $E(X)$ and $Var(X)$. This means that a $CE$ is easier to calculate when $\lambda$ is known.

Even if the linear approximation in Figure 9.5 is not appropriate, we can define $\lambda$ as the *tangent* to a utility indifference curve in Figure 9.4. The value of the coefficient of risk aversion is then reasonably valid over a restricted interval. For known functional utility forms, expressions for $\lambda$ as a function of the cash outcomes can be developed [9]. In practice, if we are not confident with assuming a single value of $\lambda$, then $\lambda$ is varied parametrically (see Appendix 11A).

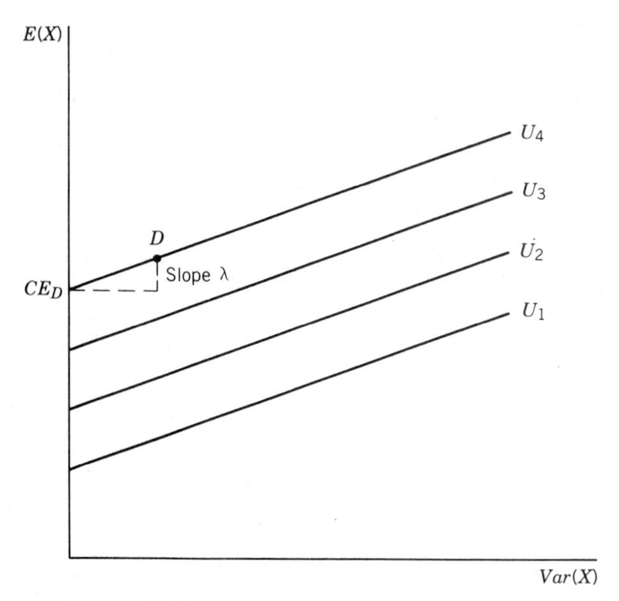


**FIGURE 9.5.** Approximation of indifference curves in Figure 9.4.

### 9.5.3 Justification of Certainty Equivalent Method

Applying Eq. 9.24 to a periodic cash flow $F_n$, which may be a random variable, we have

$$V_n = E(F_n) - \lambda \, Var(F_n) \tag{9.25}$$

For a series of cash flows from a project, we have in the simplest case, where λ is time-invariant and the $F_n$ are independent random variables,

$$\begin{aligned} PV(i) &= \sum_{n=0}^{N} \frac{V_n}{(1+i)^n} \\ &= \sum_{n=0}^{N} \frac{E(F_n) - \lambda \, Var(F_n)}{(1+i)^n} \end{aligned} \tag{9.26}$$

Here the interest rate $i$ is a *risk-free* rate, which accounts for only the time value of money. This risk-free rate can be viewed as a rate at which the individual can always invest money in some risk-free projects (such as a short-term government bond). This is the amount forgone if the project is undertaken and a net income is received from the risk-free project. Thus, having a present value of the certainty equivalents greater than zero means that the project is acceptable to this investor.

### *Example 9.5*

To illustrate the procedures involved in calculating the present value of certainty equivalents, let us examine a 5-year project with $E(F_n)$ and $Var(F_n)$ as shown in the tabulation. We assume that the λ value is known to be 0.02 for this investor. Then the certainty equivalents for the periodic random cash flows $F_n$ are

| $n$ | $E(F_n)$ | $Var(F_n)$ | $V_n$ | $PV(10\%)$ |
|---|---|---|---|---|
| 0 | −400 | 400 | −408 | −408.00 |
| 1 | 120 | 100 | 118 | 107.27 |
| 2 | 120 | 225 | 115.5 | 95.45 |
| 3 | 120 | 400 | 112 | 84.15 |
| 4 | 110 | 900 | 92 | 62.84 |
| 5 | 120 | 2500 | 70 | 43.46 |
| | | | $\sum$ = | −$14.83 |

Since the total present value of the certainty equivalents is negative, the investor would reject the project. □

Returning to the general case, let us assume that the utility function for cash flows distributed over time is

$$u = \sum_{n=0}^{N} c_n u_n \tag{9.27}$$

where $u_n$ is a utility function for the random cash flow $F_n$ occurring at time $n$ and $c_n$ is a constant. This expression implies that contributions to total utility are additive over time, and periodic utility values are multiplied by the constant $c_n$ to adjust for the time preference of the events $F_n$. The exact form of the periodic utility functions $u_n$ is not specified. In fact, $u_n$ could be different functions over time or, in the simplest case, time-invariant. For our discussion, let us assume that $u_n = u_1$ for all $n$. A Taylor expansion can be used to generate a reasonable approximation to an expected utility function [20].

$$E(u_n) = u_n[E(F_n)] + u_n^{(2)}[E(F_n)]\ Var(F_n)/2 \tag{9.28}$$

This expression is obtained by adopting "sufficient approximation" reasoning to justify ignoring the higher moments about the mean of the cash flow in the Taylor series. If the utility function is a quadratic, however, any term $u_n^{(n)}$ ($n$th derivative of $u_n$) with $n > 3$ will be zero. Thus Eq. 9.28 becomes the exact expression of the expected utility measure. Further, the term $u_n^{(2)}$ becomes a constant for the quadratic utility function. Thus, rewriting Eq. 9.28 gives us the expression

$$E(u_n) = u_n[E(F_n)] + A_n\ Var(F_n) \tag{9.29}$$

where

$$A_n = u_n^{(2)}[E(F_n)]/2 \tag{9.30}$$

Returning to the total utility function given in Eq. 9.27 and taking the expected value of each side of the equation, we obtain

$$E(u) = \sum_{n=0}^{N} c_n E(u_n) \tag{9.31}$$

Substituting Eq. 9.29 into Eq. 9.31 yields

$$E(u) = \sum_{n=0}^{N} c_n u_n \,[E(F_n)] + \sum_{n=0}^{N} c_n A_n \, Var(F_n) \tag{9.32}$$

If a certainty equivalent can be found for each time period so that

$$u_n(V_n) = u_n[E(F_n)] + A_n \, Var(F_n) \tag{9.33}$$

the present value of this set of certainty equivalents will be equal to the expected utility of the cash flows from the investment project by letting $c_n = 1/(1 + i)^n$ [20].

## 9.6 SUMMARY

Utility theory is a very important concept because it helps to reconcile real behavior with expected monetary value in decision making. The typical individual has a concave utility function, reflecting an aversion to risk, which is usually measured by the variance of the cash flow. The axioms of utility theory can be used to derive the Bernoulli hypothesis of expected utility maximization. Validation experiments reveal, however, that this hypothesis is not perfectly true.

Operationally, the utility indifference curves that relate $E(X)$ and $Var(X)$ provide the theoretical basis for the popular mean–variance analysis presented in Chapter 11. The coefficient of risk aversion, heavily used in portfolio analysis, is the slope of the indifference curve. Finally, the discounted sum of certainty equivalents is shown to be an approximation (exact for quadratic utility) to the expected utility of a random future cash flow stream. All these results will be used in later chapters.

## REFERENCES

1. Becker, J., and R. K. Sarin, "Lottery Dependent Utility," *Management Science,* Vol. 33, No. 11, pp. 1367–1382, 1987.
2. Bernhard, R. H., "Risk-Adjusted Values, Timing of Uncertainty Resolution, and the Measurement of Project Worth," *Journal of Financial and Quantitative Analysis,* Vol. 19, No. 1, pp. 83–99, 1984.
3. Bernoulli, D., "Exposition of a New Theory of the Measurement of Risk," *Econometrica,* Vol. 22, No. 1, pp. 23–36, 1954. (Accessible translation of "Specimen Theoriae Novae de Mensura Sortis," 1738.)

4. BROCKETT, P. L., and L. L. GOLDEN, "A Class of Utility Functions Containing All the Common Utility Functions," *Management Science,* Vol. 33, No. 8, pp. 955–964, 1987.
5. CURRIM, I. S., and R. K. SARIN, "Prospect Versus Utility," *Management Science,* Vol. 35, No. 1, pp. 22–41, 1989.
6. EDWARDS, E., "The Theory of Decision Making," *Psychological Bulletin,* Vol. 51, No. 4, pp. 380–417, 1954.
7. FRIEDMAN, M., and L. J. SAVAGE, "The Utility Analysis of Choices Involving Risk," *Journal of Political Economy,* Vol. 56, No. 4, pp. 279–304, 1948.
8. HIRSHLEIFER, J., "Investment Decision under Uncertainty: Choice-Theoretic Approaches," *Quarterly Journal of Economics,* Vol. 79, No. 4, pp. 509–536, 1965.
9. JEAN, W. H., *The Analytical Theory of Finance,* Holt, Rinehart and Winston, New York, 1970.
10. JOHNSON, W., *Capital Budgeting,* Wadsworth, Belmont, Calif., 1970, Ch. 5.
11. KAHNEMAN, D., and A. TVERSKY, "Prospect Theory: An Analysis of Decision under Risk," *Econometrica,* Vol. 47, pp. 263–291, 1979.
12. KEENEY, R. L., and H. RAIFFA, *Decisions with Multiple Objectives; Preferences and Value Tradeoffs,* Wiley, New York, 1976.
13. KELLER, L. R., "Testing of the 'Reduction of Compound Alternatives' Principle," *OMEGA, International Journal of Management Science,* Vol. 13, No. 4, pp. 349–358, 1985.
14. LAVALLE, I. H., and P. C. FISHBURN, "Decision Analysis under States-Additive SSB Preferences," *Operations Research,* Vol. 35, No. 5, pp. 722–735, 1987.
15. LEVY, H., and M. SARNAT, *Portfolio and Investment Selection: Theory and Practice,* Prentice–Hall, Englewood Cliffs, N.J., 1984.
16. LUCE, D. R., and H. RAIFFA, *Games and Decisions: Introduction and Critical Survey,* Wiley, New York, 1957.
17. MACHINA, M. J., "A Stronger Characterization of Declining Risk Aversion," *Econometrica,* Vol. 50, No. 4, pp. 1069–1079, 1982.
18. MACHINA, M. J., "Decision-Making in the Presence of Risk," *Science,* Vol. 236, pp. 537–543, 1 May 1987.
19. MCCORD, M., and R. DE NEUFVILLE, "'Lottery Equivalents' Reduction of the Certainty Effect Problem in Utility Assessment," *Management Science,* Vol. 32, No. 1, pp. 56–61, 1986.
20. NEUMANN, J. V., and O. MORGENSTERN, *Theory of Games and Economic Behavior,* 2nd edition, Princeton University Press, Princeton, N.J., 1947.
21. PRATT, J. W., "Risk Aversion in the Small and in the Large," *Econometrica,* Vol. 32, No. 1–2, pp. 122–136, 1964.
22. TOBIN, J., "Liquidity Preference as Behavior toward Risk," *Review of Economic Studies,* No. 67, pp. 65–85, February 1958.

## PROBLEMS

**9.1.** Consider the homeowner in Section 9.1.1 with the utility function given by Eq. 9.1. If the deductible amount on a loss is higher than $250, the homeowner might prefer not to buy fire insurance, on an *EU* basis. Using the data in Section 9.1.1 for other factors, determine the deductible amount that would make the homeowner *indifferent* about choosing between buying and not buying insurance, on an *EU* basis.

**9.2.** For an individual with zero initial wealth and a utility function

$$u(W) = 1 - e^{-0.0001W}$$

find the *CE* for each of the following alternatives (probabilities of the outcomes are given).

| | Cash Amount | | | | |
|---|---|---|---|---|---|
| Alternative | −$10,000 | 0 | $10,000 | $20,000 | $30,000 |
| 1 | 0.1 | 0.2 | 0.4 | 0.2 | 0.1 |
| 2 | 0.1 | 0.2 | 0.3 | 0.3 | 0.1 |
| 3 | 0 | 0.3 | 0.4 | 0 | 0.3 |
| 4 | 0 | 0.15 | 0.65 | 0 | 0.2 |
| 5 | 0.5 | 0 | 0 | 0 | 0.5 |

**9.3.** Solve Example 9.2 for the situation in which the individual's initial wealth is $20,000. Would you expect the probability to change as the initial wealth changes?

**9.4.** Consider a three-stage lottery. In the first stage there are a 0.2 chance of receiving $1,000 and a 0.8 chance of going on to stage 2. In stage 2 there are a 0.5 chance of receiving $2,000 and a 0.5 chance of going on to stage 3. In stage 3 there are a 0.2 chance of receiving $1,000, a 0.3 chance of receiving $2,000, and a 0.5 chance of receiving $5,000. Reduce this three-stage lottery to an equivalent one-stage lottery.

**9.5.** Construct a compound lottery and reduce it to its equivalent one-stage lottery.

**9.6.** Obtain information about a lottery. Calculate the *EMV* of the act of purchasing a ticket.

**9.7.** Derive the *CE* for an individual with initial wealth $20,000 and a quadratic utility function as given by Eq. 9.3, when facing the lottery {(0.5, $10,000), (0.5, $20,000)}. There is no cost for the lottery. Show all computations.

**9.8.** Can you specify a risk-seeking utility function with decreasing risk aversion? With constant risk aversion? With increasing risk aversion?

**9.9.** Conduct a lottery game of the type described in Section 9.4.2. Analyze the results for consistency.

**9.10.** Construct a set of lotteries, each with the same *CE* and similar to the ones in Table 9.2, to derive one of the higher utility curves in Figure 9.4.

**9.11.** Construct a set of lotteries, each with the same *CE* and similar to the ones in Table 9.2, but using the utility function given by Eq. 9.1. What is the shape of the indifference curve?

**9.12.** Using the worksheet provided, develop your utility function. In doing so, consider the following steps.

**Step 1:** Find the certainty equivalent amount *B* for a given lottery (*A* or zero with 0.5 probability each). Once the amounts *A* and *B* are specified, find the certainty equivalent amount *C* for a new lottery (*B* or zero with 0.5 probability each). Continue this procedure for the remaining lotteries. You are likely to find some inconsistencies in the certainty equivalent amounts assessed. Resolve these inconsistencies by reassessing the certainty equivalent amounts.

**Step 2:** Scale the certainty equivalent amounts (*A* through *J*) as a percentage of *A*. For example, if $A = \$1,000$ and $B = \$300$, then $A = 100\%$ of *A* and $B = 30\%$ of *A*.

**Step 3:** Plot the scaling preferences on the chart provided and smooth the curve when connecting the points plotted.

## WORKSHEET FOR DETERMINING THE UTILITY FUNCTION

*Certainty Equivalent*

| | | | |
|---|---|---|---|
| 1 | $A$ ______ | or zero | vs. $B$ ______ |
| 2 | $B$ ______ | or zero | vs. $C$ ______ |
| 3 | $C$ ______ | or zero | vs. $D$ ______ |
| 4 | $A$ ______ | or $E$ ______ | vs. zero |
| 5 | $E$ ______ | or zero | vs. $F$ ______ |
| 6 | $F$ ______ | or zero | vs. $G$ ______ |
| 7 | $A$ ______ | or $F$ ______ | vs. $H$ ______ |
| 8 | $C$ ______ | or $E$ ______ | vs. $J$ ______ |

*Scaling Preference*

| | Amount | %A | $U$ |
|---|---|---|---|
| $A$ | | | +8 |
| $B$ | | | +4 |
| $C$ | | | +2 |
| $D$ | | | +1 |
| $E$ | | | −8 |
| $F$ | | | −4 |
| $G$ | | | −2 |
| $H$ | | | +2 |
| $J$ | | | −3 |

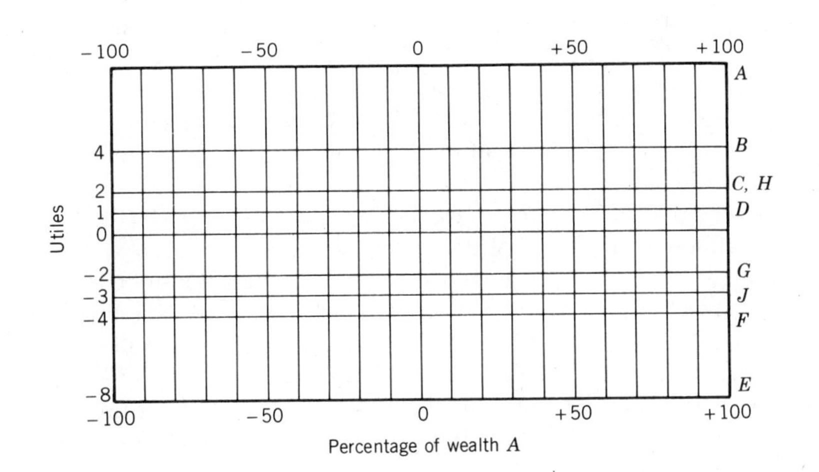

# 10

# Measures of Investment Worth under Risk—Single Project

## 10.1 INTRODUCTION

As we discussed in Chapter 9, the expected utility criterion provides an elegant and theoretically impeccable solution to the problem of investment under risk. Our general approach to investment is that "return" is to be sought and "risk" to be avoided, so utility will be an increasing function of return and a decreasing function of risk. Since risk cannot always be avoided entirely for a project with random cash flows, it was necessary to consider the trade-off between the characteristics of a project, by defining an indifference curve based on some assumptions about the investor's utility function.

In Chapter 9 we also assumed investment situations in which a utility function is known to the management, or the management is willing to develop or approximate a utility function. The approaches in this chapter will all be based on the use of probabilistic monetary values. That is, we will approach the project evaluation from the practical view that the management would have neither the time nor the motivation to participate in the task of constructing a utility function.

As seen in Part One, the principal determinants of project worth in capital budgeting models are the cash flow streams. In Part Two we assumed that management makes its investment decisions under conditions of complete certainty. However, cash flows are projected at the time the investment is first proposed, and future cash flows are subject to deviation from their expected values. Thus, probable variations in the outcomes of future events are of primary concern to most decision makers in the evaluation of investment proposals. We will define the possible variation from subjective estimates as "project risk" in investment analysis.

### 10.1.1 The Common Measures of Project Risk

The concept of risk most widely used in project evaluation is the variability of return, which is measured by variance (or standard deviation). That is, the more an investment's return varies about the expected return, the larger the investor's risk. The use of variance as a measure of risk implies that deviations below the expected value are regarded in the same way as deviations above the expected value. Even though this measure has been criticized as too conservative, since it regards all extreme returns (positive or negative) as undesirable, variance is still a popular measure of risk because of its familiarity and ease of computation [22]. The mathematical definition of this measure will be given in Section 10.2.

As an alternative, semivariance (a similar approach to risk in problems of this type) has the advantage of focusing on reduction of losses, that is, variability in negative return. When this measure is used in capital budgeting problems (for example, investment portfolio selection), full knowledge of the joint probability distribution about the projects' investment returns is required to calculate the mean, variance, and semivariance of alternatives. Unfortunately, when the decision maker is faced with a set of alternatives and each project has a large number of outcomes (for a discrete case), the development of such a distribution is usually impractical. Since the concept of semivariance is more relevant to comparing risky alternative proposals with asymmetric return distributions, its mathematical definition will be given in Section 11.2.2.

Another partial-domain approach to risk measurement is the development of Gaussian linear loss integrals [32]. The original use of loss integrals was primarily limited to Gaussian (normal) distributions and linear or quadratic relationships between the random variable and the decision criterion, but more recent developments provide a variety of extensions, including partial means [3, 37]. These partial-domain approaches describe the extent of exposure to undesired consequences (downside risk) but require complete knowledge of the probability distribution of the random variable of interest.

Another plausible measure of risk in the capital budgeting literature is the probability-of-loss criterion [2]. This measure, along with some variants of it, has become known as the safety-first rule. If risk is defined as the chance of loss, risk may be measured by the area of a probability distribution that lies below the point of profitability, the critical level. In other words, this measure treats only unfavorable returns as "loss." For example, a project with a large variation of profit may have no possibility whatever of loss. In fact, such a project would be viewed as risk-free by those who use this criterion, no matter how great the variability of potential outcomes is. For computing the probability of loss or the expected loss, however, a complete knowledge of joint probability distributions for investment proposals is required. These distributions can be difficult to obtain, and thus the measure has a rather limited application in practice. This measure may be more appropriate as a supplementary measure of risk in comparing projects. The conceptual approaches of the three common measures of risk are summarized in Figure 10.1.

### 10.1.2 How Business People Perceive Risk in Project Evaluation

There is a considerable disparity between the definitions of risk that are applied by people in business and those that are recommended by academi-

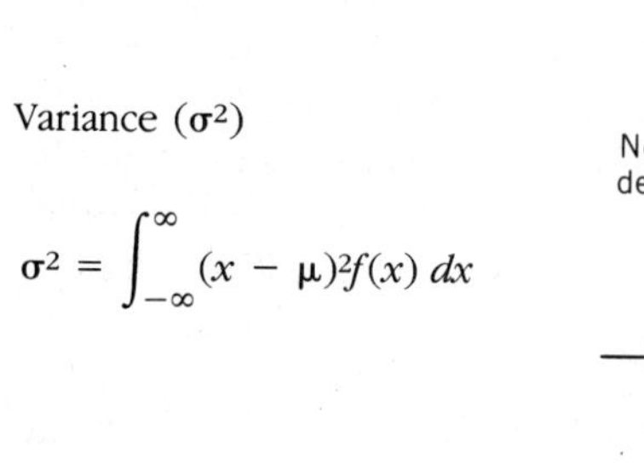

Variance ($\sigma^2$)

$$\sigma^2 = \int_{-\infty}^{\infty} (x - \mu)^2 f(x)\, dx$$

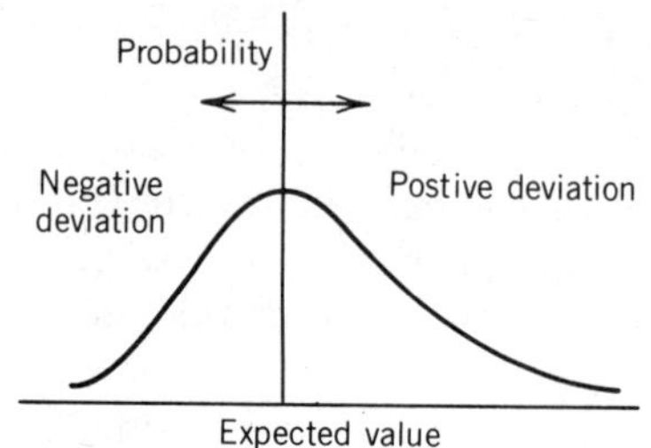


Semivariance ($S_h$)

$$S_h = \int_{-\infty}^{h} (h - x)^2 f(x)\, dx$$

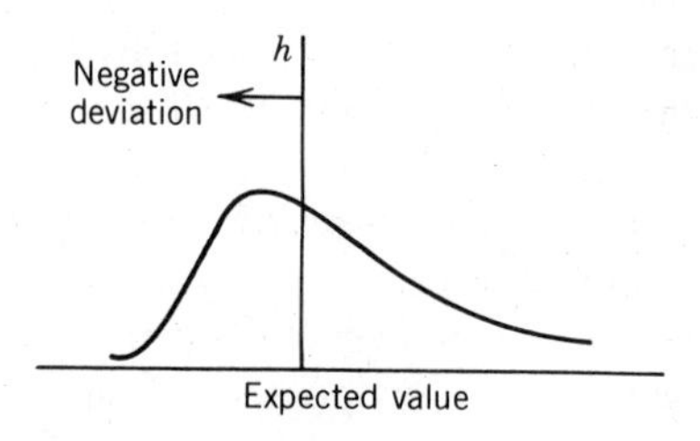


Probability-of-Loss criterion

$$L = \int_{-\infty}^{h} f(x)\, dx$$

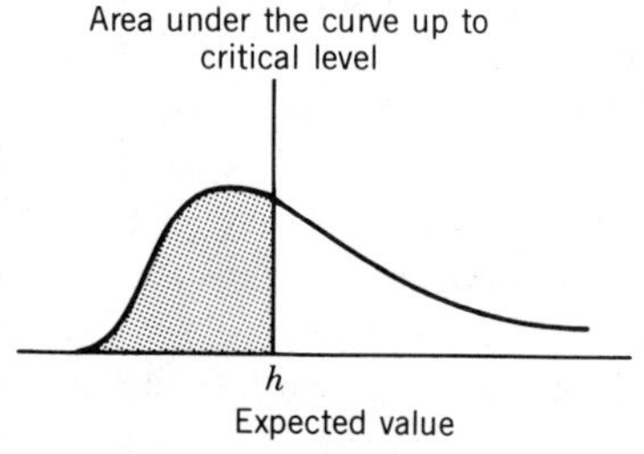


**FIGURE 10.1. Statistical measures of risk.**

cians. A survey by Petty, Scott, and Bird [30] indicates that people in business view risk as primarily associated with the probability of not achieving a target return. Almost 40% of the corporate executives interviewed described risk in this manner. In other words, management is more concerned about negative variation than about the total variation of possible investment outcomes. The second most popular definition of risk the survey found is related to variation in returns, which is equivalent to variance as a measure of risk. Only a limited number of decision makers continue to reply on the payback period as a definitive description of project riskiness. Nevertheless, it is evident that we need some sort of statistical measures to characterize the probabilistic information available in a project.

The purpose of this chapter, therefore, is twofold: (1) to extend the measure of investment worth to cover conditions of risk and (2) to demonstrate the applicability of this measure through the analysis of selected investment decisions. Before presenting analytical techniques for modeling cash flow streams under risk, we will review some fundamental methodologies for handling probabilistic information.

## 10.2 ESTIMATING VALUES IN PROBABILISTIC TERMS

This section reviews the analytic techniques that enable us to estimate the value of a variable that results from a combination of several random variables. We will consider combinations of just two random variables for most of the discussions, but if we can handle two at a time, we can easily expand our capability to any number of variables.

Random variables will be denoted with capital letters and particular values of those variables will be denoted with lowercase letters. When a random variable, say $X$, follows some probability distribution, its probability density and its cumulative distribution function are denoted respectively by $f(x)$ and $F(x)$ for a continuous case, and its probability mass and its cumulative distribution function by $p(x)$ and $P(x)$ for a discrete case.

### 10.2.1 Statistical Moments of a Single Random Variable

The moments of a distribution are the expected values of the powers of the random variable that has that distribution. In general, the $k$th moment of a random variable $X$ about a real number $C$ is defined as

$$M_k(X) = E(X - C)^k = \begin{cases} \sum_x (x - C)^k p(x) & \text{for } x \text{ discrete} \\ \int_x (x - C)^k f(x)\, dx & \text{for } x \text{ continuous} \end{cases} \quad (10.1)$$

The moments of interest in economic analysis are those about the origin ($C = 0$) and about the mean ($C = \mu$), typically for $k = 1, 2, 3$, and 4. The $k$th moments about the origin and about the mean (central moment) are denoted, respectively, by $E(X^k)$ and $\mu_k$.

$$\mu_k = E(X - \mu)^k = \begin{cases} \sum_x (x - \mu)^k p(x) & \text{for } x \text{ discrete} \\ \int_x (x - \mu)^k f(x)\, dx & \text{for } x \text{ continuous} \end{cases} \quad (10.2)$$

where $\mu = E[X]$. It can easily be shown through a binomial expansion of Eq. 10.2 that

$$\begin{aligned} \mu_1 &= E(X - \mu) = 0 \\ \mu_2 &= E(X^2) - \mu^2 = \sigma^2 \\ \mu_3 &= E(X^3) - 3\mu E(X^2) + 2\mu^3 \\ \mu_4 &= E(X^4) - 4\mu E(X^3) + 6\mu^2 E(X^2) - 3\mu^4 \end{aligned} \quad (10.3)$$

The first moment about the origin $[E(X) = \mu]$ is simply the mean, whereas the second moment about the mean is the variance ($\sigma^2$) of $X$, written as $Var(X)$.

The first moment is a measure of central tendency, a theoretical number that might never be achieved in a practical experiment. We do anticipate, however, that the average value of the random variable $X$ in a great number of trials will be somewhere near the expected value of $X$. The variance of a distribution, which is always nonnegative, is a measure of its spread or dispersion from the mean. If $f(x)$ is a curve that is sharply peaked at $\mu$, the variance is smaller than it would be were the curve flattened out. The term *standard deviation* refers to the square root of the variance,

$$\sigma = \sqrt{E(X - \mu)^2}$$

The standardized third and fourth moments of $X$ are denoted by $\alpha_3$ and $\alpha_4$, where

$$\alpha_3 = \frac{\mu_3}{\sigma^3}, \quad \text{and} \quad \alpha_4 = \frac{\mu_4}{\sigma^4} \tag{10.4}$$

Both $\mu_3$ and $\alpha_3$ measure the skewness of $X$. When $\mu_3$ (or $\alpha_3$) is greater than zero, the distribution of $X$ is skewed to the right (or positively skewed), and when $\mu_3$ (or $\alpha_3$) is less than zero, the distribution is skewed to the left (or negatively skewed). For a symmetrical distribution (such as the normal and uniform distributions), $\mu_3 = 0$ (or $\alpha_3 = 0$).

Both $\mu_4$ and $\alpha_4$ also measure the kurtosis of a distribution. Kurtosis refers to the peakedness in the middle and the thickness at the tails of a distribution. Curves such as the normal distribution, for which $\alpha_4 = 3$, are called mesokurtic. Those for which $\alpha_4 > 3$ are called leptokurtic; those for which $\alpha_4 < 3$ are called platykurtic. Tables 10.1 and 10.2 summarize the first four moments of some standard probability distributions.

***Moment-Generating Function.*** The usual definition of the moment generating function of the random variable $X$ is

$$M(\theta) = E(e^{\theta x}) = \begin{cases} \int e^{\theta x} f(x)\, dx, & \text{for } x \text{ continuous} \\ \sum e^{\theta x_j} p(x_j), & \text{for } x \text{ discrete} \end{cases} \tag{10.5}$$

The first derivative of $M(\theta)$ with respect to $\theta$ (for the continuous case) is

$$M^{(1)}(\theta) = \int x e^{\theta x} f(x)\, dx$$

Evaluation of this derivative at $\theta = 0$ yields

$$M^{(1)}(0) = \int x f(x)\, dx = E(x)$$

**Table 10.1** ***Moment Generating Functions for Discrete Probability Functions***

| Probability Density Function | Moment-Generating Function | Moments |
|---|---|---|
| Poisson distribution<br>$p(x) = \frac{\lambda^x(e^{-\lambda})}{x!}$<br>where $x = 0, 1, \ldots$ | $\exp[\lambda(e^\theta - 1)]$ | $\mu = \lambda$<br>$\mu_2 = \lambda$<br>$\mu_3 = \lambda$<br>$\mu_4 = 3\lambda^2 + \lambda$ |
| Geometric distribution<br>$p(x) = pq^{x-1}$<br>where $x = 1, 2, 3, \ldots$ | $\frac{pe^\theta}{1 - qe^\theta}$ | $\mu = \frac{1}{p}$<br>$\mu_2 = \frac{q}{p^2}$<br>$\mu_3 = \frac{(1-p)(2-p)}{p^3}$<br>$\mu_4 = \frac{(1-p)(1+9q)}{p^4}$ |
| Binomial distribution<br>$p(x) = \binom{k}{x} p^x q^{k-x}$<br>where $x = 0, 1, \ldots k$ | $(pe^\theta + q)^k$ | $\mu = kp$<br>$\mu_2 = kp(1-p)$<br>$\mu_3 = kp(1-p)(1-2p)$<br>$\mu_4 = kp(1-p)[(3p^2 - 3p)(2-k) + 1]$ |

Continuing this process through $k$ derivatives, we obtain the general expression for the $k$th moment about zero (not the central moment).

$$E(x^k) = M^{(k)}(0) \tag{10.6}$$

Thus, if we know the moment-generating function for a certain random variable, we can easily evaluate its various moments without carrying out the integration implied by the definition given in Eq. 10.1.

## *Example 10.1*

Let $X$ have a negative exponential probability density function, $f(x) = (1/b)e^{-x/b}$, for $x > 0$. The moment-generating function is

$$M(\theta) = \int_0^\infty e^{\theta x} \frac{1}{b} e^{-x/b}\, dx = (1 - b\theta)^{-1}$$

provided that $b\theta < 1$. The mean of the random variable $X$ from the moment-generating function is computed by evaluating the first derivative of $M(\theta)$ at $\theta = 0$.

$$M^{(1)}(0) = \left. \frac{b}{(1 - b\theta)^2} \right|_{\theta=0} = b$$

**Table 10.2** ***Moment-Generating Functions for Continuous Probability Functions***

| Probability Density Function | Moment-Generating Function | Moments |
|---|---|---|
| Uniform distribution<br>$\begin{cases} \dfrac{1}{b-a}, & a \le x \le b \\ 0, & \text{elsewhere} \end{cases}$ | $\dfrac{e^{\theta b} - e^{\theta a}}{(b-a)\theta}$ | $\mu = \dfrac{b+a}{2}$<br>$\mu_2 = \dfrac{(b-a)^2}{12}$<br>$\mu_3 = 0$<br>$\mu_4 = \dfrac{(b-a)^4}{80}$ |
| Exponential distribution<br>$\begin{cases} ae^{-ax}, & x > 0 \\ 0, & \text{elsewhere} \end{cases}$ | $\left(1 - \dfrac{\theta}{a}\right)^{-1}$ | $\mu = 1/a$<br>$\mu_2 = (1/a)^2$<br>$\mu_3 = 2(1/a)^3$<br>$\mu_4 = 9(1/a)^4$ |
| Beta distribution (type I)<br>$\begin{cases} \dfrac{\Gamma(a+b)}{\Gamma(a)\Gamma(b)} x^{a-1}(1-x)^{b-1}, & 0 \le x \le 1, a > 0, b > 0 \\ 0, & \text{elsewhere} \end{cases}$ | $\dfrac{\Gamma(a+b)}{\Gamma(a)} \cdot \sum_{k=0}^{\infty} \dfrac{\theta^k \Gamma(a+k)}{k!\Gamma(a+b+k)}$ | $\mu = \dfrac{a}{a+b}$<br>$\mu_2 = \dfrac{ab}{(a+b)^2(a+b+1)}$<br>$\mu_3 = \dfrac{2ab(b-a)}{(a+b)^3(a+b+1)(a+b+2)}$<br>$\mu_4 = \dfrac{3ab(a^2b + ab^2 + 2a^2 + 2b^2 - 2ab)}{(a+b)^4(a+b+1)(a+b+2)(a+b+3)}$ |
| Beta distribution (type II)<br>$\dfrac{\Gamma(a+b+2)}{\Gamma(a+1)\Gamma(b+1)} x^a(1-x)^b,$<br>$0 \le x \le 1, a > -1, b > -1$ | The MGF* and $\mu_j$'s are obtained by replacing $a$ and $b$ in the type I beta formula by $a + 1$ and $b + 1$. Type I and type II beta PDFs* are used interchangeably. | |

## Table 10.2 *(continued)*

| Probability Density Function | Moment-Generating Function | Moments |
|---|---|---|
| Symmetrical triangular distribution, $a > 0$ <br> $\begin{cases} \dfrac{x+a}{a^2}, & -a \le x \le 0 \\ \dfrac{a-x}{a^2}, & 0 \le x \le a \\ 0, & \text{elsewhere} \end{cases}$ | $\dfrac{e^{a\theta} + e^{-a\theta} - 2}{a^2\theta^2}$ | $\mu = 0$ <br> $\mu_2 = \dfrac{a^2}{6}$ <br> $\mu_3 = 0$ <br> $\mu_4 = \dfrac{a^4}{15}$ |
| Asymmetrical triangular distribution, $a, b > 0$ <br> $\begin{cases} \dfrac{2(x+a)}{a(a+b)}, & -a \le x \le 0 \\ \dfrac{2(b-x)}{b(a+b)}, & 0 \le x < b \\ 0, & \text{elsewhere} \end{cases}$ | $\dfrac{2\left(\dfrac{e^{-a\theta} - 1}{a} + \dfrac{e^{b\theta} - 1}{b}\right)}{(a+b)\theta^2}$ | $\mu = \dfrac{b-a}{3}$ <br> $\mu_2 = \dfrac{a^2 + ab + b^2}{18}$ <br> $\mu_3 = \dfrac{2b^3 + 3b^2a - 3ba^2 - 2a^3}{270}$ <br> $\mu_4 = \dfrac{a^4 + 2a^3b + 3a^2b^2 + 2ab^3 + b^4}{135}$ |
| Normal distribution <br> $\dfrac{1}{\sigma\sqrt{2\pi}} e^{-(x-\mu)^2/2\sigma^2}, \quad \infty < x < \infty$ | $e^{\mu\theta + \sigma^2\theta^2/2}$ | $\mu = \mu$ <br> $\mu_2 = \sigma^2$ <br> $\mu_3 = 0$ <br> $\mu_4 = 3\sigma^4$ |
| Gamma distribution <br> $\dfrac{a^b}{(b-1)!} x^{b-1} e^{-ax},$ <br> $0 < x < \infty, a > 0, b > 0$ | $\left(\dfrac{a}{a-\theta}\right)^b$ | $\mu = \dfrac{b}{a}$ <br> $\mu_2 = \dfrac{b}{a^2}$ <br> $\mu_3 = \dfrac{2b}{a^3}$ <br> $\mu_4 = \dfrac{3b(b+2)}{a^4}$ |

*MGF, moment-generating function; PDF, probability density function.

### 10.2.2 Statistical Moments of Linear Combinations of Random Variables

In the development of cash flow models, it is often necessary to combine several random variables in the form of sums, differences, products, or quotients. Our main interest is in characterizing the probability distribution of such combinations. When the random variables involved are statistically independent, there are means of developing exact expressions for the distribution of simple sums from a knowledge of the component distributions. Exact expressions for distributions of products and quotients of independent random variables may be possible for some simple cases [9], but in general they are not easily developed. Nor are the distributions for combinations of correlated random variables easy to come by. In many of these cases, we must be satisfied with the statistical moments (rather than the exact probability distribution) to characterize the distributions. The development of estimates for such parameters is discussed in this section.

***Covariance and Coefficient of Correlation.*** When two random variables are not independent, we need some measure of their dependence on each other before we can aggregate the variables. The parameter that tells the degree to which two variables $(X_1, X_2)$ are related is the covariance $Cov(X_1, X_2)$, denoted by $\sigma_{12}$. We define

$$\begin{aligned} \sigma_{12} &= Cov(X_1, X_2) \\ &= E\{[X_1 - E(X_1)][X_2 - E(X_2)]\} \\ &= E(X_1 X_2) - E(X_1)E(X_2) \end{aligned} \tag{10.7}$$

Note that the covariance has the same basic dimensions as the variance of a single random variable. (In fact, if $X_1$ and $X_2$ are the same random variable, the covariance reduces to the variance.)

It should be clear that if $X_1$ tends to exceed its mean whenever $X_2$ exceeds its mean, $Cov(X_1, X_2)$ will be positive. If $X_1$ tends to fall below its mean whenever $X_2$ exceeds its mean, $Cov(X_1, X_2)$ will be negative. The sign of $Cov(X_1, X_2)$, therefore, reveals whether $X_1$ and $X_2$ vary directly or inversely with one another. The coefficient of correlation is defined by

$$\rho_{12} = \frac{Cov(X_1, X_2)}{\sigma_1 \sigma_2} \tag{10.8}$$

where $\sigma_1$ and $\sigma_2$ are the standard deviations of $X_1$ and $X_2$, respectively. The value of $\rho_{12}$ can vary within the range of $-1$ and $+1$, with $\rho_{12} = -1$ indicating perfect negative correlation, $\ell_{12} = +1$ indicating perfect positive correlation. The result $\rho_{12} = 0$ implies that $Cov(X_1, X_2) = 0$ and no correlation exists between $X_1$ and $X_2$.

$$Cov(X_1, X_2) = Cov(X_2, X_1) \tag{10.9a}$$

$$Cov(aX_1 + b, cX_2 + d) = ac\, Cov(X_1, X_2) \tag{10.9b}$$

$$Cov(X_1, -X_2) = -Cov(X_1, X_2) \tag{10.9c}$$
$$Cov(X_1, X_1) = Var(X_1) \tag{10.9d}$$
$$Cov(X_1, a) = 0 \quad \text{for every constant } a \tag{10.9e}$$
$$Cov(X_1 + X_2, Y) = Cov(X_1, Y) + Cov(X_2, Y) \tag{10.9f}$$
$$Cov\left(\sum_{k=1}^{m} a_k X_k, \sum_{j=1}^{n} b_j Y_j\right) = \sum_{k=1}^{m} \sum_{j=1}^{n} a_k b_j \, Cov(X_k, X_j) \tag{10.9g}$$

***Multiplication by a Constant.*** One of the operations most often performed on random variables is multiplying them by a constant. For example, each unit of demand generates $C$ dollars of profit. If the level of demand is a random variable $X$, the profit will also be a random variable, $Y = CX$, with a mean and variance of

$$E(Y) = E(CX) = \int_x Cxf(x)\, dx = CE(X) \tag{10.10}$$

$$\begin{aligned} Var(Y) &= E\{[CX - E(Y)]^2\} \\ &= C^2 E\{[X - E(X)]^2\} \\ &= C^2\, Var(X) \end{aligned} \tag{10.11}$$

***Sums of Random Variables.*** To evaluate the mean and variance of a sum of random variables, let us consider a simple case of two random variables. Assuming that their joint probability density function is known, the random variable $Y = X_1 + X_2$ has a mean of

$$\begin{aligned} E(Y) &= E(X_1 + X_2) \\ &= \int_{x_1} \int_{x_2} (x_1 + x_2) f(x_1, x_2)\, dx_1\, dx_2 \\ &= \int_{x_1} x_1 f_1(x_1)\, dx + \int_{x_2} x_2 f_2(x_2)\, dx_2 \\ &= E(X_1) + E(X_2) \end{aligned} \tag{10.12}$$

where $f_1$ and $f_2$ are the respective marginal probability functions, without regard to the relationship of $X_1$ and $X_2$. The variance of $Y$ can be obtained in a similar manner by calculating

$$\begin{aligned} Var(Y) &= E\{[Y - E(Y)]^2\} \\ &= E\{[X_1 + X_2 - E(X_1) - E(X_2)]^2\} \\ &= E\{[X_1 - E(X_1) + X_2 - E(X_2)]^2\} \\ &= E\{[X_1 - E(X_1)]^2 + [X_2 - E(X_2)]^2 + 2[X_1 - E(X_1)]\,[X_2 - E(X_2)]\} \\ &= \sigma_1^2 + \sigma_2^2 + 2\, Cov(X_1, X_2) \end{aligned} \tag{10.13}$$

If the correlation coefficient $\rho_{12}$ is available, this relationship may be expressed as

$$Var[Y] = \sigma_1^2 + \sigma_2^2 + 2\rho_{12}\sigma_1\sigma_2 \tag{10.14}$$

Of course, if $X_1$ and $X_2$ are not correlated, their covariance and correlation coefficient are zero. Thus, this expression reduces to

$$Var[Y] = \sigma_1^2 + \sigma_2^2 \tag{10.15}$$

If we have a linear combination of two random variables, say $S_2 = C_1X_1 + C_2X_2$, we can apply the results of Eqs. 10.9g, 10.10 and 10.11 to obtain

$$E(S_2) = C_1E(X_1) + C_2E(X_2) \tag{10.16}$$

$$Var[S_2] = C_1^2\sigma_1^2 + C_2^2\sigma_2^2 + 2C_1C_2\, Cov(X_1, X_2) \tag{10.17}$$

These results can easily be extended to linear combinations of $m$ univariate random variables. Suppose that $S_m = C_1X_1 + C_2X_2 + \cdots + C_mX_m$; then

$$E(S_m) = \sum_{j=1}^{m} C_jE(X_j) \tag{10.18}$$

$$Var(S_m) = \sum_{j=1}^{m} C_j^2\sigma_j^2 + 2\sum_{j=1}^{m-1}\sum_{k=j+1}^{m} C_jC_k\, Cov(X_j, X_k) \tag{10.19}$$

As shown by Hool and Maghsoodloo [14], if all $m$ random variables are independent, the third and fourth moments of $S_m$ yield, respectively,

$$\mu_3(S_m) = \sum_{j=1}^{m} C_j^3\mu_{3j} \tag{10.20}$$

$$\mu_4(S_m) = \sum_{j=1}^{m} C_j^4\mu_{4j} + 6\sum_{j=1}^{m-1}\sum_{k=j+1}^{m} C_j^2C_k^2\sigma_j^2\sigma_k^2 \tag{10.21}$$

where $\mu_{3j}$ is the third moment and $\mu_{4j}$ is the fourth moment about the mean of random variable $j$. If all or some of these random variables are correlated, computing the higher moments such as $\mu_3(S_m)$ and $\mu_4(S_m)$ is difficult without more information about the joint probability distributions among all levels of variables involved. This, of course, complicates the problem significantly. If the $X_j$'s in Eq. 10.18 had a multivariate normal distribution (either mutually independent or dependent), then $S_m$, being a linear function of the $X$'s, would itself be normally distributed. Thus, in this situation a complete description of the distribution of $S_m$ is possible by the mean and variance alone.

***Example 10.2***

Consider the sum of two independent random variables, $S_2 = 2X_1 + X_2$, where $X_1$ is normally distributed with mean and variance $N(2, 3^2)$ and $X_2$ is uniformly distributed in a range between 4 and 6. Find the first four central moments of $S_2$.

For the normal distribution, the first four central moments from Table 10.2 would yield $E(X_1) = 2$, $\mu_2 = 9$, $\mu_3 = 0$, and $\mu_4 = 243$. For the uniform distribution, $E(X_2) = 5$, $\mu_2 = \frac{1}{3}$, $\mu_3 = 0$, and $\mu_4 = \frac{1}{5}$. Then, using Eqs. 10.18 through 10.21, we obtain

$$E(S_2) = 2(2) + 5 = 9$$

$$Var(S_2) = 4(9) + \tfrac{1}{3} = 36\tfrac{1}{3}$$

$$\mu_3(S_2) = 2^3(0) + 0 = 0$$

$$\mu_4(S_4) = 8(243) + \tfrac{1}{5} + 6(4)(1)(9)(\tfrac{1}{3}) = 2{,}016.2$$

The sum $S_2$ is certainly a symmetric distribution $[\mu_3(S_2) = 0]$, but since $\alpha_4(S_2) = 2{,}016.2/(36\frac{1}{3})^2 = 1.53 < 3$, it is not a normal distribution. □

***Differences Between Random Variables.*** The subtraction of one random variable from another can be viewed as a special case of the linear combination of the sum of random variables discussed in the previous section. That is, $Y = X_1 - X_2$ can be regarded as $Y = X_1 + C_2X_2$ where $C_2 = -1$. It follows that

$$E(Y) = E(X_1) - E(X_2) \tag{10.22}$$

$$Var(Y) = \sigma_1^2 + \sigma_2^2 - 2\,Cov(X_1, X_2) \tag{10.23}$$

Therefore, if we have a linear combination of $m$ random variables containing both addition and subtraction, the general relationships given in Eqs. 10.18 and 10.19 still hold.

### 10.2.3 Products of Random Variables

In modeling cash flow streams, we commonly have several random components contributing to the net cash flow. For example, the sales cash flow is determined by the product of unit sales price and sales volume. If these two components are random variables and we have some indication of the correlation between them, we may wish to determine the mean and variance of the corresponding sales cash flow.

Let $Z = XY$, where $X$ and $Y$ are random variables with known means and variances $(\mu_x, \sigma_x^2)$ for $X$ and $(\mu_y, \sigma_y^2)$ for $Y$. To derive the statistical moments for this type of product of two random variables, we may consider the following three cases.

**Case 1:** ***X and Y are independent of each other.***

$$E(Z) = E(XY) = \mu_x\mu_y \tag{10.24}$$

$$\begin{aligned} Var(Z) &= Var(XY) \\ &= E\{[XY - E(XY)]^2\} \\ &= E[(XY)^2] - [E(XY)]^2 \\ &= E(X^2)E(Y^2) - [E(X)E(Y)]^2 \end{aligned}$$

But from the result of Eq. 10.3, we know that

$$E(X^2) = \sigma_x^2 + \mu_x^2$$

Therefore,

$$\begin{aligned} Var(XY) &= (\sigma_x^2 + \mu_x^2)(\sigma_y^2 + \mu_y^2) - \mu_x^2\mu_y^2 \\ &= \mu_x^2\sigma_y^2 + \mu_y^2\sigma_x^2 + \sigma_x^2\sigma_y^2 \end{aligned} \tag{10.25}$$

**Case 2:** ***X and Y are dependent on each other.*** Rearranging terms in Eq. 10.8 gives us

$$E(XY) = \mu_x\mu_y + \rho_{xy}\sigma_x\sigma_y \tag{10.26}$$

and

$$\begin{aligned} Var(XY) = {} & \mu_x^2\sigma_y^2 + \mu_y^2\sigma_x^2 + 2\mu_x\mu_y\, Cov(X, Y) \\ & + 2\mu_x\, Cov(X, Y^2) + 2\mu_y\, Cov(X^2, Y) \\ & + Cov(X^2, Y^2) - [Cov(X, Y)]^2 \end{aligned} \tag{10.27}$$

where $Cov(X^k, Y^j) = E[(X - \mu_x)^k(Y - \mu_y)^j]$. This complicated formula was developed by Goodman [9], but for it to be fully operational, we need to evaluate $Cov(X^k, Y^j)$ by using a joint density function between $X^k$ and $Y^j$. Thus, we are forced to use an approximation. The approximation suggested is

$$Var(XY) = \mu_x^2\sigma_y^2 + \mu_y^2\sigma_x^2 + 2\mu_x\mu_y\, Cov(X, Y) \tag{10.28}$$

**Case 3:** ***X and Y have a joint bivariate normal distribution.*** For two normal variates $X \sim N(\mu_x, \sigma_x^2)$ and $Y \sim N(\mu_y, \sigma_y^2)$ with correlation coefficient $\rho_{xy}$, Craig [4] has shown that $E(XY)$ remains the same as in Eq. 10.26, but the variance simplifies to

$$Var(XY) = \mu_x^2\sigma_y^2 + \mu_y^2\sigma_x^2 + 2\rho_{xy}\mu_x\mu_y + \sigma_x^2\sigma_y^2(1 + \rho_{xy})^2 \tag{10.29}$$

Writing $W_j = \mu_j/\sigma_j$, $j = x, y$, we also obtain the third and fourth moments of $XY$.

$$\mu_3(xy) = \{6[(W_x^2 + W_y^2)\rho_{xy} + W_x W_y(1 + \rho_{xy}^2)] + 2\rho_{xy}(3 + \rho_{xy}^2)\}(\sigma_x\sigma_y)^3 \tag{10.30}$$

$$\mu_4(xy) = [12(W_x^2 + W_y^2)(1 + 3\rho_{xy}^2) + 24W_x W_y \rho_{xy}(3 + \rho_{xy}) + 6(1 + 6\rho_{xy} + \rho_{xy}^4)](\sigma_x\sigma_y)^4 + 3[Var(XY)]^2 \tag{10.31}$$

Thus, the mean and the second through fourth central moments of $XY$ can easily be computed.

### *Example 10.3*

Consider the break-even equation of a product

$$Z = (V - C)X - K$$

where $V$ is the unit sales price, $C$ is unit variable cost, $X$ is sales volume, $K$ is a fixed cost, and $Z$ is profit realized. Find the mean and variance of the profit to be realized, assuming that $K$ is known with certainty ($K$ = \$3,000) but $V$, $C$, and $X$ are normally distributed dependent random variables.

$$V \sim N(15, 2^2) \qquad \rho_{vc} = 0.5$$

$$C \sim N(6, 1^2) \qquad \rho_{vx} = 0.7$$

$$X \sim N(1{,}000, 100^2) \qquad \rho_{cx} = 0.1$$

Since $V$ and $C$ are normally distributed, $Y = V - C$ will also be normally distributed with a mean and variance of

$$E(Y) = 15 - 6 = 9$$

$$Var(Y) = 4 + 1 - 2(0.5)(2)(1) = 3, \qquad \text{from Eq. 10.14}$$

To find the correlation coefficient $\rho_{yx}$ between $X$ and the random variable $Y = V - C$, we use

$$Cov(Y, X) = Cov(V, X) - Cov(C, X), \qquad \text{from Eq. 10.9}$$

That is,

$$\rho_{yx}\sigma_y\sigma_x = \rho_{vx}\sigma_v\sigma_x - \rho_{cx}\sigma_c\sigma_x$$

Thus,

$$\rho_{yx} = \frac{\rho_{vx}\sigma_v - \rho_{cx}\sigma_c}{\sigma_y} = \frac{0.7(2) - 0.1(1)}{\sqrt{3}} = 0.75$$

Since $\rho_{yx} \neq 0$, Eqs. 10.26 and 10.29 give

$$E(YX) = (9)(1{,}000) + 0.75(\sqrt{3})(100) = \$9{,}130$$

$$\begin{aligned} Var(YX) &= 9^2(100^2) + 1{,}000^2(3) + 2(0.75)(9)(1{,}000) \\ &\quad + (100)^2(3)(1 + 0.75)^2 \\ &= 3{,}915{,}375 \\ \sigma_{yx} &= \$1{,}978.73 \end{aligned}$$

Finally, since $K$ is a constant,

$$E(Z) = E(YX) - K = 9{,}130 - 3{,}000 = \$6{,}130$$

$$Var(Z) = 3{,}915{,}375 = 3.92 \times 10^6 \quad \square$$

### 10.2.4 Quotients of Random Variables

We have similar difficulties with both the mean and variance of a quotient. In general, exact solutions to the mean and variance of a quotient of two random variables are not possible except in some special situations in which both random variables belong to a certain type of distribution [7]. Thus, as a practical matter, we are forced to utilize some sort of approximations for these parameters. We will again consider three cases for a random variable, $Z = X/Y$, with known means and variances for both $X$ and $Y$, $(\mu_x, \sigma_x^2)$ and $(\mu_y, \sigma_y^2)$.

Case 1: ***X and Y are independent.*** Reasonable approximations for quotients of the random variable are

$$E(Z) \cong \frac{\mu_x}{\mu_y} \tag{10.32a}$$

$$Var(Z) \cong \frac{\mu_y^2\sigma_y^2 + \mu_x^2\sigma_y^2}{\mu_y^4} \tag{10.32b}$$

Case 2: ***X and Y are correlated.*** With a known correlation coefficient between $X$ and $Y$, the mean can still be approximated as

$$E(Z) \cong \frac{\mu_x}{\mu_y} \tag{10.33a}$$

and

$$Var(Z) \cong \frac{\sigma^2 x}{(\mu_y)^2} + \left(\frac{\mu_x}{\mu_y^2}\right)^2 \sigma_y^2 - 2\rho_{xy}\left(\frac{\mu_x}{\mu_y^3}\right)\sigma_x\sigma_y \tag{10.33b}$$

The development of approximation procedures such as this will be given for a general function in Section 10.2.6.

**Case 3:** ***X and Y have a joint bivariate normal distribution.*** Given two normal variables $X \sim N(\mu_x, \sigma_x^2)$ and $Y \sim N(\mu_y, \sigma_y^2)$ with correlation coefficient $\rho$, we can completely determine the probability distribution of the quotient as shown in Hinkley [13]. The probability that $X/Y$ is less than or equal to $a$ is

$$F\left(\frac{X}{Y} \leq a\right) = L\{h, -k; r\} + L\{-h, k; r\} \tag{10.34}$$

where

$$h = \frac{\mu_x - \mu_y a}{G(a)}$$

$$k = \frac{\mu_y}{\sigma_y}$$

$$r = \frac{\sigma_y a - \rho\sigma_x}{G(a)}$$

$$G(a) = \sigma_x\sigma_y\left(\frac{a^2}{\sigma_x^2} - \frac{2\rho a}{\sigma_x\sigma_y} + \frac{1}{\sigma_y^2}\right)^{1/2}$$

and $L\{h, k; r\}$ is the standard bivariate normal integral tabulated by the National Bureau of Standards [26]. This formula will be used in Chapter 14 when we define a proabilistic benefit–cost ratio [27].

### *Example 10.4*

Suppose that both the initial investment required ($Y$) and the profit forecast ($X$) for a certain project seem rather uncertain. Let us further assume both variables to be uniformly distributed, but each with a unique range.

$$f(x) = \tfrac{1}{4}, \qquad 2 \leq x \leq 6$$

$$f(y) = \tfrac{1}{2}, \qquad 19 \leq y \leq 21 \quad (\text{units in } \$10^6)$$

On the basis of past experience, the estimate of the coefficient of correlation between profit and investment, $\rho_{xy}$, is 0.7. Then

$$\mu_x = \frac{2 + 6}{2} = 4 \qquad \mu_y = \frac{19 + 21}{2} = 20$$

$$\sigma_x^2 = \frac{(6 - 2)^2}{12} = \frac{4}{3} \qquad \sigma_y^2 = \frac{(21 - 19)^2}{12} = \frac{1}{3}$$

The accounting rate of return is defined as profit divided by investment, $Z = X/Y$. Using Eq. 10.33b, we obtain

$$E(Z) = \frac{4}{20} = 20\%$$

$$Var(Z) = \frac{4/3}{(20)^2} + \left(\frac{4}{20^2}\right)^2 \frac{1}{3} - 2(0.7)\left(\frac{4}{20^3}\right)\frac{4}{3} \cdot \frac{1}{3}$$

$$= 0.0162$$

$$\sigma_Z = \sqrt{0.0162} = 12.73\%$$

In other words, the expected accounting rate of return for the project will be about 20% with a standard deviation of 12.73%. □

### 10.2.5 Powers of Independent Random Variables[1]

It is sometimes necessary to evaluate a random variable $Y$ which is some power of a random variable $X$ in the form $Y = X^b$, or even a product of two independent powers of random variables, $Y = X_1^a X_2^b$. In this case exact solutions for the mean and variance of $Y$ are possible for some of the well-defined continuous variables. The use of the Mellin transform often makes the solution possible without complicated integration procedures [7, 28]. The Mellin transform $M_x(s)$ of a function $f(x)$, where $x$ is positive, is defined as

$$M_x(s) = \int_0^\infty x^{s-1} f(x)\, dx$$

$$= E(X^{s-1}) \tag{10.35}$$

When Eq. 10.35 is compared with Eq. 10.1, it is clear that $M_x(s)$ is the $(s - 1)$ moment of $X$ about the origin. That is, the Mellin transform provides an alternative way to find a series of moments of a distribution if $f(x)$ is viewed as a probability density function. Since Mellin transforms are available for most probability density functions of continuous random variables (such as those shown in Table 10.3), they are quite handy for finding the means and variances of such functions of random variables. In Eq. 10.35,

with $s = 1$, $\quad M_x(1) = E(X^0) = 1$

with $s = 2$, $\quad M_x(2) = E(X)$

with $s = 3$, $\quad M_x(3) = E(X^2)$

Thus, the first two statistical moments of the random variable $X$ can be stated in terms of the Mellin transform as

$$E(X) = M_x(2)$$

$$Var(X) = E(X^2) - [E(X)]^2 = M_x(3) - [M_x(2)]^2 \tag{10.36}$$

[1]This section is not used in the sequel and can be deleted without loss of continuity. The symbol $M$ is used for the Mellin transform in this section.

The relationship of Mellin transforms to expected values makes it a simple matter to establish some important operating properties of the transform in probabilistic modeling.

First, we may be faced with finding the distribution of powers of the random variable $X$. If $Y = X^b$, the expected-value argument leads to

$$\begin{aligned} M_y(s) &= E(Y^{s-1}) = E[(X^b)^{s-1}] \\ &= E[X^{b(s-1)}] = M_x(bs - b + 1) \end{aligned} \tag{10.37}$$

That is, the transform for the distribution of the $b$th power of $X$ can be obtained by replacing the $s$ argument in the transform of $f(x)$ by the expression $bs - b + 1$. For example, if $Y = X^2$, where $X$ is uniformly distributed on [0, 1], it follows that with $f(x) = 1$, $0 \leq x \leq 1$,

$$M_x(s) = \int_0^1 x^{s-1}\, dx = \frac{1}{s}$$

Then

$$M_x(2s - 2 + 1) = \frac{1}{2s - 1}$$

To find $E(Y)$ and $Var(Y)$ using the transform, we calculate

$$E(Y) = M_y(2) = \frac{1}{3}$$

$$Var(Y) = M_y(3) - [M_y(2)]^2 = \frac{1}{5} - \left(\frac{1}{3}\right)^2 = \frac{4}{45}$$

In our use of the Mellin transform, the most important property is the Mellin convolution of two functions. This convolution is defined by the integral

$$f(z) = \int_0^\infty \frac{1}{y} g\left(\frac{z}{y}\right) h(y)\, dy \tag{10.38}$$

Equation 10.38 is precisely the form of the probability density function of the random variable $Z = XY$, where $X$ and $Y$ are continuously distributed, independent random variables with probability density functions $g(x)$ and $h(y)$, respectively. The transform of this special convolution reduces to a simple product of Mellin transforms. If we define the Mellin transform of $f(z)$ as $M_z(s)$, it follows that

$$M_z(s) = M_x(s)\, M_y(s) \tag{10.39}$$

We are now in a position to develop an easy method for finding the transform of the distribution of a product of two random variables. If $Z = X^a Y^b$

**Table 10.3** ***Mellin Transforms for Selected Probability Functions***

| Probability | f(x) | | Mellin Transform |
|---|---|---|---|
| Uniform | $\frac{1}{b-a}$ | $a \leq x \leq b$ | $\frac{b^s - a^s}{s(b-a)}$ |
| Exponential | $ae^{-ax}$ | $x > 0$ | $\left(\frac{1}{a}\right)^{s-1} \Gamma(s)$ |
| Gamma | $\frac{a(ax)^{b-1}e^{-ax}}{\Gamma(b)}$ | $0 < x < \infty$<br>$b > -1$ | $\left(\frac{1}{a}\right)^{s-1} \frac{\Gamma(b+s-1)}{\Gamma(b)}$ |
| Standard beta | $\frac{\Gamma(a+b)}{\Gamma(a)\Gamma(b)} x^{a-1}(1-x)^{b-1}$ | $0 \leq x \leq 1$<br>$a > 0, b > 0$ | $\frac{\Gamma(a+b)\Gamma(a+s-1)}{\Gamma(a)\Gamma(a+b+s-1)}$ |
| Triangular | $\frac{2(x-L)}{(H-L)(M_0-L)}$ | $L \leq x \leq M_0$ | $\frac{2}{(H-L)s(s+1)}\left[\frac{H(H^s - M_0^s)}{H - M_0} - \frac{L(M_0^s - L^s)}{M_0 - L}\right]$ |
| | $\frac{2(H-x)}{(H-L)(H-M_0)}$ | $M_0 \leq x \leq H$ | |
| Generalized beta | $\frac{\Gamma(a+b)}{\Gamma(a)\Gamma(b)(H-L)^{a+b-1}} (y-L)^{a-1}(H-y)^{b-1}$ | $L \leq y \leq H$ | $\sum_{k=0}^{s-1} \binom{s-1}{k} L^{s-1-k}(H-L)^k M_x(k+1)$,<br>where $M_x(k)$ for standard beta |
| Standard normal | $\frac{1}{\sqrt{2\pi}} \exp\left(\frac{-x^2}{2}\right)$ | $-\infty < x < \infty$ | $\frac{(2)^{(s-3)/2}}{\sqrt{\pi}} \Gamma\left(\frac{s}{2}\right)$ |

(where $a$ and $b$ are constants), then $Z$ can be viewed as a product of two independent random variables $X^a$ and $Y^b$. By combining the results of Eqs. 10.37 and 10.39, we are able to find the mean and variance of $Z$.

### *Example 10.5*

Suppose that an uncertain lump-sum return $F$ is expected shortly after termination of a project. Because of current uncertain market conditions, the earning interest rate ($i$) seems to fluctuate for the foreseeable future. It is believed, however, that both the lump sum and the interest rate are uniformly distributed but each with a unique range.

$$F \sim \text{uniform } (100, 150)$$

$$i \sim \text{uniform } (9\%, 12\%)$$

Assuming that this uncertain lump sum is to be reinvested at an interest rate $i$ over the next 3 years, the future worth would be

$$Z = FV(i) = F(1 + i)^3 = FY^3$$

where $Y = 1 + i$. Since both $F$ and $Y$ are random variables, the respective Mellin transforms from Table 10.3 would be

$$M_F(s) = \frac{b^s - a^s}{(b - a)s} = \frac{150^s - 100^s}{50s}$$

$$M_y(s) = \frac{1.12^s - 1.09^s}{0.03s}\bigg|_{s=3s-3+1}$$

$$= \frac{(1.12)^{3s-2} - (1.09)^{3s-2}}{0.03(3s - 2)}$$

From Eqs. 10.37 and 10.39, the convolution yields

$$M_Z(s) = \left[\frac{150^s - 100^s}{50s}\right]\left[\frac{(1.12)^{3s-2} - (1.09)^{3s-2}}{0.03(3s - 2)}\right]$$

Then

$$E(Z) = \left[\frac{150^2 - 100^2}{50(2)}\right]\left[\frac{(1.12)^4 - (1.09)^4}{0.03(4)}\right]$$

$$= \$168.69$$

$$\sigma_Z^2 = M_Z(3) - [M_Z(2)]^2 = 395.33 = (19.88)^2$$

Thus, the expected future worth is \$168.69, with a standard deviation of \$19.88. □

### 10.2.6 General Approximation Formulas

The approximation formulas given in Eqs. 10.28, 10.32, 10.33, and 10.34 were based on Taylor's theorem. Taylor's theorem considers a function of $m$ variables, say $f(x) = f(x_1, \ldots, x_m)$, which has continuous partial deviatives of order $n$. The theorem states that the function $f(x)$ can be approximated by an $n$ th degree polynomial, commonly called a Taylor series expansion.

Suppose that $X_1, \ldots, X_m$ are random variables, and we adopt the following.

1. The mean and variance of the random variable $X_k$ is known and denoted by $\mu_k$ and $\sigma_k^2$
2. The covariance of the random variable $X_k$ and the random variable $X_j$ is known and is denoted by $\sigma_{kj}$ [or $Cov(X_k, X_j)$].
3. Let $\mu = (\mu_1, \ldots, \mu_m)$.

If the mean and variance of this approximation to $f(x)$ are computed on the basis of a Taylor's series expansion about the point $\mu$, ignoring terms of higher order than $\sigma^2$, it follows that

$$E[f(x)] \cong f(\mu) + \frac{1}{2}\sum_{k=1}^{m} \sigma_k^2 \left(\frac{\partial^2 f}{\partial x_k^2}\right)_\mu + \sum_{k<j} \sigma_{kj}\left(\frac{\partial^2 f}{\partial x_k \partial x_j}\right)_\mu \tag{10.40}$$

$$Var[f(x)] \cong \sum_{k=1}^{m} \sigma_k^2 \left[\left(\frac{\partial f}{\partial x_k}\right)_\mu\right]^2 + \sum_{k\neq j} \sigma_{kj}\left(\frac{\partial f}{\partial x_k}\right)_\mu \left(\frac{\partial f}{\partial x_j}\right)_\mu \tag{10.41}$$

The notation $(\partial f/\partial x_k)_\mu$ denotes the evaluation of the outcome of each partial derivative at its respective mean value. Of course, if the random variables are all mutually independent, the terms involving $\sigma_{kj}$ are all zero and drop out. The adequacy of the approximation provided by these formulas was investigated by Smith [33] who concluded that the approximation is close enough to suffice for most work.

### *Example 10.6*

To compare the effectiveness of these approximation formulas, let us consider Example 10.5. The function was $Z = FY^3$ where both $F$ and $Y$ are independent random variables. Further, we know that $\mu_F = 125$, $\sigma_F^2 = 208.33$, $\mu_Y = 1.105$, and $\sigma_Y^2 = 0.000075$. Then,

$$\begin{aligned} E(Z) &\cong (125)(1.105)^3 + \tfrac{1}{2}[208.33(0) + (0.000075)(6)(125)(1.105)] \\ &= 168.69 \\ Var(Z) &\cong 208.33[(1.105)^3]^2 + 0.000075[3(125)(1.105)^2]^2 \\ &= 394.97 \end{aligned}$$

Compare the results with the exact solutions in Example 10.5, and you will find that they are surprisingly close.

| Function | Exact | Approximate |
|---|---|---|
| $E(Z)$ | 168.69 | 168.69 |
| $Var(Z)$ | 395.33 | 394.97 |

The reader must be warned that the foregoing approximation formulas will not always provide such close approximations. The closeness of the approximation depends on the type of function involved. □

## 10.3 STATISTICAL MOMENTS OF DISCOUNTED CASH FLOWS

In the previous section we discussed the mechanics of handling random variables. The purpose of this section is to utilize some of the characteristics of random variables in discounted cash flow modeling [11].

### 10.3.1 Expected Net Present Value

We can treat uncertain cash flow streams as a series of random variables for the purpose of computing project present value. Let $F_n$ be a stream of random net cash flows generated by a particular project at time period $n$ ($n = 0, 1, 2, \ldots, N$). A particular cash flow $F_n$ may be positive in sign (inflow) or negative in sign (outflow), and both inflows and outflows can occur in any period. Thus, it is the net effect in a particular year that is of consequence. In the present value approach given in Chapter 6, the relation between the net present value ($PV$) of the cash flow stream and its constituent cash flow element is

$$PV(i) = \sum_{n=0}^{N} \frac{F_n}{(1+i)^n} \tag{10.42}$$

where $PV(i)$ is the net present value of the cash flow stream for the project and $i$ is the minimum attractive rate of return.

Since a random process governs the values taken by $F_n$, the relative frequencies of the random values taken by the cash flow can usually be represented by probability or density functions $f(F_n)$. This randomness can be expressed by the mean and variance of the distribution of $F_n$. Thus, the summation of the discounted random cash flows to obtain the project net present value is also a random variable. From Eq. 10.18 the random variable $PV(i)$ has a mean of

$$E[PV(i)] = \sum_{n=0}^{N} \frac{E(F_n)}{(1+i)^n} \tag{10.43}$$

### 10.3.2 Variance of Net Present Value

The value of the variance will depend on the relationship among the project cash flows. Cash flow streams are said to be completely independent if there is no causative or consequential relationship between any two cash flows in the cash flow stream. Otherwise, the cash flow streams are said to be dependent, and the degree of dependence among cash flows is then determined by the correlation coefficients.

***Independent Net Cash Flows.*** When net cash flows are independent, the variance of the project net present value is found by using the results of Eq. 10.19 and dropping the covariance terms because of the independence relationship to obtain

$$Var[PV(i)] = \sum_{n=0}^{N} \frac{\sigma_n^2}{(1+i)^{2n}} \tag{10.44}$$

where $\sigma_n^2$ is the variance of the cash flow of the project at time $n$. That is, to compute the variance of $PV$, we first multiply the variance of each $F_n$ by the square of its discount factor.

## *Example 10.7*

Let us assume that we estimate the cash flows for project A, recognizing that each annual flow can be represented by a random variable. If project A has relatively certain inflows and outflows, the variability in the flows may be due to random elements unrelated to one another. This would represent complete independence among the cash flows.

| $n$ | Expected Flow, $E(F_n)$ | Estimate of Standard Deviation, $\sigma_n$ |
|---|---|---|
| 0 | −\$400 | 20 |
| 1 | +120 | 10 |
| 2 | +120 | 15 |
| 3 | +120 | 20 |
| 4 | +110 | 30 |
| 5 | +200 | 50 |

With $i = 10\%$, the expected net present value of this project is

$$E[PV(10\%)] = -400 + \frac{120}{1.1} + \frac{120}{(1.1)^2} + \frac{120}{(1.1)^3} + \frac{110}{(1.1)^4} + \frac{200}{(1.1)^5}$$

$$= \$97.74$$

The variance of the $PV$ is obtained from Eq. 10.44 as

$$Var[PV(10\%)] = (20)^2 + \frac{(10)^2}{(1.1)^2} + \frac{(15)^2}{(1.1)^4} + \frac{(20)^2}{(1.1)^6} + \frac{(30)^2}{(1.1)^8} + \frac{(50)^2}{(1.1)^{10}}$$

$$= 2{,}247$$

and the standard deviation is

$$(2{,}247)^{0.5} = \$47 \quad \square$$

***Correlated Net Cash Flows*** For the situation in which periodic net cash flows ($F_n$'s) are somehow correlated with each other, the computation of variance will be understood better by examining the following simple situation. The cash flows for two periods are calculated from Eq. 10.42, and if $N = 2$, the present value expression reduces to

$$PV(i) = F_0 + \frac{F_1}{1+i} + \frac{F_2}{(1+i)^2}$$

Since the $F_n$'s ($n = 0, 1, 2$) are random variables, $Var[PV(i)]$ can be directly computed by utilizing the result of Eq. 10.17.

$$Var[PV(i)] = Var(F_0) + \frac{Var(F_1)}{(1+i)^2} + \frac{Var(F_2)}{(1+i)^4} + \frac{2}{1+i}\,Cov(F_0, F_1) + \frac{2}{(1+i)^2}\,Cov(F_0, F_2) + \frac{2}{(1+i)^3}\,Cov(F_1, F_2)$$

The expression of Eq. 10.8 simplifies this to

$$Var[PV(i)] = \sigma_0^2 + \frac{\sigma_1^2}{(1+i)^2} + \frac{\sigma_2^2}{(1+i)^4} + \frac{2}{1+i}\,\rho_{01}\sigma_0\sigma_1 + \frac{2}{(1+i)^2}\,\rho_{02}\sigma_0\sigma_2 + \frac{2}{(1+i)^3}\,\rho_{12}\sigma_1\sigma_2$$

where $\rho_{ks}$ is the correlation coefficient between $F_k$ and $F_s$. This variance can be generalized for the situation in which the net cash flow from an investment terminates at the end of year $N$. Basically, this is the result of Eq. 10.19.

$$Var[PV(i)] = \sum_{n=0}^{N} \frac{\sigma_n^2}{(1+i)^{2n}} + 2\sum_{n=0}^{N-1}\sum_{s=n+1}^{N} \frac{\rho_{ns}\sigma_n\sigma_s}{(1+i)^{n+s}} \tag{10.45}$$

### *Example 10.8*

Consider a project whose net cash flow streams are as follows.

| Year End, $n$ | $E(F_n)$ | $Var(F_n)$ |
|---|---|---|
| 0 | −\$1,000 | $50^2$ |
| 1 | 500 | $100^2$ |
| 2 | 800 | $200^2$ |

The correlation coefficients among the $F_n$'s are known to be $\rho_{01} = 0.2$, $\rho_{02} = 0.1$, and $\rho_{12} = 0.5$. Assuming $i = 10\%$, we obtain

$$E[PV(10\%)] = -1{,}000 + \frac{500}{1.1} + \frac{800}{(1.1)^2} = \$115.70$$

$$\begin{aligned} Var[PV(10\%)] &= \left[50^2 + \frac{100^2}{(1.1)^2} + \frac{200^2}{(1.1)^4}\right] + \left[\frac{2(0.2)(50)(100)}{1.1}\right. \\ &\quad \left. + \frac{2(0.1)(50)(200)}{(1.1)^2} + \frac{2(0.5)(100)(200)}{(1.1)^3}\right] \\ &= 56{,}582.4 \end{aligned}$$

$$\text{Standard deviation of } PV(10\%) = \$237.87 \quad \square$$

***Perfectly Correlated Cash Flows.*** Cash flows are said to be perfectly correlated if any fluctuation from the mean by one cash flow corresponds to a similar fluctuation by the other cash flows. From Eq. 10.8 it follows that two random cash flows $F_k$ and $F_s$ are perfectly correlated if and only if

$$\rho_{ks} = \frac{Cov(F_k, F_s)}{\sigma_k \sigma_s} = \pm 1$$

When more than two time periods are involved, not all cash flows can be perfectly negatively correlated with one another. This fact explains our restriction of cash flow modeling to a perfect positive correlation. With $\rho_{ks} = 1$ for all $k$ and $s$ in Eq. 10.45, the variance of the net present value becomes

$$Var[PV(i)] = \left[\sum_{n=0}^{N} \frac{\sigma_n}{(1+i)^n}\right]^2 \tag{10.46}$$

What is the economic interpretation of the assumption of perfect correlation between the periodic net cash flows? To gain insight into this question, let us consider the following.

Suppose that two random cash flows $F_k$ and $F_s$ are perfectly correlated ($\rho_{ks} = \pm 1$); then $F_k$ and $F_s$ are functionally related by a linear function of

$$F_k = mF_s + b \tag{10.47}$$

where $b$ and $m$ are constant. Further, if $\rho_{ks} = 1$, then $m$ has the same sign as $\rho_{ks}$ and is equal in absolute value to $\sigma_s/\sigma_k$. The converse is also true [21]. That is

$$F_k = \frac{\sigma_k}{\sigma_s} F_s + b$$

Solving for $F_s$ gives us

$$F_s = \frac{\sigma_s}{\sigma_k} F_k - \left(\frac{\sigma_s}{\sigma_k}\right) b \tag{10.48}$$

This functional relationship enables us to determine the entire sequence of net cash flows after we know the actual net cash flow in any one period. Suppose that $F_k$ is known to be the actual value of $\mu_k + d\,\sigma_k$. We can then compute the corresponding value of $F_s$ by substituting the value of $F_k$ into Eq. 10.48.

$$\begin{aligned} F_s &= \frac{\sigma_s}{\sigma_k}(\mu_k + d\sigma_k) + \left(\frac{-\sigma_s}{\sigma_k}\right) b \\ &= \frac{\sigma_s}{\sigma_k}(\mu_k - b) + d\sigma_s \\ &= E(F_s) + d\sigma_s \\ &= \mu_s + d\sigma_s \end{aligned} \tag{10.49}$$

It is clear that the assumption of perfect correlation between the periodic net cash flows has the following economic interpretation. If random factors cause $F_k$ to deviate from its mean value by $d$ standard deviations, the same factors will cause $F_s$ to deviate from its mean in the same direction by $d$ standard deviations. This implies also that the realization of the first random cash flow determines the other cash flows.

### 10.3.3 Mixed Net Cash Flows

Sometimes it is possible to have the initial investment of a project independent of the rest of the cash flow stream, but the remaining portions of the cash flow stream may be partially or perfectly correlated. In this case, as in the others, the mean net present value is found in exactly the same way, by adding the discounted cash flow elements. The calculation of the variance of project net present value has some minor changes, however.

If the initial investment ($F_0$) is independent and the remaining portions of the cash flow stream ($F_n$, $n > 0$) are correlated, the only difference in the calculation of the variance with respect to the general case in Eq. 10.45 is to assume $\rho_{0n} = 0$ ($n = 1, 2, \ldots, N$).

In the special case where the rest of the cash flow stream is perfectly correlated, $Var[PV(i)]$ further simplifies to

$$Var[PV(i)] = \sigma_0^2 + \left[\sum_{n=1}^{N} \frac{\sigma_n}{(1+i)^n}\right]^2 \tag{10.50}$$

### *Example 10.9*

Consider the following project whose initial investment is distributed independently of future cash flows. The future net inflows, however, are all perfectly correlated with one another. Assume again that $i = 10\%$.

| $n$ | $E(F_n)$ | $\sigma_n$ | |
|---|---|---|---|
| 0 | −\$600 | 50 | Independent |
| 1 | +50 | 50 | Perfectly correlated |
| 2 | +400 | 100 | |
| 3 | +300 | 100 | |
| 4 | +200 | 100 | |
| 5 | +200 | 100 | |

$$E[PV(10\%)] = -600 + \frac{50}{1.1} + \frac{400}{(1.1)^2} + \frac{300}{(1.1)^3} + \frac{200}{(1.1)^4} + \frac{200}{(1.1)^5} = \$262$$

For variance computation we use Eq. 10.50,

$$Var[PV(10\%)] = (50)^2 + \left[\frac{50}{1.1} + \frac{100}{(1.1)^2} + \frac{100}{(1.1)^3} + \frac{100}{(1.1)^4} + \frac{100}{(1.1)^5}\right]^2$$

$$= 113{,}500$$

and

$$\text{Standard deviation of } PV(10\%) = \$334 \quad \square$$

### 10.3.4 Net Cash Flows Consisting of Several Components

Another situation to consider is one for which the net cash flow $F_n$ can be broken down into two components (this can be any number of components) $A_n$ and $B_n$, where $A_n$ is the part of $F_n$ that varies completely independently, and $B_n$ is the part of $F_n$ that is perfectly correlated with $B_n$ in any other operating period.

To compute the variance for this situation, we first compute separately the variances of the independent components and the perfectly correlated components of the net cash flows. The sum of these subvariances is the variance of the $PV$ of the project,

$$Var[PV(i)] = \sum_{n=0}^{N} \frac{Var(A_n)}{(1+i)^{2n}} + \left[\sum_{n=0}^{N} \frac{\sqrt{Var(B_n)}}{(1+i)^n}\right]^2 \tag{10.51}$$

### *Example 10.10*

A manufacturing firm is considering the purchase of a numerically controlled machine for \$21,000. The machine has an economic service life of 5 years with a certain salvage value of $5,000. Because market conditions are uncertain, the additional annual revenues are expected to have the following normal probability distributions with mean $\mu_n$ and standard deviation $\sigma_n$. The annual operating and maintenance costs are also assumed to vary according to normal probability distributions with the following known means and standard deviations.

| Year End, ($n$) | Revenue (dollars) | Cost (dollars) |
|---|---|---|
| 1 | $N(6{,}500,\ 500^2)$ | $N(300,\ 50^2)$ |
| 2 | $N(6{,}500,\ 500^2)$ | $N(300,\ 50^2)$ |
| 3 | $N(6{,}500,\ 500^2)$ | $N(300,\ 50^2)$ |
| 4 | $N(6{,}500,\ 500^2)$ | $N(300,\ 50^2)$ |
| 5 | $N(6{,}500,\ 500^2)$ | $N(300,\ 50^2)$ |

The revenue streams are believed to be independent of the cost streams. The revenue flows themselves vary independently, whereas the cost flows are perfectly correlated with costs in any other operating period.

(a) Without considering income tax effects, but using $i = 15\%$, find the mean and variance of the net present value of the project. Let

$b_n$ = revenue at end of year $n$,

$c_n$ = cost at end of year $n$,

$F_n$ = net cash flow at end of year $n$.

Then $F_n$ yields

$$F_n = \begin{cases} -21{,}000 & \text{for } n = 0 \\ b_n - c_n & \text{for } n = 1, 2, 3, 4 \\ b_n - c_n + 5{,}000 & \text{for } n = 5 \end{cases}$$

Taking expectations, we obtain

$$E(F_n) = \begin{cases} -21{,}000 & \text{for } n = 0 \\ E(b_n) - E(c_n) = 6{,}500 - 300 = 6{,}200 & \text{for } n = 1, 2, 3, 4 \\ E(b_n) - E(c_n) + 5{,}000 = 11{,}200 & \text{for } n = 5 \end{cases}$$

The expected net present value would be

$$\begin{aligned} E[PV(15\%)] &= -21{,}000 + 6{,}200(P/A,\ 15\%,\ 5) + 5{,}000(P/F,\ 15\%,\ 1) \\ &= \$2{,}269 \end{aligned}$$

Since the initial investment and the salvage value are known with certainty, their variances will be zero. Thus, we need to consider only the variances of $b_n$ and $c_n$.

$$Var[PV(15\%)] = \left[0 + \frac{500^2}{(1.15)^2} + \frac{500^2}{(1.15)^4} + \frac{500^2}{(1.15)^6} + \frac{500^2}{(1.15)^8} + \frac{500^2}{(1.15)^{10}}\right]$$
$$+ \left[0 + \frac{50}{1.15} + \frac{50}{(1.15)^2} + \frac{50}{(1.15)^3} + \frac{50}{(1.15)^4} + \frac{50}{(1.15)^5}\right]^2$$
$$= 583{,}578 + (167.61)^2$$
$$= 611{,}671$$

Standard deviation of $PV(15\%) = \$782$

(b) Suppose both the initial investment and the salvage value are random variables governed by normal distributions of $N(\$21{,}000, 800^2)$ for the initial investment and of $N(\$5{,}000, 1{,}000^2)$ for the salvage value. Further, assuming that these realizations are mutually independent of the rest of the cash flows, the revised mean and variance of the *PV* would be as follows.

$$E[PV(15\%)] = \$2{,}269 \quad \text{(remains unchanged)}$$
$$Var[PV(15\%)] = [\text{old variance}] + 800^2 + \frac{1{,}000^2}{(1.15)^{10}}$$
$$= 1{,}498{,}856$$

Standard deviation of $PV(15\%) = \$1{,}224$ □

### 10.3.5 Cash Flows with Uncertain Timing: Continuous Case

So far we have assumed that the timing of these cash flows is known with certainty. Now we can relax this certainty assumption by allowing the timing to vary and examine how the mean and variance statistics can be determined in this situation. We will consider three plausible timings for cash flow.

***Lump-Sum Payment with Uncertain Timing.*** Suppose that a single cash flow $F$ has a known amount but that the future timing of this flow is uncertain. This situation is similar to that shown in Figure 10.2. When both $F$ and $t$ are known, the present value expression is $PV(r) = Fe^{-rt}$. It follows that the expected present value with continuous compounding at a nominal rate of $r$ is

$$E[PV(r)] = \int_0^\infty Fe^{-rt}f(t)\,dt \tag{10.52}$$
$$= F\int_0^\infty f(t)e^{-rt}\,dt \tag{10.53}$$
$$= FE(e^{-rt})$$

where $f(t)$ denotes the probability density function about the timing of $F$. From

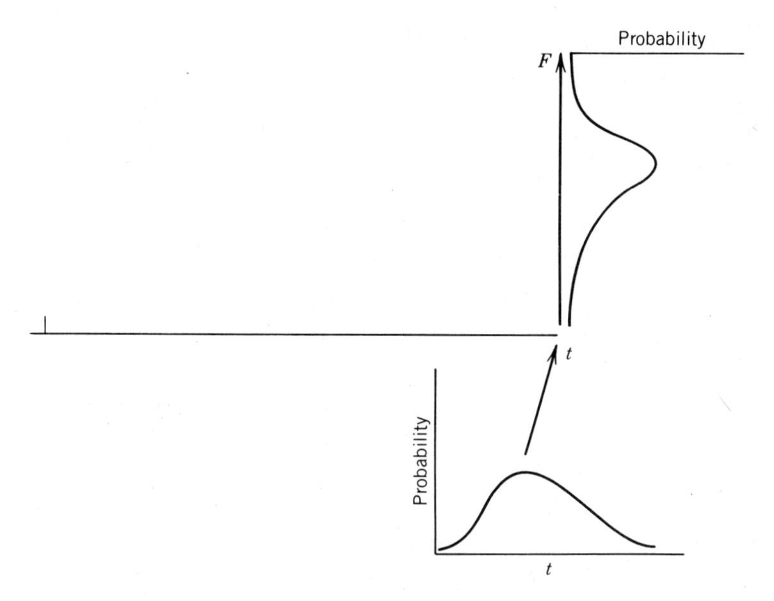


**FIGURE 10.2.** Lump-sum payment with uncertain timing and uncertain amount.

Chapter 3, the expression for $\int_0^\infty f(t)e^{-rt}\,dt$ is known as the Laplace transform of the function $f(t)$ and is denoted by $L(r)$. Since the Laplace transform of most standard forms of probability functions is known (see Table 10.4), we may easily calculate $E(e^{-rt})$. That is,

$$E(e^{-rt}) = L[f(t)]\big|_{s=r} = L(r) \tag{10.54}$$

The variance computation is

$$\begin{aligned} Var[PV(r)] &= \int_0^\infty (Fe^{-rt})^2 f(t)\,dt - [FE(e^{-rt})]^2 \\ &= F^2 \int_0^\infty f(t)\,e^{-2rt}\,dt - F^2[L(r)]^2 \\ &= F^2[L(2r) - L(r)^2] \\ &= F^2\,Var(e^{-rt}) \end{aligned} \tag{10.55}$$

or, expressed differently,

$$Var(e^{-rt}) = L(2r) - L(r)^2 \tag{10.56}$$

Both $E(e^{-rt})$ and $Var(e^{-rt})$ can be closely approximated by using a Taylor series expansion when the precise form of $f(t)$ is not known or when $f(t)$ is an arbitrary

**Table 10.4** ***Laplace Transforms for Selected Density Functions***

| Density Function | Laplace Transform, $E(e^{-rt})$ |
|---|---|
| Generalized normal | $e^{-(r\mu - r^2\sigma^2)/2}$* |
| $\frac{1}{\sigma\sqrt{2\pi}} e^{-(x-\mu)^2/2\sigma^2}, \quad -\infty < x < \infty$ | |
| Standard normal | $e^{r^2/2}$* |
| $\frac{1}{\sqrt{2\pi}} e^{-x^2/2}, \quad -\infty < x < \infty$ | |
| Uniform | |
| $\frac{1}{b-a}, \quad a \le x \le b$ | $\frac{e^{-ra} - e^{-rb}}{r(b-a)}$ |
| Gamma | |
| $\frac{a^b}{(b-1)!} x^{b-1} e^{-ax}, \quad 0 < x < \infty$ | $\left(1 + \frac{r}{a}\right)^{-b}$† |
| Exponential | |
| $ae^{-ax}, \quad 0 < x < \infty$ | $\left(1 + \frac{r}{a}\right)^{-1}$ |

*Bilateral Laplace transform that is equivalent to the MGF with changed parameter sign.
†Special cases: Erlang, $b$ = integer; exponential, $b = 1$.

probability function [38, 31]. These approximations are

$$E(e^{-rt}) \cong \left(1 + \frac{\sigma^2 r^2}{2} - \frac{\mu_3 r^3}{6}\right) e^{-r\mu} \tag{10.57}$$

$$Var(e^{rt}) \cong \sigma^2 r^2 \left(1 - \frac{\sigma^2 r^2}{4}\right) e^{-2r\mu} \tag{10.58}$$

where $\mu = E(t)$
$\sigma_2 = Var(t)$
$\mu_3 = E(t - \mu)^3$

If the amount of $F$ itself is also uncertain, we find the present value expression to be a product of two random variables. When both $E(F)$ and $Var(F)$ are given, it is a simple matter to find $E[PV(r)]$ and $Var[PV(r)]$ from the relations given by Eqs. 10.24 to 10.29.

### *Example 10.11*

Power interruptions at a certain manufacturing plant are expected to cost $30,000 each time they occur, and the interruptions are thought to occur according to a negative exponential distribution with a mean time between successive interruptions of 1.2 years. Assuming continuous compounding at $r = 10\%$, the expected present value of the cost of the next interruption would be

$$\begin{aligned} E[PV(10\%)] &= 30{,}000E(e^{-0.1t}) \\ &= 30{,}000\left[\frac{1}{(0.1)(1.2)+1}\right] \\ &= \$26{,}785.71 \end{aligned}$$

If the interruption timing is known to be precisely 1.2 years and the cost is exactly \$30,000, the present value is $30{,}000e^{-0.1(1.2)} = \$26{,}607.61$, and the difference of \$178.10 is the added cost of the uncertainty in the timing. □

***Both Project Initiation and Duration Uncertain.*** Suppose that a uniform continuous cash flow ($\bar{A}$) has an uncertain duration of $t$ time periods after starting at the uncertain time $k$. Further assume that $t$ and $k$ are statistically independent random variables with known probability functions. This situation is depicted in Figure 10.3.

We will handle this situation in two steps. First, if $k$ and $t$ are known with certainty, the present value expression at a nominal rate of interest $r$, assuming continuous compounding, would be

$$PV(r) = \bar{A}\,\frac{e^{rt}-1}{re^{rt}}\,e^{-rk} = \frac{\bar{A}}{r}\,(e^{-rk} - e^{-r(k+t)}) \tag{10.59}$$

When $k$ and $t$ are random variables, we take an expectation on both sides of Eq. 10.59 to obtain

$$E[PV(r)] = \frac{\bar{A}}{r}[E(e^{-rk}) - E(e^{-r(k+t)})] \tag{10.60}$$

Suppose that $f(k)$ and $g(t)$ stand for the probability density functions of the cash flow events $k$ and $t$, respectively. From Eq. 10.54, we define

$$E(e^{-rk}) = L[f(k)]\Big|_{s=r} = L(r)_1$$

$$E(e^{-rt}) = L[g(t)]\Big|_{s=r} = L(r)_2$$

Because these are independent, it follows that

$$\begin{aligned} E(e^{-r(k+t)}) &= E(e^{-rk}e^{-rt}) = E(e^{-rk})E(e^{-rt}) \\ &= L(r)_1 L(r)_2 \end{aligned}$$

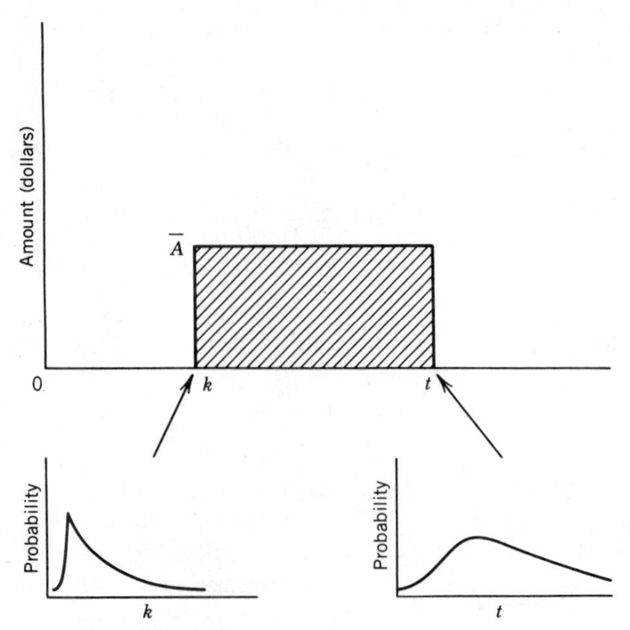


**FIGURE 10.3.** Both project initiation and duration uncertain.

Thus, Eq. 10.60 can be rewritten as

$$E[PV(r)] = \frac{\bar{A}}{r} L(r)_1[1 - L(r)_2] \tag{10.61}$$

To find the expression of variance of $PV(r)$, assuming independence, we use

$$Var\,[PV(r)] = \left(\frac{\bar{A}}{r}\right)^2 [Var\,(e^{-rk}) + Var\,(e^{-r\,(k+t)})] \tag{10.62}$$

From Eq. 10.56 it follows that

$$Var(e^{-rk}) = L(2r)_1 - L(r)_1^2$$
$$Var(e^{-rt}) = L(2r)_2 - L(r)_2^2$$

From Eq. 10.25, the variance of the product of two independent random variables is

$$\begin{aligned} Var(e^{-r(k+t)}) &= Var(e^{-rk}e^{-rt}) \\ &= L(r)_1^2\,[L(2r)_2 - L(r)_2^2] + L(r)_2^2[L(2r)_1 - L(r)_1^2] \\ &\quad + [L(2r)_1 - L(r)_1^2][L(2r)_2 - L(r)_2^2] \end{aligned} \tag{10.63}$$

Substituting Eq. 10.63 back into Eq. 10.62, we can find the variance of $PV(r)$.

### *Example 10.12*

A uniform continuous cash flow totaling $1,000 per year will begin in the near future and continue for an uncertain period of time. More precisely, the start time has an uncertain delay that is uniformly distributed between 6 months and 1 year. The project duration follows a gamma distribution with a mean of 3 years and a variance of 1. Gamma parameters yielding these statistics are $a = 3$ and $b = 9$. The nominal interest rate is 10% compounded continuously. From Table 10.4, the Laplace transforms of uniform and gamma distributions are as follows.

**1.** With $f(k) = \dfrac{1}{b-a}, \quad a \le k \le b \; (a = 0.5, b = 1),$

$$L(r)_1 = \frac{e^{-ra} - e^{-rb}}{r(b-a)} = \frac{e^{-0.1(0.5)} - e^{-0.1(1)}}{0.1(1-0.5)} = L(0.1)_1 = 0.9278$$

**2.** With $f(t) = \dfrac{a^b}{(b-1)!} t^{b-1} e^{-at}, \quad 0 < t < \infty \; (a = 3, b = 9),$

$$L(r)_2 = \left(1 + \frac{r}{a}\right)^{-b} = \left(1 + \frac{0.1}{3}\right)^{-9} = L(0.1)_2 = 0.7444$$

Thus, we also compute

$$L(2r)_1 = \frac{e^{-0.2(0.5)} - e^{-0.2(1)}}{0.2(1-0.5)} = 0.8610$$

$$L(2r)_2 = \left[1 + \frac{0.2}{3}\right]^{-9} = 0.5594$$

Finally,

$$E[PV(10\%)] = \left(\frac{1{,}000}{0.1}\right)(0.9278)(1 - 0.7444) = \$2{,}371.46$$

$$Var(e^{-rk}) = 0.8610 - (0.9278)^2 = 0.00018716$$

$$Var(e^{-rt}) = 0.5594 - (0.7444)^2 = 0.00526864$$

$$\begin{aligned} Var[PV(10\%)] &= \left(\frac{1{,}000}{0.1}\right)^2 [0.00018716 + (0.9278)^2(0.00526864) \\ &\quad + (0.7444)^2(0.00018716) \\ &\quad + (0.00018716)(0.00526864)] \\ &= \left(\frac{1{,}000}{0.1}\right)^2 (0.00482717) \\ &= 482{,}717.03 \end{aligned}$$

Standard deviation of $PV(10\%) = \$694.78$ □

***Random Timing Between Cash Inflows.*** When randomness exists in both the magnitude and the timing between cash inflows, we may develop an analytical solution to the problem with the following assumptions.

1. There is independence between cash inflow times and cash flows and between inflow times themselves.
2. Initial outlays are to take place at $t = 0$. Cash inflows occur randomly thereafter, and cash inflows are also random in magnitude.
3. The nominal discount rate $r$ is given, and continuous compounding is used.

Analytically, let $t_1, t_2, \ldots, t_n$ denote the time points at which the inflows $F_1$, $F_2, \ldots, F_n$ occur. Assume that $d_j = t_j - t_{j-1}$ $(j = 1, 2, \ldots n; t_0 = 0)$ is a random variable with a continuous distribution $G_j(t)$ for all $j$ and associated density $g(t)$. Similarly, the distribution of the $F_j$'s is $B_j(t)$ for all $j$ with $b_j(t)$ as a density function. This situation is illustrated in Figure 10.4. Then the net present value of cash flows is given by

$$PV(r) = \sum_{j=1}^{n} F_j e^{-rt_j} - F_0 = \sum_{j=1}^{n} F_j \exp\left(-r \sum_{k=1}^{j} d_k\right) - F_0 \qquad (10.64)$$

## *Example 10.13*

To illustrate the computational procedures involved, let us take a simple example. Suppose that a project has the following investment characteristics.

| Time Period | Cash Mean | Cash Variance | Distribution |
|---|---|---|---|
| $t_0$ | −\$500 | 0 | Certainty |
| $t_1$ | 400 | $(50)^2$ | Uniform |
| $t_2$ | 400 | $(50)^2$ | Uniform |

The timings between cash flows are $d_1 = t_1 - t_0$ and $d_2 = t_2 - t_1$. Let $d_1$ and $d_2$ be identical and independent exponentially distributed random variables with a mean of one year. We assume that $r = 10\%$, and since $n = 2$, we have

$$PV(r) = F_0 + F_1 e^{-rd_1} + F_2 e^{-r(d_1 + d_2)}$$

Then the expected net present value is

$$E[PV(r)] = E(F_0) + E(F_1 e^{-rd_1}) + E(F_2 e^{-r(d_1 + d_2)})$$

Because of the independence of $F$ and $d$,

$$E[PV(r)] = E(F_0) + E(F_1)E(e^{-rd_1}) + E(F_2)E(e^{-rd_1})E(e^{-rd_2})$$

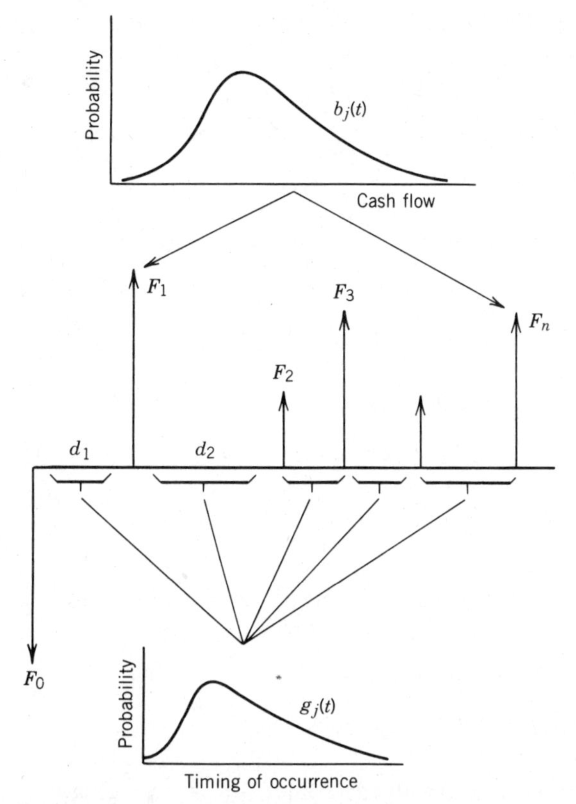


**FIGURE 10.4. Random timing between cash inflows.**

Using the Laplace transform expression given in Eq. 10.54, it follows that

$$E[PV(r)] = E(F_0) + E(F_1)L(r)_1 + E(F_2)L(r)_1L(r)_2$$

Since we assumed that $d_1$ and $d_2$ have the same exponential distribution with a mean time of one year, we have

$$L(r)_1 = L(r)_2 = \frac{1}{1 + r} \quad \text{(see Table 10.4)}$$

Thus,

$$E[PV(10\%)] = -500 + 400\left(\frac{1}{1.1}\right) + 400\left(\frac{1}{1.1}\right)^2 = \$194.21$$

The variance computation is

$$Var[PV(r)] = Var(F_1e^{-rd_1}) + Var(F_2e^{-r(d_1+d_2)})$$
$$+ 2\,Cov(F_1e^{-rd_1}, F_2e^{-r(d_1+d_2)})$$

[$Var(F_0)$ is zero in our example.]

We will compute $Var[PV(r)]$ in several steps for ease of explanation. Since $E(F_1) = E(F_2) = 400$ and

$$Var(e^{-rd_1}) = L(2r)_1 - L(r)_2^2 = \frac{1}{1+2r} - \left(\frac{1}{1+r}\right)^2$$

and

$$Var(F_1) = Var(F_2) = 2{,}500$$

Thus,

$$E(e^{-r(d_1+d_2)}) = [E(e^{-rd})]^2 = \left(\frac{1}{1+r}\right)^2$$

$$\begin{aligned} Var(e^{-r(d_1+d_2)}) &= E(e^{-2r(d_1+d_2)}) - E(e^{-r(d_1+d_2)})^2 \\ &= L(2r)^2 - L(r)^4 \\ &= \left(\frac{1}{1+2r}\right)^2 - \left(\frac{1}{1+r}\right)^4 \end{aligned}$$

Using the variance formula for the product of random variables given in Eq. 10.25, we compute

$$Var(F_1e^{-rd_1}) = 3{,}107.68$$
$$Var(F_2e^{-r(d_1+d_2)}) = 3{,}457.58$$

Since $F_1$ and $F_2$ are perfectly correlated with each other ($\rho_{12} = 1$), we have

$$\begin{aligned} Cov(F_1e^{-rd_1}, F_2e^{-r(d_1+d_2)}) &= \sqrt{(3{,}107.68)(3{,}457.58)} \\ &= 3{,}277.96 \end{aligned}$$

Thus,

$$\begin{aligned} Var[PV(10\%)] &= 3{,}107.68 + 3{,}457.58 + 2(3{,}277.96) \\ &= (114.55)^2 \quad \square \end{aligned}$$

As we see in Example 10.13, the computation becomes quite involved even though $n = 2$. For a general $n$-period problem, a closed-form formula based on the Laplace transform technique is developed by Perrakis and Henin [29]. Unfortunately, this closed-form formula requires rather extensive computations. Eventually, one has to use some sort of computer simulation as a practical method. Such a simulation technique will be discussed in Chapter 12.

### 10.3.6 Cash Flows with Uncertain Timing: Discrete Case

In this section we will consider the situation in which the cash flows occur at discrete points in time and the project life is also a discrete random variable. Two situations will be examined: (1) the random variables are independent and nonidentically distributed, and (2) the random variables are dependent and nonidentically distributed.

***Independent and Nonidentically Distributed.*** To develop the mean and variance expression of the *PV* distribution, we define the following:

$N$ = project life, a random variable

$N'$ = maximum value of $N$

$X_k$ = the present value of cash flow occurring at the end of period $k$,

$P_j$ = probability of the project terminating at the end of period $j$,

We have

$$X_k = \frac{F_k}{(1+i)^k}$$

$$Var(X_k) = \frac{Var(F_k)}{(1+i)^{2k}}$$

$$Cov(X_j, X_k) = \frac{Cov(F_j, F_k)}{(1+i)^{j+k}}$$

Then the *PV* of the project with random project life $N$ can be expressed as

$$\begin{aligned} PV(i) &= X_0 + X_1 + X_2 + \cdots + X_N \\ &= \sum_{k=0}^{N} X_k \end{aligned}$$

If $N$ is a nonnegative integer-valued random variable independent of the $X$s, the associated mean and variance of the *PV* are

$$\begin{aligned} E[PV(i)] &= E\left(\sum_{k=0}^{N} X_k\right) \\ &= \sum_{j=0}^{N'} P_j\left[\sum_{k=0}^{j} E(X_k)\right] \end{aligned} \tag{10.65}$$

and

$$Var[PV(i)] = \sum_{j=0}^{N'} P_j E\left\{\left[\sum_{k=0}^{j} (X_k)\right]^2\right\} - \left\{\sum_{j=0}^{N'} P_j\left[\sum_{k=0}^{j} E(X_k)\right]\right\}^2 \tag{10.66a}$$

If the project ends at period $j$, the present value is

$$S_j = \sum_{k=0}^{j} X_k$$

From the definition of the variance we know that

$$E(S_j^2) = Var(S_j) + [E(S_j)]^2 \tag{10.66b}$$

Substituting $S_j$ into Eq. 10.66b gives

$$E(S_j^2) = \sum_{k=0}^{j} Var(X_k) + \left[\sum_{k=0}^{j} E(X_k)\right]^2 \tag{10.66c}$$

The unconditional value is

$$E[PV(i)] = \sum_{j=0}^{N'} P_j \, E\left\{\left[\sum_{k=0}^{j} (X_k)\right]^2\right\}$$

$$= \sum_{j=0}^{N'} P_j \left\{\sum_{k=0}^{j} Var(X_k) + [\sum_{k=0}^{j} E(X_k)]^2\right\} \tag{10.66d}$$

When we substitute Eq. 10.66d into Eq. 10.66a, we obtain

$$Var[PV(i)] = \sum_{j=0}^{N'} P_j \left\{\sum_{k=0}^{j} Var(X_k) + \left[\sum_{k=0}^{j} E(X_k)\right]^2\right\}$$

$$- \left[\sum_{j=0}^{N'} P_j \sum_{k=0}^{j} E(X_k)\right]^2 \tag{10.66e}$$

### *Example 10.14*

Consider the project cash flows given in Example 10.8. Suppose that the project's life is a discrete random variable as follows:

| $j$ | $E(F_j)$ | $Var(F_j)$ | $P_j$ | $E(X_j)$ | $Var(X_j)$ |
|---|---|---|---|---|---|
| 0 | −\$1,000 | $50^2$ | 0 | −\$1,000 | 2,500.00 |
| 1 | 500 | $100^2$ | 0.30 | 454.55 | 8,264.46 |
| 2 | 800 | $200^2$ | 0.70 | 661.16 | 27,320.54 |
| | | | 1.00 | \$115.71 | 38,085.00 |

The entries for $E(X_j)$ and $Var(X_j)$ represent the means and variances of the present values of the cash flow $F_j$ occurring at the end of period $j$, at $i = 10\%$.

Assuming *independence* among the random variables, we evaluate Eq. 10.65 and Eq. 10.66e to determine the mean and variance of the $PV$ for the project.

$$E[PV(10\%)] = \sum_{j=0}^{2} P_j\left[\sum_{k=0}^{j} E(X_j)\right]$$

$$= 0(-1{,}000) + 0.30(-1{,}000 + 454.55)$$
$$+ 0.70(-1{,}000 + 454.55 + 661.16)$$
$$= -\$82.64$$

$$Var[PV(10\%)] = \sum_{j=0}^{2} P_j\left\{\sum_{k=0}^{j} Var(X_k) + \left[\sum_{k=0}^{j} E(X_k)\right]^2\right\} - \left[\sum_{j=0}^{2} P_j \sum_{k=0}^{j} E(X_k)\right]^2$$

$$= 0.30[(2{,}500 + 8{,}264.46) + (-545.45)^2]$$
$$+ 0.70[(2{,}500 + 8{,}264.46 + 27{,}320.54) + (115.71)^2]$$
$$- (-82.64)^2$$
$$= 0.30(308{,}280.16) + 0.70(51{,}473.80) - (6{,}829.37)$$
$$= 121{,}686.34$$

or

$$\text{Standard deviation of } PV(10\%) = \$348.84 \quad \square$$

***Dependent and Nonidentically Distributed.*** If we relax the independence assumption, the $E[PV(i)]$ remains unchanged but the variance expression will be

$$Var[PV(i)] = \sum_{j=0}^{N'} P_j\left\{\sum_{k=0}^{j} Var(X_k) + 2\sum_{k=0}^{j-1}\sum_{m=k+1}^{j} Cov(X_k, X_m) + \left[\sum_{k=0}^{j} E(X_k)\right]^2\right\} - \left[\sum_{j=0}^{N'} P_j \sum_{k=0}^{j} E(X_k)\right]^2 \qquad (10.66f)$$

## 10.4 STATISTICAL DISTRIBUTIONS OF NET PRESENT VALUE

Up to this point, we have been using the variance as the principal measure of risk to summarize the dispersion of a statistical distribution. Rather than use the variance or standard deviation as the single risk measure, we may compute the probability that the net present value of a project will be negative, $P\{PV \leq 0\}$—that is, that the present value of cash inflows will be less than the present value of the cost of the project. To derive this probability, we need to have at least part of the statistical distribution of the present values. In this section, we show how this information is found.

### 10.4.1 Discrete Cash Flows Described by a Probability Tree

Suppose we assume a discrete distribution of cash flows for each time period as shown in Figure 10.5. The figure shows the probabilities and the project returns for a two-period investment. This probability tree provides a useful framework for analysis. For simplicity, only two outcomes are allowed for each node. In our example the project requires an initial outlay of \$100. At $n = 1$ there are a 0.5 probability of realizing a cash inflow of \$50 and a 0.5 probability of obtaining a cash inflow of \$70. If the \$50 is realized at $n = 1$, there is a 0.3 probability of obtaining a cash inflow of \$60 at $n = 2$ and so forth. Let us assume $i = 10\%$.

We see immediately that there are four possible cash flow sequences. By first computing the net present value for every cash flow sequence and finding the joint probabilities of those cash flow sequences, we obtain Table 10.5. Collecting these *PV*s and the joint probabilities will provide the statistical distribution of *PV* for the project (as presented in Figure 10.6). From Table 10.5 it follows that

$$\begin{aligned} E[PV(10\%)] &= -4.96(0.15) + 11.57(0.35) + 0(0.20) + 38.02(0.30) \\ &= \$14.71 \\ Var[PV(10\%)] &= [(-4.96)^2(0.15) + (11.57)^2(0.35) + 0^2(0.20) \\ &\quad + (38.02)^2(0.30)] - (14.71)^2 \\ &= (16.37)^2 \end{aligned}$$

Then we compute

$$\begin{aligned} P\{PV(i) \leq 0\} &= P\{PV = -4.96\} + P\{PV = 0\} \\ &= 0.15 + 0.20 \\ &= 0.35 \end{aligned}$$

There is a 35% chance that the project may end up with no gains at all. This type of additional information is more meaningful to many people in business, because a probability requires less statistical sophistication to translate into business operational terms than does the mean or variance alone.

### 10.4.2 Use of the First Two Statistical Moments

The mean and variance of $PV(i)$ that were obtained for various cash flow series in the previous sections provide a basis for estimating the profitability of the investment in probabilistic terms. For example, if we know *only* the first two statistical moments about $PV(i)$ and have no information about the shape of the distribution, Chebyshev's inequality can be used to make probability estimates. This theorem states that

$$P\{\,|\,PV(i) - E[PV(i)]\,| \geq k\,(\sigma[PV(i)])\} \leq \frac{1}{k^2} \tag{10.67}$$

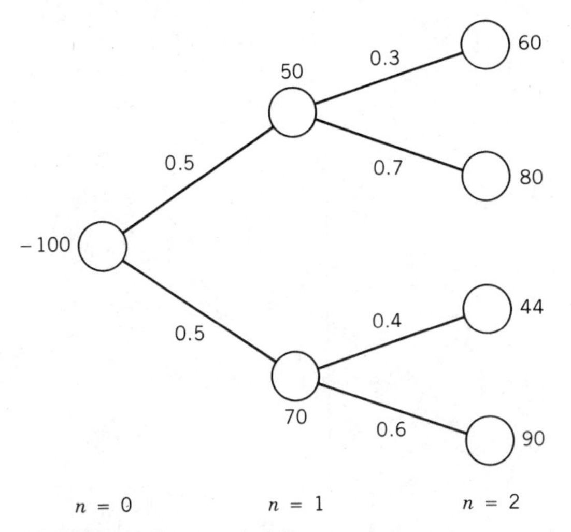


**FIGURE 10.5.** A probability tree.

or, expressed slightly differently,

$$P\{E[PV(i)] - k\sigma[PV(i)] \leq PV(i) \leq E[PV(i)] + k\sigma\,[PV(i)]\} \geq 1 - \frac{1}{k^2} \qquad (10.68)$$

This permits us to place bounds on probability statements without knowing the probability distribution. For example, if we know that $E[PV(i)] = 6$ and $Var[PV(i)] = 2^2$ for a certain project, the probability that $PV(i)$ falls between 2 and 10 (two standard deviations from the mean) is

$$P\{2 \leq PV(i) \leq 10\} = P\{6 - 2(2) \leq PV(i) \leq 6 + 2(2)\} \geq 1 - \tfrac{1}{4} = 0.75$$

If we can assume that $PV(i)$ is a normal distribution with mean $E[PV(i)]$ and variance $Var[PV(i)]$, we can make a more precise probabilistic statement of $PV$ for the project. This is possible because the normal distribution can be fully described by the first two statistical moments (this will also be true for many standard symmetrical distributions). Let

$$Z = \frac{PV(i) - E[PV(i)]}{\sqrt{Var[PV(i)]}}$$

**Table 10.5** ***Joint Probabilities***

| | $F_n$ | | | | |
|---|---|---|---|---|---|
| Cash Flow Sequence | $n = 0$ | $n = 1$ | $n = 2$ | Joint Probability | $PV(10\%)$ |
| 1 | −100 | 50 | 60 | 0.15 | −$4.96 |
| 2 | −100 | 50 | 80 | 0.35 | 11.57 |
| 3 | −100 | 70 | 44 | 0.20 | 0 |
| 4 | −100 | 70 | 90 | 0.30 | 38.02 |
| | | | | 1.00 | $14.71 |

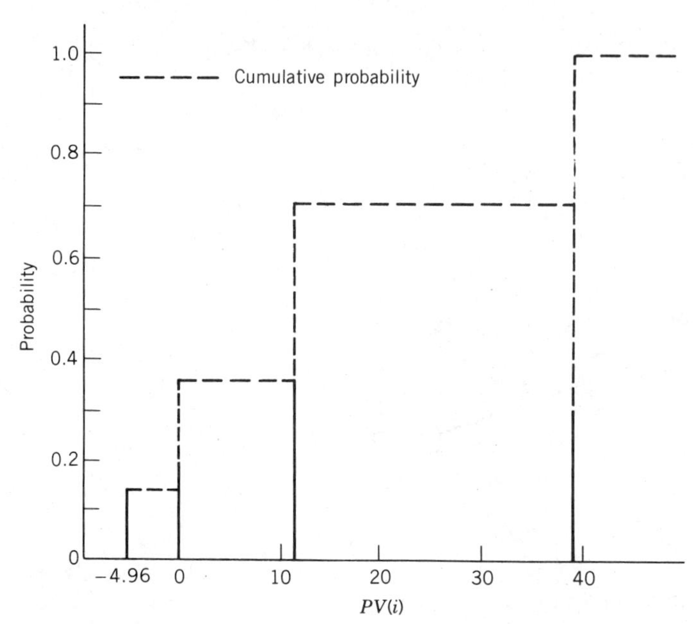


**FIGURE 10.6.** Discrete present value distribution.

then

$$P\{PV(i) \leq 0\} = P\left\{Z \leq \frac{0 - E[PV(i)]}{\sqrt{Var\,[PV(i)]}}\right\} \tag{10.69}$$

and we know that $P\{Z \leq z\}$ can be found from using the standard normal distribution table in Appendix B.

## *Example 10.15*

To illustrate the method, let us take the project in Example 10.8 and further assume that all $F_0$'s are normally distributed. Since $E[PV(10\%)] = \$115.70$ and $Var[PV(10\%)] = (237.87)^2$,

$$P[PV(.10) \leq 0] = P\left\{Z \leq \frac{0 - 115.70}{237.87}\right\} = P[Z \leq -.49] = 0.31 \quad \square$$

Some of the conditions under which $PV(i)$ will be normally distributed are examined in great detail by Hillier [12]. These are summarized as follows.

1. If $F_0, F_1, \ldots, F_n$ has a multivariate normal distribution, $PV(i)$ (being a linear function of the $F_n$'s) will itself be normally distributed. This statement is always true without regard to the relations among the $F_n$'s (either independent or dependent).
2. Since $PV(i)$ is the sum of discounted random cash flows ($F_n$'s), it follows from the central limit theorem (CLT) of statistics that, under certain condi-

tions, $PV(i)$ is also asymptotically normally distributed with $E[PV(i)]$ and $Var[PV(i)]$. The CLT holds best when $F_n$'s are identically distributed but independent random variables. For independent random variables that are not identically distributed, it is still possible to relax this requirement greatly and to assume only, in effect, that no individual cash flows ($F_n$'s) have such large variances relative to the others that their distributions dominate the sum $PV(i)$ [20].

3. A number of CLTs for *dependent* random variables have been developed (e.g., by Doob [6] and Loene [19]). In general, the dependent case is much more difficult than the independent case, so it is not possible to state a single result that completely solves this problem. The most relevant theorem for the sum of dependent random variables is that, if $F_n$'s are uniformly bounded random variables and $Var[PV(i)]$ approaches infinity as $n$ becomes large, $PV(i)$ will also be asymptotically normal.
4. An essential theoretical difficulty of assuming the CLT for $F_n$'s is that $PV(i)$ is not the direct sum of random variables but rather the discounted sum. The effect of this is that the shape of $PV(i)$ may be dominated by the early cash flows, especially at a high discount rate. The fact that $PV(i)$ is a discounted sum further implies that even if we had independently distributed cash flows continuing forever, the variance of the present value of the first $n$ cash flows would remain finite as $n$ approached infinity, and it is known in this case that $PV(i)$ will not tend to normality unless each of the net cash flows is normally distributed. In practice, however, each of the earlier net cash flows may itself be a sum of a number of variables. In some cases, therefore, it may be reasonable to assume that they are normally distributed and thus circumvent the difficulty mentioned [36].

### 10.4.3 Use of the First Four Statistical Moments

Knowledge of the third and fourth moments in addition to the first two moments can improve the accuracy of the estimates we make for various profit levels. In effect, the additional knowledge of the third and fourth moments allows us to use the Gauss–Camp–Meidall inequality for a random variable $X$ with central moments $\mu_j$ for $j = 1, 2, 3,$ and 4,

$$P[\,|X - \mu| > k\sigma] \leq \frac{4}{9}\frac{1 + S^2}{(k - S)^2} \quad \text{if } k > S \tag{10.70}$$

where

$S = |\mu - M_0|/\sigma$

$M_0$ = modal point

Assuming that the $PV(i)$ distribution can be approximated by the highly flexible Pearson system of curves, the model point $M_0$ can be computed (see Kendall and Stuart [17]) as

$$M_0 = \mu - \frac{\sigma\sqrt{\beta_1}(\beta_2 + 3)}{2(5\beta_2 - 6\beta_1 - 9)} \tag{10.71}$$

where

$$\beta_1 = \frac{\mu_3^2}{\mu_2^3} = \alpha_3^2 \quad \text{and} \quad \beta_2 = \frac{\mu_4}{\mu_2^2} = \alpha_4$$

(see Eq. 10.4).

A more relevant application of the first four statistical moments of $PV(i)$ may be in fitting a probability density function directly to the Pearson system of distributions. The Pearson system of distributions encompasses density functions of a wide variety of shapes and combinations of skewness and kurtosis. Interested readers should consult the paper by Kottas and Lau [18] on this topic.

### *Example 10.16*

Suppose that we know the following first four statistical moments for the $PV(i)$ distribution of a certain project: $\mu = \$26$, $\mu_2 = 88$, $\mu_3 = 131$, and $\mu_4 = 15{,}520$. Therefore,

$$\beta_1 = \frac{131^2}{88^3} = 0.2770 \quad \text{and} \quad \beta_2 = \frac{15{,}520}{88^2} = 0.2004$$

From these we estimate

$$M_o = 26 - \frac{\sqrt{88}\sqrt{0.2770}(3.2004)}{2[5(0.2004) - 6(0.2770) - 9]} = 25.54$$

Thus, $S = \left| 26 - 25.54 \right| / \sqrt{88} = 0.049$. Let $k = 2$, and

$$\frac{4}{9}\left[\frac{1 + 0.049^2}{(2 - 0.049)^2}\right] = 0.117$$

Thus, the probability that $PV(i)$ will be within two standard deviations is

$$P\{7.24 < PV(i) < 44.76\} \geq 1 - 0.117 = 0.883$$

If we were given only the first two moments and had to use Chebyshev's inequality, the probability we obtained would be 0.75. Thus, knowing the additional two higher moments, we can tighten the bound of the probability estimate. □

## 10.5 ESTIMATING RISKY CASH FLOWS

In the previous sections, we derived the probability distribution of the present value function and the expression of its mean and variance. These calculations, however, involve the mean and variances of the various individual cash flows and the covariances among them. Knowledge of this information is a prerequisite to our analysis. Therefore, in this section we discuss some reasonable approaches to estimating these statistical parameters.

### 10.5.1 Beta-Function Estimators for Single Cash Flows

The estimating procedure suggested by Hillier [12] is to make an "optimistic" estimate, a "pessimistic" estimate, and a "most likely" estimate for each cash flow. Hillier further defines the nature of each estimate: The *optimistic estimate* of cash flow assumes that "Everything will go as well as reasonably possible." The *pessimistic estimate* of cash flow assumes that "everything will go as poorly as reasonably possible." The third estimate considers the amount of net cash flow that is *most likely* to occur. These three estimates are used as the upper bound, the lower bound, and the mode of the corresponding cash flow probability distribution. The probability distribution itself is assumed to be a beta distribution with a standard deviation of one-sixth of the spread between the upper and lower bounds (the optimistic and pessimistic estimates).

Under the assumption that each cash flow has a beta distribution with a spread of six standard deviations between the bounds, the mean and variance become explicit functions of the bounds and mode. Given the following estimates of the bounds and mode of cash flow distribution of $F_n$ at the end of period $n$,

$$\text{Est}_\text{o}(F_n) = H \quad \text{(optimistic estimate)}$$

$$\text{Est}_\text{p}(F_n) = L \quad \text{(pessimistic estimate)}$$

$$\text{Est}_\text{m}(F_n) = M_0 \quad \text{(most likely estimate)}$$

we find

$$E(F_n) = \tfrac{1}{6}(H + 4M_0 + L) \tag{10.72}$$

and

$$Var(F_n) = \left(\frac{H - L}{6}\right)^2 \tag{10.73}$$

The model of an underlying beta distribution has been assumed primarily for convenience, but these estimating formulas are based on a system developed for the PERT (Program Evaluation and Review Technique) network planning and scheduling technique [5]. It is important to note that the difference between the approximate expected values as just calculated and those resulting from the exact formula is relatively small for a wide range of beta-distribution conditions. The greatest percentage difference between the simple PERT approximation of expected value and the exact value for the associated beta distribution is known to be 18.8% [24]. We will examine the characteristics of the beta distribution.

As seen in Table 10.2, the beta distribution for random variable $X$ is defined as

$$f(x) = \frac{\Gamma(a + b + 2)}{\Gamma(a + 1)\Gamma(b + 1)} x^a(1 - x)^b, \qquad 0 \leq x \leq 1$$

This equation becomes substantially less forbidding once we realize that the first part is merely a constant used to satisfy the basic requirement that the area under the curve be unity if it is to represent a probability distribution. Thus, for any selection of the parameters $a$ and $b$, we may rewrite the equation as

$$f(x) = \frac{1}{c} x^a (1 - x)^b, \qquad 0 \leq x \leq 1 \tag{10.74}$$

where $c$ is a constant.

It becomes clear that, for all positive values of $a$ and $b$, the curve will meet the horizontal axis at the points $x = 0$ and $x = 1$. When $a = b$, the curve will be symmetric with the mode at $M_0 = 0.5$. It should also be noted that when $a = 0$ but $b > 0$, the distribution assumes a J shape with the top of the J at $x = 0$, but when $b = 0$ and $a > 0$ the shape is reversed, with the top at $x = 1$. It is also seen that, if $a = b = 0$, Eq. 10.74 reduces to $f(x) = 1/c$, which is the probability density function for the uniform distribution. We further note that, if $a > b$, the beta distribution is skewed to the left (negative skewness). With $a < b$, the distribution is skewed to the right (positive skewness). These relationships are shown in Figure 10.7.

Notice that the standard beta distribution is bounded between 0 and 1 and the shape of the distribution is totally determined by the specification of $a$ and $b$. To use the beta distribution in some specialized form as a distribution of $PV(i)$, we need to extract the values of $a$ and $b$ from the $H$, $M_0$, and $L$ values. A transformation of these values to the standard beta yields

$$X = \frac{F_n - L}{H - L}$$

or

$$F_n = L + (H - L)X \tag{10.75}$$

This linear relationship transforms the standard beta distribution into

$$f(F_n) = \frac{\Gamma(a + b + 2)}{\Gamma(a + 1)\Gamma(b + 1)(H - L)^{a+b+1}} (F_n - L)^a (H - F_n)^b \qquad \text{for } L \leq F_n \leq H \tag{10.76}$$

Expressions for the mean, variance, and mode of this distribution are given in McBridge and McClelland [24].

$$E[F_n] = \mu = L + \frac{(H - L)(a + 1)}{a + b + 2} \tag{10.77}$$

$$Var[F_n] = \sigma^2 = \frac{(H - L)^2 (a + 1)(b + 1)}{(a + b + 2)^2 (a + b + 3)} \tag{10.78}$$

$$M_0 = \frac{Lb + Ha}{a + b} \qquad \text{for } a, b \geq -1 \tag{10.79}$$

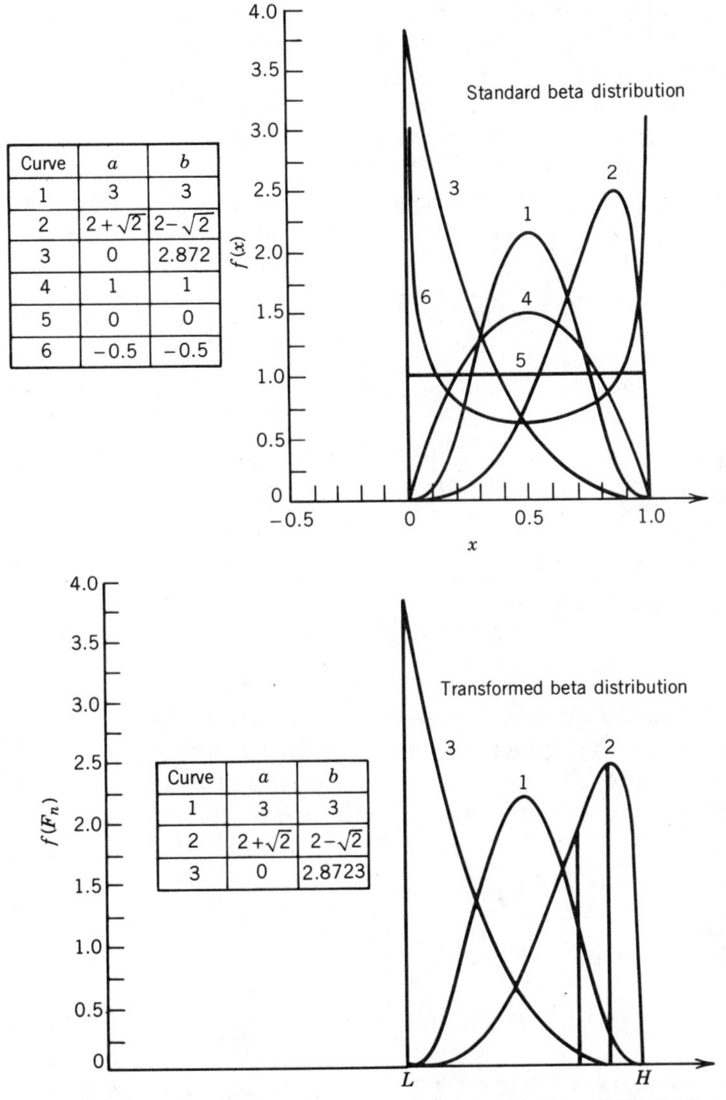


**FIGURE 10.7.** Shapes of beta distribution. Source: McBridge and McClelland [24].

Expressed in terms of beta parameters, these equations become

$$a = \frac{(\mu - L)^2 \left(\dfrac{H - \mu}{H - L}\right)}{\sigma^2} - \frac{\mu - L}{H - L} - 1 \tag{10.80}$$

$$b = \frac{\dfrac{\mu - L}{H - L}(H - \mu)^2}{\sigma^2} + \frac{\mu - L}{H - L} - 2 \tag{10.81}$$

The first task in using a beta approximation to determine the expected net present value and variance for a given cash flow is to select appropriate values for the parameters $a$ and $b$. The random variable $F_n$ in Eq. 10.75 is simply a linear function of the standard beta random variable, and it is much easier for us to work with the standard beta function to find $a$ and $b$. Once these are found, we can use Eqs. 10.77 and 10.78 to determine $\mu$ and $\sigma^2$ for the beta distribution with some arbitrary range $(L, H)$. The first step is to transform the mode $M_0$ (most likely estimate) to the mode $M$ of the standardized beta random variable.

$$M = \frac{M_0 - L}{H - L} = \frac{a}{a + b} \tag{10.82}$$

For the standardized beta distribution we know that

$$E(X) = \frac{a + 1}{a + b + 2} \tag{10.83}$$

and

$$Var(X) = \frac{(a + 1)(b + 1)}{(a + b + 3)(a + b + 2)^2} \tag{10.84}$$

Since $M$ is the only known value from Eq. 10.82, there is still not enough information to compute the values of $a$ and $b$. To solve this problem, we may assume that the curve's standard deviation is one-sixth of the range. This seems to be a reasonable approach because one-sixth of the range can be used as a rough estimate of standard deviation for most unimodal distributions. It follows that

$$Var(X) = (\tfrac{1}{6})^2 \tag{10.85}$$

Rewriting Eq. 10.82 in terms of $b$ gives us

$$b = a/M - a \tag{10.86}$$

Substituting Eqs. 10.85 and 10.86 into Eq. 10.84 and rearranging terms, we find that

$$a^3 + (7M - 36M^2 + 36M^3)a^2 - 20M^2a - 24M^3 = 0 \tag{10.87}$$

As seen in Eq. 10.82, computing $M$ from the values of $L$, $M_0$, and $H$ requires no more than simple arithmetic. Once $M$ is found, $a$ is determined from Eq. 10.87, and we return to the modal equation (Eq. 10.82) again to solve for $b$.

To reduce the effort needed to solve the cubic form of Eq. 10.87, Greer [10] developed the graph seen in Figure 10.8. As an example of its use, with cash flow estimates of $H$ = \$150, $M_0$ = \$100, and $L$ = \$70 for a given project period, we compute the standardized mode as $M = (100 - 70)/(150 - 70) = 0.375$. This value is located on the horizontal axis of the graph, and the value for $a$ is read

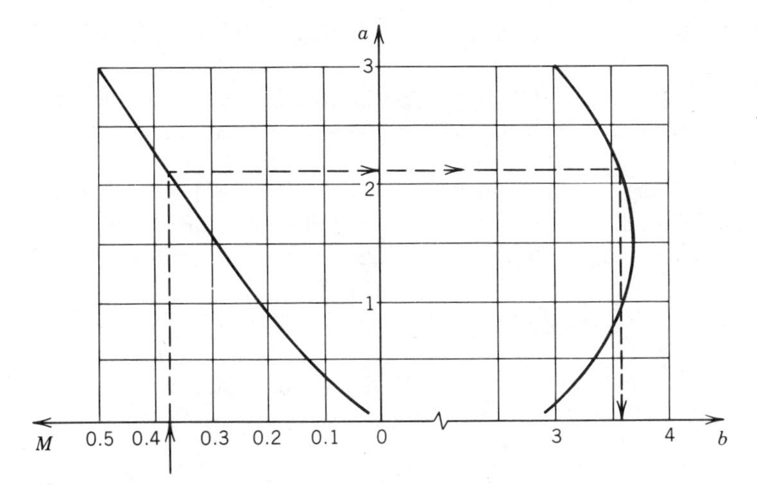


**FIGURE 10.8.** Graphic chart for finding the parameters $a$ and $b$. Source: Greer [10].

from the vertical axis. Then we return to the horizontal scale for $b$. This gives us the approximations $a = 2.14$ and $b = 3.6$. If we have to solve Eq. 10.87 for $a$ with $M = 0.375$, the exact values are $a = 2.135$ and $b = 3.559$.

It might be interesting at this point to compare the PERT-type estimates given in Eqs. 10.72 and 10.73 with the estimates obtained by assuming the analytical beta-distribution parameters. For our example the simplifying approximations made in PERT give us

$$E(F_n) = \frac{150 + 4(100) + 70}{6} = 103.33$$

$$Var(F_n) = \left(\frac{150 - 70}{6}\right)^2 = 177.7$$

Using the beta parameters $a = 2.135$ and $b = 3.559$, we obtain

$$E(F_n) = 70 + \frac{(150 - 70)(2.135 + 1)}{2.135 + 3.559 + 2} = 102.60$$

$$Var(F_n) = \frac{(150 - 70)^2(2.135 + 1)(3.559 + 1)}{(2.135 + 3.559 + 2)^2(2.135 + 3.559 + 3)} = 177.7$$

It is no coincidence that the variance figures are identical in this example. This is so because we assumed the standard deviation to be one-sixth the range to derive the parameters $a$ and $b$ [34]. [A three-point approximation utilizing the median rather than the mode has been proposed by Keefer and Bodily [16]. The mean $= \sum p_j v_j$, variance $= \sum(p_j v_j^2) - (\text{mean})^2$, with $p_1 = 0.185, p_2 = 0.630,$ $p_3 = 0.185, v_1 = x(0.05), v_2 = x(0.50), v_3 = x(0.95)$, where $x(\cdot)$ denotes the $(\cdot)$

fractile of the random variable. In terms of probability assessment, using the median rather than the mode has the advantage that the median naturally lends itself to straightforward consistency checks. On the other hand, it does require the 0.05 and 0.95 fractiles, which in practice are somewhat more difficult to assess accurately.]

One of the advantages of knowing the beta parameters is that we can make probabilistic statements about the range of cash flows. In our example we found that $a = 2.135$ and $b = 3.559$. Use of these values in Eq. 10.76 gives

$$f(F_n) = (\text{constant})(F_n - 70)^{2.135}(150 - F_n)^{3.559}$$

In principle, we may directly evaluate the following equation to find the probability of realizing a cash flow between $A$ and $B$.

$$P[A \le F_n \le B] = \int_A^B f(F_n)\, dF_n$$

Since $f(F_n)$ contains the constant, which is difficult to evaluate for noninteger values of the parameters, we need to find the same probabilities by an alternative method.

$$\int_A^B f(F_n)\, dF_n = \frac{\int_A^B (F_n - 70)^{2.135}(150 - F_n)^{3.559}\, dF_n}{\int_{70}^{150} (F_n - 70)^{2.135}(150 - F_n)^{3.559}\, dF_n} \tag{10.88}$$

Recognizing that a fairly complex set of calculations must still be performed with this equation, Greer [10] developed another graph that simplifies the process of assigning probability numbers to the cash flow intervals, as shown in Figure 10.9. Assume that we are interested in finding $P\{100 \le F_n \le 110\}$. This is equivalent to finding $P\{0.375 \le x \le 0.5\}$ for the standardized beta variable $x$. We use the graph by locating $M$ (in our example, $M = 0.375$) on the left margin. After consulting the scale at the top of the chart to locate the curve that corresponds to $x = 0.5$, we project the intersection of the $M = 0.375$ line and the $x = 0.5$ curve to the lower scale, where $P\{x \le 0.5\}$ is read as 0.707. We use the same procedure and then interpolate to locate the point $x = 0.375$, with $P\{x \le 0.375\} = 0.448$. Subtracting, we find $P\{0.375 \le x \le 0.5\} = P\{100 \le F \le 110\} = 0.259$.

### 10.5.2 Hillier's Method for Correlated Cash Flows

When cash flows are statistically correlated, the correlation must be known in order to calculate correctly the variance of the present value. In principle, the correlation could be obtained by directly estimating the correlation coefficient for each pair of random variables, but there tend to be too many pairs for this approach to be practical. To circumvent this difficulty, Hillier [12] developed a model for the pattern of correlations between cash flows that is both reasonable and simple to use, with the following restrictive assumptions.

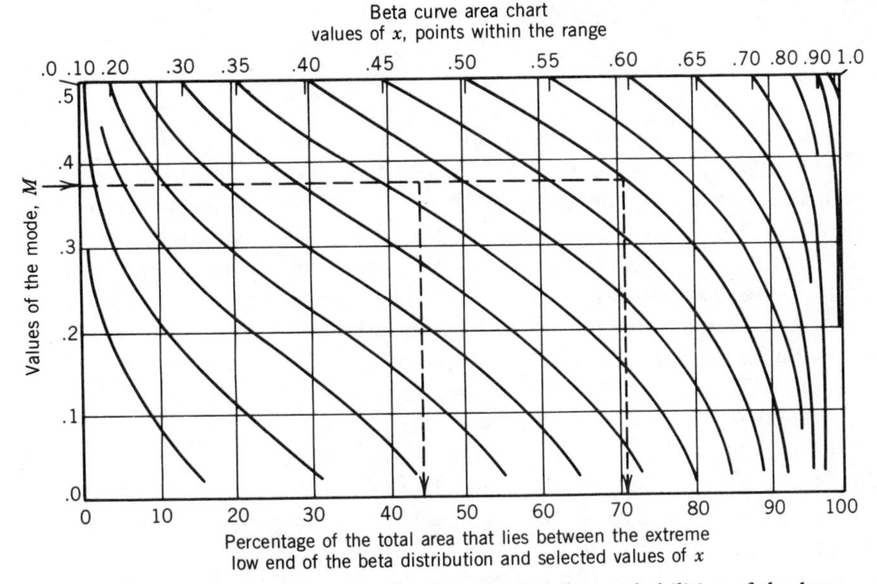


**FIGURE 10.9. Graphic chart for determining the probabilities of the beta distribution. Source: Greer [10].**

1. The random variables are Markov-dependent through time. For example, the distribution of $F_1$ depends only on $F_0$, that of $F_2$ only on $F_1$, and so forth. In other words, the partial correlation coefficient between $F_0$ and $F_2$, given $F_1$, is zero.
2. The correlation coefficient for $F_0$ and $F_1$ is the same as for $F_1$ and $F_2$, $F_2$ and $F_3$, and so forth.

Using these assumptions, we can compute the entire set of total correlation coefficients in a simple way.

$$\rho_{01} = \rho, \qquad \rho_{02} = \rho^2, \qquad \rho_{03} = \rho^3$$
$$\rho_{12} = \rho, \qquad \rho_{13} = \rho^2, \qquad \text{and so on}$$

Thus, we can express any correlation coefficient between periods $j$ and $k$ by

$$\rho_{jk} = \rho^{|j-k|} \tag{10.89}$$

The method of estimating $\rho$ itself is based on work by Mood [25]. Assume that $F_0$ and $F_1$ are correlated random variables with means $E(F_0)$ and $E(F_1)$ and variances $\sigma_0^2$ and $\sigma_1^2$, respectively, and with a correlation coefficient of $\rho_{12}$. Suppose we have a given value $x$ for $F_0$; then we can estimate the expected value of $F_1$ from

$$E(F_1 \mid F_0 = x) = E(F_1) + \rho_{01}\frac{\sigma_1}{\sigma_0}[x - E(F_0)] \tag{10.90}$$

Rewriting this expression we have

$$\frac{E(F_1 \mid F_0 = x) - E(F_1)}{\sigma_1} = \rho_{01}\frac{x - E(F_0)}{\rho_0} \tag{10.91}$$

In other words, after normalizing $F_0$ and $F_1$, the expected deviation of $F_1$ from its unconditional mean is just $\rho_{01}$ times the deviation of $F_0$ from its unconditional mean. This provides a reasonably simple basis for developing a subjective estimate of $\rho_{01}$. The foregoing expressions describe what is commonly referred to as the least square regression line of $F_1$ on $F_0$. If $F_0$ and $F_1$ are bivariate normal variables, this procedure gives the best unbiased estimate of $\rho_{01}$; otherwise, it gives the best linear estimate according to the principle of least squares [12]. Any correlation value obtained is, by definition, contained in the interval $[-1, +1]$. If the value falls outside this interval, some multiplicative effect may be present, and Hillier's model may not be appropriate.

To implement the procedure, we need to select values of $F_0$ and then estimate the expected values of $F_1$ given the $F_0$'s. It is helpful to select values for $F_0$ that are three standard deviations above and below $E(F_0)$, and we can then use the formula for estimating the mean of a beta distribution.

### *Example 10.17*

Consider a project whose cash flow estimates are given as follows. The initial investment required for the project $F_0$, is considered to be independent of future cash flows. The $F_n$'s occurring after $n = 0$, however, are partially correlated with one another. The discount rate is known to be 10%.

| $n$ | Expected Flow, $E(F_n)$ | Estimated Standard Deviation, $\sigma_n$ |
|---|---|---|
| 0 | −\$500 | \$30 |
| 1 | 100 | 20 |
| 2 | 150 | 30 |
| 3 | 200 | 40 |
| 4 | 250 | 50 |
| 5 | 300 | 60 |

Since $F_0$ was assumed to be an independent variable, the correlation coefficient $\rho_{0k}$ for $k > 0$ will be zero. To find $\rho_{12}$, however, we must estimate pessimistic, most likely, and optimistic values for $F_2$ conditional on $F_1$.

First, given $F_1 = \$160$ (three standard deviations above $E(F_1)$, an optimistic outcome), a pessimistic estimate of $F_2$ is \$140, a most likely estimate of $F_2$ is \$180, and an optimistic estimate of $F_2$ is \$210. Then

$$E(F_2|F_1 = 160) = [140 + 4(180) + 210]\tfrac{1}{6} = \$178$$

$$\frac{178 - 150}{30} = \rho_{12}\frac{160 - 100}{20}$$

Solving for $\rho_{12}$ yields 0.33.

Second, given $F_1 = \$40$ (three standard deviation below $E(F_1)$, pessimistic), a pessimistic estimate of $F_2$ is \$70, a most likely estimate of $F_2$ is \$110, and an optimistic estimate of $F_2$ is \$210. Then

$$E(F_2|F_1 = 40) = [70 + 4(110) + 210]\,\tfrac{1}{6} = \$120$$

$$\frac{120 - 150}{30} = \rho_{12}\frac{40 - 100}{20}$$

That is, $\rho_{12} = 0.30$.

We might average the two values of 0.33 and 0.30 to estimate $\rho_{12}$. We can estimate $\rho_{23}$, $\rho_{34}$, and so on and average all these values to obtain our single estimate of $\rho$. Assuming that this combined average is $\rho = 0.3$, the following matrix of correlation coefficients can be constructed.

| $j$ \ $k$ | 0 | 1 | 2 | 3 | 4 | 5 |
|---|---|---|---|---|---|---|
| 0 | 1 | 0 | 0 | 0 | 0 | 0 |
| 1 | 0 | 1 | 0.3 | 0.9 | 0.027 | 0.0081 |
| 2 | 0 | 0.3 | 1 | 0.3 | 0.09 | 0.027 |
| 3 | 0 | 0.09 | 0.3 | 1 | 0.3 | 0.09 |
| 4 | 0 | 0.027 | 0.09 | 0.3 | 1 | 0.3 |
| 5 | 0 | 0.0081 | 0.027 | 0.09 | 0.3 | 1 |

Finally, we may calculate $Var[PV(10\%)]$ for the project as

$$Var[PV(10\%)] = (\sigma_0)^2 + \sum_{n=1}^{5}\frac{\sigma_n^2}{(1.1)^{2n}} + 2\sum_{n=1}^{4}\sum_{j>n}^{5}\frac{1}{(1.1)^{n+j}}\rho_{nj}\sigma_n\sigma_j$$

Note that we have made some restrictive assumptions about $\rho$ in order to obtain an estimate of $Var[PV(i)]$ that is somewhere between the two extremes of complete independence and perfect correlation. □

## 10.6 SUMMARY

In this chapter the concept of probability was introduced as a means of describing the likelihood of the different possible investment outcomes. By computing the mean and variance of the present value of investment returns, we can summarize the investment characteristics under risk. It is theoretically desirable to derive an exact distribution of present values for a project, but practical considerations preclude this approach for all but a few simple cases. Computer simulation is suggested as an alternative means of deriving the present value distributions. (This approach will be explored in Chapter 12.) One of the most widely advocated approaches to risk in capital budgeting literature is the expected-utility criterion. The first and most difficult requirement of this criterion is to derive and specify the utility function that is to be applied. Because of its operational difficulty, the method often loses its attractiveness as a practical decision tool. In summary, the direct use of probability distributions appears to be the most feasible and practical. Our analysis in this chapter has been limited to single projects. Chapter 11 will extend this analysis to multiple investment projects.

## REFERENCES


1. BARNES, J. W., C. D. ZINN, and B. S. ELDRED, "A Methodology for Obtaining the Probability Density Function of the Present Worth of Probabilistic Cash Flow Profiles," *AIIE Transactions,* Vol. 10, No. 3, pp. 226–236, 1978.
2. BONINI, C. P., "Risk Evaluation of Investment Projects," *OMEGA,* Vol. 3, No. 6, pp. 735–750, 1975.
3. BUCK, J. R., and R. G. ASKIN, "Partial Means in the Economic Risk Analysis of Projects," *The Engineering Economist,* Vol. 31, No. 3, pp. 189–212, 1986.
4. CRAIG, C. C., "On the Frequency Function of XY," *Annals of Mathematical Statistics,* Vol. 7, pp. 1–15, 1936.
5. DONALDSON, W. A., "The Estimation of the Mean and Variance of a PERT Activity Time," *Operations Research,* Vol. 13, pp. 382–385, 1965.
6. DOOB, J. L., *Stochastic Processes,* Wiley, New York, 1953.
7. EPSTEIN, B., "Some Applications of the Mellin Transform in Statistics," *Annals of Mathematical Statistics,* Vol. 20, pp. 370–379, 1948.
8. GIFFIN, W. C., *Transform Techniques for Probability Modeling,* Academic Press, New York, 1975.
9. GOODMAN, L. A.,"On the Exact Variance of Products," *Journal of the American Statistic Association,* Vol. 55, pp. 708–713, 1960.
10. GREER, W. R., "Capital Budgeting Analysis with the Timing of Events Uncertain," *The Accounting Review,* Vol. 45, No. 1, pp. 103–114, January 1970.
11. HILLIER, F. S., "Derivation of Probabilistic Information for the Evaluation of Risky Investments," *Management Science,* Vol. 2, No. 3, pp. 485–489, 1963.
12. HILLIER, F. S., *The Evaluation of Risky Interrelated Investments,* North-Holland, Amsterdam, pp. 23–29, 1971.
13. HINKLEY, D. V., "On the Ratio of Two Correlated Normal Random Variables," *Biometrika,* Vol. 56, No. 3, p. 635, 1969.
14. HOOL, J. N., and S. MAGHSOODLOO, "Normal Approximation to Linear Combinations of Independently Distributed Random Variables," *AIIE Transactions,* Vol. 12, No. 2, pp. 140–144, 1980.
15. JEAN, W. H., *The Analytical Theory of Finance,* Holt, Rinehart and Winston, New York, 1970.
16. KEEFER, D. L., and S. E. BODILY, "Three-Point Approximations for Continuous Random Variables," *Management Science,* Vol. 29, No. 3, pp. 595–609, 1983.
17. KENDALL, M. G., and A. STUART, *The Advanced Theory of Statistics,* Vol. 1, Charles Griffin, London, 1969.
18. KOTTAS, J. F., and HON-SHIANG LAU, "A Four-Moments Alternative to Simulation for a Class of Stochastic Management Models," *Management Science,* Vol. 28, No. 7, pp. 749–758, 1982.
19. LOEVE, M., *Probability Theory,* 2nd edition, Van Nostrand, Princeton, N.J., 1960.

20. MANTELL, E. H., "A Central Limit Theorem of Present Values of Discounted Cash Flows," *Management Science,* Vol. 19, No. 3, pp. 314–318, 1972.
21. MAO, J. C. T., *Quantitative Analysis of Financial Decision,* Macmillan, New York, 1969.
22. MAO, J. C. T., "Models of Capital Budgeting, E–V vs E–S," *Journal of Financial and Quantitative Analysis,* Vol. 4, No. 5, pp. 657–675, 1970.
23. MAO, J. C. T., "Survey of Capital Budgeting: Theory and Practice," *Journal of Finance,* Vol. 25, pp. 349–360, 1970.
24. MCBRIDGE, W. J., and C. W. MCCLELLAND, "PERT and Beta Distribution," *IEEE Transactions on Engineering Management,* Vol. EM-14, No. 4, pp. 166–169, 1967.
25. MOOD, A., *Introduction to Theory of Statistics,* McGraw–Hill, New York, 1950.
26. NATIONAL BUREAU OF STANDARDS, *Tables of the Bivariate Normal Distribution Function and Related Functions,* Applied Mathematics Series, No. 50, Government Printing Office, Washington, D.C., 1971.
27. PARK, C. S., "Probabilistic Benefit–Cost Analysis," *The Engineering Economist,* Vol. 29, No. 2, pp. 83–100, 1984.
28. PARK, C. S., "The Mellin Transform in Probabilistic Cash Flow Modeling," *The Engineering Economist,* Vol. 32, No. 2, pp. 115–134, 1987.
29. PERRAKIS, S., and C. HENIN, "The Evaluation of Risky Investments with Random Timing of Cash Return," *Management Science,* Vol. 21, No. 1, pp. 79–86, 1974.
30. PETTY, W. J., D. F. SCOTT, and M. M. BIRD, "The Capital Expenditure Decision-Making Process of Large Corporations," *The Engineering Economist,* Vol. 20, No, 3, pp. 159–172, Spring 1975.
31. ROSENTHAL, R. E., "The Variance of Present Worth of Cash Flows under Uncertain Timing," *The Engineering Economist,* Vol. 23, No. 3, pp. 163–170, 1978. [See also Corrections, *The Engineering Economist,* Vol. 25, No. 3, pp. 230–233, 1980.]
32. SCHLAIFER, R., *Introduction to Statistics for Business Decisions,* McGraw–Hill, New York, 1961.
33. SMITH, D. E., "A Taylor's Theorem–Central Limit Theorem Approximation: Its Use in Obtaining the Probability Distribution of Long-Range Profit," *Management Science,* Vol. 18, No. 4, pp. 214–219, 1971.
34. SWANSON, L. A., and H. L. PAZER, "Implications of the Underlying Assumptions of PERT," *Decision Science,* Vol. 2, pp. 461–480, 1971.
35. TANCHOCO, J. M. A., J. R. BUCK, and L. C. LEUNG, "Modeling and Discounting of Continuous Cash Flows under Risk," *Engineering Costs and Production Economics,* Vol. 5, pp. 205–216, 1981.
36. WAGLE, B., "A Statistical Analysis of Risk in Capital Investment Projects," *Operational Research Quarterly,* Vol. 18, pp. 13–33, 1967.
37. WINKLER, R. L., G. M. ROODMAN, and R. R. BRITNEY, "The Determination of Partial Moments," *Management Science,* Vol. 19, No. 3, pp. 290–296, 1972.
38. YOUNG, D., and L. CONTRERAS, "Expected Present Worths of Cash Flows under Uncertain Timing," *The Engineering Economist,* Vol. 20, No. 4, pp. 257–268, 1975.

## PROBLEMS

**10.1.** Suppose that an uncertain lump-sum return $P$ is expected shortly after termination of a project. In the current uncertain market conditions, it appears that the earning interest rate ($i$) will fluctuate for the foreseeable future. It is thought, however, that both the lump sum and the interest rate are uniformly distributed but each has a unique range.

$P$: uniform ($1,000, $2,000)

$i$: uniform (9%, 11%)

Assuming that this uncertain sum is to be reinvested at an interest rate of $i$ over the next 3 years, compute the mean and variance of the total accumulated earnings.

**10.2.** An oil company drilling wells in a certain type of terrain finds that the monthly drilling cost is

$$C = 28{,}333T^{0.5} + 70{,}000T^{-0.5}$$

where $T$ is the drilling time (days). The company record indicates that $T$ has been a random variable governed by a gamma distribution with $a = 2.5$ and $b = 1$. Compute the mean and variance of the monthly drilling cost.

**10.3.** To determine the oil in place, we consider porosity, area, formation thickness, water saturation, and the formation volume factor as independent variables. We have

$$Z = \frac{7758\phi Ah(1 - W)E}{B^{1.05}}$$

where $Z$ = recoverable oil (barrels),
$\phi$ = porosity, dimensionless,
$A$ = area ($m^2$)
$h$ = thickness (m)
$W$ = water saturation, dimensionless,
$E$ = oil recovery factor, dimensionless,
$B$ = oil formation volume factor, dimensionless.

Suppose all parameters are triangularly distributed independent random variables with the following values.[2]

| Parameter | Minimum, $L$ | Most Likely, $M_0$ | Maximum, $H$ |
|---|---|---|---|
| $\phi$ | 0.125 | 0.133 | 0.140 |
| $A$ (acres) | 1,780 | 2,115 | 2,450 |
| $h$ (ft) | 225 | 250 | 275 |
| $W$ | 0.09 | 0.115 | 0.21 |
| $k = 1 - W$ | 0.79 | 0.85 | 0.91 |
| $E$ | 0.17 | 0.31 | 0.52 |
| $B$ | 1.214 | 1.240 | 1.260 |

a. Compute the mean and variance of $Z$.
b. Suppose the crude oil price per barrel fluctuates between $18.25 and $20.15. If the oil price is a uniformly distributed random variable over the bounds ($18.25 and $20.15), compute the mean and variance of the total oil revenue.

**10.4.** Consider the present value of a project portfolio consisting of two independent projects,

$$PV(i) = PV(i)_1 + PV(i)_2$$

[2]The Mellin transform will provide an easy way to compute the mean and variance of this type of equation with multiplicative random variables.

where $PV(i)_1$ is normally distributed with mean and variance $N(10, 4^2)$ and $PV(i)_2$ is uniformly distributed in a range between $-2$ and 8.

a. Find the first four central moments of $PV(i)$.
b. Approximate the probability that the $PV(i)$ will be positive.

**10.5.** Consider the break-even equation of a product

$$Z = (V - C)X - K$$

where $Z$ = profit (dollars),
$V$ = unit sales price,
$C$ = unit variable cost,
$X$ = sales volume,
$K$ = fixed cost.

Then the break-even volume is expressed by

$$X_b = \frac{K}{V - C}$$

Suppose that $K$, $V$, and $C$ are normally distributed with means and variances as follows.

$K$—$N(\$70{,}000,\ 10{,}000^2)$

$V$—$N(\$10,\ 1^2)$

$C$—$N(\$5,\ 0.5^2)$

a. Assuming that $K$, $V$, and $C$ are statistically independent random variables, determine the mean and variance of $X_b$.
b. Compute the probability that $X_b$ will be greater than 14,000 units.
c. Suppose the following correlation exists between $V$ and $C$: $\rho_{vc} = 0.3$. Determine the mean and variance of $X_b$.

**10.6.** Suppose that company XYZ wishes to predict the long-range profit (LRP) of a proposed new product. Specifically, the company wants to know the chances of LRP being above (or below) $X$ dollars over the 10-year period following the entrance of the product into the market. Assume that the equation is

$$\text{LRP} = (X_1 - X_2)[X_3 + 4(10^3)(2 - X_1)] - 10^{10}X_2[X_3 + 4(10^3)(2 - X_1)]^{-1} - X_4 - X_5 - X_6$$

where $X_1, X_2, \ldots, X_6$ denote the statistically independent variables. Suppose that the means and variances of the random variables $X_1, X_2, \ldots, X_6$ have been specified as follows.

| Component Value | Mean | Variance |
|---|---|---|
| $X_1$ | 2.0 | $(1.2)10^{-3}$ |
| $X_2$ | 1.5 | $(2.5)10^{-3}$ |
| $X_3$ | $10^6$ | $(1.6)10^9$ |
| $X_4$ | $(2.0)10^5$ | $(1.0)10^8$ |
| $X_5$ | $10^5$ | $(1.6)10^7$ |
| $X_6$ | $(8.0)10^4$ | $(9.0)10^6$ |

a. Using Eqs. 10.40 and 10.41, obtain the (approximate) values of the mean and variance of LRP.
b. If the probability distribution of LRP is assumed to be approximately normal, what is the probability that the long-range profit will exceed \$100,000?

**10.7.** Consider a one-product company whose profit function takes one of the following forms,

$$\text{Type 1:} \quad Z = a + bXY$$

$$\text{Type 2:} \quad Z = a + bX + cX^2$$

$$\text{Type 3:} \quad Z = a + bXY + cX^2$$

where $X$ and $Y$ are normally distributed random variables, not necessarily statistically independent, and $a$, $b$, and $c$ are constants. Let $\rho$ be the coefficient of correlation between $X$ and $Y$.

a. Compute the means and variances of the three types of profit equations.
b. Suppose the company estimates the following probabilistic income for a typical operating year.

- Volume ($X$)—$N(10^5, 10^8)$; that is, $X$ is normally distributed with mean $10^5$ and variance $10^8$.
- Sales price is constant at \$10 per unit. Thus, sales are $10X$.
- Unit variable manufacturing cost ($Y$)—$N(5, 0.04)$.
- Volume ($X$) and unit variable manufacturing cost ($Y$) are statistically independent. Variable manufacturing cost (VMC) = $XY$.
- Semivariable manufacturing cost has the following linear relationship to volume ($X$).

$$\text{SVM} = -\$30{,}000 + 0.5X$$

- Semivariable administrative cost (SAC) has the following quadratic relationship to volume ($X$).

$$\text{SAC} = -\$40{,}000 + 0.25X + (0.64)10^{-5}X^2$$

- Fixed costs (excluding depreciation): \$190,000.
- Depreciation: \$100,000.
- Tax rate: 40%.

Determine the mean and variance of the net after-tax cash flow for the operating year.

**10.8.** The total construction cost of a typical chemical processing system made up of three subsystems is estimated from

$$Y = \{[(1 + X_1)(1 + X_2 + X_3 + X_2X_3) + X_4 + X_5 + X_6](1 + X_7)(1 + X_8X_9)\} \times (5.7X_{10} + 9.4X_{11} + X_{12})$$

where the variables and their means, variances, and correlations are as follows.

| | $X_j$ | $\mu$ | $\sigma^2$ |
|---|---|---|---|
| $^*X_1$ | Overhead rate | 0.24 | 0.01416 |
| $^*X_2$ | Contractors' fees | 0.10 | 0.005495 |
| $^*X_3$ | Engineering fees | 0.099 | 0.00004275 |
| $^*X_4$ | Freight | 0.05 | 0.00004651 |
| $^*X_5$ | Shakedown | 0.05 | 0.00002693 |
| $^*X_6$ | Spares | 0.01 | 0.00001163 |
| $^*X_7$ | Contingency | 0.20 | 0.001163 |
| $X_8$ | Interest rate | 0.16 | 0.00009769 |
| $X_9$ | Construction time (months) | 0.375 | 0.03435 |
| $X_{10}$ | Subsystem 1 ($\$10^3$) | 146.0 | 226.0 |
| $X_{11}$ | Subsystem 2 ($\$10^3$) | 111.0 | 130.3 |
| $X_{12}$ | Subsystem 3 ($\$10^3$) | 167.0 | 295.8 |

*As a fraction of total system cost, $X_{10} + X_{11} + X_{12}$.

*Correlation coefficients:*

$\rho_{1,10} = \rho_{1,11} = \rho_{1,12} = 1.0$

$\rho_{4,10} = \rho_{4,11} = \rho_{4,12} = 0.8$

$\rho_{6,10} = \rho_{6,11} = \rho_{6,12} = 0.9$

*Covariance matrix:*

| | $X_9$ | $X_{10}$ | $X_{11}$ | $X_{12}$ |
|---|---|---|---|---|
| $X_1$ | | 1.789 | 1.358 | 2.047 |
| $X_4$ | | 0.08200 | 0.06221 | 0.09370 |
| $X_6$ | | 0.04614 | 0.03503 | 0.05279 |
| $X_8$ | −0.001099 | | | |

a. Using Eqs. 10.40 and 10.41, estimate the mean and variance of the total cost.
b. Approximate the probability distribution of the total cost.

**10.9.** Using the technique of range cost estimating (PERT), compute the expected total material cost and the variance about this expected value.

| Material | Low | Most Likely | High |
|---|---|---|---|
| 1 | $2 | $4 | $6 |
| 2 | 4 | 7 | 10 |
| 3 | 6 | 7 | 14 |
| 4 | 3 | 6 | 9 |
| 5 | 12 | 14 | 22 |

**10.10.** Consider a project whose net cash flow streams, $F_n$, are as follows.

| $n$ | $E(F_n)$ | $Var(F_n)$ |
|---|---|---|
| 0 | −$10 | 9 |
| 1 | 3 | 4 |
| 2 | 8 | 16 |
| 3 | 10 | 25 |

In addition, the correlation coefficients among the $F_n$'s are known to be $\rho_{01} = \rho_{02} = \rho_{03} = 0.5$ and $\rho_{12} = \rho_{23} = \rho_{13} = 1$. Assuming $i = 10\%$, compute the mean and variance of the *PV* for the project.

**10.11.** Consider the following investment cash flows over a 2-year life.

| | $\mu$ | $\sigma^2$ |
|---|---|---|
| $F_0$ | −$500 | 0 |
| $F_1$ | 200 | $50^2$ |
| $F_2$ | 500 | $50^2$ |

Compute the mean and variance of *PV* of this project at $i = 10\%$ if
a. $F_1$ and $F_2$ are mutually independent.
b. $F_1$ and $F_2$ are partially correlated with $\rho_{12} = 0.3$.
c. $F_1$ and $F_2$ are perfectly correlated with $\rho_{12} = 1$.
Also,
d. Compute the probability that the *PV* will be negative.

**10.12.** Consider the following risky cash flows for an investment project.

| $n$ | $E(F_n)$ | $Var(F_n)$ |
|---|---|---|
| 0 | −$500 | 0 |
| 1 | 400 | $50^2$ |
| 2 | 400 | $50^2$ |

Assume $F_1$ and $F_2$ are mutually independent. At $i = 10\%$, the mean and variance of the *PV* of the project would be

$$E[PV(10\%)] = -500 + \frac{400}{1.1} + \frac{400}{(1.1)^2} = \$194.20$$

$$Var[PV(10\%)] = 0 + \frac{2{,}500}{(1.1)^2} + \frac{2{,}500}{(1.1)^4} = (61.43)^2$$

The objective is to determine the mean and variance of the *annual equivalent value* of the project. Two engineers, John and Barbara, compute the mean and variance of the annual equivalent value as follows.

John:

$$E[AE(10\%)] = 194.20\,\overset{A/P,10\%,2}{(0.5762)} = \$111.90$$

$$Var[AE(10\%)] = (61.43)^2(0.5762)^2 = (35.40)^2$$

Barbara:

$$E[AE(10\%)] = \$111.90 \quad \text{(same as John's)}$$

$$(61.43)^2 = \frac{Var[AE(10\%)]}{(1.1)^2} + \frac{Var[AE(10\%)]}{(1.1)^4}$$

Solving for $Var[AE(10\%)]$ yields

$$Var[AE(10\%)] = (50)^2$$

Both claim that their way of computing the variance is correct.

a. Which one is correct?

b. Suggest the proper way to compute the variance.

**10.13.** Consider the following after-tax cash flows for an 8-year project life following a two-year investment period.

| $n$ | $E(F_n)$ | $Var(F_n)$ |
|---|---|---|
| 0 | −\$500 | $50^2$ |
| 1 | −200 | $30^2$ |
| 2 | 100 | $20^2$ |
| 3 | 100 | $30^2$ |
| 4 | 300 | $40^2$ |
| 5 | 300 | $40^2$ |
| 6 | 400 | $50^2$ |
| 7 | 300 | $50^2$ |
| 8 | 200 | $40^2$ |
| 9 | 100 | $30^2$ |

The investment flows ($F_0$ and $F_1$) are independent of the operating cash flows ($F_2$ through $F_9$). The firm's *MARR* is 15%.

a. If the annual operating cash flows are mutually independent, determine the mean and variance of the *PV* of the project.

b. If the annual operating cash flows are perfectly correlated (positively) with one another, determine the mean and variance of the project.

c. In part a or part b, is it meaningful to compute the mean and variance of *IRR* of this project?

d. In part b, assume that all the project cash flows (including the investment flows) are normally distributed with the means and variances specified. What is the probability that the *PV* of the project will exceed \$200?

e. Suppose that the operating cash flows are negatively correlated with one another. Assume a perfect correlation coefficient of −1. Compute the variance of the project. What can you conclude from the variance calculation?

**10.14.** Consider the following two investment projects, A and B. The company has simplified the uncertain situation somewhat and feels that it is sufficient to imagine that the proposals have the following possible returns.

| | Project A | Project B |
|---|---|---|
| First cost | \$10,000 | \$8,000 |
| Salvage | 0 | 0 |
| Life | 7 years | 5 years |
| Depreciation | 5-year MACRS | 3-year MACRS |
| Net annual receipts | | |

| Project A Prob. | Project A Receipts | Project B Prob. | Project B Receipts |
|---|---|---|---|
| 0.2 | \$2,000 | 0.4 | \$2,000 |
| 0.6 | 3,000 | 0.3 | 2,500 |
| 0.2 | 3,500 | 0.3 | 4,500 |

The firm's marginal tax rate and *MARR* are 40% and 12%, respectively.

a. Compute the mean and variance of the *PV* of project A.
b. Compute the mean and variance of the *PV* of project B.
c. Develop the *PV* distributions for both projects.
d. The company is also uncertain about the realization of the actual salvage value associated with each project, and these uncertain events are captured in terms of the following probabilities.

| Project A | | Project B | |
|---|---|---|---|
| Prob. | Salvage | Prob. | Salvage |
| 0.2 | $0 | 0.3 | $0 |
| 0.3 | $1,000 | 0.4 | $400 |
| 0.5 | $1,500 | 0.3 | 1,000 |

Compute the mean and variance of the *PV* for each project.

**10.15** Consider the following investment situations in which the present values of single payments must be computed under uncertain timing. Assume an annual interest rate of 15% or equivalent interest rate under continuous compounding (or discounting) $r = \ln(1 + i) = 0.13976$.

a. Consider the cost of an unscheduled plant shutdown caused by an unforeseen equipment failure. Each shutdown will cost about $35,000 in labor, materials, and lost profits and is known to have a negative exponential distribution. The average time between unscheduled shutdowns has been 3.5 years. Calculate the mean and variance of the present value (cost) of the next unscheduled shutdown to evaluate alternative strategies, such as changing the intervals of scheduled maintenance shutdowns.
b. Consider a $1,000 payment to be received at any time between 1 and 2 years hence (uniformly distributed). Compute the mean and variance of the present value of this receipt.
c. Consider a project whose PERT analysis shows an expected completion time of 2 years with variance $(0.3)^2$. Assume that this completion time will have a beta distribution. Upon completion of the project, the firm will receive $1,000,000. Compute the mean and variance of the present value of this sum.

**10.16.** Consider the following investment situations in which the present values of uniform cash flows must be computed under uncertain timing. Assume an annual interest rate of 15% or equivalent interest rate under continuous compounding (or discounting) $r = \ln(1 + i) = 0.13976$.

a. Consider a situation in which a $1,000/year uniform cash flow is received but will cease at any time in the next 2 years. If all cessation times in this interval are considered equally likely, compute the mean and variance of the present value of this receipt.
b. In part a, suppose that the starting time of the cash flow is also a random variable independent of the cash flow's duration. Let the starting time be uniformly distributed on the interval (0,1), so that we have a uniform continuous cash flow that can start at any time between now and one year from now and can end at any time between 0 and 2 years after it starts. Compute the mean and variance of the present value of this receipt.

**10.17.** Consider a machine that now exists in condition $j$ and generates earnings at the uniform continuous rate of $A_j$ dollars per year. If at some time $T$ its condition changes from $j$ to $k$, its earning rate will instantaneously change from $A_j$ to $A_k$. We will inspect the machine exactly one year from now. Assume the nominal interest rate of $r$ compounded continuously.

a. If the machine's condition changes to $k$ at time $T$, where $T$ is distributed uniformly on the time interval (0,1) and time is measured in years, what are the mean and variance of the present value of its earnings for the year?
b. If the machine remains in condition $j$ for the entire year, what are the mean and variance of the present value of its earnings for the year?

**10.18.** The following data are available for cost elements for operating machines A and B valued at $10,000 each. Model A has a 3-year life and model B has a 5-year life. Both have zero salvage value. Each machine will be depreciated by using a 3-year recovery property under optional MACRS (straight-line method with half-year convention). An interest rate of 12% is used, and the marginal income tax rate is 40%.

Before-Tax Operating Costs

| | Model A | | Model B | |
|---|---|---|---|---|
| Year | Mean | Variance | Mean | Variance |
| 1 | 9,000 | 250,000 | 10,000 | 360,000 |
| 2 | 13,000 | 490,000 | 12,000 | 490,000 |
| 3 | 15,000 | 1,000,000 | 12,000 | 640,000 |
| 4 | | | 13,000 | 640,000 |
| 5 | | | 13,000 | 1,000,000 |

NOTE: Yearly cash flows are normally distributed.

a. Assuming statistical independence, calculate the mean and variance for the annual equivalent cost of operating each machine.
b. From the results of part a, calculate the probability that the annual equivalent cost of operating model A will exceed the cost of operating model B.
c. Repeat the calculation of parts a and b, assuming that the correlation coefficient between operating expenses for any 2 years for model A is 0.3.
d. Suppose that costs for A are perfectly correlated, costs for B are uncorrelated, and both machines are to be leased. The receipts from each machine are $20,000 per year. Calculate the mean and variance for total present value.

**10.19.** Consider the following net present value function expressed in terms of random cash flow components $X$, $Y$, and $Z$,

$$PV(12\%) = 3.5X(Y - 45) + 2.7Y(Z - X) + 100$$

where $X$, $Y$, and $Z$ are distributed as follows.

| X | | Y | | Z | |
|---|---|---|---|---|---|
| Prob. | Event | Prob. | Event | Prob. | Event |
| 0.15 | $10 | 0.25 | $30 | 0.20 | $50 |
| 0.35 | $20 | 0.40 | $50 | 0.50 | $80 |
| 0.50 | $40 | 0.35 | $70 | 0.30 | $100 |

a. Assuming statistical independence among random variables, determine the net present value distribution.
b. Compute the mean and variance of the distribution.

**10.20.** By definition, traditional payback period analysis tries to answer the question "How long will it take for an investment outlay to be recovered by the net cash flows from operations?" Since estimates of cash flows are subject to a certain degree of risk, the probability concept can be used to improve the analysis. Suppose that three-level estimates of annual cash flow are made to determine the probability distribution of the cumulative cash flow estimates for each year. From this probability distribution information, the probability of recovering the initial investment at the end of each year can be developed. Consider the following project with a $10,000 investment over 5 years.

| | Annual Net A/T Cash Flows | | |
|---|---|---|---|
| n | Low (Pessimistic) | Most Likely | High (Optimistic) |
| 0 | −$10,000 | −$10,000 | −$10,000 |
| 1 | 2,000 | 3,000 | 4,000 |
| 2 | 2,000 | 3,000 | 4,000 |
| 3 | 3,000 | 4,000 | 5,000 |
| 4 | 2,000 | 3,000 | 3,500 |
| 5 | 2,000 | 3,000 | 3,500 |

a. Using Eqs. 10.72 and 10.73, compute $E(F_n)$ and $Var(F_n)$.
b. Find the means and variances of the cumulative cash flow for each period.
c. Compute the mean and variance of the payback period.
d. Using the result in part c, estimate the probability that the payback period will be less than or equal to 3 years.

**10.21.** Consider Example 10.20. Suppose the firm's *MARR* is known to be 12%. Repeat parts a through d for the discounted payback period.

**10.22.** Motorola Electronics Company is considering the introduction of a small heat pump unit for residential use. The project requires an initial investment of $80 million. The company has prepared the following cash flow forecasts.

| | Low | Most Likely | High |
|---|---|---|---|
| Market size (units) | 7 million | 9 million | 13 million |
| Market share | 2% | 3% | 6% |
| Unit price | $1,200 | $1,500 | $1,600 |
| Unit variable cost | $900 | $1,100 | $1,200 |
| Fixed cost | $8 million | $10 million | $11 million |
| Salvage value | $2 million | $3 million | $5 million |

The initial investment can be depreciated on a straight-line basis (with half-year convention) over 7 years, and the project is expected to have an economic service life of 8 years. The firm's marginal tax rate is 40% and its *MARR* is known to be 15%. The firm will conduct the economic analysis based on the following implicit model of cash flow.

$n$ = 0: Cash flow = −50 million.

$n$ = 1 to 8: Cash flow = (revenue − costs − depreciation)(1 − tax rate) + depreciation.

Revenues = Market size × market share × unit price.

Costs = (market size × market share × variable unit cost) + fixed cost.

$n$ = 8: Adjust the cash flow in year 8 to reflect the gains tax and the salvage value.

Net proceeds = salvage value − gains tax.

Gains tax = [salvage value − ($50 million − accumulated depreciation)](tax rate).

a. Assume that each random variable above is triangularly distributed with the parameters specified. Compute the mean and variance of PV for this heat pump project.
b. Estimate the probability that the *PV* will be negative.

**10.23.** The state of Alabama approved the sale of bonds to finance the construction of new jail systems during the 1989 legislative session. Because interest rates in the bond market are currently high, the state's finance director has decided to delay the sale of the bonds until the bond market improves. The current bond interest rate is 9% and the state will initiate the bond sale whenever the rate comes down to about 8%. The timing of the bond sale is uncertain, but the duration is expected to be a probability distribution with known mean and variance.

The original construction costs (the net proceeds from the bond sale) were estimated to be $30 million at the time of the legislature's approval but are expected to increase at the rate of 1% for each month of delay (12% over the next 12 months) inflation. The bonds will have a 20-year life and interest payable semiannually. Assume that it is now the end of the 1989 legislative session.

a. Suppose that the timing of the bond sale can be described as a gamma distribution with a mean of 6 months and a standard deviation of 12 months. What expected amount of the bond sale will cover the construction costs? Find the variance of this amount.
b. Assume that the inflation rate is also unknown but can be described as uniformly distributed from 0.5% to 1.5% per month. What is the expected amount of the bond sale in this situation? Find the variance of this amount.
c. Repeat part b when the timing of the bond sale can be described by a uniform distribution with a mean of 6 months and a standard deviation of 12 months. Compare the results with those of part b.
d. Suppose that the state's discount rate is 8% (nominal). Find the optimal timing of the bond sale, assuming the conditions given in parts a and b. You may use continuous compounding in your modeling.
e. Suppose that both the inflation rate and the timing of the bond sale are perfectly correlated with each other (correlation coefficient = 1). How would you analyze the problem?

## Acknowledgments


- Problems 10.1, 10.2, and 10.3 are from [28].
- Problem 10.6 is based on [33].
- Problem 10.7 is based on J. C. Hayya and W. L. Ferrara, "Normal Approximation of the Frequency Functions of Standard Form," *Management Science,* Vol. 19, No. 2, pp. 173–186, 1972.
- Problem 10.8 is based on P. R. Dunlap, "Interval Estimates of Projected Cost," *AIChE Symposium Series—Uncertainty Analysis for Engineers,* Vol. 78, No. 220, pp. 49–58, American Institute of Chemical Engineers, 1982.
- Problems 10.15 and 10.16 are based on Young and Contreras [38].
- Professor Don Young of Georgia Tech provided Problem 10.17.

# 11

# Methods for Comparing Risky Projects

## 11.1 INTRODUCTION

The comparison of two or more risky projects is more of a challenge than the evaluation of one project in isolation. Underlying all comparison methods are assumptions regarding the decision maker's preference or utility functions. Frequently these assumptions are not well understood. In this chapter we present some of the more important comparison methods and attempt to explain the circumstances under which they might be appropriate.

We begin with the mean–variance, mean–semivariance, and probability-of-loss criteria presented in the previous chapter. These are popular methods and relatively easy to implement. A more theoretical examination of the decision problem leads to stochastic dominance rules. Next, we present a brief introduction to portfolio theory and the capital asset pricing model. An extension of some techniques in Chapter 8 is called stochastic programming; it is designed to handle probabilistic coefficients in the mathematical programming problem. Multiperiod portfolio analysis is presented, along with some techniques for estimating beta coefficients, as a potential technique for selecting among projects in a risky environment. The last topic in the chapter is uncertainty resolution, which deals with the reduction in uncertainty about specific cash flows as time progresses.

## 11.2 COMPARATIVE MEASURES OF INVESTMENT WORTH

### 11.2.1 Mean–Variance, E–V

In Section 9.5 we introduced the mean–variance ($E$–$V$) criterion in the context of utility models. The $E$–$V$ criterion is quite popular for evaluating investments. Let us apply it to a selection problem.

### *Example 11.1*

A company has developed probability distributions for the present value (*PV*) amounts that might be obtained from each of five mutually exclusive projects. Evaluate the projects and recommend which one, if any, should be selected. All projects have the same lifetime.

| | Project | | | | |
|---|---|---|---|---|---|
| *PV* | 1 | 2 | 3 | 4 | 5 |
| −$10,000 | 0.2 | 0.1 | 0.1 | 0.0 | 0.0 |
| 0 | 0.2 | 0.2 | 0.2 | 0.3 | 0.15 |
| +10,000 | 0.2 | 0.4 | 0.3 | 0.4 | 0.65 |
| +20,000 | 0.2 | 0.2 | 0.3 | 0.0 | 0.0 |
| +30,000 | 0.2 | 0.1 | 0.1 | 0.3 | 0.2 |

For project 1 we have

$$E(PV) = [(0.2)(-10) + (0.2)(0) + (0.2)(10) + (0.2)(20) + (0.2)(30)](10^3) = (10)(10^3)$$

$$Var(PV) = [(0.2)(-10)^2 + (0.2)(0)^2 + (0.2)(10)^2 + (0.2)(20)^2 + (0.2)(30)^2](10^6) - (10)^2(10^6) = (200)(10^6)$$

In a similar manner, we compute the other values, with the results shown.

| | Project | | | | |
|---|---|---|---|---|---|
| Value | 1 | 2 | 3 | 4 | 5 |
| *E(PV)*, $\$10^3$ | 10 | 10 | 11 | 13 | 12.5 |
| *Var(PV)*, $10^6$ | 200 | 120 | 129 | 141 | 89 |

The results are also plotted in Figure 11.1.

Comparing projects 1 and 2, we see that project 2 has a much lower variance of *PV*, 120 versus 200, although projects 1 and 2 have the same mean of 10. The expected *PV* of 10 for project 2 is more certain, and we therefore prefer project 2 over project 1. We say that project 2 dominates project 1 by the first of two mean–variance rules stated below.

Comparing projects 3 and 2, we see that the increase in *E(PV)* from 10 to 11 comes at a price of a higher *Var(PV)*, from 120 to 129. We cannot choose between the two simply on the basis of *E–V*, and must resort to other information, such as a utility function. Similarly, we cannot choose between projects 2, 3, and 4 by using only *E–V*.

Examining project 5, we see that we can apply rule 1 (or rule 2) in eliminating projects 2 and 3. Project 5 promises both a higher *E(PV)* and a lower

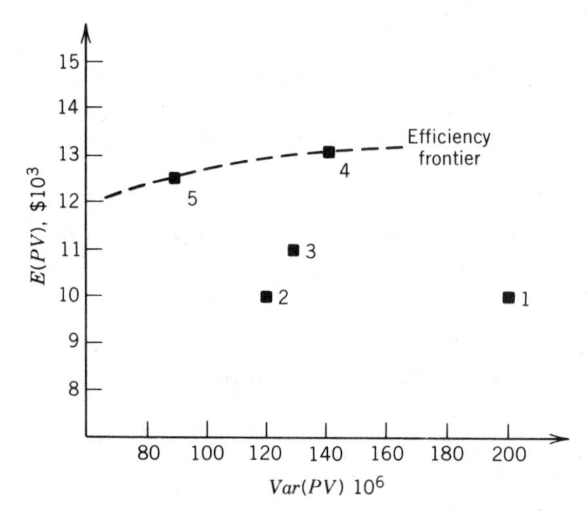


**FIGURE 11.1 Mean–variance plot, Example 11.1.**

*Var*(*PV*) than either project 2 or 3. On the other hand, we cannot say anything about project 4 versus project 5. We are thus left with projects 4 and 5 to choose from, on the basis of further information. □

We now state the dominance rules for the *E–V* criterion.

*Mean–variance rule 1.* If project A has an *E*(*PV*) the same as or higher than that of project B, and has a lower *Var*(*PV*) than B, we prefer project A.

*Mean–variance rule 2.* If project A has a *Var*(*PV*) the same as or lower than that of project B, and has a higher *E*(*PV*) than B, we prefer project A.

We may substitute other criteria such as *AE* or *FV* for *PV* in the statement of the rules, since we showed in Chapter 7 that these criteria are equivalent to *PV*.

The *E–V* criterion enables us to narrow the choice to two projects out of the five. These two, projects 4 and 5, lie on the *efficiency frontier,* shown in Figure 11.1. The efficiency frontier is a curve drawn through the points representing projects that are not dominated by some other project. Any point below and to the right of the efficiency frontier represents a project dominated by one on the frontier. (Some authors label the axes of the plot the other way, with the resulting efficiency frontier drawn through the points to the right and downward.)

The ultimate choice between projects 4 and 5 in Example 11.1 depends on the trade-off between *E*(*PV*) and *Var*(*PV*) for the decision maker. Recalling the utility concepts from Chapter 9, we can represent this as a choice among different isoutility curves, as shown in Figure 11.2. The decision problem can be solved if we can determine which of projects 4 and 5 lies on a higher utility curve. (The curves in the figure are $U_1$, $U_2$, and $U_3$ in order of increasing preference.) But as we saw in Chapter 9, determining utility curves is neither precise nor straightforward.

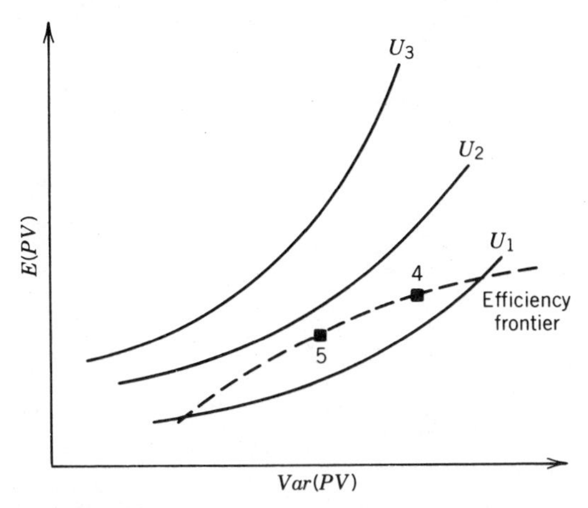


**FIGURE 11.2** Utility curves and the efficiency frontier.

### 11.2.2 Mean–Semivariance, $E\text{–}S_h$

The variability of *PV* for projects 4 and 5 in Example 11.1 comes from *PV* values that are all nonnegative. For the other three projects there are negative *PV* values. It can be argued that variability of positive *PV* values is not undesirable; only variability in the negative, or loss, region is. Consider the following extreme example.

## *Example 11.2*

The probability distributions for *PV* have been obtained for two projects, as shown. Obtain *E(PV)* and *Var(PV)* for each.

| | Project | |
|---|---|---|
| *PV* | 6 | 7 |
| −\$20,000 | 0.1 | 0.0 |
| −\$10,000 | 0.2 | 0.0 |
| 0 | 0.3 | 0.0 |
| +\$10,000 | 0.3 | 0.5 |
| +\$30,000 | 0.1 | 0.1 |
| +\$50,000 | 0.0 | 0.4 |

We obtain the following.

| | Project | |
|---|---|---|
| Value | 6 | 7 |
| *E(PV)*, $\$10^3$ | 2 | 28 |
| *Var(PV)*,. $10^6$ | 176 | 356 |

In this example project 7 has much greater variability than project 6, but all the variability is in the positive region. Figure 11.3 compares the probability

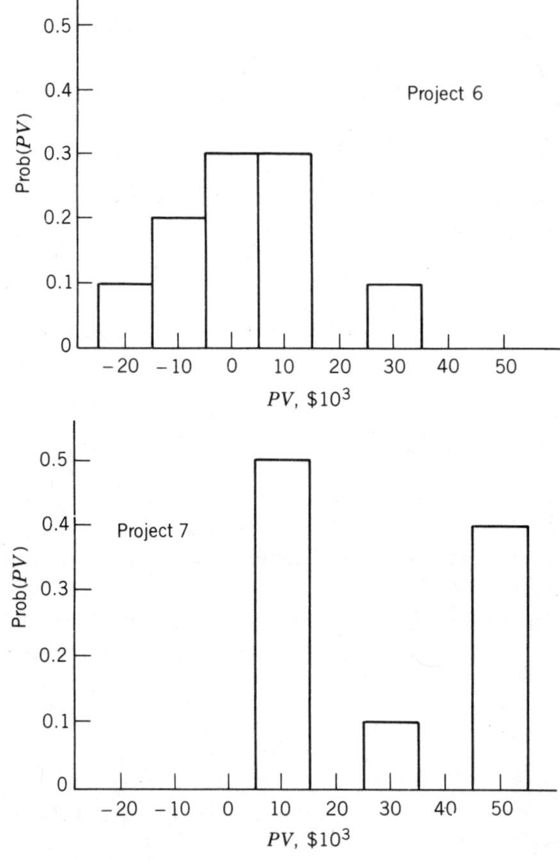


**FIGURE 11.3 Probability distributions for Example 11.2.**

distributions for the two projects. We can see that the distribution for 7 lies much farther to the right than the distribution for 6. □

A modification of the variance formula, Eq. 10.3, captures only the undesirable variation in *PV* of a project [19]. Specifically, the semivariance $S_h$ of a (discrete) probability distribution for a random variable $X$ is defined as

$$S_h = E[(X - h)^-]^2 = \sum_j [(X_j - h)^-]^2 P_j \tag{11.1}$$

where

$$\begin{aligned} (X - h)^- &= X - h \quad \text{when } X \leq h \\ &= 0 \quad \text{when } X > h \end{aligned} \tag{11.2}$$

$$P_j = \text{probability of } X_j \text{ occurring}$$

and $b$ is a reference value considered as a lower limit, below which risk is incurred. Many analysts select $b = 0$ for $E(PV)$ or $E(AE)$ analysis, but any suitable value may be specified.

Actual calculation of semivariance requires the direct form of Eq. 11.1, rather than the faster Eq. 10.3. For example, setting $b = 0$, we obtain for project 6 in Example 11.2.

$$S_b = (-10{,}000 - 0)^2(0.2) + (-20{,}000 - 0)^2(0.1)$$
$$= (60)(10^6)$$

For project 7 the $S_b = 0$. Thus, application of the $E$–$S_b$ criterion would involve examining the following numbers.

| Value | Project 6 | 7 |
|---|---|---|
| $E(PV)$, \$$10^3$ | 2 | 28 |
| $S_b(PV)$, $10^6$ | 60 | 0 |

The dominance rules in Section 11.2.1 are modified to replace $Var(PV)$ with $S_b(PV)$. In this case we can use either rule 1 or rule 2 to conclude that project 7 clearly dominates project 6.

### *Example 11.3*

Apply the $E$–$S_b$ criterion with $b$ = \$5,000 present value to the mutually exclusive projects in Example 11.1. We will demonstrate the $S_b$ computation for project 2 and present the other results. The $E(PV)$ values are the same as in Example 11.1.

For project 2,

$$S_b = (-10{,}000 - 5{,}000)^2(0.1) + (0 - 5{,}000)^2(0.2)$$
$$= (27.5)(10^6)$$

| Value | Project 1 | 2 | 3 | 4 | 5 |
|---|---|---|---|---|---|
| $E(PV)$, \$$10^3$ | 10 | 10 | 11 | 13 | 12.5 |
| $S_b(PV)$, $10^6$ | 45 | 27.5 | 27.5 | 7.5 | 3.75 |

Applying the dominance rules, we can eliminate all except projects 4 and 5. The results are plotted in Figure 11.4. ☐

### 11.2.3 Safety First

The safety-first criterion, as explained in Chapter 10, favors the alternative with the smallest probability of loss, where loss is usually defined as a negative $PV$ or $AE$ or as an $IRR$ below $MARR$.

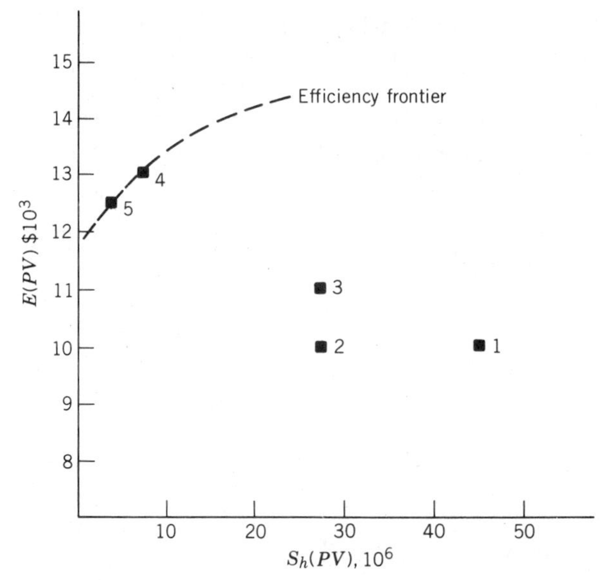


**FIGURE 11.4 Mean–semivariance plot for Example 11.3.**

## *Example 11.4*

Apply the safety-first criterion to the projects in Example 11.1, with loss defined as negative *PV*. We have

| Project | 1 | 2 | 3 | 4 | 5 |
|---|---|---|---|---|---|
| Prob( $PV < 0$ ) | 0.2 | 0.1 | 0.1 | 0.0 | 0.0 |

By this criterion, all but projects 4 and 5 are eliminated. There is no way to distinguish between projects 4 and 5 unless we use some other criterion. □

The probability-of-loss calculation obscures the magnitude of any potential losses. In this sense, the criterion provides less information than $S_h$ or the probability distribution itself.

The three criteria presented in this section are relatively easy to apply when selecting among mutually exclusive projects. When we select from investment alternatives that are combinations of projects, the procedures can become rather complicated, as shown in Section 11.5. Further, the criteria often do not indicate a clear choice among competing alternatives. This issue is the topic of the next section.

## 11.3 STOCHASTIC DOMINANCE

### 11.3.1 First-Degree Stochastic Dominance

Let us reexamine the comparison of projects 2 and 3 in Example 11.1. The probabilities for the *PV* values are shown below, along with *E*(*PV*) and *Var*(*PV*).

| | Project | |
|---|---|---|
| *PV* | 2 | 3 |
| −$10,000 | 0.1 | 0.1 |
| 0 | 0.2 | 0.2 |
| +10,000 | 0.4 | 0.3 |
| +20,000 | 0.2 | 0.3 |
| +30,000 | 0.1 | 0.1 |
| *E*(*PV*), $\$10^3$ | 10 | 11 |
| *Var* (*PV*), $10^6$ | 120 | 129 |

The probability distributions for *PV* are shown in Figure 11.5. The only difference between the two distributions is in the shaded region. Project 2 has 0.4 probability of +10 and 0.2 probability of +20, and project 3 has 0.3 probability of +10 and 0.3 probability of +20.

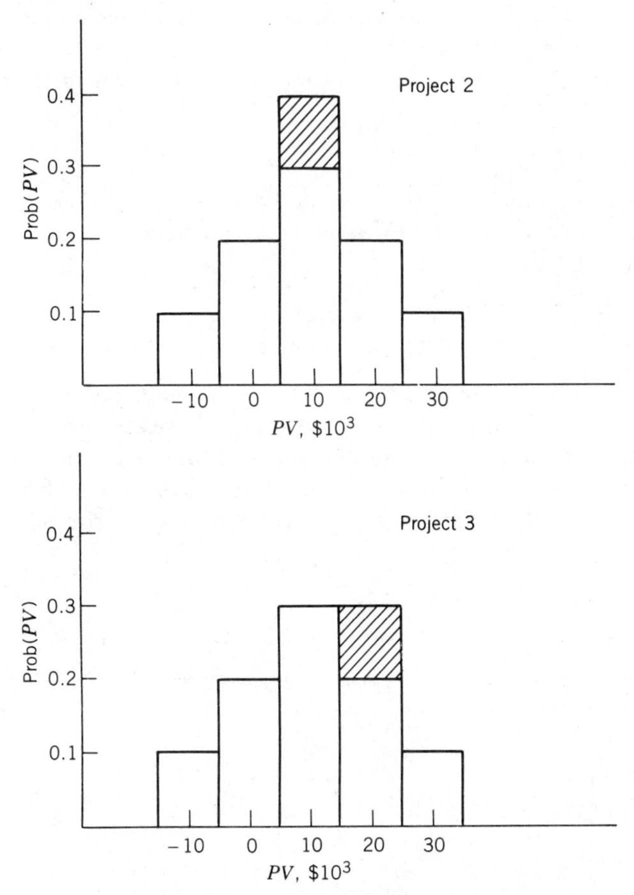


**FIGURE 11.5** Probability distributions for Example 11.1; comparison of projects 2 and 3.

- Without specifying the form of a decision maker's utility function, we can make the following statement about preferences between projects 2 and 3.

$$E[U(PV_3)] = E[U(PV_2)] + (0.1)U(\$20{,}000) - (0.1)U(\$10{,}000)$$

Then, as long as $U(\$20{,}000) > U(\$10{,}000)$, the decision maker prefers project 3 over project 2. The requirement that $U(\$20{,}000) > U(\$10{,}000)$ is satisfied by any increasing utility function. Therefore, the only type of decision maker who would *not* prefer project 3 over project 2 is one with a utility function that decreases from \$10,000 to \$20,000.

We now formally state these concepts [11].

***Definition: Dominance.*** Given two random variables $X$ and $Y$ with cumulative probability distribution functions $F(x)$ and $G(y)$, we say that $X \text{ D } Y$ ($X$ dominates $Y$) or $F \text{ D } G$ [$F(x)$ dominates $G(y)$], if

$$E[U(X)] \geq E[U(Y)]$$

for every utility function in the class of functions, and if the inequality holds strictly for at least one function in the class.

A number of authors have proved different versions of the following theorem [9,11,24].

***Theorem 11.1.*** (First-Degree Stochastic Dominance) Let $F(x)$ and $G(y)$ be cumulative distributions for random variables $X$ and $Y$. Let $U$ be any nondecreasing function with finite values for any finite $x$. A necessary and sufficient condition for $X \text{ D } Y$, or $F \text{ D } G$, is that

$$\begin{aligned} F(x) &\leq G(y) \quad \text{for every } x \\ \text{and} \quad F(x_0) &< G(y_0) \quad \text{for some } x_0 \end{aligned}$$

The theorem states that the cumulative distribution function (cdf) of $X$ must lie below that of $Y$ for at least one value and must lie nowhere above it. The theorem is equally valid for continuous and discrete probability distributions.

For selection among individual projects, first-degree stochastic dominance is easy to apply. However, it does not answer all our questions, as the next example shows.

## *Example 11.5*

Apply the first-degree dominance test to the projects in Example 11.1. We show the cdfs below and in Figure 11.6.

| | Project | | | | |
|---|---|---|---|---|---|
| PV | 1 | 2 | 3 | 4 | 5 |
| −\$10,000 | 0.2 | 0.1 | 0.1 | 0.0 | 0.00 |
| 0 | 0.4 | 0.3 | 0.3 | 0.3 | 0.15 |
| +10,000 | 0.6 | 0.7 | 0.6 | 0.7 | 0.80 |
| +20,000 | 0.8 | 0.9 | 0.9 | 0.7 | 0.80 |
| +30,000 | 1.0 | 1.0 | 1.0 | 1.0 | 1.00 |

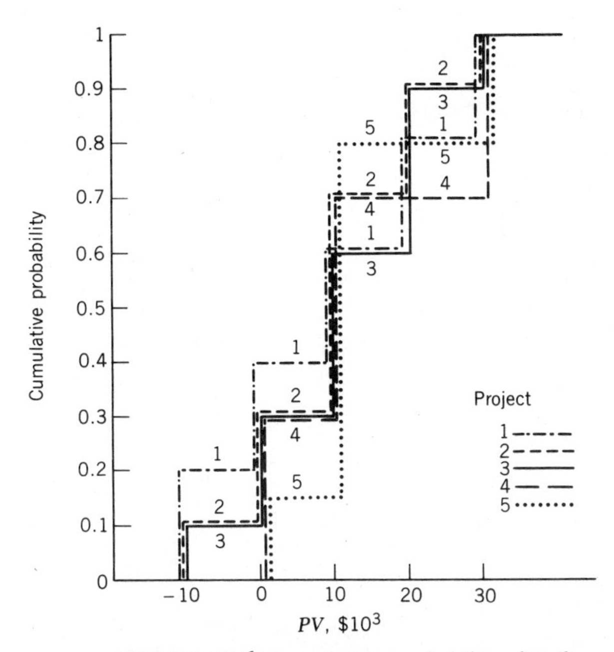


**FIGURE 11.6** Cumulative probability distributions for Example 11.1. (Lines have been shifted slightly for clarity; see Example 11.5.)

We see here that the cdf for 3 is below that for 2 in the range [10, 20) and is coincident everywhere else. In addition, the cdf for project 4 is below that for 2 in the ranges [−10, 0) and [20, 30) and is coincident everywhere else. So projects 3 and 4 are each clearly preferred over project 2, a conclusion not available from *E–V* analysis. When we search for other instances of dominance, however, we find none. □

Example 11.5 demonstrates that although the first-degree stochastic dominance test helps us decide some of the choices that are unresolved after applying *E–V*, the test may fail when *E–V* does indicate a clear choice. For example, the *E–V* criterion tells us that project 5 is preferred over projects 1, 2, and 3, but we cannot draw such a conclusion by examining the cdfs. The reason is that in Figure 11.6 the cdfs intersect one another, except for the pairs (2, 3) and (2, 4). We need to resort to a more restrictive test, as presented in the next section.

### 11.3.2 Second-Degree Stochastic Dominance

Let us examine in more detail the comparison of projects 3 and 4 in Example 11.1. Neither the first-degree test in Section 11.3.1 nor the *E–V* criterion in Section 11.2.1 enables us to make a clear-cut choice. Figure 11.7 presents the probability distributions, and Figure 11.8 shows the cdfs for an uncluttered comparison.

When we compare the probabilities for the −$10,000 *PV,* project 3 falls behind since it has 0.1 chance of obtaining that value, and project 4 has zero

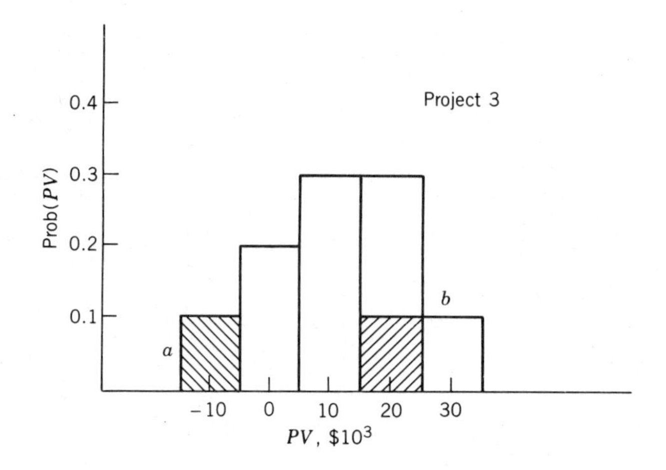


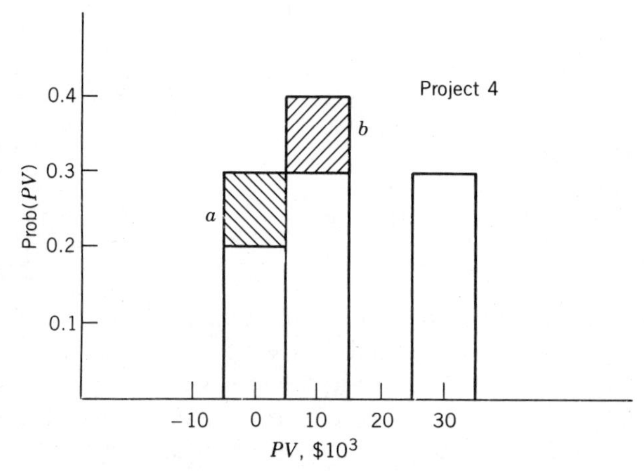


**FIGURE 11.7 Probability distributions for Example 11.1; comparison of projects 3 and 4.**

chance. At $0 *PV,* project 3 catches up, and the cdfs have the same value of 0.3. These differences are shown by the shaded regions labeled *a* in Figures 11.7 and 11.8. For $10,000 *PV,* project 4 fails behind; it has 0.4 chance of obtaining that value, versus 0.3 chance for project 3. Project 4 catches up at $20,000 *PV* (it actually overtakes project 3, but we are concerned only with the catching up). These differences are represented by the shaded regions labeled *b.* (The *c* region is discussed near the end of this section.)

The question that concerns us now is whether these two instances of falling behind and catching up balance in the expected utility calculation or whether one project is favored by them. Each instance involves a shift of 0.1 probability a distance of $10,000 along the horizontal axis. Region *a* amounts to an expected utility difference of

$$(0.1)U(\$0) - (0.1)U(-\$10{,}000)$$

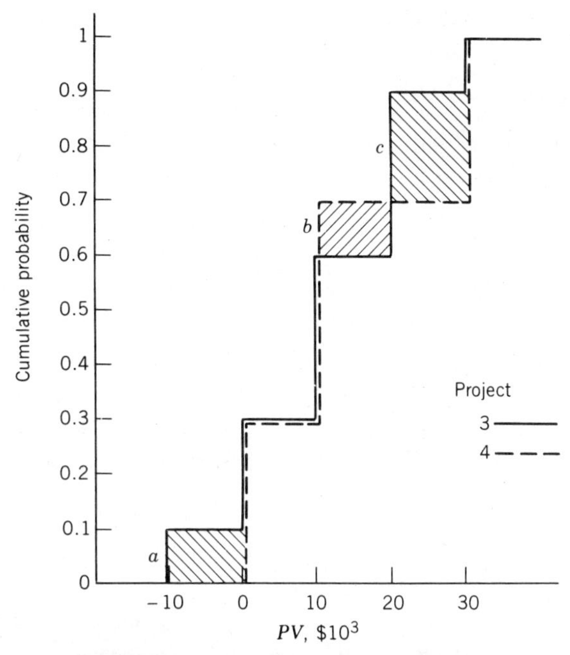


**FIGURE 11.8 Cumulative probability distributions for projects 3 and 4 in Example 11.1. (Lines have been shifted slightly for clarity; see Example 11.5.)**

in favor of project 4. Region $b$ favors project 3 by an expected utility difference of

$$(0.1)U(\$20{,}000) - (0.1)U(\$10{,}000)$$

It happens that for increasing, concave utility functions, the utility difference resulting from $a$ is greater than that from $b$. For increasing, concave functions

$$U(x + \Delta) - U(x) > U(y + \Delta) - U(y) \quad \text{for } x < y \text{ and } \Delta > 0 \quad (11.3)$$

The decreasing marginal utility of additional wealth, represented by the flattening of the utility curve (see Figure 9.1), means that shifts along the horizontal axis are more important at lower values of wealth. The 0.10 shift in probability from −\$10,000 to \$0 is more valuable than the 0.10 shift from \$10,000 to \$20,000. Considering these two shifts, therefore, we conclude that for any increasing, concave utility function, we clearly prefer project 4 over project 3. (The overtaking of 3 by 4 at \$20,000 AE simply strengthens our already clear preference.)

Our comparison of expected utility differences in this example is based on two shifts of the same probability the same distance along the wealth (horizontal) axis. In general, we may have multiple shifts of different probabilities for various distances along the wealth axis. The general case is covered by the following theorem [11].

***Theorem 11.2.*** (Second-Degree Stochastic Dominance) Let $F(x)$ and $G(y)$ be cumulative distributions for random variables $X$ and $Y$. Let $U$ be any nondecreasing, concave function. A necessary and sufficient condition for $X$ D $Y$, or $F$ D $G$, is that

$$\int_{-\infty}^{x} F(t)\, dt \leq \int_{-\infty}^{x} G(t)\, dt \quad \text{for every } x$$

and strict inequality holds at some $x_0$.

The theorem states that the *integral* of the cdf of $X$ must lie below that of $Y$ for at least one value and must lie nowhere above it. We are actually comparing the *second* integrals of the probability density (or mass) function, hence the term second-degree stochastic dominance.

## *Example 11.6*

Apply the second-degree dominance test to the projects in Example 11.1. The following table shows the values of the integrals (for the *PV* values) of the cdfs; selected curves are shown in Figure 11.9. All calculated values are in terms of $10^3.

| | Project | | | | |
|---|---|---|---|---|---|
| PV | 1 | 2 | 3 | 4 | 5 |
| −$10,000 | 0 | 0 | 0 | 0 | 0 |
| 0 | 2 | 1 | 1 | 0 | 0 |
| +10,000 | 6 | 4 | 4 | 3 | 1.5 |
| +20,000 | 12 | 11 | 10 | 10 | 9.5 |
| +30,000 | 20 | 20 | 19 | 17 | 17.5 |

The comparison of projects 3 and 4 reveals the clear preference for project 4. Notice that the curves for 4 and 5 cross, however, so we still cannot choose between the two. The table shows that

Project 2 dominates 1.

Project 3 dominates 1 and 2.

Project 4 dominates 1, 2, and 3.

Project 5 dominates 1, 2, and 3. □

If we look again at the cdfs for projects 3 and 4 in Figure 11.8, we may explain the second-degree test as follows. The two cdfs can intersect several times. All that is required for project 4 to dominate project 3 is that, for any *PV*

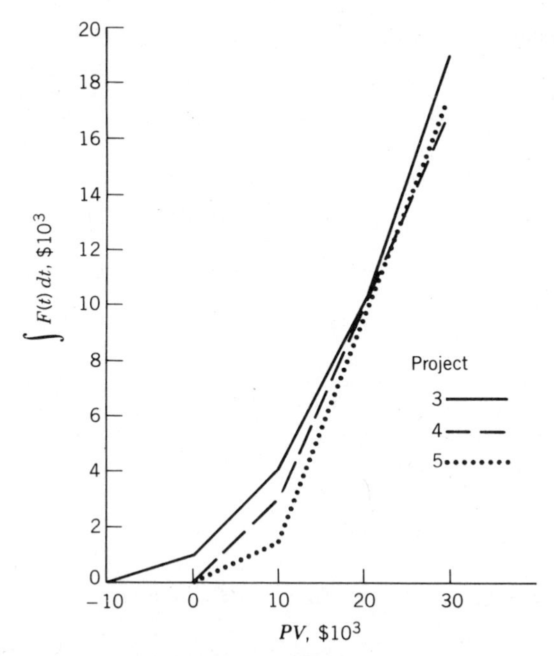


**FIGURE 11.9 Integrals of the cdfs for projects 3, 4, and 5 in Example 11.6.**

value, the accumulated shaded area in favor of 4 be greater than the accumulated shaded area in favor of 3. This guarantees the condition in Theorem 11.2, that the integral of $F$ must be less than the integral of $G$. Then for any *linear* utility function, the expected utility from $-\infty$ to any $x$ is greater for project 4. A concave utility function magnifies that preference because of the heavier weighting of the shaded regions on the left. In Figure 11.8 region $a$ represents the area in favor of project 4, $b$ the area in favor of project 3, and $c$ the area in favor of project 4. For any value $x$ between −\$10,000 and +\$30,000, the accumulated area from $-\infty$ to $x$ in favor of 4 exceeds that in favor of 3, except at +\$20,000, where they are equal. (Notice the similarity between the mathematical operation of the utility function here and that of the interest rate in choosing depreciation strategies in Section 4.3.5 and in determining the uniqueness of *IRR* using the Norstrom criterion in Section 6.3.1.)

### 11.3.3 Third-Degree Stochastic Dominance

The requirement for a nondecreasing, concave utility function to apply the second-degree test seems fairly mild from a behavioral standpoint. A more restrictive, but still fairly broad, class of utility functions is the class of nondecreasing functions with a nonnegative third derivative. Recall that in Chapter 9 we showed this class to be the class of utility functions with decreasing local risk aversion. If we are willing to ascribe such behavior to a decision maker, we can use the third-degree test for decision making [32].

***Theorem 11.3.*** (Third-Degree Stochastic Dominance) Let $F(x)$ and $G(y)$ be cumulative distributions for random variables $X$ and $Y$. Let $U$ be any nondecreasing, concave utility function with nonnegative third derivative. A necessary and sufficient condition for $X$ D $Y$, or $F$ D $G$, is that

$$\int_{-\infty}^{x}\int_{-\infty}^{t} F(w)\,dw\,dt \leq \int_{-\infty}^{x}\int_{-\infty}^{t} G(w)\,dw\,dt \quad \text{for every } x \text{ and}$$

strict inequality holds at some $x_0$ and $E(X) > E(Y)$.

### *Example 11.7*

Apply the third-degree test to projects 4 and 5 in Example 11.1. We need to separate the curves in Fig. 11.9 into segments according to slope. For project 4 the analysis proceeds as follows. In region [0, 10],

$$y = (0.3)x$$

where $y$ is the integral of the cdf (Figure 11.9) and $x$ is the *PV*. This results in

$$\int_{0}^{w} (0.3)x\,dx = 0.15w^2$$

Evaluating the integral at $PV = 10 \times 10^3$, we obtain $15 \times 10^6$. In region [10, 30],

$$y = (0.7)x - 4{,}000$$

The value of the integral, including the region [0, 10], is

$$\begin{aligned} &15 \times 10^6 + \int_{10{,}000}^{w} (0.7x - 4{,}000)\,dx \\ &= 15 \times 10^6 + [0.35x^2 - 4{,}000x]\Big|_{10{,}000}^{w} \\ &= 20 \times 10^6 - 4{,}000w + 0.35w^2 \end{aligned}$$

Evaluating at $PV = 20 \times 10^3$, we obtain $80 \times 10^6$, and at $PV = 30 \times 10^3$ we get $215 \times 10^6$.

For project 5 we proceed similarly. In region [0, 10],

$$y = (0.15)x$$

$$\int_{0}^{w} (0.15)x\,dx = 0.075w^2$$

Evaluating at $PV = 10 \times 10^3$, we obtain $7.5 \times 10^6$. In region [10, 30],

$$y = (0.8)x - 6{,}500$$

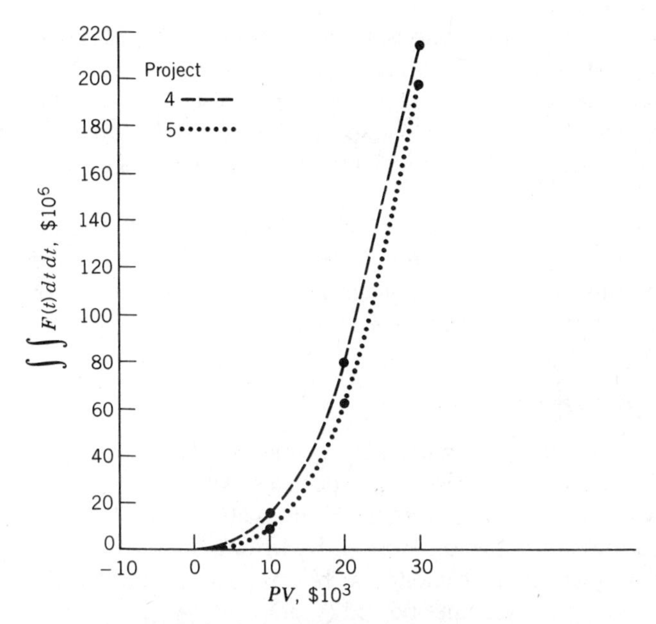


**FIGURE 11.10** Double integrals of the cdfs for projects 4 and 5 in Example 11.7.

The integral from 0 is

$$7.5 \times 10^6 + \int_{10{,}000}^{w} (0.8x - 6{,}500)\, dx$$

$$= 7.5 \times 10^6 + [0.4x^2 - 6{,}500x]\Big|_{10{,}000}^{w}$$

$$= 32.5 \times 10^6 - 6{,}500w + 0.4w^2$$

Evaluating at $PV = 20 \times 10^3$ we obtain $62.5 \times 10^6$, and at $PV = 30 \times 10^3$ we get $197.5 \times 10^6$. These results are plotted in Figure 11.10. It is clear that in the range of possible $PV$ values we prefer project 5, since the curve for it is below the curve for project 4. □

### 11.3.4 Relationship Between Dominance and Mean–Variance Criterion

Since the development of the $E$–$V$ criterion for decision making, a number of authors have studied the implications with respect to utility functions of using $E$–$V$ [12, 23, 28]. For example, it is has been shown that for a decision maker with a bounded quadratic utility function (so that we use only the increasing portion), preference by the $E$–$V$ criterion, according to either rule 1 or rule 2 in Section 11.2.1, implies dominance (according to the definition in Section 11.3.1). Another major result is stated with a condition imposed on the random variables representing wealth (*PV, AE,* etc.). If these random variables are from a restricted class of two-parameter distributions, for a decision maker preference by the $E$–$V$ criterion implies dominance with any nondecreasing utility function. One such distribution in the restricted class is the normal distribution. Hence this result is often applied to portfolio analysis (see Section 11.4). These types of relationships are quite intricate, and the interested reader is referred to other sources [11, 17].

We see that in order to implement any of the three stochastic dominance tests, we need detailed information about the probability distribution of $PV$ or $AE$. For single projects this task may be manageable. For combination alternatives the analysis can become complicated and tedious.

## 11.4 PORTFOLIO THEORY

This section contains a brief introduction to a subject that has played an important role in financial theory during the past several decades. Portfolio theory ties together ideas in utility theory and mean–variance analysis and reconciles these concepts with observed investor behavior. Most of the writing on the subject is related to investments in financial instruments, such as stocks and bonds, with an assumed planning horizon of one time period (year).

The presentation here covers only some fundamentals that are necessary for understanding the material in later sections. Procedures for solving the portfolio selection problem are mentioned in passing, except for the development of the index model in Section 11.4.4. Extensive coverage of the subject can be found [18, 26].

In this section we use the following notation.

- $a$ a fraction between 0 and 1
- $A_j$ a constant associated with the return from a share of security $j$
- $A_{J+1}$ a constant associated with the return from the market index
- $B_j$ a constant parameter that expresses the relation between security $j$ and the market index
- $\beta_j$ a measure of the relationship between variability in return from security $j$ and return from portfolio $M$, equal to $Cov(R_j, R_m)/\ Var(R_m)$
- $C_j$ an independent random variable associated with the return of a share of security $j$; it has mean zero and variance $Q_j$
- $C_{J+1}$ an independent random variable associated with the return from the market index; it has mean zero and variance $Q_{J+1}$
- $E_j$ $E(R_j)$, expected return from a share of security $j$
- $I$ a random variable representing a market index
- $J$ the number of securities being considered for a portfolio
- $\lambda$ coefficient of risk aversion when a combination of expected return and variance of return is evaluated
- $R_j$ return from a share of stock of security $j$
- $R$ return from a portfolio of securities

$R_c$ return from an investment that consists of a combination

$R_f$ known return from the risk-free asset $F$

$R_m$ return from the risky asset $M$

$\rho_{jk}$ correlation coefficient between returns from shares of securities $j$ and $k$

$\sigma_j$ standard deviation of return from a share of security $j$

$\sigma_c$ standard deviation of return from an investment combination

$\sigma_m$ standard deviation of return from portfolio $M$

$x_j$ fraction of funds invested in security $j$

### 11.4.1 Efficiency Frontier

The portfolio analysis problem is to select the amount to invest in each security, such as a stock or bond, and thereby determine a set of efficient portfolios in which no portfolio is dominated by any other portfolio. The typical formulation of the problem includes some type of constraint on how much of each security can be included in the portfolio.

To illustrate the concept, we present a short example. Note that we use *return* and *variance of return* as the two evaluation criteria for each security. For the simplified situation in which dividends are paid at year end, we define the return from a share of stock in company $j$ as

$$R_j = \frac{\text{Dividend} + \text{Share value increase}}{\text{Previous share value}} \tag{11.4}$$

For the simplified case of a bond purchased at par with one-year maturity and interest paid at year end, we have

$$R_j = \frac{\text{Interest}}{\text{Issue price}} \tag{11.5}$$

Thus, a stock that paid a \$20 dividend, sold for \$100 a year ago, and now sells for \$95 generated a return of (\$20 − 5)/100 = 15%. Similarly, a one-year bond issued at \$1,000 that paid \$120 interest yielded a return of 12%.

### *Example 11.8*

Construct portfolios by selecting from among the four securities shown in the table. Either \$100 or \$200 may be invested in any one security, except for security 1, which does not have a limit. There is an overall budget restriction of \$400.

| | | | Coefficient of Correlation of $R$ | | | |
|---|---|---|---|---|---|---|
| Security | $E(R)$ per \$100 | $Var(R)$ per \$100 | Sec. 1 | Sec. 2 | Sec. 3 | Sec. 4 |
| 1 | 10 | 0 | — | 0 | 0 | 0 |
| 2 | 30 | 25 | 0 | — | +0.5 | −0.4 |
| 3 | 30 | 100 | 0 | +0.5 | — | −0.2 |
| 4 | 20 | 49 | 0 | −0.4 | −0.2 | — |

**Table 11.1** ***Portfolios for Example 11.8***

| Portfolio Number | Amount Invested in | | | | Expected Return | Variance of Return |
|---|---|---|---|---|---|---|
| | Sec. 1 | Sec. 2 | Sec. 3 | Sec. 4 | | |
| 1 | $400 | — | — | — | $40 | 0 |
| 2 | 300 | 100 | — | — | 60 | 25 |
| 3 | 300 | — | 100 | — | 60 | 100 |
| 4 | 300 | — | — | 100 | 50 | 49 |
| 5 | 200 | 200 | — | — | 80 | 100 |
| 6 | 200 | — | 200 | — | 80 | 400 |
| 7 | 200 | — | — | 200 | 60 | 196 |
| 8 | 200 | 100 | 100 | — | 80 | 175 |
| 9 | 200 | 100 | — | 100 | 70 | 46 |
| 10 | 200 | — | 100 | 100 | 70 | 121 |
| 11 | — | 200 | 200 | — | 120 | 700 |
| 12 | — | 200 | — | 200 | 100 | 184 |
| 13 | — | — | 200 | 200 | 100 | 484 |
| 14 | 100 | 200 | 100 | — | 100 | 350 |
| 15 | 100 | 200 | — | 100 | 90 | 93 |
| 16 | — | 200 | 100 | 100 | 110 | 265 |
| 17 | 100 | 100 | 200 | — | 100 | 525 |
| 18 | 100 | — | 200 | 100 | 90 | 393 |
| 19 | — | 100 | 200 | 100 | 110 | 490 |
| 20 | 100 | 100 | — | 200 | 80 | 165 |
| 21 | 100 | — | 100 | 200 | 80 | 240 |
| 22 | — | 100 | 100 | 200 | 100 | 259 |
| 23 | 100 | 100 | 100 | 100 | 90 | 168 |

Table 11.1 and Figure 11.11 show the results for all meaningful portfolios in this example. Because of the riskless security 1, we would always invest the entire initial amount.

The curve in Figure 11.1 connects the undominated portfolios and represents the efficient frontier. Thus, we would expect an investor to select one of the following portfolios, depending on the investor's utility function: 1, 2, 9, 15, 12, 16, and 17.

The calculations of expected portfolio return and variance of portfolio return were done by using Eqs. 10.16, 10.17, and 10.8. We obtain here the result for portfolio 22 to demonstrate the use of the equations.

$$E(R) = (1)(\$30) + (1)(30) + (2)(20) = \$100$$

$$Var(R) = (1)^2(25) + (1)^2(100) + (2)^2(49) + (2)(1)(1)(5)(10)(0.5) + 2(1)(2)(5)(7)(-0.4) + 2(1)(2)(10)(7)(-0.2) = 259 \quad \square$$

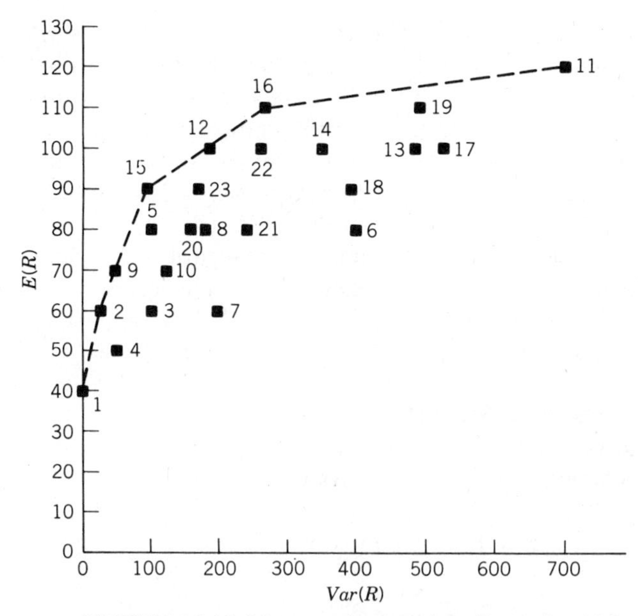


**FIGURE 11.11** Mean–variance plot, Example 11.8.

### *11.4.2 Diversification of Risk*

It is instructive to compare the results of Example 11.8 with the situation of having to invest the $400 in only *one* security.

| Security | $E(R)$ | $Var(R)$ |
|---|---|---|
| 1 | $40 | 0 |
| 2 | 120 | 400 |
| 3 | 120 | 1,600 |
| 4 | 80 | 784 |

To achieve the same expected return, investing in *one* security generally entails much greater risk, as measured by the variance of return, than spreading the $400 among several securities. For example, portfolio 11 achieves the same $E(R)$ of $120 as does investment of $400 in security 3, but with a much lower $Var(R)$ of 700, compared with 1,600.

By spreading the $400 among securities 2, 3, and 4, as in portfolio 16, we can achieve nearly the maximum $E(R)$, $110 out of a possible 120, with a relatively low $Var(R)$ of 265. If we were able to find another security similar to security 2, but with zero correlation, we could split our $400 evenly between the two with

$$E(R) = (2)(30) + (2)(30) = \$120$$

$$Var(R) = (2)^2(25) + (2)^2(25) = 200$$

Of particular interest to us is the effect of investing in securities with *negative* correlations of return. Security 4 in Example 11.8 has a negative correlation with both securities 2 and 3. The inclusion of security 4 *in combination* with either 2 or 3 will reduce the variance of portfolio return, as shown in the calculation for portfolio 22. It is noteworthy that security 4 is included in more of the efficient portfolios than security 3—four times versus two—despite the lower expected return.

Also noteworthy is the fact that the low-yield, riskless security 1 is included in four of the seven efficient portfolios. This means that some investors will be able to reach higher utility curves with a combination of risky *and* riskless securities than with only risky ones.

### 11.4.3 Full Covariance Model

The previous concepts are now presented formally. We begin by expressing the expected return and variance of return from a portfolio.

$$E(R) = E\left(\sum_j x_j R_j\right) = \sum_j x_j E(R_j) = \sum_j x_j E_j \tag{11.6}$$

$$Var(R) = \sum_j x_j^2 \sigma_j^2 + \sum_j \sum_{\substack{k \\ k \neq j}} x_j x_k \rho_{jk} \sigma_j \sigma_k \tag{11.7}$$

The coefficient of risk aversion, introduced in Section 9.5, is used to create a linear combination of portfolio return and variance, which is to be maximized.

$$\operatorname*{Max}_{x_j} E(R) - \lambda\, Var(R)$$

or

$$\text{Max} \sum_j x_j E_j - \lambda \sum_j x_j^2 \sigma_j^2 - \lambda \sum_j \sum_{\substack{k \\ k \neq j}} x_j x_k \rho_{jk} \sigma_j \sigma_k \tag{11.8}$$

subject to

$$\sum x_j = 1 \tag{11.9}$$

$$x_j \geq 0 \quad \text{for all } j \tag{11.10}$$

Equation 11.9 requires investment of all the available cash. Since one of the securities may be riskless and highly liquid, the restriction is more of a formality than anything else. Equation 11.10 prevents short sales. For $J$ securities this formulation requires $J$ parameter estimates for expected returns and $(J^2 - J)/2$ estimates for covariances, hence the designation of the model as the *full covariance model.* The earliest solution approaches used quadratic programming, which resulted in relatively long computation times [7, 14, 21]. Varying the value of the coefficient of risk aversion results in the set of efficient portfolios.

More efficient solution techniques take advantage of the mathematical structure of the problem. Specifically, any portfolio on the efficiency frontier can be represented as a nonnegative, linear combination of two "corner portfolios."

Further, both of two corner portfolios that are adjacent on the efficiency frontier will contain the same securities except for one. (The corner portfolio plays a role similar to that of a basis in linear programming.) The solution procedure essentially moves along the efficiency frontier from one corner portfolio to another [20].

### 11.4.4 Index Model

The requirement for the covariance estimates is a formidable data burden for the potential user of the full covariance model. Some restrictive assumptions about the way securities behave could greatly simplify the data collection and analysis. (Recall how the assumptions for correlated cash flows simplified the calculation of the variance of present value in Section 10.5.) The index model (sometimes called the diagonal model) achieves precisely this objective [25].

It is assumed that the returns from securities are related only through some common, underlying market factor. The return from a security is represented by

$$R_j = A_j + B_j I + C_j, \qquad j = 1, \ldots, J \tag{11.11}$$

where $A_j$ = a constant, specific to security $j$,
$I$ = a random variable representing a market index,
$B_j$ = a constant parameter that expresses the relation between security $j$ and the market index,
$C_j$ = an independent random variable with mean zero and variance $Q_j$.

Figure 11.12 shows the assumed relationship between $R_j$ and $I$. The future level of the index is expressed as

$$I = A_{J+1} + C_{J+1} \tag{11.12}$$

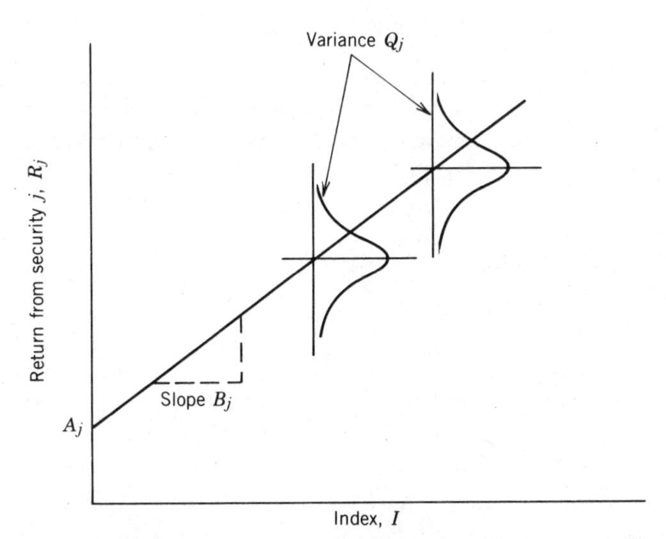


**FIGURE 11.12** Assumed relationship between return from security $j$ and the index.

where $A_{J+1}$ = a constant,

$C_{J+1}$ = an independent random variable with mean zero and variance $Q_{J+1}$.

The number of parameter estimates required for $J$ securities is $3J + 2$, considerably fewer than the $(J^2 - J)/2$ for the full covariance model.

The expressions for expected portfolio return and variance can be derived as follows. For individual securities,

$$E(R_j) = A_j + B_j A_{J+1} \tag{11.13}$$

$$\sigma_j^2 = B_j^2 Q_{J+1} + Q_j \tag{11.14}$$

$$\sigma_{jk} = B_j B_k Q_{J+1} \tag{11.15}$$

For the portfolio,

$$E(R) = \sum_{j=1}^{j} x_j E(A_j + B_j A_{J+1})$$

$$= \sum_{j=1}^{J} x_j A_j + \sum_{j=1}^{J} x_j B_j A_{J+1} \tag{11.16}$$

By letting

$$x_{J+1} = \sum_{j=1}^{J} x_j B_j \tag{11.17}$$

we can express Eq. 11.16 as

$$E(R) = \sum_{j=1}^{J+1} x_j A_j \tag{11.18}$$

Equation 11.18 can be interpreted in the following manner. Investment in the portfolio represents investment in $J$ securities *plus* an investment in the index. For portfolio variance we have from Eq. 11.7

$$Var(R) = \sum_{j=1}^{J} x_j^2 (B_j^2 Q_{J+1} + Q_j) + \sum_{j=1}^{J} \sum_{\substack{k=1 \\ k \neq j}}^{J} x_j x_k B_j B_k Q_{J+1}$$

$$= \sum_{j=1}^{J} x_j^2 Q_j + \sum_{j=1}^{J} x_j^2 B_j^2 Q_{J+1} + \sum_{j=1}^{J} \sum_{\substack{k=1 \\ k \neq j}}^{J} x_j x_k B_j B_k Q_{J+1}$$

$$= \sum_{j=1}^{J} x_j^2 Q_j + Q_{J+1} \left( \sum_{j=1}^{J} x_j B_j \right)^2 \tag{11.19}$$

Again using Eq. 11.17, we obtain

$$Var(R) = \sum_{j=1}^{J+1} x_j^2 Q_j \tag{11.20}$$

Two questions are suggested by such a formulation. (1) Does the simplified model, besides reducing the data-gathering effort, make possible faster solution techniques? (2) Is the efficiency frontier obtained from the index model reasonably consistent with that obtained from the full covariance model? The answer to the first question is a definite yes. The diagonal model can be solved by algorithms that are 50 times faster than a standard quadratic programming algorithm [25]. The diagonal model can also be approximated by a linear programming formulation [27].

The answer to the second question is not so clear. Quite a number of empirical tests have been performed in an attempt to compare the two formulations [6, 8]. On balance, the diagonal model seems to do as well or nearly as well as the full covariance model. Going one step further, some analysts have developed and applied multi-index models in an attempt to improve on the single-index model [6, 8].

### 11.4.5 Capital Market Theory

The development of portfolio theory has led to questions about the capital market. Specifically, if each investor used a portfolio model for selecting investments, what would the overall consequences be? Would some securities be favored by investors and others never be selected? Would this selection process affect the prices, and thus returns, of the securities? Capital market theory addresses these questions within the framework of a general market equilibrium. A number of assumptions are needed in order to apply this theory [4, 33].

1. All investors have the same information about security returns, variances, and covariances.
2. All investors are rational and risk-averse and select securities to maximize a utility function based on the mean and variance of return within a one-period time horizon.
3. All investors can borrow and lend without limit at a fixed interest rate.
4. Short sales are permitted.
5. There are no taxes, transaction costs, or costs for information.
6. Fractional shares are permitted; that is, investments are continuously divisible.
7. There are no monopolistic forces in the market.
8. The total quantity of any security is fixed.

This imposing list of assumptions leads to two major results that are of interest to us: the *separation theorem* and the pricing of individual securities.

Consider first an individual investor deciding between two assets, one a risk-free asset $F$ and the other a risky asset $M$ (the risky asset could be a single security or a portfolio). By combining the two, the investor will achieve an expected return for the combination $C$ as follows,

$$E(R_c) = aR_f + (1 - a)E(R_m), \qquad 0 \leq a \leq 1 \tag{11.21}$$

where $R_c$ = return for the combination $C$,
$R_f$ = known return for the risk-free asset $F$,
$R_m$ = return for the risky asset $M$.

The variance of return will be

$$Var(R_c) = a^2\, Var(R_f) + (1 - a)^2\, Var(R_m) + 2a(1 - a)\, Cov(R_f, R_m) \tag{11.22}$$

The $Var(R_f)$ is zero by definition, so that terms falls out. The last term falls out because the covariance between a random variable, $R_m$, and a constant, $R_f$, is zero. So we are left with

$$Var(R_c) = (1 - a)^2\, Var(R_m)$$

$$\sigma_c = (1 - a)\sigma_m \tag{11.23}$$

where $\sigma$ means the standard deviation of return. Solving for $a$, we have

$$a = 1 - \sigma_c/\sigma_m \tag{11.24}$$

Substituting for $a$ in Eq. 11.21 and rearranging gives

$$E(R_c) = R_f + (\sigma_c/\sigma_m)[E(R_m) - R_f] \tag{11.25}$$

For a given risky asset $M$, the values of $\sigma_m$ and $E(R_m)$ can be presumed fixed for the next time period. Equation 11.25 can thus be viewed as the equation of a straight line involving the variables $E(R_c)$ and $\sigma_c$, as shown in Figure 11.13 (the line connecting $R_f$ and $M$). Our individual investor could achieve any $E(R_c)$, $\sigma_m$ combination along this line by appropriate splitting of the investment amount between $F$ and $M$ (points along the line beyond $M$ are achieved by borrowing).

Our investor can perform the same analysis for other risky investments along the efficiency frontier, such as $K$ and $L$. The investor would prefer a point, such as $M$, which lies on a line through $R_f$ that is tangent to the efficiency frontier (see Figure 11.13). Investment (or portfolio) $M$ in combination with the risk-free asset thus dominates all other assets or portfolios for our investor.

Since all investors are assumed to have the same information, motives, and access to funds, they *all* will act similarly. All the investors will identify the *same* portfolio $M$, called the *market portfolio,* and allocate shares of their investment

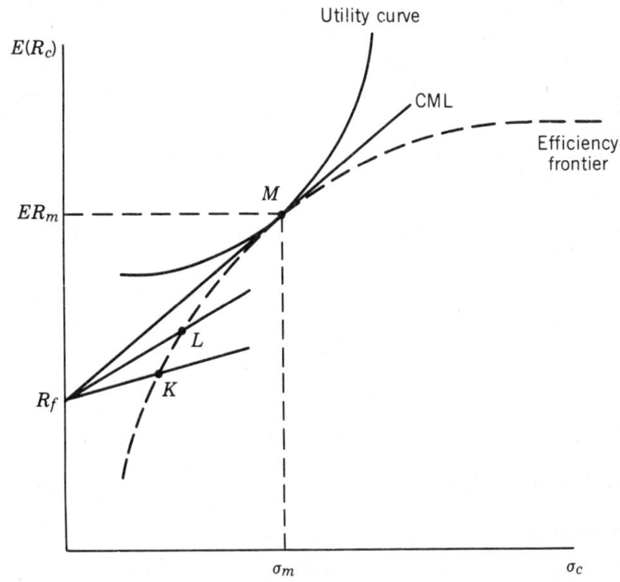


**FIGURE 11.13** **Development of the capital market line (CML).**

amounts to *M*, with the rest to be invested in the risk-free asset. The only differences in how the investors act will be in the shares they allocate to *M* and *F*, according to the differences in their particular utility curves. The line from $R_f$ through *M* in Figure 11.13 is called the *capital market line* (CML). This result is called the *separation theorem* [14].

In Eq. 11.25 for the CML the term $[E(R_m) - R_f/\sigma_m]$ represents the slope of the CML. At the same time, the term represents the additional expected return $E(R_c)$ required to compensate for additional risk as measured by $\sigma_c$, and thus it is a measure of the price of risk. This concept can be developed further to provide a way to establish the price of a risky asset under equilibrium market conditions. For a risky asset *j* that is part of the portfolio *M*, it can be shown that [4]

$$E(R_j) = R_f + \beta_j[E(R_m) - R_f] \tag{11.26}$$

where

$$\beta_j = \frac{Cov(R_j, R_m)}{Var(R_m)}$$

$$= \frac{\sigma_j \rho_{jm}}{\sigma_m} \tag{11.27}$$

Here $\rho_{jm}$ is the correlation coefficient for returns of *j* and *M*. Equations 11.26 and 11.27 represent the *capital asset pricing model*.

This concept can be applied to the evaluation of individual securities with a view to including them in the market portfolio.

### *Example 11.9*

For a CML with $E(R_m) = 11\%$, $R_f = 6\%$, and $\sigma_m = 4\%$, evaluate securities 1, 2, and 3 for inclusion in the market portfolio. Data for the securities are as follows.

| $j$ | $E(R_j)$ | $\sigma_j$ | $\rho_{jm}$ |
|---|---|---|---|
| 1 | 12% | 8% | 0.2 |
| 2 | 14% | 12% | 0.8 |
| 3 | 8% | 4% | −0.5 |

The equilibrium prices for the securities are obtained as follows.

$$j = 1: \quad 0.06 + (0.11 - 0.06)(0.08)(0.2)/0.04 = 0.08 = 8\%$$
$$j = 2: \quad 0.06 + (0.05)(0.12)(0.8)/0.04 = 0.018 = 18\%$$
$$j = 3: \quad 0.06 + (0.05)(0.04)(-0.5)/0.04 = 0.035 = 3.5\%$$

On the basis of these calculations, we would judge security 1 to be underpriced: it offers an expected return of 12% when its equilibrium price need only be 8%. The inclusion of security 1 in our market portfolio is desirable. Security 2 is overpriced, since a return of 18% is needed to compensate for the higher *volatility,* as measured by the $\beta$. Security 3 has negative correlation with the market portfolio, and the result is an equilibrium price *less* than $R_f$; here security 3 is underpriced. □

The concepts discussed here have been applied in the context of project selection for the firm [4].

## 11.5 DISCRETE CAPITAL-RATIONING MODELS UNDER RISK

The earlier sections of this chapter have dealt with risk primarily as represented by variability of cash flow. This is not the only source of variability, for each of the following factors can vary.

- amounts available for investment
- reinvestment rates
- project lifetimes
- timing of cash flows
- correlations among cash flows

In this section we address some of these issues. We first present a method for representing cash flows for correlated projects, so that we can compute $E(PV)$ for the combination. Then we examine an approach called stochastic programming, which is designed to handle directly some of the sources of variability.

### 11.5.1 Hillier's Method for Correlated Projects

The method presented in Section 10.5.2 for representing correlated cash flows can also be applied to cash flows of more than one project [15]. Let $F_{jn}$ be the random cash flow from project $j$ ($j = 1, 2, \ldots, J$) at time period $n$ ($n = 0, 1, 2, \ldots, N$). Allowing the random variables $F_{jn}$ to be correlated implies the following expression for the combined cash flow resulting from acceptance of the $J$ projects.

$$E[PV(i)] = \sum_{j=1}^{J} \sum_{n=0}^{N} \frac{E(F_{jn})}{(1+i)^n} \tag{11.28}$$

$$Var[PV(i)] = \sum_{j=1}^{J} \sum_{k=1}^{J} \sum_{n=0}^{N} \sum_{m=0}^{N} \frac{\sigma_{jn}\sigma_{km}\rho_{jn,km}}{(1+i)^{n+m}} \tag{11.29}$$

where $\sigma_{jn}^2$ = variance of cash flow $F_{jn}$,
$\rho_{jn,km}$ = correlation coefficient between $F_{jn}$ and $F_{km}$,
$i$ = interest rate for discounting.

The evaluation of Eq. 11.28 presents no unusual problems, but Eq. 11.29 requires estimating a correlation coefficient for every pair of cash flows, which would generally be impractical. Hillier's extension requires the following assumptions in addition to those stated in Section 10.5.2.

1. The partial correlation coefficient between $F_{jn}$ and $F_{km}$, $j \neq k$, $m < n$, with respect to $F_{kn}$ is zero.
2. The total correlation coefficient between $F_{jn}$ and $F_{km}$, $j \neq k$, $n$ and $m$ fixed, is a constant for all pairs $jk$.

The result of all the assumptions is that the total correlation coefficients have the following properties.

For cash flows of the same project, $j = k$,

$$\rho_{jn,jm} = \rho^{|n-m|} \tag{11.30}$$

For cash flows of different projects, $j \neq k$,

$$\rho_{jn,km} = \rho^{*}\rho^{|n-m|} \tag{11.31}$$

where $\rho$ and $\rho^*$ are constant parameters. Thus, if we have the $\sigma_{jn}$ values, we need to estimate only two additional parameters to generate the correlation matrix. It is recommended that an estimation procedure similar to that in the single-project case be used. Typical questions might be

If $F_{jn}$ is $3\sigma$ above (below) $E(F_{jn})$, what is $E(F_{j,n+1})$?
If $F_{jn}$ is $3\sigma$ above (below) $E(F_{jn})$, what is $E(F_{kn})$?

The same caution applies here as in the single-project situation. The value of the correlation coefficient must be in the interval $[-1, +1]$. If it falls outside this interval, the model presented here may not be appropriate.

### *Example 11.10*

Generate the correlation matrix when $J = 2$, $N = 3$, $\rho = 0.6$, and $\rho^* = 0.5$. The resulting (2)(3 + 1) matrix is

| | $k_m$ | | | | | | | |
|---|---|---|---|---|---|---|---|---|
| $j_n$ | 10 | 11 | 12 | 13 | 20 | 21 | 22 | 23 |
| 10 | 1 | 0.6 | 0.36 | 0.216 | 0.5 | 0.3 | 0.18 | 0.108 |
| 11 | 0.6 | 1 | 0.6 | 0.36 | 0.3 | 0.5 | 0.3 | 0.18 |
| 12 | 0.36 | 0.6 | 1 | 0.6 | 0.18 | 0.3 | 0.5 | 0.3 |
| 13 | 0.216 | 0.36 | 0.6 | 1 | 0.108 | 0.18 | 0.3 | 0.5 |
| 20 | 0.5 | 0.3 | 0.18 | 0.108 | 1 | 0.6 | 0.36 | 0.216 |
| 21 | 0.3 | 0.5 | 0.3 | 0.18 | 0.6 | 1 | 0.6 | 0.36 |
| 22 | 0.18 | 0.3 | 0.5 | 0.3 | 0.36 | 0.6 | 1 | 0.6 |
| 23 | 0.108 | 0.18 | 0.3 | 0.5 | 0.216 | 0.36 | 0.6 | 1 |

□

Although the procedure given simplifies the problem of obtaining the correlation coefficients, it has a number of disadvantages that may lead to unrealistic applications.

1. $\rho$ is the same for all projects.
2. $\rho$ is constant across time.
3. $\rho^*$ is the same for all project pairs $jk$.

In addition, the information that must be obtained concerning future cash flows is of a hypothetical nature, and we might question the validity of the resulting answers.

Hillier embedded the expressions for mean and variance of the present value of a combination of projects within an approximate linear programming approach. The approach is quite involved, and the interested reader is referred to Hillier's original work [15].

#### 11.5.2 Stochastic Programming

Consider a capital-rationing problem in which the budget limitation is a random variable,

$$\text{Max} \sum_j x_j E[PV(i)] \tag{11.32}$$

subject to

$$\text{Prob}\left[\sum_j a_j x_j \le M\right] \ge P \tag{11.33}$$

$$0 \le x_j \le 1, \qquad j = 1, \dots, n \tag{11.34}$$

where $x_j$ = project selection variable,
$a_j$ = investment needed for project $j$,
$M$ = budget limit, a random variable independent of the $PV(i)$,
$P$ = a specified probability value.

If we have probabilistic information about $M$, we can convert the *chance constraint* Eq. 11.33 to an equivalent deterministic constraint. For example, if $M$ is normally distributed with a mean of \$500,000 and a standard deviation of \$40,000 and $P = 0.8$, Eq. 11.33 can be restated as

$$\sum_j a_j x_j \leq \$500{,}000 - (0.84)(\$40{,}000)$$

or

$$\sum_j a_j x_j \leq \$466{,}400$$

Here the value 0.84 is the standard normal random deviate for a one-tail probability of 0.20.

With more than one chance constraint, questions of independence among the right-hand-side values need to be addressed. In addition, the individual probability values should be specified so that the joint probability of satisfying *all* chance constraints equals or exceeds some desired value. These issues cannot be addressed in a straightforward manner.

Many authors have treated the situation in which the investment outlays, the $a_j$ variables in Eq. 11.33, are random variables, or in the case of several constraints, $n = 0, 1, 2, \ldots, N$, some of the $a_{jn}$ are random variables. The right-hand-side coefficients are known, but the $PV(i)$ may be random variables. Such formulations have a first-stage problem, where some of the $x_j$ variables must be specified, followed by a second-stage problem, where the realizations of the random variables are known and the remaining decision variables must be specified. The typical approach is to convert the stochastic formulation into a larger, deterministic problem [5, 16, 31]. Some work is directed at solving the stochastic problem directly [13].

## 11.6 MULTIPERIOD INDEX MODEL FOR PROJECT PORTFOLIO

In this section we return to the problem of estimating the correlation coefficients among project cash flows (see Sections 11.4.3, 11.4.4, and 11.5.1). The method uses a single-index model for multiple time periods [10]. The major assumption of the model is that cash flows for project $j$ at time period $n$ are related only through common relationships with some basic underlying factor, such as gross national product or the number of housing starts.

### 11.6.1 Model Structure and Assumptions

We obtain estimates of the correlation coefficients by first estimating the parameters of a single-index model. Contained in these estimates are implied estimates of the correlation coefficients. The multiperiod index model is patterned after Sharpe's diagonal model for security portfolio analysis [25]. The underlying assumption is that the returns of various projects are related only through common relationships with some basic, common factor (index). Thus,

$$F_{jn} = A_{jn} + B_j I_n + C_{jn}, \qquad j = 1, \ldots, J \text{ and } n = 0, 1, \ldots, N \tag{11.35}$$

The notation is analogous to that in the previous sections.

- $F_{jn}$ the random cash flow for project $j$ at time period $n$
- $I_n$ the random variable representing the level of the index at time period $n$
- $A_{jn}$ a constant representing the portion of the cash flow for project $j$ at time period $n$ that is independent of the index and any inherent variability
- $B_j$ a constant parameter that expresses the responsiveness of the cash flow stream for project $j$ to changes in the index
- $C_{jn}$ an independent random variable associated with the cash flow for project $j$ at time period $n$; it has mean zero and variance $\sigma_{jn}^2$

We also define

- $A_{0n}$ a constant representing the expected value of the index at time period $n$
- $C_{0n}$ an independent random variable with mean zero and variance $\sigma_{0n}^2$, representing the variability of the index at time period $n$
- $\rho_{0,nm}$ the correlation coefficient of the index at lag$(n - m)$

This model is based on several assumptions about the random error terms, $C_{jn}$.

1. The average of each $C_{jn}$ is zero; that is, $E(C_{jn}) = 0$, $j = 1, \ldots, J$, and $n = 0, 1, \ldots, N$.
2. The variance of each project error term $\sigma_{jn}^2$ is constant; that is, $\sigma_{jn}^2 = a$, $j = 1, \ldots, J$, and $n = 0, 1, \ldots, N$.
3. The project error terms are uncorrelated with the index; that is,

$$Cov(C_{0n}, C_{jm}) = 0, \qquad j = 1, 2, \ldots, J \text{ and } n,m = 0, 1, 2, \ldots, N$$

4. The project error terms are not serially correlated; that is,

$$Cov(C_{jn}, C_{jm}) = 0, \qquad j = 1, 2, \ldots, J \text{ and } n,m = 0, 1, 2, \ldots, N, \quad n \neq m$$

5. The error terms of one project are uncorrelated with those of any other project; that is,

$$Cov(C_{jn}, C_{kn}) = 0, \qquad j,k = 1, 2, \ldots, J, \quad j \neq k, \text{ and } n = 0, 1, 2, \ldots, N$$

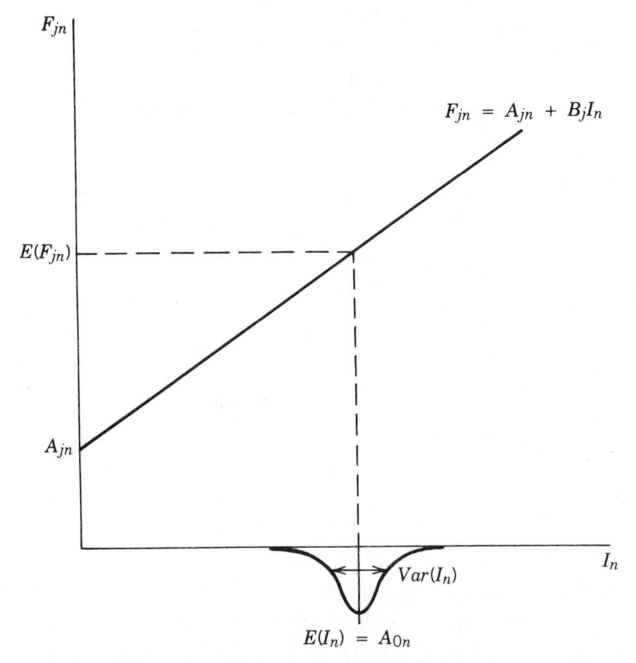


**FIGURE 11.14** The single index model.

These assumptions imply that the regression parameters $A_{jn}$ and $B_j$ are unbiased, minimum-variance linear estimates of the true parameters. Graphically, the model suggests a regression line such as the one shown in Figure 11.14.

The following relations are straightforward to obtain (see problems).

$$I_n = A_{0n} + C_{0n} \tag{11.36a}$$

$$E(F_{jn}) = A_{jn} + B_j A_{0n} \tag{11.36b}$$

$$Var(F_{jn}) = B_j^2 \sigma_{0n}^2 + \sigma_{jn}^2 \tag{11.36c}$$

An estimate of the covariance between two cash flows $F_{jn}$ and $F_{km}$ can be derived from the definition of the covariance [10].

$$\begin{aligned} Cov(F_{jn}, F_{km}) &= B_j B_k \, Cov(I_{n,m}) \\ &= B_j B_k \rho_{0,\,nm} \sigma_{0n} \sigma_{0m} \end{aligned} \tag{11.37}$$

Notice that this expression contains no terms related to project $j$ or $k$ except for $B_j$ and $B_k$, which are obtained by linear regression.

### 11.6.2 Procedure

For selecting an economic index, the stationary property of the series is critical. Many economic time series behave as though they had no fixed mean.

However, they may exhibit homogeneity in the sense that, apart from the trend, one part of the series behaves much like any other part. Series that describe such homogeneous nonstationary behavior can be differenced to obtain a stationary series. Let $\nabla^d I_n$ be the $d$th difference of the series for which the series is stationary. We may define the difference operator $\nabla$ as

$$\nabla I_n = I_n - I_{n-1} \tag{11.38}$$

Then Eq. 11.35 of the model is replaced by

$$\nabla^d F_{jn} = A_{jn} + B_j \nabla^d I_n + C_{jn} \tag{11.39}$$

For stationary series the standard deviation is the same at time $n$ as at time $m$. Then Eq. 11.37 becomes

$$Cov(F_{jn}, F_{km}) = B_j B_k \rho_{0,nm} \sigma_0^2 \tag{11.40}$$

where $\sigma_0^2$ is the variance of the index.

Most economic time series found in practice can usually be adequately reduced to a stationary series by first or second differencing. Occasionally, transformations other than differencing are useful in reducing a nonstationary time series to a stationary one.

Once the covariance terms are estimated, we could, for example, use a mean–variance criterion to make project selection decisions.

$$\begin{aligned} \underset{x_j}{\text{Max}} \sum_{j=1}^{J} \sum_{n=0}^{N} \frac{(A_{jn} + B_j A_{0n}) x_j}{(1+i)^n} &- \lambda \sum_{j=1}^{J} \sum_{n=0}^{N} \frac{\sigma_{jn}^2 x_j^2}{(1+i)^{2n}} \\ &- 2\lambda \sum_{\substack{n=0 \\ n<m}}^{N-1} \sum_{m=1}^{N} \frac{\rho_{0,nm}\sigma_0^2}{(1+i)^{n+m}} \sum_{j=1}^{J} \sum_{k=1}^{J} B_j B_k x_j x_k \end{aligned} \tag{11.41}$$

where $\lambda$ is the coefficient of risk aversion, subject to budget constraints and contingency constraints. Note that this model is a 0–1 quadratic programming model, which, although difficult, is amenable to solution by standard algorithms.

The only terms that make the problem nonlinear are the $x_j x_k$ combinations in the last term of Eq. 11.41. Accordingly, we can obtain a simplified model by deleting those terms. The resulting linear model has the same objective function as Eq. 11.41, except that the last term is replaced by

$$- 2\lambda \sum_{\substack{n=0 \\ n<m}}^{N-1} \sum_{m=1}^{N} \frac{\rho_{0,nm}\sigma_0^2}{(1+i)^{n+m}} \sum_{j=1}^{J} B_j^2 x_j^2 \tag{11.42}$$

Since the $x_j$ will be either 0 or 1, we can replace the $x_j^2$ variables in the second and third terms of Eq. 11.41 and obtain a linear model. We may distinguish between the two forms by designating Eq. 11.41 as the model with cross-correla-

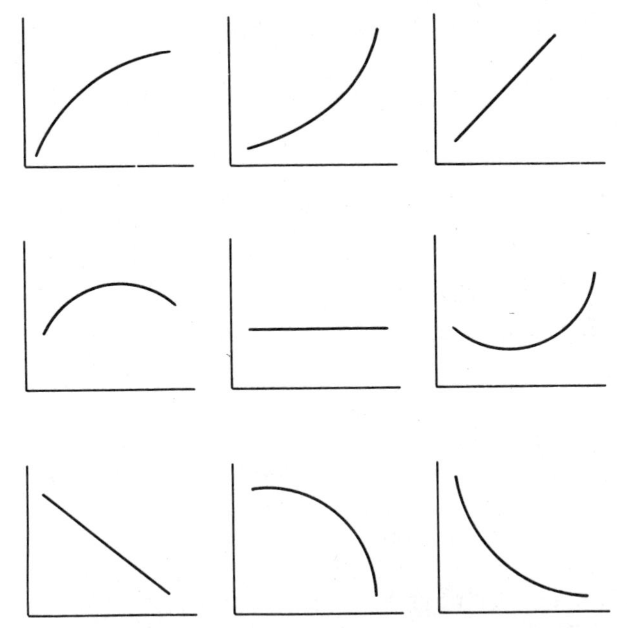

**FIGURE 11.15** **Different types of forecast patterns used for the index.**

tion (between projects) and Eq. 11.42 as the model with autocorrelation only (within projects). Note also that if we eliminate entirely the last term of Eq. 11.41, we are left with a linear model that is similar to that of Sharpe [25].

A typical procedure for using the model then consists of the following steps:

1. Identify the set of potential projects.
2. Obtain the mean and variance of each cash flow $F_{jn}$. This step can be performed by taking optimistic, most likely, and pessimistic values for the cash flows and using Eqs. 10.72 and 10.73.
3. Select an appropriate index that reflects economic circumstances common to the activities of the company. Typical indices are the gross national product, number of housing starts, automobile production, and so forth.
4. If necessary, reduce the index to a stationary series by using Eq. 11.38.
5. Obtain various forecast patterns for the index. The purpose here is not to obtain the best forecast of the index but to obtain a *variety* of plausible forecasts, or scenarios. Figure 11.15 shows some typical patterns.
6. For *each* different forecast pattern of the index, obtain the mean of the cash flow, *given* that the index forecast is correct.
7. Obtain regression coefficients $A_{jn}$ and $B_j$ by using Eq. 11.35 or 11.39, as appropriate.
8. Select the discount rate $i$.
9. Select the risk aversion factor $\lambda$. If this step poses difficulty, simply let $\lambda$ be a parameter. This will then yield the entire efficiency frontier.

10. Specify the objective function, either the quadratic model in Eq. 11.41 or the linear model in Eq. 11.42.
11. Specify budget constraints.
12. Specify contingency constraints.
13. Optimize.
14. Make a final decision based on the output results.

The case study in Appendix 11.A demonstrates the use of the procedure.

## 11.7 UNCERTAINTY RESOLUTION

When we compare projects, situations come up in which both $E(PV)$ and $Var(PV)$ are essentially the same and yet there is another characteristic of the cash flows that would lead an investor to prefer one alternative. Consider the example shown in Figure 11.16. In the example the project requires an investment of $100 at time 0, shown at node (0, 1). Between time 0 and time 1 there is a 0.5 chance of moving to node (1, 1) and obtaining $10 and a 0.5 chance of moving to node (1, 2) and obtaining $110. Having reached time 1, the investor will be faced with probabilities of reaching nodes (2, 1) and (2, 2), each with 0.5 chance, accessible from node (1, 1); *or* of reaching nodes (2, 3) and (2, 4), each with 0.5 chance, accessible from node (1, 2). Additional cash flows at time 2 will be realized, as shown in Fig. 11.16.

Examining this project, we see that there is considerable risk as measured by $Var(PV)$ or any similar measure. However, once time 1 has occurred, a large portion of the uncertainty will have been eliminated. Specifically, the investor will have learned whether the time 1 cash flow is $10 or $110. Thus, as time passes and the future cash flows are realized, the overall picture of the project is seen in better focus. This process of moving from greater uncertainty to less uncertainty is designated *uncertainty resolution.*

Van Horne developed a coefficient of variation measure for this resolution of uncertainty through time: at each time period $n$, compute the ratio of stan-

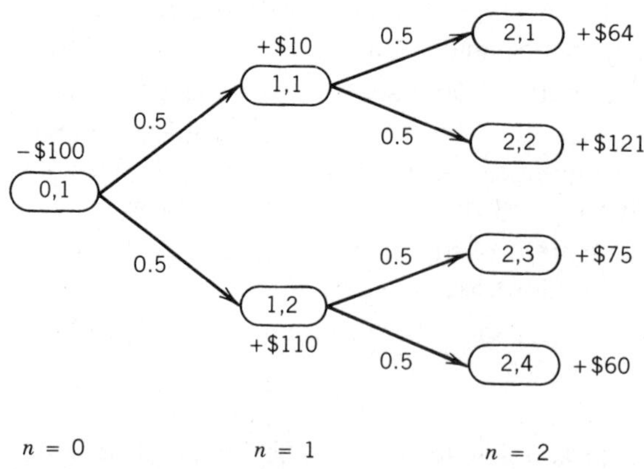


**FIGURE 11.16** Illustration of uncertainty resolution.

dard deviation of $PV$ (about the relevant conditional means) to $E(PV)$ for the project [30]. Bierman and Hausman [3] examine this measure in detail and discuss its usefulness and limitations under various circumstances.

Apart from Bierman and Hausman's criticism, the coefficient of variation has the disadvantage of treating *all* variability as undesirable (see Section 11.2.2). Moreover, the measure does not reflect other possible effects of timing and magnitude. Further, precisely defined rules for applying this measure are lacking.

To overcome these limitations, Park and Thuesen [22] developed a procedure that combines the area of negative balance ($ANB$) (see Section 6.6.1) with Baumol's expected-gain confidence limit criterion [1]. Their procedure consists of the following steps.

1. Compute the project balance at each point in time for each branch of the probability tree. (See Section 6.2.2.)
2. Compute the $ANB$ for each branch of the probability tree.
3. Compute $E(ANB)$ and $\sigma(ANB)$ for each time period by Van Horne's method.

For a risk-free discount rate of $i = 10\%$, the relevant values for steps 1 and 2 are shown in Table 11.2. The risk-free rate is used here because we are explicitly representing risk by $Var(PV)$ and $ANB$.

For step 3 we proceed as follows. For $n = 0$,

$$E(ANB)_0 = (246)(0.25) + (200)(0.25) + (100)(0.25) + (100)(0.25) = 161.5$$

$$Var(ANB)_0 = (246 - 161.5)^2(0.25) + (200 - 161.5)^2(0.25) + (100 - 161.5)^2(0.25) + (100 - 161.5)^2(0.25) = 4{,}046.75$$

$$\sigma(ANB)_0 = 63.61$$

$$EGCL(ANB)_0 = E(ANB)_0 + \delta\sigma(ANB)_0 = 161.5 + \delta(63.61)$$

For $n = 1$, first compute the conditional means based on the branches emanating from node (1, 1) and node (1, 2), respectively.

$$E(ANB)_{1,1} = (246)(0.5) + (200)(0.5) = 223$$

$$E(ANB)_{1,2} = (100)(0.5) + (100)(0.5) = 100$$

**Table 11.2** ***Data for Example in Figure 11.16***

| | Project Balance at End of Period $n$ | | | | Accumulated Value, |
|---|---|---|---|---|---|
| Branch | $n = 0$ | $n = 1$ | $n = 2$ | $ANB$ | Branch $FV$ |
| 1 | −100 | −100 | −46 | 246 | −46 |
| 2 | −100 | −100 | 11 | 200 | 11 |
| 3 | −100 | 0 | 75 | 100 | 75 |
| 4 | −100 | 0 | 60 | 100 | 60 |

Then calculate the conditional expectation of *ANB* at $n = 1$ by weighting $E(ANB)_{1,1}$ and $E(ANB)_{1,2}$ by their probability of occurrence.

$$E(ANB)_1 = (0.5)(223) + (0.5)(100) = 161.5$$

Next compute the conditional variance at node (1, 1) and node (1, 2) with respect to their conditional means, $E(ANB)_{1,1}$ and $E(ANB)_{1,2}$.

$$Var(ANB)_{1,1} = (246 - 223)^2(0.5) + (200 - 223)^2(0.5) = 529$$
$$Var(ANB)_{1,2} = 0$$

Using the conditional variances, $Var(ANB)_{1,1}$ and $Var(ANB)_{1,2}$, compute the weighted conditional variance at time $n = 1$,

$$Var(ANB)_1 = (0.5)(529) + (0.5)(0) = 264.5$$
$$\sigma(ANB)_1 = 16.26$$
$$EGCL(ANB)_1 = E(ANB)_1 + \delta\sigma(ANB)_1 = 161.5 + \delta(16.26)$$

For $n = 2$, since we are in the last time period, the weighted conditional mean is

$$E(ANB)_2 = (0.25)(246) + (0.25)(100) + (0.25)(100) + (0.25)(100) = 161.5$$

In addition, since we are in the last period, all the conditional variances are zero, so

$$EGCL(ANB)_2 = E(ANB)_2 = 161.5$$

If we select a conservative value of $\delta = 3$ for the expected-gain confidence limit criterion, we obtain the following values.

| $n$ | 0 | 1 | 2 |
|---|---|---|---|
| $EGCL(ANB)_n$ | 352.3 | 210.3 | 161.5 |

This information can then be used together with $E(PV)$ and $Var(PV)$ in evaluating and comparing investment projects [22]. In essence, we would need some trade-off mechanism, or perhaps a dominance criterion, that dealt with three attributes instead of the usual two.

## 11.8 SUMMARY

In this chapter we have attempted to treat a broad variety of techniques for comparing risky projects. The simplest techniques to understand are mean–variance, mean–semivariance, and probability of loss. These are compared with stochastic dominance criteria. For project portfolio analysis the data and computational burdens grow quickly with problem size, and the emphasis shifts to special structures and algorithms, such as Hillier's method for representing correlated cash flows and the various index models. Uncertainty resolution adds a third attribute to the usual two of $E(PV)$ and $Var(PV)$.

Despite the great variety of techniques developed for comparing risky projects, some fundamental obstacles remain. Efficient techniques to select project portfolios are still lacking for all except the simplest utility functions. Further, sources of variability related to amounts available for investment, reinvestment rates, project lifetimes, timing of cash flows, and so forth are difficult to incorporate in analytic models. A large number of researchers are addressing these issues, and the future promises more complex data structures and computational procedures.

## REFERENCES


1. Baumol, W. J., "An Expected Gain–Confidence Limit Criterion for Portfolio Selection," *Management Science,* Vol. 10, No. 1, pp. 174–182, 1963.
2. Bernhard, R. H., "Risk-Adjusted Values, Timing of Uncertainty Resolution, and the Measurement of Project Worth," *Journal of Financial and Quantitative Analysis,* Vol. 19, No. 1, pp. 83–89, 1984.
3. Bierman, H., Jr., and W. R. Hausman, "The Resolution of Investment Uncertainty through Time," *Management Science,* Vol. 18, No. 12, B:654–B:662, 1972.
4. Bussey, L. E., *The Economic Analysis of Industrial Projects,* Prentice–Hall, Englewood Cliffs, N.J., Ch. 12, 1978.
5. Charnes, A., and W. W. Cooper, "Chance Constrained Programming," *Management Science,* Vol. 6, No. 1, pp. 73–79, 1969.
6. Elton, E. J., and Gruber, M. J., Eds., *Security Evaluation and Portfolio Analysis,* Prentice–Hall, Englewood Cliffs, N.J., pp. 24–27, 491–495, 1972.
7. Farrar, D. E., *The Investment Decision under Uncertainty,* Markham, Chicago, 1967.
8. Farrell, J. L., Jr., *Guide to Portfolio Management,* McGraw–Hill, New York, 1983, pp. 43–55.
9. Feldstein, M., "Mean Variance Analysis in the Theory of Liquidity, Preference and Portfolio Selection," *Review of Economics and Statistics,* Vol. 47, pp. 5–12, 1969.
10. Guerra-Quiroga, J. E., *An Index Model for Correlated Cash Flow Streams in Capital Budgeting,* M.S. Thesis, Georgia Institute of Technology, School of Industrial and Systems Engineering, Atlanta, 1979.
11. Hanoch, G., and H. Levy, "The Efficiency Analysis of Choices Involving Risk," *Review of Economic Studies,* Vol. 36, pp. 335–346, 1969.
12. Hanoch, G., and H. Levy, "Efficient Portfolio Selection with Quadratic and Cubic Utility," *Journal of Business,* Vol. 43, pp. 181–189, 1970.
13. Hansotia, B. J., "Some Special Cases of Stochastic Programs with Recourse," *Operations Research,* Vol. 25, pp. 361–363, 1977.
14. Hester, D. R., and J. Tobin, Eds. *Risk Aversion and Portfolio Choice,* Cowles Foundation Monograph 19, Wiley, New York, 1967.

15. HILLIER, F. S., *The Evaluation of Risky Interrelated Investments,* North-Holland, Amsterdam, 1969.
16. KEOWN, J. K., and J. D. MARTIN, "A Chance Constrained Goal Programming Model for Working Capital Management," *The Engineering Economist,* Vol. 22, No. 3, pp. 153–174, 1977.
17. KIRA, D., and W. T. ZIEMBA, "Equivalence among Alternative Portfolio Selection Criteria," in H. Levy and M. Sarnat, *Financial Decision Making under Uncertainty,* Academic Press, New York, pp. 151–161, 1977.
18. LEVY, H., and M. SARNAT, *Investment and Portfolio Analysis,* Wiley, New York, 1972.
19. MAO, C. T., and J. F. BREWSTER, "An $E$–$S_b$ Model of Capital Budgeting," *The Engineering Economist,* Vol. 15, No. 2, pp. 108–121, 1970.
20. MARKOWITZ, H. M., "The Optimization of a Quadratic Function Subject to Linear Constraints," *Naval Research Logistics Quarterly,* Vol. 3, pp. 111–133, 1956.
21. MARKOWITZ, H. M., *Portfolio Selection: Efficient Diversification of Investments,* Wiley, New York, 1959.
22. PARK, C. S., and G. J. THUESEN, "Combining the Concepts of Uncertainty Resolution and Project Balance for Capital Allocation Decisions," *The Engineering Economist,* Vol. 24, No. 2, pp. 109–127, 1979.
23. PRATT, J. W., H. RAIFFA, and R. SCHLAIFER, *Introduction to Statistical Decision Theory,* McGraw–Hill, New York, 1968.
24. QUIRK, J. P., and R. SAPOSNIK, "Admissibility and Measurable Utility Functions," *Review of Economic Studies,* Vol. 29, pp. 140–146, 1962.
25. SHARPE, W. F., "A Simplified Model for Portfolio Analysis," *Management Science,* Vol. 9, No. 2, pp. 277–293, 1963.
26. SHARPE, W. F., *Portfolio Theory and Capital Markets,* McGraw–Hill, New York, 1970.
27. SHARPE, W. F., "A Linear Programming Approximation for the General Portfolio Analysis Problem," *Journal of Financial and Quantitative Analysis,* Vol. 6, pp. 1263–1275, December 1971.
28. TOBIN, J., "Liquidity Preference as Behavior towards Risk," *Review of Economic Studies,* Vol. 25, pp. 65–68, 1957–1958.
29. U.S. DEPARTMENT OF COMMERCE, *Business Statistics,* Washington, D.C., Biennal Supplement.
30. VAN HORNE, J. C., "The Analysis of Uncertainty Resolution in Capital Budgeting for New Products," *Management Science,* Vol. 15, No. 8, pp. B:376–B:386, 1969.
31. WETS, R., "Programming under Uncertainty: The Equivalent Convex Program," *SIAM Journal on Applied Mathematics,* Vol. 14, pp. 89–105, 1966.
32. WHITMORE, G. A., "Third Degree Stochastic Dominance," *American Economic Review,* Vol. 60, pp. 457–459, 1970.
33. WILKES, F. M., *Capital Budgeting Techniques,* 2nd edition, Wiley, New York, pp. 358–389, 1983.

## PROBLEMS

**11.1.** Compute $E(PV)$ and $Var(PV)$ for projects 2 through 5 in Example 11.1.

**11.2.** Compute the semivariance for the projects in Example 11.3.

**11.3.** Consider two projects with the following annual equivalent amounts. Both projects have the same lifetime. Evaluate these two projects by using (a) a linear utility function and (b) an increasing, concave utility function. Then apply the second-degree stochastic dominance test to the two projects.

| | Project | |
|---|---|---|
| *PV* | A | B |
| −$10,000 | 0.1 | 0.0 |
| 0 | 0.2 | 0.3 |
| +10,000 | 0.3 | 0.4 |
| +20,000 | 0.1 | 0.0 |
| +30,000 | 0.3 | 0.3 |

**11.4.** Consider the following two mutually exclusive investment projects.

| Project A | | | Project B | | |
|---|---|---|---|---|---|
| *PV* | Probability | Cumulative Probability | *PV* | Probability | Cumulative Probability |
| $1,000 | 4/16 | 4/16 | $500 | 3/16 | 3/16 |
| 2,000 | 4/16 | 8/16 | 1,500 | 3/16 | 6/16 |
| 9,000 | 4/16 | 12/16 | 2,500 | 4/16 | 10/16 |
| 10,000 | 4/16 | 1 | 3,500 | 3/16 | 13/16 |
| | | | 4,500 | 3/16 | 1 |

a. Compute the mean and variance of *PV* for each project.
b. Determine whether any project dominance exists, based on the *E–V* criterion.
c. Assuming that you are a risk-averse individual with decreasing marginal utility, which project would you select by applying the stochastic dominance criterion?
d. If your utility function is $u(x) = 2 - 2e^{-0.0001x}$, where $x = PV$, which project would you select based on the expected utility criterion?

**11.5.** Consider the following mutually exclusive investment proposals.

| | Present Value | |
|---|---|---|
| | Project A | Project B |
| Mean | $3,000 | $6,000 |
| Variance | $(1,000)^2$ | $(2,000)^2$ |
| *PV* distribution | Normal | Normal |

a. Compute the probability of $\{PV_A > PV_B\}$, assuming that both projects are statistically independent.
b. A plot of the *PV* distributions on the same chart is shown in the accompanying illustration. Compute the intersection point *C* and give your interpretation of the shaded area.

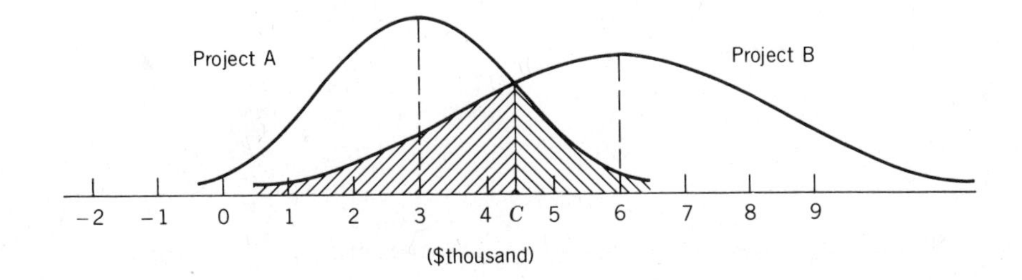

c. If you have a utility function of $u(x) = 1 - e^{-0.0001x}$, which project is more preferred, based on the expected utility maximization principle?

**11.6.** Consider the following continuous *PV* distributions. Assuming that each pair of distributions is statistically independent, determine any possible project dominance. (For notational simplicity, let $x = PV$.)

$$\text{Set 1} \begin{cases} f(x) = 1/8, & 2 \le x \le 10 \\ g(x) = 1/13, & 0 \le x \le 13 \end{cases}$$

$$\text{Set 2} \begin{cases} f(x) = 1/23, & 0 \le x \le 23 \\ g(x) = 1/10, & 5 \le x \le 15 \end{cases}$$

$$\text{Set 3} \begin{cases} f(x) = 1/8, & 1 \le x \le 9 \\ g(x) = 1/11, & 0 \le x \le 11 \end{cases}$$

$$\text{Set 4} \begin{cases} f(x) = 1/10, & 0 \le x \le 10 \\ g(x) = 1/5, & 2 \le x \le 7 \end{cases}$$

**11.7.** Consider the following sets of discrete *PV* distributions. Apply the stochastic dominance rules and identify the project that dominates in each set.

| Set | $PV(x_j)$ | Project A $p(x_j)$ | Project B $g(x_j)$ |
|---|---|---|---|
| 1 | 10 | 0.20 | 0.10 |
| | 20 | 0.50 | 0.70 |
| | 40 | 0.30 | 0.20 |
| 2 | 0 | 0.20 | 0.10 |
| | 5 | 0.20 | 0.30 |
| | 10 | 0.20 | 0.25 |
| | 15 | 0.20 | 0.30 |
| | 20 | 0.20 | 0.05 |
| 3 | 5 | 0.30 | 0.20 |
| | 10 | 0.40 | 0.60 |
| | 20 | 0.30 | 0.20 |

**11.8.** Consider the following three mutually exclusive projects. Which project would you select by applying the stochastic dominance rules?

| $PV(x_j)$ | Project A $f(x_j)$ | Project B $g(x_j)$ | Project C $z(x_j)$ |
|---|---|---|---|
| −$10,000 | 0.20 | 0 | 0 |
| 0 | 0 | 0.25 | 0.20 |
| 10,000 | 0.40 | 0.50 | 0.60 |
| 20,000 | 0.20 | 0.10 | 0.10 |
| 30,000 | 0.20 | 0.15 | 0.10 |

**11.9.** Consider the following two mutually exclusive investment projects. Assuming that they are statistically independent, which project would be selected by applying the stochastic dominance rules?

| $AE$, $x_j$ | Project A $f(x_j)$ | Project B $g(x_j)$ |
|---|---|---|
| −$200 | 0.20 | 0 |
| −100 | 0.10 | 0.30 |
| 0 | 0.10 | 0.20 |
| 100 | 0.10 | 0.20 |
| 200 | 0.30 | 0.10 |
| 300 | 0.20 | 0.20 |

**11.10.** Two projects are being compared. An analyst has obtained the following data for the projects, using $i = 10\%$.

| Project | $E(PV)$ | $Var(PV)$ |
|---|---|---|
| A | 100 | 100 |
| B | 140 | 169 |

What is your recommendation? Briefly explain the method you are using to make a decision.

**11.11.** Refer to Example 11.8, but assume that all correlation coefficients are zero. Obtain the efficient frontier for this case.

**11.12.** You have $30,000 to invest in securities. A preliminary screening has narrowed the list to three, plus a riskless security. Multiples of $10,000 may be invested in any one security. Obtain the efficient frontier of undominated portfolios.

| Security | $E(R)$ per $1,000 | $Var(R)$ per $1,000 | Coefficient of Correlation of $R$ Sec. 1 | Sec. 2 | Sec. 3 | Sec. 4 |
|---|---|---|---|---|---|---|
| 1 | 150 | 0 | — | 0 | 0 | 0 |
| 2 | 200 | 400 | 0 | — | −0.3 | −0.2 |
| 3 | 240 | 625 | 0 | −0.3 | — | 0.4 |
| 4 | 280 | 900 | 0 | −0.2 | 0.4 | — |

**11.13.** For a capital market line with $E(R_m) = 10\%$, $R_f = 5\%$, and $\sigma_m = 3\%$, evaluate these four securities for inclusion in the market portfolio.

| Security | $E(R_j)$ | $\sigma_j$ | $\rho_{jm}$ |
|---|---|---|---|
| 1 | 8% | 7% | 0.3 |
| 2 | 15% | 10% | 0.5 |
| 3 | 15% | 12% | 0.6 |
| 4 | 10% | 10% | −0.1 |

**11.14.** The Spring Valley Manufacturing Company is faced with two large mutually exclusive capital investment projects because it has sufficient funds to undertake only one or the other. There is much uncertainty associated with each project, so a capital budgeting specialist has been called in. The consultant found that the $PV$ for each project is normally distributed with mean and standard deviation as

| Project | Mean, $PV$ | Standard Deviation |
|---|---|---|
| 1 | $200,000 | $50,000 |
| 2 | $250,000 | $80,000 |

The consultant was not sure whether the two projects were statistically independent. The consultant therefore asked the management a series of questions.

Consultant: If the *PV* from project 2 is as low as \$10,000, what do you judge the most likely *PV* from project 1 to be?

Management: It is tough! But I guess it would be about \$150,000.

Consultant: If the *PV* from project 2 is as high as \$490,000, what do you judge the most likely *PV* from project 1 to be?

Management: I expect that it would be around \$300,000.

a. Can we estimate the correlation coefficient between the two *PV*s?
b. What is the probability that the *PV* of project 1 will exceed that of project 2?
c. Can we apply the set of stochastic dominance rules for the dependent projects?

**11.15.** A company is considering two projects, A and B,

| | Project A | | | Project B | | |
|---|---|---|---|---|---|---|
| $n$ | L | $M_0$ | $H$ | $L$ | $M_0$ | $H$ |
| 0 | −1,000 | −1,000 | −1,000 | −2,000 | −2,000 | −2,000 |
| 1 | 800 | 1,000 | 1,400 | 1,400 | 1,900 | 2,000 |
| 2 | 500 | 800 | 1,100 | 800 | 1,200 | 1,700 |

where $L$, $M_0$, and $H$ are pessimistic, most likely, and optimistic values for the cash flows. The pessimistic and optimistic values cover the range so that the chance of cash flow being beyond these limits is one in several hundred.

a. Estimate the mean and standard deviation of each cash flow for project A. Use estimators for the beta distribution,

$$E(\text{mean}) = \tfrac{4}{6}M_0 + \tfrac{1}{6}L + \tfrac{1}{6}H$$

and $H - L$ covers six standard deviations.

b. Assuming that the cash flows for project A are independent, determine the expected value and the variance of the present value of A, using $i = 10\%$.
c. Repeat part b, but now assume that the flows are perfectly correlated.
d. Analysts have obtained the following information concerning the dependence of the individual flows on each other.
   i. If the time 1 flow for A is 1,200, the most likely value of the time 2 flow for A is 900, the optimistic value is 1,000, and the pessimistic value is 800.
   ii. If the time 2 flow for A is 900, the most likely value of the time 2 flow for B is 1,250, the optimistic value is 1,600, and the pessimistic value is 900.

   Estimate the two-way correlations for each pair of cash flows, and place the results in a matrix. Use Hillier's technique, preferably.
e. Using the results of part d for project A only, estimate the variance of the present value of A.
f. Using the results of part d, estimate the variance of the present value of the combination of A and B. Assume that the correlation within B is the same as that within A.

**11.16.** Consider project $j$ with cash flows as follows.

| $n$ | $E(F_{jn})$ | $Var(F_{jn})$ |
|---|---|---|
| 0 | −1,000 | 1,600 |
| 1 | 400 | 2,500 |
| 2 | 400 | 3,600 |
| 3 | 400 | 4,900 |
| 4 | 400 | 6,400 |

a. With $i = 10\%$, obtain the mean and variance of $FV$ for
   i. $F_{jn}$ independent.
   ii. $F_{j0}$ independent and $F_{jn}$ perfectly correlated for $n = 1, 2, 3, 4$.

b. Now assume correlation coefficients of the type

$$\rho_{01} = 0, \qquad \rho_{02} = 0, \qquad \rho_{03} = 0, \qquad \rho_{04} = 0$$

$$\rho_{12} = \rho_{23} = \rho_{34}$$

$$\rho_{13} = \rho_{24} = \rho_{12}^2, \quad \text{etc.}$$

Obtain the variance of $FV$ for a range of values of $\rho_{12}$.

c. Find weighting factors for parts a.i and a.ii so that you obtain the same variance of $FV$ as in part b.

**11.17.** Identify 20 securities that represent a range from conservative to volatile. Obtain a performance history for each security, covering at least 15 data points. Estimate parameters for a single-period index model. Obtain the efficient frontier for the index model or a linear approximation.

**11.18.** Derive the covariance equation, Eq. 11.37.

**11.19.** Apply the multiperiod index model to a project selection problem for a firm.

**11.20.** Small-Time, Inc. is a distributor that sells a variety of home improvement products. The company is considering expansion into several product lines.

1. Fiberglass roll insulation.
2. Cellulose insulation for air-blowing installation.
3. Window and door weather stripping.
4. Storm windows.
5. Set-back energy saver thermostats.
6. Flue dampers.
7. Attic fans, motor-driven.
8. Window air conditioners for 120 volts.

The estimated cash flows for these products are as follows.

| | Product Line | | | | | | | |
|---|---|---|---|---|---|---|---|---|
| **Cash Flow Type** | 1 | 2 | 3 | 4 | 5 | 6 | 7 | 8 |
| Initial investment | | | | | | | | |
| Inventory purchase | \$100 | 150 | 60 | 190 | 30 | 40 | 30 | 120 |
| Franchise fee | \$60 | 60 | 30 | | 20 | 20 | 10 | 90 |
| Annual net sales revenue | \$110 | 120 | 80 | 175 | 50 | 70 | 35 | 130 |

The sales revenue is estimated to be good for 5 years for product lines 1, 2, and 3 and for 4 years for lines 4, 5, 6, 7, and 8, after which time the franchises expire. Product line 1 is available from Owens, Inc.; lines 2 and 3 are available from BTU Mfg. Co.; lines 7 and 8 are available from Eco-Systems, Inc.; lines 5 and 6 are available from Thermo-Save, Inc.; and line 4 is available from local manufacturers. Small-Time feels that it should either sell all the products available from a particular manufacturer, or none, for obvious reasons related to dependability of supplies, technical assistance, and so forth.

Operating expenses are expected to be negligible, since existing personnel and facilities can handle the products. The initial inventory purchases are adequate for the first year of sales. Thereafter, additional inventory purchase costs have been included in the calculation of annual net sales revenue,

$$\text{Annual net sales revenue} = \text{Gross annual sales revenue} - \text{Sales commissions} - \text{Annual inventory purchases}$$

In effect, Small-Time maintains about one year's worth of inventory. Annual net sales revenue does not get higher in the last year because gross sales are expected to drop, and some inventory must still be maintained, even if it has to be written off later. The franchise fee is typically depreciated over the franchise period, by the straight-line method.

Small-Time has a rather simple capital structure that consists of bank loans and stock.

Bank loans, 18% interest, $100,000 principal owed

Stock, 10,000 shares, market value of $10 per share

Earnings per share are stable at $1.40. The current corporate tax rate, federal and state combined, is 50%. No dividends are paid, so all earnings are available for investment. Together with cash from other sources, the amounts available for investment are

| | |
|---|---|
| Time 0 | $600 |
| Time 1 | 500 |
| Time 2 | −500 |

The other cash sources are primarily from depreciation of existing assets. The company has no problem in obtaining additional loans at 18%, provided it maintains its debt–equity ratio at 1.0. Excess cash is typically invested at 12% in U.S. Treasury notes.

Small-Time is somewhat concerned about the impact of tax changes on the home improvement activities of consumers. On the other hand, consumers have been spending more than ever on insulation, attic fans, and the like. The current prognosis is that there will be some type of energy tax credit, but the exact nature of the credit is not known at this time. The sales revenue estimates are based on a credit equal to half the old credit. The best judgment is that net sales could fluctuate either way by as much as 20%. One thing is fairly certain: if sales trend in any way, they will continue to do so. Small-Time's sales manager thinks that if first-year sales are either above or below the expected value, there is only a 50% chance the trend will revert back to the estimated value. In addition, any trend will be felt uniformly across the product lines. If one product trends up, the rest will positively follow by the same trend factor. The single exception to this phenomenon is the 120-volt window air conditioner, which seems to do exactly the opposite to all the other products. After making a final review of all the cash flow estimates before giving them to you for analysis, the sales manager tells the president of Small-Time that the net sales revenue for the two types of insulation will not hold up if Small-Time tries to sell both, since many customers who would buy one could be convinced to buy the other. The original sales estimates were based on selling only one of the products. The best guess the sales manager can make is that both net sales revenues will be reduced by 20% if both product lines are sold.

Using after-tax cash flows and assuming independent cash flows, obtain $E(PV)$ and $Var(PV)$ of each project. Then set up the correlation matrix for the projects, using the information given. Evaluate some combinations of projects.

**11.21.** Consider the case of a person contemplating life insurance. The two major types of policies are whole life and term. With term insurance the premiums increase each year. With whole life the premiums typically remain constant, but a decreasing portion of each premium is invested in the cash value portion. Term insurance returns nothing if the person survives. Whole life insurance, on the other hand, accumulates a cash value that can be withdrawn by the person.

Obtain data for a term policy and for a whole-life policy when the insurance coverage is the same for both policies. Identify a mortality curve consistent with the person's age on which the policies are based. Select a simple objective function that reflects alternative investment possibilities, the probabilities of dying, insurance benefits paid to a beneficiary, and so on. Make some specific assumptions about investment growth rates for alternative investments, planning horizons, and the like.

Determine which policy is better for the person.

# APPENDIX 11.A

# Example of Use of Multiperiod Index Model for Selecting Project Portfolio

In order to demonstrate the multiperiod index model presented in Section 11.6, a project selection problem in the Mexican leather industry is used. A set of seven projects, including the existing firm, is used. The seven projects in the data set were chosen to emphasize a variety of starting and terminating dates, cash flow streams that are a mixture of both positive and negative values, a wide range of cash flows, cash flow streams with different degrees of variability, and various degrees of correlation between the selected index and the projects. A time horizon of 14 years is used. The cash flows of projects are shown in Figure 11a.1. Project 1 refers to the existing firm. Since both inflows and outflows can occur in any period, it is the net effect in a particular year that is of consequence here. Notice that for a particular period in which both inflows and outflows occur in a given project, the variability of the cash flow is the result of both, the inflows and the outflows.

For project 4 the pessimistic, most likely, and optimistic estimates of the cash flows are given in columns 2, 3, and 4 of Table 11a.1. The means and variances of the cash flows are given in columns 5 and 6. The characteristics of each cash flow are classified by calculating the variability ratio for each cash flow in project 4, defined as

$$VR = \frac{\text{Variance of expected present value}}{\text{Expected present value}}$$

Column 7 in Table 11a.1 shows the variability ratios of the cash flows. The behavior of this ratio is as we might expect: it increases as $n$ gets larger. It is important to note that for $n = 10$ the variability ratio is the result of the uncertainty of the required future investment by the project as well as the uncertainty of the expected inflows. The variability ratio of each project is shown in Figure 11a.1.

| Time | 0 | 1 | 2 | 3 | 4 | 5 | 6 | 7 | 8 | 9 | 10 | 11 | 12 | 13 | 14 | Variability ratio |
|---|---|---|---|---|---|---|---|---|---|---|---|---|---|---|---|---|
| 1. National Market Actual line; tannery, first-class glaze leather, second-class glaze leather, goat leather (cut) | 52.4 | 56.3 | 61.5 | 66.5 | 34.9 | 74.4 | 79.1 | 82.8 | 88.5 | 46.6 | 99.9 | 100.7 | 101.6 | 102.4 | 51.46 | 0.10 |
| 2. National Market Goat leather for conversion into clothing | | | 33.6 | 36.8 | 39.7 | 41.7 | 44.4 | 47.1 | 49.4 | | 55.6 | 59.5 | 62.3 | 68.1 | 73.4 | 7.19 |
| | 123.5 | 25.0 | | | | | | | | 12.8 | | | | | | |
| 3. Export Glaze leather (goat and kid) | | 61.2 | 65.5 | 70.4 | 76.1 | 83.2 | 88.6 | 93.0 | 38.9 | | 111.7 | 120.8 | 127.2 | 135.2 | 144.8 | 2.92 |
| | 319.4 | | | | | | | | | 23.6 | | | | | | |
| 4. Exports Goat leather for conversion into clothing | | | | 44.0 | 46.1 | 47.5 | 50.9 | 53.0 | 54.9 | 55.4 | | 60.8 | 63.3 | 67.4 | 69.4 | 1.01 |
| | | 87.9 | 36.0 | | | | | | | | .17.7 | | | | | |
| 5. Exports Purses, handbags, or pocketbooks of goat and kid leather. | | | | 8.8 | 11.5 | 15.5 | 18.9 | 21.1 | | 27.0 | 30.9 | 35.1 | 36.8 | 36.1 | | 0.40 |
| | | | 55.9 | | | | | | 10.5 | | | | | | | |
| 6. Exports Leather footware for men | | | 145.8 | 150.5 | 164.2 | 179.6 | 182.6 | 178.6 | 172.0 | 165.4 | 157.8 | | | | | 2.10 |
| | 330.5 | 210.1 | | | | | | | | | | | | | | |
| 7. Exports Leather footware for women | | | | 285.3 | 304.4 | 327.3 | 349.8 | 360.3 | 350.5 | 336.5 | | | | | | 5.55 |
| | 500.0 | 240.0 | 35.7 | | | | | | | | | | | | | |
| Budget constraints | 1150 | 550 | 100 | | | | | | 25 | 25 | 25 | | | | | |

**FIGURE 11a.1** Project expected cash flows and budget constraints. All the cash flows are in thousands of dollars.

$$\text{Variability ratio} = \frac{\sum_{n=0}^{N} Var(\mu_{jn})}{\sum_{n=0}^{N} E(\mu_{jn})}$$

**Table 11a.1** ***Example of Project Cash Flows, Mean and Variance, and Variability Ratio—Project 4***

| $n$ | $Est_p^*$ ($L$) | $Est_m$ ($M_0$) | $Est_o$ ($H$) | $E$ | $Var$ | $VR$ |
|---|---|---|---|---|---|---|
| 0 | | | | | | |
| 1 | −102.62 | −87.94 | −76.49 | −88.48 | 18.97 | 0.21 |
| 2 | −52.94 | −35.98 | −22.75 | −36.60 | 25.32 | 0.69 |
| 3 | 31.94 | 43.98 | 53.37 | 43.54 | 12.76 | 0.29 |
| 4 | 32.97 | 46.07 | 56.29 | 45.59 | 15.11 | 0.33 |
| 5 | 34.01 | 47.54 | 59.21 | 47.23 | 17.64 | 0.37 |
| 6 | 35.13 | 50.86 | 62.13 | 50.12 | 20.25 | 0.40 |
| 7 | 36.26 | 52.98 | 65.05 | 52.21 | 23.01 | 0.44 |
| 8 | 37.29 | 54.91 | 67.97 | 54.15 | 26.15 | 0.48 |
| 9 | 38.32 | 55.37 | 70.89 | 55.12 | 29.47 | 0.53 |
| 10 | −39.73 | −17.69 | −0.50 | −18.50 | 42.75 | 2.31 |
| 11 | 40.48 | 60.75 | 76.73 | 60.04 | 36.50 | 0.61 |
| 12 | 41.51 | 63.26 | 79.65 | 62.37 | 40.41 | 0.65 |
| 13 | 42.54 | 67.38 | 82.57 | 65.77 | 44.51 | 0.68 |
| 14 | 42.57 | 69.39 | 85.49 | 67.60 | 51.17 | 0.76 |

*$Est_p$, pessimistic estimate; $Est_m$, most likely estimate; $Est_o$, optimistic estimate; $E$, mean; $Var$, variance; $VR$, variability ratio.

After the mean and variance of each cash flow are obtained, the analysis is concerned with identifying the best index. For testing purposes four different economic indices were studied as possible alternatives: U.S. gross national product (GNP), U.S. personal consumption expenditures, U.S. personal consumption expenditures in clothing and shoes, and U.S. disposable personal income. Personal consumption expenditures in clothing and shoes was selected as the economic index. It was found to be highly correlated with the firm's cash flows and the proposed projects. Quarterly data from 1960 to the second quarter of 1977 are used in the analysis [29].

An analysis of the selected index showed that the time series exhibits nonstationary behavior in both mean and slope. Therefore, second differencing was used to reduce the series to a stationary series. Even though the series' first differences (that is, $I_n - I_{n-1}$) are not strictly a stationary series, one may want to assume so and use first differences throughout the remainder of the procedure. On the other hand, if one decides that the small nonstationary behavior in the mean exhibited by the series is important, the second differences (that is, $I_n - 2I_{n-1} + I_{n-2}$) will be required to reduce the series to a stationary one. For testing the solution procedure, both the first and second differences of the time series are used.

Nine different patterns were chosen for the forecast of the index, as shown in Figure 11.15. For these forecasts, estimates of the $F_{jn}$ were then obtained from management personnel of a large company in the industry.

Generally, the results obtained from the regression analysis were very similar for first and second differences of the time series. The responsiveness of

**Table 11a.2** ***Results from Regression Analysis***

| | First Differences | | Second Differences | |
|---|---|---|---|---|
| Project | $B^*$ | $R^2$ | $B$ | $R^2$ |
| 1 | 0 | NA† | 0 | NA |
| 2 | 0 | NA | 0 | NA |
| 3 | −0.024 | 89.72 | −0.223 | 25.81 |
| 4 | −0.015 | 99.81 | −0.154 | 97.61 |
| 5 | 1.878 | 91.42 | 1.720 | 33.20 |
| 6 | 3.705 | 99.59 | 5.088 | 72.90 |
| 7 | 5.447 | 96.79 | 4.399 | 58.34 |

*B, measure of responsiveness of cash flows for project $j$ to changes in the index.
$R^2$, measure of percentage of variation explained by the regression.
†NA, not applicable.

projects 1 and 2 is zero in both cases, since there is no functional relationship between the index and these two projects. Projects 1 and 2 refer to the Mexican national market, and the economic index is U.S. personal consumption expenditures in clothing and shoes. Furthermore, for both the first and second differences, the same projects responded positively or negatively with the economic index. And in both cases, higher responsiveness is shown by projects 6 and 7, which correspond to exportation of leather footwear for men and women, respectively. These results are shown in Table 11a.2.

Two different programs were used for the time series analysis. Both programs are coded in FORTRAN IV for use on the CDC Cyber 74 at Georgia Institute of Technology. The identification program BOX was written by the University of Wisconsin Computing Center. This program is dimensioned to handle up to 500 observations, 150 autocorrelations and partial autocorrelations, and 2 regular differences. The "average run" for this program takes 1.21 seconds of CPU time.

The identification program TSI was written by the School of Industrial and Systems Engineering at Georgia Institute of Technology. This program is dimensioned to handle up to 500 observations. The average run for this program takes 1.45 seconds of CPU time. Although only one of the programs is needed, both were used because of the output convenience of each program.

Six different runs were made with 70, 60, and 50 observations each. It was found that the autocorrelation and partial autocorrelation functions behave in the same way no matter what the number of observations is. Consequently, the estimate of the autocorrelation function with 70 observations is considered to be adequate.

The statistical package used for the regression problem was the biomedical computer programs (BMDP) package, coded in FORTRAN IV. The average run for this program takes 15.42 seconds of CPU time. The total number of regressions solved was 45. This total number of runs is the sum of five runs for

each of the correlated projects in the first-difference case plus four runs for each correlated project in the second-difference case.

The SYMQUAD algorithm within the Multi-Purpose Optimization System (MPOS) was used to optimize the nonlinear model. The average run for this program takes 1.23 seconds of CPU time. Fourteen quadratic programming problems were run at seven different levels of the risk aversion factor for the first and second differences. Since SYMQUAD is not an integer quadratic algorithm, round-off procedures had to be used to select projects. All project variables with values greater than 0.935 were rounded to one, and those with values less than or equal to 0.935 were rounded to zero. The REGULAR algorithm from MPOS was used to make runs of the linear model. The average run for this program takes 0.61 second of CPU time. Again, as in the nonlinear model, all variables with values greater than 0.935 were rounded to one and those with lesser values to zero.

The results obtained from the runs are summarized in Figures 11a.2 and 11a.3, which show the efficiency frontiers generated by letting $\lambda$ assume different values. The following comments summarize the major findings for this example. Note that when $\lambda = 0.0$ the nonlinear model becomes a linear model, since only linear terms remain in the objective function; that is, only the expected present worths of the projects are considered in selecting the optimal portfolio composition.

In each figure the efficiency frontier is shown for the three different models, that is,

QP = nonlinear model

LP-I = linear model with autocorrelation

LP-II = linear model with no correlation

Each point in Figures 11a.2 and 11a.3 represents a specific portfolio composition. In the second-differences case the efficiency frontier is the same for the nonlinear model and the linear model with autocorrelation. Thus, in this example either model will lead to the correct efficiency frontier. However, if the linear model containing no correlation information is used by an investor who is highly averse to risk, the portfolio selection will be nonoptimal.

In the first-differences case, the shift between the efficiency frontiers is almost nonexistent. Therefore the three models will give the same efficiency frontier. The only observation is that the same portfolio composition might represent a different $\lambda$ value, depending on the model used.

Using first differences, investors who are highly risk-averse ($\lambda = 0.75, 1.0$) would choose to maintain the present firm and reject all the proposed projects. In this case the expected present value of the project portfolio is \$361,100 with a variance of only 35,670, giving a variability ratio of 0.10. On the other hand, for investors who are slightly risk-averse ($\lambda = 0.3, 0.4$), the best policy is to select projects 1, 2, 4, and 5. In this case the variability ratio is 0.38 and the net present value of the total cost is small compared with that of other project portfolios. For example, the cost of selecting projects 1, 2, 4, and 6 is two and one-half times the cost of selecting projects 1, 2, 4, and 5.

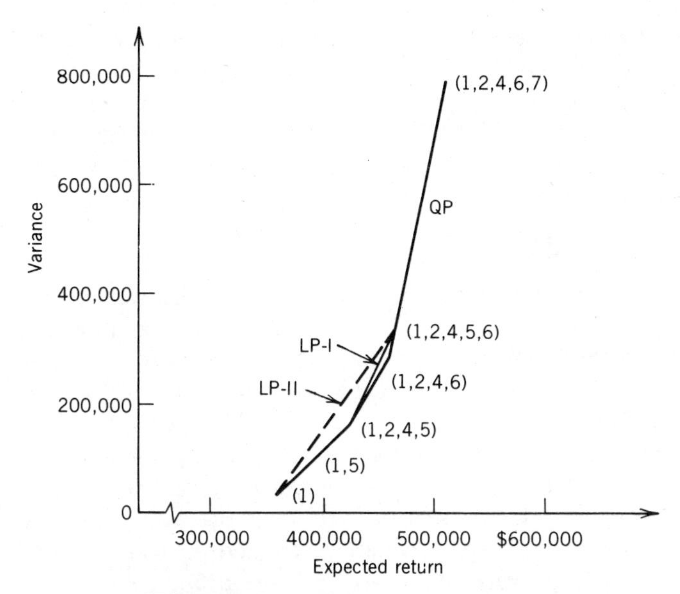


**FIGURE 11a.2** Efficiency frontiers, first-differences case. Models are QP, nonlinear model; LP-I, linear model with autocorrelation; LP-II, linear model with no correlation.

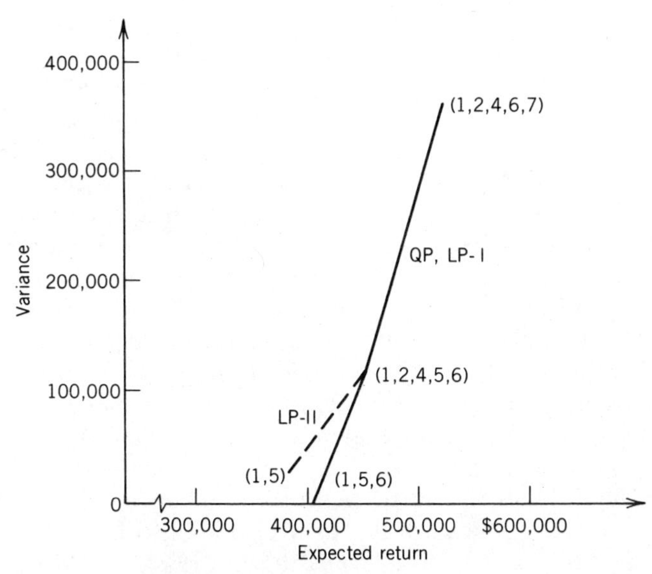


**FIGURE 11a.3** Efficiency frontiers, second-differences case. Models are QP, nonlinear model; LP-I, linear model with autocorrelation; LP-II, linear model with no correlation.

Using second differences, investors face a potential problem because the objective function is not guaranteed to be a concave function. Thus, the optimal solution might be a local maximum and not the global maximum. Solutions for all values of $\lambda$ except $\lambda = 0.10, 0.20, 0.30$ were obtained. Comparing the portfolio composition for investors who are highly risk-averse and are using first and second differences, we observe that the portfolio composition for the second-differences case includes projects 5 and 6, which are not included in the first-differences case.

Results obtained by using first and second differences are also available for the proposed linear model with autocorrelation. It is important to mention that even though the objective function of this model contains autocorrelation only among cash flows for the same project, the cross-correlation was introduced after the objective function was maximized.

In the first-differences case, for investors who are highly risk-averse, the best composition is the same as in the nonlinear model: only the existing firm will be selected. When second differences are used, investors who are highly risk-averse will select projects 1, 5, and 6, with an expected present worth of $408,110 and variance 1,440. The variability ratio for this portfolio composition is almost zero. Compared with the nonlinear models, the same results were obtained. In both cases, first and second differences, investors who are indifferent or slightly risk-averse will select the same portfolio composition. The number of projects selected begins to change at $\lambda = 0.40$. In the second-differences case the portfolio compositions are larger than in the first-differences case.

For the linear model with no correlation, since no statistical interrelationships among cash flows are considered, the same results apply for first and second differences. For investors who are highly risk-averse, the best portfolio composition will be projects 1 and 5, with an expected present value of $370,610 and variance of 48,630. The variability ratio is 0.13, and the net present value of the total cost is $41,260.

In general, we may conclude that the nonlinear model is theoretically more correct, since it considers in the optimization problem both cross-correlation and autocorrelation among cash flows. The proposed linear model considers only autocorrelation. However, for this example the differences in the portfolio compositions are almost nonexistent in the range of high risk aversion. Most of the differences in portfolio composition occur in the range of low risk aversion. Therefore, an investor who is risk-averse would probably select the same portfolio composition regardless of the model employed. Even though the linear model does not consider cross-correlation among cash flow streams while optimizing, cross-correlation should be added after obtaining the optimal solution and a sensitivity analysis should be performed. Although we were unable to define the exact level at which statistical interrelationships of cash flow streams become critical, it is clear that for utility-maximizing, risk-averse investors, assuming independent cash flows can lead to the selection of a nonoptimal portfolio.

# 12

# Risk Simulation

## 12.1 INTRODUCTION

In the previous two chapters we have shown methods for determining the distributions of performance measures (e.g., *PV* and *IRR*) analytically. All these methods essentially derive expressions for the means and standard deviations of the cash flows and the coefficients of correlation between them, assuming that aggregation of random variables results in a normal distribution. Their main limitation is that they can deal with only a restricted class of probability functions. Even with well-defined probability functions, the computational burden becomes prohibitively excessive. For example, the cash flow in a particular period is often a function of a number of other critical random variables, such as selling prices, size of the market, investment required, inflation rate, and operating costs. A number of these random variables may be correlated with one another. Consequently, the probability distribution for the *PV* is not easy to develop analytically in most real-world situations. In this chapter we will discuss an alternative approach to risk analysis, called risk simulation. Risk simulation is generally an easier way to combine the necessary probability distributions—and in some cases the only practical way. The subject of risk simulation is fairly broad and encompassing. Thus, we will not attempt to discuss in full detail every possible facet of the method. For a comprehensive treatment of this subject, see Hertz and Thomas [7]. Our discussion here will be of sufficient detail to present the major aspects of the method and how it can be applied to investment analysis.

The remainder of this chapter is divided into six sections. Section 12.2 presents an overview of Monte Carlo sampling techniques. In Section 12.3 various topics related to selecting input distributions are discussed. Section 12.4 treats sampling procedures for independent random variables, and Section 12.5 examines the three basic techniques for sampling dependent random variables. In Section 12.6 we provide statistical analyses for simulation output data and discuss the need to assess the accuracy of simulation output. Finally, Section 12.7 presents a simulation example with the aid of computer graphics.

## 12.2 AN OVERVIEW OF THE LOGIC OF SIMULATION

### 12.2.1 Monte Carlo Sampling

Before we examine the mechanical details of the Monte Carlo sampling technique, we need to review the method in relation to investment analysis. As mentioned earlier, the objective of simulation is to determine the distribution of the *PV* from a proposed project. Now we will take a step backward to see how the *PV* distribution comes into the picture in the first place. Suppose we are considering a hypothetical project in which the ultimate *PV* is a function of three cash flow components, $F_0$, $F_1$, and $F_2$:

$$PV(i) = F_0 + \frac{F_1}{1 + i} + \frac{F_2}{(1 + i)^2} \tag{12.1}$$

If we know the exact values of such variables as $F_0$, $F_1$, and $F_2$, we would merely substitute these into Eq. 12.1 and solve it directly for $PV(i)$. Of course, we usually do not know exact values of these variables. In fact, all we know about them might be the ranges of the values and the relative likelihoods of possible values for each variable. We should observe one important point, however: the actual state of nature is that all variables will have only one specific value. If we can simulate the actual state of nature for these variables, somehow the resulting $PV(i)$ can be easily obtained. This is where computer simulation enters the analysis.

If we can express the random variables by probability distributions, we can simulate the actual state of nature by sampling the state of each variable from its originally specified distribution. By collecting values of each random variable sampled from within its distribution, we can compute a single value of the *PV*. In other words, each value of *PV* computed in this manner represents one possible state of the nature or possible combination of the variables (e.g., $F_n$). These repetitive computations, or iterations, are continued until a sufficient number of *PV* values is available to tabulate or define the distribution—the objective of the analysis. This use of computer-based sampling experiments to obtain a distribution is called the *Monte Carlo method.*

Later, we will address many of the technical details, such as how to sample values of the random variables, how many replications are required, how the distributions of input variables are defined, and what happens if variables are correlated in some manner. The general logic of risk simulation, however, is simply to define the *PV* distribution (or any other measure of performance) by a series of repetitive samplings. Figure 12.1 depicts a simulation analysis of the hypothetical investment of Eq. 12.1.

### 12.2.2 Using the Simulation Output

***Single-Project Evaluation.*** When the probability distribution of the *PV* has been obtained, the crucial question is how to use this distribution in project analysis. Recall that the probability distribution provides information regarding the probability that a random variable will attain some value $x$. We can use this information to define the cumulative distribution, which expresses the probability that the random variable will attain a value smaller than or equal to some $x$, that is, $P(X \leq x)$. A common notation for the cumulative distribution is $F(x) = P(X \leq x)$. Thus, if the *PV* distribution is known, we can compute the probability

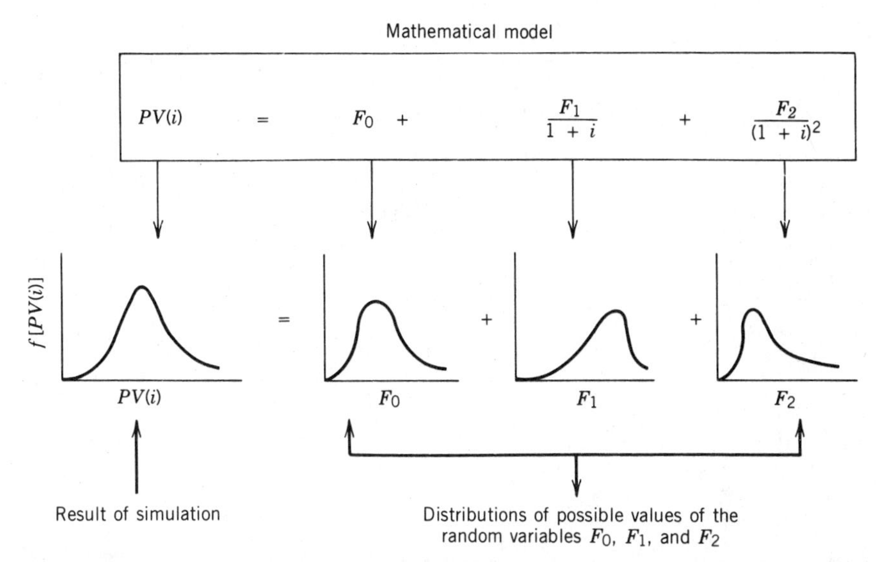


**FIGURE 12.1** Schematic of simulation analysis of the hypothetical investment of Eq. 12.1.

that the *PV* of a project will be negative. We will use this probabilistic information in judging the profitability of the project.

***Two Mutually Exclusive Projects.*** We can extend the concept of a cumulative distribution to analyze an investment choice between project A and project B. An additional computer program can be written to develop a probability distribution for the differences between any two mutually exclusive alternatives. This, in turn, enables us to develop a cumulative distribution of the difference in net present value, or

$$\{\Delta \le PV_A - PV_B\}$$

Setting $\Delta$ equal to 0 is equivalent to finding the probability that $PV_A$ will exceed $PV_B$. Then, if this probability is equal to 0 or 1, a project dominance occurs. This type of probabilistic information is useful when we have to discriminate between two competing projects. Figure 12.2 shows the process schematically.

Having discussed the general Monte Carlo method, we turn our attention to some of the technical details of carrying out risk simulation.

## 12.3 SELECTING INPUT PROBABILITY DISTRIBUTIONS

To carry out a simulation with a mathematical model (i.e., a *PV* model) whose inputs are random variables (such as cash flow elements and project life), we have to specify the probability distributions of these inputs. Then, given that the inputs to a model follow particular distributions, the simulation proceeds by generating values of these random variables from the appropriate distributions.

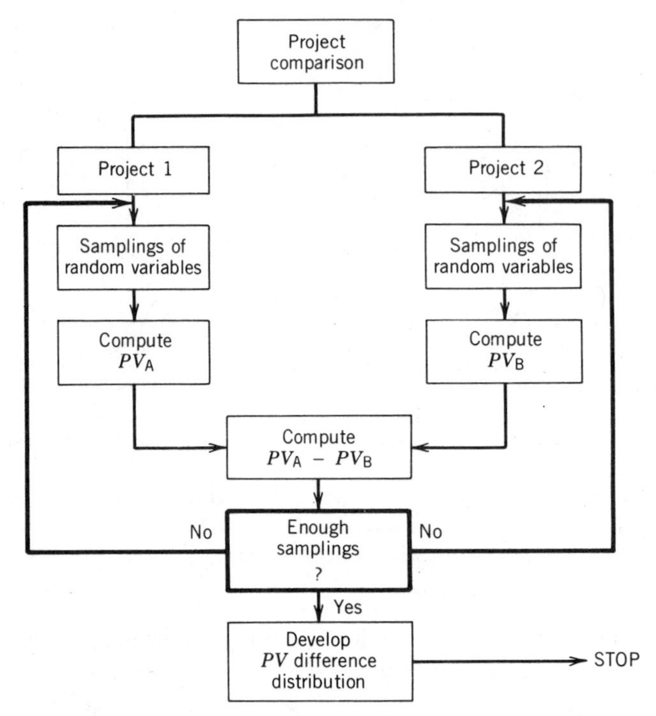


**FIGURE 12.2** Sampling process of developing a *PV* difference distribution.

Our concern in this section is how we go about specifying these input probability distributions. Throughout this section we are assuming that our data are independent observations taken from a single underlying distribution. If this is not the case, most of the techniques discussed here are not directly applicable.

### 12.3.1 Selecting a Distribution Based on Observed Data

If observed data are available, we can use them in one of two general approaches to specify a distribution. First, standard techniques of statistical inference are used to fit a theoretical distribution form (e.g., normal, lognormal, gamma, or beta) to the data and to perform hypothesis tests to determine how good the fit is. After we show that a distribution, given the appropriate parameters, has a statistically proven fit, we can use this distribution to generate the corresponding random variables during the simulation. Second, when no good fit is found, the values of the data themselves are used directly to define an empirical distribution. In the simulation we sample directly from this empirical distribution.

***Relative Frequency Distribution.*** For certain variables we may have some statistical data on which to base input distributions. From such statistical data we can always develop some form of relative frequency distribution. We can illustrate this process with a numerical example.

### *Example 12.1*

Suppose a management consultant is studying the cost of processing criminal cases through a county courthouse. The task is to collect some statistical data about the processing time. Table 12.1 shows observations of the time (in days) for 150 cases completed during the study period. The processing time $x$ appears to be a random event ranging from 6 days to 320 days.

To represent the data in a frequency distribution, we may divide the range of data into a number of cells and then tabulate the number of observations made within each cell. This can be done in two steps: sorting the observed data into increasing order and then grouping the data into specified cell intervals. The results are shown in Tables 12.2 and 12.3.

In Table 12.3 we use an interval size (cell width) of 25 and assume that the

**Table 12.1** ***Observed Processing Times for Criminal Cases (Days)***

| | | | | |
|---|---|---|---|---|
| 20 | 73 | 175 | 125 | 17 |
| 106 | 216 | 35 | 63 | 320 |
| 87 | 114 | 90 | 28 | 134 |
| 197 | 20 | 64 | 16 | 34 |
| 24 | 151 | 54 | 100 | 78 |
| 24 | 77 | 181 | 42 | 28 |
| 95 | 63 | 68 | 41 | 152 |
| 12 | 144 | 252 | 82 | 96 |
| 32 | 42 | 84 | 133 | 9 |
| 68 | 304 | 68 | 28 | 109 |
| 29 | 80 | 18 | 220 | 105 |
| 116 | 101 | 37 | 48 | 199 |
| 55 | 130 | 103 | 7 | 84 |
| 71 | 74 | 27 | 303 | 6 |
| 132 | 32 | 77 | 123 | 179 |
| 13 | 91 | 68 | 68 | 103 |
| 144 | 40 | 47 | 76 | 69 |
| 63 | 152 | 41 | 71 | 129 |
| 34 | 43 | 128 | 139 | 74 |
| 50 | 224 | 53 | 78 | 22 |
| 85 | 67 | 110 | 66 | 24 |
| 279 | 37 | 77 | 81 | 26 |
| 78 | 181 | 15 | 49 | 90 |
| 56 | 92 | 25 | 27 | 60 |
| 119 | 82 | 75 | 53 | 175 |
| 59 | 22 | 275 | 19 | 111 |
| 18 | 19 | 97 | 168 | 65 |
| 86 | 24 | 22 | 72 | 121 |
| 65 | 105 | 40 | 35 | 8 |
| 134 | 29 | 198 | 57 | 38 |

**Table 12.2** ***Sorted Processing Times (in Increasing Order)****

| | | | | |
|---|---|---|---|---|
| 6 | 7 | 8 | 9 | 12 |
| 13 | 15 | 16 | 17 | 18 |
| 18 | 19 | 19 | 20 | 20 |
| 22 | 22 | 22 | 24 | 24 |
| 24 | 24 | 25 | 26 | 27 |
| 27 | 28 | 28 | 28 | 29 |
| 29 | 32 | 32 | 34 | 34 |
| 35 | 35 | 37 | 37 | 38 |
| 40 | 40 | 41 | 41 | 42 |
| 42 | 43 | 45 | 47 | 48 |
| 50 | 53 | 53 | 54 | 55 |
| 56 | 57 | 59 | 60 | 63 |
| 63 | 63 | 64 | 65 | 65 |
| 66 | 67 | 68 | 68 | 68 |
| 68 | 68 | 69 | 71 | 71 |
| 72 | 73 | 74 | 74 | 75 |
| 76 | 77 | 77 | 77 | 78 |
| 78 | 78 | 80 | 81 | 82 |
| 82 | 84 | 84 | 85 | 86 |
| 87 | 90 | 90 | 91 | 92 |
| 95 | 96 | 97 | 100 | 101 |
| 103 | 103 | 105 | 105 | 106 |
| 109 | 110 | 111 | 114 | 116 |
| 119 | 121 | 123 | 125 | 128 |
| 129 | 130 | 132 | 133 | 134 |
| 134 | 139 | 144 | 144 | 151 |
| 152 | 152 | 168 | 175 | 175 |
| 179 | 181 | 181 | 197 | 198 |
| 199 | 216 | 220 | 224 | 252 |
| 275 | 279 | 303 | 304 | 320 |

*Read across.

data could range between 0 and 325. This cell width divides the data into 13 equal intervals, shown in column 2. The observed frequency counts are given in column 3, and column 4 shows these frequencies computed as a fraction of the total number of observations. Finally, column 5 gives the cumulative frequency values. Plotting the values in column 4 produces Figure 12.3. At this point we must emphasize that the relative frequency distribution is merely an approximation of a probability distribution with a limited set of sample data. As we take more samples, this relative frequency distribution approaches the probability distribution governing that random variable. Thus, in risk simulation, the relative frequency may be used as an approximation of an underlying probability distribution of each random variable from available data.

Plotting the values of column 5 in Table 12.3 gives us the cumulative

**Table 12.3** ***Summary of the Processing Time Distribution***

| Cell Number | Cell Interval | Frequency | Relative Frequency | Cumulative Frequency |
|---|---|---|---|---|
| 1 | $0 < x \leq 25$ | 23 | 0.153 | 0.153 |
| 2 | $25 < x \leq 50$ | 28 | 0.187 | 0.340 |
| 3 | $50 < x \leq 75$ | 29 | 0.193 | 0.533 |
| 4 | $75 < x \leq 100$ | 24 | 0.160 | 0.693 |
| 5 | $100 < x \leq 125$ | 15 | 0.100 | 0.793 |
| 6 | $125 < x \leq 150$ | 10 | 0.067 | 0.860 |
| 7 | $150 < x \leq 175$ | 6 | 0.040 | 0.900 |
| 8 | $175 < x \leq 200$ | 6 | 0.040 | 0.940 |
| 9 | $200 < x \leq 225$ | 3 | 0.020 | 0.960 |
| 10 | $225 < x \leq 250$ | 0 | 0.000 | 0.960 |
| 11 | $250 < x \leq 275$ | 2 | 0.013 | 0.973 |
| 12 | $275 < x \leq 300$ | 1 | 0.007 | 0.980 |
| 13 | $300 < x \leq 325$ | 3 | 0.020 | 1.000 |

NOTES: Sample mean = 85.75 days
Sample standard deviation = 65.16 days

frequency distribution shown in Figure 12.4. We always work with a cumulative distribution in risk simulation; the reason for this will become evident in later sections when we discuss the mechanics of Monte Carlo sampling. □

***Fitting Distributions.*** Once we have obtained a relative frequency distribution, we can determine whether the data conform to a family of well-known standard distributions (such as normal, gamma, or lognormal). The process of identifying and confirming the distribution is called *fitting a distribution.* Fitting a distribution to data is recommended in risk simulation because it can provide the

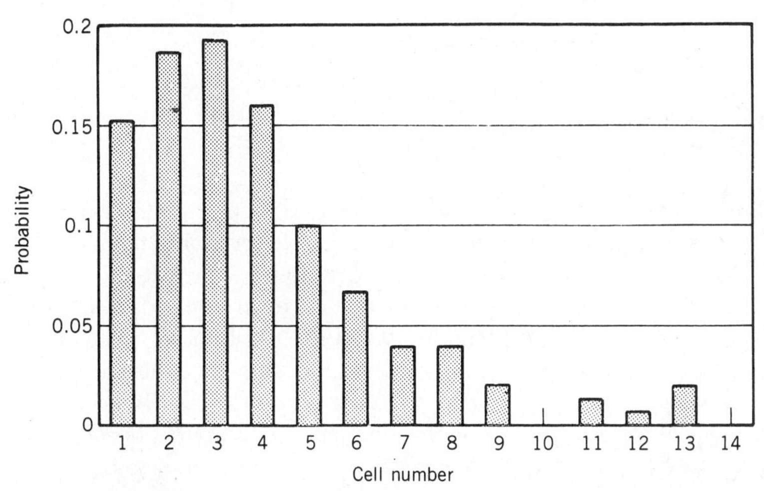


**FIGURE 12.3** Relative frequency distribution for Table 12.3

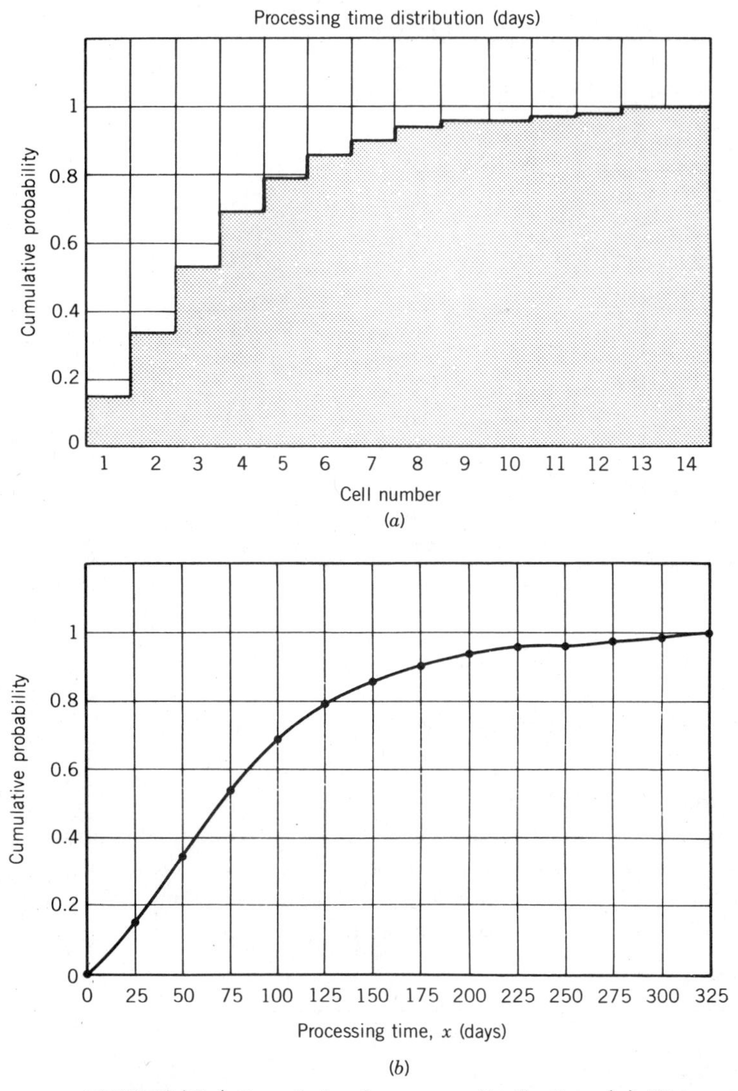


**FIGURE 12.4** Cumulative frequency distribution. (*a*) Histogram (*b*) linear fit.

analyst with several benefits. (1) The data collected are themselves random, so an empirical distribution obtained from one set of observations might differ from that obtained from another set of observations on the same variable taken at a different time. By fitting a theoretical distribution form to the data, we are able to infer the characteristics of the random variable. Distributions specified in this way should be less sensitive to the random fluctuations in the particular observations we happen to have obtained. (2) If empirical distributions are defined in the usual fashion, their use in the simulation implies that no random variables falling outside the range of the observed data will be generated. With a fitted theoretical distribution, random variables outside the range of the observed data can be generated. This is a very desirable feature in risk simulation, since many project outcomes depend heavily on the probability that an "extreme" event may occur.

Three steps are required to fit a distribution: (1) identify what type of standard distribution might fit the data by studying the shape and range of the relative frequency distribution, (2) estimate the unknown parameters that describe the distribution, and (3) test the hypothesis of the distribution assumption. We will illustrate the first two steps with the example shown in Table 12.1.

Review of the frequency distribution in Figure 12.3 indicates the possibility of a gamma-distribution fit. We will assume that the distribution has its lower limit bounded by zero but its upper limit unbounded ($0 < x < \infty$). The gamma distribution is defined as

$$f(x) = \frac{1}{\Gamma(a)b^a} x^{a-1} e^{-x/b}, \qquad 0 < x < \infty; \quad a, b > 0 \tag{12.2}$$

where $\Gamma(a)$ is the gamma function, defined by

$$\Gamma(a) = \int_0^\infty t^{a-1} e^{-t}\, dt$$

for any real number $a > 0$. Notice the property of the gamma function $\Gamma(a + 1) = a\Gamma(a)$. In Eq. 12.2, $a$ and $b$ are the shape and scale parameters, respectively.

With a distributional assumption made, the next step is to estimate the shape and scale parameters of the distribution. There are many ways to estimate the parameters of a distribution, each method having its own virtues and shortcomings [4]. Maximum likelihood estimates (MLEs) have been used for a wide range of distributions, and considerable study has been devoted to their small-sample properties. (See Banks and Carson [1] for suggested estimators for distributions often used in simulation and Law and Kelton [10] for a more advanced treatment, including the maximum likelihood equations and sampling properties of each estimator.) Two equations must be satisfied to obtain the MLE for the gamma distribution.

$$\ln \hat{b} + \frac{\Gamma'(\hat{a})}{\Gamma(\hat{a})} = \frac{1}{n} \sum_{j=1}^{n} \ln X_j \tag{12.3}$$

$$\hat{a}\hat{b} = \bar{X} \tag{12.4}$$

where $\Gamma'$ denotes the derivative $\Gamma$ and $\bar{X}$ is the sample mean. The parameters $\hat{a}$ and $\hat{b}$ can be solved for numerically. Alternatively, approximations to $\hat{a}$ and $\hat{b}$ can be obtained by letting

$$T = \left( \ln \bar{X} - \sum_{j=1}^{n} \frac{\ln X_j}{n} \right)^{-1} \tag{12.5}$$

Using a table of MLEs for the gamma distribution (see Choi and Wette [3] or [1]) to obtain $\hat{a}$ as a function of $T$, we then let $\hat{b} = \bar{X}/\hat{a}$. In Example 12.1, we compute $\bar{X} = 85.75$, and

$$\sum_{j=1}^{150} \frac{\ln X_j}{150} = 4.1567$$

Then

$$T = (\ln 85.75 - 4.1567)^{-1} = 3.39$$

Using the table in [10], we find $\hat{a} = 1.73$ for $T = 3.39$. Substituting $\hat{a}$ into Eq. 12.4 and solving for $\hat{b}$ gives us $\hat{b} = 49.51$. Now we can define the gamma function in terms of these estimated parameters.

$$f(x) = (0.0012791)x^{0.73}e^{-x/49.51} \tag{12.6}$$

(Note that $\dfrac{1}{\Gamma(1.73)(49.51)^{1.73}} = 0.0012791$.)

***Goodness-of-Fit Test.*** After we have hypothesized a distribution form for our data and estimated its parameters, we must examine whether the fitted distribution is in agreement with our observed data. This is equivalent to asking whether it is feasible to obtain our observed data by sampling from the fitted distribution. The chi-square test is used to answer this question [8]. The test is valid for a large sample size ($n > 100$) with both discrete and continuous distributional assumptions. For a small sample size and continuous distributions, the Kolmogorov–Smirnov test can be more discriminating. (See [1] for the testing procedures.)

For our example we wish to test the hypothesis ($H_0$) that a sample of $n = 150$ observations is drawn from the gamma distribution defined in Eq. 12.6. The values in column 2 of Table 12.4 are the observed frequencies for the 13 intervals from Table 12.3. From these values compute $P_j$, the theoretical hypothesized probability associated with the $j$th interval. This is equivalent to evaluating

$$P_j = \int_j (0.0012791)x^{0.73}e^{-x/49.51}\,dx \tag{12.7}$$

To evaluate Eq. 12.7, we may use a numerical integration routine available on a computer. The values of $P_j$ are shown in column 3.

The expected frequency for each interval is then computed as $E_j = nP_j$, where $n = 150$. These values appear in column 4. For applying the chi-square test, a minimum expected frequency of 5 is suggested [1]. If an $E_j$ value is too small, it has to be combined with expected frequencies in adjacent intervals. Thus we combine intervals 8 through 13. We now have only eight intervals, for which we list the chi-square statistics in column 5 through 7. Then the test value is $\chi_0^2 = 1.341$. Since we are estimating two parameters from the data, the degree

**Table 12.4** ***Chi Square ($\chi^2$) Test—Goodness of Fit***

| 1<br>Cell $j$ | 2<br>Observed | 3<br>$P_j$ | 4<br>$E_j$ | 5<br>$f_j - E_j$ | 6<br>$(f_j - E_j)^2$ | 7<br>$(f_j - E_j)^2/E_j$ |
|---|---|---|---|---|---|---|
| 1 | 23 | 0.142 | 21.277 | 1.723 | 2.967 | 0.139 |
| 2 | 28 | 0.208 | 31.267 | −3.267 | 10.629 | 0.340 |
| 3 | 29 | 0.185 | 27.692 | 1.308 | 1.711 | 0.062 |
| 4 | 24 | 0.143 | 21.456 | 2.544 | 6.470 | 0.302 |
| 5 | 15 | 0.104 | 15.592 | −0.592 | 0.351 | 0.022 |
| 6 | 10 | 0.073 | 10.910 | −0.910 | 0.828 | 0.076 |
| 7 | 6 | 0.050 | 7.446 | −1.446 | 2.090 | 0.281 |
| 8 | 6 | 0.033 | 4.992 | | | |
| 9 | 3 | 0.022 | 3.303 | | | |
| 10 | 0 | 0.014 | 2.163 | | | |
| 11 | 2 | 0.009 | 1.405 | 1.279 | 1.635 | 0.119 |
| 12 | 1 | 0.005 | 0.906 | | | |
| 13 | 3 | 0.004 | 0.581 | | | |
| 14 | 0 | 0.002 | 0.371 | Cell grouping | | |
| | 15 | 0.091 | 13.721 | | | |
| | 150 | 1.000 | | | Test statistics $\chi_0^2 =$ | 1.341 |

Degree of freedom (df) = Number of valid cell intervals − Number of parameters estimated − 1
= 8 cells after grouping − 2 parameters $(a, b)$ − 1
= 5

From Appendix B (Table B.2), $\chi^2_{5,95\%} = 11.07 > 1.341$, indicating that the gamma-distribution fit is reasonable.

of freedom (df) to use in the test is 5 (df = number of valid intervals − number of parameters estimated − 1). At the $\alpha = 0.05$ level, we read $\chi^2_{5,95\%} = 11.07$, which is not exceeded by $\chi_0^2$, so we would not reject $H_0$. (Note that we would also not reject $H_0$ for certain larger values of $\alpha$ such as 0.50.) Thus, this test gives us no reason to conclude that our data are poorly fitted by the gamma distribution. Figure 12.5 shows the distribution fitted to the data for our example.

If our hypothesized distribution for a set of data is rejected by the goodness-of-fit test, we must search for a better-fitting distribution or use the empirical distribution of the data.

### 12.3.2 Selecting a Distribution in the Absence of Data

In a more likely situation, we wish to define a distribution for a random variable but have no data available and no idea what the shape of the distribution is. In this instance many simulation practitioners use heuristic procedures

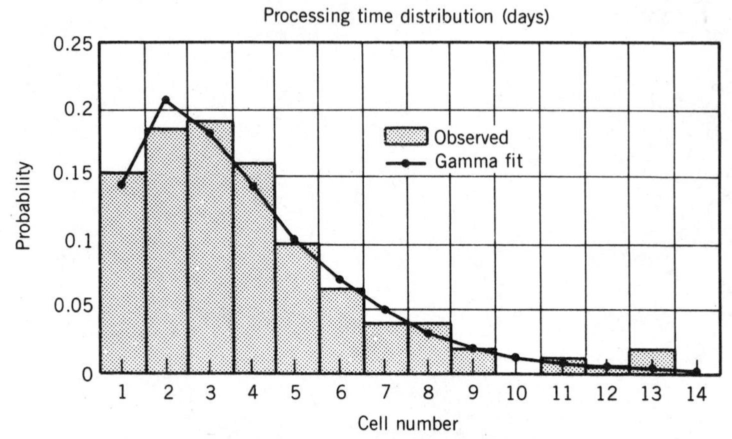


**FIGURE 12.5** Gamma distribution fit.

for choosing a distribution. Two probability distributions are often used for this purpose, the triangular and the beta distributions.

We will assume that the random quantity of interest is a continuous random variable $X$. The first step in either approach is to identify an interval $[L, H]$ within which the random variable $X$ is contained with a probability close to one. To obtain subjective estimates of $L$ and $H$, we must provide our most pessimistic ($L$) and optimistic ($H$) estimates of the random variable. Once an interval $[L, H]$ has been subjectively identified, the next step is to determine the probability distribution of $X$ over that interval.

***Triangular Distribution.*** In the triangular approach we must first identify our subjective estimate of the most likely value ($M_0$) of the random variable. This most likely value is the mode of the distribution $X$. Given a certain $L$, $H$, and $M_0$, the random variable $X$ is considered to have a triangular distribution over the interval $[L, H]$ with $M_0$. The triangular distribution is described by a continuous density function that follows the general form given in Figure 12.6. It is defined with respect to the position of $M_0$ as

$$f(x) = \begin{cases} \dfrac{2(x - L)}{(H - L)(M_0 - L)}, & L \leq x < M_0 \\ \dfrac{2(H - x)}{(H - L)(H - M_0)}, & M_0 \leq x \leq H \end{cases} \tag{12.8}$$

The cumulative probability distribution is given by

$$F(x) = \begin{cases} \dfrac{(x - L)^2}{(H - L)(M_0 - L)}, & L \leq x \leq M_0 \\ 1 - \dfrac{(H - x)^2}{(H - L)(H - M_0)}, & M_0 \leq x \leq H \end{cases} \tag{12.9}$$

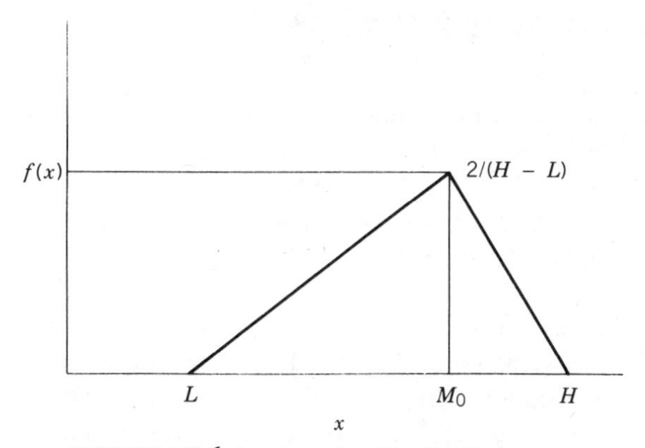


**FIGURE 12.6** A triangular distribution.

The expected value and variance of the triangular distribution are then expressed by

$$E(X) = \frac{L + M_0 + H}{3} \tag{12.10}$$

$$Var(X) = \frac{1}{18}(L^2 + M_0^2 + H^2 - LH - M_0H - M_0L) \tag{12.11}$$

***Beta Distribution.*** A second approach to determining a density function over $[L, H]$ is to assume that the random variable $X$ has a beta distribution over this interval with shape parameters $a$ and $b$ (see Section 10.5). The beta distribution offers more modeling flexibility because its properties are closer to those of random cash flows. It has a finite range ($L \leq X \leq H$), continuity, and unimodality (given that $a$ and $b$ are both greater than zero). It can also assume various degrees of skewness and peakedness, a desirable property in this application.

To estimate the shape parameters $a$ and $b$, we may use Eqs. 10.80 and 10.81. An alternative idea is to ask the analyst to estimate the average of the distribution $\mu$ in addition to $M_0$. A beta distribution has a mean $\mu$ and a mode $M_0$ given by

$$\mu = L + \frac{(a+1)(H-L)}{(a+b+2)} \quad \text{and} \quad M_0 = L + \frac{a(H-L)}{a+b}$$

Given subjective estimates of $\mu$ and $M_0$, we can solve these equations to obtain the following estimates of $a$ and $b$.

$$\hat{a} = \frac{(\mu - L)(2M_0 - L - H)}{(M_0 - \mu)(H - L)} \tag{12.12}$$

$$\hat{b} = \frac{(H - \mu)\hat{a}}{\mu - L} \tag{12.13}$$

Note that if $\mu > M_0$, the density function is skewed to the right; if $\mu < M_0$, it is skewed to the left.

## 12.4 SAMPLING PROCEDURES FOR INDEPENDENT RANDOM VARIABLES

Now that we have specified a distribution for a random variable, the next step is to seek ways to generate samples from this distribution. The basic element required is a source of independent identically distributed uniform random numbers between 0 and 1. Throughout this chapter we will use $U(0, 1)$ to denote such a statistically reliable uniform random number generator, and we will use the symbols $U_1, U_2, \ldots$ to represent uniform random numbers generated by this routine. Many computer systems provide this type of routine as a built-in function.

The general requirement in simulation is to provide a sequence of random deviates drawn from a distribution that is continuous and nonuniform. Most methods for deriving such numbers are based on the use of sequences of $U(0, 1)$. The most widely used method is based on an operation in statistics known as the probability integral transformation. The method, which is generally called the *inverse transformation method,* will be described in this section. The technique will be explained in detail for the triangular distributions and then applied to empirical distributions. It is the most straightforward technique but not always the most computationally efficient one.

Several techniques are also available for generating other common distributions [1]. We will review some of these algorithms that are useful in risk analysis. In practice, we normally use routines available in FORTRAN or the routines built into the simulation language being used. But because some computer systems do not have random-variable generators, we must sometimes construct an acceptable routine on our own.

### 12.4.1 Inverse Transformation Techniques

The probability integral transformation theorem can be stated as follows. If $F^{-1}(x)$ is the inverse of the cumulative distribution function for the random variable $X$, the random variables defined by $X_i = F^{-1}(U_i)$ are a random sample of the variable $X$. Stated differently, to produce random numbers from a given distribution, we must evaluate the inverse cumulative distribution function with a sequence of uniformly distributed numbers in the range 0 to 1.

If the density function $f(x)$ can be described mathematically, it is often possible to find an expression for the inverse of the cumulative function. To illustrate the method, we consider the triangular distribution given in Eq. 12.9.

***Triangular Random Deviate.*** We first find the mode $M$ of the standardized triangular random variable by transforming $M_0$.

$$M = \frac{M_0 - L}{H - L} \tag{12.14}$$

Then the distribution function given in Eq. 12.9 is easily inverted to obtain the relationship

$$X = \begin{cases} L + [F(x)(H - L)(M_0 - L)]^{1/2}, & 0 \leq F(x) \leq M \\ H - \{[1 - F(x)](H - L)(H - M_0)\}^{1/2}, & M \leq F(x) \leq 1 \end{cases} \tag{12.15}$$

Since we know that $0 \leq F(x) \leq 1$ and $0 \leq U \leq 1$, we can obtain the desired triangular random deviate by letting $U = F(x)$ and substituting this into Eq. 12.15. In summary, we do the following.

**Step 1:** Compute $M$.

**Step 2:** Generate a uniform random deviate $U$.

**Step 3:** If $U \leq M$, compute $X = L + [U(H - L)(M_0 - L)^{1/2}$. Otherwise, compute $X = H - [(1 - U)(H - L)(H - M_0)]^{1/2}$.

## *Example 12.2*

Consider the following triangular probability density function with $M_0 = 15$, $L = 10$, and $H = 30$.

$$f(x) = \begin{cases} \dfrac{2(x-10)}{100}, & 10 \leq x \leq 15 \\ \dfrac{30-x}{150}, & 15 \leq x \leq 30 \\ 0, & \text{otherwise} \end{cases}$$

We first compute $M = (15 - 10)/(30 - 10) = 0.25$. The cumulative function is

$$F(x) = \begin{cases} \dfrac{(x-10)^2}{100}, & 10 \leq x \leq 15 \\ 1 - \dfrac{(30-x)^2}{300}, & 15 \leq x \leq 30 \end{cases}$$

The inverted cumulative function with $M = 0.25$ is

$$X = \begin{cases} 10 + \sqrt{100\,F(x)}, & 0 \leq F(x) \leq 0.25 \\ 30 - \sqrt{[1 - F(x)](300)}, & 0.25 < F(x) \leq 1 \end{cases}$$

Suppose we generate one uniform random deviate and let this be 0.47. To find the corresponding triangular random deviate, we simply evaluate

$$X = 30 - \sqrt{(1 - 0.47)(300)} = 17.39$$

***Empirical Continuous Distribution.*** When we have to work with an empirical continuous distribution, we can apply the inverse transformation method to sample random deviates from the distribution. Numerical computation methods must be used. To illustrate the method, consider the data given in Figure 12.4. This distribution must be inverted to generate random deviates representing the processing times. The inverted graph shown in Figure 12.7 appears in the conventional form, with the input $U$ the abscissa and the output the ordinate.

Graphically, the process of generating the random deviates consists of

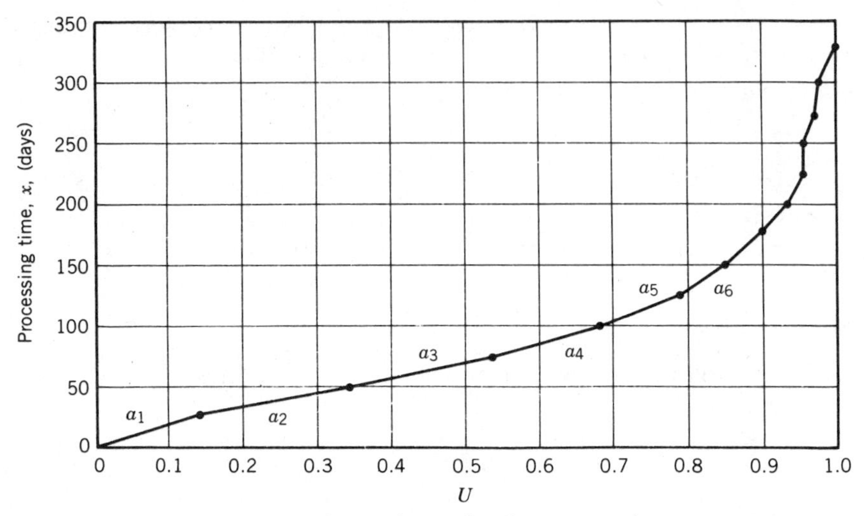


**FIGURE 12.7 Inverted cumulative distribution as a function of U.**

taking as inputs a series of uniformly distributed random numbers ($U_j$) and reading the outputs ($x_j$) from the graph. As a computer program, this process must be translated into a table search procedure. The table input entries are the cumulative distribution values (e.g., column 5 of Table 12.3), and the output entries are the processing time values at which the data were tabulated (e.g., column 2 of Table 12.3). Table 12.5 shows the data for processing time organized in this way. The second column gives the values of the cumulative distribution and is labeled $U$ because a random number between 0 and 1 supplies

**Table 12.5** ***Generation of the Processing Time Lengths***

| $j$ | $U$ | $x_j$ | $a_j$ |
|---|---|---|---|
| 0 | 0 | 0 | — |
| 1 | 0.153 | 25 | 163.40 |
| 2 | 0.340 | 50 | 133.69 |
| 3 | 0.533 | 75 | 129.53 |
| 4 | 0.693 | 100 | 156.25 |
| 5 | 0.793 | 125 | 250.00 |
| 6 | 0.860 | 150 | 373.13 |
| 7 | 0.900 | 175 | 625.00 |
| 8 | 0.940 | 200 | 625.00 |
| 9 | 0.960 | 225 | 1,250.00 |
| 10 | 0.960 | 250 | — |
| 11 | 0.973 | 275 | 1,923.08 |
| 12 | 0.980 | 300 | 3,571.43 |
| 13 | 1.000 | 325 | 1,250.00 |

the input. The third column is the output $x$, representing the generated processing time. When continuous data are represented in tabular form, all values between the tabulated points are possible. This necessitates an interpolation between these tabulated points, which is also consistent with the assumption that the cumulative distribution is approximated by straight-line segments between the tabulated points. To carry out the process of interpolation numerically, we should know the slopes of these lines. We will explain briefly how these slopes are obtained.

Note that the original cumulative function is defined at the points $x_j$ ($j = 1, \ldots, J$) to have the values $r_j$ ($j = 1, \ldots, J$). Then the inverse cumulative function in Figure 12.7 reverses the roles of input and output. The slopes, denoted by $a_j$, are computed from

$$a_j = \frac{x_j - x_{j-1}}{r_j - r_{j-1}}, \qquad j = 1, \ldots, J \tag{12.16}$$

In evaluating Eq. 12.16, we should note the relationships $r_0 = 0$ and $r_J = 1$. In our example we have $J = 13$, $x_0 = 0$, and $x_{13} = 325$. With these we compute

$$a_1 = \frac{x_1 - x_0}{r_1 - r_0} = \frac{25 - 0}{0.153 - 0} = 163.40$$

The rest of $a_j$ can be computed in this fashion, and the results are entered in Table 12.5 as column 4.

With Table 12.5, we can generate random deviates $X$ as follows.

**Step 1:** Generate $U$.

**Step 2:** Find the interval in which $U$ lies; that is, find $j$ so that $r_{j-1} \leq U \leq r_j$.

**Step 3:** Compute $X$ from

$$X = x_{j-1} + a_j(U - r_{j-1}) \tag{12.17}$$

### *Example 12.3*

If a random number $U_1 = 0.654$ is generated, $U_1$ is seen to lie in the fourth interval (between $r_3 = 0.533$ and $r_4 = 0.693$), so that by Eq. 12.17

$$\begin{aligned} X_1 &= x_3 + a_4(U_1 - r_3) \\ &= 75 + 156.25(0.654 - 0.533) \\ &= 93.91 \quad \square \end{aligned}$$

### 12.4.2 Other Frequently Used Random Deviates

We will briefly review some algorithms for generating random deviates that are frequently used in risk simulation. Excellent treatments of this topic are given in Banks and Carson [1] and Law and Kelton [10].

***Normal Distribution.*** The polar method is used to generate $N(0, 1)$ random deviates in pairs [10].

1. Generate $U_1$ and $U_2$. Let $V_j = 2U_j - 1$ for $j = 1,2$, and let $W = V_1^2 + V_2^2$.
2. If $W > 1$, go back to step 1. Otherwise, let $Y = [(-2 \ell n\, W)/W]^{1/2}$, and compute

$$X_1 = V_1 Y \quad \text{and} \quad X_2 = V_2 Y$$

   Then $X_1$ and $X_2$ are $N(0, 1)$ random variables.

To generate $N(\mu, \sigma^2)$ variates, we modify the procedure to compute

$$Z_1 = \mu + \sigma X_1 \quad \text{and} \quad Z_2 = \mu + \sigma X_2$$

Then $Z_1$ and $Z_2$ are $N(\mu, \sigma^2)$ random variables.

***Lognormal Distribution.*** A special property of the lognormal distribution [namely, that if $Y \sim N(\mu, \sigma^2)$, then $e^Y \sim LN(\mu, \sigma^2)$] is used to obtain the following algorithm.

1. Generate $Y \sim N(\mu, \sigma^2)$ by using the normal generation algorithm in the previous section.
2. Set $X = e^Y$, where $X$ is the lognormal random deviate.

***Beta Distribution.*** Several alternative algorithms can be used to generate random variables from a beta distribution. Most such algorithms are beyond the scope of this text; many of them are designed for special problem structures and require that the user have a high level of statistical sophistication. Here we will present a general method discussed in Banks and Carson [1] for generating a beta $(a, b)$ random deviate over the interval $[0, 1]$ for any $a > 0$ and $b > 0$.

The method derives from the fact that if $\gamma_1$ and $\gamma_2$ are independent random variables with $\gamma_1(a, 1)$ and $\gamma_2(b, 1)$, the ratio $\gamma_1/(\gamma_1 + \gamma_2)$ is distributed as beta $(a, b)$, where $(a, 1)$ and $(b, 1)$ are the shape parameters for the gamma distribution. More precisely, we may proceed as follows to generate a beta random deviation over the arbitrary interval $[L, H]$.

**Step 1:** Generate two gamma random deviates, $\gamma_1(a, 1)$ and $\gamma_2(b, 1)$ independent of $\gamma_1$.

**Step 2:** Set

$$Y = \frac{\gamma_1}{\gamma_1 + \gamma_2}$$

**Step 3:** Compute $X = L + (H - L)Y$.

To generate a gamma random deviate $\gamma \sim (a, 1)$, we set $d = (e + a)/e$, $e = 2.7182\ldots$, and follow one of two procedures, depending on the value of $a$ [2].

**Case I:** If $0 < a < 1$, proceed as follows.

**Step 1:** Generate $U_1 \sim U(0, 1)$ and let $Q = dU_1$. If $Q > 1$, go to step 3. Otherwise, go to step 2.

**Step 2:** Let $Z = Q^{1/a}$, and generate another uniform random deviate $U_2 \sim U(0, 1)$. If $U_2 \leq e^{-z}$, set $\gamma = Z$ and stop. Otherwise, go back to step 1.

**Step 3:** Let $Z = -\ln[(d - P)/a]$ and generate $U_2 \sim U(0, 1)$. If $U_2 \leq Z^{a-1}$, set $\gamma = Z$ and stop. Otherwise, go back to step 1.

**Case II:** If $a > 1$, proceed as follows.

**Step 1:** Compute $k_1 = (2a - 1)^{-1/2}$, $k_2 = a - \ln 4$, $k_3 = a + 1/k_1$, and $k_5 = 1 + \ln k_4$, and let $k_4 = 4.5$.

**Step 2:** Generate $U_1$ and $U_2$.

**Step 3:** Let $V = k_1 \ln[U_1/(1 - U_1)]$, $Y = ae^V$, $Z = U_1^2 U_2$, and $W = k_2 + k_3 V - Y$.

**Step 4:** If $W + k_4 - k_3 Z \geq 0$, set $X = Y$ and stop. Otherwise, go to step 5.

**Step 5:** If $W \geq \ln Z$, set $X = Y$ and stop. Otherwise, go back to step 2.

### *Example 12.4*

Generate one gamma random deviate from Eq. 12.6. Since $a = 1.73$ and $b = 49.51$, we use case II. Here we compute $k_1 = 0.1652$, $k_2 = 0.3437$, $k_3 = 7.7833$, $k_4 = 4.5$, and $k_5 = 2.5041$.

**Step 2:** Generate $U_1 = 0.8$ and $U_2 = 0.3$.

**Step 3:** Compute $V = 0.229$, $Y = 2.1752$, $Z = 0.192$, and $W = -0.0491$.

**Step 4:** Compute $W + k_4 - k_3 Z = 2.956$. Since $2.956 > 0$, set $X = Y = 2.1752$, where $X \sim \text{gamma}(a,1)$.

**Step 5:** To obtain $X' \sim \text{gamma}(a, b)$, we obtain $X' = bX$ or $X' = (49.51)(2.1752) = 107.69$. □

## 12.5 SAMPLING PROCEDURES FOR DEPENDENT RANDOM VARIABLES

Up to this point, we have essentially covered all the procedures of simulation analysis except for the important issue of dependency. We mentioned earlier that some of the random variables affecting *PV* may, in fact, be related to one another. If they are, we need to sample from distributions of the random variables in a manner that accounts for any dependency. This section will present some of the procedures for this purpose.

### 12.5.1 Assessment of Conditional Probabilities

There are two main issues at hand. (1) How do we recognize that a dependency may exist between two random variables? (2) Given that we can recognize a dependency relationship, how do we modify our sampling technique to account for the dependency?

***Cross-Plot from Observed Data.*** Various statistical measures of correlation can be used to determine whether two or more variables are dependent on one another, but an easier way to identify such a dependency is to make a cross-plot of the values of two of the variables of interest.

For example, suppose we are concerned about whether variables *A* and *B* are related to each other. We could make a cross-plot of variable *A* versus variable *B* by plotting the numerical values of *A* and *B* for each data point, as shown in Figure 12.8*a*. What would we conclude about any dependencies between the variables? With no apparent trends or patterns, we can safely conclude that there is no relationship or dependency, so we can sample from each distribution separately in the analysis.

Now suppose we make a similar cross-plot between two other random variables, *C* and *D*, and the results are as shown in Figure 12.8*b*. This example represents the other extreme of complete dependency. If we know *C*, we can

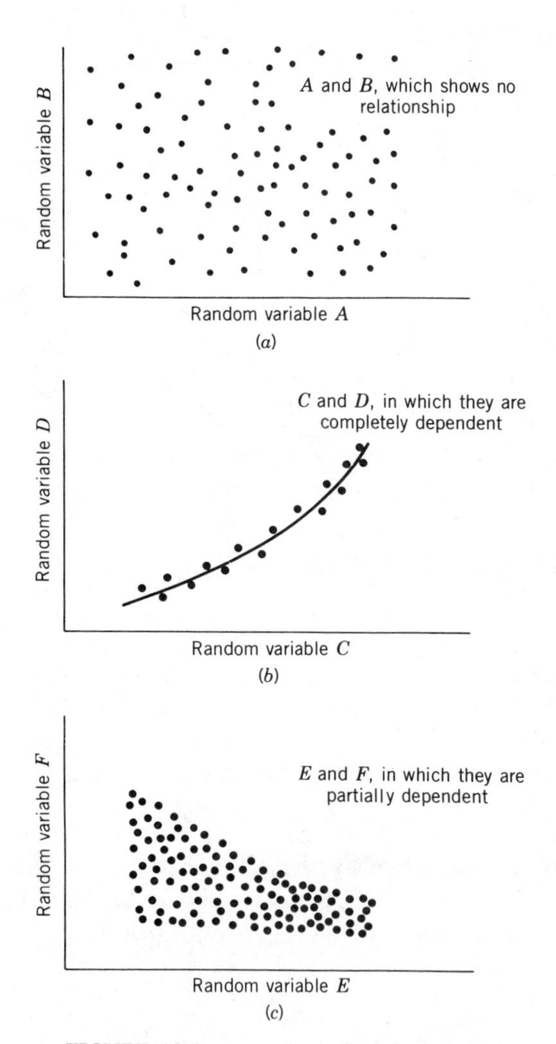


**FIGURE 12.8** Cross-plots of random variables.

predict the value of $D$ (or vice versa). Therefore, in terms of the simulation analysis, we need only specify a distribution of one of the variables. If this is done on a computer, we would determine the function (equation) of the curve and use the equation to establish the value of $D$, given $C$. For each trial, we would sample the distribution to determine its numerical value.

These two situations represent the either–or case of dependency and are simple to handle. But now let us look at the instance in which there is a partial dependency as depicted in Figure 12.8*c*. The cross-plot data are not clearly independent, but there does not appear to be an explicit dependency relationship such as in Figure 12.8*b*. In this situation we need to assess conditional probabilities between the two random variables.

***Assessment of Subjective Conditional Probabilities.*** If there are no data to use for estimating statistical dependence, we must assess the conditional probabilities subjectively. We already have shown such an example in Chapter 10. Once we assess the conditional probabilities, we can measure the degree of statistical dependence by computing the correlation coefficient.

### 12.5.2 Sampling a Pair of Dependent Random Samples

***Establishing Dependency Equation.*** To account for partial dependencies in simulation analysis, we first perform some data preparation steps.

**Step 1:** Prepare a cross-plot of available $X$–$Y$ data.

**Step 2:** Draw a boundary around the observed set of $X$–$Y$ points.

**Step 3:** Identify the conditional distribution, $f(y \mid x)$.

The first two steps are self-explanatory, and at the completion of these two steps we would have a bounded cross-plot such as in Figure 12.9. It should be noted that the boundary around the observed $X$–$Y$ plotting points does not have to be drawn in straight lines. It should, however, include all possible observed $X$–$Y$ values. If random variables are assessed on the basis of subjective probability distributions, we can skip the first two steps. In the third step we are trying to ascertain the variation of $Y$ as a function of $X$ (conditional probability distribution), which means that for a given value of $X$ we are concerned about the distribution of $Y$ values.

For the data of Figure 12.9, it appears that the variation of $Y$ as a function of $X$ is essentially uniform. That is, the conditional distribution describing the variability of $Y$ as a function of $X$ may be a uniform distribution. We have the choice of defining the variability of one variable as a function of the other with any type of distribution that approximates the data. It is simply a matter of trying to characterize the scatter of plotted $X$–$Y$ data points by a type of distribution such as the uniform or triangular distribution.

***Sampling Framework.*** Our problem of sampling $X$ and $Y$ values for each trial, as related to Figure 12.10, is that once we sample a value of $X$ (say $x_1$) from the $X$

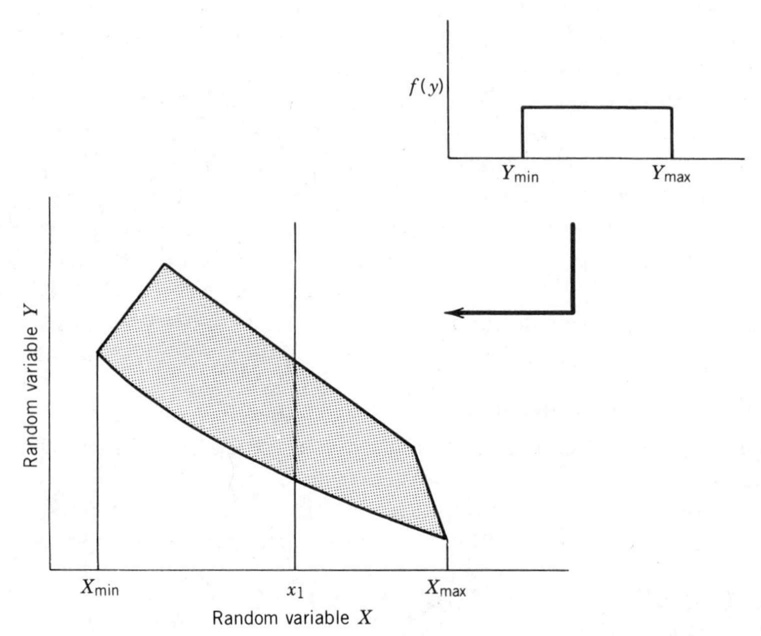


**FIGURE 12.9 Uniform variability within the bounded region of a partially dependent random variable.**

distribution, we have fixed our position on the $X$ axis of the three-dimensional (3-D) graph. To sample a value of $Y$, we would, in essence, cut the surface with a knife to look at the distribution profile along the face of the cut. This would be a uniform distribution having its walls at $Y_{\min(x1)}$and $Y_{\max(x1)}$ and a flat roof, the conditional distribution from which we would sample a value of $Y$ [11].

The step we just described implies that we need to store the conditional

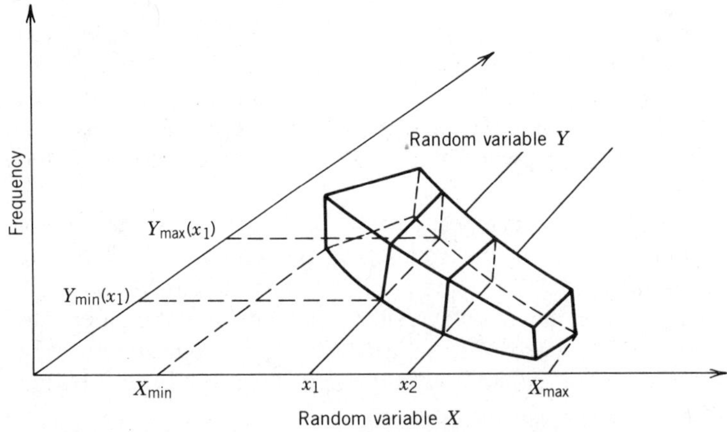


**FIGURE 12.10** Three-dimensional representation of the partial dependency of Figure 11.9, representing the case of the variability of $Y$ as a function of $X$.

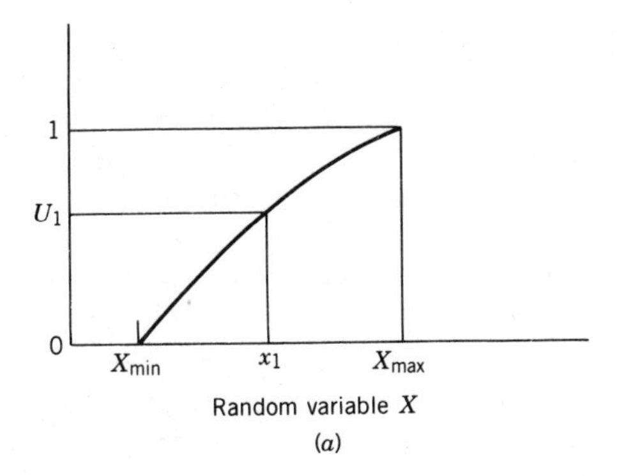


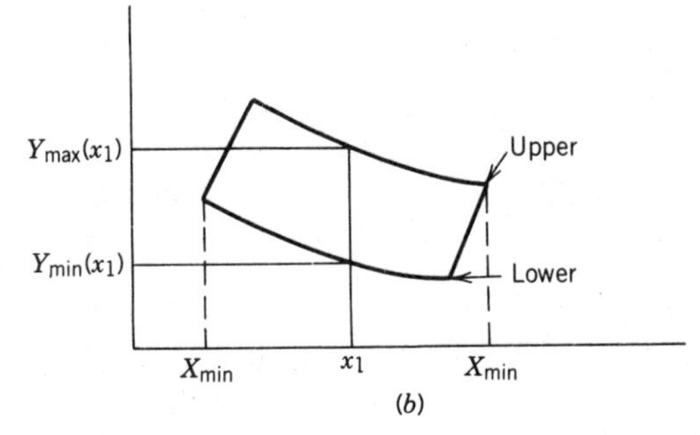


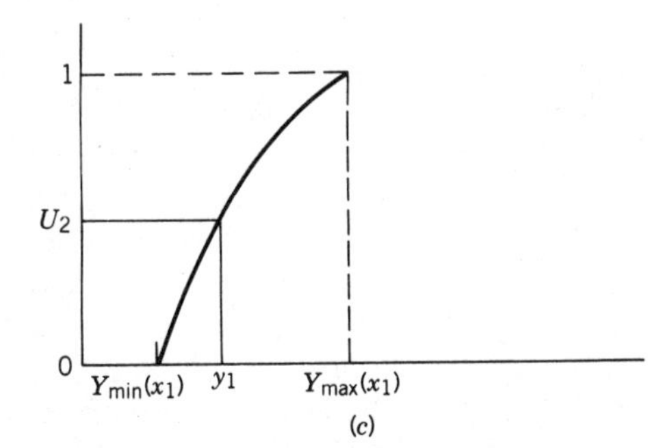


**FIGURE 12.11 Conceptual steps to sample values for each pair of partially dependent random variables $X$ and $Y$. (*a*) Sample $x_1$ from the $X$ distribution. (*b*) Locate the $Y_{min}$ and $Y_{max}$ associated with $x_1$. (*c*) Sample $y_1$ from the $f(y \mid x_1)$ distribution.**

distribution—the specific $Y$ distribution for each possible value of $X$. In this manner, once we establish a value of $X$ for each trial, we could go to the conditional cumulative frequency graph corresponding to that value of $X$ to sample the value of $Y$ for the trial. Figure 12.11 depicts the detailed simulation procedure for sampling a pair of dependent random variables.

For each dependent pair, we would need two random numbers ($U_1$ and $U_2$). We use the first random number, $U_1$, to sample the $X$ distribution and the second, $U_2$, to sample the conditional $Y$ distribution. When this is done by computer, we input to the computer the equations of the lower boundary and the upper boundary as functions of $X$ (Figure 12.11*a*). Once the value of $X$ is sampled, it is substituted into the equation of the lower boundary to compute the value of $Y_{min}(x)$, and the value of $Y_{max}(x)$ is determined for the trial by solving the equation of the upper boundary for the sampled value of $X$ (Figure 12.11*b*). From this conditional distribution bounded between $Y_{min}(x)$ and $Y_{max}(x)$, we obtain the sample value of $Y$ using $U_2$ (Figure 12.11*c*).

### 12.5.3 Sampling Based on Regression Equation

If more than one random variable appears in a simulation model as input, the relationship between these variables should be determined and taken into consideration. Regression analysis is a statistical technique for determining this relationship between the variables. We will make no attempt to describe the techniques of regression analysis fully. The complete treatment of this subject as presented in many statistical textbooks should provide an ample review for the interested reader [8]. In this section we summarize the major results of simple linear regression.

***Simple Linear Regression.*** Suppose that we want to estimate the relationship between two variables $(x, y)$. Suppose also that the true relationship between $x$ and $y$ is a linear relationship. The expected value of $y$ for a given value of $x$ is assumed to be

$$E(y \mid x) = b_0 + b_1 x \tag{12.18}$$

where $b_0$ and $b_1$ are regression coefficients. It is assumed that each observation of $y$ can be described by the model

$$\begin{aligned} y &= E(y \mid x) + \epsilon \\ &= b_0 + b_1 x + \epsilon \end{aligned} \tag{12.19}$$

where $\epsilon$ is assumed to be a random error with zero mean and constant variance $\sigma^2$. The regression model given by Eq. 12.19 involves a single variable $x$ and is thus called a simple linear regression model.

Suppose that we have $n$ pairs of observations $(y_1, x_1)$, $(y_2, x_2)$, . . . , $(y_n, x_n)$. We may estimate $b_0$ and $b_1$ in Eq. 12.19 on the basis of these observations. The method of least squares is used to form the estimates ($b_0$ and $b_1$) so that the sum of the squares of the deviations between the observations and regression line is minimized. The individual observations in Eq. 12.19 are written as

$$y_j = b_0 + b_1 x_j + \epsilon_j, \quad j = 1, 2, \ldots, n \tag{12.20}$$

where $\epsilon_1, \epsilon_2 \ldots$ are uncorrelated random variables.

Each $\epsilon_j$ in Eq. 12.19 is given by

$$\epsilon_j = y_j - b_0 - b_1 x_j \tag{12.21}$$

and represents the difference between the observed response $y_j$ and the expected response $b_0 + b_1x_j$, predicted by the model in Eq. 12.18. Then the least-square function is

$$L = \sum_{j=1}^{n} \epsilon_j^2 = \sum_{j=1}^{n} (y_j - b_0 - b_1x_j)^2 \tag{12.22}$$

Taking the partial derivatives, setting each to zero, and solving for $b_0$ and $b_1$ yields

$$\hat{b}_1 = \frac{\sum_{j=1}^{n} y_j(x_j - \bar{x})}{\sum_{j=1}^{n} (x_j - \bar{x})^2} \tag{12.23}$$

$$\hat{b}_0 = \bar{y} - \hat{b}_1\bar{x} \tag{12.24}$$

An estimate of the mean of $y$ given $x$, $E(y \mid x)$, is given by

$$\hat{y} = \hat{b}_0 + \hat{b}_1x \tag{12.25}$$

Now that we have found estimates of $b_0$ and $b_1$, the adequacy of the simple linear relationship should be tested before using the model in simulation. Several tests can be conducted to help determine model adequacy. Testing whether the order of the model tentatively assumed is correct, called the *lack-of-fit test,* is suggested. Testing for the significance of regression provides another means for assessing the adequacy of the model. The procedures for these tests are explained in [8].

***Sampling Procedure.*** A relationship such as that given by Eq. 12.25 can be used in sampling dependent random deviates, assuming that the model's premises have been tested by a lack-of-fit test and residual analysis and that the regression is significant. The simulation requires a value of $y$ given by

$$y = \hat{y} + \epsilon \tag{12.26}$$

where $\epsilon$ is normally distributed with mean zero and variance $\sigma^2$. This variance can be estimated by the mean square error (MSE),

$$\hat{\sigma}^2 = MSE = \sum_{j=1}^{n} \frac{\epsilon_j^2}{n-2} = \sum_{j=1}^{n} \frac{(y - \hat{y})^2}{n-2} \tag{12.27}$$

In the simulation, we can generate a random output $y_1$, given a random input $x_1$, as follows.

**Step 1:** Generate a random deviate $x_1$ from the marginal distribution of $x$. The marginal distribution may be a fitted or an empirical one.

**Step 2:** Generate a normal random deviate $N(0, \sigma^2)$, with $\sigma^2 = \hat{\sigma}^2$, to represent $\epsilon$.

**Step 3:** Calculate $y_1$ by using Eqs. 12.25 and 12.26.

## *Example 12.5*

Suppose a geologist is studying a new drilling prospect in an area in which 15 wells have been drilled. The unknown variables to be considered in the new prospect are net pay thickness (the thickness of the producing zone) and initial well potentials measured by barrels of oil per day (BOPD). Summarized data from the 15 wells in the field are shown in Table 12.6. Graphing these data generates Figure 12.12. Our questions are the following. (1) Is there a dependency between net pay thickness $x$ and initial potential $y$? (2) If so, how can the partial dependency be described and used in this simulation? To answer these questions, we may proceed as follows.

**Step 1:** Estimate the simple regression equation. Using Eqs. 12.23 and 12.24, we obtain

$$\hat{b}_0 = 104.762$$
$$\hat{b}_1 = 5.071$$
$$\hat{y} = \hat{b}_0 + \hat{b}_1 x = 104.76 + 5.071x$$

**Step 2:** Test the significance of the regression. Referring to Table 12.7 and noting that $F_0 = 119.5131 > F_{0.01,1,13} = 9.07$, we conclude that there is a linear relationship between $x$ and $y$. Each regression coefficient can also be shown statistically to be significant. Thus, we assume that the regression model given earlier is appropriate.

**Step 3:** Estimate $\sigma^2$ from Eq. 12.27.

$$\hat{\sigma}^2 = 421.795$$

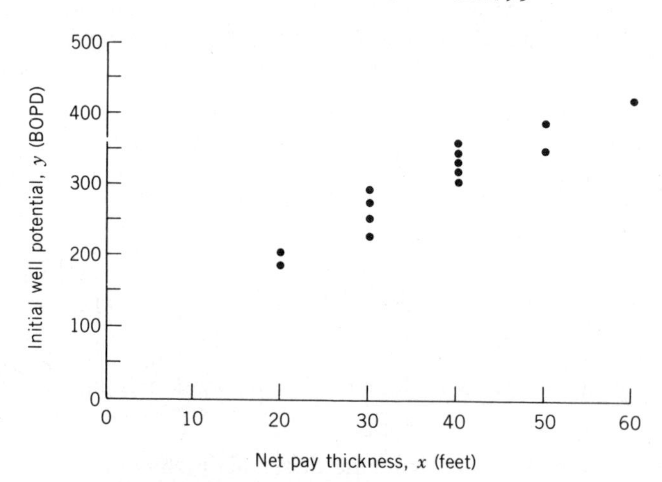


**FIGURE 12.12** Relationship between net pay thickness and initial well potential (BOPD).

**Table 12.6** ***Net Pay Thickness and Initial Potential Data from 15 Wells***

| Well Number | Net Pay Thickness, $x$ (feet) | Initial Well Potential, $y$ (BOPD)* |
|---|---|---|
| 1 | 20 | 200 |
| 2 | 30 | 240 |
| 3 | 30 | 290 |
| 4 | 50 | 340 |
| 5 | 30 | 270 |
| 6 | 40 | 330 |
| 7 | 40 | 310 |
| 8 | 60 | 390 |
| 9 | 30 | 230 |
| 10 | 20 | 180 |
| 11 | 40 | 340 |
| 12 | 40 | 320 |
| 13 | 50 | 360 |
| 14 | 20 | 210 |
| 15 | 40 | 300 |

*BOPD, Barrels of oil per day.

**Table 12.7** ***Analysis of Variance for Regression Results***

| Source of Variation | Sum of Squares | Degree of Freedom | Mean Squares | |
|---|---|---|---|---|
| Regression ($SS_R$) | 50,410.000 | 1 | *MSR* | 50,410.000 |
| Residual ($SS_E$) | 5,483.333 | 13 | *MSE* | 421.795 |
| Total ($S_{yy}$) | 55,893.333 | 14 | | |

Degree of freedom for regression = 1

Degree of freedom for residual = $n - 2$, where $n = 15$

Total degree of freedom = $1 + (n - 2) = n - 1$

- $F$ test for significance of regression.

  **Step 1:** Compute $F$ statistics ($F_0$) = 50,410.00/421.795 = 119.5131.

  **Step 2:** Choose the desired significance level; $\alpha = 0.01$.

  **Step 3:** Look up $F_{\alpha,1,n-2}$ in [8]; $F_{0.01,1,13} = 9.07$

  **Step 4:** Conclude that the regression parameters estimated are statistically significant if $F_0 > F_{\alpha,1,n-2}$; since $9.07 < 119.5131$, we conclude that there is a linear relationship between $x$ and $y$.

- The coefficient of determination ($R^2$). The quantity $R^2 = SS_R/S_{yy}$ is called the coefficient of determination, and it is often used to judge the adequacy of a regression model. We often refer loosely to $R^2$ as the amount of variability in the data accounted for by the regression model. In our example, we have $R^2 = 0.9019$; that is, 90.19% of the variability in the data is accounted for by the model.
- *MSE* ($\sigma^2$) = 421.7949.
- Regression equation: $y = 104.76 + 5.071x$.

**Step 4:** Sample one normal deviate with $N(0, 421.795)$ as shown by procedure 1 in Figure 12.13. (We assume this to be $\epsilon_1 = -8.45$.)

**Step 5:** Construct a cumulative distribution of $x$. (This is shown by procedure 2 in Figure 12.13.)

**Step 6:** Sample one random deviate of $x$ from Figure 12.13 by using Eq. 12.16. (We assume this to be $x_1 = 38$.)

**Step 7:** Calculate $y_1$ from Eqs. 12.24 and 12.25.

$$y_1 = 104.76 + (5.071)(38) - 8.45 = 289 \quad \square$$

If the simple linear regression model used is inadequate for any reason, several other possibilities exist. There may be several independent variables, for instance, in which case a multiple regression model may be employed [8]. (The use of multiple regression models in risk simulation is similar to the use of simple regression.) The regression method discussed here is workable only when statistical data are available. In the absence of such data, we need to develop a sampling scheme based on subjective estimates. This will be discussed in the next two sections.

### 12.5.4 Conditional Sampling in the Absence of Data

***Conditional Sampling.*** Conditional sampling was suggested as a method for handling correlation between variables $x$ and $y$ in the absence of data [6]. Basically, the method requires determining the marginal distribution for the independent variable $x$ and then a series of conditional subjective probability distributions for the dependent variable $y$. In simulation a value is sampled for the independent variable $x$, and this in turn determines which of the series of conditional distributions will be utilized for subsequent sampling. Such a sampling procedure is called conditional sampling.

As general as this approach may be, its practical usefulness can be quite limited. Not only is specification of the entire joint distribution required, but also derivation of many marginal and conditional distributions must be carried out. Such a level of detail is rarely obtainable in risk simulation.

***Discriminant Sampling.*** Managers may prefer to consider relationships in terms not of probability distributions, but of the range of values that one variable might assume, given the values of other related variables [5]. As methods of sampling, subjective probability distributions may be obtained for the independent variables, and then permissible ranges for dependent variables can be produced, given the possible values for the related variables. This type of sampling scheme is called *discriminant sampling*. The specific procedure for generating a correlated random pair $(x, y)$ would be as follows.

1. Divide the ranges for $x$ and $y$ into several intervals. (The more intervals we have, the more accurate the sampling scheme will be.)

Procedure 1: Generate one normal deviate with $N(0, 421.795)$ using a uniform random number $U_1$.

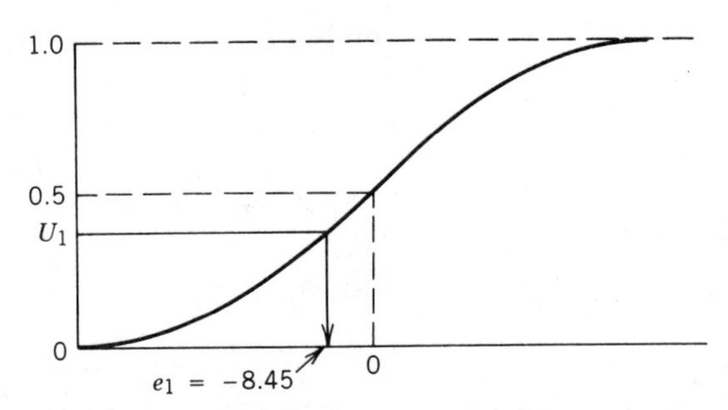


Procedure 2: Generate one random deviate of $x$ using $U_2$. First construct a cumulative distribution of $x$.

***Net Pay Thickness Distribution (x)***

| Cell | Interval | Observed Frequency | Relative Frequency | Cumulative Frequency |
|---|---|---|---|---|
| 1 | $10 < x \leq 20$ | 3 | 0.200 | 0.200 |
| 2 | $20 < x \leq 30$ | 4 | 0.267 | 0.467 |
| 3 | $30 < x \leq 40$ | 5 | 0.333 | 0.800 |
| 4 | $40 < x \leq 50$ | 2 | 0.133 | 0.933 |
| 5 | $50 < x \leq 60$ | 1 | 0.067 | 1.000 |

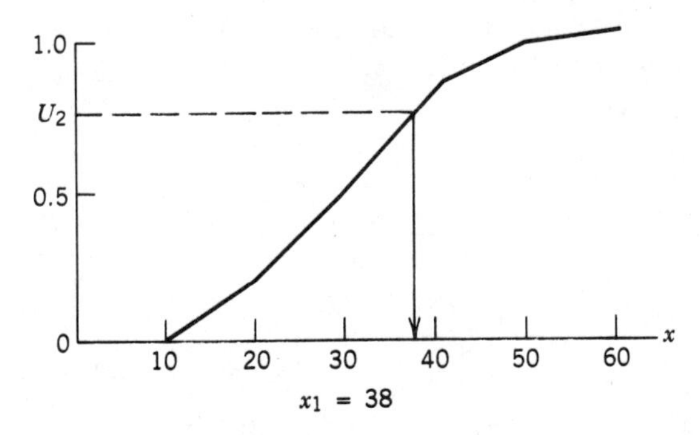


Procedure 3: Calculate $y_1$.

$$y_1 = 104.76 + (5.071)(38) - 8.45 = 289$$

**FIGURE 12.13 Procedures to generate a pair of dependent random variables.**

2. Estimate the probability for each interval of $y$ given that $x$ is in a certain interval.
3. Determine the conditional probability distributions that govern intervals of $y$.
4. Generate an independent sample from the distribution of $x$.

5. Generate a dependent sample from the conditional distribution of $y$ that governs the interval $x$.

We will explain the procedure with the following example.

### *Example 12.6*

Suppose we have two variables $x$ and $y$, where $x$ may assume values in the range 100 to 340 and $y$ may be between 0 and 6. Table 12.8 shows the probability distribution that management has decided to use when making assessments.

To assess the conditional probabilities, management then further states that if $x$ is below 145, then $y$ will probably be less than 2.5 with 0.05 probability, but if $x$ is above 145, then $y$ should be in the range 2.5 to 6.0 with 0.95 probability. Assessments similar to this can be made for other intervals, and these are summarized in Table 12.8.

Now let's consider a simple function $Z = XY$. To sample a value of $Z$, $Z_1$, the following sampling procedure is then applicable.

**Step 1:** Sample a value $x_1$ from the independent distribution of $X$.

- Generate a uniform random deviate, $U_1 = 0.25$.
- Then

$$x_1 = 190 + \left(\frac{240 - 190}{0.23}\right)(0.25 - 0.17) = 207.39$$

**Step 2:** Sample a value $y_1$ from the conditional distribution $Y$ that governs the interval containing $x_1$, that is, $145 < x_1 < 190$. The conditional distribution $(y \mid x = x_1)$ is

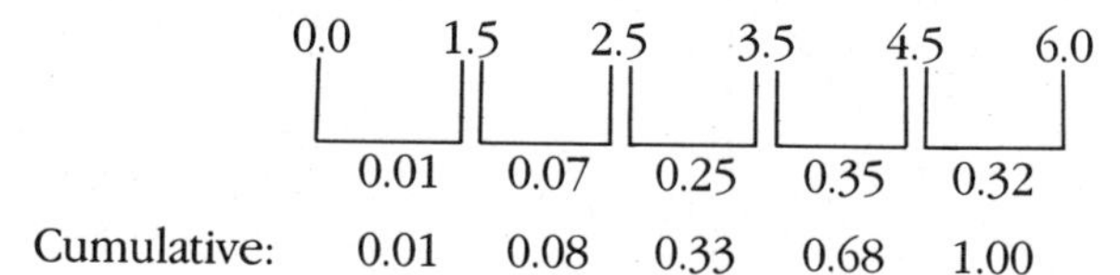


Cumulative: 0.01 0.08 0.33 0.68 1.00

- Generate a uniform random deviate, $U_2 = 0.61$. Note that $U_2$ lies between 3.5 and 4.5.
- Then,

$$y_1 = 3.5 + \left(\frac{4.5 - 3.50}{0.35}\right)(0.61 - 0.33) = 4.3$$

**Step 3:** Compute $Z$.

$$Z = (207.39)(4.3) = 891.78 \quad \square$$

As seen in this example, the essence of discriminant sampling is to restrict the range of values of the dependent variable according to the sampled value of the

**Table 12.8** ***Hypothetical Probability Assessments for Variables X and Y***

| Variable $X$ | | | Variable $Y$ | | |
|---|---|---|---|---|---|
| Range | Probability | Cumulative | Range | Probability | Cumulative |
| 100–145 | 0.05 | 0.05 | 0–1.5 | 0.15 | 0.15 |
| 145–190 | 0.12 | 0.17 | 1.5–2.5 | 0.20 | 0.35 |
| 190–240 | 0.23 | 0.40 | 2.5–3.5 | 0.30 | 0.65 |
| 240–290 | 0.40 | 0.80 | 3.5–4.5 | 0.20 | 0.85 |
| 290–340 | 0.20 | 1.00 | 4.5–6.0 | 0.15 | 1.00 |

*Conditional Probability Assessments*

| $Y$ \ $X$ | 100–145 | 145–190 | 190–240 | 240–290 | 290–340 |
|---|---|---|---|---|---|
| 0–1.5 | 0 | 0.01 | 0.04 | 0.15 | 0.40 |
| 1.5–2.5 | 0.05 | 0.07 | 0.13 | 0.25 | 0.30 |
| 2.5–3.5 | 0.20 | 0.25 | 0.33 | 0.35 | 0.23 |
| 3.5–4.5 | 0.30 | 0.35 | 0.30 | 0.45 | 0.07 |
| 4.5–6.0 | 0.45 | 0.32 | 0.20 | 0.10 | 0 |

independent variable. Furthermore, the probabilities assessed are assumed to be evenly distributed over the intervals.

### 12.5.5 Normal Transformation Method[1]

***Theory of Operation.*** In seeking a model correlation pattern by using further structural simplifications, we may use a normal transformation method [9]. To illustrate the basic idea, we again consider two correlated random variables $X$ and $Y$. Marginal distributions for these variables $f(x)$ and $g(y)$ are assessed in a manner similar to the discriminant sampling. To assess the extent of the dependence, management is asked to come up with a single estimate: "Assuming that $x = Q$, my median estimate for $y$ is $P$, where $Q$ is some value of $x$ not equal to the median of its distribution." [The term *median* is defined as the 0.50 fractile (percentile) of the distribution.] With this single estimate, it is possible to find both a set of conditional distributions for $Y$ and a correlational measure of dependence between $X$ and $Y$, where $\rho$ is the coefficient of correlation between $X$ and $Y$.

In general, $X$ and $Y$ will not be normally distributed. They can always be transformed, however, to the normal distribution $N(\mu, \sigma^2)$ on a "fractile-to-

[1]This section can be skipped without loss of continuity.

fractile" basis. In this case we define functions $z$ and $w$ so that $z(x)$ and $w(y)$ are both normal after transformation.

$$z(x) \sim N(\mu_x, \sigma_x^2) \tag{12.28a}$$

$$w(y) \sim N(\mu_y, \sigma_y^2) \tag{12.28b}$$

The distributions $f$ and $g$ are then related to

$$f(x) = \frac{dz}{dx}\,\phi_x[z(x)] \tag{12.29a}$$

$$g(y) = \frac{dw}{dy}\,\phi_y[w(y)] \tag{12.29b}$$

where $\phi_x$ and $\phi_y$ are the cumulative functions of variables $X$ and $Y$ with distribution $N(\mu_x, \sigma_x^2)$ and $N(\mu_y, \sigma_y^2)$, respectively. If $h$ is given by

$$h(y \mid x = x_1) = \frac{dw}{dy}\,\phi_h[w(y)] \tag{12.30a}$$

where $\phi_h$ is the conditional distribution with

$$N\left\{\mu_y + \rho'\frac{\sigma_y}{\sigma_x}[z(x_1) - \mu_x],\ \sigma_y^2(1 - \rho'^2)\right\} \tag{12.30b}$$

it follows from the bivariate normal distribution result given that Eq. 12.28a is satisfied. We should note, however, that $\rho'$ is the correlation coefficient between $z(x)$ and $w(y)$, not between $X$ and $Y$.

These results can be summarized as follows. Suppose $X$ and $Y$ are correlated random variables. Then, provided the marginal distributions for $X$ and $Y$ can be chosen as distributions that become normal after known transformations are applied to the variables, a conditional distribution can be determined for $Y$. The correlation coefficient $\rho'$ can be used as an *approximate* measure of the extent of the dependence between $X$ and $Y$, provided $z$ and $w$ are monotonic.

The transformations $z$ and $w$ that are chosen in a particular situation will depend on the properties that management judges the variable to have. Table 12.9 presents a list of the characteristics of these distributions for the assessment of subjective probability distributions [13]. Distributions not included in Table 12.9 can still be transformed to the normal distributions on a fractile-to-fractile basis by using

$$t(p) = q \tag{12.31}$$

where, if $p$ is the $k$th fractile of the distribution, $q$ is the $k$th fractile of the normal distribution, $N(0,1)$.

**Table 12.9** ***Characteristics of Distribution That Can Be Used for Normal Transformation***

| Type of Distribution | Transformation $t$ That Must Be Applied to Obtain a Normal Distribution $N(\mu, \sigma^2)$ | Fractiles That Must Be Assessed | Range | Other Properties |
|---|---|---|---|---|
| 1. Normal | $f(x) = x$ | 0.25, 0.75 | $[-\infty, +\infty]$ | Symmetrical |
| 2. Lognormal | $f(x) = \log x$ | 0.25, 0.75 | $[0, +\infty]$ | Positively skewed |
| 3. Three-parameter lognormal | $f(x) = \log(x - a)$ | 0, 0.25, 0.75 | $[a, +\infty]$ | Positively skewed |
| 4. Three-parameter lognormal | $f(x) = \log(a - x)$ | 0.25, 0.75, 1 | $[-\infty, a]$ | Negatively skewed |
| 5. Four-parameter lognormal | $f(x) = \log[(x - a)/(b - x)]$ | 0, 0.25, 0.75, 1 | $[a, b]$ | Any shape |

SOURCE: Schlaifer [13].

***Estimating* $\rho'$.** An estimate for $\rho'$ has to be provided to use the transformation method. An estimate for $\rho$ can be obtained if a single assessment of the form "Assuming that $x = Q$, my median estimate for $y$ is $P$" is made. To show why this is so, we note that $w(P)$ is the 0.50 fractile of the normal distribution given by Eq. 12.30b when $x = Q$, that is,

$$w(P) = \underbrace{\mu_y + \rho' \frac{\sigma_y}{\sigma_x}[z(Q) - \mu_x]}_{\text{mean value}} \tag{12.32}$$

Now solving for $\rho'$ gives

$$\rho' = \frac{[w(P) - \mu_y]\sigma_x}{[z(Q) - \mu_x]\sigma_y} \tag{12.33}$$

Generally, we may wish to obtain two or three assessments of the form "If $x = Q$, the median $y = P$" and average the estimates of $\rho'$ that are obtained. In theory, any value except the median of $X$ can be used for $Q$.

***Sampling Scheme.*** Once $\rho'$ has been obtained, the following sampling scheme is suggested.

***Step 1:*** Sample a value of $q_1$ from $N(0, 1)$.

**Step 2:** Sample a value of $q_2$ from $N(0, 1)$ for which the coefficient of correlation between $q_1$ and $q_2$ is $\rho'$. This step is equivalent to sampling $q_2$ from $N(\rho' q_1, 1 - \rho'^2)$.

**Step 3:** Calculate the fractile $k_1$ of $N(0, 1)$ to which $q_1$ corresponds and the fractile $k_2$ of $N(0, 1)$ to which $q_2$ corresponds.

**Step 4:** Put the $x$ sample equal to the $k_1$th fractile of its distribution and the $y$ sample equal to the $k_2$th fractile of its distribution.

## *Example 12.7*

The procedures we have outlined will now be illustrated with the data in Table 12.8. The variables $x$ and $y$ in this example are bounded both above and below. We assume that each can be represented by a four-parameter lognormal distribution from Table 12.9. To use this four-parameter lognormal distribution, we are required to estimate the upper (0.75) and lower (0.25) quartiles. Using Table 12.8 on the basis of linear interpolation, we compute the parameters

| Variable | Lower Bound | Upper Bound | Lower Quartile | Upper Quartile |
|---|---|---|---|---|
| $x$ | 100 | 340 | 207.4 | 283.7 |
| $y$ | 0 | 6 | 2.0 | 4.0 |

The parameters $\mu$ and $\sigma$ (equal to the means and standard deviations of the normal distributions into which the four-parameter lognormal distribution is transformed) are computed for $X$.

$$\log\left(\frac{207.4 - 100}{340 - 207.4}\right) = \mu_x - 0.6745\sigma_x$$

$$\log\left(\frac{283.7 - 100}{340 - 283.7}\right) = \mu_x + 0.6745\sigma_x$$

Solving for $\mu_x$ and $\sigma_x$ gives $\mu_x = 0.49$ and $\sigma_x = 1.03$. For the variable $y$, we evaluate

$$\log\left(\frac{2.0 - 0}{6 - 2.0}\right) = \mu_y - 0.6745\sigma_y$$

$$\log\left(\frac{4.0 - 0}{6 - 4.0}\right) = \mu_y + 0.6745\sigma_y$$

Solving for $\mu_y$ and $\sigma_y$ yields $\mu_y = 0$ and $\sigma_y = 1.03$. Suppose we make the assessment "Assuming that $x = 207.4$, my median estimate for $y$ is 3.5." Given this assessment, we compute $\rho'$ as follows. With $Q = 207.4$ and $P = 3.5$, we first evaluate

$$z(Q) = \log\left(\frac{Q - 100}{340 - Q}\right) = -0.2108$$

$$w(P) = \log\left(\frac{P - 0}{6 - P}\right) = 0.3365$$

Using Eq. 12.33, we obtain

$$\rho' = \frac{(0.3365 - 0)(1.03)}{(-0.2108 - 0.49)(1.03)} = -0.48$$

To sample a pair of correlated random deviates $(x, y)$ with $\hat{\rho} = \rho' = -0.48$, we may proceed as follows:

**Step 1:** Sample a value $q_1$ from $N(0, 1)$—let $q_1 = 0.245$.

**Step 2:** Sample one value $q_2$ from $N(0, 1)$ with $\rho' = -0.48$. Using the result in Eq. 12.28b is equivalent to sampling $q_2$ from $N(\rho q_1, 1 - \rho^2) = N(-0.11763, 0.7696)$. Let $q_2$ be $-0.145$.

**Step 3:** Compute the fractiles of $q_1$ and $q_2$.

$$q_1 = 0.245 \sim 0.5968\text{th fractile}$$
$$q_2 = -0.145 \sim 0.4876\text{th fractile}$$

**Step 4:** Compute the 0.5968th fractile of the distribution of $X$ and the 0.4876th fractile of the distribution of $Y$. From Table 12.8, they are equivalent to $x = 265$ and $y = 2.96$. These are the paired sample values. □

Getting the assessor to give a 0.50 fractile or median may be difficult in practice. In many situations it is unclear whether many assessors understand the difference between the mean, the median, and the mode. Further, the correlation obtained through the normal transformation process is *not* a true measure of dependence between the original variables, but a rough approximation at best.

## 12.6 OUTPUT DATA ANALYSIS

As we have seen, the risk simulation model contains random elements (e.g., cash flows), and outputs from the simulation are only a limited number of observed samples of these random variables. As a consequence, any decisions made on the basis of simulation results should consider the inherent variability of the simulation outputs. In this section we will discuss some statistical assessments of the accuracy of the measure of performance.

### 12.6.1 Replication and Precision of Results

In analyzing a risk simulation output, we need a method for ascertaining how close an estimator (e.g., mean value) is to the true measure. The usual approach to assessing the accuracy of an estimator is to construct a *confidence interval* for the true measure; that is, we determine an interval about the mean within which the true value may be expected to fall with a certain probability (e.g., 95%). Two approaches are possible for constructing a confidence interval.

***Fixing a Number of Replications.*** First we may fix the number of replications and then determine the confidence interval for the estimator at the specified level of precision.

Suppose that the single run produces observations $Y_1, Y_2, \ldots, Y_n$. The steady-state (or long-run) measure of performance, $\theta$, to be estimated is defined by

$$\theta = \lim_{n \to \infty} \frac{1}{n} \sum_{j=1}^{n} Y_j \tag{12.34}$$

with probability 1, where the value of $\theta$ is independent of initial conditions.

We must decide to stop the simulation after some number of observations, say $n$, have been collected. The sample size $n$ is usually a design choice based on the desired accuracy of the point estimator, as measured by an estimate of point estimator variability.

When we discuss one replication (or run), the notation $Y_1, Y_2, Y_3, \ldots, Y_n$ will be used; if several replications have been made, the output data for replication $k$ will be denoted by $Y_{k1}, Y_{k2}, \ldots, Y_{kn}$, and the replication number $m$ is to be fixed. The basic raw output data $\{Y_{kj}, k = 1, \ldots, m; j = 1, \ldots, n\}$ are exhibited in Table 12.10, where $Y_{kj}$ is an individual observation from replication $k$. In general, the total number of observations $n$ may vary from one replication to the next, but for simplicity we assume that they are constant over replications.

**Table 12.10** ***Raw Output Data from a Monte Carlo Simulation***

| Replication, $k$ | Observation, $j$ 1 | 2 | 3 | . . . | $n$ | Replication Averages |
|---|---|---|---|---|---|---|
| 1 | $Y_{1,1}$ | $Y_{1,2}$ | $Y_{1,3}$ | . . . | $Y_{1,n}$ | $\bar{Y}_1$ |
| 2 | $Y_{2,1}$ | $Y_{2,2}$ | $Y_{2,3}$ | . . . | $Y_{2,n}$ | $\bar{Y}_2$ |
| ⋮ | ⋮ | ⋮ | ⋮ | | ⋮ | ⋮ |
| $m$ | $Y_{m,1}$ | $Y_{m,2}$ | $Y_{m,3}$ | . . . | $Y_{m,n}$ | $\bar{Y}_m$ |

When the replication method is used, each replication is regarded as a single sample for the purpose of estimating $\theta$. For replication $k$, we define

$$\bar{Y}_k = \frac{1}{n}\sum_{j=1}^{n} Y_{kj} \tag{12.35}$$

as the sample mean of all observations in replication $k$. Since all replications use different random number streams and all are initialized, the replication averages $\bar{Y}_1, \bar{Y}_2, \bar{Y}_3, \ldots, \bar{Y}_m$ are independent and identically distributed random variables. Then the overall point estimator is given by

$$\bar{Y} = \frac{1}{m}\sum_{k=1}^{m} \bar{Y}_k \tag{12.36}$$

To estimate the standard error of $\bar{Y}$, we first compute the sample variance

$$S^2 = \frac{1}{m-1}\sum_{k=1}^{m} (\bar{Y}_k - \bar{Y})^2 = \frac{1}{m-1}\left(\sum_{k=1}^{m} \bar{Y}_k^2 - m\bar{Y}^2\right) \tag{12.37}$$

The standard error of $\bar{Y}_k$ is given by

$$\hat{\sigma} = \frac{S}{\sqrt{m}} \tag{12.38}$$

A $100(1 - \alpha)\%$ confidence interval for $\theta$, based on the $t$ distribution, is given by

$$\bar{Y} - t_{\alpha/2,m-1}\hat{\sigma} \leq \theta \leq \bar{Y} + t_{\alpha/2,m-1}\hat{\sigma} \tag{12.39}$$

where $t_{\alpha/2,m-1}$ is the $100(1 - \alpha/2)$ percentage point of a $t$ distribution with $m - 1$ degrees of freedom. It should be emphasized that this confidence interval is valid only if the bias of $\bar{Y}$ is approximately zero. (See Banks and Carson [1] for a statistical treatment for measuring the bias.)

### *Example 12.8*

In a risk simulation, five replications have been made with 100 present value observations in each replication. The sample means for the replications are

\$18.41, \$19.04, \$18.94, \$18.59, and \$18.89. We compute the overall sample mean $\bar{Y}$ as

$$\bar{Y} = \frac{1}{5}(18.41 + 19.04 + 18.94 + 18.59 + 18.89) = \$18.77$$

The sample variance is then computed by using Eq. 12.40.

$$\begin{aligned} S^2 &= \frac{1}{5-1}[18.41^2 + 19.04^2 + 18.94^2 + 18.59^2 + 18.89^2 - (5)(18.77)^2] \\ &= 0.0686 \end{aligned}$$

Then the standard error of estimation is

$$\hat{\sigma} = \sqrt{\frac{S^2}{5}} = 0.117$$

A 95% confidence interval for $\theta$ ($PV$) is given by

$$18.77 \pm (2.78)(0.117)$$

or

$$\$18.45 \leq \theta \leq \$19.10$$

(See [8] for a $t$-table with $m$ degrees of freedom.) □

***Confidence Intervals with Specified Accuracy.*** A second, alternative approach is to determine the number of replications required to obtain a confidence interval with a specified precision. A complete discussion of this subject is beyond the scope of this text but can be found in most standard statistics textbooks [8]. Our objective here is to remind the analyst that a few replications are not sufficient to obtain an acceptable estimate of a measure of performance.

### 12.6.2. Comparison of Two Projects

Suppose we wish to compare two mutually exclusive projects. Perhaps we may want to compare the performance of two different decision criteria in portfolio selection through computer simulation. (Many other examples of this nature can be found in risk simulation.)

The method of replication will also be used to analyze the output data. The mean performance measure for project $j$ will be denoted by $\theta_j$ ($j = 1,2$). The goal of the simulation experiments is to obtain point and interval estimates of the difference in mean performance $\Delta\theta$, namely $\theta_1 - \theta_2$.

When comparing two projects, we must decide on a number of $PV$ observations for each project, and for simplicity we will assume that the number of observations for each project during replication $k$ remains constant. The total numbers of replications to be made for the projects are also identical.

From replication $k$ of project $j$ we obtain an estimate $\bar{Y}_{kj}$ of the mean

**Table 12.11** ***Simulation Output Data and Summary Measures When Comparing Two Projects***

| | Replication | | | | Sample | Sample |
|---|---|---|---|---|---|---|
| Project | 1 | 2 | | $m$ | Mean | Variance |
| 1 | $Y_{11}$ | $Y_{21}$ | ... | $Y_{m1}$ | $\bar{Y}_1$ | $S_1^2$ |
| 2 | $Y_{12}$ | $Y_{22}$ | ... | $Y_{m2}$ | $\bar{Y}_2$ | $S_2^2$ |

performance measure, $\theta_j$. For example, $\bar{Y}_{21}$ may represent the average *PV* observed for project 1 (based on $n$ observations) during replication 2. The raw data format is shown in Table 12.11. From these we compute the sample mean $\Delta\bar{Y} = \bar{Y}_1 - \bar{Y}_2$, the sample variance $\Delta\sigma^2$, and the standard error of the specified point estimator $\Delta\theta$ ($\Delta\hat{\sigma}$) as shown in Table 12.12. Note that the value of $\Delta\hat{\sigma}$ will depend on the sampling environment and relationship between the projects. (See Banks and Carson [1] for the statistical derivations of those entries in the table.)

A two-sided $100(1 - \alpha)\%$ confidence interval for $\Delta\theta$ will always be of the form

$$\Delta\bar{Y} \pm t_{\alpha/2,v}\Delta\hat{\sigma} \tag{12.40}$$

where $v$ and $\Delta\hat{\sigma}$ are defined in Table 12.12. Then the confidence interval will lead to one of three possible conclusions, in referring to Figure 12.14.

1. If the confidence interval for $\Delta\theta$ is totally to the left of zero, we may conclude $\theta_1 < \theta_2$.
2. If the confidence interval for $\Delta\theta$ is totally to the right of zero, we may conclude $\theta_1 > \theta_2$.
3. If the confidence interval for $\Delta\theta$ contains zero, then on the basis of the data on hand, we may conclude $\theta_1 \approx \theta_2$.

## *Example 12.9*

In comparing two mutually exclusive projects, we made five replications with 100 observations in each replication for both projects. The simulation results with independent sampling are

| | Replication | | | | | Sample | Sample |
|---|---|---|---|---|---|---|---|
| Project | 1 | 2 | 3 | 4 | 5 | Mean | Variance |
| 1 | \$18.41 | 19.04 | 18.94 | 18.54 | 18.89 | 18.77 | 0.0686 |
| 2 | \$22.02 | 22.12 | 22.98 | 22.42 | 23.31 | 22.57 | 0.3108 |

We compute

$$\Delta\bar{Y} = \bar{Y}_1 - \bar{Y}_2 = 18.77 - 22.57 = -\$3.80$$

**Table 12.12** ***Standard Error ($\Delta\sigma^2$) Computation***

| Case | Computation Formula* | Remark |
|---|---|---|
| *Case I*<br>Independent sampling with equal variances ($\sigma_1^2 = \sigma_2^2$) | $\Delta\bar{Y} = \bar{Y}_1 - \bar{Y}_2$<br>$\Delta S = \sqrt{\frac{(m_1 - 1)S_1^2 + (m_2 - 1)S_2^2}{m_1 + m_2 - 2}}$<br>$\Delta\hat{\sigma} = \Delta S\sqrt{\frac{1}{m_1} + \frac{1}{m_2}}$<br>$\nu = m_1 + m_2 - 2$ | 1. Different and independent random numbers are used to simulate the two projects.<br>2. Normally $m_1 = m_2$.<br>3. Use an $F$ test to determine whether $\sigma_1^2 = \sigma_2^2$. |
| *Case II*<br>Independent sampling with unequal variances ($\sigma_1^2 \neq \sigma_2^2$) | $\Delta\bar{Y} = \bar{Y}_1 - \bar{Y}_2$<br>$\Delta\hat{\sigma} = \sqrt{\frac{S_1^2}{m_1} + \frac{S_2^2}{m_2}}$<br>$\nu_0 = \frac{(S_1^2/m_1 + S_2^2)^2}{[(S_1^2/m_1)^2/(m_1 + 1)] + [(S_2^2/m_2)^2/(m_2 + 1)]}$<br>$\nu = \nu_0 - 2$ | If $m_1 = m_2 = m$, we can use case I even if the variances ($\sigma_1^2$ and $\sigma_2^2$) are not equal. |
| *Case III*<br>Correlated sampling (or common random numbers) | $D_k = Y_{k_1} - Y_{k_2}$<br>$\bar{D} = \frac{1}{m}\sum_{k=1}^{m} D_k = \bar{Y}_1 - \bar{Y}_2$<br>$S_D^2 = \frac{1}{m-1}\sum_{k=1}^{m}(D_k - \bar{D})^2$<br>$\Delta\hat{\sigma} = S_0/\sqrt{m}$<br>$\nu = m - 1$ | 1. For each replication, the same random numbers are used to simulate the projects.<br>2. $m_1 = m_2 = m$. |

*$\nu$, degree of freedom; $m_1$ and $m_2$, number of replications for projects 1 and 2.

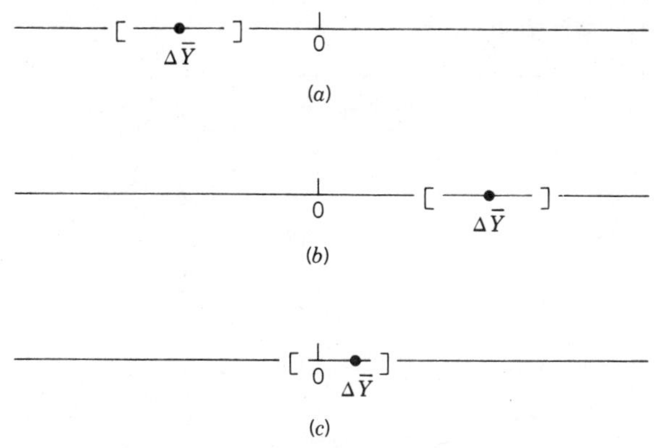


**FIGURE 12.14** Three possible confidence intervals in a comparison of two mutually exclusive projects. A project dominance is observed in either situation *a* or situation *b*.

Using Table 12.12, we also compute

$$\Delta S = \sqrt{\frac{(4)(0.0686) + (4)(0.3108)}{5 + 5 - 2}} = 0.4355$$

and

$$\Delta\hat{\sigma} = \left(\sqrt{\tfrac{1}{5} + \tfrac{1}{5}}\right) \Delta S = 0.2755$$

A 95% confidence interval is

$$-3.80 \pm t_{0.025,8}\, \Delta\hat{\sigma} = -3.80 \pm (2.31)(0.2755)$$

or

$$-\$4.43 \leq \Delta\theta \leq -\$3.16$$

The confidence interval for $\Delta\theta$ is totally to the left of zero, and we conclude that $\theta_1 < \theta_2$; project 2 is preferred over project 1. □

## 12.7 PROGRAMMING RISK SIMULATION MODEL ON COMPUTERS

So far, we have focused our attention on the analytical development of the risk simulation model. In this section we will see how the analytical relationships discussed in the previous sections can be translated into a computer model. For this purpose we will work through a simple investment problem and demonstrate the several features of risk simulation.

### 12.7.1. Decision Problem

Consider the two mutually exclusive investment proposals shown in Table 12.13. Both proposals have an economic life of three years. The size of the initial

**Table 12.13** ***Proposal Cash Flow Profiles (Lowest, Most Likely, Highest)***

| | Proposal 1, $F_{n1}$ | | | Proposal 2, $F_{n2}$ | | |
|---|---|---|---|---|---|---|
| End of Year, $n$ | $L$ | $M_0$ | $H$ | $L$ | $M_0$ | $H$ |
| 0 | (−10, | −10, | −10)* | (−15, | −15, | −15)* |
| 1 | (5, | 8, | 12) | (5, | 10, | 13) |
| 2 | (10, | 15, | 20) | (20, | 25, | 30) |
| 3 | (10, | 20, | 30) | (10, | 15, | 40) |

*When we enter a deterministic cash flow, the lowest, the most likely, and the highest estimates are identical.

investment required is known with certainty for both proposals; their respective net cash flows over the life of the investment are given in terms of three estimates (lowest possible, most likely, highest possible.) These three estimates are equivalent to the pessimistic, most likely, and optimistic estimates in a triangular distribution. The question is which alternative is more economically attractive at an interest rate of 20% (cost of capital or minimum attractive rate of return). For illustration purposes, the number of repetitions (number of Monte Carlo samplings) is set at 100.

***PV Model.*** For our simple example, the present value expression for project $j$ ($j$ = 1,2) is

$$PV(i)_j = F_{0j} + \frac{F_{1j}}{1+i} + \frac{F_{2j}}{(1+i)^2} + \frac{F_{3j}}{(1+i)^3} \tag{12.41}$$

where $F_{nj}$ = cash flow occurring at period $n$ for project $j$,
$i$ = minimum attractive rate of return.

***Interactive Simulation Model.*** Our risk simulation program is designed to run on a microcomputer, and there is an emphasis on graphic computer output. Graphics has several benefits for the risk analyst. It reduces the time required to interpret results, simplifies the communication of results to others, and enhances formal reports. Another benefit is that graphic output is much easier to use than tabular output in decision making. Visual information is presented clearly. Differences between anticipated and actual figures are evident, and discrepancies in data become obvious.

The computer program comprises three subprograms: (1) a program for entering data and simulation parameters (i.e., IN1A for cash flow inputs, IN1B for inputs to the Monte Carlo sampling process, and IN2 for the inputs to the discounting process); (2) a program for performing the Monte Carlo sampling and calculating the resulting statistics; and (3) a program for directing the next step to be performed (i.e., displaying the simulation results).

Five possible options are available to the user in the third subroutine: (1) plot the statistics, (2) make a new run with reseeded random numbers,

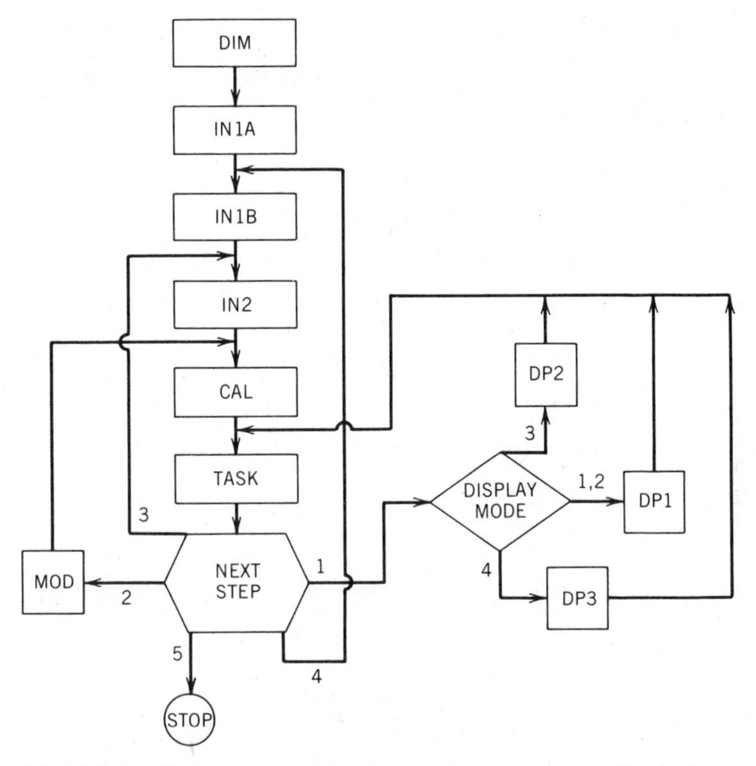


**FIGURE 12.15** Programming logic on a microcomputer. DIM: declares arrays, IN1A: reads in cash flow profiles, IN1B: reads in simulation parameters, IN2: reads in discounting parameters, CAL: calculates present value statistics, TASK: determines the next task to be performed, MOD: reseeds the random numbers, DP1: displays the distribution for a single project, DP2: displays the superimposed distribution, DP3: displays the difference distribution. (Reprinted with permission from C. S. Park, *Interactive Microcomputer Graphics,* Addison–Wesley, Reading, Mass., 1985.)

(3) change the discount rate, (4) change the number of Monte Carlo samplings (replications), and (5) exit the program.

If we select the first option, four additional options are available for displaying the statistics: (1) display the histogram for project 1, (2) display the histogram for project 2, (3) display the superimposed distributions of projects 1 and 2, and (4) display the histogram of the difference distribution (project 1 − project 2). The complete code has been provided [12]. The general programming logic is shown in Figure 12.15.

### 12.7.2 Description of the Simulation Model

***Data Entry.*** With the data given in the Table 12.13 as sample inputs, we would have

```
TYPE OF DECISION PROBLEM
1.  SINGLE PROJECT EVALUATION
2.  COMPARISON OF TWO MUTUALLY EXCLUSIVE PROJECTS
```

```
        ⟨ENTER 1 OR 2⟩? 2

        NUMBER OF MONTE CARLO SAMPLINGS (LESS THAN 500)? 100

        PROJECT LIFE FOR PROJECT 1? 3
        PROJECT LIFE FOR PROJECT 2? 3
        DISCOUNT RATE (%)? 20

        CASH FLOW PROFILE—PROJECT 1

PERIOD 0 ⟨LOW, MOST LIKELY, HIGH⟩? −10, −10, −10
PERIOD 1 ⟨LOW, MOST LIKELY, HIGH⟩? 5, 8, 12
PERIOD 2 ⟨LOW, MOST LIKELY, HIGH⟩? 10, 15, 20
PERIOD 3 ⟨LOW, MOST LIKELY, HIGH⟩? 10, 20, 30

        CASH FLOW PROFILE—PROJECT 2

PERIOD 0 ⟨LOW, MOST LIKELY, HIGH⟩? −15, −15, −15
PERIOD 1 ⟨LOW, MOST LIKELY, HIGH⟩? 5, 10, 13
PERIOD 2 ⟨LOW, MOST LIKELY, HIGH⟩? 20, 25, 30
PERIOD 3 ⟨LOW, MOST LIKELY, HIGH⟩? 10, 15, 40
```

***System Feedback.*** Because it takes time to perform the Monte Carlo samplings, the computer prints a status message on the CRT screen.

```
              Monte Carlo Sampling
    In Processing Loop 43 of 100 for Project-1
```

This message indicates that the system is busy and gives some idea of how close to completion it is.

***Statistical Summary Report.*** The first statistical summary output is

```
          ** SUMMARY OF SIMULATION RESULTS **

                 PROJECT-1 PROFILE

THE NUMBER OF ITERATIONS        =       100
THE EXPECTED VALUE              = $    18.41
THE STANDARD DEVIATION          =       2.79
THE DISCOUNT RATE USED          =       20%
THE MINIMUM VALUE OBSERVED      = $    12.41
THE MAXIMUM VALUE OBSERVED      = $    26.07
THE PROBABILITY (PV <= 0)       =        0%

                 PROJECT-2 PROFILE

THE NUMBER OF ITERATIONS        =       100
THE EXPECTED VALUE              = $    22.02
THE STANDARD DEVIATION          =       3.91
THE DISCOUNT RATE USED          =       20%
THE MINIMUM VALUE OBSERVED      = $    14.07
THE MAXIMUM VALUE OBSERVED      = $    32.09
THE PROBABILITY (PV <= 0)       =        0%

⟨ Press any key to continue . . . ⟩
```

The summary report shows that the expected *PV* of project 1 amounts to $18.41 with a standard deviation of $2.79, whereas that of project 2 is $22.02 with a standard deviation of $3.91. All the *PV*s for both projects are greater than zero, indicating that there is no possibility of realizing negative returns. Thus, if these two projects were mutually independent, both would be acceptable to the investor. Since these projects are mutually exclusive, the difference distribution provides an additional piece of information for judging the two economic choices.

The expected difference in the *PV* is −$3.61 with a standard deviation of $5.00. The probability that the *PV* of project 2 exceeds that of project 1 is calculated as 80%. This implies that there is still a 20% probability that the reverse situation may hold true. The ultimate decision is up to the investor, but this type of additional information will aid him or her in discerning the best project of a set of projects in a complex decision environment.

***Plotting Statistics.*** After displaying the present value statistics, the computer will print the "next step" options on the screen.

```
NEXT STEP TO BE PERFORMED

1. PLOT THE STATISTICS
2. MAKE NEW RUN WITH RESEEDED RANDOM NUMBERS
3. CHANGE THE DISCOUNT RATE
4. CHANGE THE # OF MONTE CARLO SAMPLINGS
5. EXIT PROGRAM

< INPUT CHOICE >?
```

With option 1, the computer puts the program in graphics mode and produces a graphical display.

```
** GRAPHICAL DISPLAY OPTIONS **

1. DISPLAY THE HISTOGRAM FOR PROJECT-1
2. DISPLAY THE HISTOGRAM FOR PROJECT-2
3. DISPLAY THE SUPERIMPOSED HISTOGRAM (PROJECTS 1 AND 2)
4. DISPLAY THE DIFFERENCE DIST. (PROJECT 1 - PROJECT 2)

    [ ENTER YOUR CHOICE ]?
```

Assume that display option 1 is selected. The computer then asks for the number of cells to use in histogram plotting.

```
        THE SUGGESTED NUMBER OF CELLS FOR DISPLAY:

          NO. OF OBSERVATIONS        NO. OF CELLS

               1 -  20                    5
              21 -  50                   10
              51 - 200                   20
                  >200                   30

    ENTER THE NUMBER OF CELLS FOR HISTOGRAM DISPLAY →?
```

Assume that we set the number of cells at 15. Figure 12.16 is then obtained as the histogram of present value for project 1, along with the cell statistics. After displaying the histogram, the program returns to the main option menu shown at the beginning of this section. We can manipulate the form of histograms by changing the number of cells; we simply choose the plotting option again, change the number of cells, and replot the statistics. Figure 12.17 can be obtained in a similar fashion.

When we choose plotting option 3, we obtain Figure 12.18, which combines and plots the two histograms in the same chart. A shade pattern is used on the histogram of project 2 to emphasize the overlapping portions of the statistics.

CELL STATISTICS (WIDTH = 0.96)

| CELL NO | OBS FREQ | REL FREQ | CUM FREQ | UPPER BOUND |
|---|---|---|---|---|
| 1 | 2 | 0.02 | 0.02 | 13.01 |
| 2 | 4 | 0.04 | 0.06 | 13.96 |
| 3 | 2 | 0.02 | 0.08 | 14.92 |
| 4 | 9 | 0.09 | 0.17 | 15.88 |
| 5 | 13 | 0.13 | 0.30 | 16.83 |
| 6 | 17 | 0.17 | 0.47 | 17.79 |
| 7 | 15 | 0.15 | 0.62 | 18.74 |
| 8 | 7 | 0.07 | 0.69 | 19.70 |
| 9 | 6 | 0.06 | 0.75 | 20.66 |
| 10 | 9 | 0.09 | 0.84 | 21.61 |
| 11 | 9 | 0.09 | 0.93 | 22.57 |
| 12 | 4 | 0.04 | 0.97 | 23.53 |
| 13 | 2 | 0.02 | 0.99 | 24.48 |
| 14 | 0 | 0.00 | 0.99 | 25.44 |
| 15 | 1 | 0.01 | 1.00 | 26.40 |

SAMPLE STATISTICS:

| | | |
|---|---|---|
| MEAN | = | 18.41 |
| STAND DEV | = | 2.79 |
| MIN VALUE | = | 12.41 |
| MAX VALUE | = | 26.07 |
| PV (≤ 0) | = | 0.00 |

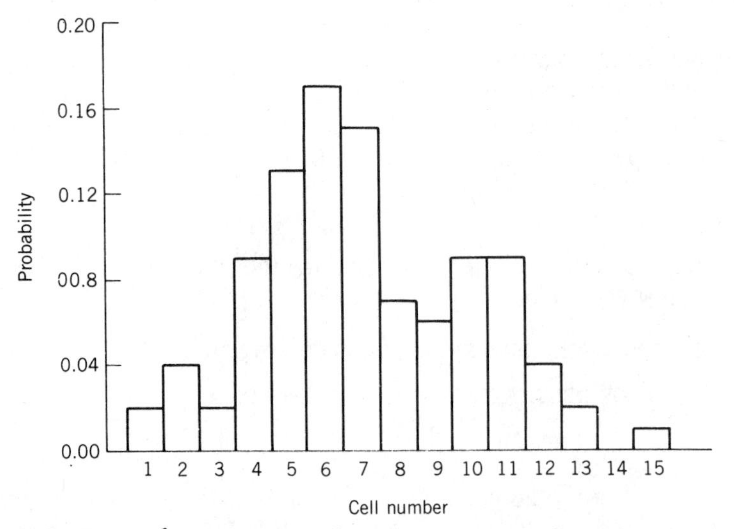


**FIGURE 12.16** Simulation result—project summary for project 1.

When we choose option 4, we obtain the histogram plot of the difference distribution (project 1 − project 2). This is shown in Figure 12.19, from which we can read the probability that the *PV* of project 2 will exceed that of project 1 for any given level.

### 12.7.3 Replication Results

Note that in our example the *PV*s range from a minimum of $12.41 to a maximum of $26.07 for project 1. If we print out all 100 observations (*PV*s), we find that most of the values are not very close to the expected value ($18.41). We clearly need a method for ascertaining how close an estimator (mean value) is to

CELL STATISTICS (WIDTH = 1.26)

| CELL NO | OBS FREQ | REL FREQ | CUM FREQ | UPPER BOUND |
|---|---|---|---|---|
| 1 | 1 | 0.01 | 0.01 | 14.86 |
| 2 | 0 | 0.00 | 0.01 | 16.12 |
| 3 | 10 | 0.10 | 0.11 | 17.38 |
| 4 | 7 | 0.07 | 0.18 | 18.64 |
| 5 | 14 | 0.14 | 0.32 | 19.90 |
| 6 | 15 | 0.15 | 0.47 | 21.17 |
| 7 | 13 | 0.13 | 0.60 | 22.43 |
| 8 | 13 | 0.13 | 0.73 | 23.69 |
| 9 | 3 | 0.03 | 0.76 | 24.95 |
| 10 | 7 | 0.07 | 0.83 | 26.21 |
| 11 | 3 | 0.03 | 0.86 | 27.47 |
| 12 | 8 | 0.08 | 0.94 | 28.74 |
| 13 | 2 | 0.02 | 0.96 | 30.00 |
| 14 | 3 | 0.03 | 0.99 | 31.26 |
| 15 | 1 | 0.01 | 1.00 | 32.52 |

SAMPLE STATISTICS:

| | | |
|---|---|---|
| MEAN | = | 22.02 |
| STAND DEV | = | 3.91 |
| MIN VALUE | = | 14.07 |
| MAX VALUE | = | 32.09 |
| PV (≤ 0) | = | 0.00 |

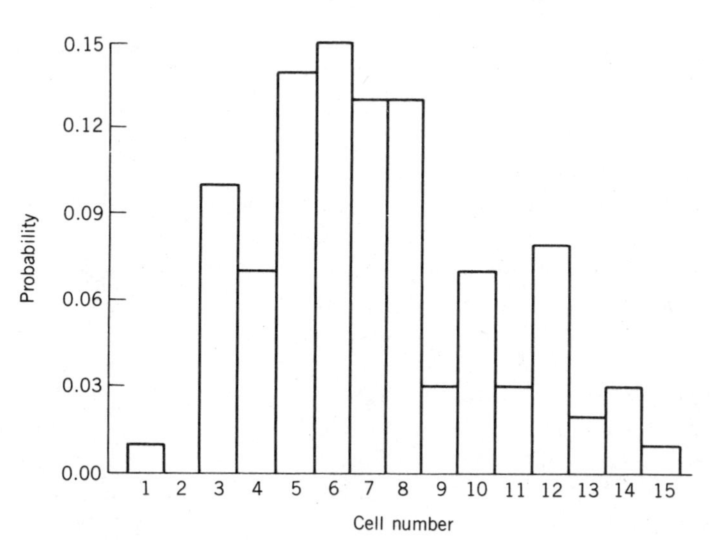


**FIGURE 12.17** Simulation result—project summary for project 2.

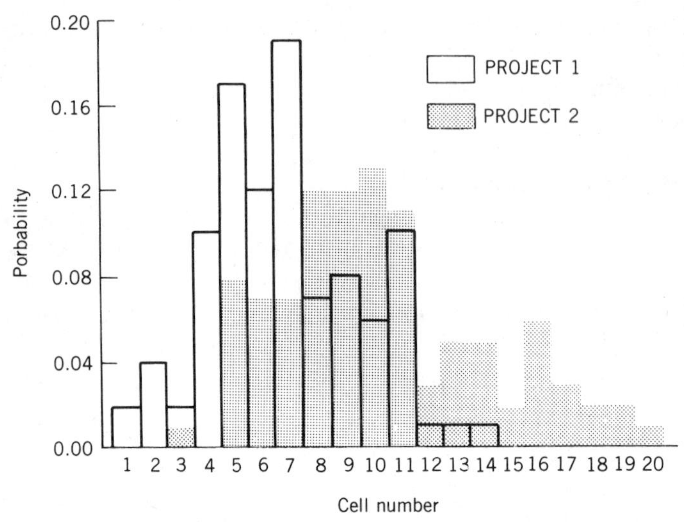


**FIGURE 12.18 Displaying the superimposed distributions.**

the true measure. The usual approach to assessing the accuracy of an estimator is to construct a confidence interval for the true measure; that is, we determine an interval about the mean within which the true value may be expected to fall with a certain probability (e.g., 95%).

Table 12.14 summarizes the simulation results obtained with five replications (100 observations in each replication). For obtaining the confidence interval for the difference, case 1 in Table 12.12 was assumed. The confidence interval for $\theta_1 - \theta_2$ is totally to the left of zero. As shown in Table 12.15, there is strong evidence for the hypothesis that $\theta_1 < \theta_2$ or, equivalently, that project 2 appears to be better than project 1. It will be clear that as we increase the number of replications, the length of the confidence interval decreases for specified precision. In other words, the relative precision of the estimator increases as the number of replicate runs increases.

## 12.8 SUMMARY

Risk analysis is a technique for economic evaluation. It provides the analyst with a convenient way to support management with reliable information about the uncertainties in alternative investment opportunities, corporate strategies, or operational procedures. Investment risk analysis requires the analyst to estimate the uncertainty associated with each factor affecting the decision. Monte Carlo methods are used to select values from probability distributions describing the likelihood of various values of the critical factors.

The issue of assessing the correlation patterns between variables has not really received adequate attention from the proponents of risk analysis approaches. The process of obtaining adequate assessments for correlation patterns is quite complicated and subject to many possible sources of bias. The

CELL STATISTICS (WIDTH = 1.91)

| CELL NO | OBS FREQ | REL FREQ | CUM FREQ | UPPER BOUND |
|---|---|---|---|---|
| 1 | 1 | 0.01 | 0.01 | −15.27 |
| 2 | 4 | 0.04 | 0.05 | −13.36 |
| 3 | 4 | 0.04 | 0.09 | −11.45 |
| 4 | 5 | 0.05 | 0.14 | −9.54 |
| 5 | 6 | 0.06 | 0.20 | −7.64 |
| 6 | 8 | 0.08 | 0.28 | −5.73 |
| 7 | 17 | 0.17 | 0.45 | −3.82 |
| 8 | 17 | 0.17 | 0.62 | −1.91 |
| 9 | 18 | 0.18 | 0.80 | −0.00 |
| 10 | 8 | 0.08 | 0.88 | 1.91 |
| 11 | 5 | 0.05 | 0.93 | 3.82 |
| 12 | 5 | 0.05 | 0.98 | 5.73 |
| 13 | 1 | 0.01 | 0.99 | 7.64 |
| 14 | 1 | 0.01 | 1.00 | 9.54 |
| 15 | 0 | 0.00 | 1.00 | 11.45 |

SAMPLE STATISTICS

| | | |
|---|---|---|
| MEAN | = | −3.61 |
| STAND DEV | = | 5.00 |
| MIN VALUE | = | −16.17 |
| MAX VALUE | = | 8.94 |
| PV (≤ 0) | = | 80.00 |

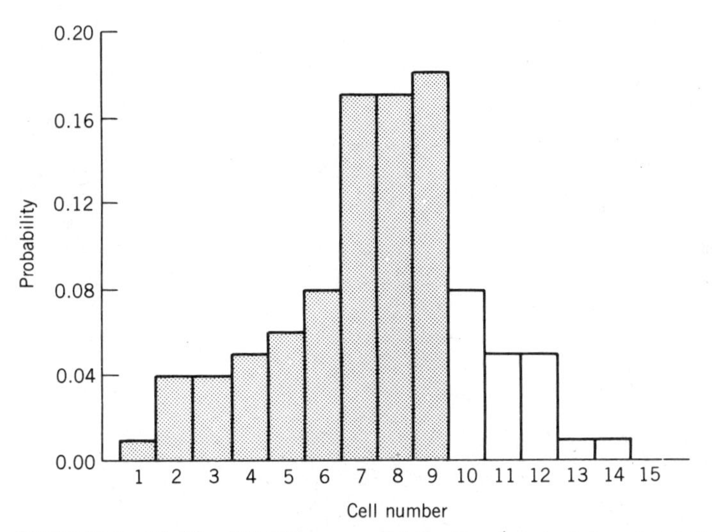


**FIGURE 12.19** **The *PV* difference distribution (project 1 − project 2).**

normal transformation approach to modeling the correlation pattern is certainly a step in the right direction, but much more research must be done in this area.

Although risk simulation has proved to be a valuable decision tool in many areas, an impediment to the use of simulation in the evaluation of risk has been the time-consuming and tedious work of writing and testing computer programs to perform the simulation. General-purpose languages, such as SIMSCRIPT, GPSS, and SLAM, have been designed to facilitate the creation of computer simulation programs, but their fixed structures and queueing-systems orienta-

**Table 12.14** *Summary of the Replication Results*

| Replication Number | Project 1 | | | Project 2 | | | Project 1 − Project 2 | | | Probability ($PV_1 < PV_2$) |
|---|---|---|---|---|---|---|---|---|---|---|
| | Mean | *SD* | Range | Mean | *SD* | Range | Mean | *SD* | Range | |
| 1 | \$18.41 | \$2.79 | \$12.41 ~ \$26.07 | \$22.02 | \$3.91 | \$14.07 ~ \$32.09 | −3.61 | 5 | −16.17 ~ 8.94 | 0.80 |
| 2 | 22.02 | 2.68 | 12.21 ~ 25.46 | 22.12 | 4.0 | 14.0 ~ 31.27 | −3.08 | 5 | −14.73 ~ 7.32 | 0.73 |
| 3 | 18.94 | 2.89 | 12.85 ~ 27.33 | 22.98 | 4.28 | 12.50 ~ 31.19 | −4.04 | 5.33 | −15.26 ~ 10.31 | 0.73 |
| 4 | 18.59 | 2.97 | 10.89 ~ 25.12 | 22.42 | 4.35 | 14.86 ~ 34.17 | −3.84 | 5.39 | −21.17 ~ 7.48 | 0.77 |
| 5 | 18.89 | 3.44 | 12.49 ~ 27.24 | 23.31 | 4.40 | 14.29 ~ 32.44 | −4.42 | 5.49 | −18.86 ~ 6.43 | 0.78 |

**Table 12.15** ***Summary of Confidence Interval Statistics***

| Project | Replication 1 | 2 | 3 | 4 | 5 | Sample Mean | Sample Variance |
|---|---|---|---|---|---|---|---|
| 1 | 18.41 | 19.04 | 18.94 | 18.59 | 18.89 | $18.77 | 0.0686 |
| 2 | 22.02 | 22.12 | 22.98 | 22.42 | 23.31 | $22.57 | 0.3108 |

Confidence intervals

Project 1: $\$18.45 < PV_1 < \$19.10$

Project 2: $\$22.18 < PV_2 < \$22.96$

Project 1 − project 2: $-\$4.43 < \Delta PV < -\$3.16$

tion render them impractical for studying risk. With the increasing use of microcomputers, several special-purpose risk simulation programs have been developed. Additional risk simulations often produce large amounts of output because users want to vary the model's constraints, relationships, or input conditions. This ability to vary conditions is precisely what makes simulation valuable.

As models become more complex and users pose more questions, however, the amount of output may become enormous and the task of interpreting the results may become overwhelming. More output may make the results more difficult to comprehend, but failure to try enough conditions may leave the analyst without an adequate understanding of the model's behavior. When one must interpret large amounts of output from a simulation, computer graphics is an important tool. In this chapter we demonstrated how interactive computer graphics can facilitate the understanding of risk simulation results.

In many simulation studies a large amount of time and money is spent on model development and programming, but little effort is made to analyze the simulation output data in an appropriate manner. For instance, the most common mode of operation is to make a single simulation run of somewhat arbitrary observations and then treat the resulting simulation estimates as being the "true" measures for the model. In this chapter we emphasized that a rigorous statistical output analysis should be made to find an acceptable estimate of the measure of performance.

## REFERENCES

1. BANKS, J., and J. S. CARSON, II, *Discrete-Event Systems Simulation,* Prentice–Hall, Englewood Cliffs, N.J., 1984, pp. 358–367.
2. CHENG, R. C. H., "The Generation of Gamma Variables," *Applied Statistician,* Vol. 26, No. 1, pp. 71–75, 1975.
3. CHOI, S. C., and R. WETTE, "Maximum Likelihood Estimation of the Parameters of the Gamma Distribution and Their Bias," *Technometrics,* Vol. 11, No. 4, pp. 683–890, 1969.
4. FISHMAN, G. S., *Principles of Discrete Event Simulation,* Wiley, New York, 1978.
5. EILON, S., and T. R. FOWKES, "Sampling Procedure for Risk Simulation," *Operations Research Quarterly,* Vol. 24, pp. 241–252, 1973.
6. HERTZ, D. B., "Risk Analysis in Capital Investment," *Harvard Business Review,* Vol. 42, pp. 95–106, 1964.

7. HERTZ, D. B., and H. THOMAS, *Risk Analysis and Its Applications,* Wiley, New York, 1983.
8. HINES, W. W., and D. C. MONTGOMERY, *Probability and Statistics in Engineering and Management Science,* 2nd edition, Wiley, New York, 1980.
9. HULL, J. C., "Dealing with Dependence in Risk Simulation," *Operations Research Quarterly,* Vol. 28, pp. 201–213, 1977.
10. LAW, A. M., and W. D. KELTON, *Simulation Modeling and Analysis,* McGraw–Hill, New York, 1982.
11. NEWENDORP, P. D., *Decision Analysis for Petroleum Exploration,* Penn Well Publishing, Tulsa, Okla., 1975.
12. PARK, C. S., *Interactive Microcomputer Graphics,* Addison–Wesley, Reading, Mass., 1985, Ch. 12.
13. SCHLAIFER, R., *Introduction to Statistics for Business Decisions,* McGraw–Hill, New York, 1961.

## PROBLEMS

**12.1.** Consider the observed processing time data for criminal cases given in Table 12.1. Knowing that the distribution has its lower limit bounded by zero but its upper limit unbounded ($0 < x < \infty$), fit the data to a lognormal distribution and examine whether the fitted distribution is in agreement with the observed data.

**12.2.** Suppose that a cash flow for a certain year $n$ ($F_n$) is not known with certainty, but three subjective estimates can be made as follows.

$$L = \$1{,}000, \qquad M_0 = \$1{,}500, \qquad H = \$1{,}800$$

If $F_n$ is considered to have a triangular distribution over the interval $[L, H]$ with $M_0$, identify the distribution parameters.

**12.3.** In Problem 12.2, if $F_n$ has a beta distribution over $[L, H]$ with shape parameters $a$ and $b$,

a. Estimate the shape parameters $a$ and $b$.
b. Compute the mean ($\mu$) of the distribution.

**12.4.** With the triangular distribution defined in Problem 12.2, generate two random deviates from the distribution when the uniform random deviates are given at $U_1 = 0.25$ and $U_2 = 0.85$, respectively.

**12.5.** Generate a beta random deviate from the beta distribution defined in Problem 12.3, using a uniform random number sequence of 0.23, 0.47, 0.13, 0.22, 0.89, 0.71, 0.57, . . . .

**12.6.** Consider Problem 10.3 and develop the oil recovery distribution ($Z$), assuming that all the random variables ($\phi, A, h, W, E,$ and $B$) have triangular distributions with the parameters specified in the problem. Set the number of iterations at 100. Compute the mean and variance of the distribution, and compare the results with the analytical mean and variance.

**12.7.** Consider Problem 10.5 and develop the break-even probability distribution by using the Monte Carlo sampling method. Compute the mean and variance of the distribution based on 200 Monte Carlo samplings.

**12.8.** Consider Problem 10.6 and develop the long-range profit (*LRP*) distribution, assuming that all random variables ($X_1, X_2, \ldots, X_6$) are normally distributed. Can you say that the *LRP* distribution is also a normal distribution? Compute the mean and variance of the *LRP* based on 200 Monte Carlo samplings.

**12.9.** Consider Problem 10.7 and develop the net cash flow (after-tax) distribution at 100 Monte Carlo samplings. What is the probability that the net cash flow would be less than \$140,000?

**12.10.** Consider Problem 10.13. Develop the *PV* distribution based on 200 Monte Carlo samplings for the following situations.

a. The annual operating cash flows are mutually independent and the $F_n$ are uniformly distributed with the means and variances specified in the problem. Can you claim that the resulting *PV* distribution is normally distributed?

b. The annual operating cash flows are perfectly correlated (positively) with one another.

**12.11.** Consider the two investment projects described in Problem 10.14. Obtain the *PV* distributions for projects A and B, respectively. Use the sample size of 200.

**12.12.** Note that the two investment projects described in Problem 10.14 have unequal investment lives (5 years and 3 years). Suppose that these two investments are mutually exclusive and you want to develop a difference distribution to determine which project is more economically attractive. Define the modeling assumptions you need before developing such a difference distribution and then develop the difference distribution.

**12.13.** Consider the *PV* expression in Problem 10.19. Develop the *PV* distribution and compute the mean and variance of the distribution based on 200 Monte Carlo samples.

**12.14.** Consider Problem 10.20. Estimate the beta-distribution parameters for the annual cash flows based on the three estimates (low, most likely, and high) specified in the problem. Develop the *PV* distribution with $i = 10\%$. Also develop the payback period distribution. Use the sample size of 200.

**12.15.** Consider Problem 10.22.

a. Assuming that the random variables (market size, market share, unit price, unit variable cost, fixed cost, and salvage value) have triangular distributions with the three estimates specified in the problem, develop the *PV* distribution for the project based on 100 Monte Carlo samples.

b. Compute the mean and variance of the *PV* in part a, and determine the 95% confidence interval for the *PV*.

**12.16.** Consider Example 12.5. Develop the initial well potential distribution based on the regression principle.

**12.17.** The classic risk analysis article by Hertz [6] contains a simulation example in which two investments, A and B, are under consideration. With the investment analysis, Hertz obtains the following tabulated and plotted data.

***Comparison of Two Investment Opportunities***

| Selected Statistics | Investment A | Investment B |
|---|---|---|
| Amount of investment | $10,000,000 | $10,000,000 |
| Life of investment (in years) | 10 | 10 |
| Expected annual net cash inflow | $ 1,300,000 | $ 1,400,000 |
| Variability of cash inflow | | |
| 1 chance in 50 of being greater than | $ 1,700,000 | $ 3,400,000 |
| 1 chance in 50 of being less than | $ 900,000 | $ (600,000) |
| Expected return on investment | 5.0% | 6.8% |
| Variability of return on investment | | |
| 1 chance in 50 of being greater than | 7.0% | 15.5% |
| 1 chance in 50 of being less than | 3.0% | (4.0%) |
| Risk of investment | | |
| Chance of a loss | Negligible | 1 in 10 |
| Expected size of loss | Negligible | $ 200,000 |

In the case of negative figures (indicated by parentheses), *less than* means worse than.

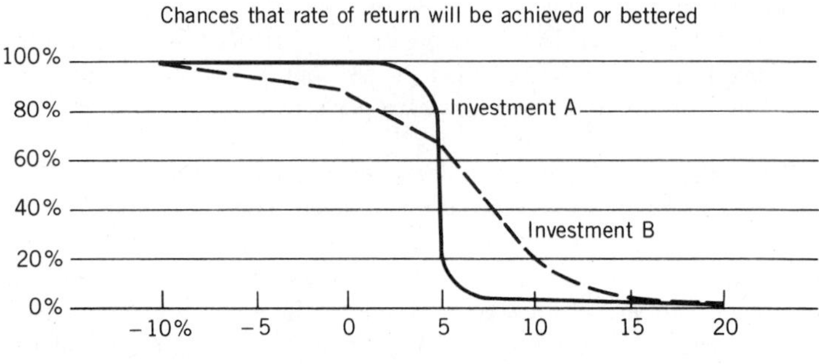


a. Discuss the validity of developing a rate of return distribution and comparing two alternatives on the basis of this criterion.
b. Suggest the correct procedures for comparing the two mutually exclusive alternatives.

**12.18.** Consider Example 12.6. Develop a $Z$ distribution by using the discriminant sampling scheme.

**12.19.** Develop a computer sampling scheme for the normalization method described in Section 12.5.5. Then generate 100 pairs of dependent variates $(x, y)$ with the data in Example 12.7.

**12.20.** Consider the two mutually exclusive alternatives in Example 11.3.
a. Generate 100 samples of the $PV$ for each alternative.
b. Replicate the simulation five times and compute the sample means and sample variances for each replication.
c. Determine the 95% confidence interval for the difference distribution and check to see whether there is a stochastic dominance.

# 13

# Decision Tree Analysis

## 13.1 INTRODUCTION

Another class of investment problems that will be described in this chapter consists of multiple-stage investment problems. This class is characterized by a series of decisions to be made, at various time intervals, with each decision influenced by the information that is available at the time it is made. Decision problems with more than a single stage introduce another source of complexity. It is the sequence of decisions and uncertain events that links the initial decision to the consequences. With this added complexity, direct choice among the initial decisions is very difficult. One popular method for dealing with such a multiple-stage problem is the decision tree method. The decision tree technique facilitates project evaluation by enabling the firm to write down all the possible future decisions, as well as their monetary outcomes, in a systematic manner.

We will take the experience of a retail chain outlet to demonstrate the selection of optimal investment strategies, given substantial market uncertainty. Section 13.2 explains a simple decision analysis model. The situation illustrated is a forecasting event, although the model could be used for any step-by-step selection of alternatives in any uncertain problem environment. Section 13.3 considers the cost of uncertainty and the value of additional information. We also discuss how sensitivity analysis may help to limit the consequences of uncertain measurements. Section 13.4 presents an example of the investment sampling process in which an investment, although initially not desirable, may appear acceptable in view of the consequent opportunity of obtaining additional information.

## 13.2 SEQUENTIAL DECISION PROCESS

A characteristic of the investment problems described up to this point has been that a single decision is made and, as a consequence of this decision, estimated revenues may be earned and estimated costs may be incurred. But in a sequential decision problem, in which the actions taken at one stage depend on actions taken in earlier stages, the evaluation of investment alternatives can become very complicated. In these situations the decision tree method provides a valuable tool for organizing the information needed to make the decision.

### 13.2.1 Structuring the Decision Tree

Before introducing a multiple-stage decision problem, we will work with a single-stage decision tree to illustrate the terminology used in decision tree analysis [4]. Then later we will expand the decision tree to include a multiple-stage decision problem. Perhaps the best way to explain the decision tree is to demonstrate its use by a specific example.

***Retail Convenience Store Problem.*** A large chain of retail convenience stores is considering the expansion of one of its Orlando outlets. Continuation of the current rate of growth will require increasing the hours of operation and paying overtime if substantial business is not to be lost. As an alternative, the company is considering enlarging the store. This expansion may be accomplished in either of two ways.

*Option 1: Expand Large.* Because there is a possibility of substantial growth in the neighborhood, the first plan is to expand the current store by leasing additional floor space and then to remodel the entire store. This would nearly double its floor space. This plan will cost $150,000 initially, consisting of the remodeling expense of $70,000 and the purchase of new equipment worth $80,000. The equipment is expected to have a residual value of $15,000 at the end of 5 years. Leasing additional floor space will cost the store $10,000 annually. This large-scale expansion requires an $18,000 investment in working capital that is recovered at the end of planning horizon.

*Option 2: Expand Small.* A less costly plan is to redo the layout of the store and make a small addition on one side. It is thought that this small expansion will be sufficient for 5 years, the usual planning period for the company, and will cost $50,000. No residual value is expected at the end of this period.

A preliminary analysis indicates the estimated incremental revenues (compared with the no-expansion alternative) affected by each of the three assumed classes of business conditions.

| Business Conditions | Estimated Probability of Occurrence | Incremental Net Revenues under Each Option: Option 1, Expand Large | Option 2, Expand Small |
|---|---|---|---|
| Good | 0.25 | $100,000 | $40,000 |
| Moderate | 0.60 | 75,000 | 30,000 |
| Poor | 0.15 | 35,000 | 10,000 |

Here the net revenue means the incremental revenue caused by expansion less regular operating and maintenance but excluding depreciation and lease expenses. For tax purposes the store will expense all remodeling costs, but the expenditures on the equipment must be capitalized. The installed equipment would be classified in the 5-year personal property class. The chain store's *MARR* is known to be 15% after tax, and the store has a marginal tax rate of 40%, which is expected to remain the same for the investment period.

***Decision Tree for the Orlando Store Expansion Problem.*** To illustrate the basic ideas, we will develop the decision tree for the Orlando store expansion problem. Recall that the decision is between investment options, and the decision point is represented by a decision node (□) in Figure 13.1. The decision alternatives are represented as branches from the decision node. The direction of the arrows refers to the time flow of the decision process.

Suppose that the store manager selects a particular alternative, say option 1. There are three chance events that can happen, each event representing a business condition that can prevail. These are shown in Figure 13.1 as branches emanating from a circle node (○). (In a decision tree analysis, we always use the convention of a square box for a decision node and a circle for an event or chance node.) Notice that these chance nodes represent events over which the store manager has no control. In our example the chance nodes represent the business conditions over the planning horizon. We must assign probabilities to each event in Figure 13.1.

Generally, traversing each branch on the decision tree will bring some reward, positive or negative, to the decision maker. At the end of each branch is the conditional profit associated with the selected action and given event. The conditional profit thus represents the profit associated with the decisions and events along the path from the first part of the tree to the end. For our example the incremental annual revenue of $100,000 in Figure 13.1 is associated with the action that leads to investment in option 1, and the store experiences a good business condition over the planning horizon.

***Relevant Net After-Tax Cash Flow.*** Once the structure of the decision tree is determined, the next step is to find the relevant cash flow (monetary value)

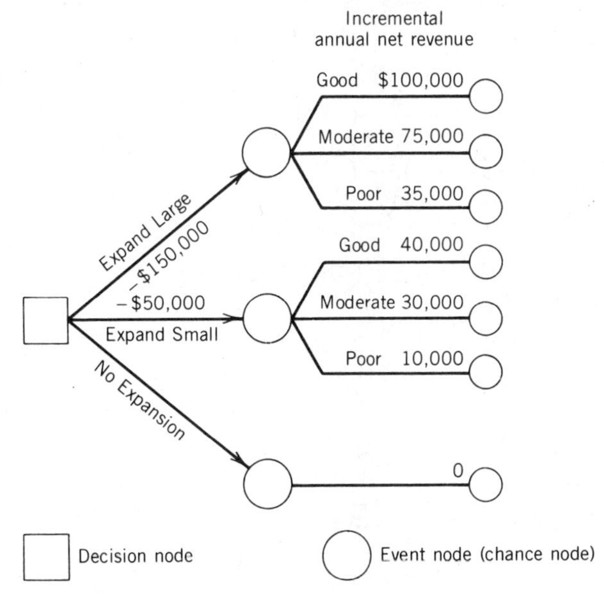


**FIGURE 13.1 The decision tree for the Orlando store expansion problem.**

associated with each of the decision alternatives and the possible chance outcomes. As we have emphasized throughout this book, the decision has to be made on an after-tax basis. Therefore, the relevant monetary value should be on an after-tax basis. Since the costs and revenues occur at different points in time over the study period (5 years), we also need to convert the various amounts on the tree's branches to their equivalent amounts (present value). For the problem being considered, the after-tax *MARR* is 15%, and Figure 13.2 shows the costs and revenues on the branches transformed to their present equivalents.

To illustrate, if the store adopts option 1 and a good business condition prevails over the planning horizon, the net cash flows after taxes are as computed in Table 13.1a. Then the present value of this branch is

$$\begin{aligned} PV(15\%) &= -140{,}000 + 60{,}400(P/F,\ 15\%,\ 1) + 64{,}240(P/F,\ 15\%,\ 2) \\ &\quad + 60{,}144(P/F,\ 15\%,\ 3) + 57{,}686(P/F,\ 15\%,\ 4) \\ &\quad + 86{,}529(P/F,\ 15\%,\ 5) \\ &= \$76{,}645 \end{aligned}$$

and the internal rate of return is 35.42%. This amount of $76,645 is entered at the end of the corresponding branch tip. This procedure is repeated for the remaining two branches associated with option 1, and the resulting amounts are shown in Figure 13.2.

For option 2 the net cash flow calculation is rather simple, as shown in Table 13.1b. Again we repeat the procedure for the remaining two branches originating from "Expand Small," and the computed *PV*s are entered at the tips of the respective branches.

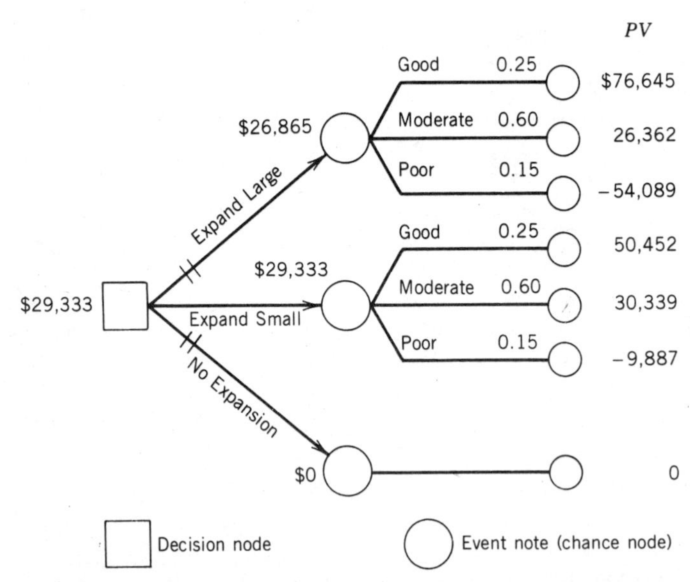


**FIGURE 13.2** Present-value amounts of after-tax cash flows for each outcome and solution of decision tree.

**Table 13.1** ***Net Cash Flows after Taxes for Decision Paths 1 and 4***

| | 0 | 1 | 2 | 3 | 4 | 5 |
|---|---|---|---|---|---|---|
| a. *Decision path 1—expand large under good business conditions* | | | | | | |
| Incremental revenue | | $100,000 | 100,000 | 100,000 | 100,000 | 100,000 |
| Lease expense | | 10,000 | 10,000 | 10,000 | 10,000 | 10,000 |
| Depreciation | | 16,000 | 25,600 | 15,360 | 9,216 | 9,216 |
| Remodeling | $70,000 | | | | | |
| Taxable income | −70,000 | 74,000 | 64,400 | 74,640 | 80,784 | 80,784 |
| Income taxes | −28,000 | 29,600 | 25,760 | 29,856 | 32,314 | 32,314 |
| Net income | −42,000 | 44,400 | 38,640 | 44,784 | 48,470 | 48,470 |
| Add depreciation | | 60,400 | 64,240 | 60,144 | 57,686 | 57,686 |
| Investment | −80,000 | | | | | |
| Salvage | | | | | | 15,000 |
| Gains tax | | | | | | −4,157 |
| Working capital | −18,000 | | | | | 18,000 |
| Net cash flow | −$140,000 | 60,400 | 64,240 | 60,144 | 57,686 | 86,529 |
| b. *Decision path 4—expand small under good business conditions* | | | | | | |
| Incremental revenue | | $40,000 | 40,000 | 40,000 | 40,000 | 40,000 |
| Remodeling | $50,000 | | | | | |
| Taxable income | −50,000 | 40,000 | 40,000 | 40,000 | 40,000 | 40,000 |
| Income taxes | −20,000 | 16,000 | 16,000 | 16,000 | 16,000 | 16,000 |
| Net income | −30,000 | 24,000 | 24,000 | 24,000 | 24,000 | 24,000 |
| Net cash flow | −$30,000 | 24,000 | 24,000 | 24,000 | 24,000 | 24,000 |

### 13.2.2 Expected Value as a Decision Criterion

At this point we can analyze the decision tree to determine which alternative should be undertaken. To analyze a decision tree, we begin at the end of the tree and work backward—the *averaging out and folding back* procedure [6]. In other words, starting at the tips of the decision tree's branches and working toward the initial node, we use the following two rules.

1. For each chance node we calculate the expected monetary value (*EMV*) by multiplying probabilities by conditional profits associated with branches emanating from that chance node and summing these conditional profits. We then place the *EMV* next to the node to indicate that it is the expected value calculated over all branches emanating from this node.
2. At each decision node we select the alternative with the highest *EMV* (or minimum cost). Then we eliminate from further consideration the decision alternatives that are not selected. On the decision tree diagram we draw a mark across the nonoptimal decision branches, indicating that they are not to be followed.

This procedure is illustrated in Figure 13.2 for our store expansion example. First, we calculate the *EMVs* for the event nodes associated with the larger-scale store expansion. For example, the *EMV* of option 1 represents the sum of the product of probabilities for good, moderate, and poor business conditions times the respective conditional profits.

$$EMV = (\$76{,}645)(0.25) + (26{,}362)(0.60) - (54{,}089)(0.15)$$
$$= \$26{,}865$$

For option 2 the *EMV* is simply

$$EMV = (\$50{,}452)(0.25) + (30{,}339)(0.60) - (9{,}887)(0.15)$$
$$= \$29{,}333$$

In Figure 13.2 the expected monetary values are shown next to the event nodes. The store manager must choose which action to take, and this would be the one with the highest *EMV,* namely option 2 (Expand Small) with *EMV* = $29,333. We indicate this expected value in the tree by putting $29,333 next to the decision node at the beginning of the tree. Notice that the decision tree uses the idea of maximizing expected monetary value that was developed in the previous section.

## 13.3 OBTAINING ADDITIONAL INFORMATION

In this section we introduce a general method for evaluating the possibility of obtaining more information. Most of the information we can obtain is imperfect in the sense that it will not tell us exactly which event will occur. Such imperfect information may have value if it improves the chances of making a correct decision, that is, if it improves the expected monetary value. The problem is whether the reduced uncertainty is valuable enough to offset its cost. The gain is in the improved efficiency of decisions that may become possible with better information.

We will use the term *experiment* in a broad sense here. An experiment may represent a market survey to predict sales volume for a typical consumer product, statistical sampling of production quality, or a seismic test to give a well-drilling firm some indications of the presence of oil [4].

### 13.3.1 The Value of Perfect Information

Let us take the prior decision of Expand Small as a starting point. How do we determine whether further strategic steps would be profitable? We could do more to obtain additional information about the future business condition, but such steps cost money. Thus, we have to balance the monetary value of reducing uncertainty with the cost of securing additional information. In general, we can evaluate the worth of a particular experiment only if we can estimate the reliability of the resulting information. In our store expansion problem an expert's opinion may be helpful in deciding whether or not to expand the store. This opinion can be of value, however, only if management can say beforehand how closely the expert can predict the future business condition. An example will make this clear.

***Orlando Retail Store Expansion Problem Revisited.*** Suppose that the store manager knows an expert who can be called in as a consultant on the store expansion described as option 1. The expert will charge a fee to provide the store with a report that the business condition is *good, moderate,* or *poor.* This expert is not infallible but can provide a market survey that is pretty reliable. From past experience management estimates that, when the business condition is relatively good (A), the survey predicts a favorable business condition (F) with probability 0.80, an inconclusive business condition (I) with probability 0.10, and an unfavorable business condition (UF) with probability 0.10. Such probabilities would reflect past experience with surveys of this nature, modified perhaps by the judgment of the store manager. The probabilities shown in Table 13.2 express the reliability or accuracy of a survey of this type.

How much are the consultant's services worth to the manager? Should the store manager hire the expert as a consultant? To answer these questions, we will first introduce the *opportunity loss* concept and illustrate how this loss concept is related to the *value of perfect information.*

***The Opportunity Loss Concept.*** The best place to start a decision improvement process is with a determination of how much we might improve incremental profit by removing uncertainty. Although we probably could not obtain perfect information, its value is worth computing as an upper bound to the value of additional information.

We can easily calculate the value of perfect information. Merely note the difference between the incremental profit from an optimal decision based on perfect information and the incremental profit from the original decision of Expand Small, made without foreknowledge of the actual business condition. We call this difference *opportunity loss,* and we must compute it for each potential degree of business condition.

For our store expansion example, the decision may hinge on the future

**Table 13.2** ***Conditional Probabilities of Survey Prediction***

| | For a Given Business Condition | | |
|---|---|---|---|
| Survey Outcome | Good (A) | Moderate (B) | Poor (C) |
| Favorable (F) | 0.80 | 0.30 | 0.10 |
| Inconclusive (I) | 0.10 | 0.40 | 0.20 |
| Unfavorable (UF) | 0.10 | 0.30 | 0.70 |
| Sum | 1.00 | 1.00 | 1.00 |

business condition. The only unknown, subject to a probability distribution, is the business condition. Recall that the business condition was assumed to be good, moderate, or poor. The opportunity loss table for this prior decision is shown in Table 13.3. For example, the conditional net present value of $76,645 is the net profit associated with option 1 should the potential business condition be good. Recall also that, without receiving any information, the indicated action (prior optimal decision) was to select option 2 (Expand Small). Under a good business condition this option will yield a net present value of $50,452. Therefore, for a good business condition with perfect information, the prior decision to Expand Small is inferior to the decision to Expand Large to the extent of $76,645 − $50,452 = $26,193. When decision strategies with and without perfect information are the same (in this case for Expand Small with a moderate business condition), the value of opportunity loss must be zero. If a poor business condition prevails, the No Expansion alternative becomes the best strategy. The opportunity loss under this situation would be $0 − (−$9,887) = $9,887.

Being reluctant to give up a chance to make $76,645, however, the store manager may wonder whether to obtain further information before action. As with the prior decision, we need a single figure to represent the expected value of perfect information (*EVPI*). Again, an average weighted by the assigned chances is used, but this time weights are applied to the set of opportunity losses (regrets) in Table 13.3.

| Business Condition | Opportunity Loss | Probability |
|---|---|---|
| Good | $26,193 | 0.25 |
| Moderate | 0 | 0.60 |
| Poor | 9,887 | 0.15 |

$$EVPI = (0.25)(26{,}193) + (0.60)(0) + (0.15)(9{,}887) = \$8{,}031$$

This figure represents the maximum expected amount that could be gained in incremental profit from perfect knowledge. This *EVPI* places an upper limit on the sum the manager would be willing to pay for additional information.

**Table 13.3** ***Conditional Value Table for Store Expansion Decison Problem***

| Business Condition | Probability | Optimal Choice with Perfect Information: Decision | Optimal Choice with Perfect Information: Outcome | Prior Decision (Expand Small) | Opportunity Loss |
|---|---|---|---|---|---|
| Good | 0.25 | Expand Large | $76,645 | $50,452 | $26,193 |
| Moderate | 0.60 | Expand Small | 30,339 | 30,339 | 0 |
| Poor | 0.15 | No Expansion | 0 | −9,887 | 9,887 |
| Expected value | | *EPPI** = | $37,365 | $29,333 (Prior optimal) | $8,031 (Expected opportunity losses) |

*EPPI = Expected profit with perfect information.

***Updating Conditional Profit.*** Thus, there is (potentially at least) some value to be obtained from additional information. The store manager can perform an experiment in this situation. Recall that the experiment takes the form of receiving a market survey from a consultant.

There are three possible outcomes from the survey: (1) the survey may predict a good business condition, (2) the survey may be inconclusive, or (3) the survey may predict a poor business condition. The manager's alternatives are to employ this service or simply to expand the store on a small scale, as previously decided.

If the store manager takes a survey before acting, the decision can be based on the survey outcomes. We can express this problem in terms of a decision tree, as shown in Figure 13.3. The upper part of the tree shows the decision process if no survey is taken. This is the same as Figure 13.2, with probabilities of 0.25, 0.60, and 0.15 for good, moderate, and poor conditions and an expected profit of $26,865 for Expand Large that is less than the expected profit from Expand Small. Thus the optimal decision had an expected value of $29,333 based on the Expand Small strategy. (Note that all of Figure 13.2 has been repeated as the top part of Figure 13.3 for completeness, but only the earlier "best decision" is actually needed at this point.)

The lower part of the tree, following the branch Take Survey, displays the results and the subsequent decision possibilities in Figure 13.3. Using a forecasting service allows for two stages of events—the survey prediction and the actual level of business condition. After each of the three possible survey outcomes, a decision about whether or not to expand the store must be made.

### 13.3.2 Determining Revised Probabilities[1]

In Figure 13.3 market survey outcomes precede actual business conditions because the former are obtained as additional information before actual business conditions are known. To complete the analysis of Figure 13.3, we need the revised probabilities for the various events after additional information has been obtained. Recall that the store manager estimates the probabilities shown in Table 13.2. These are the conditional probabilities for the various survey outcomes, given the potential business condition. For example, when the actual business is good, the survey predicts a good business condition with a probability of 0.80, the survey results are inconclusive with a probability of 0.10, and the survey predicts a poor business condition with a probability of 0.10.

In fact, the probabilities in Table 13.2 express the reliability or accuracy of the experiment. It is a question of counting the number of times the survey was "on target" or had a "complete miss" for each of the three actual levels of business condition. We may convert the counts into proportions, entered as decimal fractions (probabilities) in Figure 13.4. With these estimates the store manager can evaluate the economic worth of the service. Without these reliability estimates, no specific value can be attached to taking the survey.

[1]This discussion of revised probabilities has been considerably influenced by an excellent presentation of decisions and revision of probabilities by Bierman, Bonini, and Hausman [1].

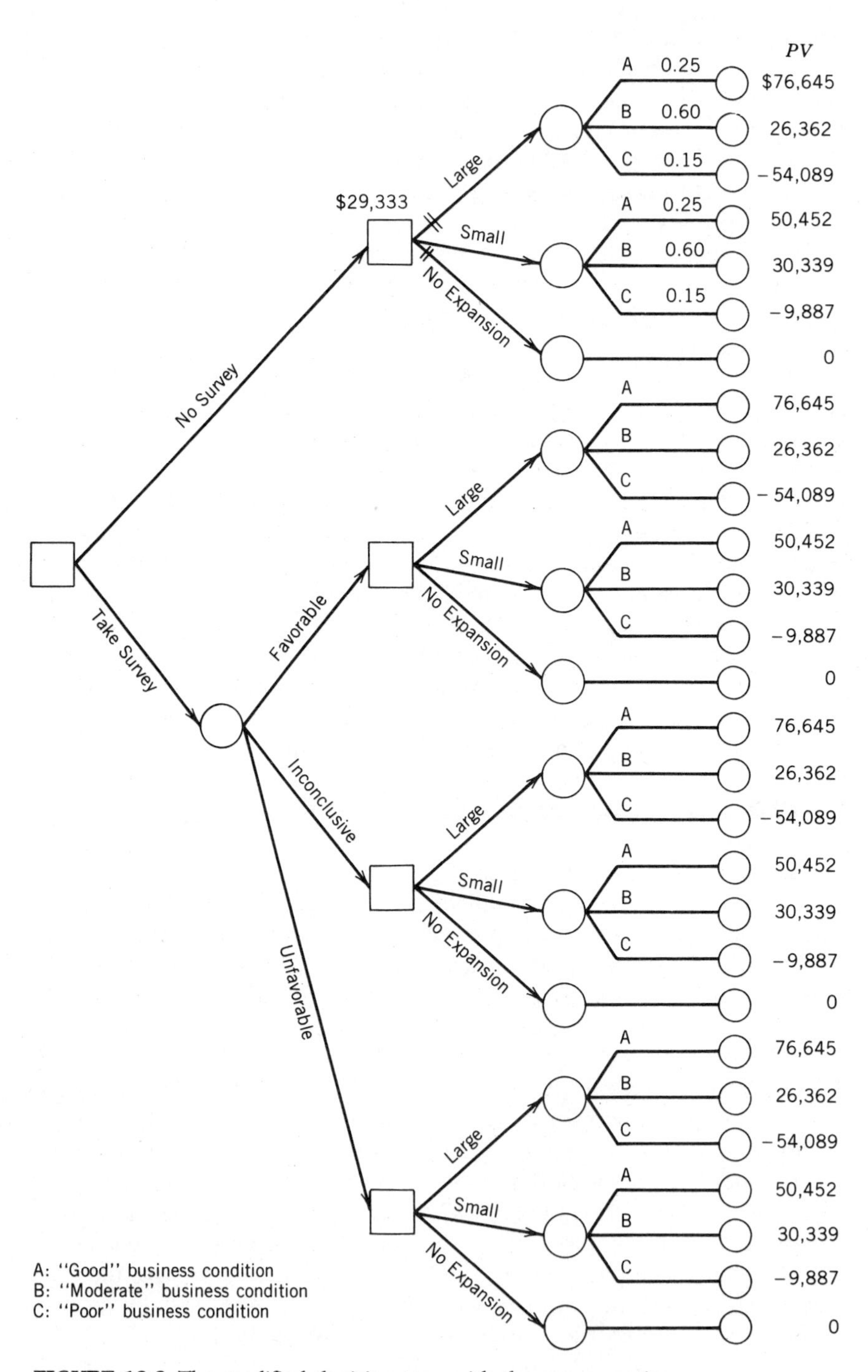


**FIGURE 13.3** The modified decision tree with the survey option.

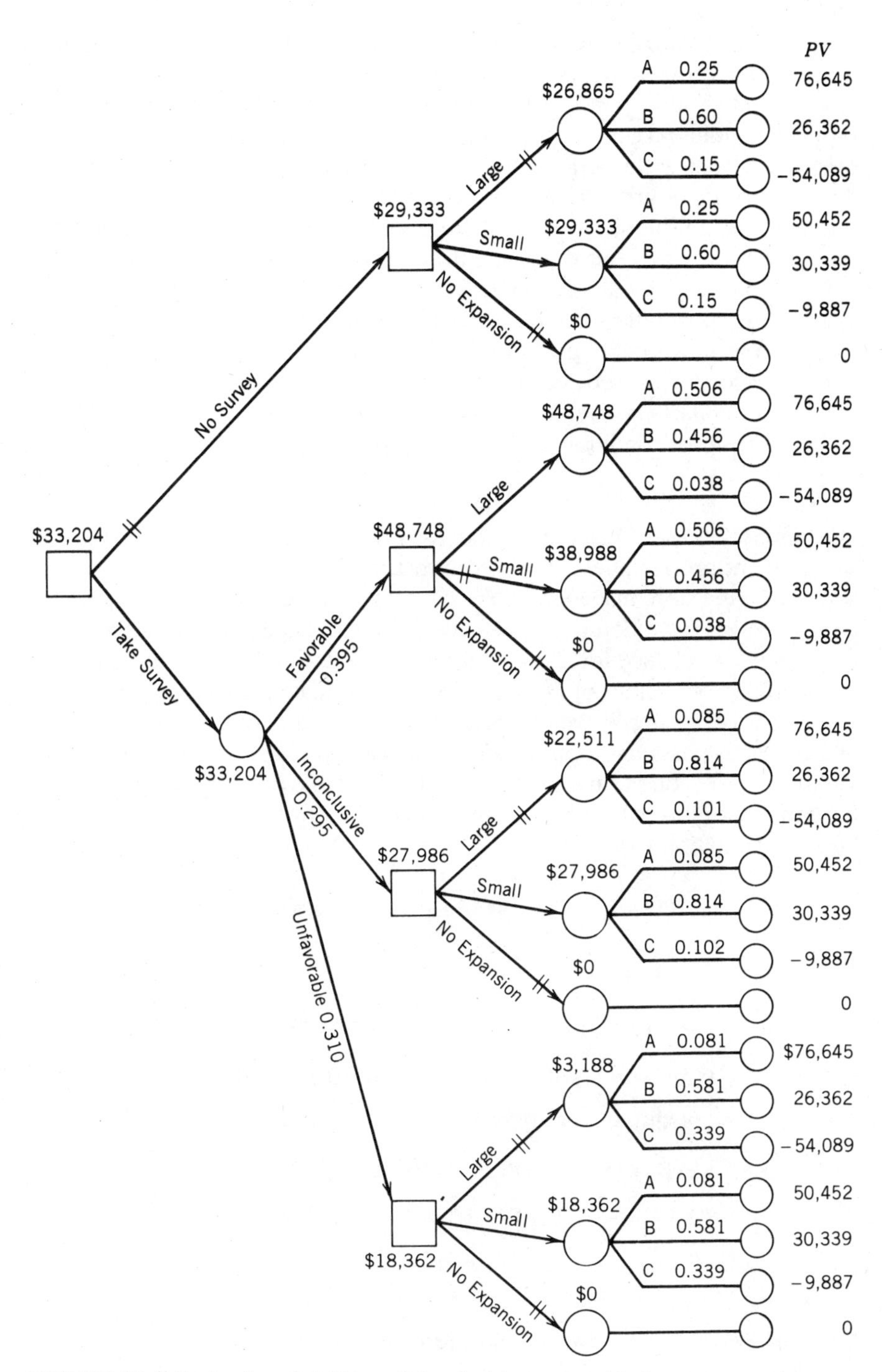


**FIGURE 13.4** Revised probabilities of the decision tree with the survey option.

Now the manager has information on (1) the chances for particular business condition outcomes before getting additional information (prior probabilities) and (2) the chances of obtaining three different forecast levels for the business condition (good, moderate, or poor), given three actual business conditions. A glance at the sequence of events in Figure 13.4, however, shows that these probabilities are not the ones required to find "expected" values of various strategies in the decision path. What are really needed are the chances of three levels of survey outcomes and the total probabilities of each of the three potential business conditions, all taking into account earlier estimates. In other words, the probabilities of Table 13.2 are not directly useful in Figure 13.4. All we have available are the probabilities of three levels of business market condition. Therefore, we need the unconditional probabilities of the three survey outcomes. Similarly, we need the conditional probabilities of a good, moderate, and poor business condition, given a prediction of good business market, and so forth.

***Joint Probabilities.*** To generate the conditional probabilities, we must construct a joint probabilities table. Recall that we have the original probabilities assessed by the store manager: a 0.25 chance that the retail store will experience a good market condition (A), a 0.60 chance for a moderate business condition (B), and a 0.15 chance for a poor business condition (C). From these and the conditional probabilities of Table 13.2, we can calculate the joint probabilities of Table 13.4. For example, we calculate the joint probability of both a prediction of a good business condition (F) and an actual good business condition (A) by multiplying the conditional probability of a favorable prediction, given a good business condition (which is 0.80 from Table 13.2), by the probability of a good business condition (A):

$$P(\mathrm{A,F}) = P(\mathrm{F} \mid \mathrm{A})P(\mathrm{A}) = (0.80)(0.25) = 0.200$$

Similarly, we calculate

$$P(\mathrm{A,I}) = P(\mathrm{I} \mid \mathrm{A})P(\mathrm{A}) = (0.10)(0.25) = 0.025$$

$$P(\mathrm{A,UF}) = P(\mathrm{UF} \mid \mathrm{A})P(\mathrm{A}) = (0.10)(0.25) = 0.025$$

$$P(\mathrm{B,F}) = P(\mathrm{F} \mid \mathrm{B})P(\mathrm{B}) = (0.30)(0.60) = 0.180$$

$$P(\mathrm{B,I}) = P(\mathrm{I} \mid \mathrm{B})P(\mathrm{B}) = (0.40)(0.60) = 0.240$$

$$P(\mathrm{B,UF}) = P(\mathrm{UF} \mid \mathrm{B})P(\mathrm{B}) = (0.30)(0.60) = 0.180$$

and so on.

***Marginal Probabilities.*** We obtain the marginal probabilities of future business conditions in Table 13.4 by summing the values across the columns. Notice that these are in fact the original probabilities for good, moderate, and poor business conditions, and they are called prior probabilities because they were assessed before any information from the survey was obtained. In interpreting

**Table 13.4** ***Probabilities for Joint Events and Revised Probabilities (Store Expansion Problem)***

| Potential Business Condition | Survey Prediction | | | Marginal Probabilities of Business Condition |
|---|---|---|---|---|
| | Favorable (F) | Inconclusive (I) | Unfavorable (UF) | |
| Good (A) | 0.200 | 0.025 | 0.025 | 0.25 |
| Moderate (B) | 0.180 | 0.240 | 0.180 | 0.60 |
| Poor (C) | 0.015 | 0.030 | 0.105 | 0.15 |
| Marginal probabilities of survey prediction | 0.395 | 0.295 | 0.310 | 1.00 |

| Given Survey Outcome | Actual Business Condition | | | Probability Sum |
|---|---|---|---|---|
| | Good (A) | Moderate (B) | Poor (C) | |
| Favorable (F) | 0.506 | 0.456 | 0.038 | 1.00 |
| Inconclusive (I) | 0.085 | 0.814 | 0.101 | 1.00 |
| Unfavorable (UF) | 0.081 | 0.581 | 0.339 | 1.00 |

the marginal probabilities, we will borrow the analogy by Bierman, Bonini, and Hausman [1].

It is useful to think of Table 13.4 as representing the results of 1,000 past situations identical to the one under consideration. The probabilities then represent the frequency with which the various outcomes occurred. For example, in 250 of the 1,000 cases, the actual business condition turned out to be good; and in these 250 good cases, the survey predicted a favorable condition in 200 instances [that is, $P(A, F) = 0.20)$], a moderate one in 25 instances, and an unfavorable one in 25 instances, respectively.

We can then interpret the marginal probabilities of survey predictions in Table 13.4 as the relative frequencies with which the survey predicted favorable, moderate, and unfavorable conditions. For example, the survey predicted favorable 395 out of 1,000 times—200 of these times when business conditions actually were good, 180 times when business conditions were moderate, and 15 times when business conditions were poor.

These marginal probabilities of survey prediction are critical in our analysis, because they provide us the probabilities associated with the information received by the store manager before the decision to invest in the store expansion is made. The probabilities are entered beside the appropriate branches in Figure 13.4.

***Posterior Probabilities.*** What we need now, after receiving the survey information, is to calculate the probabilities for the branches labeled A (good), B (moderate), and C (poor). Clearly, we cannot use the values of 0.25, 0.60, and 0.15 for these events, because these probabilities were calculated *prior* to taking the survey. The required probabilities are the conditional probabilities for the various levels of business condition given the survey result—for our example

$P(\text{A} \mid \text{F})$, the probability of a good business condition (A) given that the survey predicts a favorable market condition (F). We can easily compute this from the definition of conditional probability, using the data from Table 13.4.

$$P(\text{A} \mid \text{F}) = P(\text{A,F})/P(\text{F}) = 0.200/0.395 = 0.506$$

The probabilities of moderate and poor conditions, given a prediction of favorable condition, are

$$P(\text{B} \mid \text{F}) = P(\text{B,F})/P(\text{F}) = 0.180/0.395 = 0.456$$

$$P(\text{C} \mid \text{F}) = P(\text{C,F})/P(\text{F}) = 0.015/0.395 = 0.038$$

We call these probabilities *posterior* probabilities because they come after receiving the information from the survey. To understand the meaning of the foregoing calculations, think again of Table 13.4 as representing 1,000 past identical situations. Then, in 395 cases [since $P(\text{F}) = 0.395$], 200 actually had a good business condition. Hence, the posterior probability for good business condition is, as calculated, $200/395 = 0.506$.

The posterior probabilities after survey predictions of other situations can be calculated similarly.

$$P(\text{A} \mid \text{I}) = P(\text{A,I})/P(\text{I}) = 0.025/0.295 = 0.085$$

$$P(\text{B} \mid \text{I}) = P(\text{B,I})/P(\text{I}) = 0.240/0.295 = 0.814$$

$$P(\text{C} \mid \text{I}) = P(\text{C,I})/P(\text{I}) = 0.030/0.295 = 0.101$$

$$P(\text{A} \mid \text{UF}) = P(\text{A,UF})/P(\text{UF}) = 0.025/0.310 = 0.081$$

$$P(\text{B} \mid \text{UF}) = P(\text{B,UF})/P(\text{UF}) = 0.180/0.310 = 0.581$$

$$P(\text{C} \mid \text{UF}) = P(\text{C,UF})/P(\text{UF}) = 0.105/0.310 = 0.339$$

These values are also shown in Figure 13.4 at the appropriate points in the decision tree.

### 13.3.3 Expected Monetary Value after Receiving Sample Information

As we have all the necessary information, we can analyze Figure 13.4, starting from the right and working backward. The expected values are shown next to the circles. For example, follow the branches Take Survey, Favorable Prediction, and Expand Large. The expected value of \$48,748 shown next to the circle at the end of these branches is calculated as

$$(0.506)(\$76{,}645) + (0.456)(26{,}362) + (0.038)(-54{,}089) = \$48{,}748$$

Thus, the manager can expect a profit of \$48,748 (before survey cost) if the store is doubled in size after a prediction of favorable business market conditions is received. Since this is better than the $38,988 associated with the smaller-scale expansion, the decision to expand the store on a larger scale is made, and the Expand Small and No Expansion branches are marked to indicate that they are not optimal.

There will be an expected loss of $27,986 − $22,511 = $5,475 if the survey gives a moderate market prediction, but the manager nevertheless goes ahead with the larger store expansion. Therefore, the Expand Small option becomes a better option, and the Expand Large branch is marked out. We also observe a situation similar to this when the survey indicates an unfavorable business condition.

We now reduce the part of the decision tree related to taking the survey to three chance events ($48,748, $27,986, and $18,362). We also calculate the expected value next to the circle node as

$$(0.395)(\$48{,}748) + (0.295)(\$27{,}986) + (0.310)(\$18{,}362) = \$33{,}204$$

Thus, if the survey is taken at *no cost* and the manager acts on the basis of the information received, the expected profit is $33,204.

Suppose the market survey is available at a cost of $1,000. Since this survey cost is deductible from income as a business expense, the net after-tax survey cost is ($1,000)(1 − 0.40) = $600. Subtracting this $600 from $33,204 yields $32,604, which is still greater than the $29,333 profit that would be obtained without taking the survey, so we conclude that it is worth spending $1,000 to receive some additional information from the market survey.

### 13.3.4 Value of the Market Survey

Taking the survey in the foregoing example is a means of obtaining additional information. The information is not perfect, because the survey cannot tell exactly whether the business condition will be good, moderate, or poor. Recall that the expected profit with the survey was $33,204. In fact, this is the expected profit with *free* information. The difference between the profit with information and without it is $33,204 − $29,333 = $3,871. In this situation, the value of the sample information is simply $3,871. This implies that the market survey would be worth taking as long as its cost did not exceed this amount. (In our example, taking the survey costs $600 after taxes, which is less than $3,871, and we conclude that it is worth taking.)

Compared with the *EVPI* ($8,032), the value of the survey is substantially lower, reflecting the fact that the survey can give inconclusive or incorrect information as indicated in Table 13.4. In other words, taking a sample is a means of obtaining information, but it is imperfect, since the sample is not likely to represent exactly the population from which it is taken [1].

## 13.4 DECISION TREE AND RISK

To simplify the decision analysis in the store expansion example, we made the rather unrealistic assumption that the manager seeks to maximize the expected

*PV*, by ignoring the risk incurred in each of the possible courses of action. Conceptually, risk can be incorporated in the analysis simply by assigning a utility to each monetary outcome and then choosing the branch that maximizes the expected utility. Such a procedure provides an acceptable theoretical solution to the problem but, as we have already noted, is rather difficult to implement in practice. Alternatively, answers to "what if" questions can be obtained with relative ease by using sensitivity analysis. With this general approach we could evaluate many possible "what if" situations merely by repetitively solving the tree with different values of parameters. In this section we will illustrate both approaches, sensitivity analysis and utility theory.

### 13.4.1 Sensitivity Analysis

Sensitivity analysis deals with the consequences of incremental change. How much could the manager's subjective assessment of chances be altered before the optimal decision would shift? In our store expansion example, just how far could chances shift before Expand Small would replace Expand Large as the optimal value? Formally, this amounts to establishing the values of chances at the break-even condition. Here the expected values of the two strategies are equal because this is the crossover point of indifference between strategies.

### *Example 13.1*

Suppose the probability of a moderate business condition is held around a neutral level at 0.60 and the chances of good and poor business conditions are calculated to equate the two strategies Expand Large and Expand Small. If $p$ symbolizes the chance of good business condition, then $(1 - 0.60 - p)$ or $(0.40 - p)$ is the chance of poor demand. This follows because the chances for all possible levels of business condition must sum to 1.

From the values of strategies in Figure 13.2, we can determine the break-even probability of

$$\text{Expected value of Expand Large} = \text{expected value of Expand Small}$$

or

$$p(76{,}645) + (0.60)(26{,}362) + (0.4 - p)(-54{,}089)$$
$$= p(50{,}452) + (0.60)(30{,}339) + (0.4 - p)(-9{,}887)$$

or

$$70{,}395p = 20{,}067$$

Solving this equation for $p$ yields 0.285.

At this chance level the manager would be indifferent to choosing between Expand Large and Expand Small because the expected values of both strategies are equal. Even so, the manager might prefer one strategy over the other on the basis of other information not included in this quantitative analysis. □

### 13.4.2 Decision Based on Certainty Equivalents

Recall from the discussion of expected utility in Chapter 9 that expected monetary value is not an appropriate decision criterion when the conditional profits or losses are so large that the decision maker views the alternatives as having different amounts of risk. In such situations utility values are used in the decision tree in place of conditional monetary profits. Recall also that the preference curve takes into account an individual's attitude toward risk. The probabilities take into account the individual's beliefs about the uncertain events. The mechanics of the calculation are as follows.

1. For each branch outcome, convert the evaluation units (usually in dollars) to preference scale values (utile values).
2. Calculate the expected values of the preference numbers for each branch.
3. At each decision node, compare the expected values of the preference numbers and select the decision alternative with the largest expected utile value.
4. When the optimal decision path for the decision tree is reached, use the preference curve to obtain the *certainty equivalent amount* corresponding to the expected utile value of the optimal decision path.

#### *Example 13.2*

This example illustrates how the procedure we have discussed is used in a sequential analysis. In applying this procedure, reconsider the store expansion problem shown in Figure 13.4. Suppose that the manager's preference curve as a function of monetary outcome ($X$) can be described by

$$U(X) = 1 - e^{-X/30{,}000}$$

For example, the monetary outcome of $76,645 associated with the first branch has an equivalent preference (utility) value of

$$\begin{aligned} U(76{,}645) &= 1 - e^{-76{,}645/30{,}000} \\ &= 0.922 \end{aligned}$$

The remaining *PV*s of all branch monetary outcomes in Figure 13.4 are replaced by the preference values, and these are summarized in Figure 13.5. Then the expected utility values are calculated for event nodes. For example, the expected utility value of −0.179 at the topmost right circle in Figure 13.5 represents the sum of the products of probabilities for good (A), moderate (B), and poor (C) business conditions times the respective conditional utility values.

$$E[U(X)] = (0.25)(0.922) + (0.60)(0.585) + (0.15)(-5.068) = -0.179$$

The other expected utility values are computed similarly.

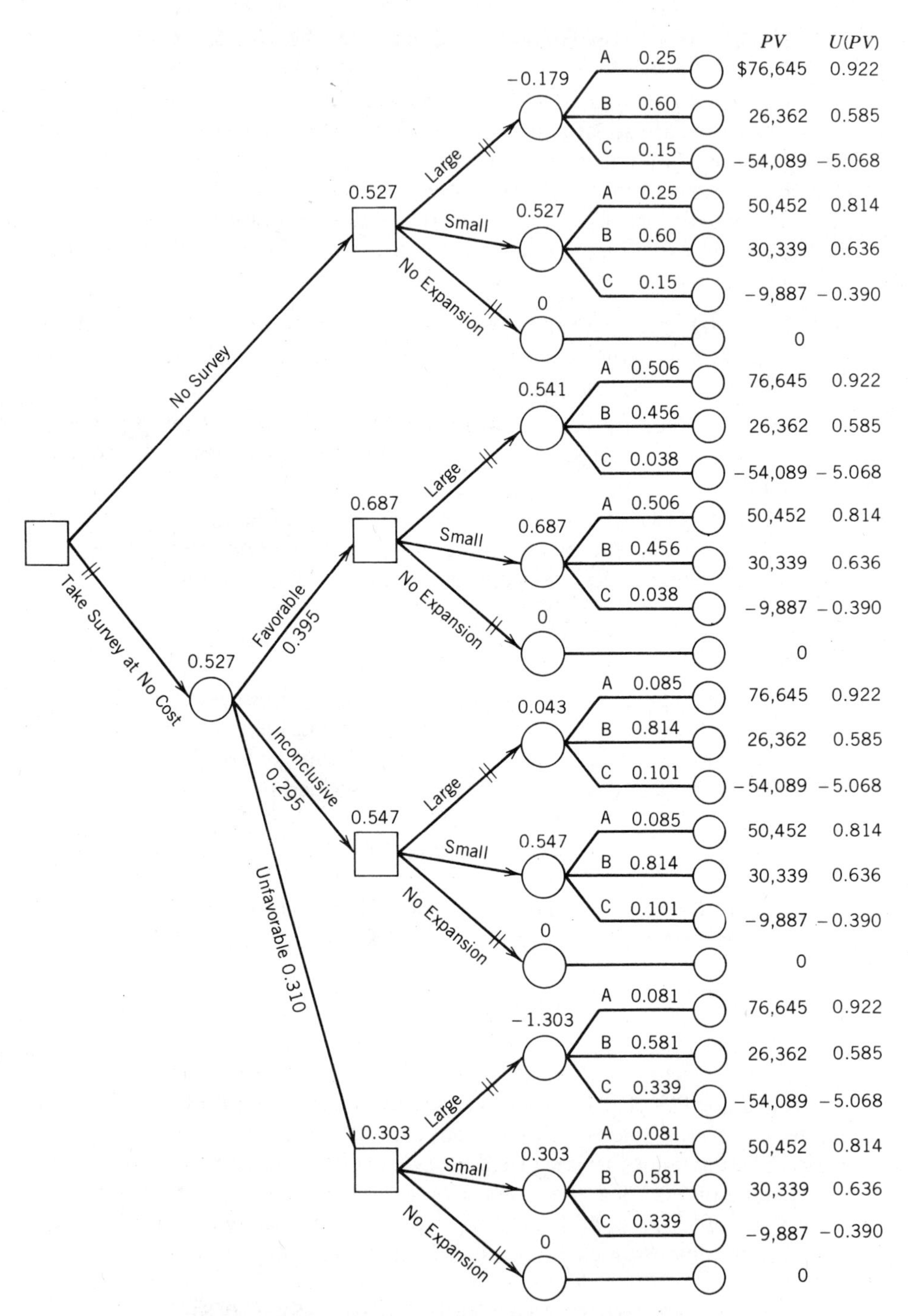


**FIGURE 13.5** Sequential decision based on preference values (utilities). The $U(X) = 1 - e^{-X/30,000}$ calculation is based on a free sample. If the sampling cost is not zero, the cost must be subtracted before computing U($X$).

Notice that the expected utility value for both options (Take Survey at No Cost versus No Survey) are the same (0.527). This is purely coincidental. Based on the expected-utility principle, the branch Take Survey at Some cost is no longer the optimal decision path. It appears that the Expand Small option becomes the ultimate choice, even if the survey is taken and the manager acts on the basis of the information received. To calculate the actual expected utility value for the Take Survey at Some Cost option, *we must subtract the sampling cost from each branch outcome before calculating the equivalent utile value for the branch.* □

## 13.5 INVESTMENT DECISIONS WITH REPLICATION OPPORTUNITIES

Consider situations in which a single type of equipment can be installed in many different locations, and it is possible to obtain additional information by trying the equipment in one location before deciding what to do about the other locations. This class of decision problem is known as investment decisions with replications [1,2]. In this section we will emphasize that an apparently good individual investment might be delayed when considered in the broader context of subsequent investments. For the same reason, an apparently poor individual investment might be good after obtaining additional information and having the opportunity to make sequential decisions.

### 13.5.1 The Opportunity to Replicate

We may observe the situations in which multiple-plant firms have an opportunity to innovate sequentially or a single plant has multiple production lines. Consider the introduction of new manufacturing technology (such as a flexible manufacturing system) in a multiple-plant company. The analysis for a single unit of the manufacturing system indicates a negative *PV*. But there is some probability that the manufacturing system would be successful and would have a positive *PV* in any subsequent use. In other words, there is uncertainty about the outcome, but there is some probability that the investment would be desirable. In such a situation the possibility that the firm may miss out on a technological breakthrough may be sufficient motivation for trying the new technology as a sample investment [3].

If one unit of a manufacturing system has a positive expected *PV*, some may argue that all the units should be installed for the entire firm. On an expected-value basis this is true. Under conditions of uncertainty and risk aversion, however, trying the investment on a small scale may help to determine whether the forecasted good result will actually occur. If the result is good, the remaining units can be installed. The cost of this policy is delay of the investments, however, which may be a disadvantage.

### 13.5.2 Experiment Leading to Perfect Information[2]

We will assume initially that undertaking one investment would allow perfect information about what could happen if all the identical investments

[2]This section is based on an article by Bierman and Rao [2].

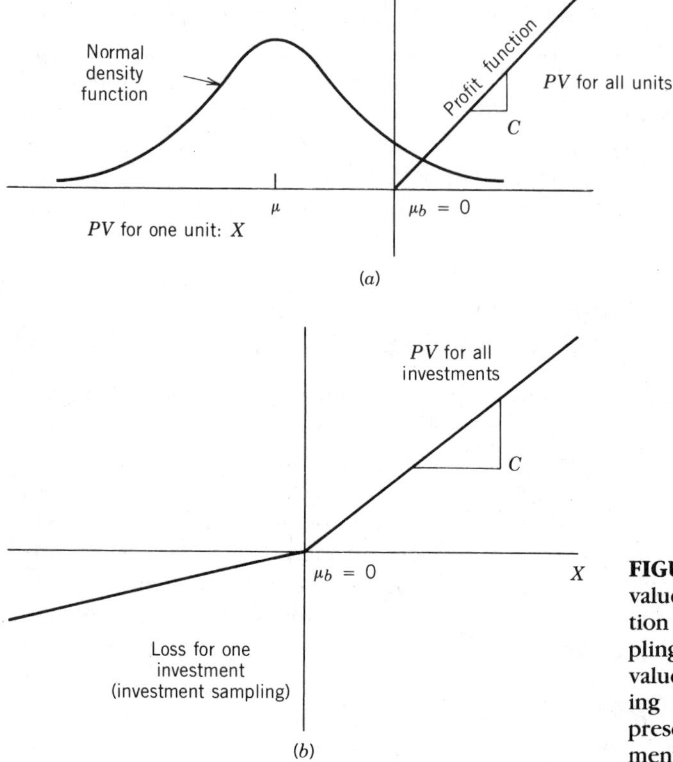


**FIGURE 13.6** Expected value of perfect information and investment sampling. (a) Net present value lost by not undertaking investment; (b) Net present value if investment is undertaken.

were undertaken. Figure 13.6*a* shows the basic model with the *PV* of the profits lost by not undertaking the investment. Here $X$ is the continuous random variable *PV* with mean $\mu$ for one unit of investment, and $\mu_b$ is the break-even present value. Under the expected value criterion, $\mu_b$ is the value when $X$ is equal to zero.

Since $\mu$ is to the left of $\mu_b$, the correct decision seems to be to reject the investment. If the investment is rejected, the present value is zero. But $X$ is a random variable with a probability density function. This is a "prior distribution." If we are certain that the *PV* of the investment is $X = \mu$ (the variance of the distribution is zero), the investment would be rejected. If there is some probability that $X > \mu_b$, then further analysis is required.

The slope $C$ of the *PV* curve for all units of investment depends on the number of units in which the firm can feasibly invest. The more units of investment, the steeper the slope. Thus, the profit potential for multiple investments, given an undesirable single investment, is a function of the slope of the *PV* line, the variance of the probability density function, and the distance between $\mu$ and $\mu_b$.

If $\mu$ is less than 0 and if the value of $X$ is positive, the expected profits are

$$C\int_0^{\infty} xf(x)\,dx$$

where $f(x)$ is the probability density function of $X$. The expected loss from undertaking one unit of investment if $X$ is less than $\mu_b$ is

$$\int_{-\infty}^{0} (-x)f(x)\,dx$$

If the expected present value is positive, the investment would be tried. If we have a discrete random variable, the integral would be replaced by the summation but the basic principle remains unchanged.

Figure 13.6*b* shows the *PV* if we undertake one investment and then we undertake the remaining investments should the actual value of $X$ be greater than $\mu_b$. If the actual value of $X$ is less than $\mu_b$, no additional investments are undertaken.

### *Example 13.3*[3]

In converting a job shop operation to a flexible cellular manufacturing operation, we often face a decision whether to convert the whole factory or convert gradually. Normally, it costs less to convert the whole factory than do the conversion partially over periods. The gradual conversion, however, is a less risky investment because we can dictate the level of automation as we see fit. Moreover, this gradual automation can serve as an investment sampling process; the investment (automating the whole factory), although initially desirable, may appear unacceptable after obtaining additional information and being able to make sequential decisions. To structure this decision problem, consider the situation in which we can convert the whole factory with six identical manufacturing cells. We may convert just one cell and see how economical it is, then convert another cell, and so forth.

To make the problem simple, we assume that the level of performance of the converted cell can be described by a discrete set of three outcomes.

Probability (excellent performance) $= P(\theta_1) = 0.5$
Probability (fair performance) $= P(\theta_2) = 0.3$
Probability (poor performance) $= P(\theta_3) = 0.2$

The cash flow information related to each performance level is summarized as follows.

Cell conversion costs (from job shop to cellular manufacturing): $1,000 per cell

[3]This problem was suggested by George Prueitt.

Annual returns after tax over 10 years,

| | |
|---|---|
| if $\theta_1$ | \$250 |
| if $\theta_2$ | 170 |
| if $\theta_3$ | 50 |

The expected *PV* at *MARR* of 10% over a planning horizon of 10 years, is

$$
\begin{aligned}
E(X) &= 0.5[-1{,}000 + 250(P/A,\ 10\%,\ 10)] \\
&\quad + 0.3[-1{,}000 + 170(P/A,\ 10\%,\ 10)] \\
&\quad + 0.2[-1{,}000 + 50(P/A,\ 10\%,\ 10)] \\
&= 0.5(536) + 0.3(45) + 0.2(-693) \\
&= \$143 > 0
\end{aligned}
$$

$$
\begin{aligned}
Var(X) &= 0.5(536 - 143)^2 + 0.3(45 - 143)^2 + 0.2(-693 - 143)^2 \\
&= (469)^2
\end{aligned}
$$

Since the expected *PV* is positive, we would accept on an expected present value basis. However, there is 0.5 probability that each unit cell will perform to produce a *PV* of \$536 or \$3,216 in total. There is 0.3 probability that each unit cell will produce \$45 per unit or \$270 in total, and there is 0.2 probability that each unit will lose \$693 or \$4,158 in total. The expected value is still positive (\$858). The variance of the total *PV* is then

$$Var\,(6X) = 36Var\,(X) = (2{,}814)^2$$

indicating a significant risk in the problem.

At this point we may compute the value of perfect information. For each possible event ($\theta_j$), the optimal act would be as follows.

| Event | Optimal Act | Based on Prior Belief | Opportunity Loss |
|---|---|---|---|
| $\theta_1$ | Convert | Convert | 0 |
| $\theta_2$ | Convert | Convert | 0 |
| $\theta_3$ | Do not convert | Convert | \$4,158 |

The expected opportunity loss would be

$$
\begin{aligned}
E(\text{loss}) &= 0.50(0) + 0.3(0) + 0.2(\$4{,}158) \\
&= \$832
\end{aligned}
$$

Therefore, the *EVPI* is \$832.

We can eliminate the uncertainty by converting just one cell for a maximum cost of \$693. We will assume that, at the end of year 1, converting one cell would allow perfect information about what could happen if all the identical cells were converted. The firm will convert the five additional cells only if event $\theta_1$ or event $\theta_2$ occurs and the process proves to be feasible. If the process is feasible, each investment adds \$536 or \$45 of net present value, depending on the performance level. The sampling procedure (trying one investment before proceeding with the remainder) will delay the other investments and adversely affect their present value if they have positive present values. Multiplying the \$536 by the five cells, we find \$2,680. Since these five cells are converted a year later, the *PV* is 2,680(*P/F*, 10%, 1) = \$2,436. Therefore, the total *PV* associated with the five-cell conversion after successful performance of the first cell is \$536 + \$2,436 = \$2,972. Similarly, if we observe $\theta_2$, we go ahead and convert the remaining five cells. This will result in \$45 + \$225(*P/F*, 10%, 1) = \$250. Therefore, the expected value corresponding to the two favorable outcomes is (0.50)(\$2,972) + (0.30)(\$250) = \$1,561.

If event $\theta_3$ is observed, the firm will make no further conversions, thereby limiting the loss to the one cell. The expected cost (in present value) of obtaining this information is equal to (0.2)(\$693) = \$254, the net expected opportunity loss associated with conversion of one cell, that will occur if the event is $\theta_3$. (If the firm returned to the job shop operation after observing the first year's operation, the net cost would depend on the disposal value of the installed cell. In this example we assume that there would be no market for this used and unsuccessful manufacturing equipment, so the abandonment cost of the already installed cell would be greater than that of the retaining option.) Figure 13.7 shows the decision tree that evolves.

Because the expected value of \$1,561 exceeds the expected cost of \$254, we advocate undertaking the single cell conversion in the hope that we will find out that either event $\theta_1$ or event $\theta_2$ is the true state of the world.

If the alternative were to make all the investments now or do nothing, the firm would convert all the cells. If the delay of other conversions is allowed, the firm certainly wants to obtain information as cheaply as possible, which is accomplished by undertaking a minimum-sized investment. The fact that the other investments will be delayed is unfortunate, but it does not affect the basic sampling strategy. Delaying the investments decreases their *PV*, but it also enables the firm to avoid investing funds in undesirable investments. On balance, it is a desirable strategy if the investment would otherwise be rejected. □

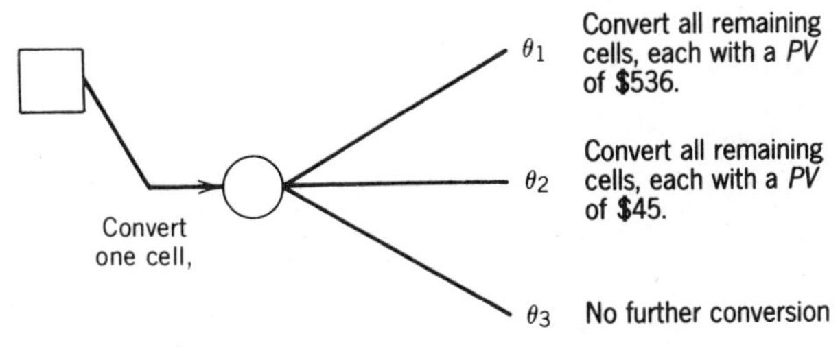


**FIGURE 13.7** Decision tree with perfect information.

Example 13.3 illustrates a situation in which the investment appeared to be acceptable on the expected-value criterion, but it was worth buying perfect information before implementing the investment. We can apply the same investment sampling process even when the investment, although initially *not* desirable, may appear acceptable after we obtain additional information and are able to make sequential decisions. In general, we will observe that the greater the uncertainty, the better it is to obtain additional information, perhaps by trying an investment if the investment can be replicated.

### 13.5.3 Sampling Leading to Imperfect Information

Now assume in Example 13.3 that after the first investment is made and the results are observed, we still cannot be certain of the desirability of the investment; that is, the information obtained is imperfect. Suppose we are able to assign a set of probabilities reflecting the reliability of the information as shown in Table 13.5a. If the conversion actually is excellent, there is still a 0.05 proba-

**Table 13.5** ***Calculation of Joint and Posterior Probabilities***

a. *Conditional Probabilities*

| | Actual State | | |
|---|---|---|---|
| Observed Event from Sample Investment | $\theta_1$ | $\theta_2$ | $\theta_3$ |
| Conversion seems to be excellent (E) | 0.80 | 0.20 | 0.10 |
| Conversion seems to be fair (F) | 0.15 | 0.60 | 0.20 |
| Conversion seems to be bad (B) | 0.05 | 0.20 | 0.70 |

b. *Joint Probabilities*

| Joint Probabilities | Sample Prediction Outcome | | | Marginal Probabilities (Performance) |
|---|---|---|---|---|
| | E | F | B | |
| $\theta_1$ | 0.400 | 0.075 | 0.025 | 0.50 |
| $\theta_2$ | 0.060 | 0.180 | 0.060 | 0.30 |
| $\theta_3$ | 0.020 | 0.040 | 0.140 | 0.20 |
| Marginal Probabilities (Sample) | 0.480 | 0.295 | 0.225 | 1.00 |

c. *Revised (Posterior) Probabilities*

| Conditonal Outcome | Posterior Probabilities for a Given Survey Prediction | | |
|---|---|---|---|
| | E | F | B |
| $\theta_1$ | 0.833 | 0.254 | 0.111 |
| $\theta_2$ | 0.125 | 0.610 | 0.267 |
| $\theta_3$ | 0.042 | 0.136 | 0.622 |

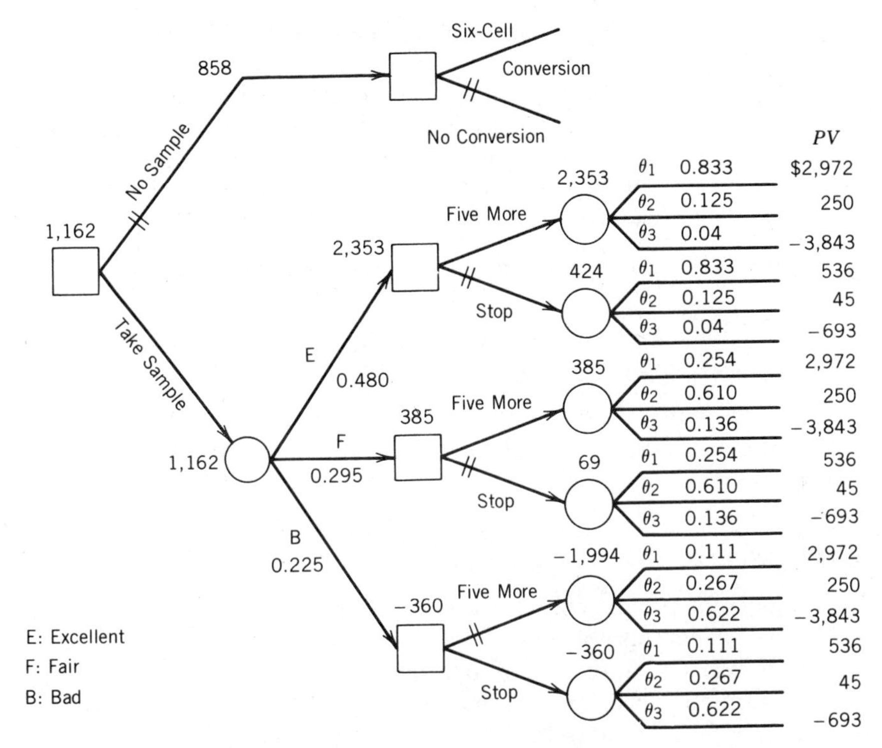


**FIGURE 13.8** Decision tree with imperfect information.

bility that it may not appear profitable. If the conversion actually is not good, there are a 0.10 probability that it may appear excellent, a 0.20 probability that it may appear fair, and so forth.

The problem that has to be solved is whether or not it is desirable to go ahead and make the initial cell conversion. The computations of the relevant revised probabilities are shown in Table 13.5c. Figure 13.8 shows the probabilities and the outcomes if one cell conversion is undertaken to obtain information and then the decision is whether or not to undertake the conversions of the remaining cells. In Figure 13.8 we find that the one-unit conversion as a means of obtaining information is still desirable, even with imperfect information. The delayed investment strategy results in an expected net present value of $1,162, which is $304 more than that of the no-delay situation.

## 13.6 SUMMARY

The decision tree is another technique that can facilitate investment decision making when uncertainty prevails, especially when the problem involves a sequence of decisions. In decision tree analysis a decision criterion is chosen—say, to maximize expected profit. If possible and feasible, an experiment is conducted. The earlier probabilities of the states of nature are revised on the basis of the experimental results. The expected profit of each possible decision is computed, and the act with the highest expected profit is chosen as the optimum action.

Before undertaking an experiment, the decision maker must determine whether the expected profit of acting after receiving the result of the experiment is sufficiently large to offset the cost of the experiment. The analysis requires finding the optimum rule and evaluating the expected profit by using this rule. In summary, for a decision tree analysis, we may follow eight steps.

1. Choose the decision criterion (we suggest the expected-value decision rule).
2. Describe the set of possible outcomes and possible decisions.
3. Assign probabilities to the possible outcomes.
4. Determine a profit function or net present value for each branch node.
5. Conduct an experiment (to obtain further information).
6. Revise the assigned probabilities.
7. Compute the expected profit (net present value) for each decision.
8. Choose the act with the highest expected profit (net present value).

The inherent project risk raises the possibility that an investment that seems acceptable may turn out to be undesirable. On the other hand, an investment that seems to be unacceptable may turn out to be desirable when additional information is obtained. Therefore, the value of the additional information must be balanced against the cost of the information.

## REFERENCES

1. BIERMAN, H., JR., C. P. BONINI, and W. H. HAUSMAN, *Quantitative Analysis for Business Decisions,* 5th edition, Richard D. Irwin, Homewood, Ill., 1977.
2. BIERMAN, H., JR., and V. R. RAO, "Investment Decisions with Sampling," *Financial Management,* Vol. 7, No. 3, pp. 19–24, Autumn 1978.
3. BIERMAN, H., JR., and S. SMIDT, *The Capital Budgeting Decision—Economic Analysis of Investment Projects,* 7th edition, Macmillan, New York, 1988.
4. MAGEE, J. F., "Decision Trees for Decision Making," *Harvard Business Review,* Vol. 42, No. 4, pp. 126–138, July–August 1964.
5. NEWENDORP, P. D., *Decision Analysis for Petroleum Exploration,* Pennwell Publishing Company, Tulsa, Okla., Chs. 4 and 10, 1975.
6. RAIFFA, H., *Decision Analysis—Introductory Lectures on Choices under Uncertainty,* Addison–Wesley, Reading, Mass., 1970.

## PROBLEMS

**13.1.** Urn 1 contains 3 red balls and 2 black balls; urn 2 contains 11 red balls and 10 black balls. You are going to be given a sample of 2 balls and on this basis decide from which urn the sample came. The sample is equally likely to come from either urn. A correct decision gives you nothing; a wrong decision costs you one dollar. You wish to choose your guess so as to minimize your expected losses.

a. You are allowed to choose whether the sample will be drawn with or without replacement. Find your better strategy and expected loss.
b. You are allowed to decide whether the sample will be drawn with or without replacement after you have seen the first ball. Find the better strategy and expected loss.

**13.2.** You must decide whether to buy or lease a car. You have gathered some data on the costs of buying, operating, and leasing a vehicle for 3 months under two circumstances.

| Use | Buy | Lease |
|---|---|---|
| Light use | $1,200 | $950 |
| Intensive use | 1,600 | 1,700 |

These figures include all quantifiable costs. The unknown factor is whether the car will have light use or intensive use. This factor is beyond your control, but you estimate a 0.6 probability of intensive use and a 0.4 probability of light use.
a. Draw a decision tree for this problem.
b. What action do you recommend? Why?
c. You think that you can determine with 95% accuracy whether your car will be subject to light or intensive use. This will require about 4 hours of work on your part. You value your time at $200 per 8-hour working day. Should you spend more time investigating the matter? Why?

**13.3.** Jones wishes to buy a loom. The dealer offers Jones a choice between a new, perfect loom for $44 and an old, possibly defective model costing $30. Jones estimates that the old loom has probability ⅓ of being defective and requiring $30 worth of repairs.
a. Which loom should Jones buy to minimize this expected cost?
b. Jones can hire a loom inspector who always passes good looms but catches bad looms with probability ¾. Jones will buy the old loom if it passes inspection and the new loom if it does not. How much are the inspector's services worth to Jones?

**13.4.** You are evaluating the development of an information system (IS) to forecast sales volumes. You will decide the staffing of a manufacturing line on the basis of this forecasted volume. The system may forecast sales as (1) increasing, (2) stable, or (3) decreasing. You respond with either (1) single-shift or (2) double-shift staff levels. Each of the six possible combinations provides an estimated dollar payoff.

| | States of Nature | | |
|---|---|---|---|
| Actions | Increasing Sales | Stable Sales | Decreasing Sales |
| Single shift | $11,000 | $10,000 | $2,000 |
| Double shift | 18,000 | 8,000 | −3,000 |

If an IS is used, it provides a sales volume forecast (prediction); then the manager chooses a staff level (strategy), and finally an actual sales level (state of nature) occurs, yielding its associated payoff. States of nature are uncontrollable but occur with a probability that can be estimated, sometimes from historical data. In our example sales historically increased 30% of the time, were stable 45% of the time, and decreased 25% of the time.

Next, the manager must estimate the accuracy of the IS in forecasting sales levels. If the IS is infallible, it is called a perfect information system. Most ISs are imperfect, and the conditional probabilities in the following table reflect this fact.

| Predicted Sales | Given Actual Sales: Increase | Stable | Decrease |
|---|---|---|---|
| Increase | 65% | 10% | 10% |
| Stable | 20 | 80 | 15 |
| Decrease | 15 | 10 | 75 |
| Historical Probability: | 30 | 45 | 25 |

The value 65% in the northwest corner means that when sales actually increase in the future period, the IS will have predicted the increase 65% of the time. Note that the sum of any column must equal 100%.

a. Is it worth using the IS system?
b. What is the expected value of perfect information without the IS?
c. What is the expected value of perfect information with the IS?
d. What is the value of the sample information?

**13.5.** Suppose you are the manager of a frozen-food factory and you have to decide on the annual production volume of brussels sprouts. The nature of the product dictates that you produce all the annual volume during a short harvest season and then store and market the frozen brussels sprouts during the entire year. Merchandise that has not been sold after a year is disposed of; on the other hand, if you are short of finished goods, you might lose customers in the long run. Since you have to contact possible vendors (farmers), you have to decide on production volume at the beginning of the year. Unfortunately, you do not know the demand for frozen brussels sprouts in the coming year. Your experience indicates that demand can be low (4,000 units of volume), medium (6,000 units), or high (8,000 units). Moreover, according to past experience, the prior probabilities of low, medium, and high demand are 0.3, 0.5, and 0.2, respectively.

***Payoff Matrix and Expected Payoffs (in PV)***

| | | Event (Actual Demand): Low 4,000 | Medium 6,000 | High 8,000 | |
|---|---|---|---|---|---|
| | Prior Probability: | 0.3 | 0.5 | 0.2 | |
| Decision (Production Volume) | | Payoff Matrix | | | Expected Payoff ($) |
| Low | 4,000 | $2,000 | 0 | −2,000 | 200 |
| Medium | 6,000 | −$8,000 | 8,000 | 6,000 | 2,800 |
| High | 8,000 | −$18,000 | −2,000 | 14,000 | −3,600 |

Suppose a market survey is available with the following reliability (values obtained from past experience where actual demand was compared with predictions made by the market survey).

| Given Actual Demand | Survey Prediction: Low | Medium | High |
|---|---|---|---|
| Low | 0.70 | 0.25 | 0.05 |
| Medium | 0.20 | 0.60 | 0.20 |
| High | 0.05 | 0.25 | 0.70 |

a. Determine the strategy that maximizes the expected payoff before taking the market survey.
b. Compute the expected value of perfect information.
c. Compute the expected value of sample information.
d. If the market survey costs $5,000, is it worth taking?

**13.6.** The Football University contemplate expanding their current 70,000-seat football stadium. The question of optimal seating capacity is being debated at a board of trustees' meeting. Mr. Hagan, the board member in charge of the university's athletic program, argues that since the demand for the season tickets is uncertain, a 5,000-seat expansion should be considered at this time. Mr. Smith, the head football coach, is in favor of adding 10,000 seats now. Mr. Smith argues that it costs less to add 10,000 seats all at once than to expand piecemeal. Both Hagan and Smith agree, however, that the future ticket demands are highly correlated with the university's postseason recruiting results and how the team performed in the previous seasons. Mr. White, the university president, expresses his opinion that, without support from student fees, no stadium expansion should be considered at this time because of difficulty in obtaining bond money at a fair market rate. "Students are already paying too much in other building programs, and they do not favor this stadium project," says the president. Mr. White feels that the university should postpone the decision for a year. If the university continues to have a strong football program for the next two seasons, he points out, then the current facility can easily be expanded to provide either 5,000 or 10,000 additional seats. The work can start right after the second season and be completed before the third season starts.

Mr. Jones, the director of the university's alumni association, has presented the results of a brief telephone survey of possible season ticket buyers. The survey shows that the probability of a light ticket demand is 0.2, of a high demand 0.4, and of a moderate demand 0.4. Depending on these demand levels, over the next 8 years the following *additional* annual receipts are expected to be received from the two alternatives that require an investment now.

| | | Additional Annual Receipts | | |
|---|---|---|---|---|
| | Investment | Light | Moderate | High |
| Add 5,000 capacity | $10* | 0 | $2 | $3 |
| Add 10,000 capacity | 18 | 0 | 3 | 5 |

*Unit: million dollars.

Take the interest rate as 6% for simplicity. If the decision to construct 10,000 capacity is postponed, the cost of expanding the stadium 2 years from now is expected to be

| | |
|---|---|
| Adding 5,000 capacity | $12 million |
| Adding 10,000 capacity | $20 million |

For example, if the university adds 5,000 seats now and adds another 5,000 at the end of 2 years, the total investment will be $22 million.

a. Draw the decision tree that describes the decision problem.
b. On the basis of expected net revenue, how should the university make the capacity expansion decision?
c. One board member has proposed that a full survey be taken to establish the actual demand to be experienced. What is the maximum value of such a survey?

**13.7.** Consider the following case story.

"How to pick a peach instead of a lemon. There is a bumper crop of used cars around, and plenty of lemons. Picking a peach instead of a lemon isn't all that difficult if you know the telltale signs of good and bad. Starting out, arm yourself with the peach picker's secret weapon: patience. . . ." (From *Motor Trend,* June 1977.) Now put yourself in the market for a used car and imagine the following conversation with a used-car dealer named John. Let's say he is an honest and good friend of yours:

John: This used-car business is a tough racket. I have a customer interested in the MODEL-X on our lot, but the practices of our business prevent me from warning him that he may get stuck if he buys it.

You: What do you mean? I came here to look at the MODEL-X, which you have been advertising for the last week.

John: I worked at a MODEL-X dealer when that car first came on the market. As you may know, the MODEL-X company made 20% of its cars in a new plant where they were still having production problems; those cars were lemons. The remaining 80% of MODEL-X for that year were made at other plants and they were good cars.

You: Oh boy! What you are saying is that you have been advertising a lemon. I can hardly believe this!

John: Well, you shouldn't feel so bad: maybe the MODEL-X on our lot doesn't have any defects, or its defects may already have been fixed.

You: Hey, John. Is there any way you can tell a peach from a lemon by a simple inspection?

John: That's the trouble. I personally don't know much about the car itself. But if you are really interested in the car, it's worth having a mechanic check what the average buyer usually can't.

You: Do you know any good mechanic from your service department who can look it over?

John: Sure I do, but you can hardly expect the mechanic to go through all the trouble of examining the car and getting dirty without some financial consideration. Furthermore, you must accept the fact that he may not provide you with perfect information about the true state of the car, even if he works on the car for 5 hours for a complete test. But I am not sure either that you need this perfect information at that kind of expense.

Let us now begin the analysis of the decision problem just described. Suppose the probability that, after a simple inspection (for 30 minutes), the mechanic will say the car is a peach, when the actual state of the car is a peach, is 0.75. The probability that he will find a lemon when the car is a lemon is 0.73. Further, he charges at the rate of $20 per hour. If you buy the car and it turns out to be a lemon, it will cost you $600 to fix the car. The asking price of the car is $2,500, which is about $300 less than the price quoted by other dealers. [The other dealers sell the cars with a guarantee (anti-lemon), so no repair cost is expected.]

a. What is the probability that the suspected car is actually a lemon, even if the mechanic tells you it is a peach?
b. If you buy the car with your own judgment but it turns out to be a lemon, it will cost you $600 to fix the car. What is the expected value of perfect information?
c. Is it worth having the mechanic give the car a simple inspection?

d. Suppose the probability that the mechanic will find the true state of the car is a linear function of the testing time. He says the car is a peach with the following probability when the car is a peach.

$$0.6 + 0.3t, \qquad 0 \leq t \leq 1$$

$$0.875 + 0.025t, \qquad 1 \leq t \leq 5$$

where $t$ is the testing time in hours. He says the car is a lemon with the following probability when it is a lemon.

$$0.7 + 0.06t, \qquad 0 \leq t \leq 5$$

How long would you let him check the car?

e. In part d, express the expected value of perfect information as a function of $t$.

## Acknowledgments


- Problem 13.4 is based on an article by G. P. Schell, "Establishing the Value of Information Systems," *Interfaces,* Vol. 16, No. 3, pp. 82–89, 1986.
- Problem 13.5 is based on an article by N. Ahituv and Y. Wand, "Information Evaluation and Decision Makers' Objectives," *Interfaces,* Vol. 11, No. 3, pp. 24–33, 1981.

## *PART FOUR*

# SPECIAL TOPICS IN ENGINEERING ECONOMIC ANALYSIS

# 14
# Evaluation of Public Investments

## 14.1 INTRODUCTION

In this chapter we will discuss investment decisions in the public sector. The public sector consists of various governmental units ranging from municipal governments to the federal government. Unlike the situation in the private sector, a basic operating principle of these governments is to serve their citizens. Thus, the question of the general welfare of the citizens should ultimately be considered in most decisions regarding government expenditures.

Divergent views of what best sustains the general welfare often lead to disagreements about the appropriateness of particular expenditures. A national defense project seemingly unjustifiable on an economic basis, for example, may be accepted as being in the best interest of the public. Conversely, such projects as synthetic fuels, power plants, hospitals, and housing loan programs may be justifiable economically but unacceptable because we consider them to be in competition with investor-owned services.

Our objective in this chapter is to present a logical sequence for analyzing these public investments. We begin by examining the nature of public activities and then asking, "Why benefit–cost analysis?" After stating the rationale for benefit–cost analysis in the public sector, we give a simple but structured overview of benefit–cost analysis. A detailed benefit and cost measurement for a mass transit system project follows the introductory discussion. Finally, decision rules, the risks, and the uncertainties of various project attributes in the public sector are discussed.

## 14.2 BENEFIT–COST ANALYSIS

### 14.2.1 The Nature of Public Activities

The nature of public projects undertaken by government agencies ranges from cultural development, protection, and economic services to management of natural resources. Cultural development projects may include public school systems, civic centers, public libraries, historic preservation, and the like. Protection projects provide services, such as those of the military, the police, and the fire department. Economic services include the postal service, rural electrification projects, transit systems, highway and bridge construction, and housing loan programs. Natural resource projects might entail the conservation of fish and wildlife, pollution control, forest management, and flood control.

These project lists are obviously incomplete, and many projects belong in more than one class. (For example, an irrigation and water supply project is considered by many to contribute to both natural resource and economic services.) Pinpointing the nature of an activity by a classification such as this is useful, however, in defining the scope of program objectives.

The scope of public program objectives also depends on the view taken in the evaluation: the local view—regional, state, county, or city—or the national view. A local government is subject to much political pressure from the taxpayers who support it but little from those living outside its jurisdiction. Therefore, the tendency is to ignore outside considerations. Higher levels of the government, on the other hand, must be responsible for making sure that local planners adequately consider project consequences affecting other areas. Local interests may try to influence federal agencies to select projects that produce local benefits, but the federal government must consider whether such projects can be justified from the national view. Ideally, decision makers weigh the national cost against the achievement of local goals.

### 14.2.2 Why Benefit–Cost Analysis?

The need for benefit–cost analysis, along the lines we are establishing, stems from two sources cited by Schultze [9]. First, as in the private sector, the resources of the government are always less than we need to accomplish all that we would like to do. Therefore, among competing claims on available resources, we must choose those that contribute most to our government objectives, the most basic of which is to maximize the general welfare of the citizenry (whatever we conceive that to be).

Second, public projects rarely have an automatic regulator to indicate when an activity has ceased to be productive or could be made more efficient and when it should be displaced by another activity. In the private sector we rely on profit and competition to furnish the needed incentives and discipline and to provide feedback on the quality of decisions. This system is imperfect, but it is basically sound in the private sector. In the public sector this type of check is virtually nonexistent. We must find another tool for making the choices that resource scarcity forces upon us. Benefit–cost analysis provides the means to improve decision-making processes for the more efficient use of government resources in the absence of competitive pressures.

### 14.2.3. The Procedure of Benefit–Cost Analysis[1]

The first step in benefit–cost analysis is a careful specification and analysis of basic program objectives—what are we really trying to accomplish? Suppose

[1]This section is based on an article by C. L. Schultze, "Why Benefit–Cost Analysis?" which appears in Hinrichs and Taylor [9].

we consider constructing a highway to connect two cities. The objective of this intercity highway project is not just to lay concrete.

Indeed, highway construction may contribute to "the good life" or to national unity (by making access easier), but to take "laying concrete" and "improving life" as our sole objectives does not tell us much, if anything, useful about the character of the highway, its location, or its relation to other elements of our transportation system. In this example we want specific objectives broader than "laying concrete" but narrower than "improving life." Highways are useful only as they serve an important goal, namely transporting people and goods safely and efficiently. Once this is accepted as an objective, we can analyze aviation, railroads, mass transit systems, and highways to determine the most effective transportation network to connect the two cities.

The second step is to analyze the output of the project in terms of the objectives specified in the first step. For our highway example, we must ask not how many miles of concrete are laid but what the project produces in terms of faster, safer, less-congested travel—how many hours of travel time are eliminated and how many accidents are prevented. Once these project attributes are specified, they often must be measured in dollar terms so that different benefits may be compared against each other and against the cost of attaining them.

The third step is to measure the total project costs, not just for one year, but over the life of the project. In deciding to build an expressway through a downtown area, for instance, we must take into account not only the cost of the expressway but also the cost of relocating the displaced residents and the effects of the freeway on the areas through which it is to run (e.g., air pollution, noise, and changes in real estate value).

The fourth and crucial step is to analyze alternatives, seeking those that have the greatest benefits in achieving the basic objectives or that achieve those objectives at the least cost. In our highway example we should be comparing the benefits of additions or improvements to highways with those of additions or improvements to aviation and railroads as alternative means of providing safe and efficient transportation. This does not mean that we pick only one. But we do need to decide, at least roughly, which combination of alternatives is the preferred one. It is this comparison of alternatives that is essential for testing the economic desirability of existing and proposed projects.

### 14.2.4 Valuation of Benefits and Costs

***Benefits, Disbenefits, and Cost.*** In a benefit–cost study of public investments, we must define precisely what constitutes the user's (public) benefits and sponsor's (government) costs. To find the user's net benefits we must identify the project outcomes that are favorable and unfavorable to the user. (The unfavorable outcomes are commonly referred to as *disbenefits*.) Then we combine these terms,

$$\text{Net user's benefit } (B) = \text{Favorable benefits} - \text{Disbenefits}$$

Similarly, to determine the net cost to the sponsor, we must identify and classify the expenditures required and any savings (or revenues) to be realized.

The elements of cost to the sponsor should include both capital investment and annual operating cost. The savings in public projects usually come from the sales of products or services as a consequence of the completion of the project. We combine these cost elements to obtain

$$\text{Net sponsor's cost} = \text{Capital cost} + \text{Operating and maintenance costs} - \text{Savings (revenues)}$$

***Primary versus Secondary Benefits.*** One of the critical issues in evaluating public projects is how to quantify benefits in dollar terms. To do so, we must question the extent of benefits to be measured by distinguishing between the benefits directly attributable to the project, which are *primary benefits,* and the benefits indirectly attributable to the project, the *secondary benefits.* For example, a port construction or expansion will bring population increases to nearby port areas. Primary benefits may include various regional economic benefits from port activities. Secondary benefits may include increasing the incomes of various producers such as vehicle mechanics through increases in population. Secondary benefits are commonly known as the multiplier effect.

A benefit–cost analysis should always consider the primary benefits and should consider the secondary benefits whenever possible. Inclusion of secondary benefits should be a function of their effect compared with that of the primary benefits and with the cost of determining them. If the primary benefits far exceed the cost of the project, we may not have to look hard into the extent of secondary benefits.

***Relating User's Benefits to the Cost of Financing.*** Proper identification of the user's benefits is important in allocating the public project cost. The procedure for dividing total financial cost among responsible parties (or users) is called *cost allocation.* For example, federal cost-sharing practice divides the burden for the project cost between payments from the beneficiaries and a subsidy from the federal tax money. The portion of the allocated project cost must be repaid by the beneficiaries. Repayment requires allocation of the total project cost among project purposes.

Many user taxes (or fees) are structured so that the more benefits a user receives, the higher taxes (or fees) the user has to pay. The gasoline tax, which is based on this operating principle, provides tax revenues in relation to the amount of public road use. (The more a person drives, the more use he or she makes of the highway system.) A view similar to this can be taken to assess the fare system in a mass transit service. Example 14.1 explains this concept.

## *Example 14.1*

Consider a mass transit system with the physical characteristics shown in Figure 14.1 and Table 14.1. There are two types of riders, core and suburban. Most

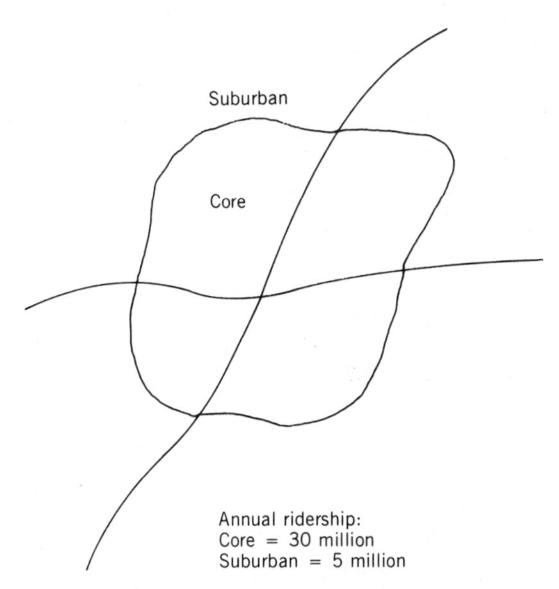


**FIGURE 14.1 Mass transit system—core and suburban riders.**

suburban riders also travel part of the core system. The system's various costs are estimated to be as follows.

| Costs | Core | Suburban | Total |
|---|---|---|---|
| Vehicle cost ($ million) | 80 | 20 | 100 |
| Annual O&M cost ($ million) | 15 | 3 | 18 |
| Construction cost ($ million) | 800 | 200 | 1,000 |

The transit authority wants to determine the fare structure for the two types of riders. Each vehicle costs $200,000 and has a 15-year service life. The track system has a 30-year service life. Assume that neither track nor vehicles will have any salvage value at the end of their service lives and that the interest rate for the transit authority is known to be 10%.

*View 1:* Allocate core cost to core riders and suburban costs to suburban riders.

$$\text{Core cost/year} = \$800(A/P,\ 10\%,\ 30) + 80(A/P,\ 10\%,\ 15) + 15 = \$110.4 \text{ million}$$

**Table 14.1** ***Physical Characteristics of the Transit System***

| Description | Core | Suburban | Total |
|---|---|---|---|
| Track length (miles) | 80 | 40 | 120 |
| Rides (millions) | 30 | 5 | 35 |
| Number of vehicles in service | 400 | 100 | 500 |

$$\text{Suburban cost/year} = \$200\,(A/P,\ 10\%,\ 30) + 20(A/P,\ 10\%,\ 15) + 3 = \$26.85 \text{ million}$$

$$\text{Core fare} = \$110.4/30 = \$3.68/\text{ride}$$

$$\text{Suburban fare} = \$26.85/5 = \$5.37/\text{ride}$$

*View 2:* Allocate core costs to all riders, since almost all use the core portion. Allocate suburban costs to suburban riders as "full system" riders.

$$\text{Core fare} = \$110.4/35 = \$3.15/\text{ride}$$

$$\text{Suburban (full system) fare} = \$3.15 + \$26.85/5 = \$8.52/\text{ride}$$

View 2 is practiced by many states for highway tax allocation among light and heavy vehicles. □

### 14.2.5 Decision Criteria

When the benefits and costs have been properly identified and quantified, the next step is to compute the benefit–cost ratio as a means of justifying the public expenditure. As presented in Section 6.4, the benefit–cost ratio can take two forms: the aggregate $B/C$ ratio, $R_A$, and the netted $B/C$ ratio, $R_N$. Although there have been many discussions of the superiority of using one $B/C$ ratio over the other [14], we have shown that both ratios can serve as a proper measure of project worth.

When comparing mutually exclusive projects, we have also shown in Chapter 7 that both $B/C$ ratios will lead to the same project selection by the *PV* criterion, as long as the *incremental analysis* is used. For this reason, we will consider only the $R_A$ criterion in subsequent discussions.

***The B/C Ratio.*** Recall that the $R_A$ criterion for project acceptance is

$$\text{If } R_A = B/(I + C') > 1, \quad \text{accept the project} \tag{14.1}$$

where $I + C' > 0$, $C = I + C'$, and

$B$ = *PV* equivalent benefits to the user computed at $i$,

$I$ = *PV* equivalent capital invested by the sponsor computed at $i$,

$C'$ = *PV* equivalent total operating and maintenance costs computed at $i$,

$i$ = interest rate used in public project evaluation.

***Incremental B/C Ratio.*** When we compare mutually exclusive alternatives A1 and A2, the correct alternative can be selected by applying the principle of incremental analysis. The additional incremental outlay is economically justifiable only if the incremental benefit realized exceeds the incremental outlay. For $R_A$ we compute the incremental differences for each term ($B$, $I$, and $C'$) and take the $B/C$ ratio based on these differences. The expression is then

$$\Delta R_A = \frac{B_1 - B_2}{(I_1 - I_2) + (C_1' - C_2')} = \frac{\Delta B}{\Delta I + \Delta C'} = \frac{\Delta B}{\Delta C} \tag{14.2}$$

where $\Delta I + \Delta C' > 0$. The decision rules without considering the do-nothing alternative are

$\Delta R_A > 1$: accept A1

$\Delta R_A < 1$: reject A1 and retain A2

When comparing multiple alternatives, we recommend that the alternatives be arranged by increasing order of their denominators $(I + C')$. Thus, the alternative with the smallest denominator should be first, the alternative with the next smallest second, and so forth. This ordering rule will always ensure that the incremental denominator $(\Delta I + \Delta C')$ is positive, so that the decision rules just described will produce a correct result.

If the do-nothing alternative is to be considered, we may first compute the $B/C$ ratio for each alternative and discard the alternatives whose $B/C$ ratios are less than or equal to 1. Then we consider only the remaining alternatives in applying the incremental analysis.

When $\Delta I + \Delta C' = 0$, it implies that both alternatives require the same initial investment and operating expenditure. If this is the case, we should not use the $B/C$ ratio. The decision to select the alternative whose $B$ is larger should be obvious, however.

***Selecting an Interest Rate.*** Before ending our examination of decision criteria, we need to discuss some ways to select an appropriate interest rate for public projects. An extensive literature on public finance addresses the question of which discount rate is appropriate to use in public sector investment decisions [2,9]. No single view is held by all authors on this subject, however. Two prevailing views are that

1. The discount rate should reflect only the prevailing government borrowing rate.
2. The discount rate should represent the rate that could have been earned had the funds not been removed from the private sector.

The first view has been adopted by the federal government with respect to water resource projects, and the second view is held by the Office of Management and Budget (OMB). Since 1972 the OMB has required that, with certain exceptions, for example, water resource projects, federal agencies must use a discount rate of 10% to evaluate federal investment decisions [6]. The prescribed discount rate of 10% represents an estimate of the average rate of return on private investment, before taxes and *after inflation*.

The interest rate to use in evaluating public projects is a matter of judgment. The rate should not be less than that paid for funds borrowed for the project. If the public project is competitive with a private project, the rate should be comparable to that used in private investment.

### *Example 14.2*

In an effort to reduce rates of recidivism (return to crime), the Department of Justice is considering three experimental prison reform programs based on rehabilitation in its correctional system. The principal focus is on offenders who were convicted for property crimes—robbery, burglary, larceny, and auto theft. The direct dollar benefits from programs designed to reduce recidivism rates are determined by combining cost elements such as the value of stolen property, loss of police services, court costs, and prison and probation costs. The dollar benefits expected for each person who does not return to crime are as follows.

***Benefits per Person of Reduced Recidivism***

| Type of Crime | Robbery | Burglary | Larceny | Auto theft |
|---|---|---|---|---|
| Benefits ($) | 10,444 | 8,596 | 7,431 | 9,130 |

The current recidivism rates for these crimes as committed by individuals in the age category 18 to 24 during the first year following release from prison are

***Selected First-Year Recidivism Rates***

| Type of Crime | Robbery | Burglary | Larceny | Auto theft |
|---|---|---|---|---|
| Recidivism Rate | 44% | 37% | 40% | 42% |

The three major programs to reduce crime are the following.

1. Improve training and education programs so that the offender can more readily compete in the marketplace and thus will be less likely to commit crimes.
2. Remove inmates from institutional settings and place them in smaller homes in the community, where access to such things as family and jobs will provide an atmosphere similar to the one the offender will return to.
3. Improve living conditions in prisons in order to reduce despair, bitterness, hatred of society, vindictiveness, and later recidivism.

The cost of each program and the overall reduction in recidivism expected to follow its application are estimated for a rehabilitation center that releases each year 3,000 inmates sentenced for property crimes.

| Program | Initial Cost | Annual Cost | Overall Reduction Rate for Each Release (%) |
|---|---|---|---|
| 1 | $700,000 | $350,000 | 5 |
| 2 | 2,200,000 | 500,000 | 10 |
| 3 | 1,200,000 | 140,000 | 4 |

The percentages of the total number of inmates now released after serving sentences for each type of crime are the following.

| Type of Crime | Robbery | Burglary | Larceny | Auto theft |
|---|---|---|---|---|
| Percentage of Total Inmates Released | 25 | 35 | 20 | 20 |

If each program lasts only 5 years and the rate of interest used is 6%, which program should be adopted?

**Benefit–cost analysis of crime reduction programs**

Program 1 (Sample Calculation)

Benefits (annual):

$$\begin{aligned}
\text{Robbery} &= 3{,}000(0.25)(0.44)(\$10{,}444)(0.05) = \$172{,}326\\
\text{Burglary} &= 3{,}000(0.35)(0.37)(\ \$8{,}596)(0.05) = \$166{,}977\\
\text{Larceny} &= 3{,}000(0.20)(0.40)(\ \$7{,}431)(0.05) = \ \ \$89{,}172\\
\text{Auto theft} &= 3{,}000(0.20)(0.42)(\ \$9{,}130)(0.05) = \underline{\$115{,}038}\\
&\qquad\qquad\qquad\qquad\qquad\qquad\qquad\quad \$543{,}513
\end{aligned}$$

Annual cost = \$350,000

Investment = \$700,000

$B = \$543{,}513(P/A, 6\%, 5) = \$2{,}289{,}476$

$C' = \$350{,}000(P/A, 6\%, 5) = \$1{,}474{,}328$

$I = \$700{,}000$

$$\frac{B}{I + C'} = 1.05 > 1 \quad \text{(acceptable)}$$

Program 2

$$B = \$4{,}578{,}954$$
$$C' = \$2{,}106{,}182$$
$$I = \$2{,}200{,}000$$
$$\frac{B}{I + C'} = 1.06 > 1 \quad \text{(acceptable)}$$

Program 3

$$B = \$1{,}831{,}582$$
$$C' = \$589{,}731$$
$$I = \$1{,}200{,}000$$
$$\frac{B}{I + C'} = 1.02 > 1 \quad \text{(acceptable)}$$

Since all programs are acceptable individually, we do not consider a do-nothing alternative. The sequence of analysis is 3, 1, and 2.

**Incremental analysis—mutually exclusive problem**

$$\text{Iteration 1: Program 1} - \text{Program 3} = \frac{\$457{,}894}{\$384{,}597} = 1.191 \quad \text{Select 1}$$

$$\text{Iteration 2: Program 2} - \text{Program 1} = \frac{\$2{,}289{,}478}{\$2{,}131{,}854} = 1.074 \quad \text{Select 2}$$

With the incremental analysis, Program 2 is selected. □

## 14.3 THE BENEFIT–COST CONCEPT APPLIED TO A MASS TRANSIT SYSTEM

Many examples can be offered to illustrate benefit–cost analysis. (See [5] and [9] for excellent collections of cases of benefit–cost studies.) In this section we present a mass transit improvement project to illustrate the benefit–cost analysis outlined in the previous sections. The material given is based on the guidelines suggested by the American Association of State Highway and Transportation Officials [1].

### 14.3.1 The Problem Statement

Consider a proposed bus transit project that entails the use of express commuter buses operating in one lane of an eight-lane urban freeway from which all other vehicles are to be banned. The commuter bus service will operate only in the peak hours of each day, and each of the two lanes to be used for buses, one going north and one going south, will be used in the peak direction of traffic for 3 hours. The extent of one part of the roadway available for other vehicles will be reduced when a bus lane goes into operation and the freeway changes from four lanes to three lanes in the peak direction. However, former users of the highway as well as new travelers (if any) will be able to use a relatively high-speed (50-mph) bus service during this time. The off-peak periods are assumed to be unaffected and are not considered.

The traffic and facility data assembled for the analysis are presented in Tables 14.2 and 14.3. The affected highway section is designated *AB* in Figure 14.2. Alternative 0 represents the existing situation, and alternative 1 represents the situation for the proposed bus lane. The drivers who are diverted to the commuter buses may or may not be changing from customary trips similar to those of other freeway users. Only one class of trip is considered in this example. We will assume that, for all the commuter bus users, the alternative before the commuter bus lane is available is a trip by car consisting, on the average, of 3.2 miles of travel on arterials at 28 mph and 10 miles of travel on the freeway under conditions noted in Table 14.2.

Bus vehicles will be rented, and no major facilities such as parking facili-

**Table 14.2** ***Highway and Transit Traffic Data***

*Highway Traffic Data*

| | | | Vehicles per Hour | | |
|---|---|---|---|---|---|
| Alternative | Study Year | Period | One-Way Volume | One-Way Capacity | $v/c$ Ratio |
| 0 | 1 | Peak hour | 6,000 | 8,000 | 0.75 |
| 1 | 1 | Peak hour | 5,000 | 6,000 | 0.83 |
| 0 | 15 | Peak hour | 7,000 | 8,000 | 0.88 |
| 1 | 15 | Peak hour | 5,400 | 6,000 | 0.90 |

*Commuter Traffic Data*

| Alternative | Study Year | Period | Hourly Patronage | Fare per Trip | In-Vehicle Time per Trip (minutes) | Wait, Walk, and Transfer Time per Trip (minutes) | Annual Bus Miles (millions) |
|---|---|---|---|---|---|---|---|
| 0 | 1 | — | — | — | — | — | — |
| 1 | 1 | Peak | 2,000 | $0.40 | 19 | 5.0 | 2.4 |
| 0 | 15 | — | — | — | — | — | — |
| 1 | 15 | Peak | 2,500 | 0.40 | 19 | 5.0 | 3.0 |

**Table 14.3** ***Highway and Transit Facility Data***

*Highway Traffic Data*

| Alternative | Section Length (miles) | Design Speed (mph) | Number of One-Way Lanes | Grade (%) | Curvature (degrees) |
|---|---|---|---|---|---|
| 0 (existing) | 10 | 70 | 4 | 0* | 0 |
| 1 (proposed) | 10 | 70 | 3 | 0 | 0 |

*Level

*Transit Facility Data*

| Service Speed | Seats per Bus | Driver's Wage |
|---|---|---|
| 40 mph | 50 | $8/hour |

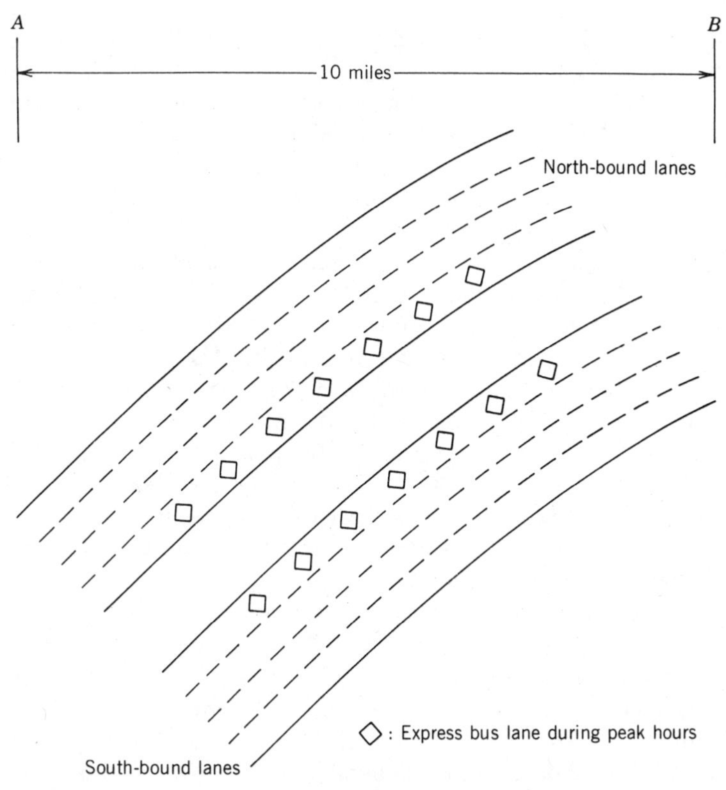


**FIGURE 14.2** Affected highway section—10-mile length.

ties, terminals, and administrative offices will be required. Enough space is available to accommodate this addition to the current transit system.

### 14.3.2 Users' Benefits and Disbenefits

In mass transit improvements, user benefits may be defined as net changes in transit user costs and highway user costs before and after the new transit service. Transit improvements that reduce users' perceived costs will often divert traffic or induce additional transit patronage above the level prevailing in the absence of the improvement. This factor should be considered in estimating traffic data. The benefit calculations may be applied to data from any specified period of time, but a year is used here for the purpose of illustration. If the analysis period spans several years, the convention is to select two representative (study) years within the analysis period for detailed study and development of user benefit and cost estimates and then extrapolate or interpolate for the other years [1]. Traffic estimates should be available or made for each study year.

***Highway User Costs.*** Unit highway user costs consist of basic section costs, accident costs, transition costs, and delay costs [1],

$$HU = (BS + A)L + T + D \tag{14.3}$$

where $HU$ = the unit highway user cost for a given section of highway (in dollars per thousand vehicles),

$BS$ = basic section costs, consisting of the unit cost (time value and vehicle running costs) associated with vehicle flow and the basic geometrics (grades and curves) of the analysis section,

$A$ = unit accident costs in the analysis section,

$L$ = length of the analysis section in miles,

$T$ = transition cost, the additional unit user time and running costs incurred through changes in speed between analysis sections,

$D$ = additional unit time and running costs caused by delays at intersections, traffic signals, or other traffic control devices.

The *Manual on User Benefit Analysis of Highway and Bus-Transit Improvements* [1] provides detailed definitions of all these cost elements, along with a variety of nomographs to aid in their estimation. (The nomographs were developed through extensive highway research.) In our example cost elements $A$ and $D$ are assumed to be negligible. (If they are not, we can estimate them according to the procedures outlined in [1].) The nomograph shown in Figure 14.3 can be used to estimate the basic section cost ($BS$) in Eq. 14.3. As seen in Figure 14.3, estimation of the transition cost ($T$) is a by-product of basic section cost calculations.

Highway user costs are calculated from the traffic and facility data in Tables 14.2 and 14.3 by using Figure 14.3. In computing the basic section cost, we must determine the unit time value for commuters, which can be a difficult task in practice. The manual [1] also describes the general procedures for obtaining the unit time value. With an estimated unit time value for commuters of \$3 per hour,

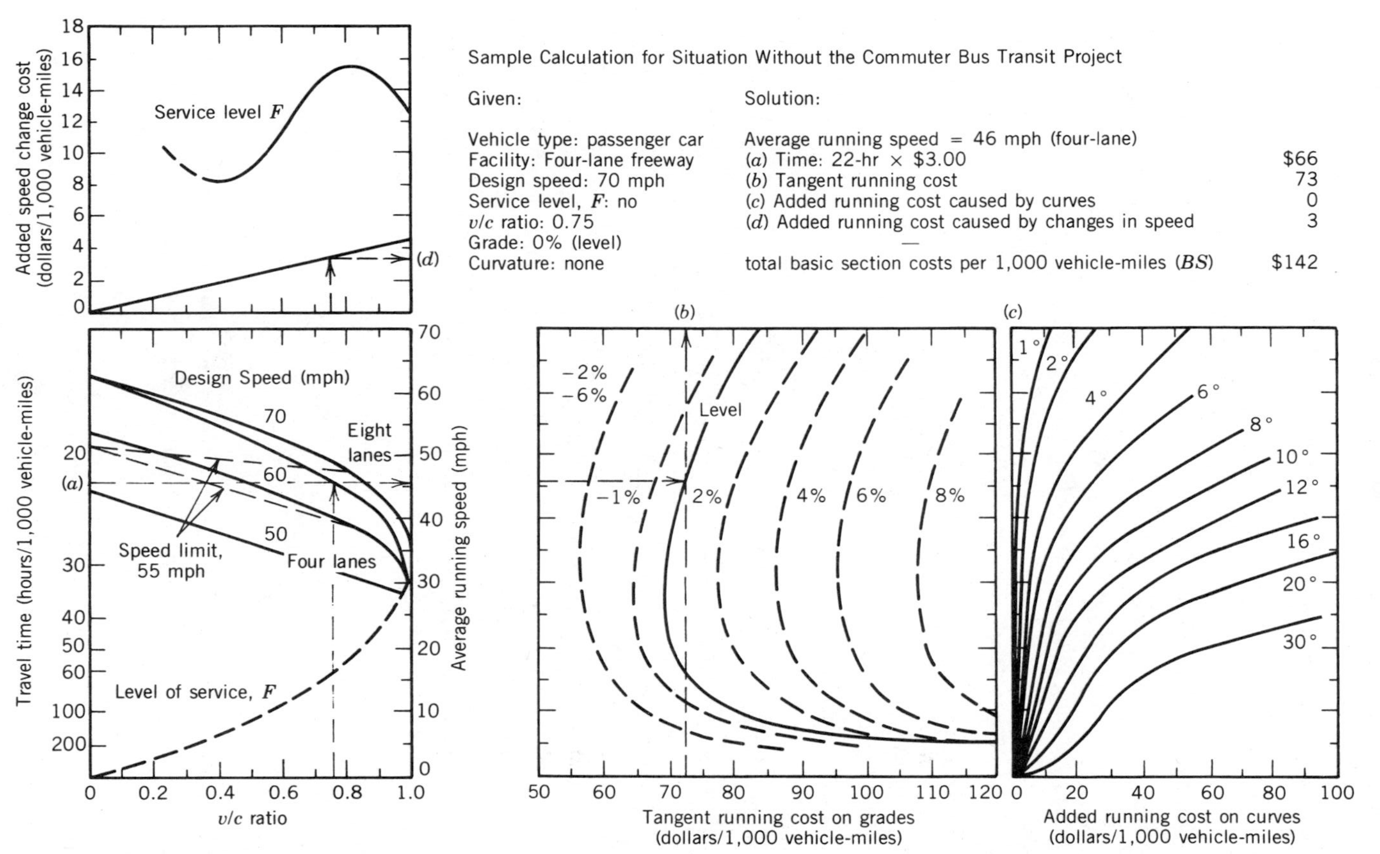


**FIGURE 14.3** Basic section (*BS*) costs for passenger cars on freeways. Source: Adapted from *A Manual on User Benefit Analysis of Highway and Bus-Transit Improvements 1977*, by the American Association of State Highway and Transportation Officials, Washington D.C., 1978.

the basic section costs (*BS*) without and with the commuter bus transit project during study year 1 are as follows.

*Without the commuter bus transit project*

Average running speed at traffic volume per capacity ($v/c$) of 0.75 = 46 mph (four lanes)

| | | |
|---|---|---|
| (*a*) Time: 22 hr × \$3.00 | = | \$66 |
| (*b*) Tangent running cost | = | 73 |
| (*c*) Added running cost caused by curves | = | 0 |
| (*d*) Added running cost caused by changes in speed | = | 3 |
| Total basic section costs per 1,000 vehicle-miles (*BS*) | = | \$142 |

*With the commuter bus transit project*

Average running speed at traffic volume per capacity ($v/c$) of 0.83 = 43 mph (three lanes)

| | | |
|---|---|---|
| (*a*) Time 23.5 hr × \$3.00 | = | \$70.5 |
| (*b*) Tangent running cost | = | 72 |
| (*c*) Added running cost caused by curves | = | 0 |
| (*d*) Added running cost caused by changes in speed | = | 3.5 |
| Total basic section costs per 1,000 vehicle-miles (*BS*) | = | \$146 |

Notice that the highway user costs increase with the commuter bus system because drivers have to travel through three-lane traffic rather than four-lane traffic, which causes more congestion. The highway user costs of the 10-mile segment for study year 1 before and after the transit service are

$$HU_0 = 142(10) = \$1{,}420/1{,}000 \text{ vehicles one way}$$
$$HU_1 = 146(10) = \$1{,}460/1{,}000 \text{ vehicles one way}$$

We can compute the highway user costs for study year 15 in a similar fashion.

$$HU_0 = 147.8(10) = \$1{,}478$$
$$HU_1 = 153(10) = \$1{,}530$$

***Commuter Bus Transit User Costs.*** Transit user costs (*TU*) consist of the value of the travel time and the money costs of a trip made in whole or in part on a bus facility. We may express them as

$$\text{TU} = 1{,}000[v(VT/60) + w(WT/60) + F] \tag{14.4}$$

where $v$ = value of in-vehicle travel time (\$),
$w$ = value of time for waiting, walking, and transferring (\$),
$VT$ = time spent in vehicles per person-trip,

$WT$ = time spent in walking, waiting, and transferring per person-trip,
$F$ = average bus fare per person-trip ($).

Since transit user costs are on a passenger trip basis (rather than a vehicle-mile basis), we must convert cost per thousand vehicles to cost per thousand person-trips. We do this by dividing by the average vehicle occupancy for a work trip. Assuming 1.22 persons per private vehicle [1] and putting the value of time at $3.00 per vehicle-hour, we find that

$$\text{Cost per person-hour} = \$3.00/1.22 = \$2.46$$

For calculating the costs to the user of the commuter transit service, the value of in-vehicle time per bus passenger is also assumed to be $2.46. The value of other transit time components (waiting, walking, and transferring) is assumed to be twice that figure, $4.92. Using the relevant bus traffic data from Tables 14.2 and 14.3, we compute one-way $TU$ per 1,000 passenger trips.

$$TU_1 = 1{,}000[2.46(19/60) + 4.92(5/60) + 0.4] = \$1{,}589$$

Since there was no commuter bus service prior to this project, we must estimate what the bus trip makers would have to pay to complete the same trips on the highway system. Recall that before this transit project, the commuter bus users were assumed to travel 3.2 miles on arterials at 28 mph and 10 miles on the freeway under conditions noted in Table 14.2. Thus, the bus user costs before the transit service consisted of the arterial user cost and the freeway user cost. Figure 14.4 will be used to estimate the cost factors for the arterial. (This figure was also developed by the American Association of State Highway and Transportation Officials [1].) The highway user cost factors for the freeway portion can be estimated from Figure 14.3 as before. For study year 1, we compute $TU_0$ as follows.

*User cost for arterial*

Average running speed at $v/c$ of 0.7 = 28 mph

| | | |
|---|---|---|
| ($a$) Time: 34 hr × $3.00 | = | $102 |
| ($b$) Tangent running cost | = | 58 |
| ($c$) Added running cost caused by curves | = | 0 |
| ($d$) Added running cost caused by changes in speed | = | 4 |
| Total user cost per 1,000 vehicle-miles | = | $164 |

The user cost for the freeway was computed previously as

$$\text{Total user cost per 1,000 vehicle-miles} = \$142$$

Thus, the total commuter bus user cost prior to the service is

$$TU_0 = 164(3.2) + 142(10) = \$1{,}944.8$$

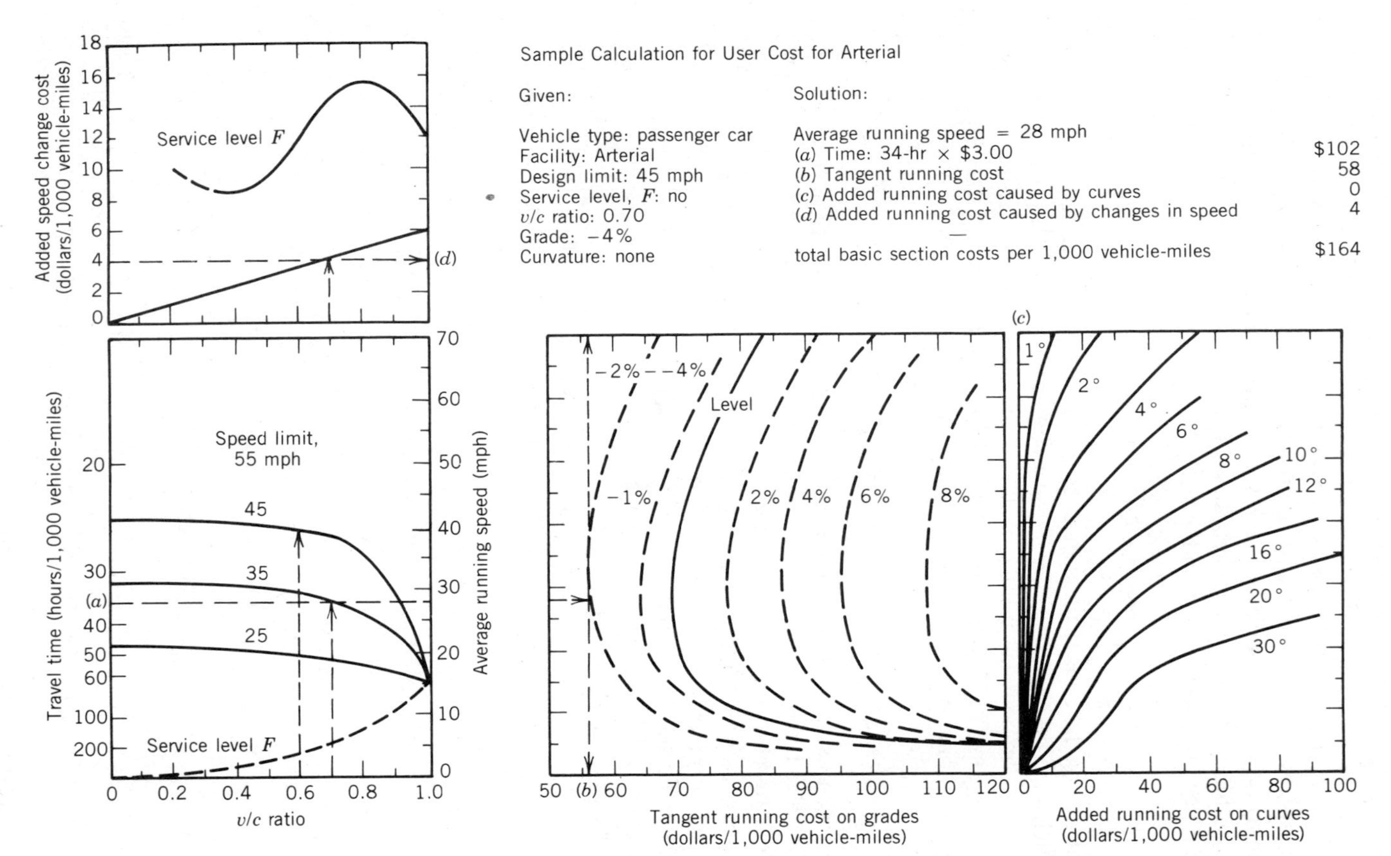


**FIGURE 14.4** Basic section (*BS*) costs for passenger cars on arterials. Source: Adapted from *A Manual on User Benefit Analysis of Highway and Bus-Transit Improvements 1977*, by the American Association of State Highway and Transportation Officials, Washington D.C., 1978.

We can compute $TU_0$ for study year 15 in a similar fashion and obtain

$$TU_0 = 164(3.2) + 147.8(10) = \$2{,}002.8$$

Table 14.4 summarizes the user cost to each group.

***User Benefit Calculation.*** Now we can express the annual user benefits of the transit improvement project by the reduction in transit user costs and highway user costs.

$$\begin{aligned}\text{Annual user benefits} &= \text{Benefits of transit service}\\ &\quad + \text{Benefits of reducing highway user costs}\end{aligned}$$

If the highway user costs are increased rather than decreased by the transit service, we treat them as disbenefits to the users. More precisely, we express annual user benefits as

$$\text{Annual user benefits} = N(TU_0 - TU_1) + V(HU_0 - HU_1) \qquad (14.5)$$

where $N$ = average number of person-trips annually via commuter bus (in thousands) for the base case ($N_0$) and transit improvement ($N_1$), $N = (N_0 + N_1)/2$,

$V$ = average traffic level on the affected highway facility annually, before and after the transit service, $V = (V_0 + V_1)/2$.

In our example the difference in user cost estimates without and with the new transit service for each study year is multiplied by the average volume to calculate user benefits. These are summarized in Table 14.5.

**Table 14.4** ***Summary of Users' Costs***

*Highway User Costs*

| Section | Alternative | Year | Period | *BS* | × Section Length, *L* | = One-Way *HU*, per Thousand Vehicles |
|---|---|---|---|---|---|---|
| *AB* | 0 | 1 | Peak hour | 142 | × 10 miles | = $1,420 |
| | 1 | 1 | Peak hour | 146 | × 10 miles | = $1,460 |
| | 0 | 15 | Peak hour | 147.8 | × 10 miles | = $1,478 |
| | 1 | 15 | Peak hour | 153 | × 10 miles | = $1,530 |

*Bus Transit User Costs*

| Alternative | Mode | Year | Period | One-Way *TU* per Thousand Passenger Trips |
|---|---|---|---|---|
| 0 | Auto | 1 | Peak hour | $1,944.8 |
| 1 | Bus | 1 | Peak hour | $1,589 |
| 0 | Auto | 15 | Peak hour | $2,002.8 |
| 1 | Bus | 15 | Peak hour | $1,589 |

**Table 14.5** ***Calculations of Users' Benefits***

***Highway User Benefits***

| Year | Period | Days/Year | × Hours/Day | × $(HU_0 - HU_1)$ | × $(V_0 + V_1)/2$ | = | Annual One-Way Benefits |
|---|---|---|---|---|---|---|---|
| 1 | Peak hours | 250 | × 3 | × (1,420 − 1,460) | × (6.0 + 5.0)/2 | = | −$165,000 |
| 15 | Peak hours | 250 | × 3 | × (1,478 − 1,530) | × (7.0 + 5.4)/2 | = | −$241,800 |

Growth rate: $-241{,}800 = -165{,}000(1 + g)^{15-1}$

$g = 2.7674\%$

***Bus Transit User Benefits***

| Year | Period | Days/Year | × Hours/Day | × $(TU_0 - TU_1)$ | × $(N_0 + N_1)/2$ | = | Annual One-Way Benefits |
|---|---|---|---|---|---|---|---|
| 1 | Peak hours | 250 | × 3 | × (1,944.8 − 1,589) | × (0 + 2.0)/2 | = | $266,850 |
| 15 | Peak hours | 250 | × 3 | × (2,002.8 − 1,589) | × (0 + 2.5)/2 | = | $387,938 |

Growth rate: $387{,}938 = 266{,}850(1 + g)^{15-1}$

$g = 2.7086\%$

### 14.3.3 Sponsor's Costs

The cost of implementing and operating a transit system are usually considered from two standpoints, (1) capital costs and (2) operating and maintenance costs.

***Capital Costs.*** Capital costs are defined as investments in fixed facilities or equipment such as the following.

1. Special busways or improvements in existing or proposed roadways necessitated by the transit service.
2. Terminals, shops, vehicle storage facilities, administrative offices, and other similar structures.
3. Benches, shelters, bus stop signs, and other route-side equipment.

Vehicle costs might be considered as capital costs, but because of the active rental and used-vehicle markets available to transit agencies, bus vehicle costs are often considered as the costs of vehicle rental or the operating costs for depreciable items [1].

In our example there will be no capital costs associated with the transit system because the new service will be operated through existing facilities.

***Operating and Maintenance Costs.*** The costs of operating a transit system vary with the level, type, and speed of bus operation and may be related to the vehicle mileage generated by the system. They basically entail the cost of

1. Driver's wages and fringe benefits.
2. Vehicle operation, including tires, gasoline, and lubricants.
3. Bus maintenance, including the cost of replacement parts.
4. Insurance and administrative labor.
5. Vehicle rental or depreciation.
6. Roadway maintenance attributable to the transit system.

We may need to estimate these cost components individually. A nomograph such as that shown in Figure 14.5 has been used, however, to estimate annual bus system operating costs as a function of four variables. This nomograph is from an empirical study performed by the Institute of Defense Analysis [1]. The study statistically examined the relationship between bus transit operating costs and several causal variables such as driver wages, vehicle size, service speed, and vehicle-miles produced. We assume that a typical statistical relationship obtained by the study is

$$OC = 2.37Q^{1.013}H^{0.785}S^{-0.862}e^{-0.002P} \qquad (14.6)$$

where $Q$ = annual bus-miles produced by the transit system,
$P$ = bus size in terms of seating capacity,
$H$ = bus driver wages (the nomograph assumes about 25% fringe benefits),
$S$ = average service speed.

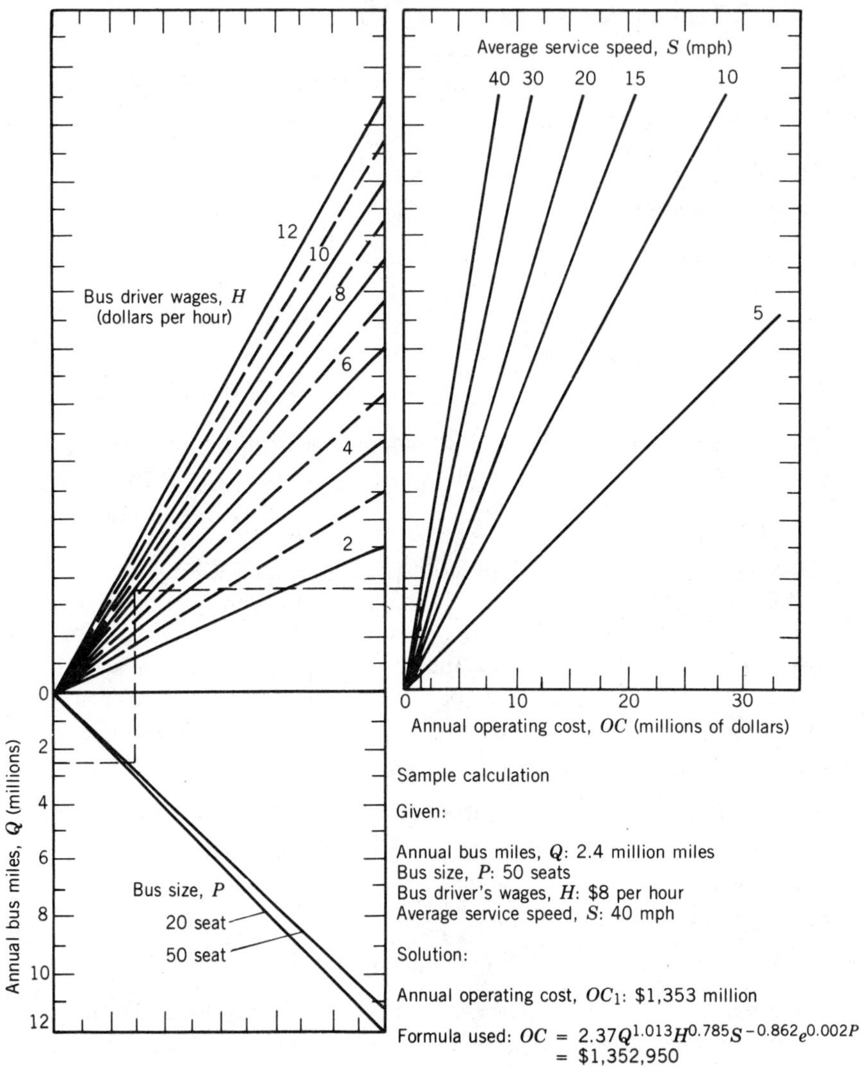


**FIGURE 14.5 Annual operating cost of a bus transit system ($OC$). Source: Adapted from *A Manual on User Benefit Analysis of Highway and Bus-Transit Improvements 1977*, by the American Association of State Highway and Transportation Officials, Washington D.C., 1978.**

In our example bus-operating costs may be calculated by using Figure 14.5 or Eq. 14.6 and the transit facility data given in Table 14.2. For study year 1, the annual operating costs are

$$Q = 2.4 \text{ million miles} \qquad P = 50 \text{ seats}$$

$$H = \$8/\text{hour} \qquad S = 40 \text{ mph}$$

Therefore, $OC_1 = \$1{,}352{,}950$. With study year 15, $Q = 3$ million miles, and the rest of the variables remain unchanged, we obtain $OC_{15} = \$1{,}696{,}101$.

The transit costs are offset to a large degree by annual transit fare revenue, which must be subtracted to avoid double counting. Annual transit revenue is obtained by computing

$$\text{Annual fare revenue} = (\text{Fare per trip})(\text{Hourly patronage})(\text{Hours per day}) \times (\text{Number of service days per year})$$

For study years 1 and 15, the annual fare revenues are

$$\text{Fare revenue for year 1} = (0.4)(2{,}000)(3\text{ hr} \times \text{two peak directions})(250)$$
$$= \$1{,}200{,}000$$

$$\text{Fare revenue for year 15} = (0.4)(2{,}500)(3\text{ hr} \times \text{two peak directions})(250)$$
$$= \$1{,}500{,}000$$

The net annual operating costs for both study years are

$$OC_1 = \$1{,}352{,}950 - \$1{,}200{,}000 = \$152{,}950$$
$$OC_{15} = \$1{,}696{,}101 - \$1{,}500{,}000 = \$196{,}101$$

Assuming an annual highway maintenance cost of $1,000 per lane-mile for the 20 lane-miles of the project gives us a total annual maintenance cost of $1,000(20) = $20,000.

### 14.3.4 Benefit–Cost Ratio for Project

The last step of the benefit–cost analysis requires calculation of the total present value of annual benefits and costs over the entire analysis period. The most accurate calculation of the present value of a series of benefits or costs requires an estimate of annual benefits or costs for each year of the analysis period. Since calculations of year-by-year values are laborious and often add little to accuracy, the suggested practice is to select only one, two, or three years of the analysis period for detailed study and extrapolate for the other years.

Assuming that benefits will grow at approximately a constant annual rate over the analysis period (resulting in a geometric progression), we first find this implied growth rate by using Eq. 2.20. For the highway user benefits, we may proceed as follows.

$$\text{Benefits for year 15} = (\text{Benefits for year 1})(1 + g)^{N-1}$$
$$-\$241{,}800 = -\$165{,}000(1 + g)^{15-1}$$

Solving for $g$, we obtain $g = 2.7674\%$. Similarly, we find the growth rate for the transit user benefits to be $g = 2.7086\%$.

Assuming annual compounding and using the present-worth factor for the geometric series in Table 2.4 for a project period of 25 years, we can compute the present value of user benefits to each group. When these are multiplied by 2, to account for symmetric two-way traffic, we obtain the total present value of benefits over 25 years, assuming a 4% inflation-free discount rate.

**1.** Present value of highway users' benefits (disbenefits),

$$PV_1 = -\$165{,}000(P/A, 2.7674\%, 4\%, 25) = -165{,}000(20.9110)$$
$$= -\$3{,}450{,}316$$

**2.** Present value of transit users' benefits,

$$PV_2 = \$266{,}850(P/A, 2.7086\%, 4\%, 25) = 266{,}850(20.7754)$$
$$= \$5{,}543{,}921$$

**3.** Total users' benefits for two-way traffic,

$$B = 2(PV_1 + PV_2) = 2(\$2{,}093{,}605) = \$4{,}187{,}210$$

Notice that although users benefit in total, highway users actually receive disbenefits because of increased congestion on the highway.

To obtain an indication of economic desirability, we must also calculate the present value of project costs. The growth factor for the *OC* based on the two one-year study periods is calculated as follows:

$$\$196{,}101 = \$152{,}950(1 + g)^{15-1}$$

Solving for $g$, we obtain

$$g = 1.7910\%.$$

**1.** Present value of operating the bus system,

$$PV_1 = 152{,}950(P/A, 1.7910\%, 4\%, 25) = 152{,}950\ (18.8023)$$
$$= \$2{,}875{,}814$$

**2.** Present value of the annual maintenance costs for the exclusive bus lane (note that there is no growth factor for the series),

$$PV_2 = 20{,}000(P/A, 4\%, 25) = \$312{,}442$$

**3.** Present value of total sponsor's costs,

$$C = PV_1 + PV_2 = \$3{,}188{,}256$$

Finally, we compute the benefit–cost ratio:

$$R_A = \frac{B}{C} = \frac{4{,}187{,}210}{3{,}188{,}256} = 1.31 > 1$$

The $B/C$ ratio greater than one indicates that the transit project should be considered unless there are offsetting environmental or other considerations not yet included in the study.

Notice that, in our study, future benefits and cost were calculated in constant dollars; the discount rate for performing present value calculations should represent the real opportunity cost. The real cost of capital has been estimated at about 4% for low-risk investments. If benefits and costs are projected in actual dollars (inflated dollars), the full current market rate of interest should be used. A range of 8 to 12% has commonly been used to represent the average long-term market interest rate in recent economic studies of public projects. (Note that the U.S. Office of Management and Budget prescribes a 10% discount rate, which represents such a market interest rate with the effects of inflation considered.) The constant-dollar approach is commonly recommended in public projects, since it avoids the need for speculation about future inflation in arriving at the economic merit of the project.

## 14.4 RISK AND UNCERTAINTY IN BENEFIT–COST ANALYSIS

Up to this point we have assumed that all the values of the economic parameters were known with certainty. In particular, correct estimates of the values for individual cash flows were assumed to be available. In this section we consider the consequences of introducing uncertainty in the benefit–cost analysis of public investments.

When the variabilities in estimates of benefits and costs are expressed in probabilistic terms, the benefit–cost ratio becomes a quotient of two different random variables (i.e., benefit–cost). The analytical procedure for evaluating such a quotient of random variables was discussed in Section 10.2.4.

In this section we first present an analytical framework that incorporates the risk elements in the benefit–cost analysis. Next, an exact probability distribution of the benefit–cost ratio for a project whose cash flow streams are normally distributed, with known means and variances, is given. Then a simulation approach is outlined for a project with nonnormal cash flows. The material in this section is based on an article by Park [13].

### 14.4.1 Exact Distribution of Benefit–Cost Ratio

The problem of the frequency distribution of ratios has been treated in [3, 11, 7, 8]. References [7] and [8] are most relevant to our model.

If both $I$ and $C'$ are normally distributed, the sum $C = I + C'$ is also normally distributed. If $B$ and $C$ are normally distributed random variables with means $\mu_j$, variances $\sigma_j^2$ ($j = 1$ for $B$ and $j = 2$ for $C$), and correlation coefficient $\rho$, the probability distribution of $R_A$ is as follows.

Let $R_A = B/C$; consequently, the probability that $R_A$ is less than or equal to $a$ is

$$\begin{aligned} P\{R_A \leq a\} &= P\left\{\frac{B}{C} \leq a\right\} \\ &= P\{B - aC \leq 0 \mid C > 0\} \cdot P\{C > 0\} \\ &\quad + P\{B - aC > 0 \mid C < 0\} \cdot P\{C < 0\} \\ &= L\{h, -k; \gamma\} + L\{-h, k; \gamma\} \end{aligned} \tag{14.7a}$$

where

$$h = \frac{\mu_1 - \mu_2 a}{G(a)} \tag{14.7b}$$

$$k = \frac{\mu_2}{\sigma_2} \tag{14.7c}$$

$$\gamma = \frac{\sigma_2 a - \rho\sigma_1}{G(a)} \tag{14.7d}$$

$$G(a) = \sigma_1\sigma_2\left\{\frac{a^2}{\sigma_1^2} - \frac{2\rho a}{\sigma_1\sigma_2} + \frac{1}{\sigma_2^2}\right\}^{1/2} \tag{14.7e}$$

and $L\{h, k; \gamma\}$ is the standard bivariate normal integral tabulated by the National Bureau of Standards [12]. Figure 14.6 shows a plot of $B$ as a function of $C$ and indicates the $P\{R_A \leq a\}$ by the two shaded areas.

We note that if $k \to \infty$, that is, as $P\{C > 0\} \to 1$, Eq. 14.7a becomes

$$P\{R_A \leq a\} \to \Phi\{-h\} \tag{14.8}$$

or with $P\{C > 0\} = 1$,

$$P\{R_A \leq a\} = \Phi\{-h\} \tag{14.9}$$

where

$$\Phi(y) = \int_{-\infty}^{y} \left(\frac{1}{\sqrt{2\pi}}\right) e^{-(1/2)u^2} \quad du$$

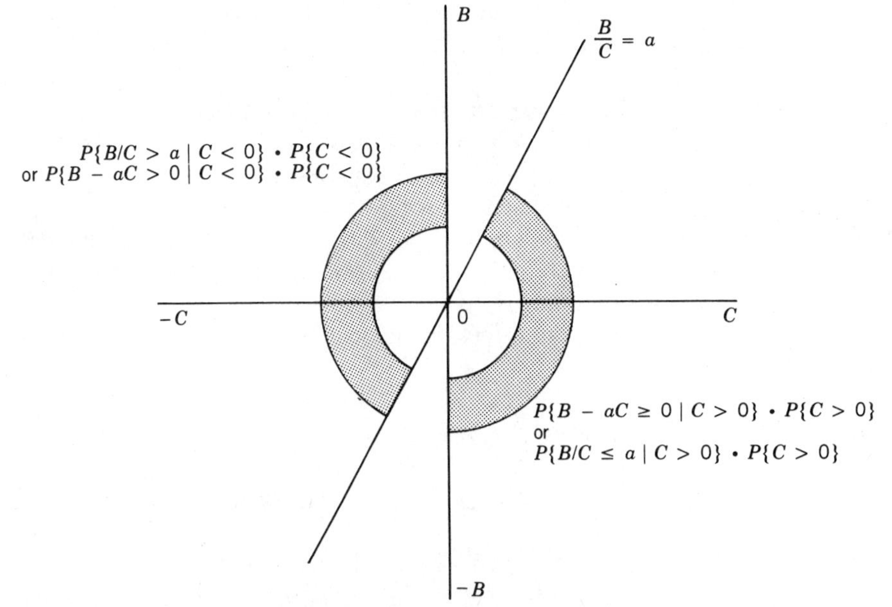


**FIGURE 14.6** Graphical illustration of P$\{B/C \leq a\}$.

This implies that, if the random variable $C$ is defined only over the nonnegative range, we can compute an exact probability from the standard normal probability table.

Recall Eq. 14.1, in which the acceptability of a project by the $R_A$ criterion was defined over $C > 0$. What if $C < 0$? Is there any economic meaning for a negative cost? Theoretically, a negative cost means a benefit to the project, which in turn has no cash outflow from the project. It is impractical, however, to conceive of such a situation in evaluating public projects. Therefore, it is natural to preclude the condition ($C < 0$), so the use of Eq. 14.9 will be sufficient for evaluating a single risky project.

### 14.4.2 Exact Distribution of Incremental Benefit–Cost Ratio

A graphic illustration will help us understand the nature of the probabilistic incremental benefit–cost ratio. Figure 14.7 shows $\Delta B$ as a function of $\Delta C$. Obviously, the line passing through the origin gives the incremental net present value equal to zero, or $\Delta B/\Delta C = 1$. For illustration we may divide the benefit–cost quadrants into six sections.

Recall Eq. 14.2 and the $\Delta R_A$ decision rules for comparing two mutually exclusive projects. That is, if we compute the incremental differences for $\Delta B$, $\Delta I$, and $\Delta C'$ based upon (project 1 − project 2), the $\Delta R_A$ criterion will select project 1 whenever the following conditions hold:

$$\Delta R_A > 1 \quad \text{with } \Delta I + \Delta C' > 0$$

or

$$\Delta R_A < 1 \quad \text{with } \Delta I + \Delta C' < 0$$

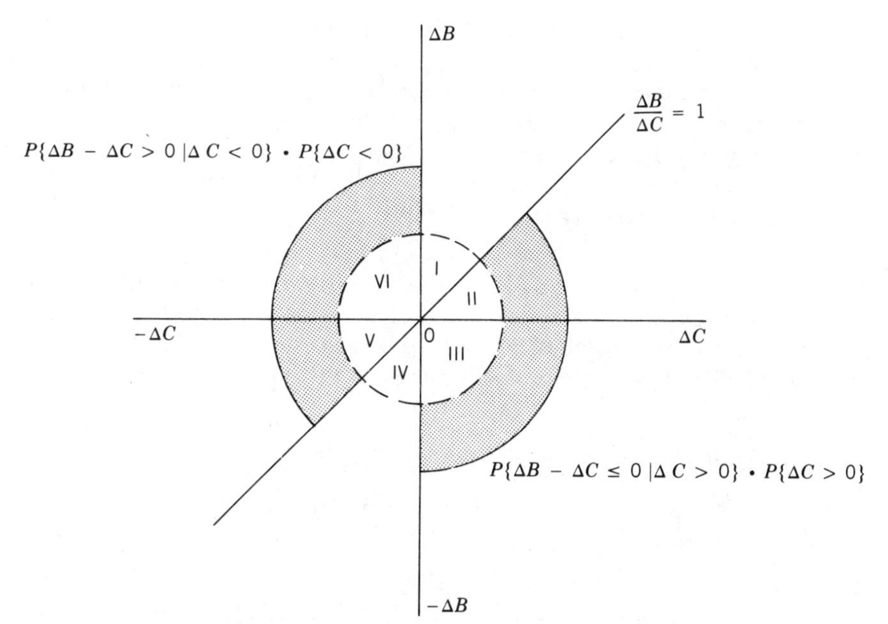


**FIGURE 14.7** Graphical illustration of P{$\Delta B/\Delta C \leq 1$}.

**Table 14.6** ***Sign Combinations and Decision Rules Based on (Project 1 − Project 2)***

| Section | Signs $\Delta B$ | Signs $\Delta C$ | Ratio, $\Delta R_A = \frac{\Delta B}{\Delta C}$ | Decision (Project 1 or 2) |
|---|---|---|---|---|
| I | + | + | >1 | 1 |
| II | + | + | <1 | 2 |
| III | − | + | <1 | 2 |
| IV | − | − | <1 | 2 |
| V | − | − | >1 | 1 |
| VI | + | − | >1 | 1 |

In referring to Figure 14.7, we can determine all sign combinations of $\Delta B$ and $\Delta C$ to form a magnitude of the ratio $\Delta B/\Delta C$ that leads to selecting project 1. We will not consider the do-nothing alternative. Table 14.6 summarizes all the possible sign combinations and decision rules to apply for each combination.

Clearly, we will select project 2 if and only if the $\Delta R_A$ falls in section II, III, or IV. To compute the probability of preferring project 2 over project 1, we may still use Eq. 14.7a by simply replacing the $(B, C, a)$ with the $(\Delta B, \Delta C, 1)$. That is,

$$P\{\text{project 2 is preferred over project 1}\}$$
$$= P(\Delta B, \Delta C \in \text{II,III}) + P(\Delta B, \Delta C \in \text{IV})$$

or

$$P\{\text{project 1} < \text{project 2}\} = P\{\Delta B - \Delta C \le 0 \mid \Delta C > 0\} \cdot P\{\Delta C > 0\} + P\{\Delta B - \Delta C < 0 \mid \Delta C < 0\} \cdot P\{\Delta C < 0\} \quad (14.10)$$

Since we have assumed normal distributions for both $\Delta B$ and $\Delta C$ with means $\mu_j$, variances $\sigma_j^2$ ($j = 1$ for $\Delta B$ and $j = 2$ for $\Delta C$), $a = 1$, and correlation coefficient $\rho$ if any, Eq. 14.10 becomes

$$\begin{aligned} P\{\text{project 1} < \text{project 2}\} &= L\{h, -k; \gamma\} + [P\{\Delta C < 0\} - L\{-h, k; \gamma\}] \\ &= L\{h, -k; \gamma\} + [\Phi(-k) - L\{-h, k; \gamma\}] \end{aligned} \quad (14.11)$$

We can easily verify the result of Eq. 14.11 to be consistent with the *PV* criterion. If we use the *PV* criterion, the probability that the *PV* of project 2 exceeds that of project 1 is

$$\begin{aligned} P\{PV_1 \le PV_2\} &= P\{B_1 - C_1 \le B_2 - C_2\} \\ &= P\{(B_1 - B_2) - (C_1 - C_2) \le 0\} \\ &= P\{\Delta B - \Delta C \le 0\} \\ &= \Phi\left(\frac{0 - (\mu_1 - \mu_2)}{\sqrt{\sigma_1^2 + \sigma_2^2 - 2\rho\sigma_1\sigma_2}}\right) \end{aligned} \quad (14.12)$$

### *Example 14.3*

Consider the two mutually exclusive projects whose capital outlays, benefits, and costs are predicted to be normally distributed with means and variances given in Table 14.7. For simplicity, we will assume statistical independence among the cash flows and between the projects ($\rho = 0$).

At a discount rate of 15%, we compute the *PV*s of *I, B,* and $C'$ to be normally distributed with the following means and variances.

$$I_1 = N(3{,}388.47,\ 629.26^2) \qquad I_2 = N(2{,}899.81,\ 461.49^2)$$

$$B_1 = N(12{,}673.56,\ 715.93^2) \qquad B_2 = N(10{,}645.79,\ 762.22^2)$$

$$C_1' = N(1{,}140.62,\ 127.08^2) \qquad C_2' = N(1{,}013.88,\ 115.53^2)$$

Then the *PV*s on incremental capital outlays, benefits, and costs are also normally distributed as

$$\Delta B = B_1 - B_2 = N(2{,}027.77,\ 1045.72^2) = N(\mu_1, \sigma_1^2)$$

$$\Delta I = I_1 - I_2 = N(488.66,\ 780.35^2)$$

$$\Delta C' = C_1' - C_2' = N(126.74,\ 171.74^2)$$

$$\Delta C = \Delta I + \Delta C = N(615.39,\ 799.02^2) = N(\mu_2, \sigma_2^2)$$

Thus, the incremental $\Delta R_A$ becomes the ratio of two normal variables.

Now from Eq. 14.7 with $\rho = 0$ and $a = 1$, we obtain

$$h = \frac{2{,}027.77 - (615.39)(1)}{1{,}316.043} = 1.0732$$

$$k = \frac{615.39}{799.02} = 0.7702$$

$$\gamma = \frac{799.02}{1{,}316.043} = 0.6071$$

$$G(1) = (1{,}045.72)(799.02)\left[\frac{1}{(1{,}045.72)^2} + \frac{1}{(799.02)^2}\right]^{1/2}$$

Substituting these values into Eq. 14.7a and utilizing the bivariate normal table in [14] yields

$$\begin{aligned} P\{\Delta R_A \le 1\} &= L\{1.0732, -0.7702; 0.6071\} \\ &\quad + L\{-1.0732, 0.7702; 0.6071\} \\ &= 0.13802 + 0.21744 = 0.35546 \end{aligned}$$

The value 0.35546 is simply the probability of $\Delta R_A$ being less than or equal to 1; it does not tell us the probability of project 2 being more economically attractive than project 1. To find this probability, we use Eq. 14.11 with $a = 1$.

**Table 14.7** ***Input Data***

| | Project 1 | | | Project 2 | | |
|---|---|---|---|---|---|---|
| Period $n$ | $I$ (mean, SD) | $B$ (mean, SD) | $C'$ (mean, SD) | $I$ (mean, SD) | $B$ (mean, SD) | $C'$ (mean, SD) |
| 0 | (2,500, 600) | | | (2,000, 400) | | |
| 1 | (500, 200) | | | (600, 250) | | |
| 2 | (600, 100) | | | (500, 100) | | |
| 3 | | (5,000, 500) | (450, 110) | | (4,200, 500) | (400, 100) |
| 4 | | (5,000, 500) | (450, 110) | | (4,200, 600) | (400, 100) |
| 5 | | (5,000, 500) | (450, 110) | | (4,200, 700) | (400, 100) |
| 6 | | (5,000, 800) | (450, 110) | | (4,200, 800) | (400, 100) |
| 7 | | (5,000, 1,000) | (450, 110) | | (4,200, 900) | (400, 100) |

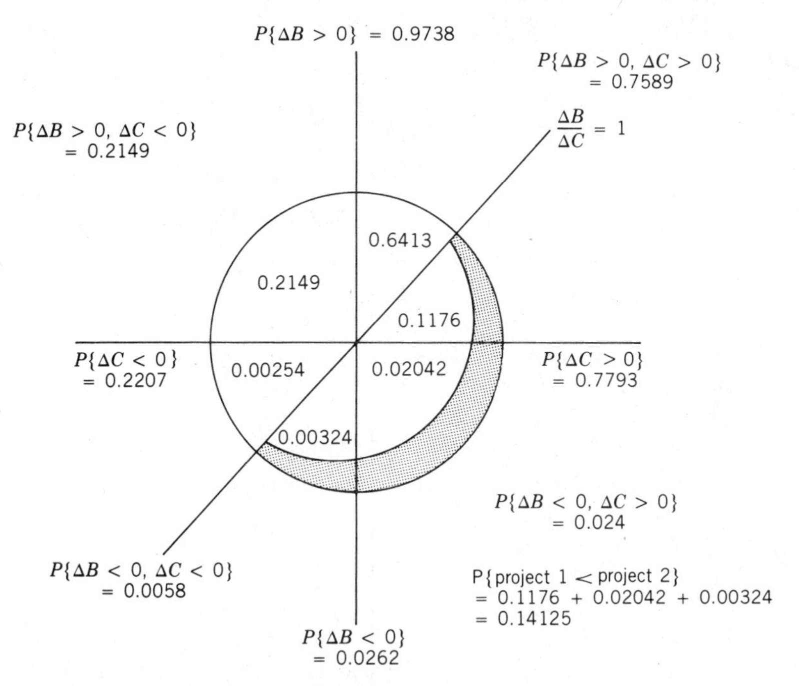


**FIGURE 14.8 Graphical illustration of computing P{project 1 < project 2}.**

$$\begin{aligned} P\{\text{project 1} < \text{project 2}\} &= L\{1.0732, -0.7702; 0.6071\} \\ &\quad + \Phi(-0.7702) - L\{-1.0732, -0.7702; 0.6071\} \\ &= 0.13802 + (0.2207 - 0.21744) = 0.1412 \end{aligned}$$

Figure 14.8 illustrates the specific probability figures applicable to each possible combination of $\Delta B$ and $\Delta C$. By varying the value of $a$, we can similarly describe the complete distribution picture for the incremental benefit–cost profile.

We will confirm the foregoing result by evaluating Eq. 14.12 to obtain

$$\Phi\left(\frac{0 - (2027.77 - 615.39)}{\sqrt{1045.72^2 + 799.02^2}}\right) = \Phi\{-1.0732\} = 0.1407$$

The slight difference between the two results (0.1407 versus 0.1412) is due to rounding errors associated with interpolating the bivariate normal table entries. □

***Computer Simulation Approach.*** If benefits and costs are not normally distributed, we may obtain a $B/C$ ratio distribution through computer simulation.

## Table 14.8 *Simulation Results*

| Observed Frequency | Relative Frequency | Cumulative Frequency | Upper Cell Limit ($B/C$ Ratio) |
|---|---|---|---|
| a. *B/C Ratio (project 1 − project 2)* | | | |
| 1 | 0.001 | 0.001 | −20 |
| 0 | 0.000 | 0.001 | −19 |
| 0 | 0.000 | 0.001 | −18 |
| 0 | 0.000 | 0.001 | −17 |
| 0 | 0.000 | 0.001 | −16 |
| 0 | 0.000 | 0.001 | −15 |
| 0 | 0.000 | 0.001 | −14 |
| 0 | 0.000 | 0.001 | −13 |
| 0 | 0.000 | 0.001 | −12 |
| 0 | 0.000 | 0.001 | −11 |
| 0 | 0.000 | 0.001 | −10 |
| 0 | 0.000 | 0.001 | −9 |
| 1 | 0.001 | 0.001 | −8 |
| 0 | 0.000 | 0.001 | −7 |
| 0 | 0.000 | 0.001 | −6 |
| 0 | 0.000 | 0.001 | −5 |
| 0 | 0.000 | 0.001 | −4 |
| 1 | 0.001 | 0.002 | −3 |
| 2 | 0.001 | 0.003 | −2 |
| 6 | 0.004 | 0.007 | −1 |
| 29 | 0.019 | 0.026 | 0 |
| 232 | 0.151 | 0.177 | 1 |
| 432 | 0.281 | 0.457 | 2 |
| 259 | 0.168 | 0.625 | 3 |
| 158 | 0.103 | 0.728 | 4 |
| 86 | 0.056 | 0.784 | 5 |
| 53 | 0.034 | 0.818 | 6 |
| 43 | 0.028 | 0.846 | 7 |
| 31 | 0.020 | 0.866 | 8 |
| 19 | 0.012 | 0.879 | 9 |
| 23 | 0.015 | 0.894 | 10 |
| 15 | 0.010 | 0.903 | 11 |
| 12 | 0.008 | 0.911 | 12 |
| 9 | 0.006 | 0.917 | 13 |
| 13 | 0.008 | 0.925 | 14 |
| 8 | 0.005 | 0.931 | 15 |
| 7 | 0.005 | 0.935 | 16 |
| 7 | 0.005 | 0.940 | 17 |
| 6 | 0.004 | 0.944 | 18 |
| 6 | 0.004 | 0.947 | 19 |
| 11 | 0.007 | 0.955 | 20 |
| 70 | 0.045 | 1.000 | INF |
| $N_1 = 1540$ | | | |

$M_1 = 272$ (rows −20 through 1)

*(continued)*

**Table 14.8** *(continued)*

| Observed Frequency | Relative Frequency | Cumulative Frequency | Upper Cell Limit ($B/C$ Ratio) |
|---|---|---|---|
| ***b. B/C Ratio (project 2 − project 1) with*** $\Delta I + \Delta C' < 0$ | | | |
| 80 | 0.174 | 0.174 | −19 |
| 0 | 0.0 | 0.174 | −18.5 |
| 2 | 0.004 | 0.178 | −18 |
| 2 | 0.004 | 0.183 | −17.5 |
| 1 | 0.002 | 0.185 | −17 |
| 2 | 0.004 | 0.189 | −16.5 |
| 0 | 0 | 0.189 | −16 |
| 1 | 0.002 | 0.191 | −15.5 |
| 1 | 0.002 | 0.193 | −15 |
| 4 | 0.009 | 0.202 | −14.5 |
| 5 | 0.011 | 0.213 | −14 |
| 4 | 0.009 | 0.222 | −13.5 |
| 5 | 0.011 | 0.233 | −13 |
| 6 | 0.013 | 0.246 | −12.5 |
| 4 | 0.009 | 0.254 | −12 |
| 5 | 0.011 | 0.265 | −11.5 |
| 8 | 0.017 | 0.283 | −11 |
| 6 | 0.013 | 0.296 | −10.5 |
| 7 | 0.015 | 0.311 | −10 |
| 10 | 0.022 | 0.333 | −9.5 |
| 4 | 0.009 | 0.341 | −9 |
| 1 | 0.002 | 0.343 | −8.5 |
| 8 | 0.017 | 0.361 | −8 |
| 6 | 0.013 | 0.374 | −7.5 |
| 12 | 0.026 | 0.400 | −7 |
| 11 | 0.024 | 0.424 | −6.5 |
| 3 | 0.007 | 0.430 | −6 |
| 11 | 0.024 | 0.454 | −5.5 |
| 21 | 0.046 | 0.500 | −5 |
| 14 | 0.030 | 0.530 | −4.5 |
| 24 | 0.052 | 0.583 | −4 |
| 25 | 0.054 | 0.637 | −3.5 |
| 24 | 0.052 | 0.689 | −3 |
| 26 | 0.057 | 0.746 | −2.5 |
| 28 | 0.061 | 0.807 | −2 |
| 24 | 0.052 | 0.859 | −1.5 |
| 22 | 0.048 | 0.907 | −1 |
| 19 | 0.041 | 0.948 | −0.5 |
| 10 | 0.022 | 0.970 | 0 |
| 4 | 0.009 | 0.978 | 0.5 |
| 0 | 0.000 | 0.978 | 1 ] $M_2 = 10$ |
| 10 | 0.022 | 1.000 | INF ] |
| $N_2 =$ 460 | | | |

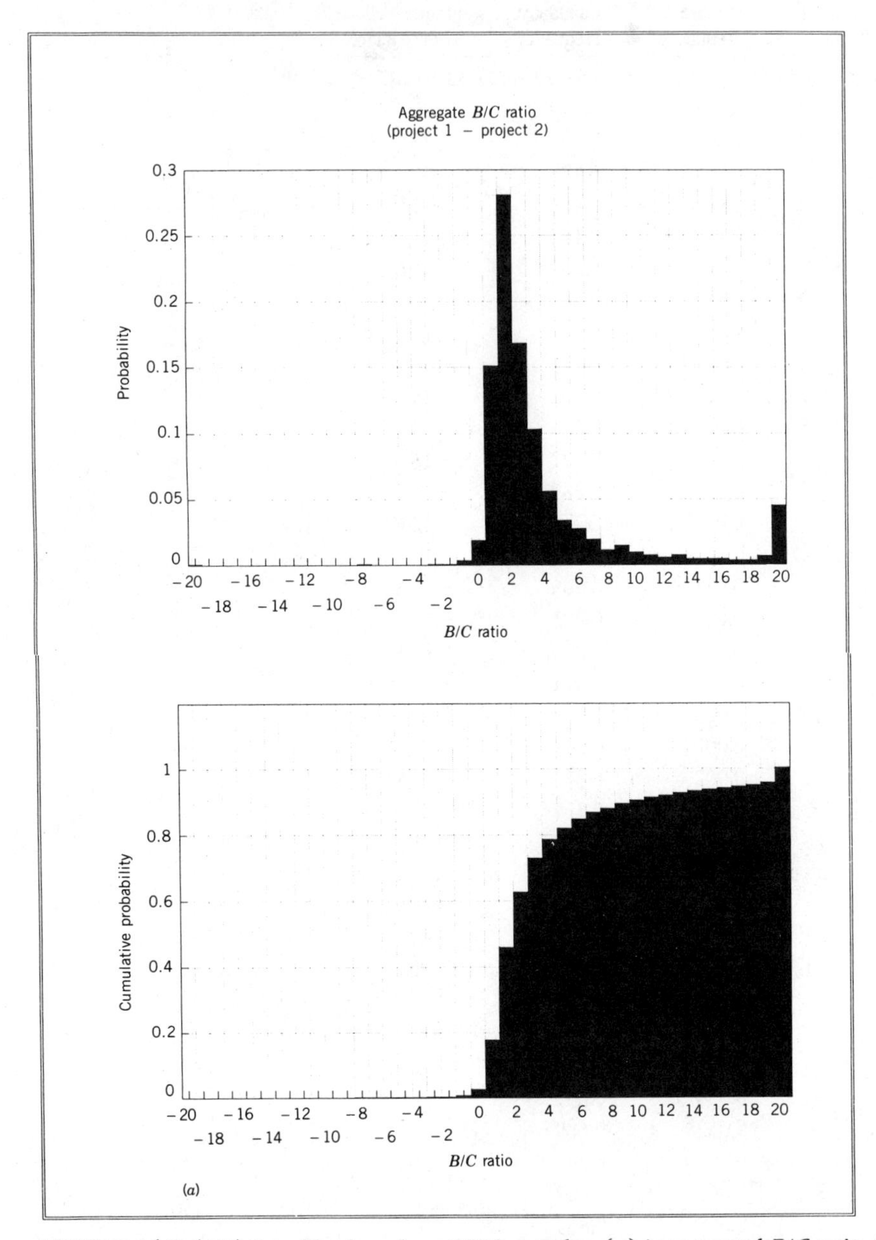


**FIGURE 14.9** $\Delta B/\Delta C$ **profiles based on 2,000 samples: (*a*) incremental** $B/C$ **ratio distributions based on (project 1 − project 2); (*b*) incremental** $B/C$ **ratio distributions based on (project 2 − project 1).**

(See Chapter 12.) A proper procedure for creating a $B/C$ ratio profile (incremental) is as follows.

1. Generate $N$ observations of $B$, $I$, $C'$ for each project.
2. Compute $\Delta B$, $\Delta I$, and $\Delta C'$ based on (project 1 − project 2) from the

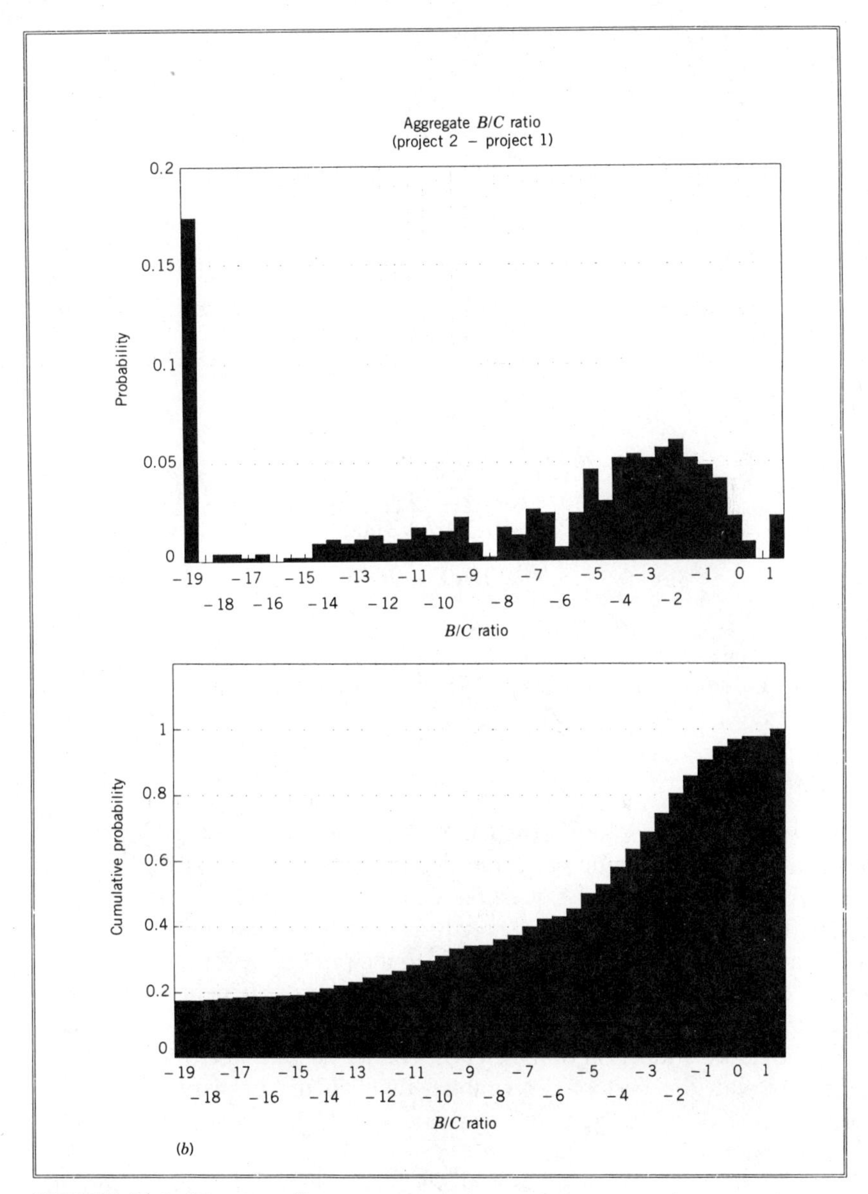


**FIGURE 14.9 (Continued)**

computer samples. Count how many times you observe $\Delta I + \Delta C' > 0$ and let this be $N_1$. (See Table 14.8a.)

3. Count how many times you observe $\Delta R_A < 1$ with $\Delta I + \Delta C' > 0$. Let this be $M_1$. (See Table 14.8a.)
4. Compute $\Delta B$, $\Delta I$, and $\Delta C'$ based on (project 2 − project 1) from the same computer sample and count how many times you observe $\Delta I + \Delta C' < 0$. Let this be $N_2$ (note that $N = N_1 + N_2$). (See Table 14.8b.)

5. Count how many times you observe $\Delta R_A > 1$ with $\Delta I + \Delta C' < 0$ and let this be $M_2$. (See Table 14.8b.)
6. Finally, compute

$$P\{\text{project 2} > \text{project 1}\} = \left(\frac{M_1}{N_1}\right)\left(\frac{N_1}{N}\right) + \left(\frac{M_2}{N_2}\right)\left(\frac{N_2}{N}\right).$$

### *Example 14.4*

Let us use a simulation approach to compute $P$ {project 2 > project 1} for the data in Example 14.3 with $N = 2{,}000$. We count

$$M_1 = 272, \qquad N_1 = 1{,}540$$

$$M_2 = 10, \qquad N_2 = 460$$

$$\begin{aligned} P\{\text{project 2} > \text{project 1}\} &= \left(\frac{272}{1{,}540}\right)\left(\frac{1{,}540}{2{,}000}\right) + \left(\frac{10}{460}\right)\left(\frac{460}{2{,}000}\right) \\ &= 0.1360 + 0.005 \\ &= 0.1410 \end{aligned}$$

The results are summarized in Table 14.8 and Figure 14.9. □

## 14.5 SUMMARY

In this chapter we have examined the nature of public investment activities. One of the critical issues in public project evaluation is how to quantify benefits in dollar terms. To do so, we must question the extent of benefits to be measured by distinguishing between benefits directly attributable to the project (primary benefits) and benefits indirectly attributable to the project (secondary benefits). A benefit–cost analysis should always consider the primary benefits and should consider the secondary benefits whenever appropriate. Inclusion of secondary benefits should be a function of their effect compared with that of the primary benefits and with the cost of determining them. If the primary benefits far exceed the cost of the project, we may not have to look into the extent of secondary benefits.

Two different benefit–cost ratios, the aggregate benefit–cost ratio and the netted benefit–cost ratio, have been widely used in evaluation of public investment proposals. Even though there has been much discussion of the superiority of one ratio over the other, both ratios can serve properly as a measure of investment worth. In particular, as long as the incremental analysis approach is used, either criterion will always result in the same project selection as the present value criterion.

Introducing the risk element in computing the benefit–cost ratio complicates the decision process. But it was possible to develop an exact probability distribution of the benefit–cost ratio by assuming that the cash flow streams are normally distributed. Thus, the exact probability can be computed with straightforward formulas, circumventing the need to use computer simulation to obtain similar probabilistic information. If a normal distribution cannot be assumed, computer simulation to obtain the $B/C$ ratio distribution becomes inevitable.

## REFERENCES

1. American Association of State Highway and Transportation Officials, *A Manual on User Benefit Analysis of Highway and Bus-Transit Improvements,* 444 North Capital Street, N.W., Washington, D.C. 20001, 1978.
2. Baumol, W. J., "On the Social Rate of Discount," *American Economic Review,* Vol. 58, pp. 788–802, 1968.
3. Curtis, J. H., "On the Distribution of the Quotient of Two Chance Variables," *Annals of Mathematical Statistics,* Vol. 12, pp. 409–421, 1941.
4. Dasgupta, A. K., and D. W. Pearce, *Cost–Benefit Analysis: Theory and Practice,* Barnes and Noble, New York, 1972.
5. Drake, A. W., R. L. Keeney, and P. M. Morse, editors, *Analysis of Public Systems,* MIT Press, Cambridge, Mass., 1972.
6. "Discount Rates to Be Used in Evaluating Time-Distributed Costs and Benefits," Circular A-94, Revised, Executive Office of the President, Bureau of the Budget, Washington, D.C., March 1972.
7. Hayya, J., D. Armstrong, and N. Gressis, "Note on the Ratio of Two Normally Distributed Variables," *Management Science,* Vol. 21, No. 11, pp. 1338–1341, July 1975.
8. Hinkley, D. V., "On the Ratio of Two Correlated Normal Random Variables," *Biometrika,* Vol. 56, No. 3, p. 635, 1969.
9. Hinrichs, H. H., and G. M. Taylor, editors, *Program Budgeting and Benefit–Cost Analysis—Cases, Text and Readings,* Goodyear Publishing Company, 1969.
10. Lutz, R. P., and H. A. Cowles, "Estimation Deviations: Their Effect upon the Benefit–Cost Ratio," *The Engineering Economist,* Vol. 16, No. 1, pp. 21–42, Fall 1971.
11. Marsaglia, G., "Ratios of Normal Variables and Ratios of Sums of Uniform Variables," *Journal of American Statistical Association,* Vol. 60, No. 309, pp. 193–204, 1965.
12. National Bureau of Standards, *Tables of the Bivariate Normal Distribution Function and Related Functions,* Applied Mathematics Series, No. 50, U.S. Government Printing Office, Washington, D.C., 1955.
13. Park, C. S., "Probabilistic Benefit–Cost Analysis," *The Engineering Economist,* Vol. 29, No. 2, pp. 83–100, 1984.
14. Schwab, B., and P. Lusztig, "A Comparative Analysis of the Net Present Value and the Benefit–Cost Ratio as Measure of Economic Desirability of Investment," *Journal of Finance,* Vol. 24, pp. 507–516, 1969.
15. Tohagan, J. K., *Quantitative Analysis for Public Policy,* McGraw–Hill, New York, 1980, Ch. 13.
16. Winfrey, R., *Economic Analysis of Highways,* International Textbook Company, Scranton, Pa., 1969.

## PROBLEMS

**14.1.** The U.S. Postal Service (USPS) recently decided to continue the nine-digit zip code (ZIP + 4) for first-class business mailers and to purchase additional capital equipment as part of postal automation. Prior to this decision, the Office of Technology

Assessment (OTA) and its contractors performed technical and economic analyses and evaluations of the options available to the U.S. Congress and the Postal Service. The optical character readers (OCRs) that the USPS had purchased in the past and proposed to purchase in the future were single-line OCRs. Single-line OCRs read the last line of an address, which usually consists of the city, state, and ZIP code. An alternative technology, the multiple-line OCR, is capable of reading up to four lines of an address. Investment in single-line OCRs and associated site preparation and program contingency costs would occur in year 0 through year 2. At the same time, research costs would be incurred to develop single-line to multiple-line conversion equipment. Conversion equipment would be procured and installed in year 3 through year 5. Maintenance, spare parts, and address information costs would occur throughout the entire horizon of the analysis, years 0 to 14. Revenue will be lost because a rate reduction will be granted to mailers who use ZIP + 4. The planned reduction is 0.5 cents per item of presorted first-class mail. The General Accounting Office (GAO) estimated an annual cost of $140 million at 90% usage. The economic justification was the expected savings in labor through reduced sorting.

| | Year | | | | | | |
|---|---|---|---|---|---|---|---|
| Costs and Savings | 0 | 1 | 2 | 3 | 4 | 5 | 6...14 |
| Single-line equipment | (140)* | (140) | (113) | | | | |
| Site preparation and contingency | (20) | (20) | (16) | | | | |
| Research | (5) | (5) | (5) | | | | |
| Conversion equipment | | | | (44) | (44) | (44) | |
| Maintenance and spares† | (28) | (34) | (37) | (43) | (58) | (99) | (86)...(142) |
| Address information† | (32) | (31) | (16) | (17) | (18) | (19) | (21)...(34) |
| Rate reduction | (58) | (89) | (117) | (136) | (140) | (140) | (140)...(140) |
| Clerk and carrier savings† | 161 | 269 | 466 | 712 | 827 | 888 | 1,016...1,676 |

*Thousands of dollars.

†Linear growth was assumed during the years between year 6 and 14.

a. Identify the values of $I$, $C$, $C'$, and $B$, assuming an interest rate of 10%.
b. Determine the user's benefits and the sponsor's costs.
c. Compute the $B/C$ ratio and recommend the best course of action.

**14.2.** A state is considering two different types of bridge design.

| | Design Type | |
|---|---|---|
| Basic Data | A | B |
| Initial construction cost | $200,000 | $250,000 |
| Renewal cost, end-of-service life | 100,000 | 125,000 |
| Annual maintenance cost | 1,000 | 1,500 |
| Periodic repairs every 5 years | 5,000 | 2,000 |
| Salvage value at end-of-service life | 10,000 | 15,000 |
| Service life (years) | 20 | 30 |

If the state's interest rate is 6%, compute the $B/C$ ratio for each design. Assuming a planning horizon of 60 years, which alternative design should the state select?

**14.3.** A state highway department is considering two alternative routes to relieve ever-growing traffic congestion on the present highway.

| Basic Data | Present Highway | Route 1 | Route 2 |
|---|---|---|---|
| Total construction costs | — | $6,500,000 | $8,200,000 |
| Annual maintenance costs | $200,000 | 400,000 | 300,000 |
| Equivalent annual vehicle-miles of travel | 35,860,000 | 22,125,000 | 20,368,000 |
| Total user cost per vehicle-mile | $0.223 | $0.205 | $0.191 |

The equivalent annual maintenance costs include required maintenance for sections of existing highway to be kept in service. Using an analysis period of 20 years and assuming a zero salvage value, determine which route should be constructed at $i = 6\%$.

**14.4.** A problem that frequently comes up in the public transportation sector is the selection of accident prevention countermeasures on state public highways and bridges. Suppose a state is considering the following set of projects recommended for evaluation at four different locations and assume the budget is $9,000.

| Location | Alternative | Benefit | Cost | $B/C$ Ratio |
|---|---|---|---|---|
| I | I-A | $40,000 | $11,000 | 3.64 |
| | I-B | 32,000 | 9,000 | 3.56 |
| | I-C | 10,000 | 2,500 | 4.00 |
| II | II-A | 35,000 | 5,200 | 6.73 |
| | II-B | 20,000 | 3,010 | 6.64 |
| III | III-A | 10,000 | 1,000 | 10.00 |
| | III-B | 30,000 | 4,600 | 6.52 |
| IV | IV-A | 5,000 | 490 | 10.20 |
| | IV-B | 12,000 | 1,200 | 10.00 |

Using the principle of benefit–cost analysis, determine the best combination of projects within the budget constraint.

**14.5.** The following five independent program elements within a restructured Department of Interior are being considered for funding and have annual costs as indicated. The department's budget for new project expenditures is $60 million per year. The objective of the department is to maximize the total utility.

| Project Number | Utility Value | Project Nature | Expenditure Year 1 | Year 2 | Year 3 |
|---|---|---|---|---|---|
| 1 | 30 | Acquiring new national forests | 20 | 10 | 5 |
| 2 | 40 | Constructing new national parks | 10 | 40 | 20 |
| 3 | 70 | Building a new dam on the Colorado River | 10 | 60 | 95 |
| 4 | 60 | Financing a 5-year oil exploration project for Indian-owned land | 20 | 20 | 20 |
| 5 | 50 | R&D to develop cheap methods of treating polluted streams | 5 | 10 | 20 |

Pressure from Congress requires that project 3 or 4 be funded. Project 2 is applicable only to project 1. Project 5 is contingent on implementation of project 3. The department can transfer any unspent funds from one fiscal year to another. Each proposal is considered to be an indivisible unit, and it is not possible to undertake "multiples" of any public project.

a. Formulate an integer programming model for the problem.

b. Solve the problem on a computer and identify the projects that should be funded.

**14.6.** A metropolitan city has come up with several water conservation measures that are not mutually dependent but can work together to save consumer and taxpayer dollars.

| Measure | Description | Benefit | Cost |
|---|---|---|---|
| 1 | Realistic rate structure that leads to a sensible pricing policy | $8* | $3 |
| 2 | Universal metering of all water users | $2 | $1 |
| 3 | Improved landscape practices to reduce outdoor water use | $1.5 | $1 |
| 4 | New building code requiring efficient plumbing equipment | $3.5 | $2 |
| 6 | Public education on television and radio | $2.3 | $1.8 |
| 7 | Provision of an accelerated leak detection program to the public | $1.8 | $1.3 |
| 8 | Coordination of water operations with neighboring cities | $5.7 | $4.2 |

*Millions of dollars.

The city has a special water conservation fund of $5 million. Select the best conservation measures based on benefit–cost analysis.

**14.7.** Consider the two mutually exclusive projects whose capital outlays, benefits, and costs are predicted to be normally distributed with means and variances given in the following.

| | Project 1 | | | Project 2 | | |
|---|---|---|---|---|---|---|
| Period $n$ | $c_n$ | $b_n$ | $c_n$ | $c_n$ | $b_n$ | $c_n$ |
| 0 | $(10, 2^2)$ | | | $(12, 3^2)$ | | |
| 1 | $(5, 1^2)$ | | | $(8, 2^2)$ | | |
| 2 | | $(9, 2^2)$ | $(4, 1^2)$ | | $(12, 3^2)$ | $(5, 1^2)$ |
| 3 | | $(13, 3^2)$ | $(6, 1^2)$ | | $(18, 3^2)$ | $(8, 1^2)$ |
| 4 | | $(16, 3^2)$ | $(8, 2^2)$ | | $(22, 3^2)$ | $(10, 2^2)$ |
| 5 | | $(12, 3^2)$ | $(7, 2^2)$ | | $(14, 3^2)$ | $(5, 1^2)$ |

Assume that the benefit–cost flows between the periods and between the projects are statistically independent. The period of investment is $m = 1$.

a. For each project, compute the means and variances of $I$, $B$, and $C'$ at $i = 6\%$.

b. Determine the $B/C$ probability distribution for each project.

c. Determine the incremental $B/C$ distribution (project 2 − project 1).

d. Determine the probability of preferring project 2 over project 1.

**14.8.** Consider the two mutually exclusive projects whose capital outlays, benefits, and costs are predicted in the following.

| | Project 1 | | | Project 2 | | |
|---|---|---|---|---|---|---|
| Period $n$ | $c_n$ | $b_n$ | $c_n$ | $c_n$ | $b_n$ | $c_n$ |
| 0 | (8, 10, 12) | | | (10, 12, 15) | | |
| 1 | (4, 5, 7) | | | (6, 8, 9) | | |
| 2 | | $(9, 2^2)$ | (3, 5) | | $(12, 3^2)$ | (4, 6) |
| 3 | | $(13, 3^2)$ | (5, 7) | | $(18, 3^2)$ | (6, 10) |
| 4 | | $(16, 3^2)$ | (6, 10) | | $(22, 3^2)$ | (8, 14) |
| 5 | | $(12, 3^2)$ | (5, 10) | | $(14, 3^2)$ | (5, 7) |

Assume that the benefit–cost flows between the periods and between the projects are statistically independent.

Note 1: For $n = 0,1$ the $c_n$ have *triangular* distributions with the three estimates (low, most likely, high) specified. The period of investment is $m = 1$.

Note 2: $b_n$ are *normally* distributed with means and variances specified.

Note 3: For $n = 2$ throught 5 the $c_n$ are *uniformly* distributed with the lower and upper bounds specified.

a. Using computer simulation, develop a $B/C$ ratio distribution for each project. (Assume 100 Monte Carlo samplings.)
b. Determine $P\{\text{project } 2 > \text{project } 1\}$ through computer simulation.

## Acknowledgments

- Problem 14.1 is based on an article by J. W. Ulvila, "Postal Automation (ZIP + 4) Technology: A Decision Analysis," *Interfaces,* Vol. 17, No. 2, pp. 1–12, 1987.
- Data in Problem 14.2 were extracted from a paper by D. T. Phillips and W. F. McFarland, "Optimal Economic Selection of Highway Accident Prevention Countermeasures," Proceedings of 1979 Spring Annual Institute of Industrial Engineers Conference.

# 15

# Economic Analysis in Public Utilities

## 15.1 INTRODUCTION

In this chapter we will consider a method commonly used by public utilities (power, gas, and telephone companies) to evaluate capital investments. This method is often referred to as the revenue requirement (*RR*). Although traditionally used for public utilities, the method can be applied as well to the private sector.

Before we explain the revenue requirement, we need to understand the nature of public utilities. Because of the capital-intensive nature of the business and for economic efficiency, public utility industries in the United States and elsewhere are granted monopoly privileges. To ensure that the public interest is best served by such firms, various government agencies and commissions regulate them. If a utility operates wholly within one state, its primary regulatory agency will be the utility commission or agency in that state. Interstate utilities are subject to regulation by agencies in the several states in which they operate.

One of the principal public interests is to regulate the utilities' profits. Thus, the rates charged by utilities are normally subject to review and approval by the regulatory agency. On the other hand, utilities are allowed to price their services (or products) within the reasonable limit that all costs are recovered, including a fair return on the rate base. There can be considerable variation in the interpretation of the term "fair return." In our presentation we will follow the definition given in Barish and Kaplan [2].

***Definition.*** "A fair return should be one which is sufficiently high so as to permit the utility to render high quality service to existing customers, to allow funds to be raised at reasonable interest rates for expansion to meet demands of new customers, to permit the timely replacement of used or worn out equipment, and to permit research into future improvement in the quality and costs of service."

In practice, the level of return on their base that constitutes a fair return to the utilities is often a political issue. Both parties (regulatory agency and utilities) normally do not agree on this issue. Since the regulatory agencies act on behalf of the public interest, they tend to allow utilities the lowest possible return within the reasonable limit. On the other hand, utilities seek a higher return so that they can maintain competitiveness with other nonregulated industries in attracting new investment funds, by providing investors with a higher return. (We will not dwell on this issue, but interested readers can find more in Jeynes [3].)

In Section 15.2 we review the concepts underlying the costs of the various types of capital invested by utilities. Precise definitions of these costs are essential to the development of the revenue requirement method. In Section 15.3 we formally define the elements of the revenue requirement and the relationships between cost of capital, cash flow definitions, and the levelized annual revenue requirement. We also present the technique for handling inflation in revenue requirement analysis. In Section 15.4 we show the equivalence between the *RR* method and conventional methods such as the *PV* and *IRR*. This equivalence assures us of the usefulness of the *RR* method in analyzing nonregulated industry. In Section 15.5 we extend the *RR* method to handle the capitalized interest during the construction period, the investment tax credit if allowed, and an accelerated depreciation schedule such as MACRS. Two methods are introduced, the flow-through and normalizing methods.

## 15.2 CAPITAL COSTS

### 15.2.1 Debt and Equity Financing for Public Utilities

It is important to understand the sources of investment funds for utilities. Broadly speaking, in public utilities two types of financing are common, debt and equity.

Debt financing includes both short-term borrowing from financial institutions and the sale of long-term bonds; bonds borrow money from investors for a fixed period. With debt financing, the interest paid on the bonds is treated as an expense for income tax purposes. As we will see in a later section, such interest charges on a debt are often referred to as the return on debt or debt costs and are usually considered separately from other tax-deductible expenses of the utility in income tax calculations.

Equity financing consists of funds invested by current and new owners of the utility. These funds come from the sale of common stock, from current earnings, and from retained earnings. Equity cost is the opportunity cost associated with equity capital, in the sense that this capital could be invested elsewhere if not used for the project. The equity cost is frequently referred to as the return on equity.

Another variation of equity financing is through the sale of preferred stock. Preferred stock is a hybrid form of financing, combining features of debt and common stock. In the event of liquidation, the claim of a holder of preferred stock on assets comes after that of creditors but before that of holders of common stock. Preferred stock carries a stipulated dividend, but the actual payment of a dividend is a discretionary obligation, although many utilities regard the obligation as fixed. Thus, the dividend payments are viewed as the financing cost but, like equity cost, are not treated as tax-deductible expenses. For this reason issuing preferred stocks is an expensive financing method compared with debt financing, because debt cost is a tax-deductible expense.

One reason for the use of preferred stock by utilities is that the Securities and Exchange Commission stated in 1952 that the capital structure of an electrical utility should not exceed a 60% debt and that common stock should not be less than 30%. Thus the 10% residual could be filled by preferred stock. Another reason is that a public utility can pass on the cost of preferred stock in the rates it charges. In other words, public utilities commissions allow utilities to base their rates on their overall measured cost of capital, including the preferred stock.

### 15.2.2 Weighted After-Tax Cost of Capital

When future investments are pooled from several financing sources at different costs, the weighted cost of capital is used as a composite index that reflects the cost of raising funds from different sources. Although we detailed the concept of weighted cost of capital in Chapter 5, we will present it in the context of use in public utilities. As we did in the previous chapters, we will use the symbols B/T and A/T to denote before tax and after tax.

The following example shows the computations needed to find the weighted cost of capital. The estimates assumed for future financing ratios and interest costs are as follows.

| Source of Financing | Percent of Total Funds | | B/T Cost | A/T Cost |
|---|---|---|---|---|
| Debt | | 57% | | |
| Short term | 0% | | 15% | |
| Long term | 57% | | 14% | |
| Equity | | 43% | | |
| Common stock | 32% | | | 30% |
| Preferred stock | 11% | | | 14% |

Note that in these data the equity cost is expressed in terms of A/T cost. The reason is that any return either to holders of common stock or to holders of preferred stock is made after payment of income taxes. Before we compute the A/T cost of capital, we must calculate the composite return on debt, $i_b$, and return on equity, $i_e$.

$$i_b = (0/0.57)0.15 + (0.57/0.57)0.14 = 0.14$$

$$i_e = (0.32/0.43)0.30 + (0.11/0.43)0.14 = 0.2591$$

Assuming an income tax rate of 40%, the A/T cost of capital, $k$, for this example is

$$k = (1 - 0.40)(0.57)(0.14) + (0.43)0.2591 = 0.1593$$

The 15.93% is the A/T cost of capital that a utility with the financial structure illustrated would expect to pay to its investors.

Although most firms try to keep the different types of capital roughly in some proportion in the balance sheet, at any particular time one type or another will be used to finance a set of projects. The traditional approach separates the investment from the financing decision in that no specific assumption is made about the exact instruments that will be used to finance a particular project. The one assumption made is that over time the capital structure proportions will be maintained. To illustrate the potential problems associated with matching the investment decision with a specific financing source, consider the following two investment situations.

- This year we will finance all projects with a bond issue at an after-tax cost of $14\%(1 - 0.40) = 8.4\%$. Suppose project X has an expected return of 9%. With 8.4% money financing, project X returns more than it costs. Is it acceptable?
- Next year we will finance all projects with a stock issue at an after-tax cost of 30%. Suppose project Y has an expected return of 25%. Financed with 30% money, project Y costs more than it returns. Is it acceptable?

If we used the cost of specific funds acquired to finance a project as the discount rate, we would accept X and reject Y. Does it make any sense to accept a 9% project and then reject a 25% project? The problem with the preceding approach is a failure to recognize that, over the long run, the *mix* of financing sources will be maintained. Projects must earn enough to cover a weighted average of various costs of capital, so particular sources *cannot* be linked to specific uses when assessing economic viability. This is a very important assumption used in evaluating public utility projects.

If this weighted average cost (15.93%) is used as the project discount rate, project X will be rejected and project Y will be accepted. The A/T cost of capital has other names in the literature: tax-adjusted cost of capital and tax-adjusted weighted average cost of capital. As we will see in a later section, the appropriate cost of capital for computing the equivalent annual revenue requirement is the A/T cost of capital. Now that we can see why it is necessary to use a weighted average cost of all the forms of capital we intend to use over the long run, we can turn to the issue of calculating the required return (weighted by the proportion of each form we wish to maintain) to investors from each financing source.

### 15.2.3 Capital Recovery Cost Based on Book Depreciation Schedule

In developing the *RR* method, the concept of capital recovery cost plays a critical role. The capital recovery cost was briefly introduced in Chapter 2 in the context of computing the cash flow equivalence. We now present the concept here in the context of computing revenue requirements.

The capital recovery cost at any period consists of two elements: capital repayment (principal) amount and interest cost incurred on the unpaid balance at the beginning of the period. To illustrate the concept, consider a $1,000 investment committed at time 0 with a negligible salvage value and a 4-year life. Suppose that the capital (principal) will be repaid in equal annual installments of $250. This type of accounting for a capital repayment schedule is called straight-line book depreciation. (The reader should not confuse this with the tax depreciation discussed in Chapter 4.) Book depreciation simply refers to how the original investment principal is to be repaid. We assume that the designated

**Table 15.1** *Capital Recovery Costs*

| Year | Beginning Unrecovered Balance | Return on Invested Capital | Capital (Principal) Repayment | Capital Recovery Cost |
|---|---|---|---|---|
| 1 | 1,000 | 120 | 250 | 370 |
| 2 | 750 | 90 | 250 | 340 |
| 3 | 500 | 60 | 250 | 310 |
| 4 | 250 | 30 | 250 | 280 |

NOTE: $PV(\text{Capital recovery costs}) = 370(P/F, 12\%, 1) + 340(P/F, 12\%, 2) + 310(P/F, 12\%, 3) + 280(P/F, 12\%, 4) = \$1{,}000$

portion of the capital is repaid at each installment, and that interest costs are incurred at the end of year but only on the amount left unpaid at the beginning of that year. The annual capital recovery costs with an interest rate of 12% are shown in Table 15.1. The present value of the capital recovery cost is $1,000 at the interest rate of 12%. This was found by adding return (interest cost) to each capital repayment amount and then discounting each combined amount at 12% interest. This example is just another demonstration of the equivalence shown in Chapter 2 by the relation

$$PV(\text{Capital recovery costs}) = PV(\text{First cost}) \quad (15.1)$$

The example of the $1,000 investment can be extended to include salvage value (either positive or negative). Suppose we expect a salvage value of $200 at the end of the investment's 4-year life. This implies that only $800 ($1,000 − $200) must be repaid over the project life. Again assuming a straight-line book depreciation, the annual equal installments would be $200 ($800/4). Table 15.2 illustrates the capital recovery cost calculations. Note also the relationship

$$PV\ (\text{Capital recovery costs}) = PV(\text{First cost}) - PV(\text{Salvage value}) \quad (15.2)$$

When the principal payments are made each year, the question is which financing source of capital is repaid. A common assumption in public utility economic analysis is that the principal payments are a constant percentage of the book depreciation, so that the debt–equity ratio remains unchanged over the planning horizon.

## 15.3 THE REVENUE REQUIREMENT METHOD

In this section we present one common method used to determine revenue requirements. The reader should recognize, however, that different regulatory agencies may require different approaches and assumptions in deriving the revenue requirements. There are also some variations in defining the methodology such as the minimum revenue requirement discipline [3].

**Table 15.2** *Capital Recovery Costs with Salvage Value*

| Year | Beginning Unrecovered Balance | Return on Invested Capital | Capital (Principal) Repayment | Capital Recovery Cost |
|---|---|---|---|---|
| 1 | 1,000 | 120 | 200 | 320 |
| 2 | 800 | 96 | 200 | 296 |
| 3 | 600 | 72 | 200 | 272 |
| 4 | 400 | 48 | 200 | 248 |

NOTE: $PV(\text{capital recovery costs}) = 320(P/F, 12\%, 1) + 296(P/F, 12\%, 2) + 272(P/F, 12\%, 3) + 248(P/F, 12\%, 4)$

$= \$873$

$PV(\text{salvage value}) = 200(P/F, 12\%, 4) = \$127$

$PV(\text{capital recovery costs}) = PV(\text{first cost}) - PV(\text{salvage value})$

$= 1{,}000 - 127$

$= \$873$

### 15.3.1 Assumptions of the Revenue Requirement Method

With public utilities a common method of comparison is to determine the equivalent annual revenue requirements for each alternative and choose the alternative with the minimum requirements. The term "annual revenue requirements" commonly means the total amount of funds needed for capital recovery, income taxes, operating expenses, and other allowed expenses such as property taxes, all reduced to an annual basis. If the actual annual equivalent revenue from the project exceeds the minimum revenue requirement, the project is considered "profitable." The underlying philosophy in this approach is that the charges are to be recovered from revenues paid by the utility's customers. Therefore, the alternative that requires the customer costs to be minimized is preferred in the absence of intangible factors.

The following assumptions are also made throughout the remainder of this chapter [7].

1. Unrecovered investment in an asset during any one year will be equal to its book value at the beginning of that year.
2. Debt capital invested in an asset during any one year will be a fraction of its book value during that year, and this fraction will remain constant throughout the asset's life.
3. Equity, preferred stock, and debt capital give fixed rates of return to the investors throughout the life of the project.
4. Book depreciation charges are used to retire stock issues and bond issues each year in proportion to the debt–equity financing used.
5. The effective income tax rate is constant over the life of the project.

### 15.3.2 Determination of Annual Revenue Requirements

***Required Revenues.*** The first component of the annual revenue requirement is the capital recovery cost resulting from capital investment. As we explained before, the capital recovery cost includes book depreciation (capital repayment amount) and return to investors (interest expenses).

The second component is income tax. When we combine these two components (capital recovery cost + income tax), the resulting costs are called *carrying charges.* The following equation is used to find the annual carrying charges at the end of year $n$,

$$CC_n = D'_n + RI_n + T_n \tag{15.3a}$$

where $D'_n$ = book depreciation in year $n$,
$RI_n$ = return to investors in year $n$,
$T_n$ = income taxes in year $n$.

The return to investors, $RI_n$, can be broken into two components,

$$RI_n = RD_n + RE_n \tag{15.3b}$$

$$RD_n = d(i_b U_{n-1}) \tag{15.3c}$$

$$RE_n = (1 - d)(i_e U_{n-1}) \tag{15.3d}$$

where $RD_n$ = return on debt in year $n$,
$RE_n$ = return on equity in year $n$,
$d$ = proportion of debt financing,
$U_{n-1}$ = unrecovered investment (book balance) at end of year $n-1$,
$i_b$ = interest rate on borrowed capital,
$i_e$ = rate of return on equity capital.

The revenue requirement in year $n$ is defined as

$$RR_n = CC_n + C_n \tag{15.4}$$

where $RR_n$ = revenue requirement in year $n$,
$C_n$ = summation of all recurring annual expenses (including property taxes) in year $n$.

Figure 15.1 illustrates the elements of the revenue requirement and their relationships to one another.

***Income Tax Computation.*** Income tax treatment of a public utility is no different from that of any other profit-making corporation. A public utility pays taxes on the basis of taxable income. Its tax payments are affected by the method of depreciation used and the current tax law. Since tax depreciation and interest paid on debt are tax-deductible, income taxes in any particular year can be estimated with the equation

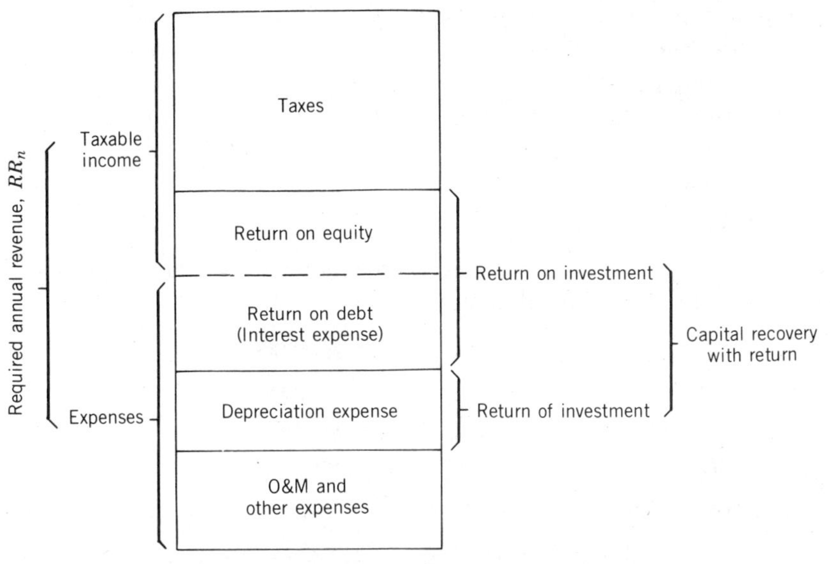


**FIGURE 15.1** Elements of revenue requirements.

$$\begin{aligned} T_n &= (\text{Taxable income})t_m \\ &= (\text{Revenue} - \text{Expenses} - \text{Debt interest} - \text{Tax depreciation})t_m \\ &= (RR_n - C_n - RD_n - D_n)t_m \\ &= (CC_n - RD_n - D_n)t_m \end{aligned} \tag{15.5}$$

where $D_n$ = tax depreciation in year $n$,
$t_m$ = marginal income tax rate in year $n$ (constant over study period).
Substituting Eq. 15.3 for $CC_n$ into Eq. 15.5 and solving for $T_n$, we obtain

$$T_n = \frac{t_m}{1 - t_m}(RE_n + D'_n - D_n) \tag{15.6}$$

***Levelized Annual Revenue Requirement.*** Once we obtain the revenue requirements over the project life, the next step is to find the equivalent annual revenue requirement, often referred to in the literature as the levelized annual revenue requirement. In doing so, we may first find the present value of the revenue requirement series and then convert this present value into the annual equivalent as follows.

$$\text{Levelized revenue requirement} = PV(\text{Annual revenue requirements})(A/P, k, N)$$

$$LRR(k) = \sum_{n=0}^{N} RR_n(1 + k)^{-n}(A/P, k, N) \tag{15.7}$$

where LRR($k$) = levelized annual revenue requirement computed at $k$,
$k$ = A/T cost of capital.

Note that we use $k$ to find the *LRR* in Eq. 15.7. The reader should know that there is considerable controversy over which discount rate to use in finding the *LRR*. (See for example, [1] and [3].) Oso [4] has proved mathematically, however, that the A/T cost of capital, $k$, is the proper discount rate in revenue requirement analysis.

### *Example 15.1*

An investment project being considered by a utility has the following characteristics (a flow-through accounting procedure is used).

| | |
|---|---|
| Required total capital ($V$) | = \$100,000 |
| Project life (= tax life) ($N$) | = 5 years |
| Salvage value ($F$) | = \$10,000 |
| Tax rate ($t_m$) | = 45% |
| Book depreciation method ($D'_n$) | = SL |
| Tax depreciation method ($D_n$) | = SOYD |
| Annual total operating costs ($C_n$) | = \$10,000 |
| Debt financing ratio ($d$) | = 60% |
|   Debt (short-term) = 0% | |
|   Debt (long-term bonds) = 60% | |
| Equity financing ratio $(1 - d)$ | = 40% |
|   Common stock = 30% | |
|   Preferred stock = 10% | |
| Return to investors ($RI_n$) | |
|   Debt cost before tax | = 15% |
|   Common stock after tax | = 32% |
|   Preferred stock after tax | = 14% |

First we compute both composite rates of return to investors and the A/T cost of capital.

$$i_b = 0.15$$
$$i_e = (0.3/0.40)(0.32) + (0.10/0.40)(0.14) = 0.275$$
$$\text{A/T cost of capital} = (1 - 0.45)(0.60)(0.15) + (0.4)(0.275) = 0.1595$$

The remaining calculations for this example are summarized in Table 15.3. The table columns (with entries for year 1) are described as follows.

Col. 1: Project life = 5 years

Col. 2: Unrecovered investment at the beginning of year 1,

$$U_0 = \text{initial investment } (P) = \$100{,}000$$
$$U_1 = U_0 - D'_1 = \$100{,}000 - \$18{,}000 = \$82{,}000$$

**Table 15.3** ***Revenue Requirements for Example 15.1***

*Debt to Equity: 60% to 40%*

| 1<br>End of Year, $n$ | 2<br>Unrecovered Balance, $U_n$ | 3<br>Book Depreciation, $D'_n$ | 4<br>Return on Debt, $RD_n$ | 5<br>Return on Equity, $RE_n$ | 6<br>Total Return to Investors, $RI_n$ | 7<br>Tax Depreciation, $D_n$ | 8<br>Taxes, $T_n$ | 9<br>Carrying Charges, $CC_n$ | 10<br>O & M Costs, $C_n$ | 11<br>Required Revenue, $RR_n$ |
|---|---|---|---|---|---|---|---|---|---|---|
| 0 | 100,000 | | | | | | | | | |
| 1 | 82,000 | 18,000 | 9,000 | 11,000 | 20,000 | 30,000 | −818 | 37,182 | 10,000 | 47,182 |
| 2 | 64,000 | 18,000 | 7,380 | 9,020 | 16,400 | 24,000 | 2,471 | 36,871 | 10,000 | 46,871 |
| 3 | 46,000 | 18,000 | 5,760 | 7,040 | 12,800 | 18,000 | 5,760 | 36,560 | 10,000 | 46,560 |
| 4 | 28,000 | 18,000 | 4,140 | 5,060 | 9,200 | 12,000 | 9,049 | 36,249 | 10,000 | 46,249 |
| 5 | 10,000 | 18,000 | 2,520 | 3,080 | 5,600 | 6,000 | 12,338 | 35,938 | 10,000 | 45,938 |

NOTE: Col. 3 = (100,000 − 10,000)/5
Col. 4 = $U_{n-1}(d)(i_b)$, where $i_b$ = 15%
Col. 5 = $U_{n-1}(1 - d)(i_e)$, where $i_e$ = 27.5%
Col. 6 = Col. 4 + Col. 5
Col. 7 = SOYD, assumed a flow-through accounting procedure that disallows MACRS tax depreciation.
Col. 8 = [0.45/(1 − 0.45)] (Col. 3 + Col. 5 − Col. 7)
Col. 9 = Col. 3 + Col. 6 + Col. 8
Col. 11 = Col. 9 + Col. 10

$$PV(15.95\%) = 47{,}182(P/F, 15.95\%, 1) + 46{,}871(P/F, 15.95\%, 2) + \cdots + 45{,}938(P/F, 15.95\%, 5) = \$152{,}928$$

$$LRR(15.95\%) = 156{,}181(A/P, 15.95\%, 5) = \$46{,}651$$

Col. 3: Book depreciation, assuming the SL method,

$$D'_1 = (100{,}000 - 10{,}000)/5 = \$18{,}000$$

Col. 4: Return on debt, $RD_1 = U_0(d)(i_b)$,

$$RD_1 = (100{,}000)(0.60)(0.15) = \$9{,}000$$

Col. 5: Return on equity, $RE_1 = U_0(1 - d)(i_e)$,

$$RE_1 = (100{,}000)(1 - 0.60)(0.275) = \$11{,}000$$

Col. 6: Total return on investment, Col. 4 + Col. 5,

$$RI_1 = 9{,}000 + 11{,}000 = \$20{,}000$$

Col. 7: Depreciation for tax purposes, $D_1$, assuming the SOYD,

$$D_1 = (100{,}000 - 10{,}000)(5/15) = \$30{,}000$$

Col. 8: Taxes for year 1, $T_1$, using Eq. 15.6,

$$T_1 = (0.45/0.55)(11{,}000 + 18{,}000 - 30{,}000) = -\$818$$

Col. 9: Carrying charges, $CC_1$, Col. 2 + Col. 6 + Col. 8,

$$CC_1 = 18{,}000 + 20{,}000 - 818 = \$37{,}182$$

Col. 10: Total annual operating costs, $C_1$,

$$C_1 = \$10{,}000$$

Col. 11: Revenue requirement, $RR_n$, Col. 9 + Col. 10,

$$RR_1 = 37{,}182 + 10{,}000 = \$47{,}182$$

Finally, the levelized annual revenue requirement (*LRR*) can be obtained by using Eq. 15.7.

$$\begin{aligned} PV(15.95\%) &= \$47{,}182(P/F,\ 15.95\%,\ 1) + 46{,}871(P/F,\ 15.95\%,\ 2) \\ &\quad + \cdots + 45{,}938(P/F,\ 15.95\%,\ 5) = \$152{,}928 \\ LRR(15.95\%) &= 152{,}928(A/P,\ 15.95\%,\ 5) = \$46{,}651 \quad \square \end{aligned}$$

### 15.3.3 Effect of Inflation in Revenue Requirements

When there is inflation, we must identify precisely the cost elements that are responsive to inflation and those that are not responsive. For the inflation-responsive elements, the likely price escalation rates must be estimated. The cost of capital to use in computing the levelized annual revenue requirement should also reflect the effect of inflation. This is done as follows.

Recall that, in Chapter 2, the market interest rate was defined as

$$i = (1 + i')(1 + f) - 1 = i' + f + i'f \tag{15.8}$$

where $i$ = market interest rate,
$i'$ = inflation-free interest rate,
$f$ = general inflation rate.

Note that $i = i'$ in the absence of inflation ($f = 0$). We will use this relationship in deriving the inflation-adjusted A/T cost of capital, $k$. For the returns on debt and equity, we may introduce the effect of inflation in the following fashion,

$$i_b = (1 + i_b')(1 + f) - 1 = i_b' + f + i_b' f \tag{15.9}$$

$$i_e = (1 + i_e')(1 + f) - 1 = i_e' + f + i_e' f \tag{15.10}$$

where $i_b$ = inflation-adjusted B/T cost of borrowed capital,
$i_b'$ = real B/T cost of borrowed capital,
$i_e$ = inflation-adjusted A/T cost of equity capital,
$i_e'$ = real A/T cost of equity capital.
Then the inflation-adjusted A/T cost of capital is also defined as

$$k = (1 + k')(1 + f) - 1 \tag{15.11a}$$

where $k'$ is the real A/T cost of capital. Using the inflation-adjusted $i_b$ and $i_e$ in Eqs. 15.9 and 15.10, we can also express $k$ as

$$\begin{aligned} k &= d(1 - t_m)i_b + (1 - d)i_e \\ &= d(1 - t_m)(i_b' + f + i_b' f) + (1 - d)(i_e' + f + i_e' f) \end{aligned} \tag{15.11b}$$

Consequently, we can solve Eq. 15.11a for $k'$. This gives us the expression

$$k' = \frac{1 + k}{1 + f} - 1 \tag{15.12}$$

Substituting Eq. 15.11b into Eq. 15.12 for $k$ and simplifying the equation yields

$$k' = d(1 - t_m)i_b' + (1 - d)i_e' - \frac{dft_m}{1 + f} \tag{15.13}$$

Since we know the relationship between $k$ and $k'$, we can handle the effect of inflation explicitly. If the revenue requirements are expressed in terms of actual dollars, we use $k$ to find the *LRR*. On the other hand, if the revenue requirements are given in constant dollars, we use $k'$ to find the *LRR*. There can be various practical difficulties when we must consider the effects of extreme price escalation in revenue requirement studies. This problem has been addressed in Sullivan, Ward, and Lee [6].

### *Example 15.2*

Reconsider the investment situation in Example 15.1. Suppose that the general inflation rate is 8% but that the annual operating cost ($C_n$) escalates at the annual rate of 10%. Before the firm experienced inflation, its debt and equity costs were known to be

Real B/T cost of borrowed capital ($i_b'$) = 10%
Real A/T cost of equity capital ($i_e'$) = 25%

Since the revenue requirements are given in actual dollars, we must use the inflation-adjusted A/T cost of capital in finding the *LRR*. We first compute $i_b$ and $i_e$ with Eqs. 15.9 and 15.10.

$$i_b = 0.10 + 0.08 + 0.008 = 18.8\%$$
$$i_e = 0.25 + 0.08 + 0.020 = 35.0\%$$

The inflation-adjusted A/T cost of capital is then

$$k = 0.6(1 - 0.45)0.188 + 0.4(0.35) = 20.20\%$$

Using Eq. 15.12, we find the real A/T cost of capital to be

$$k' = \frac{1 + 0.2020}{1 + 0.08} - 1 = 11.30\%$$

With the remaining parameters unchanged, the revenue requirements in actual dollars are as shown in Table 15.4. Finally, the *LRR* is obtained.

$$\begin{aligned} PV(20.20\%) &= \$55{,}916(P/F,\ 20.20\%,\ 1) + 55{,}314(P/F,\ 20.20\%,\ 2) + \cdots \\ &\quad + 54{,}208(P/F,\ 20.20\%,\ 5) \\ &= \$164{,}058 \\ LRR(20.20\%) &= \$164{,}058(A/P,\ 20.20\%,\ 5) \\ &= \$55{,}099 \quad \square \end{aligned}$$

## 15.4 EQUIVALENCE OF THE PRESENT VALUE AND REVENUE REQUIREMENT METHODS

As we mentioned briefly at the beginning of this chapter, the *RR* method can provide a viable decision criterion for analysis of investments in private industries. In this section we show that the *RR* method is directly related to the conventional approaches used in private industries, such as the *PV* and *IRR* methods. Recall that these conventional methods depend on the determination of the A/T cash flows, usually the equity cash flows, from which the appropriate figure of merit is computed. We will first reexamine Example 15.1 to illustrate these relationships. Then we will generalize the equivalence of the *PV* (or *IRR*) and *RR* methods.

### 15.4.1 The A/T Equity Cash Flows and Revenue Requirement Series

The definition of the A/T cash flow ($Y_n$) was given in Chapter 4, but we will redefine it here using the symbols introduced in the *RR* analysis.

$$Y_n = \begin{cases} -(U_0 - B), & n = 0 \\ R_n - C_n - (RD_n + dD'_n) - T_n, & 1 \le n \le N - 1 \\ R_n - C_n - (RD_n + dD'_n) - T_n - (1 - d)F, & n = N \end{cases} \qquad (15.14)$$

where $R_n$ is the gross revenue in year $n$.

**Table 15.4** ***Effects of Inflation on Revenue Requirements for Example 15.1***

*Debt to Equity: 60% to 40%*

| 1<br>End of Year, $n$ | 2<br>Unrecovered Balance, $U_n$ | 3<br>Book Depreciation, $D'_n$ | 4<br>Return on Debt, $RD_n$ | 5<br>Return on Equity, $RE_n$ | 6<br>Total Return to Investors, $RI_n$ | 7<br>Tax Depreciation, $D_n$ | 8<br>Taxes, $T_n$ | 9<br>Carrying Charges, $CC_n$ | 10<br>O & M Costs, $C_n$ | 11<br>Required Revenue, $RR_n$ |
|---|---|---|---|---|---|---|---|---|---|---|
| 0 | 100,000 | | | | | | | | | |
| 1 | 82,000 | 18,000 | 11,280 | 14,000 | 25,280 | 30,000 | 1,636 | 44,916 | 11,000 | 55,916 |
| 2 | 64,000 | 18,000 | 9,250 | 11,480 | 20,730 | 24,000 | 4,484 | 43,214 | 12,100 | 55,314 |
| 3 | 46,000 | 18,000 | 7,219 | 8,960 | 16,179 | 18,000 | 7,331 | 41,510 | 13,310 | 54,820 |
| 4 | 28,000 | 18,000 | 5,189 | 6,440 | 11,629 | 12,000 | 10,178 | 39,807 | 14,641 | 54,448 |
| 5 | 10,000 | 18,000 | 3,158 | 3,920 | 7,078 | 6,000 | 13,025 | 38,103 | 16,105 | 54,208 |

NOTE: Col. 3 = (100,000 − 10,000)/5
Col. 4 = $U_{n-1}(d)(i_b)$, where $i_b$ = 18.8%
Col. 5 = $U_{n-1}(1 - d)(i_e)$, where $i_e$ = 35%
Col. 6 = Col. 4 + Col. 5
Col. 7 = SOYD, assumed a flow-through accounting procedure in which no MACRS tax depreciation is allowed.
Col. 8 = [0.45/(1 − 0.45)] (Col. 3 + Col. 5 − Col. 7)
Col. 9 = Col. 3 + Col. 6 + Col. 8
Col. 11 = Col. 9 + Col. 10

$$PV(20.20\%) = 55{,}916(P/F, 20.20\%, 1) + 55{,}314(P/F, 20.20\%, 2) + \cdots + 54{,}208(P/F, 20.20\%, 5) = \$164{,}058$$

$$LRR(20.20\%) = 168{,}149(A/P, 20.20\%, 5) = \$55{,}099$$

In Eq. 15.14 it is assumed that the year $N$ salvage value, $F$, equals the book value for tax purposes so that there is no gain or loss on disposal. Note that this salvage value is weighted by $(1 - d)$ to reflect only the portion that is equity capital, to be consistent with the set of assumptions in Section 15.3.1. The net cash flows defined by Eq. 15.14 are referred to as the equity cash flows; that is, these cash flows are the yearly returns of the equity stockholders' capital, since debt payments are explicitly taken into account.

The relevant A/T cash flow calculations for the investment situation in Example 15.1 are summarized in Table 15.5. Notice that $R_n$ is the gross revenue in year $n$ that must be determined. Now we set the value of $R_n$ equal to the value of the minimum revenue requirement, $RR_n$, for each year. The equity cash flow series shown in Table 15.5 then yields an internal rate of return of 27.5%. This is the return on equity capital specified in the problem! Stated differently, the *PV* of all future $Y_n$, when discounted at $i_e = 27.5\%$, is equal to the capital investment by owners,

$$-\$40{,}000 + 4{,}000(P/F, 27.5\%, 5) = \$18{,}200 - 1{,}980(P/G, 27.5\%, 5)$$

This is an expected result inasmuch as the revenue requirements have been determined so that they provide the 27.5% return to owners. If the utility's actual revenues exceed the permitted revenues ($R_n > RR_n$), the *IRR* of the resulting A/T equity cash flows will exceed the specified return on equity, leading to selection of the project. This return on equity serves as the minimum attractive rate of return to the utility. (See Section 5.4.1.) When the *PV* of such cash flows is computed at $i_e$, it will always be positive.

### 15.4.2 Important Results Regarding the Equivalence of the PV and RR Methods

The situation just observed is not an isolated one, but represents a precise relation between the *PV* and *RR* methods. Before presenting some mathematical results that relate the two methods, we will vary only the debt ratio to see how this might affect the outcome. Two additional calculations are made, 100% debt financing and 100% equity financing. The revenue requirements and the corresponding A/T equity cash flows for each case are summarized in Table 15.6. These additional examples further confirm the findings in the previous section and assure us of the exact equivalence between the two methods. We will present some mathematical results that establish this equivalence without proofs under the assumptions of Section 15.3.1. (See Oso [4] and Ward and Sullivan [7] for proofs.)

***Theorem 15.1.*** Given a revenue sequence $\{R_n\}$ that is just enough to meet all capital obligations and operating costs of a project, the *PV* of the sequence discounted at the A/T cost of capital, $k$, is given by the expression

$$PV(R_n; k) = \frac{U_0 - F(1 + k)^{-N}}{1 - t_m} - \frac{t_m}{1 - t_m} PV(D_n; k) + PV(C_n; k) \quad (15.15)$$

**Table 15.5** ***After-Tax Equity Cash Flows for Example 15.1***

*Debt to Equity: 60% to 40%*

| 1 | 2 | 3 | 4 | 5 | 6 | 7 | 8 | 9 | 10 | 11 | 12 |
|---|---|---|---|---|---|---|---|---|---|---|---|
| | | Financing | | | | Debt Repayment | | | | | |
| End of Year, $n$ | Capital Investment, $V$ | Equity, $S$ | Debt, $B$ | Gross Revenue, $R_n = RR_n$ | O&M Costs, $C_n$ | Principal, $DP_n$ | Interest, $IP_n$ | Tax Depreciation, $D_n$ | Taxable Income, $TI_n$ | Taxes (Savings), $T_n$ | A/T Cash Flow, $Y_n$ |
| 0 | −100,000 | 40,000 | 60,000 | | | | | | | | −40,000 |
| 1 | | | | 47,182 | 10,000 | 10,800 | 9,000 | 30,000 | −1,818 | −818 | 18,200 |
| 2 | | | | 46,871 | 10,000 | 10,800 | 7,380 | 24,000 | 5,491 | 2,471 | 16,220 |
| 3 | | | | 46,560 | 10,000 | 10,800 | 5,760 | 18,000 | 12,800 | 5,760 | 14,240 |
| 4 | | | | 46,249 | 10,000 | 10,800 | 4,140 | 12,000 | 20,109 | 9,049 | 12,260 |
| 5 | | | | 45,938 | 10,000 | 10,800 | 2,520 | 6,000 | 27,418 | 12,338 | 10,280 |
| 5 | 10,000 | | | | | 6,000 | | | | | 4,000 |

NOTE: Col. 7 = 60,000/5
Col. 10 = Col. 5 − Col. 6 − Col. 8 − Col. 9
Col. 11 = (Col. 10)(0.45)
Col. 12 = Col. 5 − Col. 6 − Col. 7 − Col. 8 − Col. 9 − Col. 11

$$PV(15.95\%) = -40{,}000 + 18{,}200(P/F, 15.95\%, 1) + 16{,}220(P/F, 15.95\%, 2) + \cdots + (10{,}280 + 4{,}000)(P/F, 15.95\%, 5) = \$10{,}492$$

$$PV(27.5\%) = 0 \rightarrow IRR \text{ on equity cash flows} = 27.5\% = i_e$$

**Table 15.6** ***A/T Equity Cash Flows with Revenues Set Equal to $RR_n$ under Various Financing Mixes***

| End of Year, $n$ | Financing Mixes | | | | | |
|---|---|---|---|---|---|---|
| | $d$ = 100% | | $d$ = 60% | | $d$ = 0% | |
| | Required Revenue, $RR_n$ | A/T Equity Cash Flow, $Y_n$ | Required Revenue, $RR_n$ | A/T Equity Cash Flow, $Y_n$ | Required Revenue, $RR_n$ | A/T Equity Cash Flow, $Y_n$ |
| 0 | | 0 | | −40,000 | | −100,000 |
| 1 | 33,182 | 0 | 47,182 | 18,200 | 68,182 | 45,500 |
| 2 | 35,391 | 0 | 46,871 | 16,220 | 64,091 | 40,550 |
| 3 | 37,600 | 0 | 46,560 | 14,240 | 60,000 | 35,600 |
| 4 | 39,809 | 0 | 46,249 | 12,260 | 55,909 | 30,650 |
| 5 | 42,018 | 0 | 45,938 | 10,280 | 51,818 | 25,700 |
| 5* | | 0 | | 4,000 | | 10,000 |
| | | | | $IRR$ = 27.5% | | $IRR$ = 27.5% |

*Accountable salvage value, which is proportional to financing ratio.

Then

$$LRR(k) = PV(R_n; k)(A/P, k, N) \tag{15.16}$$

If the values from Example 15.1 with $d = 0.6$ are substituted into Eq. 15.15, the *PV* of revenue requirements is found to be \$152,928 and the *LRR* is \$46,651. These are the same as the results obtained from Eq. 15.7.

***Theorem 15.2.*** Suppose we are given a project and the set of all possible revenue sequences that provide for all project obligations to be just met. Then there exists one and only one rate that discounts every member of the revenue sequence to the same value, and this rate is the A/T cost of capital, $k$.

***Theorem 15.3.*** Assume a minimum *RR* sequence ($R_n = RR_n$) and the A/T cash flow model in Eq. 15.14. Then

$$PV(Y_n; i_e) = 0 \tag{15.17}$$

## 15.5 FLOW-THROUGH AND NORMALIZATION ACCOUNTING

In this section we consider some extensions of the *RR* method and introduce such items as (1) capitalized interest during the construction period, (2) the investment tax credit, and (3) accelerated depreciation schedules such as MACRS. In practice, the procedure for computing revenue requirements is often dictated by many of the accounting conventions and procedures required by regulatory agencies in establishing utility rates. Two methods are allowed, the *flow-through* and *normalizing* methods. The procedure for computing the revenue requirements discussed in the previous sections is, in fact, based on the flow-through accounting principle. The two methods differ only in handling the tax and depreciation credits and interest expenses during the construction period.

### 15.5.1 Flow-Through Method

The flow-through method for calculating revenue requirements allocates credits and costs in the year in which they occur. More precisely,

1. The investment tax credit (if any) is not amortized. Note that the Tax Reform Act of 1986 repealed the regular investment credit for property placed in service after 1986.
2. The interest paid during construction is not capitalized but is taken as an expense in the first year.
3. Under the Tax Reform Act of 1986, no MACRS depreciation is allowed when the utility elects to use the flow-through accounting principle.

Since the investment tax credit reduces tax obligation, we need to adjust the way to compute the tax amount for the year in which the credit is given.

Generally, if the investment tax credit remains in effect, it is allowed for the year in which we place the property in service. This is the earlier of (1) the first tax year for which depreciation can be taken or (2) the tax year in which the property is ready to be used for its intended purpose. We will assume the first situation, year 1. In figuring $T_1$, we modify Eq. 15.5 to reflect the investment tax credit.

$$\begin{aligned} T_1 &= (\text{Taxable income})t_m - ITC \\ &= (RR_1 - C_1 - RD_1 - D_1)t_m - ITC \\ &= (D_1' + RE_1 + T_1 - D_1)t_m - ITC \end{aligned} \tag{15.18}$$

where $ITC$ is the investment tax allowed. Solving for $T_n$ gives

$$T_1 = \frac{t_m}{1 - t_m}(D_1' + RE_1 - D_1) - \frac{ITC}{1 - t_m} \tag{15.19}$$

When no investment tax credit is allowed, we simply drop the $ITC$ term in Eqs. 15.18 and 15.19. Then to compute the $RR_n$, we can use Eq. 15.4, where $T_n$ is replaced by Eq. 15.19. The interest paid during the construction period, $I_n$, is treated as part of the operating expenses so that $RR_1 = CC_1 + C_1 + I_1$.

## *Example 15.3*

Generate the annual revenue requirements by using the flow-through method and the following data. Assume during that the asset is placed in service during 1990.

Total required investment capital ($V$) = \$120,000

Salvage value ($F$) = \$30,000

Project life (= tax life or book life) = 5 years

Debt ratio ($d$) = 0.60

Cost of debt capital ($i_b$) = 15%

Required return on equity ($i_e$) = 30%

Marginal tax rate ($t_m$) = 0.45

Book depreciation ($D_n'$) = SL

A/T cost of capital ($k$) = (0.60)(1 − 0.45)(0.15) + (0.40)(0.30) = 16.95%

Tax depreciation ($D_n$) = SOYD

Investment tax credit allowed for non-MACRS property ($ITC$) = 0%

Interest expense during construction as percentage of the total capital = 1%

The annual revenue requirements and associated data for this problem are given in Table 15.7. The steps required to obtain the table entries are as follows.

## Table 15.7 *Revenue Requirements under the Flow-Through Method*

| 1<br>End of Year, $n$ | 2<br>Unrecovered Book Balance, $U_n$ | 3<br>Book Depreciation, $D'_n$ | 4<br>Tax Depreciation, $D_n$ | 5<br>Return on Debt, $RD_n$ | 6<br>Return on Equity, $RE_n$ | 7<br>Investment Tax Credit, $ITC$ | 8<br>Interest During Construction, $I_n$ | 9<br>Taxes, $T_n$ | 10<br>O&M Costs, $C_n$ | 11<br>Required Revenue, $RR_n$ |
|---|---|---|---|---|---|---|---|---|---|---|
| 0 | 120,000 | | | | | | | | | |
| 1 | 102,000 | 18,000 | 30,000 | 10,800 | 14,400 | 0 | 1,200 | 1,964 | 10,000 | 56,364 |
| 2 | 84,000 | 18,000 | 24,000 | 9,180 | 12,240 | | | 5,150 | 10,000 | 54,570 |
| 3 | 66,000 | 18,000 | 18,000 | 7,560 | 10,080 | | | 8,247 | 10,000 | 53,887 |
| 4 | 48,000 | 18,000 | 12,000 | 5,940 | 7,920 | | | 11,389 | 10,000 | 53,249 |
| 5 | 30,000 | 18,000 | 6,000 | 4,320 | 5,760 | | | 14,531 | 10,000 | 52,611 |

NOTE 1: No investment tax credits are allowed for assets placed in service after 1986.

NOTE 2: Col. 3 = (120,000 − 30,000)/5 = 18,000

Col. 4 = SOYD

Col. 5 = $U_{n-1}\,(d)\,(i_b)$, where $i_b = 15\%$

Col. 6 = $U_{n-1}\,(1 - d)\,(i_e)$, where $i_e = 30\%$

Col. 9 = $[0.45/(1 - 0.45)](D'_n + RE_n - D_n) - ITC/(1 - 0.45)$

Col. 11 = Col. 3 + Col. 5 + Col. 6 + Col. 9 + Col. 10

$$PV(16.95\%) = 56{,}364(P/F, 16.95\%, 1) + 54{,}570(P/F, 16.95\%, 2) + \cdots + 52{,}611(P/F, 16.95\%, 5) = \$174{,}295$$

$$LRR(16.95\%) = 174{,}295(A/P, 16.95\%, 5) = \$54{,}416$$

**Step 1:** Determine the construction interest. The amount of interest is

$$(0.01)(120{,}000) = \$1{,}200$$

and will be covered in the first year of the project.

**Step 2:** Calculate the depreciation schedules.

**Step 3:** Compute the investment tax credit.

$$(0\%)(120{,}000) = \$0$$

**Step 4:** Compute the unrecovered book balances.

$$U_0 = \$120{,}000$$
$$U_1 = 120{,}000 - 18{,}000 = \$102{,}000$$

**Step 5:** Calculate the debt interests.

$$RD_1 = (0.6)(120{,}000)(0.15) = \$10{,}800$$

$$RD_2 = (0.6)(102{,}000)(0.15) = \$9{,}180$$

**Step 6:** Calculate the return on the equity capital.

$$RE_1 = (0.4)(120{,}000)(0.30) = \$14{,}400$$

$$RE_2 = (0.4)(102{,}000)(0.30) = \$12{,}240$$

**Step 7:** Calculate taxes using Eq. 15.19.

$$T_1 = \frac{0.45}{1 - 0.45}(14{,}400 + 18{,}000 - 30{,}000) - \frac{0}{1 - 0.45}$$
$$= \$1{,}964$$

$$T_2 = \frac{0.45}{1 - 0.45}(12{,}240 + 18{,}000 - 24{,}000) = \$5{,}150$$

**Step 8:** Calculate the annual revenue requirements.

$$RR_1 = 18{,}000 + 10{,}800 + 14{,}400 + 1{,}200 + 1{,}964 + 10{,}000$$
$$= \$56{,}364$$

$$RR_2 = 18{,}000 + 9{,}180 + 12{,}240 + 0 + 5{,}150 + 10{,}000$$
$$= \$54{,}570$$

**Step 9:** Compute the *PV* of the revenue requirements and the *LRR*.

$$PV(16.95\%) = \$174{,}295$$

$$LRR(16.95\%) = \$54{,}416$$ □

### 15.5.2 Normalizing Method

The normalizing method is frequently used by public utilities because it includes many of the accounting conventions and procedures required by regulatory agencies in establishing utility rates. Public utility property placed in service during 1987 and later is eligible for the MACRS only if the tax benefits of the MACRS are normalized. Normalization is used to take shorter periods and accelerated recovery into account. It is done (1) by using the same depreciation for income tax that is used for book accounting or (2), if different methods are used, by creating and making adjustments to a reserve to reflect the deferral of taxes that results from using different depreciation methods [1]. Other differences in accounting procedure compared with the flow-through method are the following.

1. The investment tax credit (if any) is amortized over the project life. Note again that the Tax Reform Act of 1986 repealed the regular investment tax credit for property placed in service after 1986.
2. The interest paid on money during the construction phase of the project is capitalized.
3. The tax depreciation can be based on the MACRS (see Chapter 4).
4. The deferred tax reserve account must be established.

We will first explain the special features of the normalizing method by using the data in Example 15.3; only the tax depreciation method (MACRS) is different.

***Capitalized Investment.*** The total capitalized investment cost, $K_n$, is obtained from

$$K_0 = U_0 + I_0 \tag{15.20}$$

where $I_0$ is the capitalized interest expense. In our example,

$$K_0 = 120{,}000 + (0.01)(120{,}000) = \$121{,}200$$

Usually, the interest rate that is used in determining the capitalized interest is set by regulatory agencies. The underlying philosophy of the utility companies in capitalizing the interest expense is that, in this way, present customers are not penalized for the interest cost of new construction. Instead, by spreading it over the project life, the utility company assesses this interest to both present and future customers [5]. The book depreciation is based on this total capitalized cost as shown in Table 15.8.

***Deferred Taxes.*** The deferred taxes are those that are delayed because an accelerated depreciation model (MACRS in this example) is used instead of the SL. There is a reluctance to pass these "savings" on to customers since tax rates and laws may change in the future [5]. Therefore, the normalizing procedure treats the deferred taxes as an expense in the early years and as a credit in later years. In determining the deferred taxes, we need to calculate the depreciation schedules. There are three depreciation schedules.

$D'_n$ = straight-line depreciation for book purposes

$D''_n$ = straight-line depreciation for tax purposes

$D_n$ = accelerated depreciation (MACRS) for tax purposes

**Table 15.8** ***Revenue Requirements Using the Normalizing Method***

| 1 | 2 | 3 | 4 | 5 | 6 | 7 | 8 | 9 | 10 | 11 | 12 | 13 | 14 |
|---|---|---|---|---|---|---|---|---|---|---|---|---|---|
| End of Year, $n$ | Unrecovered Book Balance, $U_n$ | Book Depreciation, $D'_n$ | Tax Depreciation (MACRS), $D_n$ | Deferred Tax, $DT_n$ | $ITC$ | Amortized Tax Credit, $AITC_n$ | Capitalized Interest, $I_c$ | Chargeable Investment, $K_n$ | Return on Debt, $RD_n$ | Return on Equity, $RE_n$ | Tax, $T_n$ | O&M Costs, $C_n$ | Required Revenue, $RR_n$ |
| 0 | 120,000 + 1,200 | | | | | | | 121,200 | | | | | |
| 1 | 102,960 | 18,240 | 24,000 | 2,700 | 0 | 0 | 1,200 | 100,260 | 10,908 | 14,544 | 8,414 | 10,000 | 64,806 |
| 2 | 84,720 | 18,240 | 38,400 | 9,180 | | | | 72,840 | 9,023 | 12,031 | 860 | 10,000 | 59,334 |
| 3 | 66,480 | 18,240 | 23,040 | 2,268 | | | | 52,332 | 6,556 | 8,741 | 5,080 | 10,000 | 50,885 |
| 4 | 48,240 | 18,240 | 13,824 | −1,879 | | | | 35,971 | 4,710 | 6,280 | 7,214 | 10,000 | 44,565 |
| 5 | 30,000 | 18,240 | 13,824 | −1,879 | | | | 19,610 | 3,237 | 4,317 | 5,608 | 10,000 | 39,523 9,850* |

*Gains tax: Accumulated tax depreciation = 24,000 + 38,400 + 23,040 + 13,814 + 13,814 = $113,088
Book value = 121,200 − 113,088 = $8,112
Taxable gains = 30,000 − 8,112 = $21,888
Gains tax = 21,888(0.45) = $9,850

NOTE: Col. 3 = (121,200 − 30,000)/5 = 18,240
Col. 4 = MACRS 5-year property with depreciation percentages (20%, 32%, 19.2%, 11.52%, 11.52%)
Depreciation base = $120,000 for tax purpose (not including the capitalized interest, $1,200)
Col. 5 = $(D_n - D'_n)(0.45)$
Col. 6, Col. 7 = No investment tax credit is allowed for the asset purchased after 1986.
Col. 9 = $K_{n-1} - D'_n - ITC - DT_n + AITC_n$
Col. 10 = $K_{n-1}(d)(i_b)$, where $i_b = 15\%$
Col. 11 = $K_{n-1}(1 - d)(i_e)$, where $i_e = 30\%$
Col. 12 = $[0.45/(1 - 0.45)](D'_n - D_n + DT_n + ITC - AITC_n - I_c)$
Col. 13 = Col. 3 + Col. 5 + Col. 10 + Col. 11 + Col. 12 + Col. 13

$PV(16.95\%) = \$176{,}997$; $LRR(16.95\%) = 176{,}997(A/P, 16.95\%, 5) = \$55{,}259$

These three depreciation models require different depreciation bases because of the capitalized interest. Assuming that the asset placed in service falls into the 5-year MACRS property class, the depreciation bases for these three depreciation methods are

| | $D'_n$ | $D''_n$ | $D_n$ |
|---|---|---|---|
| Depreciation base | \$121,200 | \$120,000 | \$120,000 |
| Salvage value | 30,000 | 30,000 | 0 |
| Depreciation life | 5 years | 5 years | $5\frac{1}{2}$ years |

Note that the interest on money borrowed during construction is "depreciated" (capitalized) over the life of the project in the case of the book depreciation schedule. The tax depreciation schedule is based on the initial cost that excludes the capitalized interest, however.

Now, to calculate the deferred taxes for each year, we define

$$DT_n = (D_n - D''_n)t_m \tag{15.21}$$

In our example, the deferred tax amounts are

$$DT_1 = (24{,}000 - 18{,}000)(0.45) = \$2{,}700$$

$$DT_2 = (38{,}400 - 18{,}000)(0.45) = \$9{,}180$$

***Amortized Investment Tax Credit.*** The normalizing procedure treats the investment tax credit (if allowed) similarly to deferred taxes. That is, credit is not given for the entire amount of tax savings in the year in which they occur. Instead, the tax credit is spread over the life of the project. The amortized investment tax credit is

$$AITC_n = ITC/N \tag{15.22}$$

In our example, *AITC* for a MACRS asset is

$$AITC_1 = (0)(120{,}000)/5 = 0$$

***Chargeable Investment.*** The chargeable investment, $K_n$, for year $n$ is

$$\begin{aligned} K_n = {} & \text{Chargeable investment in preceding year} \\ & - \text{Book depreciation} - \text{Investment tax credit if allowed} \\ & - \text{Deferred taxes} + \text{Amortized investment tax credit} \end{aligned}$$

Using the symbols defined previously, we have

$$K_n = K_{n-1} - D'_n - ITC - DT_n + AITC_n \tag{15.23}$$

In our example, the chargeable investments are

$$K_0 = 121{,}200$$
$$K_1 = 121{,}200 - 18{,}240 - 0 - 2{,}700 + 0 = \$100{,}260$$
$$K_2 = 95{,}280 - 18{,}240 - 0 - 9{,}180 + 0 = 72{,}840$$
$$K_3 = 72{,}840 - 18{,}240 - 0 - 2{,}268 + 0 = 52{,}332$$
$$K_4 = 52{,}332 - 18{,}240 - 0 + 1{,}879 + 0 = 35{,}971$$
$$K_5 = 35{,}971 - 18{,}240 - 0 + 1{,}879 + 0 = 19{,}610$$

***Return to Investors.*** The debt and equity interests are based on the chargeable investments computed in Eq. 15.23.

$$RD_n = d(i_b)K_{n-1} \tag{15.24}$$

$$RE_n = (1 - d)(i_e)K_{n-1} \tag{15.25}$$

For our example, the first 2 years are

$$RD_1 = (0.6)(0.15)(121{,}200) = \$10{,}908$$
$$RD_2 = (0.6)(0.15)(100{,}260) = 9{,}023$$
$$RE_1 = (0.4)(0.30)(121{,}200) = \$14{,}544$$
$$RE_2 = (0.4)(0.30)(100{,}260) = 12{,}031$$

***Income Taxes.*** The taxes are computed using the relationship

$$T_n = \frac{t_m}{1 - t_m}(RE_n + D'_n + DT_n + ITC - AITC_n - D_n - I_n) - \frac{ITC}{1 - t_m} \tag{15.26}$$

The derivation of this formula is left to the student. The tax equation can be verified, however, by calculating the taxes with the A/T equity cash flow relationship $T_n = (RR_n - C_n - D_n - RD_n)t_m - ITC$. The same values should be obtained. For our example,

$$T_1 = (0.45/0.55)(14{,}544 + 18{,}240 + 2{,}700 + 0 - 0 - 24{,}000 - 1{,}200) - 0/0.55 = \$8{,}414$$

$$T_2 = (0.45/0.55)(12{,}031 + 18{,}240 + 9{,}180 + 0 - 0 - 38{,}400 - 0) = \$860$$

***Required Revenues.*** Finally, the revenue requirements are computed as

$$RR_n = D'_n + DT_n + ITC - AITC_n + RD_n + RE_n + T_n + C_n \tag{15.27}$$

For example,

$$RR_1 = 18{,}240 + 2{,}700 + 0 - 0 + 10{,}908 + 14{,}544 + 8{,}414 + 10{,}000$$
$$= \$64{,}806$$
$$RR_2 = 18{,}240 + 9{,}180 + 0 - 0 + 9{,}023 + 12{,}031 + 860 + 10{,}000$$
$$= \$59{,}334$$

The *PV* of the revenue requirements and the *LRR* are determined to be

$$PV(16.95\%) = \$176{,}997 \quad \text{and} \quad LRR(16.95\%) = \$55{,}259$$

Some interesting points can be observed by comparing the results obtained with the two methods. First, the *LRR* is less for the flow-through method. This result is not a pure coincidence but can be generalized. That is, for a given investment, the flow-through method will always give a smaller *LRR*. In addition, the flow-through method requires that income tax savings resulting from accelerated depreciation are passed on to customers in the year in which they occur. This practice is a closer representation of actual cash flows in engineering economy studies of projects in unregulated firms. Because of the way utility rates are set, SL book depreciation in combination with accelerated tax depreciation and flow-through would have resulted in lower rates and a decreased after-tax rate of return [5]. These points, however, should not be interpreted as meaning that the flow-through method is better. Selection between the methods is often dictated by regulatory agencies and the firm's accounting procedures on which utility rates are based.

## 15.6 SUMMARY

In this chapter we introduced a new decision criterion frequently used in public utilities, the minimum revenue requirement. Although there are many ways of defining the revenue requirement method, the concept explained in this chapter is used by many utility firms. We have also shown that a conventional present value (or internal rate of return) of A/T cash flows is equivalent to a popular variant of the minimum revenue requirement method discussed in this chapter. Equivalence is established when the A/T cost of capital is utilized as the *MARR* in present value studies and when the same A/T cost of capital is used in finding the *LRR*. This statement is valid only when the assumptions underlying the application of the two methods are the same.

Finally, we introduced two accounting conventions in generating the minimum annual revenue requirements, the flow-through and normalizing methods. The flow-through method gives values that are a closer representation of the actual cash flows because credit and costs are taken in the year in which they occur. This principle is more in line with the definition of cash flows in engineering economy studies. Most public utilities use the normalizing method, however, because it agrees with their accounting procedures dictated by regulatory agencies, and it is on values given by the accounting procedures that utility rates are argued.

## REFERENCES

1. AMERICAN TELEPHONE AND TELEGRAPH COMPANY, *Engineering Economy: A Manager's Guide to Decision Making,* 3rd edition, McGraw–Hill, New York, 1977.
2. BARISH, N. N., and S. KAPLAN, *Economic Analysis for Engineering and Managerial Decision Making,* 2nd edition, McGraw–Hill, New York, p. 265, 1978.
3. JEYNES, P. H., *Profitability and Economic Choice,* Iowa State University Press, Ames, 1968.
4. OSO, J. B., "The Proper Role of the Tax-Adjusted Cost of Capital in Present Value Studies," *The Engineering Economist,* Vol. 24, No. 1, pp. 1–12, 1978.
5. STEVENS, G. T., JR., *Economic and Financial Analysis of Capital Investment,* Wiley, New York, 1979, Ch. 7.
6. SULLIVAN, W. G., T. L. WARD, and L. S. LEE, "Analysis of Revenue Requirements during Extreme Price Escalation," *The Engineering Economist,* Vol. 29, No. 3, pp. 161–180, 1984.
7. WARD, T. L., and W. G. SULLIVAN, "Equivalence of the Present Worth and Revenue Requirement Methods of Capital Investment Analysis," *AIIE Transactions,* Vol. 13, No. 1, pp. 29–40, 1981.

## PROBLEMS

**15.1.** A solid-waste-fired plant is considering two types of incinerator design. Both alternative designs will have a life of 20 years and a zero salvage value. The advanced-design incinerator (with energy recovery system) will cost $50,000 more than the simple-design incinerator. The advanced design will reduce fuel costs by $10,000 a year, however. The company depreciates assets of this type by the sum-of-the-years'-digits method, using the flow-through accounting procedure. Its book depreciation method is the straight-line method. The company's marginal tax rate is 40%, its debt ratio is 50%, its B/T cost of debt is 8%, and its A/T cost of equity is 15%.

a. Using revenue requirement analysis, show whether or not to install the advanced design.
b. Compute the internal rate of return on this incremental investment required by the advanced design, and recommend which design to select by using the rate of return principle.

**15.2.** A public utility is considering an investment project with the following characteristics.

Required total capital ($V$) = $300,000

Project life (= tax life) ($N$) = 6 years

Salvage value ($F$) = $30,000

Tax rate ($t_m$) = 40%

Book depreciation method = straight line

Tax depreciation method = SOYD

Annual total operating costs = $50,000

Debt financing ratio = 50%

Return to investors

Debt cost before tax = 12%

Common stock after tax = 25%

Preferred stock after tax = 15%

Accounting method = flow-through

Capitalized interest = 0%

a. Compute the composite rates of return to investors and the A/T cost of capital.
b. Determine the required revenues for each year.
c. Determine the levelized annual revenue requirement (*LRR*).

**15.3.** Repeat Problem 15.2, assuming that the capitalized interest is 1%.

**15.4.** In Problem 15.2, suppose that the general inflation rate (average inflation rate, $\bar{f}$) is expected to be 6% over the project period but the annual operating cost is expected to escalate at the annual rate of 8% during the project period. Before the firm's debt and equity costs were affected by inflation, they were known to be

Real B/T cost of borrowed capital ($i_b'$) = 12%

Real A/T cost of equity capital ($i_e'$) = 20%

a. Compute the inflation-adjusted A/T cost of capital.
b. Compute the required revenues (in actual dollars) for each year.
c. Compute the levelized revenue requirement.

**15.5.** A public utility is considering whether to replace or repair a failed transformer that has been in service for a number of years. The following data are available for each option.

*Option 1: Replace by a new transformer*

First cost = $65,000

Service life = 15 years

Salvage value = $5,000

Depreciation method = SL

*Option 2: Repair the old transformer*

Repair cost (treated as expense) = $15,000

Current market value = $4,000

Current book value = $0

Extended service life with repair = 4 years

Expected salvage value (at the end of 4 years) = $2,000

*Other investment data*

Debt financing ratio = 50%

Tax rate = 40%

B/T cost of borrowing = 10%

A/T cost of equity = 20%

Accounting method = flow-through

Other than those described, there are no significant differences between the alternatives. (The operating costs will be about the same with each.) If the option is to repair the old transformer now, the repaired transformer will be replaced at the end of 4 years by a new transformer identical (in cost and function) to the current new transformer. The transformer will be needed for an indefinite period.

a. Compute the composite A/T cost of capital.
b. Select the option based on the revenue requirement principle.
c. Show that you will select the same option if the present value criterion is applied instead of the revenue requirement principle.

**15.6.** Consider Problem 15.1. Suppose that the incinerators have a 10-year MACRS class life. Using the normalized accounting procedure, determine whether or not to install the advanced design for the following two situations.

a. No investment tax credits are allowed.
b. An energy tax credit of 10% is allowed and must be amortized over the life of the asset.

**15.7.** Consider Problem 15.2. Assume that the asset has a 5-year MACRS class life. Using the normalizing method, compute the required annual revenues.

**15.8.** Using the minimal annual revenue requirements as the gross revenues and the other results obtained in the solution to Problem 15.1, determine the cash flows for each year and the internal rate of return.

**15.9.** Using the minimum annual revenue requirements as the gross revenues and the other results obtained in the solution to Problem 15.2, determine the cash flows for each year and the internal rate of return.

**15.10.** In a move to expand its international presence, the U.S. Communication Company has decided to enter the international long-distance market. The cable called PTAT-1, which cost about $450 million to construct, is scheduled to go into service 2 years from now. The PTAT-1 cable will span 4,600 miles and carry voice, video, and computer data on beams of laser light. The new cable can carry 80,000 simultaneous telephone calls. It is estimated that the expansion will require the following capital expenditures.

| End of Year | Asset | Capital Expenditure | Service Life | Salvage | MACRS Class |
|---|---|---|---|---|---|
| 0 | Land | $10,000,000 | — | — | — |
| 1 | Building | 40,000,000 | 30 years | 4,000,000 | 31.5 Real |
| 1 | Equipment and cable | 400,000,000 | 10 years | 40,000,000 | 10 |

In addition, it is estimated that annual gross revenues and operating costs will increase by the amounts shown below.

| End of Year | Gross Revenue | Operating Costs |
|---|---|---|
| 0 | — | — |
| 1 | — | — |
| 2 | $150,000,000 | $50,000,000 |
| 3 | 200,000,000 | 80,000,000 |
| 4 | 250,000,000 | 100,000,000 |
| 5 | 300,000,000 | 150,000,000 |
| 6 | 350,000,000 | 150,000,000 |
| 7 | 400,000,000 | 200,000,000 |
| 8 | 400,000,000 | 200,000,000 |
| 9 | 400,000,000 | 200,000,000 |
| 10 | 400,000,000 | 200,000,000 |

The following investment characteristics are assumed.

- The project life is 10 years.
- The debt-to-equity ratio is 50%.
- B/T cost of borrowing is 12%.
- A/T cost of equity is 15%.
- The land will be sold at the end of 10 years at an estimated value of $20,000,000.
- The building will be depreciated using the straight-line method and is expected to sell at $15,000,000 at the end of 10 years.
- The equipment and cable will be depreciated by the SOYD over 10 years.
- The flow-through method is used.
- The capitalized interest during the construction period is 1%.
- The tax rate is 45%.

a. Under the flow-through method, determine whether this expansion is economically justified.
b. Under the normalizing method, determine whether this expansion is economically justified.
c. Using the MACRS and the present value criterion, determine whether this expansion is economically justified.

# 16

# Procedures for Replacement Analysis

## 16.1 INTRODUCTION

In principle, we can analyze an equipment replacement problem by applying the techniques of Chapter 7 to select between two alternatives or from several alternatives. Section 7.5 contains a discussion of how alternatives with unequal lives, especially service projects, can be compared. Typically, an existing asset, the defender, has a remaining life that is shorter than that of a new, potential replacement asset, the challenger. In practice, equipment replacement problems have characteristics that enable us to use specialized concepts and analysis techniques.

In this chapter we present some of these techniques. First, we discuss obsolescence and deterioration, two major reasons for considering the replacement of an existing asset. In Section 16.2 we present the basic concepts of sunk costs, the outsider point of view, and the economic life of an asset. Section 16.3 contains methods for infinite planning periods, and Section 16.4 has methods for finite planning periods. Most of the analysis techniques assume a fixed *MARR* and complete certainty. Uncertainty would be handled by sensitivity analysis. In Section 16.5 the data base requirements for conducting systematic replacement analyses are presented.

### 16.1.1 Quantifying Obsolescence and Deterioration

There are two basic reasons for considering the replacement of a physical asset.

1. Physical impairment—changes in the condition of the asset itself.
2. Obsolescence—changes in the environment external to the asset.

Examples of physical impairment attributable to age are

- Decline in output—in horsepower, in speed, in the number of units passing quality control.

- Increase in operating costs—with more fuel needed, more material waste, and more consumable parts worn out.
- Increase in maintenance costs—for spares and for maintenance labor.
- More frequent and longer unplanned breakdowns.
- Decline in the value of service—such as fewer first-quality runs from a textile loom, or greater dimensional variance of machined surfaces.

Examples of obsolescence are

- The service provided by the defender is no longer required.
- It becomes difficult to find skilled operators for the defender.
- The challenger's capacity better matches present needs (e.g., a 250-seat, 1,000-mile commercial jet).
- The challenger generates more sales (e.g., new rental cars, new fleet of aircraft).
- The challenger has a longer service life.
- The challenger has inherently lower operating and maintenance costs.
- A higher output is available from the challenger.
- The challenger has inherently fewer unplanned breakdowns.
- The challenger is more effective (e.g., a missile cruiser replacing a battleship).

Physical impairment and obsolescence may occur independently, or they may occur jointly with respect to a particular asset. No "standard" methods can be applied in quantifying these effects. Instead, each situation requires careful collection of data and analysis. Characteristics of certain types of products may follow a particular functional form that can be applied to the entire product class, such as failure probabilities of electronic components [12]. Manufacturing learning curves may apply to individual products or product classes [15]. On the other hand, the curve describing unit manufacturing costs may have a limiting value, beyond which there may or may not be a technological breakthrough to the "next generation" of the particular type of equipment.

Physical impairment (deterioration) is typically quantified by examining historical data for the asset type and making some projection for the portion of the asset's service life for which no data exist. The projection is often made by fitting to the historical data one of several common functional forms.

**1.** Straight line,

$$y(t) = a + bt$$

**2.** Exponential,

$$y(t) = ab^{ct}, \qquad a, b, c > 0$$

**3.** Negative exponential,

$$y(t) = ab^{-ct}, \qquad a, b, c > 0$$

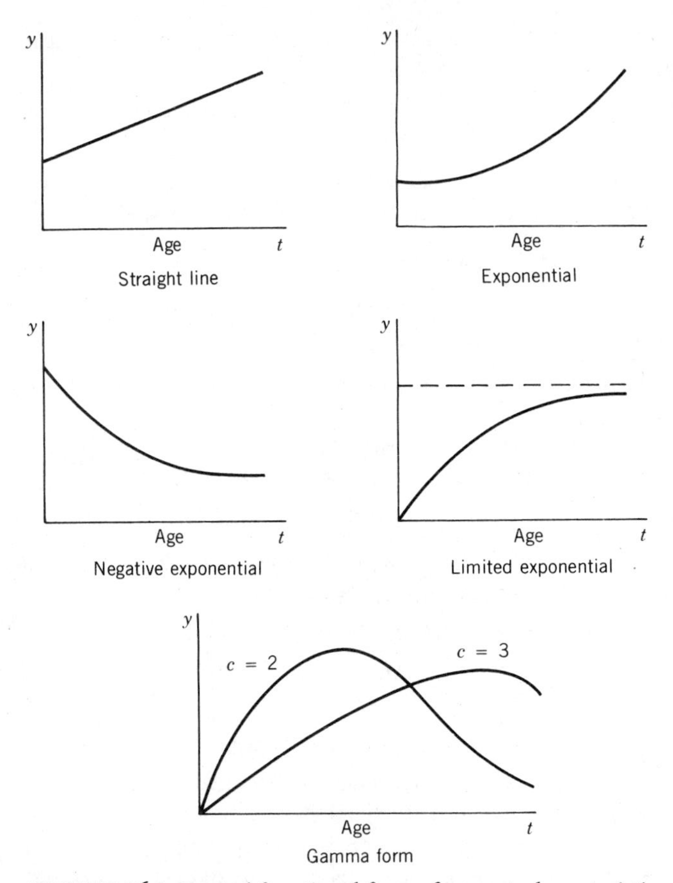


**FIGURE 16.1** Typical functional forms for asset characteristics.

4. Limited exponential,

$$y(t) = a(1 - e^{-ct}), \qquad a, c > 0$$

5. Gamma form,

$$y(t) = a(bt)^{c-1}e^{-bt}$$

Figure 16.1 shows typical (continuous) curves corresponding to the five forms.

The following examples represent real situations in which assets are impaired and technological advancements are made.

### *Example 16.1*

An important factor in evaluating transit buses is the probability of unexpected breakdown while in service. Data collected by a transit operator in a large metropolitan area showed that the mean time between failures (MTBF) varied from 2,000 to 1,200 hours and followed roughly an exponential growth curve, as shown in Figure 16.2.

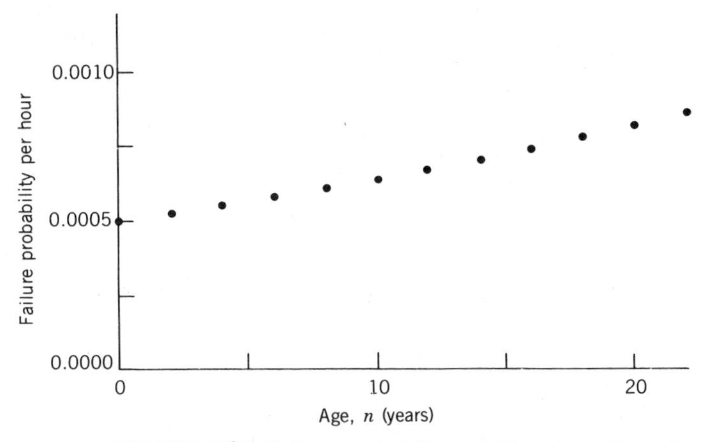


**FIGURE 16.2 Failure probability as a function of age, Example 16.1.**

$$y(n) = \text{failure probability per hour}$$
$$= (0.005)e^{0.025n}$$

where $n$ is the age of the vehicle in years. □

### *Example 16.2*

A public utility analyzed maintenance costs for a fleet of aerial basket trucks. Because the data covered a period of high inflation, an industry-specific price index for auto and truck parts was used to convert actual dollars to constant dollars. Disparities in the hours a vehicle was used as a function of age were also observed: new vehicles were used less than those 2 to 5 years old, and as vehicle age approached 7 years, there was again less utilization. A typical small aerial basket truck has as a function of age the maintenance costs shown in Table 16.1.

**Table 16.1** ***Truck Maintenance Costs in 1981 Dollars, Example 16.2***

| Age of Truck | Maintenance Labor Cost | Maintenance Parts Cost | Total Maintenance Cost | Hours Used per Year | Total Maintenance Cost per 2,000 hours |
|---|---|---|---|---|---|
| 1 | $ 630 | $ 932 | $1,562 | 1,210 | $2,582 |
| 2 | 1,519 | 2,750 | 4,269 | 2,598 | 3,286 |
| 3 | 1,330 | 2,716 | 4,046 | 2,411 | 3,356 |
| 4 | 1,288 | 4,106 | 5,394 | 2,300 | 4,690 |
| 5 | 1,316 | 3,117 | 4,433 | 2,112 | 4,198 |
| 6 | 1,372 | 3,298 | 4,670 | 1,793 | 5,209 |
| 7 | 1,099 | 2,008 | 3,107 | 1,469 | 4,230 |
| 8 | 784 | 984 | 1,768 | 1,133 | 3,120 |
| 9 | 427 | 824 | 1,251 | 817 | 3,062 |

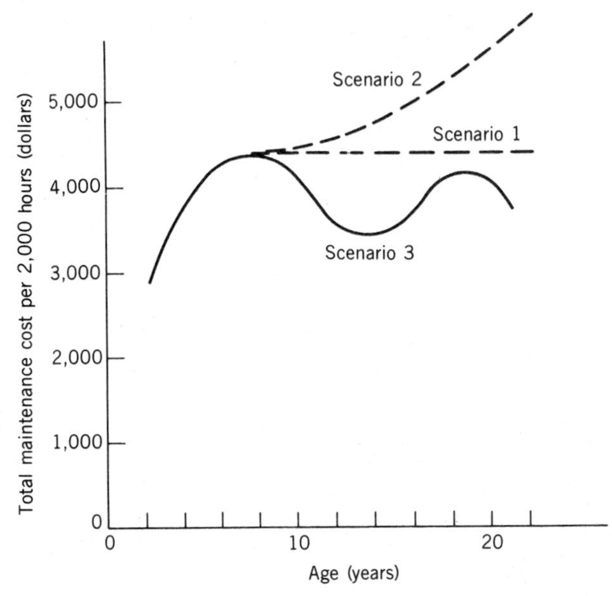


**FIGURE 16.3 Truck maintenance costs for three scenarios, Example 16.2.**

The costs are given in 1981 dollars. The resulting total maintenance cost curve appears to follow a gamma form. The high costs during years 4 through 6 were explained by "burnout" of certain components and the warranty-required replacement of hydraulic components. It was suspected that major repairs were not performed on the older vehicles, in anticipation of their retirement.

An interesting aspect of this problem became apparent when forecasts were made of maintenance costs beyond age 9. Three scenarios were envisioned.

1. Costs would remain constant from age 6 on (this was considered the last year of normal utilization).
2. Costs would increase continuously from age 6.
3. Following a major overhaul at age 7 or 8, costs would follow another gamma form, seen as a double-humped curve. (The overhaul cost would have to be capitalized and depreciated.)

The curves of these three scenarios are given in Figure 16.3. □

## *Example 16.3*

Fuel consumption improvements had been observed in the small aerial basket trucks as a function of the year they were built. These miles-per-gallon (mpg) increases were brought about by small yearly improvements in the basket truck model and by two major changes—shifting from gasoline to diesel engines and installing a battery pack for the basket movement. Older models operated the

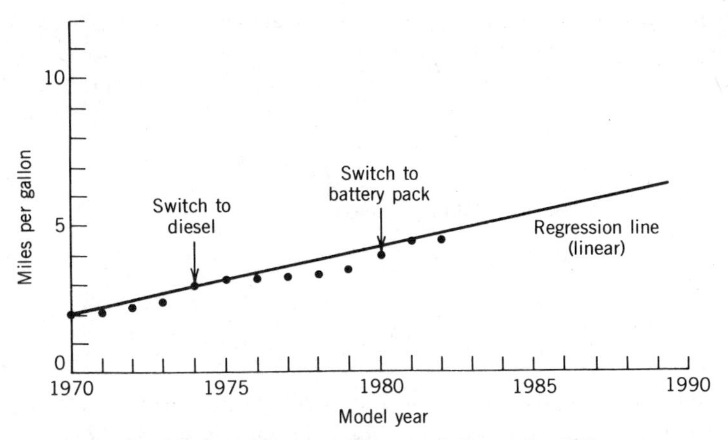


**FIGURE 16.4 Miles-per-gallon trend, Example 16.3.**

main engine 6 to 8 hours a day while the truck was parked at a work site. It was anticipated that future improvements in mileage would continue to come from incremental yearly improvements in the model, but that technological advancements would be less frequent. Accordingly, miles per gallon were forecast to increase linearly, as shown in Figure 16.4. It should be noted that the linear trend in miles-per-gallon increases means that the fuel consumption trend will follow a negative exponential function. □

### *Example 16.4*

The newer models of aerial basket trucks are fitted with small derricks, which suffice for lifting operations at a large fraction of the job sites. Previously, a derrick truck accompanied an aerial basket truck on jobs requiring any lifting. (About 45% of the jobs for aerial basket trucks require some lifting.) A survey of job records indicates that the small derrick is adequate in about 65% of the situations requiring lifting. Accordingly, when the older aerial basket trucks are compared with newer models, a figure equal to 29% (45% × 65%) of the *AE* of a derrick truck is subtracted from the annual equivalent cost of the new model. Alternatively, we could add 29% of the *AE* of a derrick truck to the *AE* of the old model aerial basket truck. □

### *Example 16.5*

Discussions with supervisors revealed that the public utility designated the lifting capacity of derrick trucks to be one-half that given in the manufacturer's rating. The manufacturer's rating was based on the boom being set at a maximum angle from the ground (minimum offset from the vertical), but the derrick crew was judged unable to perform calculations (or follow tabular guides) for other situations. A new model series will soon be available that incorporates a microprocessor and a set of strain gauges, so that the derrick operator can read the lift capacity for the boom angle being used and the actual load being hoisted. A survey of job records (with accompanying site visits) shows that the following fleet compositions would be needed to perform the typical work load during a year.

| Rated Capacity | Old Model | New Model |
|---|---|---|
| 10 tons | 23 | 29 |
| 25 tons | 11 | 7 |
| 50 tons | 6 | 4 |

Thus, even though the new models are more expensive, their versatility could enable the company to operate with a lower total annual cost for the fleet. □

### 16.1.2 Forecasting Future Data

The functional forms presented in the previous section may also be applied when forecasting the capabilities of future replacements and their costs. In Example 16.3 we saw how the miles per gallon for a type of truck was forecast to increase linearly, creating a negative exponential trend in fuel consumption as a function of the model year. In the data analysis for this example, little or no change in fuel consumption was observed as a particular vehicle aged.

The major elements that must be estimated are

- Purchase cost—often assumed to follow an exponential growth (positive or negative) when measured in constant dollars.
- Salvage values—often assumed to follow a negative exponential curve based on the initial cost of *that* year's model, or based on the initial cost of the *current* year's model.
- First-year operating and maintenance costs—often assumed to follow a negative exponential curve based on those of the current year's model. The operating and maintenance (O&M) costs of future models are usually assumed to follow the same pattern as that of the O&M costs of the current year's model.

The implicit assumption in the use of curves, as we have described, is that technological changes will be gradual. This is often a good assumption and has been verified in a wide variety of situations.

Major technological breakthroughs are more difficult to predict far in advance, say 20 years. They usually are known 5 years in advance, however. For example, magnetic levitation has been tested in the Federal Republic of Germany for more than 10 years and is now available on a limited basis [19]. The procurement lead time of fuel-efficient commercial jets is approaching 4 years [3].

Although we may know of such future challengers 5 years in advance, there still remains the difficulty of predicting costs and performance. Analysts familiar with a particular industry will often take the view of what the market will bear. By projecting costs and performance of the new model, they will calculate the likely purchase cost of the new model. Such an analysis requires intimate knowledge of the industry, and the procedures are difficult to generalize.

Learning curves are a specific type of negative exponential curve applicable to situations in which manufacturing efficiencies continue over time at a known rate [15]. The general form is

$$P_n = P_0(n)^a \tag{16.1}$$

$$a = \frac{\ln b}{\ln 2} \tag{16.2}$$

where $P_n$ = manufacturing cost of the $n$th production item,
$P_0$ = manufacturing cost of the first production item,
$b$ = slope parameter, percentage reduction in unit cost for doubled cumulative production units.

### *Example 16.6*

Costs of new assets have been observed to decline with cumulative sales as follows.

| Units | Average Unit Cost |
|---|---|
| 1–100 | $27,000 |
| 101–200 | 21,500 |
| 201–300 | 19,000 |
| 301–400 | 17,000 |

Analyzing the reduction from the first 100 units to the second 100 units, we observe a decline of 0.204, or a factor of $1 - 0.204 = 0.796$. The reduction from the second 100 units to the fourth 100 units (doubled cumulative production units) is 0.209, or a factor of 0.791. We select 0.79 as the slope factor and compute

$$a = \frac{\ln 0.79}{\ln 2} = -0.34$$

Thus, we predict the cost of future units to be

$$P_n = (27{,}000)(n)^{-0.34}$$

where $n$ represents the cumulative production in hundreds. Average unit costs for the fifth 100 units are estimated to be

$$P_5 = (27{,}000)(5)^{-0.34} = \$15{,}621 \quad \square$$

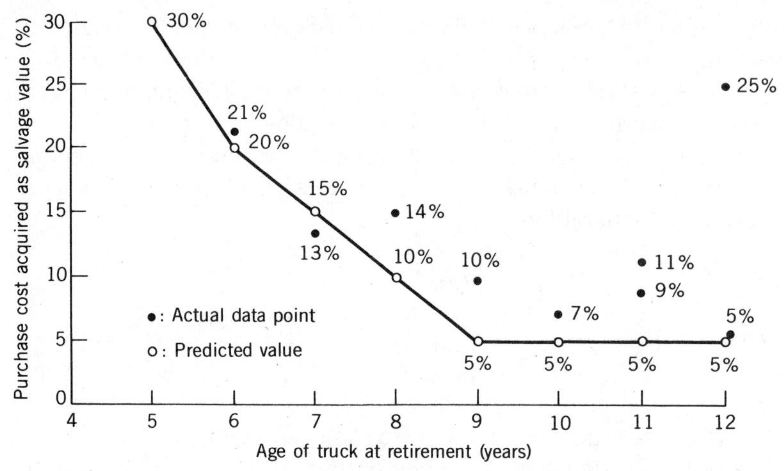


**FIGURE 16.5 Salvage value estimates, Example 16.7.**

### *Example 16.7*

Figure 16.5 represents the salvage values for a class of trucks as a function of vehicle age. The salvage values is a percentage of purchase cost, and the analysis is performed in constant dollars. The actual data points show dispersion about any hypothetical curve, including some outliers. The curve that was fitted for analysis purposes is a simple, piecewise linear curve. □

Analysis of a real situation usually requires a combination of estimating functional forms, extrapolating to estimate future data, and using expert judgment to predict major changes in technology. Uncertainties about costs and technology often complicate matters; this issue is addressed in Section 16.4. Another common difficulty in performing the analysis is unavailability of data for specific model years in constant dollars; this issue is addressed in Section 16.5.

## 16.2 BASIC CONCEPTS IN REPLACEMENT ANALYSIS

There are three basic concepts in replacement analysis.

1. Sunk costs.
2. The outsider point of view.
3. The economic life of an asset.

The next three subsections clarify when and how to use these concepts.

### 16.2.1 Sunk Costs

The concept of sunk costs relates to past expenditures for an asset. Past expenditures, no matter how great and whether recovered or not, are not relevant when considering whether to keep or replace the asset. Only current and future cash flows should be considered. Sometimes there is a tendency to charge a potential new asset, a challenger, with the unrecovered cost of an existing asset, the defender. Such allocation is improper and should be avoided.

There are situations in which the unrecovered cost of a defender will lead to a tax loss on disposal of the asset and a subsequent tax credit. The inclusion of the tax credit in the analysis is proper. The next example demonstrates how to compute the tax credit.

### *Example 16.8*

An asset that was purchased less than 4 years ago for $800,000 is being considered for replacement. Because of extensive customizing, only $60,000 salvage value is expected. The depreciation book value upon disposal will be $500,000. Determine the cash flow from disposal (after-tax salvage value) when the marginal tax rate is (a) 40%, (b) 0%.

There will be a loss on disposal of

| | |
|---|---|
| Salvage proceeds | $60,000 |
| Book value | −500,000 |
| Gain (loss) | −$440,000 |

With a marginal tax rate of 40%, there will be a tax reduction of (440,000)(0.4) = $176,000. The cash flow from disposal will be $60,000 + 176,000 = $236,000. With a marginal tax rate of 0%, the cash flow from disposal will consist only of the $60,000. □

In either situation in Example 16.8, most of the expenditure on the asset should be considered a *sunk cost,* since only a fraction of the initial cost (and of the book value) will be recovered upon disposal.

### 16.2.2 Outsider Point of View

Another way to regard the $236,000 or the $60,000 after-tax salvage value in Example 16.8 is to consider it as the *opportunity cost* of *keeping* the asset. We are giving up $236,000 (with a marginal tax rate of 40%) or $60,000 (with a marginal tax rate of 0%) in cash flow as a consequence of the decision not to sell the asset. This is the point of view of an outsider who owns neither asset. It is also the point of view of the corporate treasurer.

The outsider point of view differs from the cash flow approach followed in earlier chapters. The cash flow approach can be used in analyzing replacement problems provided the study period is the same for all alternatives (see Section 7.5). The following examples illustrate both approaches, as well as the rationale for making *AE* comparisons under various circumstances.

### *Example 16.9*

An existing asset A (the defender), which was purchased 2 years ago for $8,000, has a current salvage value of $3,000. The remaining life of the defender is 8

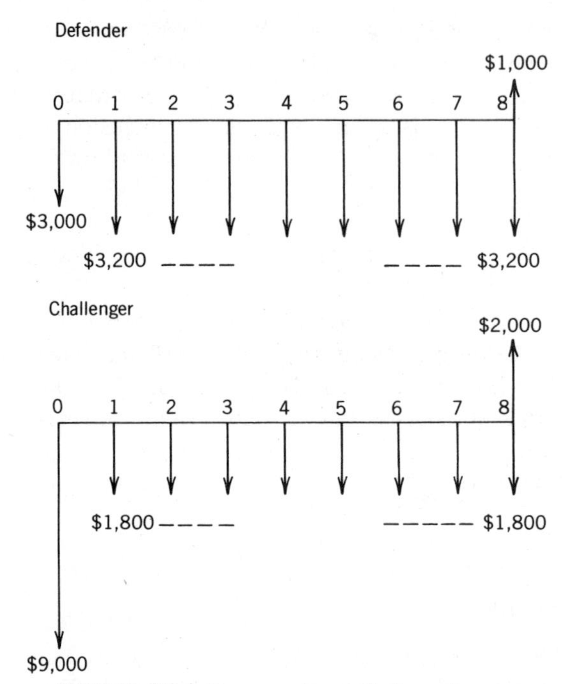


**FIGURE 16.6 Opportunity cash flow diagrams, Example 16.9.**

years, at which time the salvage value will be $1,000. Annual operating and maintenance costs are $3,200. A new asset B (the challenger) that would provide the same function is available at a purchase price of $9,000. It has a life of 8 years, a salvage value of $2,000, and annual costs of $1,800. Decide whether replacement is justified, using a before-tax comparison with $i = 10\%$ and the outsider point of view.

Since the lifetimes are the same, we will use *AE*. Figure 16.6 shows the opportunity cash flow diagrams. We will use an opposite sign convention for convenience, with outflows being positive and inflows negative.

$$\begin{aligned} AE(10\%)_A &= (3{,}000)(A/P,\ 10\%\ 8) - (1{,}000)(A/F,\ 10\%,\ 8) + 3{,}200 \\ &= (3{,}000)(0.18744) - (1{,}000)(0.08744) + 3{,}200 \\ &= 562.3 - 87.4 + 3{,}200 \\ &= 3{,}674.9 = \$3{,}675 \quad \text{for the defender} \end{aligned}$$

$$\begin{aligned} AE(10\%)_B &= (9{,}000)(A/P,\ 10\%,\ 8) - (2{,}000)(A/F,\ 10\%,\ 8) + 1{,}800 \\ &= (9{,}000)(0.18744) - (2{,}000)(0.08744) + 1{,}800 \\ &= 1{,}687.0 - 174.9 + 1{,}800 \\ &= 3{,}312.1 = \$3{,}312 \quad \text{for the challenger} \end{aligned}$$

There is a difference of $363 *AE* in favor of the challenger, and the replacement should be made. □

In Example 16.9 the outsider point of view treats the $3,000 current salvage of the defender as a cost that is incurred if the decision is to keep the defender. The current book value is likely to be greater, but, following the principle of ignoring sunk costs, this is not a factor in this before-tax analysis.

### *Example 16.10*

Rework Example 16.9 by using a cash flow approach. The actual cash flow diagrams are shown in Figure 16.7. The difference from Figure 16.6 is that the $3,000 salvage value of the defender is now credited against the $9,000 purchase price of the challenger, and there is no initial outlay for the decision to keep the defender.

$$\begin{aligned} AE(10\%)_A &= (-1{,}000)(A/F,\ 10\%,\ 8) + 3{,}200 \\ &= (-1{,}000)(0.08744) + 3{,}200 \\ &= -\ 87.4 + 3{,}200 \\ &= 3{,}012.6 = \$3{,}113 \quad \text{for the defender} \end{aligned}$$

$$\begin{aligned} AE(10\%)_B &= (9{,}000 - 3{,}000)(A/P,\ 10\%,\ 8) - 2{,}000\ (A/F,\ 10\%,\ 8) + 1{,}800 \\ &= (6{,}000)(0.18744) - (2{,}000)(0.08744) + 1{,}800 \\ &= 1{,}124.6 - 174.9 + 1{,}800 \\ &= 2{,}749.7 = \$2{,}750 \quad \text{for the challenger} \end{aligned}$$

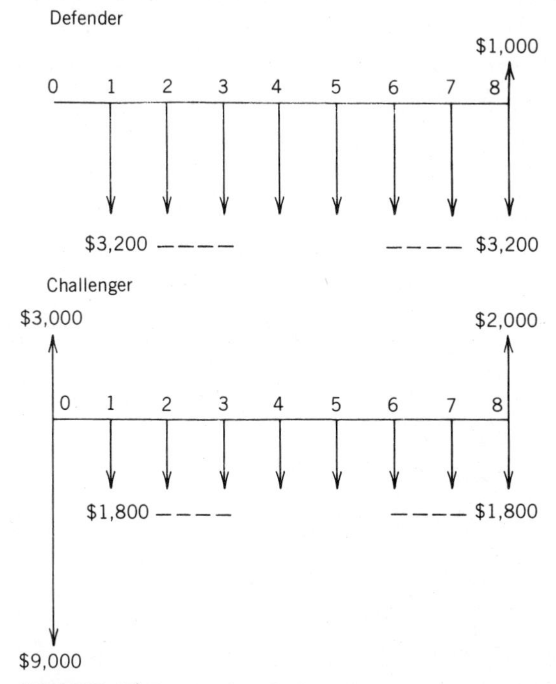


**FIGURE 16.7** Actual cash flow diagrams, Example 16.10.

There is again a difference of \$363 *AE* in favor of the challenger. This is not surprising, since the only difference between the two examples is the addition of \$3,000 at time 0 to both defender and challenger flows (compare Figures 16.6 and 16.7). We should note here that PV analysis would give results consistent with *AE* analysis. □

### *Example 16.11*

Rework Example 16.9 with the following changes: the remaining life of asset A (defender) is 3 years, the salvage value is \$1,428, and annual costs are \$2,900. Use the outsider point of view. The opportunity cash flow diagrams are shown in Figure 16.8.

$$\begin{aligned} AE(10\%)_A &= 3{,}000\ (A/P,\ 10\%,\ 3) - 1{,}428\ (A/F,\ 10\%,\ 3) + 2{,}900 \\ &= (3{,}000)(0.40211) - (1{,}428)(0.30211) + 2{,}900 \\ &= 1{,}206.3 - 431.4 + 2{,}900 \\ &= 3{,}674.9 = \$3{,}675 \quad \text{for the defender} \end{aligned}$$

From Example 16.9 $AE(10\%)_B = \$3{,}312$ for the challenger. Again, the challenger is favored by \$363 *AE*. □

We may note two things in Example 16.11. First, the costs for the defender have been contrived so that the *AE* is the same \$3,675 as in Example 16.9, despite

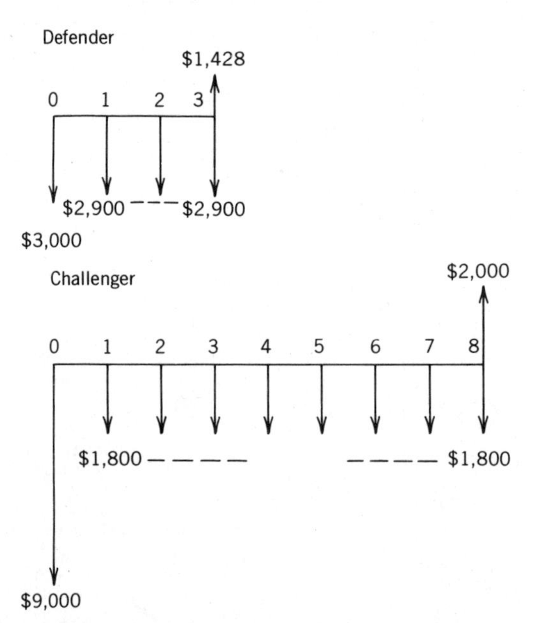


**FIGURE 16.8 Opportunity cash flow diagrams, Example 16.11.**

the shorter remaining life of 3 years compared with 8. This was done to allow easy comparison of the several examples. Second, and more important, is the question of how we can perform *AE* analysis with unequal lives. Recall from Section 7.5 that the use of *AE* with unequal lives is justified in some circumstances. We have here another special situation in which *AE* with unequal lives gives valid results, but only when using the outsider point of view.

The implicit assumption made in using *AE* when the defender's remaining life is shorter than the challenger's life is that *after the initial decision we make perpetual replacements with assets similar to the challenger.* In Example 16.11 this means that if we decide to keep A, we will replace it at time 3 by an asset similar to B. This asset will in turn be replaced 8 years later, at time 11, by another asset B. There are two implied infinite sequences in Example 16.11.

*Sequence 1:* Keep defender (A), buy a challenger (B) at time 3, buy a challenger (B) at time 11, buy a challenger (B) at time 19, . . . .

*Sequence 2:* Buy challenger (B) at time 0, buy a challenger (B) at time 8, buy a challenger (B) at time 16, . . . .

It is clear that the *AE* cost for either sequence of assets is the same after the remaining life of the defender. In Example 16.11 we have *AE* costs of \$3,312 for year 4 and after, no matter which decision we make at time 0. Thus, to make the decision we can use *AE,* \$3,675 for the remaining life of the defender and \$3,312 for the challenger, provided the implicit assumption is reasonable.

## Example 16.12

Rework Example 16.11 using *PV* analysis of the cash flows for an implied infinite sequence of replacements.

The defender, years 1 through 3:

$$\begin{aligned} PV(10\%)_{A1} &= \$2{,}900(P/A,\ 10\%,\ 3) - 1{,}428(P/F,\ 10\%,\ 3) \\ &= (2{,}900)(2.48685) - (1{,}428)(0.75131) \\ &= \$6{,}138.99 \end{aligned}$$

The defender, after year 3:

$$\begin{aligned} PV(10\%)_{A2} &= \frac{\$3{,}312.1}{0.1}(P/F,\ 10\%,\ 3) \\ &= (33{,}121)(0.75131) \\ &= \$24{,}884.14 \end{aligned}$$

The \$3,312.1 is the *AE* of the challenger, as calculated in Example 16.9. The defender, total:

$$PV(10\%)_{A} = \$6{,}138.99 + 24{,}884.14 = \$31{,}023$$

The challenger, years 1 through 8:

$$PV(10\%)_{B1} = \$6{,}000 + 1{,}800(P/A,\ 10\%,\ 8) - 2{,}000(P/F,\ 10\%,\ 8)$$
$$= 6{,}000 + (1{,}800)(5.33493) - (2{,}000)(0.46651)$$
$$= \$14{,}669.86$$

The challenger, after year 8:

$$PV(10\%)_{B2} = \frac{\$3{,}312.1}{0.1}(P/F,\ 10\%,\ 8)$$
$$= (33{,}121)(0.46651)$$
$$= \$15{,}451.28$$

The challenger, total:

$$PV(10\%)_{B} = \$14{,}669.86 + 15{,}451.28 = \$30{,}121$$

The *PV* difference of \$902 in favor of the challenger is exactly equal to the *AE* difference for 3 years,

$$\$363\ (P/A,\ 10\%,\ 3) = (363)(2.48685) = \$903 \approx 902 \quad \square$$

The next example illustrates why the cash flow approach with *AE* for the respective lives can give false results when there is an infinite sequence of replacements.

### *Example 16.13*

Rework Example 16.11 by using the cash flow approach with *AE* for the respective lives.

The defender:

$$AE(10\%)_{A} = -\$1{,}428\ (A/F,\ 10\%,\ 3) + 2{,}900$$
$$= -(1{,}428)(0.30211) + 2{,}900$$
$$= -\ 431.4 + 2{,}900$$
$$= 2{,}468.6 = \$2{,}469$$

The challenger, from Example 16.10:

$$AE(10\%)_{B} = \$2{,}750$$

Now the defender is favored by \$281 *AE*. What is wrong in this approach is that the \$3,000 current salvage value of the defender is spread over the 8-year life of the challenger. Since the remaining life of the defender is only 3 years, there is a distortion of the *AE* values.

One way to correct the distortion is to credit the $3,000 salvage value of the defender against the purchase price of the challenger, *but* spread the $3,000 over *3* years. Recalculating the *AE* would yield (see Example 16.10):

$$\begin{aligned} AE(10\%)_B &= \$9{,}000\ (A/P,\ 10\%,\ 8) - 3{,}000\ (A/P,\ 10\%,\ 3) \\ &\quad - 2{,}000\ (A/F,\ 10\%,\ 8) + 1{,}800 \\ &= (9{,}000)(0.18744) - (3{,}000)(0.40211) \\ &\quad -(2{,}000)(0.08744) + 1{,}800 \\ &= 1{,}687.0 - 1{,}206.3 - 174.9 + 1{,}800 \\ &= \$2{,}105.8 = \$2{,}106 \quad \text{for the challenger during years 1 to 3} \end{aligned}$$

Now the challenger is favored again and by the same difference of $363 *AE*. □

When the defender and challenger have unequal lifetimes, we must make an assumption in order to obtain a common planning period. A typical assumption is that after the initial decision we make perpetual replacements with assets similar to the challenger. In this situation the *AE* criterion with the outsider point of view is easy to apply. We need to evaluate only the opportunity cash flows for the respective lives of the two assets. Certainly, we can use *PV* with actual cash flows, but this requires evaluation of infinite cash flow streams. *AE* with actual cash flows is difficult to apply correctly and should be avoided.

### 16.2.3 Economic Life of an Asset

The concept of an infinite sequence of replacements can be generalized to the situation in which the life of an asset is a decision variable. A common example of this type of problem is deciding on the replacement interval for an automobile. With proper maintenance, a car can function for more than 10 years. And yet for various reasons, including economic efficiency, many individuals and corporations replace cars after as few as 3 years.

In the previous examples the lives of the defender and the challenger were given. In this section we take the approach that *the life to be assumed for the new asset (the challenger) should be the one most favorable to the asset.* This life, called the *economic life of an asset,* is the *life that results in the minimum AE cost of owning and operating the asset.* The following example demonstrates the concept.

#### *Example 16.14*

A municipality is trying to establish a policy on how long to keep economy passenger cars that are used for official business. The purchase cost of a car at fleet discount is $6,000. Annual costs for insurance, oil, and repairs for 20,000 miles of annual use are $1,500 for a new car and increase by $100/year with the age of the car. Fuel consumption is not perceived to change with age and is therefore excluded from the analysis. It is recognized that as a vehicle becomes older it suffers more breakdowns, causing employees to become less productive. The average loss of employee productivity is estimated to be 4 days/year for a new car and to increase by 2 days/year with the age of the car. The average payroll cost of an employee likely to use the car is $100/day. After 8 years spare parts become difficult to obtain, and the downtimes of the vehicle become unacceptably long and frequent. Salvage values as a function of age up to 8 years are given in column 2 of Table 16.2. Using $i = 10\%$, find the economic life of such a car. Ignore income taxes, inflation, and technological improvements.

If we replace cars after one year of use, the relevant flows are

| | | |
|---|---|---|
| Time 0: | Purchase cost | $6,000 |
| Time 1: | Operating cost | $1,500 |
| | Lost productivity cost | $400 |
| | Salvage value | $4,500 |

**Table 16.2** ***Economic-Life Calculations, Example 16.14***

| 1 | 2 | 3 | 4 | 5 | 6 | 7 | 8 | 9 | 10 | 11 |
|---|---|---|---|---|---|---|---|---|---|---|
| | | | | | Col. 3 × Col. 5 | | Col. 6 + Col. 7 | Col. 4 × 300 | | Col. 8 + Col. 10 |
| *N* | *S* | *A/P* | *A/G* | *P* − *S* | | *Si* | | | | |
| 1 | $4,500 | 1.1000 | 0 | $1,500 | $1,650 | $450 | $2,100 | $ 0 | $1,900 | $4,000 |
| 2 | 3,375 | 0.5762 | .4762 | 2,625 | 1,513 | 338 | 1,851 | 143 | 2,043 | 3,894 |
| 3 | 2,530 | 0.4021 | .9365 | 3,470 | 1,395 | 253 | 1,648 | 281 | 2,181 | 3,829 |
| 4 | 1,900 | 0.3155 | 1.3811 | 4,100 | 1,293 | 190 | 1,483 | 414 | 2,314 | 3,797 |
| 5 | 1,425 | 0.2638 | 1.8101 | 4,575 | 1,207 | 143 | 1,350 | 543 | 2,443 | 3,793 |
| 6 | 1,070 | 0.2296 | 2.2236 | 4,930 | 1,132 | 107 | 1,239 | 667 | 2,567 | 3,806 |
| 7 | 800 | 0.2054 | 2.6217 | 5,200 | 1,068 | 80 | 1,148 | 787 | 2,687 | 3,835 |
| 8 | 600 | 0.1874 | 3.0044 | 5,400 | 1,012 | 60 | 1,072 | 901 | 2,801 | 3,873 |

Explanation of columns:

1. Age, also the replacement cycle.
2. Salvage value as a function of age.
3. (*A/P*, 10%, *N*) interest factor.
4. (*A/G*, 10%, *N*) interest factor.
5. Purchase cost minus salvage value.
6. $(P - S)(A/P, 10\%, N)$, first component of ownership costs.
7. $(S)(0.1)$, second component of ownership costs.
8. *AE* ownership costs, as a function of the replacement cycle.
9. *AE* of changing portion of costs of operation and lost productivity, as a function of the replacement cycle.
10. Col. 9 + $1,500 + $400, *AE* of costs of operation and lost productivity, as a function of the replacement cycle.
11. Total *AE* costs, as a function of the replacement cycle.

Converting to *AE* for the one-year life results in

$$\begin{aligned} AE(10\%)_1 &= (\$6{,}000 - 4{,}500)(A/P, 10\%, 1) + (\$4{,}500)(0.1) + \$1{,}900 \\ &= 1{,}650 + 450 + 1{,}900 \\ &= 2{,}100 + 1{,}900 = \$4{,}000 \\ &= \text{ownership} + \text{operating} = \text{total costs} \end{aligned}$$

The \$4,000 consists of two components: ownership costs of \$2,100 and operating and lost productivity costs of \$1,900.

If we replace cars after 2 years of use, the relevant flows are

| | |
|---|---|
| Time 0: Purchase cost | \$6,000 |
| Time 1: Operating and lost productivity | \$1,900 |
| Time 2: Operating and lost productivity | \$2,200 |
| Salvage value | \$3,375 |

Computing *AE* for the 2-year life gives

$$\begin{aligned} AE(10\%)_2 &= (6{,}000 - 3{,}375)(A/P, 10\%, 2) + (3{,}375)(0.1) \\ &\quad + 1{,}900 + 300\,(A/G, 10\%, 2) \\ &= 1{,}513 + 338 + 1{,}900 + 143 \\ &= 1{,}851 + 2{,}043 = \$3{,}894 \\ &= \text{ownership} + \text{operating} = \text{total costs} \end{aligned}$$

The ownership costs per year have declined, but the operating costs per year have increased.

Repeating this process for longer lives yields the figures shown in Table 16.2. We see that the minimum *AE* of \$3,793 is obtained with a 5-year life. Thus, the economic life of an economy passenger car, as operated by the municipality, is 5 years. Figure 16.9 shows the curves for ownership costs, operating (and lost productivity) costs, and total costs.

The rationale for finding the economic life of a new asset is that by replacing perpetually according to the economic life, we obtain the minimum infinite *AE* cost stream. Figure 16.10 illustrates this concept. Of course, we should be envisioning a long period of required service of the asset. □

The concept of economic life can be applied to an existing asset (defender). In some special cases, however, the optimum retention time for a defender is different from its economic life (see Problem 16.6 at the end of the chapter). For this reason ***marginal analysis*** is proposed when the economic life of a defender makes its *AE* cost lower than that of a challenger. This analysis is presented in the next section.

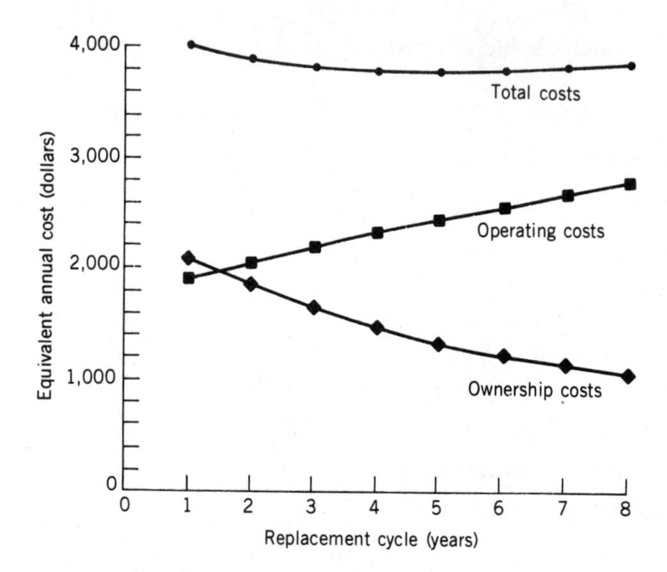


**FIGURE 16.9** Ownership, operating, and total costs for an asset, Example 16.14.

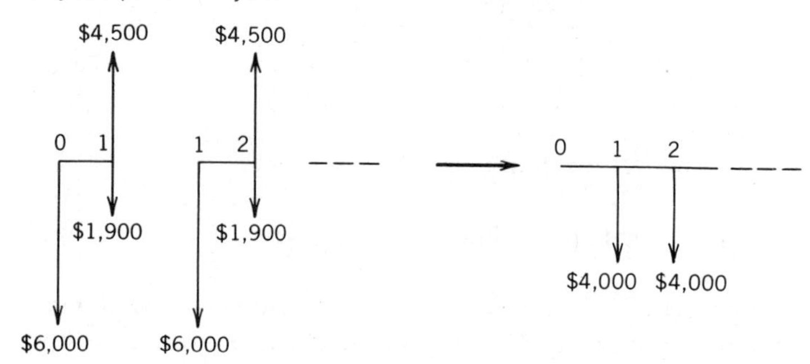


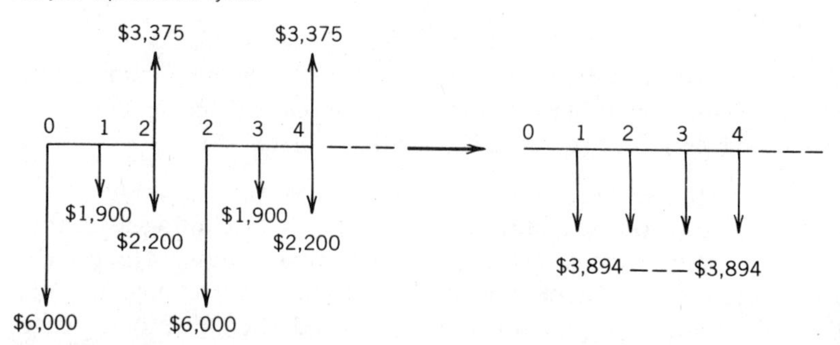


**FIGURE 16.10** Conversion of replacement cycles to infinite *AE* cost streams, Example 16.14.

## 16.3 INFINITE PLANNING PERIOD METHODS

By now the concept of an infinite planning period, implying an infinite series of replacements, should be familiar. If the concept can be applied to the decision problem, the relatively simple solution techniques presented here can be used. In the first subsection we will use the economic-life concept in an after-tax analysis. The second subsection applies *PV* analysis.

### 16.3.1 No Technology or Cost Changes, AE Method

If the decision is simply how long to retain a new asset, we should use the economic-life concept presented in Section 16.2.3. If the choice is between a defender and a challenger, the analysis procedure is as follows.

1. Find the economic life of the new asset (challenger).
2. Find the economic life of the old asset (defender), using the current salvage value as the purchase cost. In many situations the economic life of the old asset will be one year.
3. Compare *AE*s for the respective economic lives using the outsider point of view.
4. a. If the *AE* cost of the challenger is lower, replace the defender.
   b. If the *AE* cost of the defender is lower, keep it for one year and then perform another analysis.

The marginal analysis method avoids some of the pitfalls introduced when the current salvage value of the defender is zero or nearly zero (see Problem 16.6 at the end of the chapter). The implicit assumption in this method is that replacements, of either the defender or the challenger, will be by models identical to the current challenger. We also assume revenue does not depend on which asset is used.

### *Example 16.15*

An 8-year-old defender has a current salvage value of $4,500, with O&M costs of $1,500 this year and increasing by $1,500 per year (see Table 16.3, column 3). Future salvage values are expected to decline by $750 per year (see Table 16.3, column 2). The cost of the defender is fully depreciated. The challenger costs $9,000, falls in the 5-year property class, and has O&M costs of $1,393 per year, which increase (see Table 16.4, column 3). Salvage values are expected to be $5,018 after one year and to decline (see Table 16.4, column 2). Depreciation is by the MACRS SL method, but the half-year convention does not apply. The firm is a large, profitable corporation with a combined income tax rate of 40% and a *MARR* of 15%. What is your recommendation?

We will use the generalized cash flow method (see Chapter 4) to examine each asset.

**Defender:** Since the asset is fully depreciated, all salvage values are treated as depreciation recapture and taxed at 40%, leaving 60%. Using a tabular approach,

**Table 16.3** ***Economic-Life Calculations for the Defender, Example 16.15***

| 1<br>$N$ | 2<br>$S$ | 3<br>B/T<br>O&M | 4<br>$A/P$ | 5<br>$A/G$ | 6<br>A/T $S$ | 7<br>$AE$ of<br>$(P - S)$ | 8<br>A/T<br>O&M | 9<br>$AE$ of<br>Col. 8 | 10<br>Col. 7<br>+ Col. 9 |
|---|---|---|---|---|---|---|---|---|---|
| 1 | $3,750 | $1,500 | 1.1500 | 0 | $2,250 | $855 | $ 900 | $ 900 | $1,755 |
| 2 | 3,000 | 3,000 | 0.6151 | 0.4651 | 1,800 | 824 | 1,800 | 1,319 | 2,143 |
| 3 | 2,250 | 4,500 | 0.4380 | 0.9071 | 1,350 | 794 | 2,700 | 1,716 | 2,510 |
| 4 | 1,500 | 6,000 | 0.3503 | 1.3263 | 900 | 766 | 3,600 | 2,094 | 2,860 |

Explanation of columns:

6. (Col. 2) (1 − 0.4), *A/T* salvage proceeds.
7. ($2,700 − Col. 6) (*A/P*, 15%, *N*) + (0.15) (Col. 6)
8. (Col. 3) (1 − 0.4).
9. $900 + 900 (*A/G*, 15%, *N*).
10. Total *AE* costs, as a function of the retention period.

we obtain the results given in Table 16.3. The remaining economic life of the defender is obviously one year.

**Challenger:** The uneven increases in the O&M costs prevent us from using the *A/G* factor, so we must use the more tedious *PV* of O&M approach. Table 16.4 shows preliminary calculations. Some of the columns require explanation. Since book value exceeds salvage value during the early years, a sale would result in a loss and subsequent tax credit. A sale of the asset after one year gives a loss of $7,200 − 5,018 = $2,182 and a tax credit of ($2,182)(0.4) = $873. The after-tax salvage proceeds, given in column 6, are $5,018 + $873 = $5,891. The deprecia-

**Table 16.4** ***Preliminary Calculations for the Economic Life of the Challenger, Example 16.15***

| 1<br>$N$ | 2<br>$S$ | 3<br>B/T<br>O&M | 4<br>Depreciation | 5<br>Book<br>Value | 6<br>A/T $S$ | 7<br>Depreciation<br>Tax Shield | 8<br>A/T<br>O&M | 9<br>A/T<br>Annual |
|---|---|---|---|---|---|---|---|---|
| 1 | $5,018 | $1,393 | $1,800 | $7,200 | $5,891 | $720 | $ 836 | $ 116 |
| 2 | 4,262 | 1,533 | 1,800 | 5,400 | 4,717 | 720 | 920 | 200 |
| 3 | 3,628 | 2,372 | 1,800 | 3,600 | 3,617 | 720 | 1,423 | 703 |
| 4 | 3,030 | 3,122 | 1,800 | 1,800 | 2,538 | 720 | 1,873 | 1,153 |
| 5 | 2,460 | 3,870 | 1,800 | 0 | 1,476 | 720 | 2,323 | 1,603 |
| 6 | 2,215 | 5,250 | 0 | 0 | 1,329 | 0 | 3,150 | 3,150 |
| 7 | 1,993 | 6,000 | 0 | 0 | 1,196 | 0 | 3,600 | 3,600 |

Explanation of columns:

4. $9,000/5 for first 5 years.
5. $9,000 − (1,800) (*N*) until *BV* reaches 0.
6. Col. 2 − (Col. 2 − Col. 5) (0.4).
7. (Col. 4) (0.4), represents a cash inflow.
8. (Col. 3) (1 − 0.4)
9. Col. 8 − Col. 7

tion tax shield during the first 5 years is ($1,800)(0.4) = $720. This amount represents a cash inflow since it reduces taxes owed. Column 9 is the sum of the after-tax O&M costs minus the depreciation tax shield (we are using the opposite sign convention, with positive values representing cash outflows).

With the preliminary data calculated in Table 16.4, we are ready to perform the economic-life calculations on an after-tax basis, as shown in Table 16.5. Column 6 gives the *AE* ownership costs (excluding depreciation, which is combined with O&M). The *AE* O&M costs are obtained by summing, for each replacement cycle, the *PV* of yearly O&M costs, shown in columns 7 and 8. This sum is then multiplied by the *A/P* factor to give *AE* O&M in column 9. The total *AE* costs in column 10 show the economic life of the challenger to be 5 years, with an *AE* of $3,113. This life is no doubt heavily influenced by the tax allowances for depreciation.

The entries in Table 16.5 must be interpreted with caution. Some, such as those in columns 3 and 7, refer to costs that occur during the years of ownership. Others, as in column 2, refer to one-time flows at the end of an ownership period. Still others, like those in columns 6, 9, and 10, refer to annual equivalent costs for *each* year of an ownership period. The index *N* in column 1 is used for the individual year *and* the replacement cycle.

Since the *AE* cost for the defender's remaining economic life (one year) is $1,755, which is less than $3,113, the decision will be to keep the defender for at least one more year.

Looking ahead one year, we might try to determine how the analysis would then be performed. We could perform another economic life computa-

**Table 16.5** ***Economic-Life Calculations for the Challenger, Example 16.15***

| 1 | 2 | 3 | 4 | 5 | 6 | 7 | 8 | 9 | 10 |
|---|---|---|---|---|---|---|---|---|---|
| *N* | A/T *S* | A/T Annual | *P/F* | *A/P* | *AE*, (*P* − *S*) | *PV*, Annual | Sum *PV*, Annual | *AE*, Annual | Total |
| 1 | $5,891 | $ 116 | 0.8696 | 1.1500 | $4,459 | $ 101 | $ 101 | $ 116 | $4,575 |
| 2 | 4,717 | 200 | 0.7562 | 0.6151 | 3,342 | 151 | 252 | 155 | 3,497 |
| 3 | 3,617 | 703 | 0.6575 | 0.4380 | 2,900 | 462 | 714 | 313 | 3,213 |
| 4 | 2,538 | 1,153 | 0.5718 | 0.3503 | 2,644 | 659 | 1,373 | 481 | 3,125 |
| 5 | 1,476 | 1,603 | 0.4972 | 0.2983 | 2,466 | 797 | 2,170 | 647 | 3,113 |
| 6 | 1,329 | 3,150 | 0.4323 | 0.2642 | 2,226 | 1,362 | 3,532 | 933 | 3,159 |
| 7 | 1,196 | 3,600 | 0.3759 | 0.2404 | 2,055 | 1,353 | 4,885 | 1,174 | 3,229 |

Explanation of columns:

2. From Table 16.4, Col. 6.
3. From Table 16.4, Col. 9.
6. ($9,000 − Col. 2) (*A/P*, 15%, *N*) + (Col. 2) (0.15), *AE* ownership costs as a function of the replacement cycle.
7. (Col. 3) (*P/F*, 15%, *N*)
8. Cumulative sum of Col. 7.
9. (Col. 8) (*A/P*, 15%, *N*), *AE* O&M costs including the value of the depreciation tax shield.
10. Col. 6 + Col. 9, total *AE* costs, as a function of the replacement cycle.

tion for the defender, but in our example it is simpler to do marginal analysis (since the economic life will be one year). The costs incurred in owning and operating the defender from time 1 to time 2, expressed at time 1, are

$$(\$2{,}250 - 1{,}800)(1.15) + (1{,}800)\overset{A/P,\ 15\%,\ 1}{(0.15)} + 1{,}800 = \$2{,}588$$

This value is still lower than $3,113, so we would retain the defender at least until time 2. Examining the period from time 2 to time 3, we have costs at time 2 of

$$(\$1{,}800 - 1{,}350)(1.15) + (1{,}350)(0.15) + 2{,}700 = \$3{,}420$$

This value is greater than $3,113, so we would retire the defender at time 2 and acquire the challenger, which would be replaced at 5-year intervals in our infinite planning period. □

When we performed the analysis at time 0 in Example 16.15, the defender's economic life was one year, and thus it is necessary to reexamine it at time 1 and so on. If the remaining economic life had been 2 years, we could have made a decision to keep the defender for 2 years before reexamining it. In that case, had we performed marginal analysis at time 1, we would have decided to keep the defender another year.

Some reasons for performing the analysis for a defender each year are

- Changes in O&M costs.
- Changes in salvage values.
- Changes in the purchase cost of the challenger.
- Changes in tax law.

Although we assumed an infinite planning period in Example 16.15, we would perform exactly the same analysis by using the concept of recognition of unused value (see Section 7.5.2). The crucial assumption that must be satisfied in such an approach is that any asset with remaining economic life can be transferred to equally productive use, with no change in tax treatment, after the study period.

The economic-life concept presented in this section is based on the assumption of a unique minimum for the *AE* as a function of the retention period. When more than one minimum value exists, as in the situation of two or more local minima, we must proceed more cautiously. One approach is to evaluate every possible combination of retention period of the defender followed by replacement by the challenger. This approach requires *PV* analysis of infinite replacement sequences. Some computational short cuts are possible, as shown by Matsuo [13].

### 16.3.2 Geometric Changes in Purchase Costs and O&M Costs, PV Method

In this section we allow gradual changes in asset costs. The analysis method is *PV* for an infinite planning period [1, 4].

### *Example 16.16*

A machine costs $10,000 new, has operating costs of $3,000 in the first year, and has a salvage value of $5,000 at the end of the first year. Operating costs increase each year by 5%. Similarly, salvage value declines each year by 10%. New models of the machine are introduced annually with the following cost advantages: a reduction of 3% in the purchase price from that of the previous year's model and a first-year operating cost reduction of 4% from that of the previous year's model. Subsequent-year operating costs always increase by 5%. The end-of-year-1 salvage value is always 50% of the purchase price, and the salvage value declines each year by 10%. Using B/T analysis, find the best replacement cycle for $i' = 10\%$. The solution to Example 16.16 follows in the text. □

Example 16.16 is getting closer to reality. Before stating a problem as in Example 16.16, we must spend a considerable amount of time performing the data analysis for projections in price, operating costs, salvage values, and so forth. This type of analysis is best done in constant dollars. We will assume that constant dollars have been used and that the 10% is an inflation-free interest rate.

Before we solve Example 16.16, we must make two statements, one about infinite planning periods and one about replacement cycles. First, the infinite planning period is just a concept that allows us to perform some convenient mathematical tricks. For interest rates of 10% or higher, the solution for an infinite planning period will rarely differ from that for a 50-year period, but usually it is easier to obtain. The dynamic programming technique in the next section often shows that a planning period of one and a half to two times the economic life of the current challenger will give the same current decision—to keep or reject the defender—as an infinite planning period. [2]

Second, in the solution to Example 16.16 we assume a constant replacement cycle. This may lead us to a suboptimal solution—for example, to replace every 5 years, when the optimal solution could be to replace at 5, 5, 5, 6, 6, . . . years. Given the uncertainty in the projections, we will not worry about the suboptimality. The effect of the discounting will tend to give us an optimal *current* decision, even if subsequent decisions are not optimal.

We will solve Example 16.16 by obtaining the *PV* of all costs as a function of $n$, the replacement cycle. We first define some terms.

- $P$ purchase price of the asset at time 0
- $P_n$ purchase price of the asset at time $n$
- $a$ annual multiplier to calculate purchase price
- $i'$ inflation-free (real) interest rate
- $S_n$ salvage value of the asset at time $n$
- $b$ fraction to calculate end-of-year-1 salvage value, assumed to be less than 1
- $c$ fraction to calculate subsequent-year salvage values, assumed to be less than 1

- $A$ first-year O&M costs for asset purchased at time 0
- $p$ annual multiplier to calculate O&M costs for a given asset, assumed to be greater than 1
- $q$ annual multiplier to calculate first-year O&M costs for an asset purchased after time 0

We assume certain relationships among the terms. Let us consider purchase cost. An asset bought at time 1 costs

$$P_1 = aP$$

where

$$\frac{a}{1 + i'} < 1 \tag{16.1}$$

An asset bought at time 2 costs

$$P_2 = aP_1 = a^2P$$

In general,

$$P_n = a^nP \tag{16.2}$$

If we consider a sequence of purchase costs every $n$ years, the contribution to $PV$ of costs is

$$P1 = P + \frac{a^nP}{(1 + i')^n} + \frac{a^{2n}P}{(1 + i')^{2n}} + \cdots = P\left[\frac{1}{1 - \left(\frac{a}{1 + i'}\right)^n}\right] \tag{16.3}$$

Consider a sequence of salvage value cash flows. If we sell the asset purchased at time 0 after one year, we would receive

$$S_1 = bP$$

If we sell the asset after two years of use, we would receive

$$S_2 = bcP$$

If we consider a sequence of asset retirements every $n$ years, the first salvage value will be

$$S_n = bc^{n-1}P \tag{16.4}$$

The second salvage value will be

$$S_{2n} = bc^{n-1}P_n = bc^{n-1}a^nP \tag{16.5}$$

and so forth. The contribution to *PV* of costs from a sequence of asset retirements every $n$ years is

$$P2 = -P\left[\frac{bc^{n-1}}{(1+i')^n} + \frac{bc^{n-1}a^n}{(1+i')^{2n}} + \frac{bc^{n-1}a^{2n}}{(1+i')^{3n}} + \cdots\right]$$

$$= \frac{-Pbc^{n-1}}{(1+i')^n}\left[\frac{1}{1-\left(\dfrac{a}{1+i'}\right)^n}\right] \tag{16.6}$$

The expression for the operating costs is more tedious, but it follows along similar lines. Each replacement cycle contributes $n$ terms, with each term showing a higher cost than that of the previous one. The first term of each cycle reflects the improvements in first-year costs. The first asset has operating costs of $A$ in year 1, $pA$ in year 2, $p^2A$ in year 3, and so forth. The second asset has costs of $q^nA$ in year $n+1$, $pq^nA$ in year $n+2$, $p^2q^nA$ in year $n+3$, and so forth. The third asset has costs of $q^{2n}A$ in year $2n+1$, $pq^{2n}A$ in year $2n+2$, $p^2q^{2n}A$ in year $2n+3$, and so forth. The closed-form expression for the *PV* of the operating costs is

$$P3 = \frac{\dfrac{A}{p}\displaystyle\sum_{k=1}^{n}\left(\frac{p}{1+i'}\right)^k}{1-\left(\dfrac{q}{1+i'}\right)^n} \tag{16.7}$$

The derivation of Eq. 16.7 is left as an exercise (see Problem 16.28). If $n$ is likely to be large, we can also save computations by substituting for the finite series the term representing its sum.

To find the best replacement cycle, we simply evaluate $(P1 + P2 + P3)$ for different values of $n$. We have the following parameter values for Example 16.16.

$P = \$10{,}000$ $\quad c = 0.9$

$a = 0.97$ $\quad A = \$3{,}000$

$i' = 10\%$ $\quad p = 1.05$

$b = 0.5$ $\quad q = 0.96$

The closed-form expressions for the components of present value are (see Problem 16.28)

$$P1 = \frac{\$10{,}000}{1-0.8818^n}$$

$$P2 = \frac{-\$5{,}556(0.81818)^n}{1-0.8818^n}$$

$$P3 = \frac{\$2{,}857\displaystyle\sum_{k=1}^{n}(0.9546)^k}{1-0.8727^n}$$

**Table 16.6** ***Replacement Cycle Calculations for Geometric Cost Changes, Example 16.16***

| *n* | *P*1 | *P*2 | *P*3 | Total *PV* |
|---|---|---|---|---|
| 1 | 84,615 | −38,462 | 21,429 | 67,582 |
| 2 | 44,965 | −16,722 | 22,365 | 50,608 |
| 3 | 31,817 | −9,681 | 23,310 | 45,446 |
| 4 | 25,295 | −6,297 | 24,263 | 43,261 |
| 5 | 21,423 | −4,364 | 25,220 | 42,279 |
| 6 | 18,875 | −3,146 | 26,181 | 41,910 |
| 7 | 17,083 | −2,329 | 27,143 | 41,897* |
| 8 | 15,763 | −1,759 | 28,103 | 42,107 |
| 9 | 14,758 | −1,347 | 29,060 | 42,471 |
| 10 | 13,973 | −1,044 | 30,012 | 42,941 |

*The best replacement cycle.

The results are shown in Table 16.6 (nine-digit accuracy is used in computations on original data; e.g., 0.97/1.1 = 0.881818182). The *PV* is minimized with a 7-year replacement cycle.

The computations for this method are easily done on a calculator or on a desktop computer. The method can be generalized to a wide variety of situations.

## 16.4 FINITE PLANNING PERIOD METHODS

Present value methods are applicable to a more general class of replacement decisions than *AE* or other methods. As indicated in Section 7.5, the *PV* method can be applied in the following situations.

- Service projects (revenues unknown or unchanged) with repeatability—use a common planning period.
- Service projects without repeatability—use recognition of the unused value or estimate explicitly a salvage value.
- Revenue projects (all benefits and costs known) with repeatability—use a common planning period.
- Revenue projects without repeatability—compute *PV* of each alternative for its own (best) lifetime.

*PV* methods can also be applied to infinite planning periods as shown in Section 16.3.2; they allow for certain types of convenient sensitivity analysis, and they can take advantage of optimization techniques. The second and third features are illustrated by examples in this section.

### 16.4.1 Sensitivity Analysis of PV with Respect to Inflation

A typical question faced by management is what the effect of inflation is on a replacement decision. By considering separately the A/T effects from purchasing and operating an asset, we can conveniently perform a sensitivity analysis of the A/T *PV* with respect to one or more factors. For example, many organizations are in a position to increase their revenues to account for inflation in labor, material, and money interest costs. The depreciation tax shield, however, is based on actual historical cost, and in times of high inflation it loses much of its intended value. We will use generalized cash flows (see Section 4.4.1) in the next example to illustrate the loss in value of the depreciation tax shield.

## *Example 16.17*

A piece of equipment costing $10,000 is expected to produce an annual net benefit before taxes of $2,900 in constant dollars over the next five years, after which the salvage value will be zero. Depreciation to be used is SL without the half-life convention, and the tax rate is 34%. The after-tax inflation-free *MARR* is $i' = 8\%$. Determine whether the investment is worthwhile for a range of inflation rates from 0% to 10%. We will use Eq. 16.8 to calculate the market *MARR*,

$$i = (1 + i')(1 + f) - 1 \qquad (16.8)$$

where $i$ = market interest rate,
$i'$ = inflation-free interest rate,
$f$ = inflation rate.

The components of *PV* in this example are

1. Investment of $10,000 at time 0, in constant dollars.
2. Annual net benefit, in constant dollars, equal to $2,900(1 − 0.34).
3. Depreciation tax shield, in actual dollars, equal to ($10,000/5)(0.34).

The first two components will be discounted at $i'$, the inflation-free interest rate, and the third at $i$, the market interest rate.

$$PV = -\$10{,}000 + (2{,}900)(1 - 0.34)(P/A, i', N) + (2{,}000)(0.34)(P/A, i, N)$$

The results of the computations are given in Table 16.7. It is seen that for inflation rates of 5% or lower the investment is worthwhile, but not for higher inflation rates. The alternative way to work this problem is to convert all flows into actual dollars, perform the tax calculation, obtain A/T cash flow, and then reconvert (optional) into constant dollars. Table 16.8 shows this for one value of $f = 5\%$. The *PV* is computed at 8% since we are discounting constant dollars. It is clear that in using the long (income statement) approach we do more computations for *one* value of $f$ than we do when using the shortcut approach for the entire *range* of $f$ values. The reason is that in the long approach we are inflating constant dollars to actual dollars and then deflating them. We can see that if most of the cash flow types increase with inflation, we are better off doing our computations in constant dollars, even if depreciation follows an irregular pattern. □

**Table 16.7** ***PV as a Function of Inflation, Example 16.17***

| $i'$ (%) | $f$ (%) | $i$ (%) | $(P/A, i, N)$ | $PV$ |
|---|---|---|---|---|
| 8 | 0 | 8 | 3.9927 | \$357 |
| 8 | 2 | 10.16 | 3.7753 | 209 |
| 8 | 3 | 11.24 | 3.6737 | 140 |
| 8 | 4 | 12.32 | 3.5764 | 74 |
| 8 | 5 | 13.40 | 3.4832 | 11 |
| 8 | 6 | 14.48 | 3.3938 | −50 |
| 8 | 8 | 16.64 | 3.2260 | −164 |
| 8 | 10 | 18.80 | 3.0713 | −269 |

**Table 16.8** ***Constant-Dollar Analysis for Example 16.17, $f$ = 5%***

| 1<br>$N$ | 2<br>Constant-Dollar Flow B/T | 3<br>Actual-Dollar Flow B/T | 4<br>Depreciation | 5<br>Taxable Income | 6<br>Tax | 7<br>Actual-Dollar Flow A/T | 8<br>Constant-Dollar Flow A/T |
|---|---|---|---|---|---|---|---|
| 0 | −10,000 | −10,000 | — | — | — | −10,000 | −10,000 |
| 1 | 2,900 | 3,045 | 2,000 | 1,045 | 355 | 2,690 | 2,562 |
| 2 | 2,900 | 3,197 | 2,000 | 1,197 | 407 | 2,790 | 2,531 |
| 3 | 2,900 | 3,357 | 2,000 | 1,357 | 461 | 2,896 | 2,502 |
| 4 | 2,900 | 3,525 | 2,000 | 1,525 | 519 | 3,006 | 2,473 |
| 5 | 2,900 | 3,701 | 2,000 | 1,701 | 578 | 3,123 | 2,447 |

Explanation of columns:

3. (Col. 2) $(1.05)^N$
5. Col. 3 − Col. 4
6. (Col. 5) (0.34)
7. Col. 3 − Col. 6
8. (Col. 7) $(1.05)^{-N}$

$$\begin{aligned} PV(8\%) &= -\$10{,}000 + (2{,}562)(0.9259) + (2{,}531)(0.8573) \\ &\quad + (2{,}502)(0.7938) + (2{,}473)(0.7350) + (2{,}447)(0.6806) \\ &= \$11 \end{aligned}$$

### 16.4.2 Dynamic Programming Method

If the planning period is infinite and changes in asset costs are geometric, the method in Section 16.3.2 can be applied. However, if there are jumps in the costs from one year's model to the next, the method tends to break down. If the planning period is *finite* (e.g., 20 years), that method does not apply. The *AE* methods may also break down, especially if technological changes occur rapidly or if recognition of unused value does not apply (asset with remaining economic life cannot be transferred to other productive use).

A general type of replacement problem can be described as having the following characteristics.

- The planning period is finite.
- Technological changes are made in an irregular manner.
- Cost changes are made in an irregular manner.
- Assets cannot be transferred to other productive use after the end of the planning period.

The procedure for solving such a problem is to establish all "reasonable" replacement patterns and then use *PV* for the planning period to select the most economical pattern. For example, if the planning period is 20 years and the economic life of the current challenger is 8 years, we can have the following patterns for the situation in which we replace a defender at time 0:

| Pattern | 1 | 2 | 3 | 4 | 5 | etc. |
|---|---|---|---|---|---|---|
| Asset 1 Life | 8 | 8 | 8 | 7 | 10 | |
| Asset 2 Life | 8 | 7 | 6 | 7 | 10 | |
| Asset 3 Life | 4 | 5 | 6 | 6 | — | |

Dynamic programming is an optimization technique that can be applied to such a situation [22]. If the planning period is finite, we can use either backward or forward recursion. If the planning period is infinite, we must use forward recursion (we can use backward recursion, but it is too much work). Therefore, we will demonstrate forward recursion. We will use the data from Example 16.15 and assume that there are no changes in technology, costs, or tax law during the planning period. The maximum remaining life of the defender will be limited to 4 years. The maximum life of any challenger will be limited to 7 years, and we will assume a planning period of 10 years, with all assets to be retired at that time.

The sequence of possible decisions can be represented by a diagram as shown in Figure 16.11. At time 0 we are in state D8; the defender is 8 years old. We will determine the best decision as a function of the state at time 1. There are two possible states at time 1: D9, the defender is 9 years old and C1, the challenger is 1 year old. But there is only one way to reach each state.

| State $N = 1$ | Decision $N = 0$ | State $N = 0$ | *PV* costs |
|---|---|---|---|
| D9 | *K def** | D8 | \$2,700 + (900 − 2,250)(0.8696) = \$1,526 |
| C1 | *R def** | D8 | \$9,000 + (116 − 5,891)(0.8696) = \$3,978 |

Here

$K$ = keep

$R$ = replace

*def* = defender

*ch* = challenger

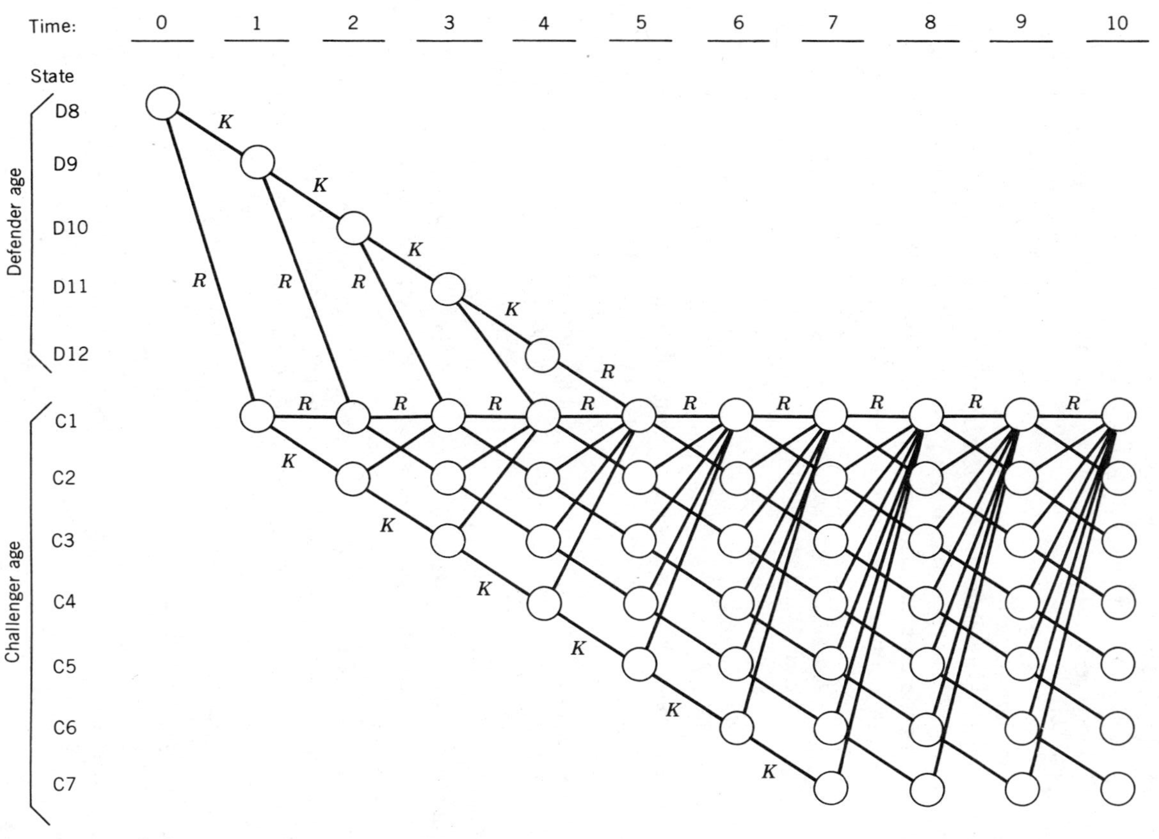


**FIGURE 16.11** Diagram for the dynamic programming method, Example 16.15. All arrows move toward the right, to a later time period. Arrows headed into the nodes on the C1 line imply replacement; all other arrows imply retention of asset.

We are basically computing incremental annual costs. For *R def,* we have

Purchase cost

$+$ (first-year A/T O&M)$(P/F,\ i,\ 1)$

$-$ (first-year salvage A/T)$(P/F,\ i,\ 1)$

For D9, the decision *K def* is optimal, since we have only one way to reach D9. Because *R def* is the only way to reach C1, it too is optimal. We use an asterisk to denote an optimal decision on how to reach a state. Notice that the *PV* costs agree with the *AE* costs for $N = 1$ shifted back to time 0. That is, \$1,526 = \$1,755/1.15, and \$3,978 = \$4,575/1.15. From an overall view, the best way to reach *any* state at time 1 is *K def.*

Now we move to time 1 and look ahead to the states at time 2. We can reach D10 only from D9, so *K def* will be optimal for D10, but we will add the cost of owning and operating the defender for the year. We can reach C1 from either D9 or C1 at time 1, and we select the decision that gives us lower cost. State C2 can be reached only from C1 (since we cannot purchase a used asset in this example).

| State $N = 2$ | Decision $N = 1$ | State $N = 1$ | *PV* to 1 | *PV* 1 to 2 | *PV* to 2 |
|---|---|---|---|---|---|
| D10 | *K def** | D9 | \$1,526 | \$1,957 | \$3,483 |
| C1 | *R def** | D9 | 1,526 | 3,459 | 4,985 |
| | *R ch* | C1 | 3,978 | 3,459 | 7,437 |
| C2 | *K ch** | C1 | 3,978 | 1,707 | 5,685 |

The computations for *PV* 1 to 2 are as follows.

$$\$2{,}250(0.8696) + (1{,}800 - 1{,}800)(0.7562) = \$1{,}957$$

$$\$9{,}000(0.8696) + (116 - 5{,}891)(0.7562) = \$3{,}459$$

$$\$5{,}891(0.8696) + (200 - 4{,}717)(0.7562) = \$1{,}707$$

The *PV* to 2 figure is the sum of *PV* to 1 plus *PV* 1 to 2. It is cheaper to reach C1 at time 2 by *R def* at time 1, which implies *K def* at time 0. The *PV* to 2 for this decision is \$4,985, compared with \$7,437 for *R ch* at time 1, which implies *R def* at time 0. (We will leave aside the question of why we might want to reach C1 at time 2.)

Notice that we have broken down the costs of owning and operating in such a way that we can analyze the situation on a year-by-year basis. To show the validity of this approach, for the simple case of buying the challenger at time 0 and keeping it 2 years, we can convert the *PV* to time 2 of reaching C2 via C1.

$$(\$5{,}685)(A/P,\ 15\%,\ 2) = (5{,}685)(0.6151) = \$3{,}497$$

This figure agrees with the *AE* for $N = 2$ as obtained in Table 16.5.

We can also take advantage of the fact that the *PV* cost of *R ch* at time 1 is equal to the *PV* cost of *R def* at time 0 multiplied by 1/1.15. The results for time 2 agree with the *AE* analysis for Example 16.15 in Section 16.3.1, which showed that $AE(N = 2) =$ \$2,143 for the defender and \$3,497 for the challenger.

| State $N = 3$ | Decision $N = 2$ | State $N = 2$ | *PV* to 2 | *PV* 2 to 3 | *PV* to 3 |
|---|---|---|---|---|---|
| D11 | *K def** | D10 | $3,483 | $2,249 | $5,732 |
| C1 | *R def** | D10 | 3,483 | 3,009 | 6,492 |
| | *R ch* | C1 | 4,985 | 3,009 | 7,994 |
| | *R ch* | C2 | 5,685 | 3,009 | 8,694 |
| C2 | *K ch** | C1 | 4,985 | 1,485 | 6,470 |
| C3 | *K ch** | C2 | 5,685 | 1,651 | 7,336 |

Continuing to time 3, we see that it is better to reach C1 at time 3 via D10 at time 2. Note that in calculating the cost of reaching C1 at time 3 via C1 at time 2, we take advantage of the fact that we know the best way to reach C1 at time 2, since we solved that problem earlier.

Table 16.9 gives the next two tables, those showing computations for time 4 and time 5. The results of these computations imply that for a 5-year planning period it would be best to keep the defender for 2 years and then replace it with

**Table 16.9** ***Tables Showing Computations for Time 4 and Time 5***

| State $N = 4$ | Decision $N = 3$ | State $N = 3$ | *PV* to 3 | *PV* 3 to 4 | *PV* to 4 |
|---|---|---|---|---|---|
| D12 | *K def** | D11 | $5,732 | $2,431 | $8,163 |
| C1 | *R def** | D11 | 5,732 | 2,615 | 8,347 |
| | *R ch* | C1 | 6,492 | 2,615 | 9,107 |
| | *R ch* | C2 | 6,470 | 2,615 | 9,085 |
| | *R ch* | C3 | 7,336 | 2,615 | 9,951 |
| C2 | *K ch** | C1 | 6,492 | 1,291 | 7,783 |
| C3 | *K ch** | C2 | 6,470 | 1,435 | 7,905 |
| C4 | *K ch** | C3 | 7,336 | 1,586 | 8,922 |

| State $N = 5$ | Decision $N = 4$ | State $N = 4$ | *PV* to 4 | *PV* 4 to 5 | *PV* to 5 |
|---|---|---|---|---|---|
| C1 | *R def* | D12 | $8,163 | $2,275 | $10,438 |
| | *R ch* | C1 | 8,347 | 2,275 | 10,622 |
| | *R ch** | C2 | 7,783 | 2,275 | 10,058 |
| | *R ch* | C3 | 7,905 | 2,275 | 10,180 |
| | *R ch* | C4 | 8,922 | 2,275 | 11,179 |
| C2 | *K ch** | C1 | 8,347 | 1,123 | 9,470 |
| C3 | *K ch** | C2 | 7,783 | 1,248 | 9,031 |
| C4 | *K ch** | C3 | 7,905 | 1,380 | 9,285 |
| C5 | *K ch** | C4 | 8,922 | 1,514 | 10,436 |

a challenger for 3 years. Continuing, we have tables for times 6 through 10, as shown in Table 16.10.

The last table indicates that the best way to reach *any* state at time 10 is via C3 at time 9. This is shown by the double asterisk. Tracing back through the tables indicates that this implies the following.

- Keep the defender for 2 years.
- Buy a challenger at time 2 and keep it for 4 years.
- Buy a challenger at time 6 and keep it for 4 years.

This pattern takes advantage of the low costs of the defender for the next 2 years and the near-optimal *AE* cost of the challenger over a 4-year life.

The sequence of optimal decisions at each time period is shown in Figure 16.12. Examination of this diagram shows that if we trace back from *any* state at time 10 to D8 at time 0, the path will include the decision to keep the defender at time 0. For that matter, *all paths to time 10 states* include keeping the defender for 2 years. This means that even if we increase our planning period, our initial decision *will remain the same.* The ability of the dynamic programming algorithm to give us this type of information makes it a powerful tool for decision and analysis. In effect, we can begin an infinite planning period problem, with changes in costs, technology, and tax law, and stop when all paths to any time period include the same initial decision [14].

The selection of the best path to time 10 can also be solved on the diagram as a shortest-path problem: build a shortest-path tree from D8.

## 16.5 BUILDING A DATA BASE

Replacement analysis requires a variety of data inputs, many of which are not readily available to an analyst who just "walks into" an organization, either to make an individual decision or to study a class of assets. Before deciding what elements should be in a data base, let us first examine the types of analyses to be performed by a replacement analyst.

1. Determine the best lifetime for replacement.
2. Obtain *PV* for a specified replacement cycle.
3. Obtain *PV* for specified acquisitions and retirements by year.
4. Keep or replace a given individual asset.
5. Switch from a current asset model to a different model.
6. Consider the option to rebuild (overhaul) the asset.
7. Determine investment, cash flow, and energy consumption by year for one of situations 1 to 6.
8. Perform sensitivity analysis on selected factors for any one of situations 1 to 6.

Two general types of data are needed for analysis: economic data and data on assets. The economic data are usually readily available in a large, established

**Table 16.10** ***Tables Showing Computations for Times 6 Through 10***

| State $N = 6$ | Decision $N = 5$ | State $N = 5$ | *PV* to 5 | *PV* 5 to 6 | *PV* to 6 |
|---|---|---|---|---|---|
| C1 | *R* | C1 | $10,058 | $1,978 | $12,036 |
| | *R* | C2 | 9,470 | 1,978 | 11,448 |
| | *R** | C3 | 9,031 | 1,978 | 11,009 |
| | *R* | C4 | 9,285 | 1,978 | 11,263 |
| | *R* | C5 | 10,436 | 1,978 | 12,414 |
| C2 | *K** | C1 | 10,058 | 976 | 11,034 |
| C3 | *K** | C2 | 9,470 | 1,086 | 10,556 |
| C4 | *K** | C3 | 9,031 | 1,200 | 10,231 |
| C5 | *K** | C4 | 9,285 | 1,317 | 10,602 |
| C6 | *K** | C5 | 10,436 | 1,521 | 11,957 |

| State $N = 7$ | Decision $N = 6$ | State $N = 6$ | *PV* to 6 | *PV* 6 to 7 | *PV* to 7 |
|---|---|---|---|---|---|
| C1 | *R* | C1 | $11,009 | $1,720 | $12,729 |
| | *R* | C2 | 11,034 | 1,720 | 12,754 |
| | *R* | C3 | 10,556 | 1,720 | 12,276 |
| | *R** | C4 | 10,231 | 1,720 | 11,951 |
| | *R* | C5 | 10,602 | 1,720 | 12,322 |
| | *R* | C6 | 11,957 | 1,720 | 13,677 |
| C2 | *K** | C1 | 11,009 | 845 | 11,858 |
| C3 | *K** | C2 | 11,034 | 944 | 11,978 |
| C4 | *K** | C3 | 10,556 | 1,043 | 11,599 |
| C5 | *K** | C4 | 10,231 | 1,145 | 11,376 |
| C6 | *K** | C5 | 10,602 | 1,323 | 11,925 |
| C7 | *K** | C6 | 11,957 | 1,478 | 13,435 |

| State $N = 8$ | Decision $N = 7$ | State $N = 7$ | *PV* to 7 | *PV* 7 to 8 | *PV* to 8 |
|---|---|---|---|---|---|
| C1 | *R* | C1 | $11,951 | $1,495 | $13,446 |
| | *R* | C2 | 11,858 | 1,495 | 13,353 |
| | *R* | C3 | 11,978 | 1,495 | 13,473 |
| | *R* | C4 | 11,599 | 1,495 | 13,094 |
| | *R** | C5 | 11,376 | 1,495 | 12,871 |
| | *R* | C6 | 11,925 | 1,495 | 13,420 |
| | *R* | C7 | 13,435 | 1,495 | 14,930 |
| C2 | *K** | C1 | 11,951 | 738 | 12,689 |
| C3 | *K** | C2 | 11,858 | 821 | 12,679 |

*(continued)*

**Table 16.10** *(continued)*

| State<br>$N = 6$ | Decision<br>$N = 5$ | State<br>$N = 5$ | *PV*<br>to 5 | *PV*<br>5 to 6 | *PV*<br>to 6 |
|---|---|---|---|---|---|
| C4 | $K^*$ | C3 | 11,978 | 907 | 12,885 |
| C5 | $K^*$ | C4 | 11,599 | 996 | 12,595 |
| C6 | $K^*$ | C5 | 11,376 | 1,150 | 12,526 |
| C7 | $K^*$ | C6 | 11,925 | 1,285 | 13,210 |

| State<br>$N = 9$ | Decision<br>$N = 8$ | State<br>$N = 8$ | *PV*<br>to 8 | *PV*<br>8 to 9 | *PV*<br>to 9 |
|---|---|---|---|---|---|
| C1 | $R$ | C1 | $12,871 | $1,300 | $14,171 |
| | $R$ | C2 | 12,689 | 1,300 | 13,989 |
| | $R$ | C3 | 12,679 | 1,300 | 13,979 |
| | $R$ | C4 | 12,885 | 1,300 | 14,185 |
| | $R$ | C5 | 12,595 | 1,300 | 13,895 |
| | $R^*$ | C6 | 12,526 | 1,300 | 13,826 |
| | $R$ | C7 | 13,210 | 1,300 | 14,510 |
| C2 | $K^*$ | C1 | 12,871 | 642 | 13,513 |
| C3 | $K^*$ | C2 | 12,689 | 714 | 13,403 |
| C4 | $K^*$ | C3 | 12,679 | 789 | 13,468 |
| C5 | $K^*$ | C4 | 12,885 | 866 | 13,751 |
| C6 | $K^*$ | C5 | 12,595 | 1,000 | 13,595 |
| C7 | $K^*$ | C6 | 12,526 | 1,118 | 13,644 |

| State<br>$N = 10$ | Decision<br>$N = 9$ | State<br>$N = 9$ | *PV*<br>to 9 | *PV*<br>9 to 10 | *PV*<br>to 10 |
|---|---|---|---|---|---|
| C1 | $R$ | C1 | $13,826 | $1,131 | $14,957 |
| | $R$ | C2 | 13,513 | 1,131 | 14,644 |
| | $R^*$ | C3 | 13,403 | 1,131 | 14,534 |
| | $R$ | C4 | 13,468 | 1,131 | 14,599 |
| | $R$ | C5 | 13,751 | 1,131 | 14,882 |
| | $R$ | C6 | 13,595 | 1,131 | 14,726 |
| | $R$ | C7 | 13,644 | 1,131 | 14,775 |
| C2 | $K^*$ | C1 | 13,826 | 558 | 14,384 |
| C3 | $K^*$ | C2 | 13,513 | 621 | 14,134 |
| C4 | $K^{**}$ | C3 | 13,403 | 686 | 14,089** |
| C5 | $K^*$ | C4 | 13,468 | 753 | 14,221 |
| C6 | $K^*$ | C5 | 13,751 | 870 | 14,621 |
| C7 | $K^*$ | C6 | 13,595 | 972 | 14,567 |

** The best way reach time 10.

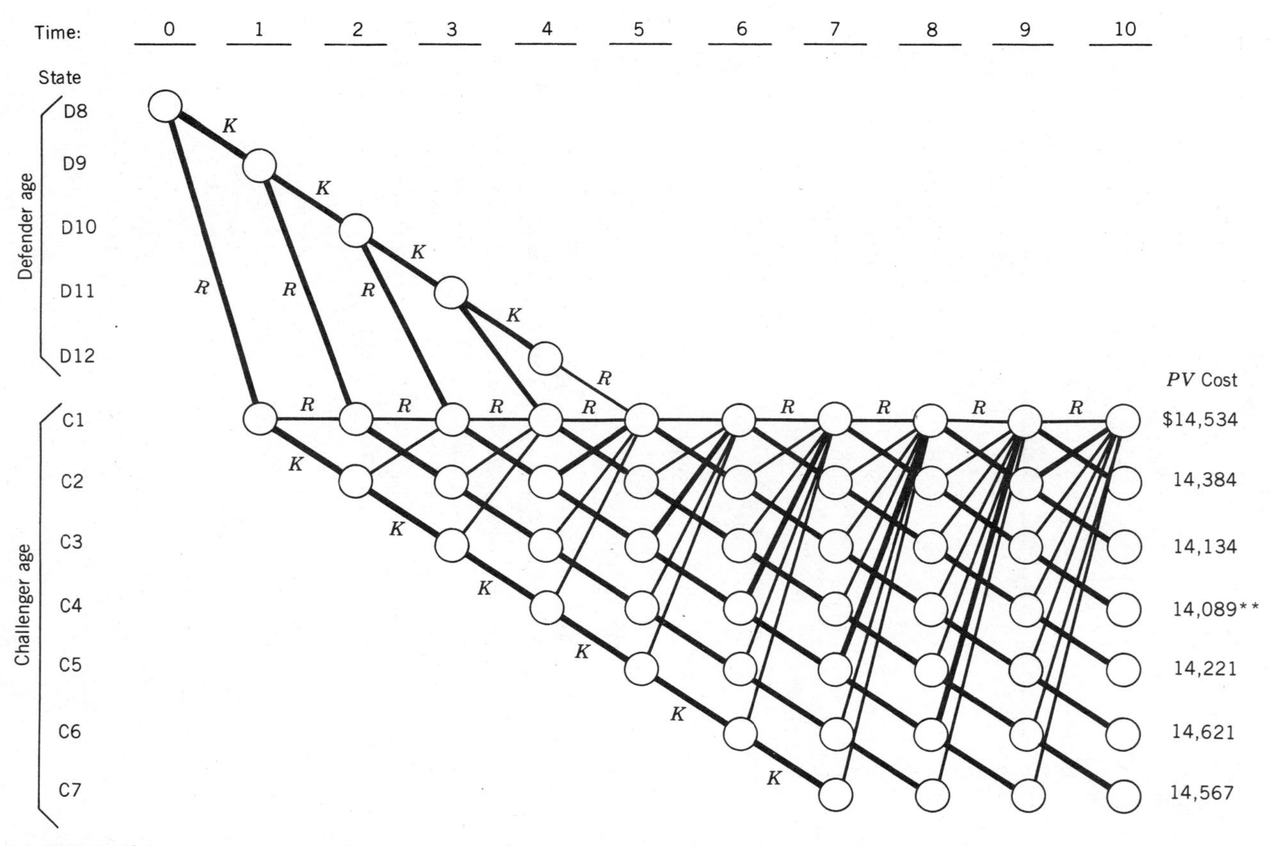


**FIGURE 16.12** Results for the dynamic programming method, Example 16.15. All arrows move toward the right, to a later time period. Arrows headed into the nodes on the C1 line imply replacement; all other arrows imply retention of asset. A heavy line represents an optimal path to a node.

firm, but they may need to be determined in consultation with management in a new or small firm. The economic data are

- Minimum attractive rate of return, specified as either a market rate or an inflation-free rate.
- Inflation forecast; if *MARR* is a market rate, check for consistency.
- Depreciation schedules and tax allowances.
- Marginal tax rate (combined federal plus state and local rate).
- Property tax rates on assets.
- Overhead rates on maintenance parts and labor, if performed in house. Overhead rates should be realistic and reflect variable overhead costs only, unless the analysis concerns large groups of assets.
- Old price indices, for curve fitting of historical maintenance costs.

The asset data are

- Purchase cost, by model year.
- Salvage values, by retirement age and model year.
- Maintenance costs (parts and labor, or contract amount) by age and model year.
- Energy and material consumption by model year.
- Utilization by age and model year.
- Downtime by age and model year.
- Overhaul costs.

These data needs generally imply a data base within the organization's management information system that tracks each model year by age. This must be an ongoing effort, similar to that for accounting or productivity data. It is generally too difficult to obtain the detailed data months or years later. The reason for the level of detail, of course, is to determine whether new models do show O&M reductions and so forth. We have seen too many instances in which commonly assumed trends were not true, for a variety of valid reasons. Above all, it is important to keep historical data by *age and model year* in order to convert to constant dollars.

## 16.6 SUMMARY

We have covered a variety of analysis techniques applicable to replacement problems. These problems are usually special cases of decision problems in which the alternatives have unequal lives. After discussing ways to quantify obsolescence and forecast future data, we presented the basic concepts of sunk costs, the outsider point of view, and the economic life of an asset. The second two are useful when we consider an infinite planning period. The outsider point of view allows the analyst to perform comparisons quickly by using *AE* for the respective lives of the defender and challenger. The cash flow method, on the other hand, requires the analyst to use *PV* and to perform more extensive computations.

A popular textbook assumption for the context of asset replacement decisions is an infinite planning period. We presented a rather long example in which we obtained the after-tax life of a challenger and then used marginal analysis to determine when the defender should be replaced. We presented another example in which costs for new asset models change geometrically.

Most texts ignore the problem of replacing equipment within a finite planning period. Our treatment includes a forward recursion dynamic programming algorithm. By this method the costs of owning and operating are broken down so that the analysis can be done on a year-by-year, after-tax basis. This powerful technique allows us to be somewhat imprecise about the planning horizon.

Most of the analysis techniques in the chapter lend themselves readily to sensitivity analysis, and we worked one example in this way, using generalized cash flows. Finally, we included a brief discussion of the types of data that must be maintained in a corporate management information system if a replacement analyst is to be in a position to perform various types of analyses.

There are many different scenarios for which asset replacement decisions can be considered. The analysis technique that should be used is the one that is most appropriate for the scenario and best satisfies the objective of performing the analysis.

## REFERENCES


1. Alchian, A., *Economic Replacement Policy,* Report R-224, RAND Corporation, Santa Monica, Calif., 1952.
2. Bean, J. C., J. R. Lohmann, and R. L. Smith, "A Dynamic Infinite Horizon Replacement Economy Decision Model," *The Engineering Economist,* Vol. 30, No. 2, pp. 99–120, 1985.
3. "Boeing's Gamble on the 757 May Finally Pay Off," *Atlanta Constitution,* May 28, 1988, p. 1B.
4. Dreyfus, S. E., "A Generalized Equipment Replacement Study," *Journal of the Society for Industrial and Applied Mathematics,* Vol. 8, No. 3, pp. 425–435, 1960.
5. Ghare, P. M., and P. E. Torgersen, "The Effect of Inflation and Increased Productivity on Machine Replacement," *Journal of Industrial Engineering,* Vol. 15, No. 4, pp. 201–207, July–August 1964.
6. Goldstein, T., S. P. Ladany, and A. Mehrez, "A Dual Replacement Model: A Note on Planning Horizon Procedures for Machine Replacements," *Operations Research,* Vol. 34, No. 6, pp. 938–941, 1986.
7. Herroelen, W., Z. Degreave, and M. Lambrecht, "Justifying CIM: Quantitative Analysis Tool," *CIM Review,* Vol. 3, No. 1, pp. 33–43, Fall 1986.
8. Kulonda, D. J., "Replacement Analysis with Unequal Lives—the Study Period Method," *The Engineering Economist,* Vol. 23, No. 3, pp. 171–179, 1978.
9. Leung, L. C., and J. M. A. Tanchoco, "Multiple Machine Replacement within an Integrated Systems Framework," *The Engineering Economist,* Vol. 32, No. 2, pp. 89–114, 1987.

10. LOHMANN, J. R., "A Stochastic Replacement Economic Decision Model," *IIE Transactions,* Vol. 18, No. 2, pp. 182–194, 1986.
11. LOHMANN, J. R., E. W. FOSTER, and D. J. LAYMAN, "A Comparative Analysis of the Effect of ACRS on Replacement Economy Decisions," *The Engineering Economist,* Vol. 27, No. 4, pp. 247–260, 1982.
12. MARSTON, A., R. WINFREY, and J. C. HEMPSTEAD, *Engineering Valuation and Depreciation,* 2nd edition, Iowa State University Press, Ames, 1953.
13. MATSUO, H., "A Modified Approach to the Replacement of an Existing Asset," *The Engineering Economist,* Vol. 33, No. 2, pp. 109–120, 1988.
14. OAKFORD, R. V., J. R. LOHMANN, and A. SALAZAR, "A Dynamic Replacement Economy Decision Model," *IIE Transactions,* Vol. 16, No. 1, pp. 65–72, 1984.
15. OSTWALD, P. F., *Cost Estimating,* 2nd edition, Prentice–Hall, Englewood Cliffs, N.J., pp. 207–216, 1984.
16. PARK, C. S., "Case Study: Buy vs. Lease Decision of Automobiles," *The Engineering Economist,* Vol. 26, No. 1, pp. 53–74, 1980.
17. PARK, C. S., and Y. SON, "An Economic Evaluation Model for Advanced Manufacturing Systems," *The Engineering Economist,* Vol. 34, No. 1, pp. 1–26, 1988.
18. RIEDER, M. L., R. R. WAGNER, and D. S. REMER, "An Economic Model for Using an Airplane via Lease-Back, Ownership, or Rental Arrangements," *The Engineering Economist,* Vol. 28, No. 2, pp. 101–130, 1982.
19. SCHROEDER, R. G., "Tech Update," *Mass Transit,* Vol. 16, No. 1/2, p. 20, January–February 1989.
20. SETHI, S., and S. CHAND, "Planning Horizon Procedures for Machine Replacement Problems," *Management Science,* Vol. 25, No. 2, pp. 140–151, 1979.
21. THUESEN, G. J., and G. P. SHARP, "Replacement Analysis in the 1980's: Obsolescence, New Tax Laws, and Inflation," School of Industrial and Systems Engineering, Georgia Institute of Technology Atlanta, May 1983.
22. WAGNER, H. W., *Principles of Operations Research,* Prentice–Hall, Englewood Cliffs, N.J., pp. 340–342, 1969.
23. WESTWOOD, J., "An Empirical Excursion into the Quicksand of Vehicle Replacement," *OMEGA, International Journal of Management Science,* Vol. 9, No. 2, pp. 195–202, 1981.

## PROBLEMS

**16.1.** Operating and maintenance (O&M) expenses for a truck increase by $400/year for the first 5 years of operation. First-year O&M expenses are $2,400. The initial cost of a truck is $8,400, estimated salvage after 4 years is $2,400, and salvage after 5 years is estimated to be $1,500. Using $i = 10\%$, compare the annual equivalent costs of a truck kept for 4 years and one kept for 5 years, on a before-tax basis.

**16.2.** Your company must decide whether conversion to solar energy (with natural gas backup) is economical. The solar conversion costs $180,000; in addition, the natural gas backup system costs $50,000 and is expected to use $15,000 of gas per year, in time 0 dollars, inflating at 6% per year. If you do nothing to change equipment, your current petroleum system will cost $30,000/year, in time 0 dollars, inflating at 8% per year. Your current petroleum system has no salvage value and zero book value. Maintenance costs for the two alternatives are expected to be the same. Use a planning period of 15 years and a real interest rate of $i' = 10\%$ to make the decision. Assume zero salvage value for the solar–natural gas system. The solar system can be depreciated over 5 years, the natural gas system over 10 years, both using MACRS SL; there is no investment tax credit, nor is there any energy tax credit. The marginal tax rate is 40%.

**16.3.** For an asset, obtain data on operating and maintenance costs and lost productivity as a function of asset age. Plot the data, and fit functional forms if possible.

**16.4.** Obtain data on technological improvements over a period of at least 10 years. Translate the data into a convenient form, such as dollars/output unit, material/output unit, labor cost/output unit, or output units/year, that would simplify comparison of assets with different capabilities.

**16.5.** Fit an exponential curve to the raw data on salvage values in Example 16.7 and Figure 16.5.

**16.6.** Fit a learning curve to the following data.

| Period | Average Price | Units Sold/Period |
|---|---|---|
| 1st quarter 88 | $3,500 | 1,000 |
| 2nd quarter 88 | 3,000 | 1,500 |
| 3rd quarter 88 | 2,700 | 2,300 |
| 4th quarter 88 | 2,000 | 2,700 |
| 1st quarter 89 | 1,800 | 3,000 |

a. What statistical and other problems come up in this type of situation?
b. What other factors might contribute to the reduction in average price?

**16.7.** Besides the factors listed in Section 16.1.1, what other factors could lead to (rational) replacement of assets?

**16.8.** A manufacturer is considering the purchase of a multipurpose machine which costs $800,000. The machine has a useful life of 6 years, with an estimated salvage value of $150,000. The manufacturer needs the machine for only 3 years but computes the annual equivalent by using the 6-year life. What is the implied salvage value after 3 years of use? Depreciation is 5-year SL with the half-year convention. *MARR* = 20%. The marginal tax rate is (a) zero, (b) 34%.

**16.9.** A manufacturer owns a multipurpose machine which 4 years ago cost $800,000. It is being depreciated by the SL method with the half-year convention, in the 5-year property class. The current salvage value of the machine is $400,000. The marginal tax rate is 34%. What is the opportunity cost for keeping the machine?

**16.10.** An asset in the 10-year property class was purchased on June 10, 1987 for $200,000. Depreciation is according to the MACRS percentages method. If the market salvage value on December 10, 1990 is $100,000 and the company's marginal tax rate is 40%, what is the opportunity cost, on December 10, 1990, of keeping the asset?

**16.11.** A defender has a current salvage value of $15,000, a remaining life of 5 years with zero salvage thereafter, and O&M costs of $26,000/year. The challenger costs $50,000, has a life of 12 years with $5,000 salvage value thereafter, and O&M costs of $20,000/year. *MARR* = 15%.

a. Use *AE* with the outsider point of view to make a decision.
b. Use *AE* with the cash flow approach to make a decision.
c. Use *PV* with the outsider point of view to make a decision.
d. Use *PV* with the cash flow approach to make a decision.
e. Reconcile any conflicting answers.
(Hint: What is the study period?)

**16.12.** An existing asset was purchased 5 years ago for $80,000. Depreciation is SL with the half-year convention, in the 8-year property class. O&M costs are $25,000 this year, $26,000 next year, $27,000 the third year, etc. The current salvage value of the asset is $20,000; salvage after one year is estimated to be $10,000, after 2 years $9,000, after 3 years $8,000, after 4 years $7,000, etc. With a marginal tax rate of 45% and *MARR* = 12%, find the remaining economic life of the asset. (Make a simplifying assumption about depreciation for the period prior to sale.)

**16.13.** A new asset is available for $150,000. Depreciation would be SL with the half-year convention, in the 8-year property class. O&M costs are $15,000 each year for the first 5 years, $18,000 in year 6, $21,000 in year 7, $24,000 in year 8, etc. Salvage values are estimated to be $131,250 after 1 year, $113,750 after 2 years, $97,500 after 3 years (SOYD pattern with $N$ = 15). With a marginal tax rate of 45% and *MARR* = 12%, find the economic life of the asset.

**16.14.** An old, fully depreciated asset (defender) has O&M costs of $1,500 this year, and these will increase by $1,500 each year. The current salvage value of the asset is $4,500, and this is expected to decline by $750 each year. The new asset (challenger) costs $9,000; falls in the 5-year property class; has O&M costs of $1,400/year, increasing by $500 each year; and has a salvage value of $5,000 after one year, declining by 10% each year. Depreciation is by the MACRS SL method. The firm is a large, profitable corporation with a combined income tax rate of 40% and a *MARR* of 15%. What is your recommendation?

**16.15.** An asset costs $20,000 and is in the 5-year property class. O&M costs are $2,000 the first year, $3,000 the second year, $4,000 the third year, etc. Salvage values are $12,000 after one year, $9,000 after 2 years, $6,000 after 3 years, and $4,000 thereafter. Depreciation is by SL, *without* the half-year convention, and *MARR* = 10%; the marginal tax rate is 40%. Use the *tabular* approach to find the after-tax economic life.

**16.16.** Marginal analysis involves examining the costs of a defender for the next year and making a comparison with an alternative, such as the annual equivalent cost for a challenger based on the challenger's economic life.

a. Describe a situation in which marginal analysis will lead to the correct decision about keeping or replacing the defender.

b. Describe a situation in which the remaining economic life of the defender will lead to an incorrect decision about keeping or replacing the defender.

c. Describe a situation in which marginal analysis might lead to an incorrect decision about keeping or replacing the defender. (See reference [13]).

**16.17.** Rework Example 16.13 so that the depreciation tax shield is included in the ownership cost.

**16.18.** Rework Example 16.12 with the following change. *Used* vehicles may be purchased from a dealer at the following prices. The salvage values given in the original problem statement apply to both vehicles purchased new and those purchased used.

| **Age** | 1 | 2 | 3 | 4 | 5 | 6 | 7 |
|---|---|---|---|---|---|---|---|
| **Price (dollars)** | 5,500 | 4,000 | 2,950 | 2,200 | 1,650 | 1,250 | 900 |

**16.19.** If we were to use the common service period approach to compare two new assets, one with a life of 7 years and the other with a life of 13 years, what would the implied study period be?

**16.20.** If we were to use annual equivalent to compare an existing asset (defender) with a remaining life of 3 years and a new asset (challenger) with a life of 7 years, what would the implied study period be?

**16.21.** An automated guided vehicle (AGV) currently costs $80,000. New technology is expected to yield models with more flexibility each year, but the purchase costs are expected to increase 6% each year. Because of the extensive customizing of AGVs, salvage values of used AGVs are relatively low, about 25% of the price paid for the vehicle, *regardless* of age. It is anticipated that the company will use AGVs in its manufacturing operations for a long time, more than 40 years. *MARR* = 20%.

a. Write an expression that gives the *PV* of ownership costs (or approximates it) for the planning period of 40 years or longer, as a function of the replacement cycle.
b. Are 5-year replacement cycles better than 10-year replacement cycles? Explain why.
c. Briefly discuss some of the practical limitations of the methodology you are using in parts a and b.

**16.22.** Obtain data for an asset type and represent purchase, salvage, and O&M costs by geometric patterns. Find the best replacement cycle for an infinite planning period.

**16.23.** XYZ Corporation was just awarded a contract to perform tunneling operations in a mountainous developing country. The contract is for a period of 14 years. One of the provisions of the contract is that XYZ will establish and train a local firm, which will continue operations after XYZ's contract has expired; thus, there is little chance for follow-on work for XYZ. In addition, when the local firm begins operation, it is to be provided with a *new* tunneling machine reflecting the latest technology.

The problem facing XYZ is the scheduling of asset (tunneling machine) purchases for the contract period. The company will need one machine throughout the contract period. Considering the usage expected, the *economic* life of an asset is about 8 years. After 8 years, maintenance costs make it cheaper to replace the machine. An important cost factor is the major overhaul performed after a machine is 4 years old. This overhaul is almost mandatory in order to keep good relations with the equipment manufacturer. Although a similar overhaul after 8 years of use would prolong the life of the asset, the manufacturer does not urge users to perform it, since it is expected that an 8-year-old asset will be retired soon.

a. *Specify* no more than *eight* schedules (patterns) for asset purchases for XYZ for the 14-year contract. (A schedule shows when assets are bought and sold.)
b. *Explain* why each of your eight schedules might be a best schedule.

**16.24.** Construct data for an asset so that it has an economic life of 8 years, in the context of Problem 16.23. Evaluate the *PV* of each of your eight schedules obtained in Problem 16.23.

**16.25.** An asset costing $200,000 is expected to generate additional revenues before taxes of $40,000/year, in constant dollars, for 15 years. Depreciation is MACRS SL for the 10-year property class; salvage value is expected to be $20,000. With a marginal tax rate of 40% and an after-tax real *MARR* of $i' = 10\%$, obtain the following.

a. *PV* with zero inflation.
b. *PV* with inflation ranging from 0 to 10%.
c. The loss in *PV* caused by the loss in value of the depreciation tax shield, for each inflation rate in part b.

**16.26.** Assume that the decision is between the defender in Problem 16.12 and the challenger in Problem 16.13. Set up the dynamic programming (DP) problem for such a decision. Assume that new models of the challenger are available each year, with the same costs. Assume that models similar to the defender are no longer available. Draw a decision network corresponding to the DP problem and show the transition costs on each arc of the network. Solve the DP problem for a 20-year planning period.

**16.27.** In the closed-form solution approach with geometric cost changes, is it possible for the *PV* of the purchase costs to be infinite (divergent series) but for the *PV* of total costs to be finite?

**16.28.** Derive Eq. 16.7 and the expressions for the components of present value in Example 16.16.

# APPENDIX A: Discrete Interest Compounding Tables

## Appendix A 0.25% Interest Rate Factors

| | Single Payment | | Equal-Payment Series | | | | Uniform |
|---|---|---|---|---|---|---|---|
| N | Compound Amount Factor, (F/P, i, N) | Present-Worth Factor, (P/F, i, N) | Compound Amount Factor, (F/A, i, N) | Sinking-Fund Factor, (A/F, i, N) | Present-Worth Factor, (P/A, i, N) | Capital Recovery Factor, (A/P, i, N) | Gradient Series Factor, (A/G, i, N) |
| 1 | 1.00250 | 0.9975062 | 1.00000 | 1.0000000 | 0.9975062 | 1.0025000 | 0.0000000 |
| 2 | 1.00501 | 0.9950187 | 2.00250 | 0.4993758 | 1.9925249 | 0.5018758 | 0.4993758 |
| 3 | 1.00752 | 0.9925373 | 3.00751 | 0.3325014 | 2.9850623 | 0.3350014 | 0.9983354 |
| 4 | 1.01004 | 0.9900622 | 4.01503 | 0.2490645 | 3.9751245 | 0.2515645 | 1.4968789 |
| 5 | 1.01256 | 0.9875932 | 5.02506 | 0.1990025 | 4.9627177 | 0.2015025 | 1.9950063 |
| 6 | 1.01509 | 0.9851304 | 6.03763 | 0.1656280 | 5.9478480 | 0.1681280 | 2.4927175 |
| 7 | 1.01763 | 0.9826737 | 7.05272 | 0.1417893 | 6.9305217 | 0.1442893 | 2.9900125 |
| 8 | 1.02018 | 0.9802231 | 8.07035 | 0.1239103 | 7.9107449 | 0.1264103 | 3.4868915 |
| 9 | 1.02273 | 0.9777787 | 9.09053 | 0.1100046 | 8.8885236 | 0.1125046 | 3.9833543 |
| 10 | 1.02528 | 0.9753403 | 10.11325 | 0.0988801 | 9.8638639 | 0.1013801 | 4.4794010 |
| 11 | 1.02785 | 0.9729081 | 11.13854 | 0.0897784 | 10.8367720 | 0.0922784 | 4.9750315 |
| 12 | 1.03042 | 0.9704819 | 12.16638 | 0.0821937 | 11.8072538 | 0.0846937 | 5.4702460 |
| 13 | 1.03299 | 0.9680617 | 13.19680 | 0.0757760 | 12.7753156 | 0.0782760 | 5.9650443 |
| 14 | 1.03557 | 0.9656476 | 14.22979 | 0.0702751 | 13.7409631 | 0.0727751 | 6.4594265 |
| 15 | 1.03816 | 0.9632395 | 15.26537 | 0.0655078 | 14.7042026 | 0.0680078 | 6.9533927 |
| 16 | 1.04076 | 0.9608374 | 16.30353 | 0.0613364 | 15.6650400 | 0.0638364 | 7.4469427 |
| 17 | 1.04336 | 0.9584413 | 17.34429 | 0.0576559 | 16.6234813 | 0.0601559 | 7.9400767 |
| 18 | 1.04597 | 0.9560512 | 18.38765 | 0.0543843 | 17.5795325 | 0.0568843 | 8.4327946 |
| 19 | 1.04858 | 0.9536670 | 19.43362 | 0.0514572 | 18.5331995 | 0.0539572 | 8.9250964 |
| 20 | 1.05121 | 0.9512888 | 20.48220 | 0.0488229 | 19.4844883 | 0.0513229 | 9.4169822 |
| 21 | 1.05383 | 0.9489165 | 21.53341 | 0.0464395 | 20.4334048 | 0.0489395 | 9.9084519 |
| 22 | 1.05647 | 0.9465501 | 22.58724 | 0.0442728 | 21.3799549 | 0.0467728 | 10.3995056 |
| 23 | 1.05911 | 0.9441896 | 23.64371 | 0.0422945 | 22.3241445 | 0.0447945 | 10.8901433 |
| 24 | 1.06176 | 0.9418351 | 24.70282 | 0.0404812 | 23.2659796 | 0.0429812 | 11.3803650 |
| 25 | 1.06441 | 0.9394863 | 25.76457 | 0.0388130 | 24.2054659 | 0.0413130 | 11.8701707 |
| 26 | 1.06707 | 0.9371435 | 26.82899 | 0.0372731 | 25.1426094 | 0.0397731 | 12.3595604 |
| 27 | 1.06974 | 0.9348065 | 27.89606 | 0.0358474 | 26.0774158 | 0.0383474 | 12.8485341 |
| 28 | 1.07241 | 0.9324753 | 28.96580 | 0.0345235 | 27.0098911 | 0.0370235 | 13.3370919 |
| 29 | 1.07510 | 0.9301499 | 30.03821 | 0.0332909 | 27.9400410 | 0.0357909 | 13.8252337 |
| 30 | 1.07778 | 0.9278303 | 31.11331 | 0.0321406 | 28.8678713 | 0.0346406 | 14.3129596 |
| 31 | 1.08048 | 0.9255165 | 32.19109 | 0.0310645 | 29.7933879 | 0.0335645 | 14.8002695 |
| 32 | 1.08318 | 0.9232085 | 33.27157 | 0.0300557 | 30.7165964 | 0.0325557 | 15.2871636 |
| 33 | 1.08589 | 0.9209062 | 34.35475 | 0.0291081 | 31.6375026 | 0.0316081 | 15.7736418 |
| 34 | 1.08860 | 0.9186097 | 35.44064 | 0.0282162 | 32.5561123 | 0.0307162 | 16.2597042 |
| 35 | 1.09132 | 0.9163189 | 36.52924 | 0.0273753 | 33.4724313 | 0.0298753 | 16.7453507 |
| 36 | 1.09405 | 0.9140338 | 37.62056 | 0.0265812 | 34.3864651 | 0.0290812 | 17.2305813 |
| 37 | 1.09679 | 0.9117545 | 38.71461 | 0.0258300 | 35.2982196 | 0.0283300 | 17.7153962 |
| 38 | 1.09953 | 0.9094807 | 39.81140 | 0.0251184 | 36.2077003 | 0.0276184 | 18.1997952 |
| 39 | 1.10228 | 0.9072127 | 40.91093 | 0.0244433 | 37.1149130 | 0.0269433 | 18.6837785 |
| 40 | 1.10503 | 0.9049503 | 42.01320 | 0.0238020 | 38.0198634 | 0.0263020 | 19.1673460 |
| 42 | 1.11057 | 0.9004425 | 44.22603 | 0.0226111 | 39.8229995 | 0.0251111 | 20.1332339 |
| 48 | 1.12733 | 0.8870533 | 50.93121 | 0.0196343 | 45.1786946 | 0.0221343 | 23.0209218 |
| 50 | 1.13297 | 0.8826346 | 53.18868 | 0.0188010 | 46.9461704 | 0.0213010 | 23.9801598 |
| 60 | 1.16162 | 0.8608691 | 64.64671 | 0.0154687 | 55.6523577 | 0.0179687 | 28.7514241 |
| 70 | 1.19099 | 0.8396404 | 76.39444 | 0.0130900 | 64.1438534 | 0.0155900 | 33.4811674 |
| 72 | 1.19695 | 0.8354579 | 78.77939 | 0.0126937 | 65.8168577 | 0.0151937 | 34.4221364 |
| 75 | 1.20595 | 0.8292231 | 82.37922 | 0.0121390 | 68.3107515 | 0.0146390 | 35.8304790 |
| 80 | 1.22110 | 0.8189351 | 88.43918 | 0.0113072 | 72.4259517 | 0.0138072 | 38.1694234 |
| 90 | 1.25197 | 0.7987405 | 100.78845 | 0.0099218 | 80.5038163 | 0.0124218 | 42.8162307 |
| 100 | 1.28362 | 0.7790438 | 113.44996 | 0.0088145 | 88.3824835 | 0.0113145 | 47.4216334 |

## Appendix A *0.5% Interest Rate Factors*

| | Single Payment | | Equal-Payment Series | | | | Uniform Gradient Series |
|---|---|---|---|---|---|---|---|
| N | Compound Amount Factor, (F/P, i, N) | Present-Worth Factor, (P/F, i, N) | Compound Amount Factor, (F/A, i, N) | Sinking-Fund Factor, (A/F, i, N) | Present-Worth Factor, (P/A, i, N) | Capital Recovery Factor, (A/P, i, N) | Factor, (A/G, i, N) |
| 1 | 1.00500 | 0.9950249 | 1.00000 | 1.0000000 | 0.9950249 | 1.0050000 | 0.0000000 |
| 2 | 1.01003 | 0.9900745 | 2.00500 | 0.4987531 | 1.9850994 | 0.5037531 | 0.4987531 |
| 3 | 1.01508 | 0.9851488 | 3.01502 | 0.3316722 | 2.9702481 | 0.3366722 | 0.9966750 |
| 4 | 1.02015 | 0.9802475 | 4.03010 | 0.2481328 | 3.9504957 | 0.2531328 | 1.4937656 |
| 5 | 1.02525 | 0.9753707 | 5.05025 | 0.1980100 | 4.9258663 | 0.2030100 | 1.9900250 |
| 6 | 1.03038 | 0.9705181 | 6.07550 | 0.1645955 | 5.8963844 | 0.1695955 | 2.4854532 |
| 7 | 1.03553 | 0.9656896 | 7.10588 | 0.1407285 | 6.8620740 | 0.1457285 | 2.9800502 |
| 8 | 1.04071 | 0.9608852 | 8.14141 | 0.1228289 | 7.8229592 | 0.1278289 | 3.4738161 |
| 9 | 1.04591 | 0.9561047 | 9.18212 | 0.1089074 | 8.7790639 | 0.1139074 | 3.9667509 |
| 10 | 1.05114 | 0.9513479 | 10.22803 | 0.0977706 | 9.7304119 | 0.1027706 | 4.4588545 |
| 11 | 1.05640 | 0.9466149 | 11.27917 | 0.0886590 | 10.6770267 | 0.0936590 | 4.9501271 |
| 12 | 1.06168 | 0.9419053 | 12.33556 | 0.0810664 | 11.6189321 | 0.0860664 | 5.4405687 |
| 13 | 1.06699 | 0.9372192 | 13.39724 | 0.0746422 | 12.5561513 | 0.0796422 | 5.9301793 |
| 14 | 1.07232 | 0.9325565 | 14.46423 | 0.0691361 | 13.4887078 | 0.0741361 | 6.4189591 |
| 15 | 1.07768 | 0.9279169 | 15.53655 | 0.0643644 | 14.4166246 | 0.0693644 | 6.9069079 |
| 16 | 1.08307 | 0.9233004 | 16.61423 | 0.0601894 | 15.3399250 | 0.0651894 | 7.3940260 |
| 17 | 1.08849 | 0.9187068 | 17.69730 | 0.0565058 | 16.2586319 | 0.0615058 | 7.8803134 |
| 18 | 1.09393 | 0.9141362 | 18.78579 | 0.0532317 | 17.1727680 | 0.0582317 | 8.3657701 |
| 19 | 1.09940 | 0.9095882 | 19.87972 | 0.0503025 | 18.0823562 | 0.0553025 | 8.8503962 |
| 20 | 1.10490 | 0.9050629 | 20.97912 | 0.0476665 | 18.9874191 | 0.0526665 | 9.3341918 |
| 21 | 1.11042 | 0.9005601 | 22.08401 | 0.0452816 | 19.8879793 | 0.0502816 | 9.8171570 |
| 22 | 1.11597 | 0.8960797 | 23.19443 | 0.0431138 | 20.7840590 | 0.0481138 | 10.2992918 |
| 23 | 1.12155 | 0.8916216 | 24.31040 | 0.0411347 | 21.6756806 | 0.0461347 | 10.7805964 |
| 24 | 1.12716 | 0.8871857 | 25.43196 | 0.0393206 | 22.5628662 | 0.0443206 | 11.2610708 |
| 25 | 1.13280 | 0.8827718 | 26.55912 | 0.0376519 | 23.4456380 | 0.0426519 | 11.7407151 |
| 26 | 1.13846 | 0.8783799 | 27.69191 | 0.0361116 | 24.3240179 | 0.0411116 | 12.2195295 |
| 27 | 1.14415 | 0.8740099 | 28.83037 | 0.0346856 | 25.1980278 | 0.0396856 | 12.6975140 |
| 28 | 1.14987 | 0.8696616 | 29.97452 | 0.0333617 | 26.0676894 | 0.0383617 | 13.1746688 |
| 29 | 1.15562 | 0.8653349 | 31.12439 | 0.0321291 | 26.9330242 | 0.0371291 | 13.6509939 |
| 30 | 1.16140 | 0.8610297 | 32.28002 | 0.0309789 | 27.7940540 | 0.0359789 | 14.1264895 |
| 31 | 1.16721 | 0.8567460 | 33.44142 | 0.0299030 | 28.6508000 | 0.0349030 | 14.6011557 |
| 32 | 1.17304 | 0.8524836 | 34.60862 | 0.0288945 | 29.5032835 | 0.0338945 | 15.0749927 |
| 33 | 1.17891 | 0.8482424 | 35.78167 | 0.0279473 | 30.3515259 | 0.0329473 | 15.5480005 |
| 34 | 1.18480 | 0.8440223 | 36.96058 | 0.0270559 | 31.1955482 | 0.0320559 | 16.0201792 |
| 35 | 1.19073 | 0.8398231 | 38.14538 | 0.0262155 | 32.0353713 | 0.0312155 | 16.4915292 |
| 36 | 1.19668 | 0.8356449 | 39.33610 | 0.0254219 | 32.8710162 | 0.0304219 | 16.9620503 |
| 37 | 1.20266 | 0.8314875 | 40.53279 | 0.0246714 | 33.7025037 | 0.0296714 | 17.4317430 |
| 38 | 1.20868 | 0.8273507 | 41.73545 | 0.0239604 | 34.5298544 | 0.0289604 | 17.9006071 |
| 39 | 1.21472 | 0.8232346 | 42.94413 | 0.0232861 | 35.3530890 | 0.0282861 | 18.3686430 |
| 40 | 1.22079 | 0.8191389 | 44.15885 | 0.0226455 | 36.1722279 | 0.0276455 | 18.8358508 |
| 42 | 1.23303 | 0.8110085 | 46.60654 | 0.0214562 | 37.7982999 | 0.0264562 | 19.7677827 |
| 48 | 1.27049 | 0.7870984 | 54.09783 | 0.0184850 | 42.5803178 | 0.0234850 | 22.5437211 |
| 50 | 1.28323 | 0.7792861 | 56.64516 | 0.0176538 | 44.1427863 | 0.0226538 | 23.4624199 |
| 60 | 1.34885 | 0.7413722 | 69.77003 | 0.0143328 | 51.7255608 | 0.0193328 | 28.0063816 |
| 70 | 1.41783 | 0.7053029 | 83.56611 | 0.0119666 | 58.9394176 | 0.0169666 | 32.4679615 |
| 72 | 1.43204 | 0.6983024 | 86.40886 | 0.0115729 | 60.3395139 | 0.0165729 | 33.3504143 |
| 75 | 1.45363 | 0.6879318 | 90.72650 | 0.0110221 | 62.4136454 | 0.0160221 | 34.6679396 |
| 80 | 1.49034 | 0.6709885 | 98.06771 | 0.0101970 | 65.8023054 | 0.0151970 | 36.8474249 |
| 90 | 1.56655 | 0.6383435 | 113.31094 | 0.0088253 | 72.3312996 | 0.0138253 | 41.1450768 |
| 100 | 1.64667 | 0.6072868 | 129.33370 | 0.0077319 | 78.5426448 | 0.0127319 | 45.3612613 |

## Appendix A *0.75% Interest Rate Factors*

| | Single Payment | | Equal-Payment Series | | | | Uniform Gradient Series |
|---|---|---|---|---|---|---|---|
| *N* | Compound Amount Factor, (*F*/*P*, *i*, *N*) | Present-Worth Factor, (*P*/*F*, *i*, *N*) | Compound Amount Factor, (*F*/*A*, *i*, *N*) | Sinking-Fund Factor, (*A*/*F*, *i*, *N*) | Present-Worth Factor, (*P*/*A*, *i*, *N*) | Capital Recovery Factor, (*A*/*P*, *i*, *N*) | Uniform Gradient Series Factor, (*A*/*G*, *i*, *N*) |
| 1 | 1.00750 | 0.9925558 | 1.00000 | 1.0000000 | 0.9925558 | 1.0075000 | 0.0000000 |
| 2 | 1.01506 | 0.9851671 | 2.00750 | 0.4981320 | 1.9777229 | 0.5056320 | 0.4981320 |
| 3 | 1.02267 | 0.9778333 | 3.02256 | 0.3308458 | 2.9555562 | 0.3383458 | 0.9950187 |
| 4 | 1.03034 | 0.9705542 | 4.04523 | 0.2472050 | 3.9261104 | 0.2547050 | 1.4906601 |
| 5 | 1.03807 | 0.9633292 | 5.07556 | 0.1970224 | 4.8894396 | 0.2045224 | 1.9850563 |
| 6 | 1.04585 | 0.9561580 | 6.11363 | 0.1635689 | 5.8455976 | 0.1710689 | 2.4782074 |
| 7 | 1.05370 | 0.9490402 | 7.15948 | 0.1396749 | 6.7946378 | 0.1471749 | 2.9701133 |
| 8 | 1.06160 | 0.9419754 | 8.21318 | 0.1217555 | 7.7366132 | 0.1292555 | 3.4607743 |
| 9 | 1.06956 | 0.9349632 | 9.27478 | 0.1078193 | 8.6715764 | 0.1153193 | 3.9501904 |
| 10 | 1.07758 | 0.9280032 | 10.34434 | 0.0966712 | 9.5995796 | 0.1041712 | 4.4383617 |
| 11 | 1.08566 | 0.9210949 | 11.42192 | 0.0875509 | 10.5206745 | 0.0950509 | 4.9252883 |
| 12 | 1.09381 | 0.9142382 | 12.50759 | 0.0799515 | 11.4349127 | 0.0874515 | 5.4109705 |
| 13 | 1.10201 | 0.9074324 | 13.60139 | 0.0735219 | 12.3423451 | 0.0810219 | 5.8954083 |
| 14 | 1.11028 | 0.9006773 | 14.70340 | 0.0680115 | 13.2430224 | 0.0755115 | 6.3786020 |
| 15 | 1.11860 | 0.8939725 | 15.81368 | 0.0632364 | 14.1369950 | 0.0707364 | 6.8605517 |
| 16 | 1.12699 | 0.8873177 | 16.93228 | 0.0590588 | 15.0243126 | 0.0665588 | 7.3412576 |
| 17 | 1.13544 | 0.8807123 | 18.05927 | 0.0553732 | 15.9050249 | 0.0628732 | 7.8207200 |
| 18 | 1.14396 | 0.8741561 | 19.19472 | 0.0520977 | 16.7791811 | 0.0595977 | 8.2989391 |
| 19 | 1.15254 | 0.8676488 | 20.33868 | 0.0491674 | 17.6468298 | 0.0566674 | 8.7759150 |
| 20 | 1.16118 | 0.8611899 | 21.49122 | 0.0465306 | 18.5080197 | 0.0540306 | 9.2516482 |
| 21 | 1.16989 | 0.8547790 | 22.65240 | 0.0441454 | 19.3627987 | 0.0516454 | 9.7261387 |
| 22 | 1.17867 | 0.8484159 | 23.82230 | 0.0419775 | 20.2112146 | 0.0494775 | 10.1993870 |
| 23 | 1.18751 | 0.8421001 | 25.00096 | 0.0399985 | 21.0533147 | 0.0474985 | 10.6713934 |
| 24 | 1.19641 | 0.8358314 | 26.18847 | 0.0381847 | 21.8891461 | 0.0456847 | 11.1421580 |
| 25 | 1.20539 | 0.8296093 | 27.38488 | 0.0365165 | 22.7187555 | 0.0440165 | 11.6116814 |
| 26 | 1.21443 | 0.8234336 | 28.59027 | 0.0349769 | 23.5421891 | 0.0424769 | 12.0799637 |
| 27 | 1.22354 | 0.8173038 | 29.80470 | 0.0335518 | 24.3594929 | 0.0410518 | 12.5470054 |
| 28 | 1.23271 | 0.8112197 | 31.02823 | 0.0322287 | 25.1707125 | 0.0397287 | 13.0128068 |
| 29 | 1.24196 | 0.8051808 | 32.26094 | 0.0309972 | 25.9758933 | 0.0384972 | 13.4773683 |
| 30 | 1.25127 | 0.7991869 | 33.50290 | 0.0298482 | 26.7750802 | 0.0373482 | 13.9406903 |
| 31 | 1.26066 | 0.7932376 | 34.75417 | 0.0287735 | 27.5683178 | 0.0362735 | 14.4027732 |
| 32 | 1.27011 | 0.7873326 | 36.01483 | 0.0277663 | 28.3556504 | 0.0352663 | 14.8636175 |
| 33 | 1.27964 | 0.7814716 | 37.28494 | 0.0268205 | 29.1371220 | 0.0343205 | 15.3232235 |
| 34 | 1.28923 | 0.7756542 | 38.56458 | 0.0259305 | 29.9127762 | 0.0334305 | 15.7815917 |
| 35 | 1.29890 | 0.7698801 | 39.85381 | 0.0250917 | 30.6826563 | 0.0325917 | 16.2387225 |
| 36 | 1.30865 | 0.7641490 | 41.15272 | 0.0242997 | 31.4468053 | 0.0317997 | 16.6946166 |
| 37 | 1.31846 | 0.7584605 | 42.46136 | 0.0235508 | 32.2052658 | 0.0310508 | 17.1492742 |
| 38 | 1.32835 | 0.7528144 | 43.77982 | 0.0228416 | 32.9580802 | 0.0303416 | 17.6026960 |
| 39 | 1.33831 | 0.7472103 | 45.10817 | 0.0221689 | 33.7052905 | 0.0296689 | 18.0548825 |
| 40 | 1.34835 | 0.7416480 | 46.44648 | 0.0215302 | 34.4469384 | 0.0290302 | 18.5058341 |
| 42 | 1.36865 | 0.7306472 | 49.15329 | 0.0203445 | 35.9137126 | 0.0278445 | 19.4040352 |
| 48 | 1.43141 | 0.6986141 | 57.52071 | 0.0173850 | 40.1847819 | 0.0248850 | 22.0690621 |
| 50 | 1.45296 | 0.6882516 | 60.39426 | 0.0165579 | 41.5664471 | 0.0240579 | 22.9475622 |
| 60 | 1.56568 | 0.6386997 | 75.42414 | 0.0132584 | 48.1733735 | 0.0207584 | 27.2664915 |
| 70 | 1.68715 | 0.5927153 | 91.62007 | 0.0109146 | 54.3046221 | 0.0184146 | 31.4633714 |
| 72 | 1.71255 | 0.5839236 | 95.00703 | 0.0105255 | 55.4768488 | 0.0180255 | 32.2881765 |
| 75 | 1.75137 | 0.5709800 | 100.18331 | 0.0099817 | 57.2026679 | 0.0174817 | 33.5163124 |
| 80 | 1.81804 | 0.5500417 | 109.07253 | 0.0091682 | 59.9944401 | 0.0166682 | 35.5390803 |
| 90 | 1.95909 | 0.5104404 | 127.87899 | 0.0078199 | 65.2746092 | 0.0153199 | 39.4946225 |
| 100 | 2.11108 | 0.4736903 | 148.14451 | 0.0067502 | 70.1746227 | 0.0142502 | 43.3311243 |

## Appendix A *1% Interest Rate Factors*

| | Single Payment | | Equal-Payment Series | | | | Uniform Gradient Series |
|---|---|---|---|---|---|---|---|
| *N* | Compound Amount Factor, (*F/P, i, N*) | Present-Worth Factor, (*P/F, i, N*) | Compound Amount Factor, (*F/A, i, N*) | Sinking-Fund Factor, (*A/F, i, N*) | Present-Worth Factor, (*P/A, i, N*) | Capital Recovery Factor, (*A/P, i, N*) | Factor, (*A/G, i, N*) |
| 1 | 1.01000 | 0.9900990 | 1.00000 | 1.0000000 | 0.9900990 | 1.0100000 | 0.0000000 |
| 2 | 1.02010 | 0.9802960 | 2.01000 | 0.4975124 | 1.9703951 | 0.5075124 | 0.4975124 |
| 3 | 1.03030 | 0.9705901 | 3.03010 | 0.3300221 | 2.9409852 | 0.3400221 | 0.9933666 |
| 4 | 1.04060 | 0.9609803 | 4.06040 | 0.2462811 | 3.9019656 | 0.2562811 | 1.4875624 |
| 5 | 1.05101 | 0.9514657 | 5.10101 | 0.1960398 | 4.8534312 | 0.2060398 | 1.9801002 |
| 6 | 1.06152 | 0.9420452 | 6.15202 | 0.1625484 | 5.7954765 | 0.1725484 | 2.4709800 |
| 7 | 1.07214 | 0.9327181 | 7.21354 | 0.1386283 | 6.7281945 | 0.1486283 | 2.9602020 |
| 8 | 1.08286 | 0.9234832 | 8.28567 | 0.1206903 | 7.6516778 | 0.1306903 | 3.4477664 |
| 9 | 1.09369 | 0.9143398 | 9.36853 | 0.1067404 | 8.5660176 | 0.1167404 | 3.9336734 |
| 10 | 1.10462 | 0.9052870 | 10.46221 | 0.0955821 | 9.4713045 | 0.1055821 | 4.4179234 |
| 11 | 1.11567 | 0.8963237 | 11.56683 | 0.0864541 | 10.3676282 | 0.0964541 | 4.9005167 |
| 12 | 1.12683 | 0.8874492 | 12.68250 | 0.0788488 | 11.2550775 | 0.0888488 | 5.3814536 |
| 13 | 1.13809 | 0.8786626 | 13.80933 | 0.0724148 | 12.1337401 | 0.0824148 | 5.8607344 |
| 14 | 1.14947 | 0.8699630 | 14.94742 | 0.0669012 | 13.0037030 | 0.0769012 | 6.3383597 |
| 15 | 1.16097 | 0.8613495 | 16.09690 | 0.0621238 | 13.8650525 | 0.0721238 | 6.8143297 |
| 16 | 1.17258 | 0.8528213 | 17.25786 | 0.0579446 | 14.7178738 | 0.0679446 | 7.2886451 |
| 17 | 1.18430 | 0.8443775 | 18.43044 | 0.0542581 | 15.5622513 | 0.0642581 | 7.7613063 |
| 18 | 1.19615 | 0.8360173 | 19.61475 | 0.0509820 | 16.3982686 | 0.0609820 | 8.2323138 |
| 19 | 1.20811 | 0.8277399 | 20.81090 | 0.0480518 | 17.2260085 | 0.0580518 | 8.7016682 |
| 20 | 1.22019 | 0.8195445 | 22.01900 | 0.0454153 | 18.0455530 | 0.0554153 | 9.1693702 |
| 21 | 1.23239 | 0.8114302 | 23.23919 | 0.0430308 | 18.8569831 | 0.0530308 | 9.6354204 |
| 22 | 1.24472 | 0.8033962 | 24.47159 | 0.0408637 | 19.6603793 | 0.0508637 | 10.0998193 |
| 23 | 1.25716 | 0.7954418 | 25.71630 | 0.0388858 | 20.4558211 | 0.0488858 | 10.5625679 |
| 24 | 1.26973 | 0.7875661 | 26.97346 | 0.0370735 | 21.2433873 | 0.0470735 | 11.0236667 |
| 25 | 1.28243 | 0.7797684 | 28.24320 | 0.0354068 | 22.0231557 | 0.0454068 | 11.4831165 |
| 26 | 1.29526 | 0.7720480 | 29.52563 | 0.0338689 | 22.7952037 | 0.0438689 | 11.9409182 |
| 27 | 1.30821 | 0.7644039 | 30.82089 | 0.0324455 | 23.5596076 | 0.0424455 | 12.3970725 |
| 28 | 1.32129 | 0.7568356 | 32.12910 | 0.0311244 | 24.3164432 | 0.0411244 | 12.8515804 |
| 29 | 1.33450 | 0.7493421 | 33.45039 | 0.0298950 | 25.0657853 | 0.0398950 | 13.3044427 |
| 30 | 1.34785 | 0.7419229 | 34.78489 | 0.0287481 | 25.8077082 | 0.0387481 | 13.7556604 |
| 31 | 1.36133 | 0.7345771 | 36.13274 | 0.0276757 | 26.5422854 | 0.0376757 | 14.2052343 |
| 32 | 1.37494 | 0.7273041 | 37.49407 | 0.0266709 | 27.2695895 | 0.0366709 | 14.6531656 |
| 33 | 1.38869 | 0.7201031 | 38.86901 | 0.0257274 | 27.9896925 | 0.0357274 | 15.0994552 |
| 34 | 1.40258 | 0.7129733 | 40.25770 | 0.0248400 | 28.7026659 | 0.0348400 | 15.5441042 |
| 35 | 1.41660 | 0.7059142 | 41.66028 | 0.0240037 | 29.4085801 | 0.0340037 | 15.9871136 |
| 36 | 1.43077 | 0.6989249 | 43.07688 | 0.0232143 | 30.1075050 | 0.0332143 | 16.4284847 |
| 37 | 1.44508 | 0.6920049 | 44.50765 | 0.0224680 | 30.7995099 | 0.0324680 | 16.8682184 |
| 38 | 1.45953 | 0.6851534 | 45.95272 | 0.0217615 | 31.4846633 | 0.0317615 | 17.3063161 |
| 39 | 1.47412 | 0.6783697 | 47.41225 | 0.0210916 | 32.1630330 | 0.0310916 | 17.7427789 |
| 40 | 1.48886 | 0.6716531 | 48.88637 | 0.0204556 | 32.8346861 | 0.0304556 | 18.1776081 |
| 42 | 1.51879 | 0.6584189 | 51.87899 | 0.0192756 | 34.1581081 | 0.0292756 | 19.0423707 |
| 48 | 1.61223 | 0.6202604 | 61.22261 | 0.0163338 | 37.9739595 | 0.0263338 | 21.5975899 |
| 50 | 1.64463 | 0.6080388 | 64.46318 | 0.0155127 | 39.1961175 | 0.0255127 | 22.4363454 |
| 60 | 1.81670 | 0.5504496 | 81.66967 | 0.0122444 | 44.9550384 | 0.0222444 | 26.5333139 |
| 70 | 2.00676 | 0.4983149 | 100.67634 | 0.0099328 | 50.1685143 | 0.0199328 | 30.4702553 |
| 72 | 2.04710 | 0.4884961 | 104.70993 | 0.0095502 | 51.1503915 | 0.0195502 | 31.2386140 |
| 75 | 2.10913 | 0.4741295 | 110.91285 | 0.0090161 | 52.5870512 | 0.0190161 | 32.3793391 |
| 80 | 2.21672 | 0.4511179 | 121.67152 | 0.0082189 | 54.8882061 | 0.0182189 | 34.2491991 |
| 90 | 2.44863 | 0.4083912 | 144.86327 | 0.0069031 | 59.1608815 | 0.0169031 | 37.8724493 |
| 100 | 2.70481 | 0.3697112 | 170.48138 | 0.0058657 | 63.0288788 | 0.0158657 | 41.3425687 |

## Appendix A 2% *Interest Rate Factors*

| | Single Payment | | Equal-Payment Series | | | | Uniform Gradient Series |
|---|---|---|---|---|---|---|---|
| N | Compound Amount Factor, (F/P, i, N) | Present-Worth Factor, (P/F, i, N) | Compound Amount Factor, (F/A, i, N) | Sinking-Fund Factor, (A/F, i, N) | Present-Worth Factor, (P/A, i, N) | Capital Recovery Factor, (A/P, i, N) | Factor, (A/G, i, N) |
| 1 | 1.02000 | 0.9803922 | 1.00000 | 1.0000000 | 0.9803922 | 1.0200000 | 0.0000000 |
| 2 | 1.04040 | 0.9611688 | 2.02000 | 0.4950495 | 1.9415609 | 0.5150495 | 0.4950495 |
| 3 | 1.06121 | 0.9423223 | 3.06040 | 0.3267547 | 2.8838833 | 0.3467547 | 0.9867991 |
| 4 | 1.08243 | 0.9238454 | 4.12161 | 0.2426238 | 3.8077287 | 0.2626238 | 1.4752495 |
| 5 | 1.10408 | 0.9057308 | 5.20404 | 0.1921584 | 4.7134595 | 0.2121584 | 1.9604015 |
| 6 | 1.12616 | 0.8879714 | 6.30812 | 0.1585258 | 5.6014309 | 0.1785258 | 2.4422563 |
| 7 | 1.14869 | 0.8705602 | 7.43428 | 0.1345120 | 6.4719911 | 0.1545120 | 2.9208154 |
| 8 | 1.17166 | 0.8534904 | 8.58297 | 0.1165098 | 7.3254814 | 0.1365098 | 3.3960803 |
| 9 | 1.19509 | 0.8367553 | 9.75463 | 0.1025154 | 8.1622367 | 0.1225154 | 3.8680532 |
| 10 | 1.21899 | 0.8203483 | 10.94972 | 0.0913265 | 8.9825850 | 0.1113265 | 4.3367361 |
| 11 | 1.24337 | 0.8042630 | 12.16872 | 0.0821779 | 9.7868480 | 0.1021779 | 4.8021314 |
| 12 | 1.26824 | 0.7884932 | 13.41209 | 0.0745596 | 10.5753412 | 0.0945596 | 5.2642420 |
| 13 | 1.29361 | 0.7730325 | 14.68033 | 0.0681184 | 11.3483737 | 0.0881184 | 5.7230708 |
| 14 | 1.31948 | 0.7578750 | 15.97394 | 0.0626020 | 12.1062488 | 0.0826020 | 6.1786209 |
| 15 | 1.34587 | 0.7430147 | 17.29342 | 0.0578255 | 12.8492635 | 0.0778255 | 6.6308958 |
| 16 | 1.37279 | 0.7284458 | 18.63929 | 0.0536501 | 13.5777093 | 0.0736501 | 7.0798993 |
| 17 | 1.40024 | 0.7141626 | 20.01207 | 0.0499698 | 14.2918719 | 0.0699698 | 7.5256353 |
| 18 | 1.42825 | 0.7001594 | 21.41231 | 0.0467021 | 14.9920313 | 0.0667021 | 7.9681081 |
| 19 | 1.45681 | 0.6864308 | 22.84056 | 0.0437818 | 15.6784620 | 0.0637818 | 8.4073220 |
| 20 | 1.48595 | 0.6729713 | 24.29737 | 0.0411567 | 16.3514333 | 0.0611567 | 8.8432819 |
| 21 | 1.51567 | 0.6597758 | 25.78332 | 0.0387848 | 17.0112092 | 0.0587848 | 9.2759926 |
| 22 | 1.54598 | 0.6468390 | 27.29898 | 0.0366314 | 17.6580482 | 0.0566314 | 9.7054594 |
| 23 | 1.57690 | 0.6341559 | 28.84496 | 0.0346681 | 18.2922041 | 0.0546681 | 10.1316878 |
| 24 | 1.60844 | 0.6217215 | 30.42186 | 0.0328711 | 18.9139256 | 0.0528711 | 10.5546833 |
| 25 | 1.64061 | 0.6095309 | 32.03030 | 0.0312204 | 19.5234565 | 0.0512204 | 10.9744520 |
| 26 | 1.67342 | 0.5975793 | 33.67091 | 0.0296992 | 20.1210358 | 0.0496992 | 11.3910000 |
| 27 | 1.70689 | 0.5858620 | 35.34432 | 0.0282931 | 20.7068978 | 0.0482931 | 11.8043337 |
| 28 | 1.74102 | 0.5743746 | 37.05121 | 0.0269897 | 21.2812724 | 0.0469897 | 12.2144597 |
| 29 | 1.77584 | 0.5631123 | 38.79223 | 0.0257784 | 21.8443847 | 0.0457784 | 12.6213850 |
| 30 | 1.81136 | 0.5520709 | 40.56808 | 0.0246499 | 22.3964556 | 0.0446499 | 13.0251166 |
| 31 | 1.84759 | 0.5412460 | 42.37944 | 0.0235963 | 22.9377015 | 0.0435963 | 13.4256618 |
| 32 | 1.88454 | 0.5306333 | 44.22703 | 0.0226106 | 23.4683348 | 0.0426106 | 13.8230283 |
| 33 | 1.92223 | 0.5202287 | 46.11157 | 0.0216865 | 23.9885636 | 0.0416865 | 14.2172237 |
| 34 | 1.96068 | 0.5100282 | 48.03380 | 0.0208187 | 24.4985917 | 0.0408187 | 14.6082562 |
| 35 | 1.99989 | 0.5000276 | 49.99448 | 0.0200022 | 24.9986193 | 0.0400022 | 14.9961339 |
| 36 | 2.03989 | 0.4902232 | 51.99437 | 0.0192329 | 25.4888425 | 0.0392329 | 15.3808653 |
| 37 | 2.08069 | 0.4806109 | 54.03425 | 0.0185068 | 25.9694534 | 0.0385068 | 15.7624591 |
| 38 | 2.12230 | 0.4711872 | 56.11494 | 0.0178206 | 26.4406406 | 0.0378206 | 16.1409241 |
| 39 | 2.16474 | 0.4619482 | 58.23724 | 0.0171711 | 26.9025888 | 0.0371711 | 16.5162694 |
| 40 | 2.20804 | 0.4528904 | 60.40198 | 0.0165557 | 27.3554792 | 0.0365557 | 16.8885044 |
| 42 | 2.29724 | 0.4353041 | 64.86222 | 0.0154173 | 28.2347936 | 0.0354173 | 17.6236815 |
| 48 | 2.58707 | 0.3865376 | 79.35352 | 0.0126018 | 30.6731196 | 0.0326018 | 19.7555947 |
| 50 | 2.69159 | 0.3715279 | 84.57940 | 0.0118232 | 31.4236059 | 0.0318232 | 20.4419757 |
| 60 | 3.28103 | 0.3047823 | 114.05154 | 0.0087680 | 34.7608867 | 0.0287680 | 23.6961025 |
| 70 | 3.99956 | 0.2500276 | 149.97791 | 0.0066676 | 37.4986193 | 0.0266676 | 26.6632301 |
| 72 | 4.16114 | 0.2403187 | 158.05702 | 0.0063268 | 37.9840631 | 0.0263268 | 27.2234094 |
| 75 | 4.41584 | 0.2264577 | 170.79177 | 0.0058551 | 38.6771143 | 0.0258551 | 28.0434389 |
| 80 | 4.87544 | 0.2051097 | 193.77196 | 0.0051607 | 39.7445136 | 0.0251607 | 29.3571782 |
| 90 | 5.94313 | 0.1682614 | 247.15666 | 0.0040460 | 41.5869292 | 0.0240460 | 31.7929241 |
| 100 | 7.24465 | 0.1380330 | 312.23231 | 0.0032027 | 43.0983516 | 0.0232027 | 33.9862823 |

## Appendix A *3% Interest Rate Factors*

| | Single Payment | | Equal-Payment Series | | | | Uniform Gradient Series |
|---|---|---|---|---|---|---|---|
| $N$ | Compound Amount Factor, $(F/P, i, N)$ | Present-Worth Factor, $(P/F, i, N)$ | Compound Amount Factor, $(F/A, i, N)$ | Sinking-Fund Factor, $(A/F, i, N)$ | Present-Worth Factor, $(P/A, i, N)$ | Capital Recovery Factor, $(A/P, i, N)$ | Factor, $(A/G, i, N)$ |
| 1 | 1.03000 | 0.9708738 | 1.00000 | 1.0000000 | 0.9708738 | 1.0300000 | 0.0000000 |
| 2 | 1.06090 | 0.9425959 | 2.03000 | 0.4926108 | 1.9134697 | 0.5226108 | 0.4926108 |
| 3 | 1.09273 | 0.9151417 | 3.09090 | 0.3235304 | 2.8286114 | 0.3535304 | 0.9802970 |
| 4 | 1.12551 | 0.8884870 | 4.18363 | 0.2390270 | 3.7170984 | 0.2690270 | 1.4630606 |
| 5 | 1.15927 | 0.8626088 | 5.30914 | 0.1883546 | 4.5797072 | 0.2183546 | 1.9409048 |
| 6 | 1.19405 | 0.8374843 | 6.46841 | 0.1545975 | 5.4171914 | 0.1845975 | 2.4138332 |
| 7 | 1.22987 | 0.8130915 | 7.66246 | 0.1305064 | 6.2302830 | 0.1605064 | 2.8818508 |
| 8 | 1.26677 | 0.7894092 | 8.89234 | 0.1124564 | 7.0196922 | 0.1424564 | 3.3449630 |
| 9 | 1.30477 | 0.7664167 | 10.15911 | 0.0984339 | 7.7861089 | 0.1284339 | 3.8031762 |
| 10 | 1.34392 | 0.7440939 | 11.46388 | 0.0872305 | 8.5302028 | 0.1172305 | 4.2564978 |
| 11 | 1.38423 | 0.7224213 | 12.80780 | 0.0780774 | 9.2526241 | 0.1080774 | 4.7049358 |
| 12 | 1.42576 | 0.7013799 | 14.19203 | 0.0704621 | 9.9540040 | 0.1004621 | 5.1484991 |
| 13 | 1.46853 | 0.6809513 | 15.61779 | 0.0640295 | 10.6349553 | 0.0940295 | 5.5871976 |
| 14 | 1.51259 | 0.6611178 | 17.08632 | 0.0585263 | 11.2960731 | 0.0885263 | 6.0210418 |
| 15 | 1.55797 | 0.6418619 | 18.59891 | 0.0537666 | 11.9379351 | 0.0837666 | 6.4500431 |
| 16 | 1.60471 | 0.6231669 | 20.15688 | 0.0496108 | 12.5611020 | 0.0796108 | 6.8742137 |
| 17 | 1.65285 | 0.6050164 | 21.76159 | 0.0459525 | 13.1661185 | 0.0759525 | 7.2935667 |
| 18 | 1.70243 | 0.5873946 | 23.41444 | 0.0427087 | 13.7535131 | 0.0727087 | 7.7081158 |
| 19 | 1.75351 | 0.5702860 | 25.11687 | 0.0398139 | 14.3237991 | 0.0698139 | 8.1178756 |
| 20 | 1.80611 | 0.5536758 | 26.87037 | 0.0372157 | 14.8774749 | 0.0672157 | 8.5228616 |
| 21 | 1.86029 | 0.5375493 | 28.67649 | 0.0348718 | 15.4150241 | 0.0648718 | 8.9230898 |
| 22 | 1.91610 | 0.5218925 | 30.53678 | 0.0327474 | 15.9369166 | 0.0627474 | 9.3185772 |
| 23 | 1.97359 | 0.5066917 | 32.45288 | 0.0308139 | 16.4436084 | 0.0608139 | 9.7093413 |
| 24 | 2.03279 | 0.4919337 | 34.42647 | 0.0290474 | 16.9355421 | 0.0590474 | 10.0954006 |
| 25 | 2.09378 | 0.4776056 | 36.45926 | 0.0274279 | 17.4131477 | 0.0574279 | 10.4767741 |
| 26 | 2.15659 | 0.4636947 | 38.55304 | 0.0259383 | 17.8768424 | 0.0559383 | 10.8534818 |
| 27 | 2.22129 | 0.4501891 | 40.70963 | 0.0245642 | 18.3270315 | 0.0545642 | 11.2255440 |
| 28 | 2.28793 | 0.4370768 | 42.93092 | 0.0232932 | 18.7641082 | 0.0532932 | 11.5929821 |
| 29 | 2.35657 | 0.4243464 | 45.21885 | 0.0221147 | 19.1884546 | 0.0521147 | 11.9558179 |
| 30 | 2.42726 | 0.4119868 | 47.57542 | 0.0210193 | 19.6004413 | 0.0510193 | 12.3140740 |
| 31 | 2.50008 | 0.3999871 | 50.00268 | 0.0199989 | 20.0004285 | 0.0499989 | 12.6677736 |
| 32 | 2.57508 | 0.3883370 | 52.50276 | 0.0190466 | 20.3887655 | 0.0490466 | 13.0169405 |
| 33 | 2.65234 | 0.3770262 | 55.07784 | 0.0181561 | 20.7657918 | 0.0481561 | 13.3615992 |
| 34 | 2.73191 | 0.3660449 | 57.73018 | 0.0173220 | 21.1318367 | 0.0473220 | 13.7017749 |
| 35 | 2.81386 | 0.3553834 | 60.46208 | 0.0165393 | 21.4872201 | 0.0465393 | 14.0374932 |
| 36 | 2.89828 | 0.3450324 | 63.27594 | 0.0158038 | 21.8322525 | 0.0458038 | 14.3687803 |
| 37 | 2.98523 | 0.3349829 | 66.17422 | 0.0151116 | 22.1672354 | 0.0451116 | 14.6956632 |
| 38 | 3.07478 | 0.3252262 | 69.15945 | 0.0144593 | 22.4924616 | 0.0444593 | 15.0181692 |
| 39 | 3.16703 | 0.3157535 | 72.23423 | 0.0138439 | 22.8082151 | 0.0438439 | 15.3363262 |
| 40 | 3.26204 | 0.3065568 | 75.40126 | 0.0132624 | 23.1147720 | 0.0432624 | 15.6501628 |
| 42 | 3.46070 | 0.2889592 | 82.02320 | 0.0121917 | 23.7013592 | 0.0421917 | 16.2649910 |
| 48 | 4.13225 | 0.2419988 | 104.40840 | 0.0095778 | 25.2667066 | 0.0395778 | 18.0088952 |
| 50 | 4.38391 | 0.2281071 | 112.79687 | 0.0088655 | 25.7297640 | 0.0388655 | 18.5575093 |
| 60 | 5.89160 | 0.1697331 | 163.05344 | 0.0061330 | 27.6755637 | 0.0361330 | 21.0674159 |
| 70 | 7.91782 | 0.1262974 | 230.59406 | 0.0043366 | 29.1234214 | 0.0343366 | 23.2145415 |
| 72 | 8.40002 | 0.1190474 | 246.66724 | 0.0040540 | 29.3650875 | 0.0340540 | 23.6036263 |
| 75 | 9.17893 | 0.1089452 | 272.63086 | 0.0036680 | 29.7018263 | 0.0336680 | 24.1634248 |
| 80 | 10.64089 | 0.0939771 | 321.36302 | 0.0031117 | 30.2007634 | 0.0331117 | 25.0353447 |
| 90 | 14.30047 | 0.0699278 | 443.34890 | 0.0022556 | 31.0024071 | 0.0322556 | 26.5666537 |
| 100 | 19.21863 | 0.0520328 | 607.28773 | 0.0016467 | 31.5989053 | 0.0316467 | 27.8444470 |

## Appendix A 4% *Interest Rate Factors*

| | Single Payment | | Equal-Payment Series | | | | Uniform Gradient Series |
|---|---|---|---|---|---|---|---|
| N | Compound Amount Factor, (F/P, i, N) | Present-Worth Factor, (P/F, i, N) | Compound Amount Factor, (F/A, i, N) | Sinking-Fund Factor, (A/F, i, N) | Present-Worth Factor, (P/A, i, N) | Capital Recovery Factor, (A/P, i, N) | Factor, (A/G, i, N) |
| 1 | 1.04000 | 0.9615385 | 1.00000 | 1.0000000 | 0.9615385 | 1.0400000 | 0.0000000 |
| 2 | 1.08160 | 0.9245562 | 2.04000 | 0.4901961 | 1.8860947 | 0.5301961 | 0.4901961 |
| 3 | 1.12486 | 0.8889964 | 3.12160 | 0.3203485 | 2.7750910 | 0.3603485 | 0.9738596 |
| 4 | 1.16986 | 0.8548042 | 4.24646 | 0.2354900 | 3.6298952 | 0.2754900 | 1.4509955 |
| 5 | 1.21665 | 0.8219271 | 5.41632 | 0.1846271 | 4.4518223 | 0.2246271 | 1.9216108 |
| 6 | 1.26532 | 0.7903145 | 6.63298 | 0.1507619 | 5.2421369 | 0.1907619 | 2.3857146 |
| 7 | 1.31593 | 0.7599178 | 7.89829 | 0.1266096 | 6.0020547 | 0.1666096 | 2.8433179 |
| 8 | 1.36857 | 0.7306902 | 9.21423 | 0.1085278 | 6.7327449 | 0.1485278 | 3.2944336 |
| 9 | 1.42331 | 0.7025867 | 10.58280 | 0.0944930 | 7.4353316 | 0.1344930 | 3.7390766 |
| 10 | 1.48024 | 0.6755642 | 12.00611 | 0.0832909 | 8.1108958 | 0.1232909 | 4.1772639 |
| 11 | 1.53945 | 0.6495809 | 13.48635 | 0.0741490 | 8.7604767 | 0.1141490 | 4.6090142 |
| 12 | 1.60103 | 0.6245970 | 15.02581 | 0.0665522 | 9.3850738 | 0.1065522 | 5.0343482 |
| 13 | 1.66507 | 0.6005741 | 16.62684 | 0.0601437 | 9.9856478 | 0.1001437 | 5.4532885 |
| 14 | 1.73168 | 0.5774751 | 18.29191 | 0.0546690 | 10.5631229 | 0.0946690 | 5.8658594 |
| 15 | 1.80094 | 0.5552645 | 20.02359 | 0.0499411 | 11.1183874 | 0.0899411 | 6.2720874 |
| 16 | 1.87298 | 0.5339082 | 21.82453 | 0.0458200 | 11.6522956 | 0.0858200 | 6.6720003 |
| 17 | 1.94790 | 0.5133732 | 23.69751 | 0.0421985 | 12.1656689 | 0.0821985 | 7.0656281 |
| 18 | 2.02582 | 0.4936281 | 25.64541 | 0.0389933 | 12.6592970 | 0.0789933 | 7.4530023 |
| 19 | 2.10685 | 0.4746424 | 27.67123 | 0.0361386 | 13.1339394 | 0.0761386 | 7.8341563 |
| 20 | 2.19112 | 0.4563869 | 29.77808 | 0.0335818 | 13.5903263 | 0.0735818 | 8.2091248 |
| 21 | 2.27877 | 0.4388336 | 31.96920 | 0.0312801 | 14.0291599 | 0.0712801 | 8.5779447 |
| 22 | 2.36992 | 0.4219554 | 34.24797 | 0.0291988 | 14.4511153 | 0.0691988 | 8.9406539 |
| 23 | 2.46472 | 0.4057263 | 36.61789 | 0.0273091 | 14.8568417 | 0.0673091 | 9.2972923 |
| 24 | 2.56330 | 0.3901215 | 39.08260 | 0.0255868 | 15.2469631 | 0.0655868 | 9.6479012 |
| 25 | 2.66584 | 0.3751168 | 41.64591 | 0.0240120 | 15.6220799 | 0.0640120 | 9.9925233 |
| 26 | 2.77247 | 0.3606892 | 44.31174 | 0.0225674 | 15.9827692 | 0.0625674 | 10.3312027 |
| 27 | 2.88337 | 0.3468166 | 47.08421 | 0.0212385 | 16.3295857 | 0.0612385 | 10.6639851 |
| 28 | 2.99870 | 0.3334775 | 49.96758 | 0.0200130 | 16.6630632 | 0.0600130 | 10.9909173 |
| 29 | 3.11865 | 0.3206514 | 52.96629 | 0.0188799 | 16.9837146 | 0.0588799 | 11.3120477 |
| 30 | 3.24340 | 0.3083187 | 56.08494 | 0.0178301 | 17.2920333 | 0.0578301 | 11.6274256 |
| 31 | 3.37313 | 0.2964603 | 59.32834 | 0.0168554 | 17.5884936 | 0.0568554 | 11.9371019 |
| 32 | 3.50806 | 0.2850579 | 62.70147 | 0.0159486 | 17.8735515 | 0.0559486 | 12.2411282 |
| 33 | 3.64838 | 0.2740942 | 66.20953 | 0.0151036 | 18.1476457 | 0.0551036 | 12.5395576 |
| 34 | 3.79432 | 0.2635521 | 69.85791 | 0.0143148 | 18.4111978 | 0.0543148 | 12.8324442 |
| 35 | 3.94609 | 0.2534155 | 73.65222 | 0.0135773 | 18.6646132 | 0.0535773 | 13.1198429 |
| 36 | 4.10393 | 0.2436687 | 77.59831 | 0.0128869 | 18.9082820 | 0.0528869 | 13.4018098 |
| 37 | 4.26809 | 0.2342968 | 81.70225 | 0.0122396 | 19.1425788 | 0.0522396 | 13.6784019 |
| 38 | 4.43881 | 0.2252854 | 85.97034 | 0.0116319 | 19.3678642 | 0.0516319 | 13.9496768 |
| 39 | 4.61637 | 0.2166206 | 90.40915 | 0.0110608 | 19.5844848 | 0.0510608 | 14.2156933 |
| 40 | 4.80102 | 0.2082890 | 95.02552 | 0.0105235 | 19.7927739 | 0.0505235 | 14.4765107 |
| 42 | 5.19278 | 0.1925749 | 104.81960 | 0.0095402 | 20.1856267 | 0.0495402 | 14.9827893 |
| 48 | 6.57053 | 0.1521948 | 139.26321 | 0.0071806 | 21.1951309 | 0.0471806 | 16.3832229 |
| 50 | 7.10668 | 0.1407126 | 152.66708 | 0.0065502 | 21.4821846 | 0.0465502 | 16.8122494 |
| 60 | 10.51963 | 0.0950604 | 237.99069 | 0.0042018 | 22.6234900 | 0.0442018 | 18.6972323 |
| 70 | 15.57162 | 0.0642194 | 364.29046 | 0.0027451 | 23.3945150 | 0.0427451 | 20.1961410 |
| 72 | 16.84226 | 0.0593744 | 396.05656 | 0.0025249 | 23.5156388 | 0.0425249 | 20.4551946 |
| 75 | 18.94525 | 0.0527837 | 448.63137 | 0.0022290 | 23.6804083 | 0.0422290 | 20.8206221 |
| 80 | 23.04980 | 0.0433843 | 551.24498 | 0.0018141 | 23.9153918 | 0.0418141 | 21.3718490 |
| 90 | 34.11933 | 0.0293089 | 827.98333 | 0.0012078 | 24.2672776 | 0.0412078 | 22.2825540 |
| 100 | 50.50495 | 0.0198000 | 1237.62370 | 0.0008080 | 24.5049990 | 0.0408080 | 22.9799999 |

## Appendix A *5% Interest Rate Factors*

| | Single Payment | | Equal-Payment Series | | | | Uniform |
|---|---|---|---|---|---|---|---|
| N | Compound Amount Factor, (F/P, i, N) | Present-Worth Factor, (P/F, i, N) | Compound Amount Factor, (F/A, i, N) | Sinking-Fund Factor, (A/F, i, N) | Present-Worth Factor, (P/A, i, N) | Capital Recovery Factor, (A/P, i, N) | Gradient Series Factor, (A/G, i, N) |
| 1 | 1.05000 | 0.9523810 | 1.00000 | 1.0000000 | 0.9523810 | 1.0500000 | 0.0000000 |
| 2 | 1.10250 | 0.9070295 | 2.05000 | 0.4878049 | 1.8594104 | 0.5378049 | 0.4878049 |
| 3 | 1.15763 | 0.8638376 | 3.15250 | 0.3172086 | 2.7232480 | 0.3672086 | 0.9674861 |
| 4 | 1.21551 | 0.8227025 | 4.31013 | 0.2320118 | 3.5459505 | 0.2820118 | 1.4390534 |
| 5 | 1.27628 | 0.7835262 | 5.52563 | 0.1809748 | 4.3294767 | 0.2309748 | 1.9025202 |
| 6 | 1.34010 | 0.7462154 | 6.80191 | 0.1470175 | 5.0756921 | 0.1970175 | 2.3579038 |
| 7 | 1.40710 | 0.7106813 | 8.14201 | 0.1228198 | 5.7863734 | 0.1728198 | 2.8052254 |
| 8 | 1.47746 | 0.6768394 | 9.54911 | 0.1047218 | 6.4632128 | 0.1547218 | 3.2445098 |
| 9 | 1.55133 | 0.6446089 | 11.02656 | 0.0906901 | 7.1078217 | 0.1406901 | 3.6757856 |
| 10 | 1.62889 | 0.6139133 | 12.57789 | 0.0795046 | 7.7217349 | 0.1295046 | 4.0990850 |
| 11 | 1.71034 | 0.5846793 | 14.20679 | 0.0703889 | 8.3064142 | 0.1203889 | 4.5144439 |
| 12 | 1.79586 | 0.5568374 | 15.91713 | 0.0628254 | 8.8632516 | 0.1128254 | 4.9219016 |
| 13 | 1.88565 | 0.5303214 | 17.71298 | 0.0564558 | 9.3935730 | 0.1064558 | 5.3215011 |
| 14 | 1.97993 | 0.5050680 | 19.59863 | 0.0510240 | 9.8986409 | 0.1010240 | 5.7132886 |
| 15 | 2.07893 | 0.4810171 | 21.57856 | 0.0463423 | 10.3796580 | 0.0963423 | 6.0973137 |
| 16 | 2.18287 | 0.4581115 | 23.65749 | 0.0422699 | 10.8377696 | 0.0922699 | 6.4736294 |
| 17 | 2.29202 | 0.4362967 | 25.84037 | 0.0386991 | 11.2740662 | 0.0886991 | 6.8422918 |
| 18 | 2.40662 | 0.4155207 | 28.13238 | 0.0355462 | 11.6895869 | 0.0855462 | 7.2033600 |
| 19 | 2.52695 | 0.3957340 | 30.53900 | 0.0327450 | 12.0853209 | 0.0827450 | 7.5568961 |
| 20 | 2.65330 | 0.3768895 | 33.06595 | 0.0302426 | 12.4622103 | 0.0802426 | 7.9029651 |
| 21 | 2.78596 | 0.3589424 | 35.71925 | 0.0279961 | 12.8211527 | 0.0779961 | 8.2416350 |
| 22 | 2.92526 | 0.3418499 | 38.50521 | 0.0259705 | 13.1630026 | 0.0759705 | 8.5729762 |
| 23 | 3.07152 | 0.3255713 | 41.43048 | 0.0241368 | 13.4885739 | 0.0741368 | 8.8970619 |
| 24 | 3.22510 | 0.3100679 | 44.50200 | 0.0224709 | 13.7986418 | 0.0724709 | 9.2139676 |
| 25 | 3.38635 | 0.2953028 | 47.72710 | 0.0209525 | 14.0939446 | 0.0709525 | 9.5237714 |
| 26 | 3.55567 | 0.2812407 | 51.11345 | 0.0195643 | 14.3751853 | 0.0695643 | 9.8265533 |
| 27 | 3.73346 | 0.2678483 | 54.66913 | 0.0182919 | 14.6430336 | 0.0682919 | 10.1223957 |
| 28 | 3.92013 | 0.2550936 | 58.40258 | 0.0171225 | 14.8981273 | 0.0671225 | 10.4113830 |
| 29 | 4.11614 | 0.2429463 | 62.32271 | 0.0160455 | 15.1410736 | 0.0660455 | 10.6936014 |
| 30 | 4.32194 | 0.2313774 | 66.43885 | 0.0150514 | 15.3724510 | 0.0650514 | 10.9691390 |
| 31 | 4.53804 | 0.2203595 | 70.76079 | 0.0141321 | 15.5928105 | 0.0641321 | 11.2380854 |
| 32 | 4.76494 | 0.2098662 | 75.29883 | 0.0132804 | 15.8026767 | 0.0632804 | 11.5005319 |
| 33 | 5.00319 | 0.1998725 | 80.06377 | 0.0124900 | 16.0025492 | 0.0624900 | 11.7565711 |
| 34 | 5.25335 | 0.1903548 | 85.06696 | 0.0117554 | 16.1929040 | 0.0617554 | 12.0062971 |
| 35 | 5.51602 | 0.1812903 | 90.32031 | 0.0110717 | 16.3741943 | 0.0610717 | 12.2498049 |
| 36 | 5.79182 | 0.1726574 | 95.83632 | 0.0104345 | 16.5468517 | 0.0604345 | 12.4871909 |
| 37 | 6.08141 | 0.1644356 | 101.62814 | 0.0098398 | 16.7112873 | 0.0598398 | 12.7185521 |
| 38 | 6.38548 | 0.1566054 | 107.70955 | 0.0092842 | 16.8678927 | 0.0592842 | 12.9439866 |
| 39 | 6.70475 | 0.1491480 | 114.09502 | 0.0087646 | 17.0170407 | 0.0587646 | 13.1635931 |
| 40 | 7.03999 | 0.1420457 | 120.79977 | 0.0082782 | 17.1590864 | 0.0582782 | 13.3774711 |
| 42 | 7.76159 | 0.1288396 | 135.23175 | 0.0073947 | 17.4232076 | 0.0573947 | 13.7884410 |
| 48 | 10.40127 | 0.0961421 | 188.02539 | 0.0053184 | 18.0771578 | 0.0553184 | 14.8943066 |
| 50 | 11.46740 | 0.0872037 | 209.34800 | 0.0047767 | 18.2559255 | 0.0547767 | 15.2232645 |
| 60 | 18.67919 | 0.0535355 | 353.58372 | 0.0028282 | 18.9292895 | 0.0528282 | 16.6061786 |
| 70 | 30.42643 | 0.0328662 | 588.52851 | 0.0016992 | 19.3426766 | 0.0516992 | 17.6211858 |
| 72 | 33.54513 | 0.0298106 | 650.90268 | 0.0015363 | 19.4037883 | 0.0515363 | 17.7876877 |
| 75 | 38.83269 | 0.0257515 | 756.65372 | 0.0013216 | 19.4849700 | 0.0513216 | 18.0175872 |
| 80 | 49.56144 | 0.0201770 | 971.22882 | 0.0010296 | 19.5964605 | 0.0510296 | 18.3526024 |
| 90 | 80.73037 | 0.0123869 | 1594.60730 | 0.0006271 | 19.7522617 | 0.0506271 | 18.8711954 |
| 100 | 131.50126 | 0.0076045 | 2610.02516 | 0.0003831 | 19.8479102 | 0.0503831 | 19.2337239 |

## Appendix A *6% Interest Rate Factors*

| | Single Payment | | Equal-Payment Series | | | | Uniform Gradient Series |
|---|---|---|---|---|---|---|---|
| *N* | Compound Amount Factor, (*F/P, i, N*) | Present-Worth Factor, (*P/F, i, N*) | Compound Amount Factor, (*F/A, i, N*) | Sinking-Fund Factor, (*A/F, i, N*) | Present-Worth Factor, (*P/A, i, N*) | Capital Recovery Factor, (*A/P, i, N*) | Factor, (*A/G, i, N*) |
| 1 | 1.06000 | 0.9433962 | 1.00000 | 1.0000000 | 0.9433962 | 1.0600000 | 0.0000000 |
| 2 | 1.12360 | 0.8899964 | 2.06000 | 0.4854369 | 1.8333927 | 0.5454369 | 0.4854369 |
| 3 | 1.19102 | 0.8396193 | 3.18360 | 0.3141098 | 2.6730119 | 0.3741098 | 0.9611760 |
| 4 | 1.26248 | 0.7920937 | 4.37462 | 0.2285915 | 3.4651056 | 0.2885915 | 1.4272338 |
| 5 | 1.33823 | 0.7472582 | 5.63709 | 0.1773964 | 4.2123638 | 0.2373964 | 1.8836333 |
| 6 | 1.41852 | 0.7049605 | 6.97532 | 0.1433626 | 4.9173243 | 0.2033626 | 2.3304038 |
| 7 | 1.50363 | 0.6650571 | 8.39384 | 0.1191350 | 5.5823814 | 0.1791350 | 2.7675812 |
| 8 | 1.59385 | 0.6274124 | 9.89747 | 0.1010359 | 6.2097938 | 0.1610359 | 3.1952076 |
| 9 | 1.68948 | 0.5918985 | 11.49132 | 0.0870222 | 6.8016923 | 0.1470222 | 3.6133314 |
| 10 | 1.79085 | 0.5583948 | 13.18079 | 0.0758680 | 7.3600871 | 0.1358680 | 4.0220070 |
| 11 | 1.89830 | 0.5267875 | 14.97164 | 0.0667929 | 7.8868746 | 0.1267929 | 4.4212947 |
| 12 | 2.01220 | 0.4969694 | 16.86994 | 0.0592770 | 8.3838439 | 0.1192770 | 4.8112608 |
| 13 | 2.13293 | 0.4688390 | 18.88214 | 0.0529601 | 8.8526830 | 0.1129601 | 5.1919772 |
| 14 | 2.26090 | 0.4423010 | 21.01507 | 0.0475849 | 9.2949839 | 0.1075849 | 5.5635212 |
| 15 | 2.39656 | 0.4172651 | 23.27597 | 0.0429628 | 9.7122490 | 0.1029628 | 5.9259757 |
| 16 | 2.54035 | 0.3936463 | 25.67253 | 0.0389521 | 10.1058953 | 0.0989521 | 6.2794284 |
| 17 | 2.69277 | 0.3713644 | 28.21288 | 0.0354448 | 10.4772597 | 0.0954448 | 6.6239721 |
| 18 | 2.85434 | 0.3503438 | 30.90565 | 0.0323565 | 10.8276035 | 0.0923565 | 6.9597045 |
| 19 | 3.02560 | 0.3305130 | 33.75999 | 0.0296209 | 11.1581165 | 0.0896209 | 7.2867276 |
| 20 | 3.20714 | 0.3118047 | 36.78559 | 0.0271846 | 11.4699212 | 0.0871846 | 7.6051477 |
| 21 | 3.39956 | 0.2941554 | 39.99273 | 0.0250045 | 11.7640766 | 0.0850045 | 7.9150753 |
| 22 | 3.60354 | 0.2775051 | 43.39229 | 0.0230456 | 12.0415817 | 0.0830456 | 8.2166249 |
| 23 | 3.81975 | 0.2617973 | 46.99583 | 0.0212785 | 12.3033790 | 0.0812785 | 8.5099142 |
| 24 | 4.04893 | 0.2469785 | 50.81558 | 0.0196790 | 12.5503575 | 0.0796790 | 8.7950647 |
| 25 | 4.29187 | 0.2329986 | 54.86451 | 0.0182267 | 12.7833562 | 0.0782267 | 9.0722007 |
| 26 | 4.54938 | 0.2198100 | 59.15638 | 0.0169043 | 13.0031662 | 0.0769043 | 9.3414498 |
| 27 | 4.82235 | 0.2073680 | 63.70577 | 0.0156972 | 13.2105341 | 0.0756972 | 9.6029418 |
| 28 | 5.11169 | 0.1956301 | 68.52811 | 0.0145926 | 13.4061643 | 0.0745926 | 9.8568093 |
| 29 | 5.41839 | 0.1845567 | 73.63980 | 0.0135796 | 13.5907210 | 0.0735796 | 10.1031868 |
| 30 | 5.74349 | 0.1741101 | 79.05819 | 0.0126489 | 13.7648312 | 0.0726489 | 10.3422109 |
| 31 | 6.08810 | 0.1642548 | 84.80168 | 0.0117922 | 13.9290860 | 0.0717922 | 10.5740199 |
| 32 | 6.45339 | 0.1549574 | 90.88978 | 0.0110023 | 14.0840434 | 0.0710023 | 10.7987534 |
| 33 | 6.84059 | 0.1461862 | 97.34316 | 0.0102729 | 14.2302296 | 0.0702729 | 11.0165524 |
| 34 | 7.25103 | 0.1379115 | 104.18375 | 0.0095984 | 14.3681411 | 0.0695984 | 11.2275589 |
| 35 | 7.68609 | 0.1301052 | 111.43478 | 0.0089739 | 14.4982464 | 0.0689739 | 11.4319156 |
| 36 | 8.14725 | 0.1227408 | 119.12087 | 0.0083948 | 14.6209871 | 0.0683948 | 11.6297658 |
| 37 | 8.63609 | 0.1157932 | 127.26812 | 0.0078574 | 14.7367803 | 0.0678574 | 11.8212531 |
| 38 | 9.15425 | 0.1092389 | 135.90421 | 0.0073581 | 14.8460192 | 0.0673581 | 12.0065215 |
| 39 | 9.70351 | 0.1030555 | 145.05846 | 0.0068938 | 14.9490747 | 0.0668938 | 12.1857146 |
| 40 | 10.28572 | 0.0972222 | 154.76197 | 0.0064615 | 15.0462969 | 0.0664615 | 12.3589761 |
| 42 | 11.55703 | 0.0865274 | 175.95054 | 0.0056834 | 15.2245433 | 0.0656834 | 12.6882760 |
| 48 | 16.39387 | 0.0609984 | 256.56453 | 0.0038977 | 15.6500266 | 0.0638977 | 13.5485427 |
| 50 | 18.42015 | 0.0542884 | 290.33590 | 0.0034443 | 15.7618606 | 0.0634443 | 13.7964280 |
| 60 | 32.98769 | 0.0303143 | 533.12818 | 0.0018757 | 16.1614277 | 0.0618757 | 14.7909452 |
| 70 | 59.07593 | 0.0169274 | 967.93217 | 0.0010331 | 16.3845439 | 0.0610331 | 15.4613480 |
| 72 | 66.37772 | 0.0150653 | 1089.62859 | 0.0009177 | 16.4155784 | 0.0609177 | 15.5653740 |
| 75 | 79.05692 | 0.0126491 | 1300.94868 | 0.0007687 | 16.4558481 | 0.0607687 | 15.7058294 |
| 80 | 105.79599 | 0.0094522 | 1746.59989 | 0.0005725 | 16.5091308 | 0.0605725 | 15.9032787 |
| 90 | 189.46451 | 0.0052780 | 3141.07519 | 0.0003184 | 16.5786994 | 0.0603184 | 16.1891232 |
| 100 | 339.30208 | 0.0029472 | 5638.36806 | 0.0001774 | 16.6175462 | 0.0601774 | 16.3710729 |

# Appendix A 7% *Interest Rate Factors*

| | Single Payment | | Equal-Payment Series | | | | Uniform Gradient Series |
|---|---|---|---|---|---|---|---|
| N | Compound Amount Factor, (F/P, i, N) | Present-Worth Factor, (P/F, i, N) | Compound Amount Factor, (F/A, i, N) | Sinking-Fund Factor, (A/F, i, N) | Present-Worth Factor, (P/A, i, N) | Capital Recovery Factor, (A/P, i, N) | Factor, (A/G, i, N) |
| 1 | 1.07000 | 0.9345794 | 1.00000 | 1.0000000 | 0.9345794 | 1.0700000 | 0.0000000 |
| 2 | 1.14490 | 0.8734387 | 2.07000 | 0.4830918 | 1.8080182 | 0.5530918 | 0.4830918 |
| 3 | 1.22504 | 0.8162979 | 3.21490 | 0.3110517 | 2.6243160 | 0.3810517 | 0.9549286 |
| 4 | 1.31080 | 0.7628952 | 4.43994 | 0.2252281 | 3.3872113 | 0.2952281 | 1.4155362 |
| 5 | 1.40255 | 0.7129862 | 5.75074 | 0.1738907 | 4.1001974 | 0.2438907 | 1.8649504 |
| 6 | 1.50073 | 0.6663422 | 7.15329 | 0.1397958 | 4.7665397 | 0.2097958 | 2.3032172 |
| 7 | 1.60578 | 0.6227497 | 8.65402 | 0.1155532 | 5.3892894 | 0.1855532 | 2.7303923 |
| 8 | 1.71819 | 0.5820091 | 10.25980 | 0.0974678 | 5.9712985 | 0.1674678 | 3.1465414 |
| 9 | 1.83846 | 0.5439337 | 11.97799 | 0.0834865 | 6.5152322 | 0.1534865 | 3.5517396 |
| 10 | 1.96715 | 0.5083493 | 13.81645 | 0.0723775 | 7.0235815 | 0.1423775 | 3.9460710 |
| 11 | 2.10485 | 0.4750928 | 15.78360 | 0.0633569 | 7.4986743 | 0.1333569 | 4.3296292 |
| 12 | 2.25219 | 0.4440120 | 17.88845 | 0.0559020 | 7.9426863 | 0.1259020 | 4.7025162 |
| 13 | 2.40985 | 0.4149644 | 20.14064 | 0.0496508 | 8.3576507 | 0.1196508 | 5.0648425 |
| 14 | 2.57853 | 0.3878172 | 22.55049 | 0.0443449 | 8.7454680 | 0.1143449 | 5.4167266 |
| 15 | 2.75903 | 0.3624460 | 25.12902 | 0.0397946 | 9.1079140 | 0.1097946 | 5.7582947 |
| 16 | 2.95216 | 0.3387346 | 27.88805 | 0.0358576 | 9.4466486 | 0.1058576 | 6.0896805 |
| 17 | 3.15882 | 0.3165744 | 30.84022 | 0.0324252 | 9.7632230 | 0.1024252 | 6.4110245 |
| 18 | 3.37993 | 0.2958639 | 33.99903 | 0.0294126 | 10.0590869 | 0.0994126 | 6.7224739 |
| 19 | 3.61653 | 0.2765083 | 37.37896 | 0.0267530 | 10.3355952 | 0.0967530 | 7.0241817 |
| 20 | 3.86968 | 0.2584190 | 40.99549 | 0.0243929 | 10.5940142 | 0.0943929 | 7.3163069 |
| 21 | 4.14056 | 0.2415131 | 44.86518 | 0.0222890 | 10.8355273 | 0.0922890 | 7.5990138 |
| 22 | 4.43040 | 0.2257132 | 49.00574 | 0.0204058 | 11.0612405 | 0.0904058 | 7.8724713 |
| 23 | 4.74053 | 0.2109469 | 53.43614 | 0.0187139 | 11.2721874 | 0.0887139 | 8.1368528 |
| 24 | 5.07237 | 0.1971466 | 58.17667 | 0.0171890 | 11.4693340 | 0.0871890 | 8.3923357 |
| 25 | 5.42743 | 0.1842492 | 63.24904 | 0.0158105 | 11.6535832 | 0.0858105 | 8.6391010 |
| 26 | 5.80735 | 0.1721955 | 68.67647 | 0.0145610 | 11.8257787 | 0.0845610 | 8.8773325 |
| 27 | 6.21387 | 0.1609304 | 74.48382 | 0.0134257 | 11.9867090 | 0.0834257 | 9.1072169 |
| 28 | 6.64884 | 0.1504022 | 80.69769 | 0.0123919 | 12.1371113 | 0.0823919 | 9.3289430 |
| 29 | 7.11426 | 0.1405628 | 87.34653 | 0.0114487 | 12.2776741 | 0.0814487 | 9.5427014 |
| 30 | 7.61226 | 0.1313671 | 94.46079 | 0.0105864 | 12.4090412 | 0.0805864 | 9.7486842 |
| 31 | 8.14511 | 0.1227730 | 102.07304 | 0.0097969 | 12.5318142 | 0.0797969 | 9.9470844 |
| 32 | 8.71527 | 0.1147411 | 110.21815 | 0.0090729 | 12.6465553 | 0.0790729 | 10.1380958 |
| 33 | 9.32534 | 0.1072347 | 118.93343 | 0.0084081 | 12.7537900 | 0.0784081 | 10.3219121 |
| 34 | 9.97811 | 0.1002193 | 128.25876 | 0.0077967 | 12.8540094 | 0.0777967 | 10.4987272 |
| 35 | 10.67658 | 0.0936629 | 138.23688 | 0.0072340 | 12.9476723 | 0.0772340 | 10.6687345 |
| 36 | 11.42394 | 0.0875355 | 148.91346 | 0.0067153 | 13.0352078 | 0.0767153 | 10.8321264 |
| 37 | 12.22362 | 0.0818088 | 160.33740 | 0.0062368 | 13.1170166 | 0.0762368 | 10.9890946 |
| 38 | 13.07927 | 0.0764569 | 172.56102 | 0.0057951 | 13.1934735 | 0.0757951 | 11.1398292 |
| 39 | 13.99482 | 0.0714550 | 185.64029 | 0.0053868 | 13.2649285 | 0.0753868 | 11.2845185 |
| 40 | 14.97446 | 0.0667804 | 199.63511 | 0.0050091 | 13.3317088 | 0.0750091 | 11.4233492 |
| 42 | 17.14426 | 0.0583286 | 230.63224 | 0.0043359 | 13.4524490 | 0.0743359 | 11.6841699 |
| 48 | 25.72891 | 0.0388668 | 353.27009 | 0.0028307 | 13.7304744 | 0.0728307 | 12.3446661 |
| 50 | 29.45703 | 0.0339478 | 406.52893 | 0.0024598 | 13.8007463 | 0.0724598 | 12.5286789 |
| 60 | 57.94643 | 0.0172573 | 813.52038 | 0.0012292 | 14.0391812 | 0.0712292 | 13.2320924 |
| 70 | 113.98939 | 0.0087727 | 1614.13417 | 0.0006195 | 14.1603893 | 0.0706195 | 13.6661871 |
| 72 | 130.50646 | 0.0076625 | 1850.09222 | 0.0005405 | 14.1762506 | 0.0705405 | 13.7297574 |
| 75 | 159.87602 | 0.0062548 | 2269.65742 | 0.0004406 | 14.1963593 | 0.0704406 | 13.8136481 |
| 80 | 224.23439 | 0.0044596 | 3189.06268 | 0.0003136 | 14.2220054 | 0.0703136 | 13.9273466 |
| 90 | 441.10298 | 0.0022670 | 6287.18543 | 0.0001591 | 14.2533279 | 0.0701591 | 14.0812167 |
| 100 | 867.71633 | 0.0011525 | 12381.66179 | 0.0000808 | 14.2692507 | 0.0700808 | 14.1703363 |

# Appendix A *8% Interest Rate Factors*

| | Single Payment | | Equal-Payment Series | | | | Uniform |
|---|---|---|---|---|---|---|---|
| N | Compound Amount Factor, (F/P, i, N) | Present-Worth Factor, (P/F, i, N) | Compound Amount Factor, (F/A, i, N) | Sinking-Fund Factor, (A/F, i, N) | Present-Worth Factor, (P/A, i, N) | Capital Recovery Factor, (A/P, i, N) | Gradient Series Factor, (A/G, i, N) |
| 1 | 1.08000 | 0.9259259 | 1.00000 | 1.0000000 | 0.9259259 | 1.0800000 | 0.0000000 |
| 2 | 1.16640 | 0.8573388 | 2.08000 | 0.4807692 | 1.7832647 | 0.5607692 | 0.4807692 |
| 3 | 1.25971 | 0.7938322 | 3.24640 | 0.3080335 | 2.5770970 | 0.3880335 | 0.9487432 |
| 4 | 1.36049 | 0.7350299 | 4.50611 | 0.2219208 | 3.3121268 | 0.3019208 | 1.4039598 |
| 5 | 1.46933 | 0.6805832 | 5.86660 | 0.1704565 | 3.9927100 | 0.2504565 | 1.8464716 |
| 6 | 1.58687 | 0.6301696 | 7.33593 | 0.1363154 | 4.6228797 | 0.2163154 | 2.2763460 |
| 7 | 1.71382 | 0.5834904 | 8.92280 | 0.1120724 | 5.2063701 | 0.1920724 | 2.6936649 |
| 8 | 1.85093 | 0.5402689 | 10.63663 | 0.0940148 | 5.7466389 | 0.1740148 | 3.0985239 |
| 9 | 1.99900 | 0.5002490 | 12.48756 | 0.0800797 | 6.2468879 | 0.1600797 | 3.4910327 |
| 10 | 2.15892 | 0.4631935 | 14.48656 | 0.0690295 | 6.7100814 | 0.1490295 | 3.8713139 |
| 11 | 2.33164 | 0.4288829 | 16.64549 | 0.0600763 | 7.1389643 | 0.1400763 | 4.2395030 |
| 12 | 2.51817 | 0.3971138 | 18.97713 | 0.0526950 | 7.5360780 | 0.1326950 | 4.5957475 |
| 13 | 2.71962 | 0.3676979 | 21.49530 | 0.0465218 | 7.9037759 | 0.1265218 | 4.9402067 |
| 14 | 2.93719 | 0.3404610 | 24.21492 | 0.0412969 | 8.2442370 | 0.1212969 | 5.2730508 |
| 15 | 3.17217 | 0.3152417 | 27.15211 | 0.0368295 | 8.5594787 | 0.1168295 | 5.5944603 |
| 16 | 3.42594 | 0.2918905 | 30.32428 | 0.0329769 | 8.8513692 | 0.1129769 | 5.9046256 |
| 17 | 3.70002 | 0.2702690 | 33.75023 | 0.0296294 | 9.1216381 | 0.1096294 | 6.2037458 |
| 18 | 3.99602 | 0.2502490 | 37.45024 | 0.0267021 | 9.3718871 | 0.1067021 | 6.4920284 |
| 19 | 4.31570 | 0.2317121 | 41.44626 | 0.0241276 | 9.6035992 | 0.1041276 | 6.7696885 |
| 20 | 4.66096 | 0.2145482 | 45.76196 | 0.0218522 | 9.8181474 | 0.1018522 | 7.0369478 |
| 21 | 5.03383 | 0.1986557 | 50.42292 | 0.0198323 | 10.0168032 | 0.0998323 | 7.2940343 |
| 22 | 5.43654 | 0.1839405 | 55.45676 | 0.0180321 | 10.2007437 | 0.0980321 | 7.5411812 |
| 23 | 5.87146 | 0.1703153 | 60.89330 | 0.0164222 | 10.3710589 | 0.0964222 | 7.7786264 |
| 24 | 6.34118 | 0.1576993 | 66.76476 | 0.0149780 | 10.5287583 | 0.0949780 | 8.0066115 |
| 25 | 6.84848 | 0.1460179 | 73.10594 | 0.0136788 | 10.6747762 | 0.0936788 | 8.2253815 |
| 26 | 7.39635 | 0.1352018 | 79.95442 | 0.0125071 | 10.8099780 | 0.0925071 | 8.4351838 |
| 27 | 7.98806 | 0.1251868 | 87.35077 | 0.0114481 | 10.9351648 | 0.0914481 | 8.6362675 |
| 28 | 8.62711 | 0.1159137 | 95.33883 | 0.0104889 | 11.0510785 | 0.0904889 | 8.8288830 |
| 29 | 9.31727 | 0.1073275 | 103.96594 | 0.0096185 | 11.1584060 | 0.0896185 | 9.0132810 |
| 30 | 10.06266 | 0.0993773 | 113.28321 | 0.0088274 | 11.2577833 | 0.0888274 | 9.1897125 |
| 31 | 10.86767 | 0.0920160 | 123.34587 | 0.0081073 | 11.3497994 | 0.0881073 | 9.3584274 |
| 32 | 11.73708 | 0.0852000 | 134.21354 | 0.0074508 | 11.4349994 | 0.0874508 | 9.5196747 |
| 33 | 12.67605 | 0.0788889 | 145.95062 | 0.0068516 | 11.5138884 | 0.0868516 | 9.6737016 |
| 34 | 13.69013 | 0.0730453 | 158.62667 | 0.0063041 | 11.5869337 | 0.0863041 | 9.8207532 |
| 35 | 14.78534 | 0.0676345 | 172.31680 | 0.0058033 | 11.6545682 | 0.0858033 | 9.9610718 |
| 36 | 15.96817 | 0.0626246 | 187.10215 | 0.0053447 | 11.7171928 | 0.0853447 | 10.0948967 |
| 37 | 17.24563 | 0.0579857 | 203.07032 | 0.0049244 | 11.7751785 | 0.0849244 | 10.2224638 |
| 38 | 18.62528 | 0.0536905 | 220.31595 | 0.0045389 | 11.8288690 | 0.0845389 | 10.3440053 |
| 39 | 20.11530 | 0.0497134 | 238.94122 | 0.0041851 | 11.8785824 | 0.0841851 | 10.4597493 |
| 40 | 21.72452 | 0.0460309 | 259.05652 | 0.0038602 | 11.9246133 | 0.0838602 | 10.5699192 |
| 42 | 25.33948 | 0.0394641 | 304.24352 | 0.0032868 | 12.0066987 | 0.0832868 | 10.7744086 |
| 48 | 40.21057 | 0.0248691 | 490.13216 | 0.0020403 | 12.1891365 | 0.0820403 | 11.2758404 |
| 50 | 46.90161 | 0.0213212 | 573.77016 | 0.0017429 | 12.2334846 | 0.0817429 | 11.4107136 |
| 60 | 101.25706 | 0.0098759 | 1253.21330 | 0.0007979 | 12.3765518 | 0.0807979 | 11.9015384 |
| 70 | 218.60641 | 0.0045744 | 2720.08007 | 0.0003676 | 12.4428196 | 0.0803676 | 12.1783183 |
| 72 | 254.98251 | 0.0039218 | 3174.78140 | 0.0003150 | 12.4509770 | 0.0803150 | 12.2165159 |
| 75 | 321.20453 | 0.0031133 | 4002.55662 | 0.0002498 | 12.4610840 | 0.0802498 | 12.2657747 |
| 80 | 471.95483 | 0.0021188 | 5886.93543 | 0.0001699 | 12.4735144 | 0.0801699 | 12.3301323 |
| 90 | 1018.91509 | 0.0009814 | 12723.93862 | 0.0000786 | 12.4877320 | 0.0800786 | 12.4115840 |
| 100 | 2199.76126 | 0.0004546 | 27484.51570 | 0.0000364 | 12.4943176 | 0.0800364 | 12.4545198 |

## Appendix A *9% Interest Rate Factors*

| | Single Payment | | Equal-Payment Series | | | | Uniform Gradient Series |
|---|---|---|---|---|---|---|---|
| N | Compound Amount Factor, (F/P, i, N) | Present-Worth Factor, (P/F, i, N) | Compound Amount Factor, (F/A, i, N) | Sinking-Fund Factor, (A/F, i, N) | Present-Worth Factor, (P/A, i, N) | Capital Recovery Factor, (A/P, i, N) | Factor, (A/G, i, N) |
| 1 | 1.09000 | 0.9174312 | 1.00000 | 1.0000000 | 0.9174312 | 1.0900000 | 0.0000000 |
| 2 | 1.18810 | 0.8416800 | 2.09000 | 0.4784689 | 1.7591112 | 0.5684689 | 0.4784689 |
| 3 | 1.29503 | 0.7721835 | 3.27810 | 0.3050548 | 2.5312947 | 0.3950548 | 0.9426192 |
| 4 | 1.41158 | 0.7084252 | 4.57313 | 0.2186687 | 3.2397199 | 0.3086687 | 1.3925039 |
| 5 | 1.53862 | 0.6499314 | 5.98471 | 0.1670925 | 3.8896513 | 0.2570925 | 1.8281968 |
| 6 | 1.67710 | 0.5962673 | 7.52333 | 0.1329198 | 4.4859186 | 0.2229198 | 2.2497922 |
| 7 | 1.82804 | 0.5470342 | 9.20043 | 0.1086905 | 5.0329528 | 0.1986905 | 2.6574042 |
| 8 | 1.99256 | 0.5018663 | 11.02847 | 0.0906744 | 5.5348191 | 0.1806744 | 3.0511664 |
| 9 | 2.17189 | 0.4604278 | 13.02104 | 0.0767988 | 5.9952469 | 0.1667988 | 3.4312309 |
| 10 | 2.36736 | 0.4224108 | 15.19293 | 0.0658201 | 6.4176577 | 0.1558201 | 3.7977678 |
| 11 | 2.58043 | 0.3875329 | 17.56029 | 0.0569467 | 6.8051906 | 0.1469467 | 4.1509642 |
| 12 | 2.81266 | 0.3555347 | 20.14072 | 0.0496507 | 7.1607253 | 0.1396507 | 4.4910233 |
| 13 | 3.06580 | 0.3261786 | 22.95338 | 0.0435666 | 7.4869039 | 0.1335666 | 4.8181636 |
| 14 | 3.34173 | 0.2992465 | 26.01919 | 0.0384332 | 7.7861504 | 0.1284332 | 5.1326175 |
| 15 | 3.64248 | 0.2745380 | 29.36092 | 0.0340589 | 8.0606884 | 0.1240589 | 5.4346307 |
| 16 | 3.97031 | 0.2518698 | 33.00340 | 0.0302999 | 8.3125582 | 0.1202999 | 5.7244605 |
| 17 | 4.32763 | 0.2310732 | 36.97370 | 0.0270462 | 8.5436314 | 0.1170462 | 6.0023753 |
| 18 | 4.71712 | 0.2119937 | 41.30134 | 0.0242123 | 8.7556251 | 0.1142123 | 6.2686530 |
| 19 | 5.14166 | 0.1944897 | 46.01846 | 0.0217304 | 8.9501148 | 0.1117304 | 6.5235800 |
| 20 | 5.60441 | 0.1784309 | 51.16012 | 0.0195465 | 9.1285457 | 0.1095465 | 6.7674500 |
| 21 | 6.10881 | 0.1636981 | 56.76453 | 0.0176166 | 9.2922437 | 0.1076166 | 7.0005630 |
| 22 | 6.65860 | 0.1501817 | 62.87334 | 0.0159050 | 9.4424254 | 0.1059050 | 7.2232239 |
| 23 | 7.25787 | 0.1377814 | 69.53194 | 0.0143819 | 9.5802068 | 0.1043819 | 7.4357418 |
| 24 | 7.91108 | 0.1264049 | 76.78981 | 0.0130226 | 9.7066118 | 0.1030226 | 7.6384283 |
| 25 | 8.62308 | 0.1159678 | 84.70090 | 0.0118063 | 9.8225796 | 0.1018063 | 7.8315971 |
| 26 | 9.39916 | 0.1063925 | 93.32398 | 0.0107154 | 9.9289721 | 0.1007154 | 8.0155627 |
| 27 | 10.24508 | 0.0976078 | 102.72313 | 0.0097349 | 10.0265799 | 0.0997349 | 8.1906395 |
| 28 | 11.16714 | 0.0895484 | 112.96822 | 0.0088520 | 10.1161284 | 0.0988520 | 8.3571408 |
| 29 | 12.17218 | 0.0821545 | 124.13536 | 0.0080557 | 10.1982829 | 0.0980557 | 8.5153783 |
| 30 | 13.26768 | 0.0753711 | 136.30754 | 0.0073364 | 10.2736540 | 0.0973364 | 8.6656606 |
| 31 | 14.46177 | 0.0691478 | 149.57522 | 0.0066856 | 10.3428019 | 0.0966856 | 8.8082935 |
| 32 | 15.76333 | 0.0634384 | 164.03699 | 0.0060962 | 10.4062403 | 0.0960962 | 8.9435783 |
| 33 | 17.18203 | 0.0582003 | 179.80032 | 0.0055617 | 10.4644406 | 0.0955617 | 9.0718118 |
| 34 | 18.72841 | 0.0533948 | 196.98234 | 0.0050766 | 10.5178354 | 0.0950766 | 9.1932855 |
| 35 | 20.41397 | 0.0489861 | 215.71075 | 0.0046358 | 10.5668215 | 0.0946358 | 9.3082854 |
| 36 | 22.25123 | 0.0449413 | 236.12472 | 0.0042350 | 10.6117628 | 0.0942350 | 9.4170911 |
| 37 | 24.25384 | 0.0412306 | 258.37595 | 0.0038703 | 10.6529934 | 0.0938703 | 9.5199757 |
| 38 | 26.43668 | 0.0378262 | 282.62978 | 0.0035382 | 10.6908196 | 0.0935382 | 9.6172055 |
| 39 | 28.81598 | 0.0347030 | 309.06646 | C.0032356 | 10.7255226 | 0.0932356 | 9.7090394 |
| 40 | 31.40942 | 0.0318376 | 337.88245 | 0.0029596 | 10.7573602 | 0.0929596 | 9.7957292 |
| 42 | 37.31753 | 0.0267971 | 403.52813 | 0.0024781 | 10.8133660 | 0.0924781 | 9.9546449 |
| 48 | 62.58524 | 0.0159782 | 684.28041 | 0.0014614 | 10.9335755 | 0.0914614 | 10.3317035 |
| 50 | 74.35752 | 0.0134485 | 815.08356 | 0.0012269 | 10.9616829 | 0.0912269 | 10.4295177 |
| 60 | 176.03129 | 0.0056808 | 1944.79213 | 0.0005142 | 11.0479910 | 0.0905142 | 10.7683153 |
| 70 | 416.73009 | 0.0023996 | 4619.22318 | 0.0002165 | 11.0844485 | 0.0902165 | 10.9427326 |
| 72 | 495.11702 | 0.0020197 | 5490.18906 | 0.0001821 | 11.0886697 | 0.0901821 | 10.9653966 |
| 75 | 641.19089 | 0.0015596 | 7113.23215 | 0.0001406 | 11.0937822 | 0.0901406 | 10.9939586 |
| 80 | 986.55167 | 0.0010136 | 10950.57409 | 0.0000913 | 11.0998485 | 0.0900913 | 11.0299383 |
| 90 | 2335.52658 | 0.0004282 | 25939.18425 | 0.0000386 | 11.1063537 | 0.0900386 | 11.0725594 |
| 100 | 5529.04079 | 0.0001809 | 61422.67546 | 0.0000163 | 11.1091015 | 0.0900163 | 11.0930215 |

## Appendix A *10% Interest Rate Factors*

| | Single Payment | | Equal-Payment Series | | | | Uniform Gradient Series |
|---|---|---|---|---|---|---|---|
| N | Compound Amount Factor, (F/P, i, N) | Present-Worth Factor, (P/F, i, N) | Compound Amount Factor, (F/A, i, N) | Sinking-Fund Factor, (A/F, i, N) | Present-Worth Factor, (P/A, i, N) | Capital Recovery Factor, (A/P, i, N) | Gradient Series Factor, (A/G, i, N) |
| 1 | 1.10000 | 0.9090909 | 1.00000 | 1.0000000 | 0.9090909 | 1.1000000 | 0.0000000 |
| 2 | 1.21000 | 0.8264463 | 2.10000 | 0.4761905 | 1.7355372 | 0.5761905 | 0.4761905 |
| 3 | 1.33100 | 0.7513148 | 3.31000 | 0.3021148 | 2.4868520 | 0.4021148 | 0.9365559 |
| 4 | 1.46410 | 0.6830135 | 4.64100 | 0.2154708 | 3.1698654 | 0.3154708 | 1.3811679 |
| 5 | 1.61051 | 0.6209213 | 6.10510 | 0.1637975 | 3.7907868 | 0.2637975 | 1.8101260 |
| 6 | 1.77156 | 0.5644739 | 7.71561 | 0.1296074 | 4.3552607 | 0.2296074 | 2.2235572 |
| 7 | 1.94872 | 0.5131581 | 9.48717 | 0.1054055 | 4.8684188 | 0.2054055 | 2.6216150 |
| 8 | 2.14359 | 0.4665074 | 11.43589 | 0.0874440 | 5.3349262 | 0.1874440 | 3.0044786 |
| 9 | 2.35795 | 0.4240976 | 13.57948 | 0.0736405 | 5.7590238 | 0.1736405 | 3.3723515 |
| 10 | 2.59374 | 0.3855433 | 15.93742 | 0.0627454 | 6.1445671 | 0.1627454 | 3.7254605 |
| 11 | 2.85312 | 0.3504939 | 18.53117 | 0.0539631 | 6.4950610 | 0.1539631 | 4.0640544 |
| 12 | 3.13843 | 0.3186308 | 21.38428 | 0.0467633 | 6.8136918 | 0.1467633 | 4.3884022 |
| 13 | 3.45227 | 0.2896644 | 24.52271 | 0.0407785 | 7.1033562 | 0.1407785 | 4.6987919 |
| 14 | 3.79750 | 0.2633313 | 27.97498 | 0.0357462 | 7.3666875 | 0.1357462 | 4.9955287 |
| 15 | 4.17725 | 0.2393920 | 31.77248 | 0.0314738 | 7.6060795 | 0.1314738 | 5.2789335 |
| 16 | 4.59497 | 0.2176291 | 35.94973 | 0.0278166 | 7.8237086 | 0.1278166 | 5.5493407 |
| 17 | 5.05447 | 0.1978447 | 40.54470 | 0.0246641 | 8.0215533 | 0.1246641 | 5.8070972 |
| 18 | 5.55992 | 0.1798588 | 45.59917 | 0.0219302 | 8.2014121 | 0.1219302 | 6.0525600 |
| 19 | 6.11591 | 0.1635080 | 51.15909 | 0.0195469 | 8.3649201 | 0.1195469 | 6.2860950 |
| 20 | 6.72750 | 0.1486436 | 57.27500 | 0.0174596 | 8.5135637 | 0.1174596 | 6.5080750 |
| 21 | 7.40025 | 0.1351306 | 64.00250 | 0.0156244 | 8.6486943 | 0.1156244 | 6.7188781 |
| 22 | 8.14027 | 0.1228460 | 71.40275 | 0.0140051 | 8.7715403 | 0.1140051 | 6.9188862 |
| 23 | 8.95430 | 0.1116782 | 79.54302 | 0.0125718 | 8.8832184 | 0.1125718 | 7.1084831 |
| 24 | 9.84973 | 0.1015256 | 88.49733 | 0.0112998 | 8.9847440 | 0.1112998 | 7.2880537 |
| 25 | 10.83471 | 0.0922960 | 98.34706 | 0.0101681 | 9.0770400 | 0.1101681 | 7.4579820 |
| 26 | 11.91818 | 0.0839055 | 109.18177 | 0.0091590 | 9.1609455 | 0.1091590 | 7.6186500 |
| 27 | 13.10999 | 0.0762777 | 121.09994 | 0.0082576 | 9.2372232 | 0.1082576 | 7.7704366 |
| 28 | 14.42099 | 0.0693433 | 134.20994 | 0.0074510 | 9.3065665 | 0.1074510 | 7.9137163 |
| 29 | 15.86309 | 0.0630394 | 148.63093 | 0.0067281 | 9.3696059 | 0.1067281 | 8.0488583 |
| 30 | 17.44940 | 0.0573086 | 164.49402 | 0.0060792 | 9.4269145 | 0.1060792 | 8.1762255 |
| 31 | 19.19434 | 0.0520987 | 181.94342 | 0.0054962 | 9.4790132 | 0.1054962 | 8.2961737 |
| 32 | 21.11378 | 0.0473624 | 201.13777 | 0.0049717 | 9.5263756 | 0.1049717 | 8.4090507 |
| 33 | 23.22515 | 0.0430568 | 222.25154 | 0.0044994 | 9.5694324 | 0.1044994 | 8.5151959 |
| 34 | 25.54767 | 0.0391425 | 245.47670 | 0.0040737 | 9.6085749 | 0.1040737 | 8.6149398 |
| 35 | 28.10244 | 0.0355841 | 271.02437 | 0.0036897 | 9.6441590 | 0.1036897 | 8.7086032 |
| 36 | 30.91268 | 0.0323492 | 299.12681 | 0.0033431 | 9.6765082 | 0.1033431 | 8.7964970 |
| 37 | 34.00395 | 0.0294083 | 330.03949 | 0.0030299 | 9.7059165 | 0.1030299 | 8.8789220 |
| 38 | 37.40434 | 0.0267349 | 364.04343 | 0.0027469 | 9.7326514 | 0.1027469 | 8.9561685 |
| 39 | 41.14478 | 0.0243044 | 401.44778 | 0.0024910 | 9.7569558 | 0.1024910 | 9.0285162 |
| 40 | 45.25926 | 0.0220949 | 442.59256 | 0.0022594 | 9.7790507 | 0.1022594 | 9.0962342 |
| 42 | 54.76370 | 0.0182603 | 537.63699 | 0.0018600 | 9.8173973 | 0.1018600 | 9.2188038 |
| 48 | 97.01723 | 0.0103074 | 960.17234 | 0.0010415 | 9.8969255 | 0.1010415 | 9.5000897 |
| 50 | 117.39085 | 0.0085186 | 1163.90853 | 0.0008592 | 9.9148145 | 0.1008592 | 9.5704130 |
| 60 | 304.48164 | 0.0032843 | 3034.81640 | 0.0003295 | 9.9671573 | 0.1003295 | 9.8022945 |
| 70 | 789.74696 | 0.0012662 | 7887.46957 | 0.0001268 | 9.9873377 | 0.1001268 | 9.9112516 |
| 72 | 955.59382 | 0.0010465 | 9545.93818 | 0.0001048 | 9.9895353 | 0.1001048 | 9.9245753 |
| 75 | 1271.89537 | 0.0007862 | 12708.95371 | 0.0000787 | 9.9921377 | 0.1000787 | 9.9409865 |
| 80 | 2048.40021 | 0.0004882 | 20474.00215 | 0.0000488 | 9.9951181 | 0.1000488 | 9.9609261 |
| 90 | 5313.02261 | 0.0001882 | 53120.22612 | 0.0000188 | 9.9981178 | 0.1000188 | 9.9830573 |
| 100 | 13780.61234 | 0.0000726 | 137796.12340 | 0.0000073 | 9.9992743 | 0.1000073 | 9.9927429 |

## Appendix A *11% Interest Rate Factors*

| | Single Payment | | Equal-Payment Series | | | | Uniform Gradient Series |
|---|---|---|---|---|---|---|---|
| N | Compound Amount Factor, (F/P, i, N) | Present-Worth Factor, (P/F, i, N) | Compound Amount Factor, (F/A, i, N) | Sinking-Fund Factor, (A/F, i, N) | Present-Worth Factor, (P/A, i, N) | Capital Recovery Factor, (A/P, i, N) | Factor, (A/G, i, N) |
| 1 | 1.11000 | 0.9009009 | 1.00000 | 1.0000000 | 0.9009009 | 1.1100000 | 0.0000000 |
| 2 | 1.23210 | 0.8116224 | 2.11000 | 0.4739336 | 1.7125233 | 0.5839336 | 0.4739336 |
| 3 | 1.36763 | 0.7311914 | 3.34210 | 0.2992131 | 2.4437147 | 0.4092131 | 0.9305526 |
| 4 | 1.51807 | 0.6587310 | 4.70973 | 0.2123264 | 3.1024457 | 0.3223264 | 1.3699509 |
| 5 | 1.68506 | 0.5934513 | 6.22780 | 0.1605703 | 3.6958970 | 0.2705703 | 1.7922587 |
| 6 | 1.87041 | 0.5346408 | 7.91286 | 0.1263766 | 4.2305379 | 0.2363766 | 2.1976420 |
| 7 | 2.07616 | 0.4816584 | 9.78327 | 0.1022153 | 4.7121963 | 0.2122153 | 2.5863010 |
| 8 | 2.30454 | 0.4339265 | 11.85943 | 0.0843211 | 5.1461228 | 0.1943211 | 2.9584688 |
| 9 | 2.55804 | 0.3909248 | 14.16397 | 0.0706017 | 5.5370475 | 0.1806017 | 3.3144093 |
| 10 | 2.83942 | 0.3521845 | 16.72201 | 0.0598014 | 5.8892320 | 0.1698014 | 3.6544157 |
| 11 | 3.15176 | 0.3172833 | 19.56143 | 0.0511210 | 6.2065153 | 0.1611210 | 3.9788084 |
| 12 | 3.49845 | 0.2858408 | 22.71319 | 0.0440273 | 6.4923561 | 0.1540273 | 4.2879324 |
| 13 | 3.88328 | 0.2575143 | 26.21164 | 0.0381510 | 6.7498704 | 0.1481510 | 4.5821554 |
| 14 | 4.31044 | 0.2319948 | 30.09492 | 0.0332282 | 6.9818652 | 0.1432282 | 4.8618653 |
| 15 | 4.78459 | 0.2090043 | 34.40536 | 0.0290652 | 7.1908696 | 0.1390652 | 5.1274673 |
| 16 | 5.31089 | 0.1882922 | 39.18995 | 0.0255167 | 7.3791618 | 0.1355167 | 5.3793823 |
| 17 | 5.89509 | 0.1696326 | 44.50084 | 0.0224715 | 7.5487944 | 0.1324715 | 5.6180433 |
| 18 | 6.54355 | 0.1528222 | 50.39594 | 0.0198429 | 7.7016166 | 0.1298429 | 5.8438940 |
| 19 | 7.26334 | 0.1376776 | 56.93949 | 0.0175625 | 7.8392942 | 0.1275625 | 6.0573857 |
| 20 | 8.06231 | 0.1240339 | 64.20283 | 0.0155756 | 7.9633281 | 0.1255756 | 6.2589751 |
| 21 | 8.94917 | 0.1117423 | 72.26514 | 0.0138379 | 8.0750704 | 0.1238379 | 6.4491225 |
| 22 | 9.93357 | 0.1006687 | 81.21431 | 0.0123131 | 8.1757391 | 0.1223131 | 6.6282889 |
| 23 | 11.02627 | 0.0906925 | 91.14788 | 0.0109712 | 8.2664316 | 0.1209712 | 6.7969347 |
| 24 | 12.23916 | 0.0817050 | 102.17415 | 0.0097872 | 8.3481366 | 0.1197872 | 6.9555175 |
| 25 | 13.58546 | 0.0736081 | 114.41331 | 0.0087402 | 8.4217447 | 0.1187402 | 7.1044904 |
| 26 | 15.07986 | 0.0663136 | 127.99877 | 0.0078126 | 8.4880583 | 0.1178126 | 7.2443005 |
| 27 | 16.73865 | 0.0597420 | 143.07864 | 0.0069892 | 8.5478002 | 0.1169892 | 7.3753871 |
| 28 | 18.57990 | 0.0538216 | 159.81729 | 0.0062571 | 8.6016218 | 0.1162571 | 7.4981812 |
| 29 | 20.62369 | 0.0484879 | 178.39719 | 0.0056055 | 8.6501098 | 0.1156055 | 7.6131035 |
| 30 | 22.89230 | 0.0436828 | 199.02088 | 0.0050246 | 8.6937926 | 0.1150246 | 7.7205641 |
| 31 | 25.41045 | 0.0393539 | 221.91317 | 0.0045063 | 8.7331465 | 0.1145063 | 7.8209611 |
| 32 | 28.20560 | 0.0354540 | 247.32362 | 0.0040433 | 8.7686004 | 0.1140433 | 7.9146806 |
| 33 | 31.30821 | 0.0319405 | 275.52922 | 0.0036294 | 8.8005409 | 0.1136294 | 8.0020954 |
| 34 | 34.75212 | 0.0287752 | 306.83744 | 0.0032591 | 8.8293161 | 0.1132591 | 8.0835649 |
| 35 | 38.57485 | 0.0259236 | 341.58955 | 0.0029275 | 8.8552398 | 0.1129275 | 8.1594350 |
| 36 | 42.81808 | 0.0233546 | 380.16441 | 0.0026304 | 8.8785944 | 0.1126304 | 8.2300375 |
| 37 | 47.52807 | 0.0210402 | 422.98249 | 0.0023642 | 8.8996346 | 0.1123642 | 8.2956903 |
| 38 | 52.75616 | 0.0189551 | 470.51056 | 0.0021254 | 8.9185897 | 0.1121254 | 8.3566970 |
| 39 | 58.55934 | 0.0170767 | 523.26673 | 0.0019111 | 8.9356664 | 0.1119111 | 8.4133475 |
| 40 | 65.00087 | 0.0153844 | 581.82607 | 0.0017187 | 8.9510508 | 0.1117187 | 8.4659176 |
| 42 | 80.08757 | 0.0124863 | 718.97790 | 0.0013909 | 8.9773970 | 0.1113909 | 8.5598522 |
| 48 | 149.79695 | 0.0066757 | 1352.69958 | 0.0007393 | 9.0302209 | 0.1107393 | 8.7683218 |
| 50 | 184.56483 | 0.0054182 | 1668.77115 | 0.0005992 | 9.0416532 | 0.1105992 | 8.8185258 |
| 60 | 524.05724 | 0.0019082 | 4755.06584 | 0.0002103 | 9.0735619 | 0.1102103 | 8.9761989 |
| 70 | 1488.01913 | 0.0006720 | 13518.35574 | 0.0000740 | 9.0847997 | 0.1100740 | 9.0438350 |
| 72 | 1833.38837 | 0.0005454 | 16658.07611 | 0.0000600 | 9.0859506 | 0.1100600 | 9.0516161 |
| 75 | 2507.39877 | 0.0003988 | 22785.44339 | 0.0000439 | 9.0872835 | 0.1100439 | 9.0609857 |
| 80 | 4225.11275 | 0.0002367 | 38401.02500 | 0.0000260 | 9.0887575 | 0.1100260 | 9.0719702 |
| 90 | 11996.87381 | 0.0000834 | 109053.39829 | 0.0000092 | 9.0901513 | 0.1100092 | 9.0834065 |
| 100 | 34064.17527 | 0.0000294 | 309665.22972 | 0.0000032 | 9.0906422 | 0.1100032 | 9.0879734 |

## Appendix A *12% Interest Rate Factors*

| N | Single Payment | | Equal-Payment Series | | | | Uniform Gradient Series |
|---|---|---|---|---|---|---|---|
| | Compound Amount Factor, (F/P, i, N) | Present-Worth Factor, (P/F, i, N) | Compound Amount Factor, (F/A, i, N) | Sinking-Fund Factor, (A/F, i, N) | Present-Worth Factor, (P/A, i, N) | Capital Recovery Factor, (A/P, i, N) | Factor, (A/G, i, N) |
| 1 | 1.12000 | 0.8928571 | 1.00000 | 1.0000000 | 0.8928571 | 1.1200000 | 0.0000000 |
| 2 | 1.25440 | 0.7971939 | 2.12000 | 0.4716981 | 1.6900510 | 0.5916981 | 0.4716981 |
| 3 | 1.40493 | 0.7117802 | 3.37440 | 0.2963490 | 2.4018313 | 0.4163490 | 0.9246088 |
| 4 | 1.57352 | 0.6355181 | 4.77933 | 0.2092344 | 3.0373493 | 0.3292344 | 1.3588521 |
| 5 | 1.76234 | 0.5674269 | 6.35285 | 0.1574097 | 3.6047762 | 0.2774097 | 1.7745945 |
| 6 | 1.97382 | 0.5066311 | 8.11519 | 0.1232257 | 4.1114073 | 0.2432257 | 2.1720474 |
| 7 | 2.21068 | 0.4523492 | 10.08901 | 0.0991177 | 4.5637565 | 0.2191177 | 2.5514654 |
| 8 | 2.47596 | 0.4038832 | 12.29969 | 0.0813028 | 4.9676398 | 0.2013028 | 2.9131439 |
| 9 | 2.77308 | 0.3606100 | 14.77566 | 0.0676789 | 5.3282498 | 0.1876789 | 3.2574167 |
| 10 | 3.10585 | 0.3219732 | 17.54874 | 0.0569842 | 5.6502230 | 0.1769842 | 3.5846530 |
| 11 | 3.47855 | 0.2874761 | 20.65458 | 0.0484154 | 5.9376991 | 0.1684154 | 3.8952546 |
| 12 | 3.89598 | 0.2566751 | 24.13313 | 0.0414368 | 6.1943742 | 0.1614368 | 4.1896526 |
| 13 | 4.36349 | 0.2291742 | 28.02911 | 0.0356772 | 6.4235484 | 0.1556772 | 4.4683039 |
| 14 | 4.88711 | 0.2046198 | 32.39260 | 0.0308712 | 6.6281682 | 0.1508712 | 4.7316880 |
| 15 | 5.47357 | 0.1826963 | 37.27971 | 0.0268242 | 6.8108645 | 0.1468242 | 4.9803034 |
| 16 | 6.13039 | 0.1631217 | 42.75328 | 0.0233900 | 6.9739862 | 0.1433900 | 5.2146643 |
| 17 | 6.86604 | 0.1456443 | 48.88367 | 0.0204567 | 7.1196305 | 0.1404567 | 5.4352969 |
| 18 | 7.68997 | 0.1300396 | 55.74971 | 0.0179373 | 7.2496701 | 0.1379373 | 5.6427366 |
| 19 | 8.61276 | 0.1161068 | 63.43968 | 0.0157630 | 7.3657769 | 0.1357630 | 5.8375242 |
| 20 | 9.64629 | 0.1036668 | 72.05244 | 0.0138788 | 7.4694436 | 0.1338788 | 6.0202033 |
| 21 | 10.80385 | 0.0925596 | 81.69874 | 0.0122401 | 7.5620032 | 0.1322401 | 6.1913173 |
| 22 | 12.10031 | 0.0826425 | 92.50258 | 0.0108105 | 7.6446457 | 0.1308105 | 6.3514067 |
| 23 | 13.55235 | 0.0737880 | 104.60289 | 0.0095600 | 7.7184337 | 0.1295600 | 6.5010067 |
| 24 | 15.17863 | 0.0658821 | 118.15524 | 0.0084634 | 7.7843158 | 0.1284634 | 6.6406450 |
| 25 | 17.00006 | 0.0588233 | 133.33387 | 0.0075000 | 7.8431391 | 0.1275000 | 6.7708396 |
| 26 | 19.04007 | 0.0525208 | 150.33393 | 0.0066519 | 7.8956599 | 0.1266519 | 6.8920974 |
| 27 | 21.32488 | 0.0468936 | 169.37401 | 0.0059041 | 7.9425535 | 0.1259041 | 7.0049123 |
| 28 | 23.88387 | 0.0418693 | 190.69889 | 0.0052439 | 7.9844228 | 0.1252439 | 7.1097639 |
| 29 | 26.74993 | 0.0373833 | 214.58275 | 0.0046602 | 8.0218060 | 0.1246602 | 7.2071167 |
| 30 | 29.95992 | 0.0333779 | 241.33268 | 0.0041437 | 8.0551840 | 0.1241437 | 7.2974189 |
| 31 | 33.55511 | 0.0298017 | 271.29261 | 0.0036861 | 8.0849857 | 0.1236861 | 7.3811020 |
| 32 | 37.58173 | 0.0266087 | 304.84772 | 0.0032803 | 8.1115944 | 0.1232803 | 7.4585796 |
| 33 | 42.09153 | 0.0237577 | 342.42945 | 0.0029203 | 8.1353521 | 0.1229203 | 7.5302482 |
| 34 | 47.14252 | 0.0212123 | 384.52098 | 0.0026006 | 8.1565644 | 0.1226006 | 7.5964858 |
| 35 | 52.79962 | 0.0189395 | 431.66350 | 0.0023166 | 8.1755039 | 0.1223166 | 7.6576527 |
| 36 | 59.13557 | 0.0169103 | 484.46312 | 0.0020641 | 8.1924142 | 0.1220641 | 7.7140911 |
| 37 | 66.23184 | 0.0150985 | 543.59869 | 0.0018396 | 8.2075127 | 0.1218396 | 7.7661257 |
| 38 | 74.17966 | 0.0134808 | 609.83053 | 0.0016398 | 8.2209935 | 0.1216398 | 7.8140634 |
| 39 | 83.08122 | 0.0120364 | 684.01020 | 0.0014620 | 8.2330299 | 0.1214620 | 7.8581942 |
| 40 | 93.05097 | 0.0107468 | 767.09142 | 0.0013036 | 8.2437767 | 0.1213036 | 7.8987915 |
| 42 | 116.72314 | 0.0085673 | 964.35948 | 0.0010370 | 8.2619393 | 0.1210370 | 7.9703981 |
| 48 | 230.39078 | 0.0043405 | 1911.58980 | 0.0005231 | 8.2971629 | 0.1205231 | 8.1240834 |
| 50 | 289.00219 | 0.0034602 | 2400.01825 | 0.0004167 | 8.3044985 | 0.1204167 | 8.1597235 |
| 60 | 897.59693 | 0.0011141 | 7471.64111 | 0.0001338 | 8.3240493 | 0.1201338 | 8.2664136 |
| 70 | 2787.79983 | 0.0003587 | 23223.33190 | 0.0000431 | 8.3303441 | 0.1200431 | 8.3082149 |
| 72 | 3497.01610 | 0.0002860 | 29133.46753 | 0.0000343 | 8.3309503 | 0.1200343 | 8.3127385 |
| 75 | 4913.05584 | 0.0002035 | 40933.79867 | 0.0000244 | 8.3316372 | 0.1200244 | 8.3180648 |
| 80 | 8658.48310 | 0.0001155 | 72145.69250 | 0.0000139 | 8.3323709 | 0.1200139 | 8.3240928 |
| 90 | 26891.93422 | 0.0000372 | 224091.11853 | 0.0000045 | 8.3330235 | 0.1200045 | 8.3299865 |
| 100 | 83522.26573 | 0.0000120 | 696010.54772 | 0.0000014 | 8.3332336 | 0.1200014 | 8.3321360 |

## Appendix A *13% Interest Rate Factors*

| | Single Payment | | Equal-Payment Series | | | | Uniform Gradient Series |
|---|---|---|---|---|---|---|---|
| N | Compound Amount Factor, (F/P, i, N) | Present-Worth Factor, (P/F, i, N) | Compound Amount Factor, (F/A, i, N) | Sinking-Fund Factor, (A/F, i, N) | Present-Worth Factor, (P/A, i, N) | Capital Recovery Factor, (A/P, i, N) | Factor, (A/G, i, N) |
| 1 | 1.13000 | 0.8849558 | 1.00000 | 1.0000000 | 0.8849558 | 1.1300000 | 0.0000000 |
| 2 | 1.27690 | 0.7831467 | 2.13000 | 0.4694836 | 1.6681024 | 0.5994836 | 0.4694836 |
| 3 | 1.44290 | 0.6930502 | 3.40690 | 0.2935220 | 2.3611526 | 0.4235220 | 0.9187238 |
| 4 | 1.63047 | 0.6133187 | 4.84980 | 0.2061942 | 2.9744713 | 0.3361942 | 1.3478708 |
| 5 | 1.84244 | 0.5427599 | 6.48027 | 0.1543145 | 3.5172313 | 0.2843145 | 1.7571329 |
| 6 | 2.08195 | 0.4803185 | 8.32271 | 0.1201532 | 3.9975498 | 0.2501532 | 2.1467739 |
| 7 | 2.35261 | 0.4250606 | 10.40466 | 0.0961108 | 4.4226104 | 0.2261108 | 2.5171106 |
| 8 | 2.65844 | 0.3761599 | 12.75726 | 0.0783867 | 4.7987703 | 0.2083867 | 2.8685096 |
| 9 | 3.00404 | 0.3328848 | 15.41571 | 0.0648689 | 5.1316551 | 0.1948689 | 3.2013837 |
| 10 | 3.39457 | 0.2945883 | 18.41975 | 0.0542896 | 5.4262435 | 0.1842896 | 3.5161880 |
| 11 | 3.83586 | 0.2606977 | 21.81432 | 0.0458415 | 5.6869411 | 0.1758415 | 3.8134154 |
| 12 | 4.33452 | 0.2307059 | 25.65018 | 0.0389861 | 5.9176470 | 0.1689861 | 4.0935922 |
| 13 | 4.89801 | 0.2041645 | 29.98470 | 0.0333503 | 6.1218115 | 0.1633503 | 4.3572736 |
| 14 | 5.53475 | 0.1806766 | 34.88271 | 0.0286675 | 6.3024881 | 0.1586675 | 4.6050389 |
| 15 | 6.25427 | 0.1598908 | 40.41746 | 0.0247418 | 6.4623788 | 0.1547418 | 4.8374870 |
| 16 | 7.06733 | 0.1414962 | 46.67173 | 0.0214262 | 6.6038751 | 0.1514262 | 5.0552315 |
| 17 | 7.98608 | 0.1252179 | 53.73906 | 0.0186084 | 6.7290930 | 0.1486084 | 5.2588965 |
| 18 | 9.02427 | 0.1108123 | 61.72514 | 0.0162009 | 6.8399053 | 0.1462009 | 5.4491124 |
| 19 | 10.19742 | 0.0980640 | 70.74941 | 0.0141344 | 6.9379693 | 0.1441344 | 5.6265116 |
| 20 | 11.52309 | 0.0867823 | 80.94683 | 0.0123538 | 7.0247516 | 0.1423538 | 5.7917249 |
| 21 | 13.02109 | 0.0767985 | 92.46992 | 0.0108143 | 7.1015501 | 0.1408143 | 5.9453778 |
| 22 | 14.71383 | 0.0679633 | 105.49101 | 0.0094795 | 7.1695133 | 0.1394795 | 6.0880878 |
| 23 | 16.62663 | 0.0601445 | 120.20484 | 0.0083191 | 7.2296578 | 0.1383191 | 6.2204611 |
| 24 | 18.78809 | 0.0532252 | 136.83147 | 0.0073083 | 7.2828830 | 0.1373083 | 6.3430904 |
| 25 | 21.23054 | 0.0471020 | 155.61956 | 0.0064259 | 7.3299850 | 0.1364259 | 6.4565524 |
| 26 | 23.99051 | 0.0416831 | 176.85010 | 0.0056545 | 7.3716681 | 0.1356545 | 6.5614064 |
| 27 | 27.10928 | 0.0368877 | 200.84061 | 0.0049791 | 7.4085559 | 0.1349791 | 6.6581926 |
| 28 | 30.63349 | 0.0326440 | 227.94989 | 0.0043869 | 7.4411999 | 0.1343869 | 6.7474307 |
| 29 | 34.61584 | 0.0288885 | 258.58338 | 0.0038672 | 7.4700884 | 0.1338672 | 6.8296191 |
| 30 | 39.11590 | 0.0255651 | 293.19922 | 0.0034107 | 7.4956534 | 0.1334107 | 6.9052345 |
| 31 | 44.20096 | 0.0226239 | 332.31511 | 0.0030092 | 7.5182774 | 0.1330092 | 6.9747311 |
| 32 | 49.94709 | 0.0200212 | 376.51608 | 0.0026559 | 7.5382986 | 0.1326559 | 7.0385405 |
| 33 | 56.44021 | 0.0177179 | 426.46317 | 0.0023449 | 7.5560164 | 0.1323449 | 7.0970719 |
| 34 | 63.77744 | 0.0156795 | 482.90338 | 0.0020708 | 7.5716960 | 0.1320708 | 7.1507119 |
| 35 | 72.06851 | 0.0138757 | 546.68082 | 0.0018292 | 7.5855716 | 0.1318292 | 7.1998251 |
| 36 | 81.43741 | 0.0122794 | 618.74933 | 0.0016162 | 7.5978510 | 0.1316162 | 7.2447548 |
| 37 | 92.02428 | 0.0108667 | 700.18674 | 0.0014282 | 7.6087177 | 0.1314282 | 7.2858227 |
| 38 | 103.98743 | 0.0096165 | 792.21101 | 0.0012623 | 7.6183343 | 0.1312623 | 7.3233306 |
| 39 | 117.50580 | 0.0085102 | 896.19845 | 0.0011158 | 7.6268445 | 0.1311158 | 7.3575604 |
| 40 | 132.78155 | 0.0075312 | 1013.70424 | 0.0009865 | 7.6343756 | 0.1309865 | 7.3887751 |
| 42 | 169.54876 | 0.0058980 | 1296.52895 | 0.0007713 | 7.6469384 | 0.1307713 | 7.4431216 |
| 48 | 352.99234 | 0.0028329 | 2707.63342 | 0.0003693 | 7.6705160 | 0.1303693 | 7.5559411 |
| 50 | 450.73593 | 0.0022186 | 3459.50712 | 0.0002891 | 7.6752416 | 0.1302891 | 7.5811313 |

## Appendix A *14% Interest Rate Factors*

| | Single Payment | | Equal-Payment Series | | | | Uniform |
|---|---|---|---|---|---|---|---|
| *N* | Compound Amount Factor, (*F*/*P*, *i*, *N*) | Present-Worth Factor, (*P*/*F*, *i*, *N*) | Compound Amount Factor, (*F*/*A*, *i*, *N*) | Sinking-Fund Factor, (*A*/*F*, *i*, *N*) | Present-Worth Factor, (*P*/*A*, *i*, *N*) | Capital Recovery Factor, (*A*/*P*, *i*, *N*) | Gradient Series Factor, (*A*/*G*, *i*, *N*) |
| 1 | 1.14000 | 0.8771930 | 1.00000 | 1.0000000 | 0.8771930 | 1.1400000 | 0.0000000 |
| 2 | 1.29960 | 0.7694675 | 2.14000 | 0.4672897 | 1.6466605 | 0.6072897 | 0.4672897 |
| 3 | 1.48154 | 0.6749715 | 3.43960 | 0.2907315 | 2.3216320 | 0.4307315 | 0.9128968 |
| 4 | 1.68896 | 0.5920803 | 4.92114 | 0.2032048 | 2.9137123 | 0.3432048 | 1.3370062 |
| 5 | 1.92541 | 0.5193687 | 6.61010 | 0.1512835 | 3.4330810 | 0.2912835 | 1.7398733 |
| 6 | 2.19497 | 0.4555865 | 8.53552 | 0.1171575 | 3.8886675 | 0.2571575 | 2.1218216 |
| 7 | 2.50227 | 0.3996373 | 10.73049 | 0.0931924 | 4.2883048 | 0.2331924 | 2.4832383 |
| 8 | 2.85259 | 0.3505591 | 13.23276 | 0.0755700 | 4.6388639 | 0.2155700 | 2.8245701 |
| 9 | 3.25195 | 0.3075079 | 16.08535 | 0.0621684 | 4.9463718 | 0.2021684 | 3.1463182 |
| 10 | 3.70722 | 0.2697438 | 19.33730 | 0.0517135 | 5.2161156 | 0.1917135 | 3.4490328 |
| 11 | 4.22623 | 0.2366174 | 23.04452 | 0.0433943 | 5.4527330 | 0.1833943 | 3.7333072 |
| 12 | 4.81790 | 0.2075591 | 27.27075 | 0.0366693 | 5.6602921 | 0.1766693 | 3.9997720 |
| 13 | 5.49241 | 0.1820694 | 32.08865 | 0.0311637 | 5.8423615 | 0.1711637 | 4.2490884 |
| 14 | 6.26135 | 0.1597100 | 37.58107 | 0.0266091 | 6.0020715 | 0.1666091 | 4.4819427 |
| 15 | 7.13794 | 0.1400965 | 43.84241 | 0.0228090 | 6.1421680 | 0.1628090 | 4.6990397 |
| 16 | 8.13725 | 0.1228917 | 50.98035 | 0.0196154 | 6.2650596 | 0.1596154 | 4.9010971 |
| 17 | 9.27646 | 0.1077997 | 59.11760 | 0.0169154 | 6.3728593 | 0.1569154 | 5.0888399 |
| 18 | 10.57517 | 0.0945611 | 68.39407 | 0.0146212 | 6.4674205 | 0.1546212 | 5.2629948 |
| 19 | 12.05569 | 0.0829484 | 78.96923 | 0.0126632 | 6.5503688 | 0.1526632 | 5.4242855 |
| 20 | 13.74349 | 0.0727617 | 91.02493 | 0.0109860 | 6.6231306 | 0.1509860 | 5.5734283 |
| 21 | 15.66758 | 0.0638261 | 104.76842 | 0.0095449 | 6.6869566 | 0.1495449 | 5.7111280 |
| 22 | 17.86104 | 0.0559878 | 120.43600 | 0.0083032 | 6.7429444 | 0.1483032 | 5.8380740 |
| 23 | 20.36158 | 0.0491121 | 138.29704 | 0.0072308 | 6.7920565 | 0.1472308 | 5.9549379 |
| 24 | 23.21221 | 0.0430808 | 158.65862 | 0.0063028 | 6.8351373 | 0.1463028 | 6.0623702 |
| 25 | 26.46192 | 0.0377902 | 181.87083 | 0.0054984 | 6.8729274 | 0.1454984 | 6.1609986 |
| 26 | 30.16658 | 0.0331493 | 208.33274 | 0.0048000 | 6.9060767 | 0.1448000 | 6.2514260 |
| 27 | 34.38991 | 0.0290783 | 238.49933 | 0.0041929 | 6.9351550 | 0.1441929 | 6.3342295 |
| 28 | 39.20449 | 0.0255073 | 272.88923 | 0.0036645 | 6.9606623 | 0.1436645 | 6.4099590 |
| 29 | 44.69312 | 0.0223748 | 312.09373 | 0.0032042 | 6.9830371 | 0.1432042 | 6.4791371 |
| 30 | 50.95016 | 0.0196270 | 356.78685 | 0.0028028 | 7.0026641 | 0.1428028 | 6.5422584 |
| 31 | 58.08318 | 0.0172167 | 407.73701 | 0.0024526 | 7.0198808 | 0.1424526 | 6.5997900 |
| 32 | 66.21483 | 0.0151024 | 465.82019 | 0.0021468 | 7.0349832 | 0.1421468 | 6.6521712 |
| 33 | 75.48490 | 0.0132477 | 532.03501 | 0.0018796 | 7.0482308 | 0.1418796 | 6.6998143 |
| 34 | 86.05279 | 0.0116208 | 607.51991 | 0.0016460 | 7.0598516 | 0.1416460 | 6.7431054 |
| 35 | 98.10018 | 0.0101937 | 693.57270 | 0.0014418 | 7.0700453 | 0.1414418 | 6.7824047 |
| 36 | 111.83420 | 0.0089418 | 791.67288 | 0.0012631 | 7.0789871 | 0.1412631 | 6.8180477 |
| 37 | 127.49099 | 0.0078437 | 903.50708 | 0.0011068 | 7.0868308 | 0.1411068 | 6.8503462 |
| 38 | 145.33973 | 0.0068804 | 1030.99808 | 0.0009699 | 7.0937112 | 0.1409699 | 6.8795894 |
| 39 | 165.68729 | 0.0060355 | 1176.33781 | 0.0008501 | 7.0997467 | 0.1408501 | 6.9060447 |
| 40 | 188.88351 | 0.0052943 | 1342.02510 | 0.0007451 | 7.1050409 | 0.1407451 | 6.9299593 |
| 42 | 245.47301 | 0.0040738 | 1746.23582 | 0.0005727 | 7.1137588 | 0.1405727 | 6.9710590 |
| 48 | 538.80655 | 0.0018560 | 3841.47534 | 0.0002603 | 7.1296003 | 0.1402603 | 7.0536057 |
| 50 | 700.23299 | 0.0014281 | 4994.52135 | 0.0002002 | 7.1326565 | 0.1402002 | 7.0713502 |

## Appendix A *15% Interest Rate Factors*

| | Single Payment | | Equal-Payment Series | | | | Uniform Gradient Series |
|---|---|---|---|---|---|---|---|
| N | Compound Amount Factor, (F/P, i, N) | Present-Worth Factor, (P/F, i, N) | Compound Amount Factor, (F/A, i, N) | Sinking-Fund Factor, (A/F, i, N) | Present-Worth Factor, (P/A, i, N) | Capital Recovery Factor, (A/P, i, N) | Factor, (A/G, i, N) |
| 1 | 1.15000 | 0.8695652 | 1.00000 | 1.0000000 | 0.8695652 | 1.1500000 | 0.0000000 |
| 2 | 1.32250 | 0.7561437 | 2.15000 | 0.4651163 | 1.6257089 | 0.6151163 | 0.4651163 |
| 3 | 1.52088 | 0.6575162 | 3.47250 | 0.2879770 | 2.2832251 | 0.4379770 | 0.9071274 |
| 4 | 1.74901 | 0.5717532 | 4.99338 | 0.2002654 | 2.8549784 | 0.3502654 | 1.3262573 |
| 5 | 2.01136 | 0.4971767 | 6.74238 | 0.1483156 | 3.3521551 | 0.2983156 | 1.7228149 |
| 6 | 2.31306 | 0.4323276 | 8.75374 | 0.1142369 | 3.7844827 | 0.2642369 | 2.0971904 |
| 7 | 2.66002 | 0.3759370 | 11.06680 | 0.0903604 | 4.1604197 | 0.2403604 | 2.4498497 |
| 8 | 3.05902 | 0.3269018 | 13.72682 | 0.0728501 | 4.4873215 | 0.2228501 | 2.7813286 |
| 9 | 3.51788 | 0.2842624 | 16.78584 | 0.0595740 | 4.7715839 | 0.2095740 | 3.0922258 |
| 10 | 4.04556 | 0.2471847 | 20.30372 | 0.0492521 | 5.0187686 | 0.1992521 | 3.3831958 |
| 11 | 4.65239 | 0.2149432 | 24.34928 | 0.0410690 | 5.2337118 | 0.1910690 | 3.6549412 |
| 12 | 5.35025 | 0.1869072 | 29.00167 | 0.0344808 | 5.4206190 | 0.1844808 | 3.9082046 |
| 13 | 6.15279 | 0.1625280 | 34.35192 | 0.0291105 | 5.5831470 | 0.1791105 | 4.1437604 |
| 14 | 7.07571 | 0.1413287 | 40.50471 | 0.0246885 | 5.7244756 | 0.1746885 | 4.3624076 |
| 15 | 8.13706 | 0.1228945 | 47.58041 | 0.0210171 | 5.8473701 | 0.1710171 | 4.5649614 |
| 16 | 9.35762 | 0.1068648 | 55.71747 | 0.0179477 | 5.9542349 | 0.1679477 | 4.7522463 |
| 17 | 10.76126 | 0.0929259 | 65.07509 | 0.0153669 | 6.0471608 | 0.1653669 | 4.9250889 |
| 18 | 12.37545 | 0.0808051 | 75.83636 | 0.0131863 | 6.1279659 | 0.1631863 | 5.0843122 |
| 19 | 14.23177 | 0.0702653 | 88.21181 | 0.0113364 | 6.1982312 | 0.1613364 | 5.2307289 |
| 20 | 16.36654 | 0.0611003 | 102.44358 | 0.0097615 | 6.2593315 | 0.1597615 | 5.3651373 |
| 21 | 18.82152 | 0.0531307 | 118.81012 | 0.0084168 | 6.3124622 | 0.1584168 | 5.4883159 |
| 22 | 21.64475 | 0.0462006 | 137.63164 | 0.0072658 | 6.3586627 | 0.1572658 | 5.6010202 |
| 23 | 24.89146 | 0.0401744 | 159.27638 | 0.0062784 | 6.3988372 | 0.1562784 | 5.7039795 |
| 24 | 28.62518 | 0.0349343 | 184.16784 | 0.0054298 | 6.4337714 | 0.1554298 | 5.7978939 |
| 25 | 32.91895 | 0.0303776 | 212.79302 | 0.0046994 | 6.4641491 | 0.1546994 | 5.8834329 |
| 26 | 37.85680 | 0.0264153 | 245.71197 | 0.0040698 | 6.4905644 | 0.1540698 | 5.9612337 |
| 27 | 43.53531 | 0.0229699 | 283.56877 | 0.0035265 | 6.5135343 | 0.1535265 | 6.0319000 |
| 28 | 50.06561 | 0.0199738 | 327.10408 | 0.0030571 | 6.5335081 | 0.1530571 | 6.0960022 |
| 29 | 57.57545 | 0.0173685 | 377.16969 | 0.0026513 | 6.5508766 | 0.1526513 | 6.1540769 |
| 30 | 66.21177 | 0.0151031 | 434.74515 | 0.0023002 | 6.5659796 | 0.1523002 | 6.2066270 |
| 31 | 76.14354 | 0.0131331 | 500.95692 | 0.0019962 | 6.5791127 | 0.1519962 | 6.2541229 |
| 32 | 87.56507 | 0.0114201 | 577.10046 | 0.0017328 | 6.5905328 | 0.1517328 | 6.2970025 |
| 33 | 100.69983 | 0.0099305 | 664.66552 | 0.0015045 | 6.6004633 | 0.1515045 | 6.3356731 |
| 34 | 115.80480 | 0.0086352 | 765.36535 | 0.0013066 | 6.6090985 | 0.1513066 | 6.3705118 |
| 35 | 133.17552 | 0.0075089 | 881.17016 | 0.0011349 | 6.6166074 | 0.1511349 | 6.4018673 |
| 36 | 153.15185 | 0.0065295 | 1014.34568 | 0.0009859 | 6.6231369 | 0.1509859 | 6.4300609 |
| 37 | 176.12463 | 0.0056778 | 1167.49753 | 0.0008565 | 6.6288147 | 0.1508565 | 6.4553886 |
| 38 | 202.54332 | 0.0049372 | 1343.62216 | 0.0007443 | 6.6337519 | 0.1507443 | 6.4781216 |
| 39 | 232.92482 | 0.0042932 | 1546.16549 | 0.0006468 | 6.6380451 | 0.1506468 | 6.4985087 |
| 40 | 267.86355 | 0.0037332 | 1779.09031 | 0.0005621 | 6.6417784 | 0.1505621 | 6.5167773 |
| 42 | 354.24954 | 0.0028229 | 2354.99693 | 0.0004246 | 6.6478475 | 0.1504246 | 6.5477705 |
| 48 | 819.40071 | 0.0012204 | 5456.00475 | 0.0001833 | 6.6585306 | 0.1501833 | 6.6080157 |
| 50 | 1083.65744 | 0.0009228 | 7217.71628 | 0.0001385 | 6.6605147 | 0.1501385 | 6.6204840 |

## Appendix A *16% Interest Rate Factors*

| | Single Payment | | Equal-Payment Series | | | | Uniform |
|---|---|---|---|---|---|---|---|
| *N* | Compound Amount Factor, $(F/P, i, N)$ | Present-Worth Factor, $(P/F, i, N)$ | Compound Amount Factor, $(F/A, i, N)$ | Sinking-Fund Factor, $(A/F, i, N)$ | Present-Worth Factor, $(P/A, i, N)$ | Capital Recovery Factor, $(A/P, i, N)$ | Gradient Series Factor, $(A/G, i, N)$ |
| 1 | 1.16000 | 0.8620690 | 1.00000 | 1.0000000 | 0.8620690 | 1.1600000 | 0.0000000 |
| 2 | 1.34560 | 0.7431629 | 2.16000 | 0.4629630 | 1.6052319 | 0.6229630 | 0.4629630 |
| 3 | 1.56090 | 0.6406577 | 3.50560 | 0.2852579 | 2.2458895 | 0.4452579 | 0.9014149 |
| 4 | 1.81064 | 0.5522911 | 5.06650 | 0.1973751 | 2.7981806 | 0.3573751 | 1.3156233 |
| 5 | 2.10034 | 0.4761130 | 6.87714 | 0.1454094 | 3.2742937 | 0.3054094 | 1.7059568 |
| 6 | 2.43640 | 0.4104423 | 8.97748 | 0.1113899 | 3.6847359 | 0.2713899 | 2.0728799 |
| 7 | 2.82622 | 0.3538295 | 11.41387 | 0.0876127 | 4.0385654 | 0.2476127 | 2.4169454 |
| 8 | 3.27841 | 0.3050255 | 14.24009 | 0.0702243 | 4.3435909 | 0.2302243 | 2.7387870 |
| 9 | 3.80296 | 0.2629530 | 17.51851 | 0.0570825 | 4.6065439 | 0.2170825 | 3.0391101 |
| 10 | 4.41144 | 0.2266836 | 21.32147 | 0.0469011 | 4.8332275 | 0.2069011 | 3.3186823 |
| 11 | 5.11726 | 0.1954169 | 25.73290 | 0.0388608 | 5.0286444 | 0.1988608 | 3.5783233 |
| 12 | 5.93603 | 0.1684628 | 30.85017 | 0.0324147 | 5.1971072 | 0.1924147 | 3.8188950 |
| 13 | 6.88579 | 0.1452266 | 36.78620 | 0.0271841 | 5.3423338 | 0.1871841 | 4.0412911 |
| 14 | 7.98752 | 0.1251953 | 43.67199 | 0.0228980 | 5.4675291 | 0.1828980 | 4.2464273 |
| 15 | 9.26552 | 0.1079270 | 51.65951 | 0.0193575 | 5.5754562 | 0.1793575 | 4.4352323 |
| 16 | 10.74800 | 0.0930405 | 60.92503 | 0.0164136 | 5.6684967 | 0.1764136 | 4.6086384 |
| 17 | 12.46768 | 0.0802074 | 71.67303 | 0.0139522 | 5.7487040 | 0.1739522 | 4.7675735 |
| 18 | 14.46251 | 0.0691443 | 84.14072 | 0.0118849 | 5.8178483 | 0.1718849 | 4.9129541 |
| 19 | 16.77652 | 0.0596071 | 98.60323 | 0.0101417 | 5.8774554 | 0.1701417 | 5.0456784 |
| 20 | 19.46076 | 0.0513855 | 115.37975 | 0.0086670 | 5.9288409 | 0.1686670 | 5.1666210 |
| 21 | 22.57448 | 0.0442978 | 134.84051 | 0.0074162 | 5.9731387 | 0.1674162 | 5.2766278 |
| 22 | 26.18640 | 0.0381878 | 157.41499 | 0.0063526 | 6.0113265 | 0.1663526 | 5.3765126 |
| 23 | 30.37622 | 0.0329205 | 183.60138 | 0.0054466 | 6.0442470 | 0.1654466 | 5.4670538 |
| 24 | 35.23642 | 0.0283797 | 213.97761 | 0.0046734 | 6.0726267 | 0.1646734 | 5.5489921 |
| 25 | 40.87424 | 0.0244653 | 249.21402 | 0.0040126 | 6.0970920 | 0.1640126 | 5.6230289 |
| 26 | 47.41412 | 0.0210908 | 290.08827 | 0.0034472 | 6.1181827 | 0.1634472 | 5.6898257 |
| 27 | 55.00038 | 0.0181817 | 337.50239 | 0.0029629 | 6.1363644 | 0.1629629 | 5.7500035 |
| 28 | 63.80044 | 0.0156739 | 392.50277 | 0.0025478 | 6.1520383 | 0.1625478 | 5.8041433 |
| 29 | 74.00851 | 0.0135120 | 456.30322 | 0.0021915 | 6.1655503 | 0.1621915 | 5.8527861 |
| 30 | 85.84988 | 0.0116482 | 530.31173 | 0.0018857 | 6.1771985 | 0.1618857 | 5.8964344 |
| 31 | 99.58586 | 0.0100416 | 616.16161 | 0.0016230 | 6.1872401 | 0.1616230 | 5.9355533 |
| 32 | 115.51959 | 0.0086565 | 715.74746 | 0.0013971 | 6.1958966 | 0.1613971 | 5.9705718 |
| 33 | 134.00273 | 0.0074625 | 831.26706 | 0.0012030 | 6.2033592 | 0.1612030 | 6.0018848 |
| 34 | 155.44317 | 0.0064332 | 965.26979 | 0.0010360 | 6.2097924 | 0.1610360 | 6.0298543 |
| 35 | 180.31407 | 0.0055459 | 1120.71295 | 0.0008923 | 6.2153383 | 0.1608923 | 6.0548118 |
| 36 | 209.16432 | 0.0047809 | 1301.02703 | 0.0007686 | 6.2201192 | 0.1607686 | 6.0770597 |
| 37 | 242.63062 | 0.0041215 | 1510.19135 | 0.0006622 | 6.2242407 | 0.1606622 | 6.0968737 |
| 38 | 281.45151 | 0.0035530 | 1752.82197 | 0.0005705 | 6.2277937 | 0.1605705 | 6.1145042 |
| 39 | 326.48376 | 0.0030629 | 2034.27348 | 0.0004916 | 6.2308566 | 0.1604916 | 6.1301784 |
| 40 | 378.72116 | 0.0026405 | 2360.75724 | 0.0004236 | 6.2334971 | 0.1604236 | 6.1441018 |
| 42 | 509.60719 | 0.0019623 | 3178.79494 | 0.0003146 | 6.2377357 | 0.1603146 | 6.1674215 |
| 48 | 1241.60509 | 0.0008054 | 7753.78179 | 0.0001290 | 6.2449662 | 0.1601290 | 6.2113092 |
| 50 | 1670.70380 | 0.0005986 | 10435.64877 | 0.0000958 | 6.2462591 | 0.1600958 | 6.2200546 |

## Appendix A *17% Interest Rate Factors*

| | Single Payment | | Equal-Payment Series | | | | Uniform Gradient Series |
|---|---|---|---|---|---|---|---|
| *N* | Compound Amount Factor, (*F/P, i, N*) | Present-Worth Factor, (*P/F, i, N*) | Compound Amount Factor, (*F/A, i, N*) | Sinking-Fund Factor, (*A/F, i, N*) | Present-Worth Factor, (*P/A, i, N*) | Capital Recovery Factor, (*A/P, i, N*) | Factor, (*A/G, i, N*) |
| 1 | 1.17000 | 0.8547009 | 1.00000 | 1.0000000 | 0.8547009 | 1.1700000 | 0.0000000 |
| 2 | 1.36890 | 0.7305136 | 2.17000 | 0.4608295 | 1.5852144 | 0.6308295 | 0.4608295 |
| 3 | 1.60161 | 0.6243706 | 3.53890 | 0.2825737 | 2.2095850 | 0.4525737 | 0.8957586 |
| 4 | 1.87389 | 0.5336500 | 5.14051 | 0.1945331 | 2.7432350 | 0.3645331 | 1.3051032 |
| 5 | 2.19245 | 0.4561112 | 7.01440 | 0.1425639 | 3.1993462 | 0.3125639 | 1.6892981 |
| 6 | 2.56516 | 0.3898386 | 9.20685 | 0.1086148 | 3.5891848 | 0.2786148 | 2.0488893 |
| 7 | 3.00124 | 0.3331954 | 11.77201 | 0.0849472 | 3.9223801 | 0.2549472 | 2.3845253 |
| 8 | 3.51145 | 0.2847824 | 14.77325 | 0.0676899 | 4.2071625 | 0.2376899 | 2.6969463 |
| 9 | 4.10840 | 0.2434037 | 18.28471 | 0.0546905 | 4.4505662 | 0.2246905 | 2.9869730 |
| 10 | 4.80683 | 0.2080374 | 22.39311 | 0.0446566 | 4.6586036 | 0.2146566 | 3.2554943 |
| 11 | 5.62399 | 0.1778097 | 27.19994 | 0.0367648 | 4.8364134 | 0.2067648 | 3.5034547 |
| 12 | 6.58007 | 0.1519741 | 32.82393 | 0.0304656 | 4.9883875 | 0.2004656 | 3.7318413 |
| 13 | 7.69868 | 0.1298924 | 39.40399 | 0.0253781 | 5.1182799 | 0.1953781 | 3.9416718 |
| 14 | 9.00745 | 0.1110192 | 47.10267 | 0.0212302 | 5.2292991 | 0.1912302 | 4.1339820 |
| 15 | 10.53872 | 0.0948882 | 56.11013 | 0.0178221 | 5.3241872 | 0.1878221 | 4.3098151 |
| 16 | 12.33030 | 0.0811010 | 66.64885 | 0.0150040 | 5.4052882 | 0.1850040 | 4.4702108 |
| 17 | 14.42646 | 0.0693171 | 78.97915 | 0.0126616 | 5.4746053 | 0.1826616 | 4.6161960 |
| 18 | 16.87895 | 0.0592454 | 93.40561 | 0.0107060 | 5.5338507 | 0.1807060 | 4.7487770 |
| 19 | 19.74838 | 0.0506371 | 110.28456 | 0.0090675 | 5.5844878 | 0.1790675 | 4.8689318 |
| 20 | 23.10560 | 0.0432796 | 130.03294 | 0.0076904 | 5.6277673 | 0.1776904 | 4.9776048 |
| 21 | 27.03355 | 0.0369911 | 153.13854 | 0.0065300 | 5.6647584 | 0.1765300 | 5.0757016 |
| 22 | 31.62925 | 0.0316163 | 180.17209 | 0.0055502 | 5.6963747 | 0.1755502 | 5.1640854 |
| 23 | 37.00623 | 0.0270225 | 211.80134 | 0.0047214 | 5.7233972 | 0.1747214 | 5.2435746 |
| 24 | 43.29729 | 0.0230961 | 248.80757 | 0.0040192 | 5.7464933 | 0.1740192 | 5.3149407 |
| 25 | 50.65783 | 0.0197403 | 292.10486 | 0.0034234 | 5.7662336 | 0.1734234 | 5.3789076 |
| 26 | 59.26966 | 0.0168720 | 342.76268 | 0.0029175 | 5.7831056 | 0.1729175 | 5.4361516 |
| 27 | 69.34550 | 0.0144205 | 402.03234 | 0.0024874 | 5.7975262 | 0.1724874 | 5.4873013 |
| 28 | 81.13423 | 0.0123253 | 471.37783 | 0.0021214 | 5.8098514 | 0.1721214 | 5.5329392 |
| 29 | 94.92705 | 0.0105344 | 552.51207 | 0.0018099 | 5.8203859 | 0.1718099 | 5.5736027 |
| 30 | 111.06465 | 0.0090038 | 647.43912 | 0.0015445 | 5.8293896 | 0.1715445 | 5.6097859 |
| 31 | 129.94564 | 0.0076955 | 758.50377 | 0.0013184 | 5.8370851 | 0.1713184 | 5.6419416 |
| 32 | 152.03640 | 0.0065774 | 888.44941 | 0.0011256 | 5.8436625 | 0.1711256 | 5.6704835 |
| 33 | 177.88259 | 0.0056217 | 1040.48581 | 0.0009611 | 5.8492842 | 0.1709611 | 5.6957885 |
| 34 | 208.12263 | 0.0048049 | 1218.36839 | 0.0008208 | 5.8540891 | 0.1708208 | 5.7181990 |
| 35 | 243.50347 | 0.0041067 | 1426.49102 | 0.0007010 | 5.8581958 | 0.1707010 | 5.7380251 |
| 36 | 284.89906 | 0.0035100 | 1669.99450 | 0.0005988 | 5.8617058 | 0.1705988 | 5.7555473 |
| 37 | 333.33191 | 0.0030000 | 1954.89356 | 0.0005115 | 5.8647058 | 0.1705115 | 5.7710185 |
| 38 | 389.99833 | 0.0025641 | 2288.22547 | 0.0004370 | 5.8672699 | 0.1704370 | 5.7846661 |
| 39 | 456.29805 | 0.0021916 | 2678.22379 | 0.0003734 | 5.8694615 | 0.1703734 | 5.7966948 |
| 40 | 533.86871 | 0.0018731 | 3134.52184 | 0.0003190 | 5.8713346 | 0.1703190 | 5.8072875 |
| 42 | 730.81288 | 0.0013683 | 4293.01695 | 0.0002329 | 5.8743039 | 0.1702329 | 5.8248039 |
| 48 | 1874.65504 | 0.0005334 | 11021.50024 | 0.0000907 | 5.8792151 | 0.1700907 | 5.8567346 |
| 50 | 2566.21528 | 0.0003897 | 15089.50167 | 0.0000663 | 5.8800607 | 0.1700663 | 5.8628614 |

## Appendix A *18% Interest Rate Factors*

| | Single Payment | | Equal-Payment Series | | | | Uniform |
|---|---|---|---|---|---|---|---|
| N | Compound Amount Factor, (F/P, i, N) | Present-Worth Factor, (P/F, i, N) | Compound Amount Factor, (F/A, i, N) | Sinking-Fund Factor, (A/F, i, N) | Present-Worth Factor, (P/A, i, N) | Capital Recovery Factor, (A/P, i, N) | Gradient Series Factor, (A/G, i, N) |
| 1 | 1.18000 | 0.8474576 | 1.00000 | 1.0000000 | 0.8474576 | 1.1800000 | 0.0000000 |
| 2 | 1.39240 | 0.7181844 | 2.18000 | 0.4587156 | 1.5656421 | 0.6387156 | 0.4587156 |
| 3 | 1.64303 | 0.6086309 | 3.57240 | 0.2799239 | 2.1742729 | 0.4599239 | 0.8901579 |
| 4 | 1.93878 | 0.5157889 | 5.21543 | 0.1917387 | 2.6900618 | 0.3717387 | 1.2946962 |
| 5 | 2.28776 | 0.4371092 | 7.15421 | 0.1397778 | 3.1271710 | 0.3197778 | 1.6728377 |
| 6 | 2.69955 | 0.3704315 | 9.44197 | 0.1059101 | 3.4976026 | 0.2859101 | 2.0252179 |
| 7 | 3.18547 | 0.3139250 | 12.14152 | 0.0823620 | 3.8115276 | 0.2623620 | 2.3525889 |
| 8 | 3.75886 | 0.2660382 | 15.32700 | 0.0652444 | 4.0775658 | 0.2452444 | 2.6558063 |
| 9 | 4.43545 | 0.2254561 | 19.08585 | 0.0523948 | 4.3030218 | 0.2323948 | 2.9358144 |
| 10 | 5.23384 | 0.1910645 | 23.52131 | 0.0425146 | 4.4940863 | 0.2225146 | 3.1936310 |
| 11 | 6.17593 | 0.1619190 | 28.75514 | 0.0347764 | 4.6560053 | 0.2147764 | 3.4303320 |
| 12 | 7.28759 | 0.1372195 | 34.93107 | 0.0286278 | 4.7932249 | 0.2086278 | 3.6470350 |
| 13 | 8.59936 | 0.1162877 | 42.21866 | 0.0236862 | 4.9095126 | 0.2036862 | 3.8448850 |
| 14 | 10.14724 | 0.0985489 | 50.81802 | 0.0196781 | 5.0080615 | 0.1996781 | 4.0250399 |
| 15 | 11.97375 | 0.0835160 | 60.96527 | 0.0164028 | 5.0915776 | 0.1964028 | 4.1886570 |
| 16 | 14.12902 | 0.0707763 | 72.93901 | 0.0137101 | 5.1623539 | 0.1937101 | 4.3368814 |
| 17 | 16.67225 | 0.0599799 | 87.06804 | 0.0114853 | 5.2223338 | 0.1914853 | 4.4708355 |
| 18 | 19.67325 | 0.0508304 | 103.74028 | 0.0096395 | 5.2731642 | 0.1896395 | 4.5916099 |
| 19 | 23.21444 | 0.0430766 | 123.41353 | 0.0081028 | 5.3162409 | 0.1881028 | 4.7002559 |
| 20 | 27.39303 | 0.0365056 | 146.62797 | 0.0068200 | 5.3527465 | 0.1868200 | 4.7977799 |
| 21 | 32.32378 | 0.0309370 | 174.02100 | 0.0057464 | 5.3836835 | 0.1857464 | 4.8851384 |
| 22 | 38.14206 | 0.0262178 | 206.34479 | 0.0048463 | 5.4099012 | 0.1848463 | 4.9632352 |
| 23 | 45.00763 | 0.0222185 | 244.48685 | 0.0040902 | 5.4321197 | 0.1840902 | 5.0329189 |
| 24 | 53.10901 | 0.0188292 | 289.49448 | 0.0034543 | 5.4509489 | 0.1834543 | 5.0949826 |
| 25 | 62.66863 | 0.0159569 | 342.60349 | 0.0029188 | 5.4669058 | 0.1829188 | 5.1501630 |
| 26 | 73.94898 | 0.0135228 | 405.27211 | 0.0024675 | 5.4804287 | 0.1824675 | 5.1991421 |
| 27 | 87.25980 | 0.0114600 | 479.22109 | 0.0020867 | 5.4918887 | 0.1820867 | 5.2425476 |
| 28 | 102.96656 | 0.0097119 | 566.48089 | 0.0017653 | 5.5016006 | 0.1817653 | 5.2809557 |
| 29 | 121.50054 | 0.0082304 | 669.44745 | 0.0014938 | 5.5098310 | 0.1814938 | 5.3148927 |
| 30 | 143.37064 | 0.0069749 | 790.94799 | 0.0012643 | 5.5168060 | 0.1812643 | 5.3448380 |
| 31 | 169.17735 | 0.0059110 | 934.31863 | 0.0010703 | 5.5227169 | 0.1810703 | 5.3712263 |
| 32 | 199.62928 | 0.0050093 | 1103.49598 | 0.0009062 | 5.5277262 | 0.1809062 | 5.3944514 |
| 33 | 235.56255 | 0.0042452 | 1303.12526 | 0.0007674 | 5.5319713 | 0.1807674 | 5.4148681 |
| 34 | 277.96381 | 0.0035976 | 1538.68781 | 0.0006499 | 5.5355689 | 0.1806499 | 5.4327958 |
| 35 | 327.99729 | 0.0030488 | 1816.65161 | 0.0005505 | 5.5386177 | 0.1805505 | 5.4485210 |
| 36 | 387.03680 | 0.0025837 | 2144.64890 | 0.0004663 | 5.5412015 | 0.1804663 | 5.4623002 |
| 37 | 456.70343 | 0.0021896 | 2531.68570 | 0.0003950 | 5.5433911 | 0.1803950 | 5.4743624 |
| 38 | 538.91004 | 0.0018556 | 2988.38913 | 0.0003346 | 5.5452467 | 0.1803346 | 5.4849118 |
| 39 | 635.91385 | 0.0015725 | 3527.29918 | 0.0002835 | 5.5468192 | 0.1802835 | 5.4941299 |
| 40 | 750.37834 | 0.0013327 | 4163.21303 | 0.0002402 | 5.5481519 | 0.1802402 | 5.5021780 |
| 42 | 1044.82681 | 0.0009571 | 5799.03782 | 0.0001724 | 5.5502384 | 0.1801724 | 5.5153190 |
| 48 | 2820.56655 | 0.0003545 | 15664.25859 | 0.0000638 | 5.5535859 | 0.1800638 | 5.5385317 |
| 50 | 3927.35686 | 0.0002546 | 21813.09367 | 0.0000458 | 5.5541410 | 0.1800458 | 5.5428211 |

## Appendix A *19% Interest Rate Factors*

| | Single Payment | | Equal-Payment Series | | | | Uniform Gradient Series |
|---|---|---|---|---|---|---|---|
| $N$ | Compound Amount Factor, $(F/P, i, N)$ | Present-Worth Factor, $(P/F, i, N)$ | Compound Amount Factor, $(F/A, i, N)$ | Sinking-Fund Factor, $(A/F, i, N)$ | Present-Worth Factor, $(P/A, i, N)$ | Capital Recovery Factor, $(A/P, i, N)$ | Factor, $(A/G, i, N)$ |
| 1 | 1.19000 | 0.8403361 | 1.00000 | 1.0000000 | 0.8403361 | 1.1900000 | 0.0000000 |
| 2 | 1.41610 | 0.7061648 | 2.19000 | 0.4566210 | 1.5465010 | 0.6466210 | 0.4566210 |
| 3 | 1.68516 | 0.5934158 | 3.60610 | 0.2773079 | 2.1399168 | 0.4673079 | 0.8846122 |
| 4 | 2.00534 | 0.4986688 | 5.29126 | 0.1889909 | 2.6385855 | 0.3789909 | 1.2844013 |
| 5 | 2.38635 | 0.4190494 | 7.29660 | 0.1370502 | 3.0576349 | 0.3270502 | 1.6565746 |
| 6 | 2.83976 | 0.3521423 | 9.68295 | 0.1032743 | 3.4097772 | 0.2932743 | 2.0018645 |
| 7 | 3.37932 | 0.2959179 | 12.52271 | 0.0798549 | 3.7056951 | 0.2698549 | 2.3211352 |
| 8 | 4.02139 | 0.2486705 | 15.90203 | 0.0628851 | 3.9543657 | 0.2528851 | 2.6153659 |
| 9 | 4.78545 | 0.2089668 | 19.92341 | 0.0501922 | 4.1633325 | 0.2401922 | 2.8856325 |
| 10 | 5.69468 | 0.1756024 | 24.70886 | 0.0404713 | 4.3389349 | 0.2304713 | 3.1330890 |
| 11 | 6.77667 | 0.1475650 | 30.40355 | 0.0328909 | 4.4864999 | 0.2228909 | 3.3589479 |
| 12 | 8.06424 | 0.1240042 | 37.18022 | 0.0268960 | 4.6105041 | 0.2168960 | 3.5644618 |
| 13 | 9.59645 | 0.1042052 | 45.24446 | 0.0221022 | 4.7147093 | 0.2121022 | 3.7509053 |
| 14 | 11.41977 | 0.0875674 | 54.84091 | 0.0182346 | 4.8022768 | 0.2082346 | 3.9195585 |
| 15 | 13.58953 | 0.0735861 | 66.26068 | 0.0150919 | 4.8758628 | 0.2050919 | 4.0716916 |
| 16 | 16.17154 | 0.0618370 | 79.85021 | 0.0125234 | 4.9376998 | 0.2025234 | 4.2085517 |
| 17 | 19.24413 | 0.0519639 | 96.02175 | 0.0104143 | 4.9896637 | 0.2004143 | 4.3313515 |
| 18 | 22.90052 | 0.0436671 | 115.26588 | 0.0086756 | 5.0333309 | 0.1986756 | 4.4412595 |
| 19 | 27.25162 | 0.0366951 | 138.16640 | 0.0072376 | 5.0700259 | 0.1972376 | 4.5393929 |
| 20 | 32.42942 | 0.0308362 | 165.41802 | 0.0060453 | 5.1008621 | 0.1960453 | 4.6268115 |
| 21 | 38.59101 | 0.0259128 | 197.84744 | 0.0050544 | 5.1267749 | 0.1950544 | 4.7045137 |
| 22 | 45.92331 | 0.0217754 | 236.43846 | 0.0042294 | 5.1485503 | 0.1942294 | 4.7734344 |
| 23 | 54.64873 | 0.0182987 | 282.36176 | 0.0035416 | 5.1668490 | 0.1935416 | 4.8344432 |
| 24 | 65.03199 | 0.0153770 | 337.01050 | 0.0029673 | 5.1822261 | 0.1929673 | 4.8883453 |
| 25 | 77.38807 | 0.0129219 | 402.04249 | 0.0024873 | 5.1951480 | 0.1924873 | 4.9358817 |
| 26 | 92.09181 | 0.0108587 | 479.43056 | 0.0020858 | 5.2060067 | 0.1920858 | 4.9777316 |
| 27 | 109.58925 | 0.0091250 | 571.52237 | 0.0017497 | 5.2151317 | 0.1917497 | 5.0145145 |
| 28 | 130.41121 | 0.0076681 | 681.11162 | 0.0014682 | 5.2227997 | 0.1914682 | 5.0467933 |
| 29 | 155.18934 | 0.0064437 | 811.52283 | 0.0012323 | 5.2292435 | 0.1912323 | 5.0750774 |
| 30 | 184.67531 | 0.0054149 | 966.71217 | 0.0010344 | 5.2346584 | 0.1910344 | 5.0998262 |
| 31 | 219.76362 | 0.0045503 | 1151.38748 | 0.0008685 | 5.2392087 | 0.1908685 | 5.1214524 |
| 32 | 261.51871 | 0.0038238 | 1371.15110 | 0.0007293 | 5.2430325 | 0.1907293 | 5.1403260 |
| 33 | 311.20726 | 0.0032133 | 1632.66981 | 0.0006125 | 5.2462458 | 0.1906125 | 5.1567774 |
| 34 | 370.33664 | 0.0027002 | 1943.87708 | 0.0005144 | 5.2489461 | 0.1905144 | 5.1711010 |
| 35 | 440.70061 | 0.0022691 | 2314.21372 | 0.0004321 | 5.2512152 | 0.1904321 | 5.1835583 |
| 36 | 524.43372 | 0.0019068 | 2754.91433 | 0.0003630 | 5.2531220 | 0.1903630 | 5.1943813 |
| 37 | 624.07613 | 0.0016024 | 3279.34805 | 0.0003049 | 5.2547244 | 0.1903049 | 5.2037751 |
| 38 | 742.65059 | 0.0013465 | 3903.42418 | 0.0002562 | 5.2560709 | 0.1902562 | 5.2119208 |
| 39 | 883.75421 | 0.0011315 | 4646.07477 | 0.0002152 | 5.2572024 | 0.1902152 | 5.2189780 |
| 40 | 1051.66751 | 0.0009509 | 5529.82898 | 0.0001808 | 5.2581533 | 0.1901808 | 5.2250869 |
| 42 | 1489.26636 | 0.0006715 | 7832.98082 | 0.0001277 | 5.2596238 | 0.1901277 | 5.2349371 |
| 48 | 4229.16030 | 0.0002365 | 22253.47527 | 0.0000449 | 5.2619134 | 0.1900449 | 5.2518054 |
| 50 | 5988.91390 | 0.0001670 | 31515.33633 | 0.0000317 | 5.2622791 | 0.1900317 | 5.2548077 |

## Appendix A *20% Interest Rate Factors*

| | Single Payment | | Equal-Payment Series | | | | Uniform |
|---|---|---|---|---|---|---|---|
| *N* | Compound Amount Factor, (*F/P, i, N*) | Present-Worth Factor, (*P/F, i, N*) | Compound Amount Factor, (*F/A, i, N*) | Sinking-Fund Factor, (*A/F, i, N*) | Present-Worth Factor, (*P/A, i, N*) | Capital Recovery Factor, (*A/P, i, N*) | Gradient Series Factor, (*A/G, i, N*) |
| 1 | 1.20000 | 0.8333333 | 1.00000 | 1.0000000 | 0.8333333 | 1.2000000 | 0.0000000 |
| 2 | 1.44000 | 0.6944444 | 2.20000 | 0.4545455 | 1.5277778 | 0.6545455 | 0.4545455 |
| 3 | 1.72800 | 0.5787037 | 3.64000 | 0.2747253 | 2.1064815 | 0.4747253 | 0.8791209 |
| 4 | 2.07360 | 0.4822531 | 5.36800 | 0.1862891 | 2.5887346 | 0.3862891 | 1.2742176 |
| 5 | 2.48832 | 0.4018776 | 7.44160 | 0.1343797 | 2.9906121 | 0.3343797 | 1.6405074 |
| 6 | 2.98598 | 0.3348980 | 9.92992 | 0.1007057 | 3.3255101 | 0.3007057 | 1.9788276 |
| 7 | 3.58318 | 0.2790816 | 12.91590 | 0.0774239 | 3.6045918 | 0.2774239 | 2.2901626 |
| 8 | 4.29982 | 0.2325680 | 16.49908 | 0.0606094 | 3.8371598 | 0.2606094 | 2.5756231 |
| 9 | 5.15978 | 0.1938067 | 20.79890 | 0.0480795 | 4.0309665 | 0.2480795 | 2.8364242 |
| 10 | 6.19174 | 0.1615056 | 25.95868 | 0.0385228 | 4.1924721 | 0.2385228 | 3.0738622 |
| 11 | 7.43008 | 0.1345880 | 32.15042 | 0.0311038 | 4.3270601 | 0.2311038 | 3.2892913 |
| 12 | 8.91610 | 0.1121567 | 39.58050 | 0.0252650 | 4.4392167 | 0.2252650 | 3.4841021 |
| 13 | 10.69932 | 0.0934639 | 48.49660 | 0.0206200 | 4.5326806 | 0.2206200 | 3.6596999 |
| 14 | 12.83918 | 0.0778866 | 59.19592 | 0.0168931 | 4.6105672 | 0.2168931 | 3.8174861 |
| 15 | 15.40702 | 0.0649055 | 72.03511 | 0.0138821 | 4.6754726 | 0.2138821 | 3.9588410 |
| 16 | 18.48843 | 0.0540879 | 87.44213 | 0.0114361 | 4.7295605 | 0.2114361 | 4.0851092 |
| 17 | 22.18611 | 0.0450732 | 105.93056 | 0.0094401 | 4.7746338 | 0.2094401 | 4.1975875 |
| 18 | 26.62333 | 0.0375610 | 128.11667 | 0.0078054 | 4.8121948 | 0.2078054 | 4.2975153 |
| 19 | 31.94800 | 0.0313009 | 154.74000 | 0.0064625 | 4.8434957 | 0.2064625 | 4.3860669 |
| 20 | 38.33760 | 0.0260841 | 186.68800 | 0.0053565 | 4.8695797 | 0.2053565 | 4.4643469 |
| 21 | 46.00512 | 0.0217367 | 225.02560 | 0.0044439 | 4.8913164 | 0.2044439 | 4.5333864 |
| 22 | 55.20614 | 0.0181139 | 271.03072 | 0.0036896 | 4.9094304 | 0.2036896 | 4.5941419 |
| 23 | 66.24737 | 0.0150949 | 326.23686 | 0.0030653 | 4.9245253 | 0.2030653 | 4.6474954 |
| 24 | 79.49685 | 0.0125791 | 392.48424 | 0.0025479 | 4.9371044 | 0.2025479 | 4.6942552 |
| 25 | 95.39622 | 0.0104826 | 471.98108 | 0.0021187 | 4.9475870 | 0.2021187 | 4.7351589 |
| 26 | 114.47546 | 0.0087355 | 567.37730 | 0.0017625 | 4.9563225 | 0.2017625 | 4.7708756 |
| 27 | 137.37055 | 0.0072796 | 681.85276 | 0.0014666 | 4.9636021 | 0.2014666 | 4.8020100 |
| 28 | 164.84466 | 0.0060663 | 819.22331 | 0.0012207 | 4.9696684 | 0.2012207 | 4.8291064 |
| 29 | 197.81359 | 0.0050553 | 984.06797 | 0.0010162 | 4.9747237 | 0.2010162 | 4.8526525 |
| 30 | 237.37631 | 0.0042127 | 1181.88157 | 0.0008461 | 4.9789364 | 0.2008461 | 4.8730837 |
| 31 | 284.85158 | 0.0035106 | 1419.25788 | 0.0007046 | 4.9824470 | 0.2007046 | 4.8907880 |
| 32 | 341.82189 | 0.0029255 | 1704.10946 | 0.0005868 | 4.9853725 | 0.2005868 | 4.9061093 |
| 33 | 410.18627 | 0.0024379 | 2045.93135 | 0.0004888 | 4.9878104 | 0.2004888 | 4.9193521 |
| 34 | 492.22352 | 0.0020316 | 2456.11762 | 0.0004071 | 4.9898420 | 0.2004071 | 4.9307851 |
| 35 | 590.66823 | 0.0016930 | 2948.34115 | 0.0003392 | 4.9915350 | 0.2003392 | 4.9406446 |
| 36 | 708.80187 | 0.0014108 | 3539.00937 | 0.0002826 | 4.9929458 | 0.2002826 | 4.9491383 |
| 37 | 850.56225 | 0.0011757 | 4247.81125 | 0.0002354 | 4.9941215 | 0.2002354 | 4.9564482 |
| 38 | 1020.67470 | 0.0009797 | 5098.37350 | 0.0001961 | 4.9951013 | 0.2001961 | 4.9627332 |
| 39 | 1224.80964 | 0.0008165 | 6119.04820 | 0.0001634 | 4.9959177 | 0.2001634 | 4.9681323 |
| 40 | 1469.77157 | 0.0006804 | 7343.85784 | 0.0001362 | 4.9965981 | 0.2001362 | 4.9727664 |
| 42 | 2116.47106 | 0.0004725 | 10577.35529 | 0.0000945 | 4.9976376 | 0.2000945 | 4.9801463 |
| 48 | 6319.74872 | 0.0001582 | 31593.74358 | 0.0000317 | 4.9992088 | 0.2000317 | 4.9924036 |
| 50 | 9100.43815 | 0.0001099 | 45497.19075 | 0.0000220 | 4.9994506 | 0.2000220 | 4.9945052 |

## Appendix A *25% Interest Rate Factors*

| | Single Payment | | Equal-Payment Series | | | | Uniform Gradient Series |
|---|---|---|---|---|---|---|---|
| N | Compound Amount Factor, (F/P, i, N) | Present-Worth Factor, (P/F, i, N) | Compound Amount Factor, (F/A, i, N) | Sinking-Fund Factor, (A/F, i, N) | Present-Worth Factor, (P/A, i, N) | Capital Recovery Factor, (A/P, i, N) | Uniform Gradient Series Factor, (A/G, i, N) |
| 1 | 1.25000 | 0.8000000 | 1.00000 | 1.0000000 | 0.8000000 | 1.2500000 | 0.0000000 |
| 2 | 1.56250 | 0.6400000 | 2.25000 | 0.4444444 | 1.4400000 | 0.6944444 | 0.4444444 |
| 3 | 1.95313 | 0.5120000 | 3.81250 | 0.2622951 | 1.9520000 | 0.5122951 | 0.8524590 |
| 4 | 2.44141 | 0.4096000 | 5.76563 | 0.1734417 | 2.3616000 | 0.4234417 | 1.2249322 |
| 5 | 3.05176 | 0.3276800 | 8.20703 | 0.1218467 | 2.6892800 | 0.3718467 | 1.5630652 |
| 6 | 3.81470 | 0.2621440 | 11.25879 | 0.0888195 | 2.9514240 | 0.3388195 | 1.8683320 |
| 7 | 4.76837 | 0.2097152 | 15.07349 | 0.0663417 | 3.1611392 | 0.3163417 | 2.1424337 |
| 8 | 5.96046 | 0.1677722 | 19.84186 | 0.0503985 | 3.3289114 | 0.3003985 | 2.3872478 |
| 9 | 7.45058 | 0.1342177 | 25.80232 | 0.0387562 | 3.4631291 | 0.2887562 | 2.6047768 |
| 10 | 9.31323 | 0.1073742 | 33.25290 | 0.0300726 | 3.5705033 | 0.2800726 | 2.7970975 |
| 11 | 11.64153 | 0.0858993 | 42.56613 | 0.0234929 | 3.6564026 | 0.2734929 | 2.9663143 |
| 12 | 14.55192 | 0.0687195 | 54.20766 | 0.0184476 | 3.7251221 | 0.2684476 | 3.1145163 |
| 13 | 18.18989 | 0.0549756 | 68.75958 | 0.0145434 | 3.7800977 | 0.2645434 | 3.2437417 |
| 14 | 22.73737 | 0.0439805 | 86.94947 | 0.0115009 | 3.8240781 | 0.2615009 | 3.3559478 |
| 15 | 28.42171 | 0.0351844 | 109.68684 | 0.0091169 | 3.8592625 | 0.2591169 | 3.4529882 |
| 16 | 35.52714 | 0.0281475 | 138.10855 | 0.0072407 | 3.8874100 | 0.2572407 | 3.5365964 |
| 17 | 44.40892 | 0.0225180 | 173.63568 | 0.0057592 | 3.9099280 | 0.2557592 | 3.6083754 |
| 18 | 55.51115 | 0.0180144 | 218.04460 | 0.0045862 | 3.9279424 | 0.2545862 | 3.6697923 |
| 19 | 69.38894 | 0.0144115 | 273.55576 | 0.0036556 | 3.9423539 | 0.2536556 | 3.7221773 |
| 20 | 86.73617 | 0.0115292 | 342.94470 | 0.0029159 | 3.9538831 | 0.2529159 | 3.7667262 |
| 21 | 108.42022 | 0.0092234 | 429.68087 | 0.0023273 | 3.9631065 | 0.2523273 | 3.8045061 |
| 22 | 135.52527 | 0.0073787 | 538.10109 | 0.0018584 | 3.9704852 | 0.2518584 | 3.8364620 |
| 23 | 169.40659 | 0.0059030 | 673.62636 | 0.0014845 | 3.9763882 | 0.2514845 | 3.8634258 |
| 24 | 211.75824 | 0.0047224 | 843.03295 | 0.0011862 | 3.9811105 | 0.2511862 | 3.8861254 |
| 25 | 264.69780 | 0.0037779 | 1054.79118 | 0.0009481 | 3.9848884 | 0.2509481 | 3.9051945 |
| 26 | 330.87225 | 0.0030223 | 1319.48898 | 0.0007579 | 3.9879107 | 0.2507579 | 3.9211816 |
| 27 | 413.59031 | 0.0024179 | 1650.36123 | 0.0006059 | 3.9903286 | 0.2506059 | 3.9345598 |
| 28 | 516.98788 | 0.0019343 | 2063.95153 | 0.0004845 | 3.9922629 | 0.2504845 | 3.9457352 |
| 29 | 646.23485 | 0.0015474 | 2580.93941 | 0.0003875 | 3.9938103 | 0.2503875 | 3.9550551 |
| 30 | 807.79357 | 0.0012379 | 3227.17427 | 0.0003099 | 3.9950482 | 0.2503099 | 3.9628158 |
| 31 | 1009.74196 | 0.0009904 | 4034.96783 | 0.0002478 | 3.9960386 | 0.2502478 | 3.9692687 |
| 32 | 1262.17745 | 0.0007923 | 5044.70979 | 0.0001982 | 3.9968309 | 0.2501982 | 3.9746269 |
| 33 | 1577.72181 | 0.0006338 | 6306.88724 | 0.0001586 | 3.9974647 | 0.2501586 | 3.9790705 |
| 34 | 1972.15226 | 0.0005071 | 7884.60905 | 0.0001268 | 3.9979718 | 0.2501268 | 3.9827512 |
| 35 | 2465.19033 | 0.0004056 | 9856.76132 | 0.0001015 | 3.9983774 | 0.2501015 | 3.9857966 |
| 36 | 3081.48791 | 0.0003245 | 12321.95164 | 0.0000812 | 3.9987019 | 0.2500812 | 3.9883135 |
| 37 | 3851.85989 | 0.0002596 | 15403.43956 | 0.0000649 | 3.9989615 | 0.2500649 | 3.9903918 |
| 38 | 4814.82486 | 0.0002077 | 19255.29944 | 0.0000519 | 3.9991692 | 0.2500519 | 3.9921061 |
| 39 | 6018.53108 | 0.0001662 | 24070.12430 | 0.0000415 | 3.9993354 | 0.2500415 | 3.9935189 |
| 40 | 7523.16385 | 0.0001329 | 30088.65538 | 0.0000332 | 3.9994683 | 0.2500332 | 3.9946824 |
| 42 | 11754.94351 | 0.0000851 | 47015.77403 | 0.0000213 | 3.9996597 | 0.2500213 | 3.9964267 |
| 48 | 44841.55086 | 0.0000223 | 179362.20343 | 0.0000056 | 3.9999108 | 0.2500056 | 3.9989295 |
| 50 | 70064.92322 | 0.0000143 | 280255.69286 | 0.0000036 | 3.9999429 | 0.2500036 | 3.9992864 |

## Appendix A *30% Interest Rate Factors*

| | Single Payment | | Equal-Payment Series | | | | Uniform Gradient Series |
|---|---|---|---|---|---|---|---|
| $N$ | Compound Amount Factor, $(F/P, i, N)$ | Present-Worth Factor, $(P/F, i, N)$ | Compound Amount Factor, $(F/A, i, N)$ | Sinking-Fund Factor, $(A/F, i, N)$ | Present-Worth Factor, $(P/A, i, N)$ | Capital Recovery Factor, $(A/P, i, N)$ | Factor, $(A/G, i, N)$ |
| 1 | 1.30000 | 0.7692308 | 1.00000 | 1.0000000 | 0.7692308 | 1.3000000 | 0.0000000 |
| 2 | 1.69000 | 0.5917160 | 2.30000 | 0.4347826 | 1.3609467 | 0.7347826 | 0.4347826 |
| 3 | 2.19700 | 0.4551661 | 3.99000 | 0.2506266 | 1.8161129 | 0.5506266 | 0.8270677 |
| 4 | 2.85610 | 0.3501278 | 6.18700 | 0.1616292 | 2.1662407 | 0.4616292 | 1.1782770 |
| 5 | 3.71293 | 0.2693291 | 9.04310 | 0.1105815 | 2.4355698 | 0.4105815 | 1.4903075 |
| 6 | 4.82681 | 0.2071762 | 12.75603 | 0.0783943 | 2.6427460 | 0.3783943 | 1.7654474 |
| 7 | 6.27485 | 0.1593663 | 17.58284 | 0.0568736 | 2.8021123 | 0.3568736 | 2.0062818 |
| 8 | 8.15731 | 0.1225895 | 23.85769 | 0.0419152 | 2.9247018 | 0.3419152 | 2.2155945 |
| 9 | 10.60450 | 0.0942996 | 32.01500 | 0.0312354 | 3.0190013 | 0.3312354 | 2.3962725 |
| 10 | 13.78585 | 0.0725382 | 42.61950 | 0.0234634 | 3.0915395 | 0.3234634 | 2.5512187 |
| 11 | 17.92160 | 0.0557986 | 56.40535 | 0.0177288 | 3.1473381 | 0.3177288 | 2.6832768 |
| 12 | 23.29809 | 0.0429220 | 74.32695 | 0.0134541 | 3.1902601 | 0.3134541 | 2.7951705 |
| 13 | 30.28751 | 0.0330169 | 97.62504 | 0.0102433 | 3.2232770 | 0.3102433 | 2.8894581 |
| 14 | 39.37376 | 0.0253976 | 127.91255 | 0.0078178 | 3.2486746 | 0.3078178 | 2.9685007 |
| 15 | 51.18589 | 0.0195366 | 167.28631 | 0.0059778 | 3.2682112 | 0.3059778 | 3.0344446 |
| 16 | 66.54166 | 0.0150282 | 218.47220 | 0.0045772 | 3.2832394 | 0.3045772 | 3.0892138 |
| 17 | 86.50416 | 0.0115601 | 285.01386 | 0.0035086 | 3.2947995 | 0.3035086 | 3.1345126 |
| 18 | 112.45541 | 0.0088924 | 371.51802 | 0.0026917 | 3.3036920 | 0.3026917 | 3.1718338 |
| 19 | 146.19203 | 0.0068403 | 483.97343 | 0.0020662 | 3.3105323 | 0.3020662 | 3.2024722 |
| 20 | 190.04964 | 0.0052618 | 630.16546 | 0.0015869 | 3.3157941 | 0.3015869 | 3.2275410 |
| 21 | 247.06453 | 0.0040475 | 820.21510 | 0.0012192 | 3.3198416 | 0.3012192 | 3.2479899 |
| 22 | 321.18389 | 0.0031135 | 1067.27963 | 0.0009370 | 3.3229551 | 0.3009370 | 3.2646228 |
| 23 | 417.53905 | 0.0023950 | 1388.46351 | 0.0007202 | 3.3253500 | 0.3007202 | 3.2781164 |
| 24 | 542.80077 | 0.0018423 | 1806.00257 | 0.0005537 | 3.3271923 | 0.3005537 | 3.2890366 |
| 25 | 705.64100 | 0.0014172 | 2348.80334 | 0.0004257 | 3.3286095 | 0.3004257 | 3.2978543 |
| 26 | 917.33330 | 0.0010901 | 3054.44434 | 0.0003274 | 3.3296996 | 0.3003274 | 3.3049594 |
| 27 | 1192.53329 | 0.0008386 | 3971.77764 | 0.0002518 | 3.3305382 | 0.3002518 | 3.3106735 |
| 28 | 1550.29328 | 0.0006450 | 5164.31093 | 0.0001936 | 3.3311832 | 0.3001936 | 3.3152606 |
| 29 | 2015.38126 | 0.0004962 | 6714.60421 | 0.0001489 | 3.3316794 | 0.3001489 | 3.3189369 |
| 30 | 2619.99564 | 0.0003817 | 8729.98548 | 0.0001145 | 3.3320611 | 0.3001145 | 3.3218786 |
| 31 | 3405.99434 | 0.0002936 | 11349.98112 | 0.0000881 | 3.3323547 | 0.3000881 | 3.3242291 |
| 32 | 4427.79264 | 0.0002258 | 14755.97546 | 0.0000678 | 3.3325805 | 0.3000678 | 3.3261046 |
| 33 | 5756.13043 | 0.0001737 | 19183.76810 | 0.0000521 | 3.3327542 | 0.3000521 | 3.3275993 |
| 34 | 7482.96956 | 0.0001336 | 24939.89853 | 0.0000401 | 3.3328879 | 0.3000401 | 3.3287891 |
| 35 | 9727.86043 | 0.0001028 | 32422.86808 | 0.0000308 | 3.3329907 | 0.3000308 | 3.3297350 |
| 36 | 12646.21855 | 0.0000791 | 42150.72851 | 0.0000237 | 3.3330697 | 0.3000237 | 3.3304864 |
| 37 | 16440.08412 | 0.0000608 | 54796.94706 | 0.0000182 | 3.3331306 | 0.3000182 | 3.3310826 |
| 38 | 21372.10935 | 0.0000468 | 71237.03118 | 0.0000140 | 3.3331774 | 0.3000140 | 3.3315552 |
| 39 | 27783.74216 | 0.0000360 | 92609.14053 | 0.0000108 | 3.3332134 | 0.3000108 | 3.3319296 |
| 40 | 36118.86481 | 0.0000277 | 120392.88269 | 0.0000083 | 3.3332410 | 0.3000083 | 3.3322258 |

## Appendix A *40% Interest Rate Factors*

| | Single Payment | | Equal-Payment Series | | | | Uniform |
|---|---|---|---|---|---|---|---|
| *N* | Compound Amount Factor, (*F*/*P*, *i*, *N*) | Present-Worth Factor, (*P*/*F*, *i*, *N*) | Compound Amount Factor, (*F*/*A*, *i*, *N*) | Sinking-Fund Factor, (*A*/*F*, *i*, *N*) | Present-Worth Factor, (*P*/*A*, *i*, *N*) | Capital Recovery Factor, (*A*/*P*, *i*, *N*) | Gradient Series Factor, (*A*/*G*, *i*, *N*) |
| 1 | 1.40000 | 0.7142857 | 1.00000 | 1.0000000 | 0.7142857 | 1.4000000 | 0.0000000 |
| 2 | 1.96000 | 0.5102041 | 2.40000 | 0.4166667 | 1.2244898 | 0.8166667 | 0.4166667 |
| 3 | 2.74400 | 0.3644315 | 4.36000 | 0.2293578 | 1.5889213 | 0.6293578 | 0.7798165 |
| 4 | 3.84160 | 0.2603082 | 7.10400 | 0.1407658 | 1.8492295 | 0.5407658 | 1.0923423 |
| 5 | 5.37824 | 0.1859344 | 10.94560 | 0.0913609 | 2.0351639 | 0.4913609 | 1.3579886 |
| 6 | 7.52954 | 0.1328103 | 16.32384 | 0.0612601 | 2.1679742 | 0.4612601 | 1.5810986 |
| 7 | 10.54135 | 0.0948645 | 23.85338 | 0.0419228 | 2.2628387 | 0.4419228 | 1.7663512 |
| 8 | 14.75789 | 0.0677604 | 34.39473 | 0.0290742 | 2.3305991 | 0.4290742 | 1.9185155 |
| 9 | 20.66105 | 0.0484003 | 49.15262 | 0.0203448 | 2.3789994 | 0.4203448 | 2.0422421 |
| 10 | 28.92547 | 0.0345716 | 69.81366 | 0.0143238 | 2.4135710 | 0.4143238 | 2.1419039 |
| 11 | 40.49565 | 0.0246940 | 98.73913 | 0.0101277 | 2.4382650 | 0.4101277 | 2.2214883 |
| 12 | 56.69391 | 0.0176386 | 139.23478 | 0.0071821 | 2.4559036 | 0.4071821 | 2.2845366 |
| 13 | 79.37148 | 0.0125990 | 195.92869 | 0.0051039 | 2.4685025 | 0.4051039 | 2.3341233 |
| 14 | 111.12007 | 0.0089993 | 275.30017 | 0.0036324 | 2.4775018 | 0.4036324 | 2.3728660 |
| 15 | 155.56810 | 0.0064281 | 386.42024 | 0.0025879 | 2.4839299 | 0.4025879 | 2.4029554 |
| 16 | 217.79533 | 0.0045915 | 541.98833 | 0.0018451 | 2.4885213 | 0.4018451 | 2.4261977 |
| 17 | 304.91347 | 0.0032796 | 759.78367 | 0.0013162 | 2.4918010 | 0.4013162 | 2.4440630 |
| 18 | 426.87885 | 0.0023426 | 1064.69714 | 0.0009392 | 2.4941435 | 0.4009392 | 2.4577345 |
| 19 | 597.63040 | 0.0016733 | 1491.57599 | 0.0006704 | 2.4958168 | 0.4006704 | 2.4681545 |
| 20 | 836.68255 | 0.0011952 | 2089.20639 | 0.0004787 | 2.4970120 | 0.4004787 | 2.4760675 |
| 21 | 1171.35558 | 0.0008537 | 2925.88894 | 0.0003418 | 2.4978657 | 0.4003418 | 2.4820567 |
| 22 | 1639.89781 | 0.0006098 | 4097.24452 | 0.0002441 | 2.4984755 | 0.4002441 | 2.4865763 |
| 23 | 2295.85693 | 0.0004356 | 5737.14232 | 0.0001743 | 2.4989111 | 0.4001743 | 2.4899776 |
| 24 | 3214.19970 | 0.0003111 | 8032.99925 | 0.0001245 | 2.4992222 | 0.4001245 | 2.4925308 |
| 25 | 4499.87958 | 0.0002222 | 11247.19895 | 0.0000889 | 2.4994444 | 0.4000889 | 2.4944431 |
| 26 | 6299.83141 | 0.0001587 | 15747.07853 | 0.0000635 | 2.4996032 | 0.4000635 | 2.4958723 |
| 27 | 8819.76398 | 0.0001134 | 22046.90994 | 0.0000454 | 2.4997165 | 0.4000454 | 2.4969383 |
| 28 | 12347.66957 | 0.0000810 | 30866.67392 | 0.0000324 | 2.4997975 | 0.4000324 | 2.4977322 |
| 29 | 17286.73740 | 0.0000578 | 43214.34349 | 0.0000231 | 2.4998554 | 0.4000231 | 2.4983223 |
| 30 | 24201.43236 | 0.0000413 | 60501.08089 | 0.0000165 | 2.4998967 | 0.4000165 | 2.4987604 |
| 31 | 33882.00530 | 0.0000295 | 84702.51324 | 0.0000118 | 2.4999262 | 0.4000118 | 2.4990850 |
| 32 | 47434.80742 | 0.0000211 | 118584.51854 | 0.0000084 | 2.4999473 | 0.4000084 | 2.4993254 |
| 33 | 66408.73038 | 0.0000151 | 166019.32596 | 0.0000060 | 2.4999624 | 0.4000060 | 2.4995031 |
| 34 | 92972.22254 | 0.0000108 | 232428.05634 | 0.0000043 | 2.4999731 | 0.4000043 | 2.4996343 |
| 35 | 130161.11155 | 0.0000077 | 325400.27888 | 0.0000031 | 2.4999808 | 0.4000031 | 2.4997311 |

## Appendix A *50% Interest Rate Factors*

| | Single Payment | | Equal-Payment Series | | | | Uniform |
|---|---|---|---|---|---|---|---|
| N | Compound Amount Factor, (F/P, i, N) | Present-Worth Factor, (P/F, i, N) | Compound Amount Factor, (F/A, i, N) | Sinking-Fund Factor, (A/F, i, N) | Present-Worth Factor, (P/A, i, N) | Capital Recovery Factor, (A/P, i, N) | Gradient Series Factor, (A/G, i, N) |
| 1 | 1.50000 | 0.6666667 | 1.00000 | 1.0000000 | 0.6666667 | 1.5000000 | 0.0000000 |
| 2 | 2.25000 | 0.4444444 | 2.50000 | 0.4000000 | 1.1111111 | 0.9000000 | 0.4000000 |
| 3 | 3.37500 | 0.2962963 | 4.75000 | 0.2105263 | 1.4074074 | 0.7105263 | 0.7368421 |
| 4 | 5.06250 | 0.1975309 | 8.12500 | 0.1230769 | 1.6049383 | 0.6230769 | 1.0153846 |
| 5 | 7.59375 | 0.1316872 | 13.18750 | 0.0758294 | 1.7366255 | 0.5758294 | 1.2417062 |
| 6 | 11.39063 | 0.0877915 | 20.78125 | 0.0481203 | 1.8244170 | 0.5481203 | 1.4225564 |
| 7 | 17.08594 | 0.0585277 | 32.17188 | 0.0310831 | 1.8829447 | 0.5310831 | 1.5648373 |
| 8 | 25.62891 | 0.0390184 | 49.25781 | 0.0203013 | 1.9219631 | 0.5203013 | 1.6751784 |
| 9 | 38.44336 | 0.0260123 | 74.88672 | 0.0133535 | 1.9479754 | 0.5133535 | 1.7596370 |
| 10 | 57.66504 | 0.0173415 | 113.33008 | 0.0088238 | 1.9653169 | 0.5088238 | 1.8235243 |
| 11 | 86.49756 | 0.0115610 | 170.99512 | 0.0058481 | 1.9768780 | 0.5058481 | 1.8713414 |
| 12 | 129.74634 | 0.0077073 | 257.49268 | 0.0038836 | 1.9845853 | 0.5038836 | 1.9067935 |
| 13 | 194.61951 | 0.0051382 | 387.23901 | 0.0025824 | 1.9897235 | 0.5025824 | 1.9328580 |
| 14 | 291.92926 | 0.0034255 | 581.85852 | 0.0017186 | 1.9931490 | 0.5017186 | 1.9518783 |
| 15 | 437.89389 | 0.0022837 | 873.78778 | 0.0011444 | 1.9954327 | 0.5011444 | 1.9656667 |
| 16 | 656.84084 | 0.0015224 | 1311.68167 | 0.0007624 | 1.9969551 | 0.5007624 | 1.9756038 |
| 17 | 985.26125 | 0.0010150 | 1968.52251 | 0.0005080 | 1.9979701 | 0.5005080 | 1.9827282 |
| 18 | 1477.89188 | 0.0006766 | 2953.78376 | 0.0003385 | 1.9986467 | 0.5003385 | 1.9878122 |
| 19 | 2216.83782 | 0.0004511 | 4431.67564 | 0.0002256 | 1.9990978 | 0.5002256 | 1.9914254 |
| 20 | 3325.25673 | 0.0003007 | 6648.51346 | 0.0001504 | 1.9993985 | 0.5001504 | 1.9939836 |
| 21 | 4987.88510 | 0.0002005 | 9973.77019 | 0.0001003 | 1.9995990 | 0.5001003 | 1.9957890 |
| 22 | 7481.82764 | 0.0001337 | 14961.65529 | 0.0000668 | 1.9997327 | 0.5000668 | 1.9970591 |
| 23 | 11222.74146 | 0.0000891 | 22443.48293 | 0.0000446 | 1.9998218 | 0.5000446 | 1.9979504 |
| 24 | 16834.11220 | 0.0000594 | 33666.22439 | 0.0000297 | 1.9998812 | 0.5000297 | 1.9985742 |
| 25 | 25251.16829 | 0.0000396 | 50500.33659 | 0.0000198 | 1.9999208 | 0.5000198 | 1.9990099 |
| 26 | 37876.75244 | 0.0000264 | 75751.50488 | 0.0000132 | 1.9999472 | 0.5000132 | 1.9993135 |
| 27 | 56815.12866 | 0.0000176 | 113628.25732 | 0.0000088 | 1.9999648 | 0.5000088 | 1.9995248 |
| 28 | 85222.69299 | 0.0000117 | 170443.38598 | 0.0000059 | 1.9999765 | 0.5000059 | 1.9996714 |
| 29 | 127834.03949 | 0.0000078 | 255666.07898 | 0.0000039 | 1.9999844 | 0.5000039 | 1.9997731 |
| 30 | 191751.05923 | 0.0000052 | 383500.11847 | 0.0000026 | 1.9999896 | 0.5000026 | 1.9998435 |

# APPENDIX B: Statistical Tables

**Table B.1** ***Cumulative Standard Normal Distribution***

$$\phi(z) = \int_{-\infty}^{z} \frac{1}{\sqrt{2\pi}} e^{-\mu^2/2}\, du$$

| z | .00 | .01 | .02 | .03 | .04 | .05 | .06 | .07 | .08 | .09 | z |
|---|---|---|---|---|---|---|---|---|---|---|---|
| .0 | .500 00 | .503 99 | .507 98 | .511 97 | .515 95 | .519 94 | .523 92 | .527 90 | .531 88 | .535 86 | .0 |
| .1 | .539 83 | .543 79 | .547 76 | .551 72 | .555 67 | .559 62 | .563 56 | .567 49 | .571 42 | .575 34 | .1 |
| .2 | .579 26 | .583 17 | .587 06 | .590 95 | .594 83 | .598 71 | .602 57 | .606 42 | .610 26 | .614 09 | .2 |
| .3 | .617 91 | .621 72 | .625 51 | .629 30 | .633 07 | .636 83 | .640 58 | .644 31 | .648 03 | .651 73 | .3 |
| .4 | .655 42 | .659 10 | .662 76 | .666 40 | .670 03 | .673 64 | .677 24 | .680 82 | .684 38 | .687 93 | .4 |
| .5 | .691 46 | .694 97 | .698 47 | .701 94 | .705 40 | .708 84 | .712 26 | .715 66 | .719 04 | .722 40 | .5 |
| .6 | .725 75 | .729 07 | .732 37 | .735 65 | .738 91 | .742 15 | .745 37 | .748 57 | .751 75 | .754 90 | .6 |
| .7 | .758 03 | .761 15 | .764 24 | .767 30 | .770 35 | .773 37 | .776 37 | .779 35 | .782 30 | .785 23 | .7 |
| .8 | .788 14 | .791 03 | .793 89 | .796 73 | .799 54 | .802 34 | .805 10 | .807 85 | .810 57 | .813 27 | .8 |
| .9 | .815 94 | .818 59 | .821 21 | .823 81 | .826 39 | .828 94 | .831 47 | .833 97 | .836 46 | .838 91 | .9 |
| 1.0 | .841 34 | .843 75 | .846 13 | .848 49 | .850 83 | .853 14 | .855 43 | .857 69 | .859 93 | .862 14 | 1.0 |
| 1.1 | .864 33 | .866 50 | .868 64 | .870 76 | .872 85 | .874 93 | .876 97 | .879 00 | .881 00 | .882 97 | 1.1 |
| 1.2 | .884 93 | .886 86 | .888 77 | .890 65 | .892 51 | .894 35 | .896 16 | .897 96 | .899 73 | .901 47 | 1.2 |
| 1.3 | .903 20 | .904 90 | .906 58 | .908 24 | .909 88 | .911 49 | .913 08 | .914 65 | .916 21 | .917 73 | 1.3 |
| 1.4 | .919 24 | .920 73 | .922 19 | .923 64 | .925 06 | .926 47 | .927 85 | .929 22 | .930 56 | .931 89 | 1.4 |
| 1.5 | .933 19 | .934 48 | .935 74 | .936 99 | .938 22 | .939 43 | .940 62 | .941 79 | .942 95 | .944 08 | 1.5 |
| 1.6 | .945 20 | .946 30 | .947 38 | .948 45 | .949 50 | .950 53 | .951 54 | .952 54 | .953 52 | .954 48 | 1.6 |
| 1.7 | .955 43 | .956 37 | .957 28 | .958 18 | .959 07 | .959 94 | .960 80 | .961 64 | .962 46 | .963 27 | 1.7 |
| 1.8 | .964 07 | .964 85 | .965 62 | .966 37 | .967 11 | .967 84 | .968 56 | .969 26 | .969 95 | .970 62 | 1.8 |
| 1.9 | .971 28 | .971 93 | .972 57 | .973 20 | .973 81 | .974 41 | .975 00 | .975 58 | .976 15 | .976 70 | 1.9 |

(continued)

**Table B.1** *(Continued)*

| $z$ | .00 | .01 | .02 | .03 | .04 | .05 | .06 | .07 | .08 | .09 | $z$ |
|---|---|---|---|---|---|---|---|---|---|---|---|
| 2.0 | .977 25 | .977 78 | .978 31 | .978 82 | .979 32 | .979 82 | .980 30 | .980 77 | .981 24 | .981 69 | 2.0 |
| 2.1 | .982 14 | .982 57 | .983 00 | .983 41 | .983 82 | .984 22 | .984 61 | .985 00 | .985 37 | .985 74 | 2.1 |
| 2.2 | .986 10 | .986 45 | .986 79 | .987 13 | .987 45 | .987 78 | .988 09 | .988 40 | .988 70 | .988 99 | 2.2 |
| 2.3 | .989 28 | .989 56 | .989 83 | .990 10 | .990 36 | .990 61 | .990 86 | .991 11 | .991 34 | .991 58 | 2.3 |
| 2.4 | .991 80 | .992 02 | .992 24 | .992 45 | .992 66 | .992 86 | .993 05 | .993 24 | .993 43 | .993 61 | 2.4 |
| 2.5 | .993 79 | .993 96 | .994 13 | .994 30 | .994 46 | .994 61 | .994 77 | .994 92 | .995 06 | .995 20 | 2.5 |
| 2.6 | .995 34 | .995 47 | .995 60 | .995 73 | .995 85 | .995 98 | .996 09 | .996 21 | .996 32 | .996 43 | 2.6 |
| 2.7 | .996 53 | .996 64 | .996 74 | .996 83 | .996 93 | .997 02 | .997 11 | .997 20 | .997 28 | .997 36 | 2.7 |
| 2.8 | .997 44 | .997 52 | .997 60 | .997 67 | .997 74 | .997 81 | .997 88 | .997 95 | .998 01 | .998 07 | 2.8 |
| 2.9 | .998 13 | .998 19 | .998 25 | .998 31 | .998 36 | .998 41 | .998 46 | .998 51 | .998 56 | .998 61 | 2.9 |
| 3.0 | .998 65 | .998 69 | .998 74 | .998 78 | .998 82 | .998 86 | .998 89 | .998 93 | .998 97 | .999 00 | 3.0 |
| 3.1 | .999 03 | .999 06 | .999 10 | .999 13 | .999 16 | .999 18 | .999 21 | .999 24 | .999 26 | .999 29 | 3.1 |
| 3.2 | .999 31 | .999 34 | .999 36 | .999 38 | .999 40 | .999 42 | .999 44 | .999 46 | .999 48 | .999 50 | 3.2 |
| 3.3 | .999 52 | .999 53 | .999 55 | .999 57 | .999 58 | .999 60 | .999 61 | .999 62 | .999 64 | .999 65 | 3.3 |
| 3.4 | .999 66 | .999 68 | .999 69 | .999 70 | .999 71 | .999 72 | .999 73 | .999 74 | .999 75 | .999 76 | 3.4 |
| 3.5 | .999 77 | .999 78 | .999 78 | .999 79 | .999 80 | .999 81 | .999 81 | .999 82 | .999 83 | .999 83 | 3.5 |
| 3.6 | .999 84 | .999 85 | .999 85 | .999 86 | .999 86 | .999 87 | .999 87 | .999 88 | .999 88 | .999 89 | 3.6 |
| 3.7 | .999 89 | .999 90 | .999 90 | .999 90 | .999 91 | .999 91 | .999 92 | .999 92 | .999 92 | .999 92 | 3.7 |
| 3.8 | .999 93 | .999 93 | .999 93 | .999 94 | .999 94 | .999 94 | .999 94 | .999 95 | .999 95 | .999 95 | 3.8 |
| 3.9 | .999 95 | .999 95 | .999 96 | .999 96 | .999 96 | .999 96 | .999 96 | .999 96 | .999 97 | .999 97 | 3.9 |

SOURCE: This table is adapted from W. W. Hines and D. C. Montgomery, *Probability and Statistics in Engineering and Management Science,* 2nd edition, Wiley, New York, 1980.

**Table B.2** *Percentage Points of the $\chi^2$ Distribution*

| $\nu^*$ \ $\alpha$ | .995 | .990 | .975 | .950 | .900 | .500 | .100 | .050 | .025 | .010 | .005 |
|---|---|---|---|---|---|---|---|---|---|---|---|
| 1 | .00+ | .00+ | .00+ | .00+ | .02 | .45 | 2.71 | 3.84 | 5.02 | 6.63 | 7.88 |
| 2 | .01 | .02 | .05 | .10 | .21 | 1.39 | 4.61 | 5.99 | 7.38 | 9.21 | 10.60 |
| 3 | .07 | .11 | .22 | .35 | .58 | 2.37 | 6.25 | 7.81 | 9.35 | 11.34 | 12.84 |
| 4 | .21 | .30 | .48 | .71 | 1.06 | 3.36 | 7.78 | 9.49 | 11.14 | 13.28 | 14.86 |
| 5 | .41 | .55 | .83 | 1.15 | 1.61 | 4.35 | 9.24 | 11.07 | 12.83 | 15.09 | 16.75 |
| 6 | .68 | .87 | 1.24 | 1.64 | 2.20 | 5.35 | 10.65 | 12.59 | 14.45 | 16.81 | 18.55 |
| 7 | .99 | 1.24 | 1.69 | 2.17 | 2.83 | 6.35 | 12.02 | 14.07 | 16.01 | 18.48 | 20.28 |
| 8 | 1.34 | 1.65 | 2.18 | 2.73 | 3.49 | 7.34 | 13.36 | 15.51 | 17.53 | 20.09 | 21.96 |
| 9 | 1.73 | 2.09 | 2.70 | 3.33 | 4.17 | 8.34 | 14.68 | 16.92 | 19.02 | 21.67 | 23.59 |
| 10 | 2.16 | 2.56 | 3.25 | 3.94 | 4.87 | 9.34 | 15.99 | 18.31 | 20.48 | 23.21 | 25.19 |
| 11 | 2.60 | 3.05 | 3.82 | 4.57 | 5.58 | 10.34 | 17.28 | 19.68 | 21.92 | 24.72 | 26.76 |
| 12 | 3.07 | 3.57 | 4.40 | 5.23 | 6.30 | 11.34 | 18.55 | 21.03 | 23.34 | 26.22 | 28.30 |
| 13 | 3.57 | 4.11 | 5.01 | 5.89 | 7.04 | 12.34 | 19.81 | 22.36 | 24.74 | 27.69 | 29.82 |
| 14 | 4.07 | 4.66 | 5.63 | 6.57 | 7.79 | 13.34 | 21.06 | 23.68 | 26.12 | 29.14 | 31.32 |
| 15 | 4.60 | 5.23 | 6.27 | 7.26 | 8.55 | 14.34 | 22.31 | 25.00 | 27.49 | 30.58 | 32.80 |
| 16 | 5.14 | 5.81 | 6.91 | 7.96 | 9.31 | 15.34 | 23.54 | 26.30 | 28.85 | 32.00 | 34.27 |
| 17 | 5.70 | 6.41 | 7.56 | 8.67 | 10.09 | 16.34 | 24.77 | 27.59 | 30.19 | 33.41 | 35.72 |
| 18 | 6.26 | 7.01 | 8.23 | 9.39 | 10.87 | 17.34 | 25.99 | 28.87 | 31.53 | 34.81 | 37.16 |
| 19 | 6.84 | 7.63 | 8.91 | 10.12 | 11.65 | 18.34 | 27.20 | 30.14 | 32.85 | 36.19 | 38.58 |
| 20 | 7.43 | 8.26 | 9.59 | 10.85 | 12.44 | 19.34 | 28.41 | 31.41 | 34.17 | 37.57 | 40.00 |

(continued)

**Table B.2** *(Continued)*

| $\nu^*$ \ $\alpha$ | .995 | .990 | .975 | .950 | .900 | .500 | .100 | .050 | .025 | .010 | .005 |
|---|---|---|---|---|---|---|---|---|---|---|---|
| 21 | 8.03 | 8.90 | 10.28 | 11.59 | 13.24 | 20.34 | 29.62 | 32.67 | 35.48 | 38.93 | 41.40 |
| 22 | 8.64 | 9.54 | 10.98 | 12.34 | 14.04 | 21.34 | 30.81 | 33.92 | 36.78 | 40.29 | 42.80 |
| 23 | 9.26 | 10.20 | 11.69 | 13.09 | 14.85 | 22.34 | 32.01 | 35.17 | 38.08 | 41.64 | 44.18 |
| 24 | 9.89 | 10.86 | 12.40 | 13.85 | 15.66 | 23.34 | 33.20 | 36.42 | 39.36 | 42.98 | 45.56 |
| 25 | 10.52 | 11.52 | 13.12 | 14.61 | 16.47 | 24.34 | 34.28 | 37.65 | 40.65 | 44.31 | 46.93 |
| 26 | 11.16 | 12.20 | 13.84 | 15.38 | 17.29 | 25.34 | 35.56 | 38.89 | 41.92 | 45.64 | 48.29 |
| 27 | 11.81 | 12.88 | 14.57 | 16.15 | 18.11 | 26.34 | 36.74 | 40.11 | 43.19 | 46.96 | 49.65 |
| 28 | 12.46 | 13.57 | 15.31 | 16.93 | 18.94 | 27.34 | 37.92 | 41.34 | 44.46 | 48.28 | 50.99 |
| 29 | 13.12 | 14.26 | 16.05 | 17.71 | 19.77 | 28.34 | 39.09 | 42.56 | 45.72 | 49.59 | 52.34 |
| 30 | 13.79 | 14.95 | 16.79 | 18.49 | 20.60 | 29.34 | 40.26 | 43.77 | 46.98 | 50.89 | 53.67 |
| 40 | 20.71 | 22.16 | 24.43 | 26.51 | 29.05 | 39.34 | 51.81 | 55.76 | 59.34 | 63.69 | 66.77 |
| 50 | 27.99 | 29.71 | 32.36 | 34.76 | 37.69 | 49.33 | 63.17 | 67.50 | 71.42 | 76.15 | 79.49 |
| 60 | 35.53 | 37.48 | 40.48 | 43.19 | 46.46 | 59.33 | 74.40 | 79.08 | 83.30 | 88.38 | 91.95 |
| 70 | 43.28 | 45.44 | 48.76 | 51.74 | 55.33 | 69.33 | 85.53 | 90.53 | 95.02 | 100.42 | 104.22 |
| 80 | 51.17 | 53.54 | 57.15 | 60.39 | 64.28 | 79.33 | 96.58 | 101.88 | 106.63 | 112.33 | 116.32 |
| 90 | 59.20 | 61.75 | 65.65 | 69.13 | 73.29 | 89.33 | 107.57 | 113.14 | 118.14 | 124.12 | 128.30 |
| 100 | 67.33 | 70.06 | 74.22 | 77.93 | 82.36 | 99.33 | 118.50 | 124.34 | 129.56 | 135.81 | 140.17 |

*$\nu$ = degrees of freedom.

SOURCE: This table is adapted from W. W. Hines and D. C. Montgomery, *Probability and Statistics in Engineering and Management Science,* 2nd edition, Wiley, New York, 1980.

# INDEX